BET1808

S0-AHG-462

Biomechanics and Neural Control of Posture and Movement

Springer

New York
Berlin
Heidelberg
Barcelona
Hong Kong
London
Milan
Paris
Singapore
Tokyo

Jack M. Winters   Patrick E. Crago
Editors

Biomechanics and Neural Control of Posture and Movement

With 277 Figures

Springer

Jack M. Winters
Biomedical Engineering Department
Marquette University
Milwaukee, WI 53201, USA
jack.winters@marquette.edu

Patrick E. Crago
Department of Biomedical Engineering
Case Western Reserve University
Cleveland, OH 44106, USA
pec3@po.cwru.edu

QP
303
. B5684
2000

Library of Congress Cataloging-in-Publication Data
Biomechanics and neural control of posture and movement / edited by
  Jack M. Winters, Patrick E. Crago.
       p.   cm.
    Includes bibliographical references and index.
    ISBN 0-387-94974-7 (alk. paper)
    1. Human locomotion—Mathematical models.   2. Posture—
Mathematical models.   3. Human mechanics—Mathematical models.
4. Afferent pathways. I. Winters, Jack M., 1957–   . II. Crago,
Patrick E.
  QP303.B5684   2000
  612.7'6—dc21   99-16554

Printed on acid-free paper.

© 2000 Springer-Verlag New York, Inc.
All rights reserved. This work may not be translated or copied in whole or in part without the
written permission of the publisher (Springer-Verlag New York, Inc., 175 Fifth Avenue, New
York, NY 10010, USA), except for brief excerpts in connection with reviews or scholarly
analysis. Use in connection with any form of information storage and retrieval, electronic adaptation, computer software, or by similar or dissimilar methodology now known or hereafter
developed is forbidden.
The use of general descriptive names, trade names, trademarks, etc., in this publication, even
if the former are not especially identified, is not to be taken as a sign that such names, as understood by the Trade Marks and Merchandise Marks Act, may accordingly be used freely by
anyone.

Production coordinated by Chernow Editorial Services, Inc., and managed by Francine
McNeill; manufacturing supervised by Erica Bresler.
Typeset by Matrix Publishing Services, York, PA.
Printed and bound by Edwards Brothers, Inc., Ann Arbor, MI.
Printed in the United States of America.

9 8 7 6 5 4 3 2 1

ISBN 0-387-94974-7 Springer-Verlag New York Berlin Heidelberg   SPIN 10556948

Preface

•

Most routine motor tasks are complex, involving load transmission throughout the body, intricate balance, and eye-head-shoulder-hand-torso-leg coordination. The quest toward understanding how we perform such tasks with skill and grace, often in the presence of unpredictable perturbations, has a long history. This book arose from the Ninth Engineering Foundation Conference on Biomechanics and Neural Control of Movement, held in Deer Creek, Ohio, in June 1996. This unique conference, which has met every 2 to 4 years since the late 1960s, is well known for its informal format that promotes high-level, up-to-date discussions on the key issues in the field. The intent is to capture the high quality of the knowledge and discourse that is an integral part of this conference series.

The book is organized into ten sections. Section I provides a brief introduction to the terminology and conceptual foundations of the field of movement science; it is intended primarily for students. All but two of the remaining nine sections share a common format: (1) a designated section editor; (2) an introductory didactic chapter, solicited from recognized leaders; and (3) three to six state-of-the-art perspective chapters. Some perspective chapters are followed by commentaries by selected experts that provide balance and insight. Section VI is the largest section, and it consists of nine perspective chapters without commentaries. Sections II through IX reflect the themes from the eight sessions of the EFC, slightly reorganized. Section IX adds an additional important topic, neuroprostheses (the focus of a related Engineering Foundation Conference), and synthesizes it with the rest of the book.

The editors are indebted to the authors and commentators for their high-quality contributions and for their patience in seeing the final result. We are also indebted to Srinath Jayasundera for help with the web-based posting of early working documents, and to Laura Polacek for assistance in gathering the miscellaneous missing pieces from around the globe.

Jack M. Winters
Patrick E. Crago

Contents

Section III

Section IV

Section V

Section VII

Section VIII

Section IX

Section X

Contributors

James J. Abbas
Center for Biomedical Engineering, University of Kentucky, Wenner-Gren Research Laboratory, Lexington, KY 40506-0070, USA

Allison S. Arnold
Department of Mechanical Engineering, Division of Biomechanical Engineering, Stanford University, Stanford, CA 94035-3030, USA

Guido Baroni
Dipartimento di Bioingegneria, Centro di Bioingegneria, Fondazione Don Gnocchi, Politecnico di Milano, Milan, Italy

Christian Bartling
Faculty of Biology, University of Bielefeld, D-33501 Bielefeld, Germany

Randall D. Beer
Department of Electrical Engineering and Computer Science, Case Western Reserve University, Cleveland, OH 44106, USA

Marc D. Binder
Department of Physiology and Biophysics, University of Washington School of Medicine, Seattle, WA 98195, USA

Guido G. Brouwn
Man-Machine Systems Group, Department of Mechanical Engineering, Delft University of Technology, 2628 BL Delft, The Netherlands

Ian E. Brown
Division of Biology, California Institute of Technology, Pasadena, CA 91125, USA

Anne Burleigh-Jacobs
Intelligent Inference Systems Corporation, 333 West Maude Avenue, Sunnyvale, CA 94086, USA

Timothy W. Cacciatore
Department of Biology, University of California at San Diego, La Jolla, CA 92093, USA

Hillel J. Chiel
Departments of Biology, Neuroscience, and Biomedical Engineering, Case
Western Reserve University, Cleveland, OH 44106, USA

François Clarac
CNRS-NBM, Marseille Cedex 20, France

Patrick E. Crago
Department of Biomedical Engineering, Case Western Reserve University,
Cleveland, OH 44106, USA

Holk Cruse
Faculty of Biology, University of Bielefeld, D-33501 Bielefeld, Germany

Michelle Davies
Department of Neurobiology and Anatomy, Hahnemann Allegheny University of the Health Sciences, Medical College of Pennsylvania, Philadelphia, PA 19129, USA

Jeffrey Dean
Faculty of Biology, University of Bielefeld, D-33501 Bielefeld, Germany

Scott L. Delp
Department of Mechanical Engineering, Stanford University, Stanford, CA
94305, USA

Natalia V. Dounskaia
Motor Control Laboratory, Arizona State University, Tempe, AZ 85287,
USA

William K. Durfee
Department of Mechanical Engineering, University of Minnesota, Minneapolis, MN 55455, USA

F. James Eisenhart
Department of Biology, University of California at San Diego, La Jolla, CA
92093, USA

Örjan Ekeberg
Department of Numerical Analysis and Computing Science, Royal Institute
of Technology, S-100 44 Stockholm, Sweden

Claire T. Farley
Department of Integrative Biology, University of California at Berkeley,
Berkeley, CA 94720-3140, USA

Giancarlo Ferrigno
Dipartimento di Bioingegneria, Centro di Bioingegneria, Fondazione Don
Gnocchi, Politecnico di Milano, Milan, Italy

Jan Fridén
Hand Surgery Unit, Sahlgrenska University Hospital, S-901 Göteborg,
Sweden

Robert J. Full
Department of Integrative Biology, University of California at Berkeley,
Berkeley, CA 94720-3140, USA

Zoubin Ghahramani
Sobell Department of Neurophysiology, University College London, London WC1N 3BG, UK

Simon Giszter
Department of Neurobiology and Anatomy, Hahnemann Allegheny University of the Health Sciences, Medical College of Pennsylvania, Philadelphia, PA 19129, USA

Ary L. Goldberger
Department of Medicine, Beth Israel Hospital and Harvard Medical School, Boston, MA 02215, USA

Gerald L. Gottlieb
NeuroMuscular Research Center, Boston University, Boston, MA 02215, USA

Aram Z. Hajian
Exponent, 21 Strathmore Road, Natick, MA 01760, USA

Thomas M. Hamm
Division of Neurobiology, Barrow Neurological Institute, St. Joseph's Hospital, Phoenix, AZ 85013, USA

Herbert Hatze
Department of Biomechanics, University of Vienna, A-1150 Wien, Austria

Morten Haugland
Center for Sensorimotor Interaction, Aalborg University, DK-9220 Aalborg, Denmark

Jeffrey M. Hausdorff
Department of Medicine, Beth Israel Hospital and Harvard Medical School, Boston, MA 02215, USA

Frans C.T. van der Helm
Department of Mechanical Engineering, Man-Machine Systems Group, Delft University of Technology, 2628 BL Delft, The Netherlands

Neville Hogan
Department of Mechanical Engineering, Massachusetts Institute of Technology, Cambridge, MA 02139, USA

Fay Horak
Neurological Sciences Institute of Oregon Health Sciences University, Portland, OR 97209, USA

Robert D. Howe
Division of Engineering and Applied Sciences, Harvard University, Cambridge, MA 02138, USA

Peter A. Huijing
Instituut voor Fundamentele, Klinische Bewegingsweterschappen, Faculteit Bewegingsweterschappen, Vrije Universiteit, 1081 BT Amsterdam, The Netherlands

D. Keoki Jackson
Department of Aeronautics and Astronautics, Massachusetts Institute of Technology, Cambridge, MA 02139, USA

Ron Jacobs
Intelligent Inference Systems Corporation, 333 West Maude Avenue, Sunnyvale, CA 94086, USA

Marc Jamon
CNRS-NBM, Marseille Cedex 20, France

Srinath P. Jayasundera
Department of Biomedical Engineering, Catholic University of America, Washington, DC 20064, USA

Ranu Jung
Center for Biomedical Engineering, University of Kentucky, Lexington, KY 40506-0070, USA

Bart L. Kaptein
Department of Electrical Engineering, Information Theory Group, Delft University of Technology, 2600 GA Delft, The Netherlands

William Kargo
Department of Neurobiology and Anatomy, Hahnemann Allegheny University of the Health Sciences, Medical College of Pennsylvania, Philadelphia, PA 19129, USA

Mitsuo Kawato
ATR Human Information Processing Labs, Kyoto 619-02, Japan

Robert E. Kearney
Department of Biomedical Engineering, McGill University, Montreal, Quebec H3A 2B4, Canada

Tom M. Kepple
Biomechanics Laboratory, National Institutes of Health, Bethesda, MD 20892-1604, USA

Kevin Kilgore
Rehabilitation Engineering Center, MetroHealth Medical Center, Cleveland, OH 44109, USA

Thomas Kindermann
Faculty of Biology, University of Bielefeld, D-33501 Bielefeld, Germany

Robert F. Kirsch
Rehabilitation Engineering Center, MetroHealth Medical Center, Cleveland, OH 44109, USA

Yasuharu Koike
Future Project, Bio Research Lab, Toyota Motor Corporation, Aichi 471-71, Japan

Erik Kouwenhoven
Perceptual Motor Integration Group, Helmholtz Instituut, University of Utrecht, 3508 TA Utrecht, The Netherlands

Hermano I. Krebs
Department of Mechanical Engineering, Massachusetts Institute of Technology, Cambridge, MA 02139, USA

William B. Kristan, Jr.
Department of Biology, University of California at San Diego, La Jolla, CA 92093, USA

Art Kuo
Department of Mechanical Engineering, University of Michigan, Ann Arbor, MI 48109-2125, USA

Corinna Lathan
Biomedical Engineering Program, Catholic University of America, Washington, DC 20064, USA

Steven L. Lehman
Department of Integrative Biology, University of California at Berkeley, Berkeley, CA 94720-3140, USA

Richard L. Lieber
Department of Orthopaedics, University of California San Diego School of Medicine and VA Medical Center, San Diego, CA 92161, USA

Gerald E. Loeb
Bio-Medical Engineering, University of Southern California, Los Angeles, CA 90089, USA

Peter S. Lum
Rehabilitation R&D Center, VA Palo Alto Health Care System, Palo Alto, CA 94304, USA

Mitchell G. Maltenfort
Division of Neurobiology, Barrow Neurological Institute, St. Joseph's Hospital, Phoenix, AZ 85013, USA

James A. Murray
Department of Biology, University of California at San Diego, La Jolla, CA 92093, USA

Wendy M. Murray
Rehabilitation Engineering Center, MetroHealth Medical Center, Cleveland, OH 44109, USA

Ferdinando A. Mussa-Ivaldi
Departments of Physiology and Physical Medicine and Rehabilitation, Northwestern University, Chicago, IL 60611-3008, USA

Dava J. Newman
Department of Aeronautics and Astronautics, Massachusetts Institute of Technology, Cambridge, MA 02139, USA

Evert-Jan Nijhof
Perceptual Motor Integration Group, Helmholtz Instituut, University of Utrecht, 3508 TA Utrecht, The Netherlands

E. Otten
Department of Medical Physiology, University of Groningen Bloemsingel 10, 9712 KZ Groningen, The Netherlands

William H. Paloski
Life Sciences Research Laboratories, NASA/Johnson Space Center, Houston, TX 77058, USA

Antonio Pedotti
Dipartimento di Bioingegneria, Centro di Bioingegneria, Fondazione Don Gnocchi, Politecnico di Milano, Milan, Italy

C.K. Peng
Rey Laboratory for Nonlinear Dynamics in Medicine, Beth Israel Hospital and Harvard Medical School, Boston, MA 02215, USA

Robert J. Peterka
R.S. Dow Neurologial Sciences Institute, Portland, OR 97209, USA

Stephen J. Piazza
Center for Locomotion Studies, Penn State University, University Park, PA 16802, USA

Tanja M. Pieters
Biomedical Engineering Program, Catholic University of America, Washington, DC 20064, USA

Dejan Popovic
Center for Sensorimotor Interaction, Aalborg University, DK-9220 Aalborg, Denmark

Mirjana Popovic
Institute for Medical Research, 11000 Belgrade, Yugoslavia

Jochen Quintern
Neurological Clinic, Klinikum Grosshadern, University of Munich, D-81377 Munich, Germany

Tariq Rahman
Extended Manipulation Laboratory, DuPont Hospital for Children, Wilmington, DE 19899, USA

David J. Reinkensmeyer
Department of Mechanical and Aerospace Engineering, University of California at Irvine, Irvine, CA 92697-3975, USA

Robert Riener
Lehrstuhl für Steuerungs- und Regelungstechnik, Technical University of Munich, D-80333 Munich, Germany

Leonard A. Rozendaal
Department of Mechanical Engineering, Man-Machine Systems Group, Delft University of Technology, 2628 CD Delft, The Netherlands

Andy L. Ruina
Theoretical and Applied Mechanics, Cornell University, Ithaca, NY 14853, USA

Josef Schmitz
Faculty of Biology, University of Bielefeld, D-33501 Bielefeld, Germany

Michael Schumm
Faculty of Biology, University of Bielefeld, D-33501 Bielefeld, Germany

Reza Shadmehr
Department of Biomedical Engineering, The Johns Hopkins University, Baltimore, MD 21205, USA

Thomas Sinkjær
Center for Sensorimotor Interaction, Aalborg University, DK-9220 Aalborg, Denmark

Richard Skalak
Department of Bioengineering, University of California at San Diego, La Jolla, CA 92093, USA

Boguslaw A. Skierczynski
Department of Bioengineering, University of California at San Diego, La Jolla, CA 92093, USA

Michael P. Slawnych
Department of Biomedical Engineering, McGill University, Montreal, Quebec H3A 2B4, Canada

A.J. (Knoek) van Soest
Faculty of Human Movement Sciences, Vrije University, NL 1081 BT Amsterdam, The Netherlands

Steven J. Stanhope
Biomechanics Laboratory, National Institutes of Health, Bethesda, MD 20892-1604, USA

Richard B. Stein
Division of Neuroscience, University of Alberta, Edmonton, Alberta T6G 2S2, Canada

Sybert Stroeve
Department of Mechanical Engineering, Delft University of Technology, 2628 CD Delft, The Netherlands

Sujat Sukthankar
Biomedical Engineering Program, Catholic University of America, Washington, DC 20064, USA

Stephan P. Swinnen
Laboratory of Motor Control, Department of Kinesiology, K.U. Leuven, 3001 Heverlee, Belgium

Kurt Thoroughman
Department of Biomedical Engineering, The Johns Hopkins University, Baltimore, MD 21205, USA

Rajko Tomovic
Faculty of Electrical Engineering, University of Belgrade, 11000 Belgrade, Yugoslavia

Ronald J. Triolo
Rehabilitation Engineering Center, MetroHealth Medical Center, Cleveland, OH 44109, USA

Carole A. Tucker
Sargent College of Health and Rehabilitation Sciences, Boston University, Boston, MA 02215, USA

Peter H. Veltink
Faculty of Electrical Engineering, University of Twente, 7500 AE Enschede, The Netherlands

Charles B. Walter
Laboratory of Motor Control, Department of Kinesiology, K.U. Leuven, 3001 Heverlee, Belgium

Jeanne Y. Wei
Department of Medicine, Beth Israel Hospital and Harvard Medical School, Boston, MA 02215, USA

Richard J.A. Wilson
Department of Biology, University of California at San Diego, La Jolla, CA 92093, USA

David A. Winter
Department of Kinesiology, University of Waterloo, Waterloo, Ontario N2L 3G1, Canada

Jack M. Winters
Biomedical Engineering Department, Marquette University, Milwaukee, WI 53201, USA

Daniel M. Wolpert
Sobell Department of Neurophysiology, University College London, London WC1N 3BG, UK

Gary T. Yamaguchi
Department of Chemical, Biological and Materials Engineering, Arizona State University, Tempe, AZ 85287-6006, USA

E. Paul Zehr
Division of Neuroscience, University of Alberta, Edmonton, Alberta T6G 2S2, Canada

Section I

1
Terminology and Foundations of Movement Science

Jack M. Winters

1 Introduction

The purpose of this chapter is to provide a technical foundation in movement science. This involves addressing five areas: (1) key terminology (i.e., the "language" of movement science) (Sections 2–3); (2) key aspects of neuromotor physiology, with a primary focus on the interaction between neural circuits and muscles (Section 4); (3) key terminology associated with making measurements and processing experimental data, especially as related to motion analysis systems (Section 5); (4) key concepts in mechanics and control that impact on the study of movement science (Section 6); and (5) basic implications of living tissues, such as tissue remodeling, adaptation, and sensorimotor learning (Section 7).

2 Terminology from a "Systems" Perspective

2.1 Basics: System, Input, Output

The scientist or engineer who studies human or animal movement typically starts by defining, either explicitly or implicitly, a *system*. As shown in Figure 1.1, a system has conceptually closed boundaries. Arrows represent unidirectional flow of information (*signals*), with arrows entering the system called *inputs* and those leaving the system called *outputs*. Notice the inherent cause–effect relationship between inputs and outputs, with the identified system inbetween. Sometimes such information flow is conceptual (e.g., a brain path-

way), but often it represents quantities with known units (e.g., length, force) that typically change as a function of time. *State variables* are used to capture the internal state of the system.

The systems framework can apply equally to both experiments and computer simulation. For instance, the experimentalist normally applies some type of change (inputs) to the subject or preparation, and then measures subsequent responses (outputs). In computer simulation, models are normally represented by a set of *state equations* of the form:

$$\frac{dx(t)}{dt} = F(x(t),u(t)); \qquad y(t) = G(x(t),u(t)) \quad (1.1)$$

where u is the input vector, y is the output vector, x is the state variable vector, and t is time.

A system with a single input and single output is called an *SISO system* (u and y are then scalars), one with multiple inputs yet one output a *MISO system* (u is a vector, y a scalar), and one with multiple inputs and outputs a *MIMO system*. For musculoskeletal models, we are typically interested in MIMO systems, because more than one control signal (e.g., drives to each muscle) is usually required to cause most movements (e.g., see Chapters 7, 10, 11, 19, 28, 29, 32–36, 43, 46). For experimental studies, certain inputs are often purposely held constant while one is changed, so that SISO-type analysis can be utilized for a MIMO system (see also Chapter 9).

The identification of a "system" is itself an exercise in modeling because a *set of assumptions* had to be made. Classic assumptions involve, for instance, treating a whole muscle as a single entity, assuming each muscle has a single line of action, as-

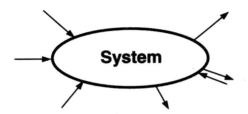

FIGURE 1.1. A "system."

suming a single common neuromotor drive to a muscle, assuming the existence of a certain type of central pattern generator (neural oscillator circuitry), assuming an idealized joint (e.g., hinge), or idealizing contact between a person and the environment as occurring at a single point. Although some assumptions may seem obvious, it is generally good practice to state one's assumptions as explicitly as possible. For instance, in Chapters 5 to 6, Huijing questions the common assumption of treating a whole muscle as if it were a "big sarcomere."

2.2 Example: Muscle Mechanics and the Hill Model

This section develops an example of systems analysis that has its roots in the classic physiological studies of A.V. Hill and colleagues during the 1920s and 1930s (Hill 1938) that were continued through the 1960s (Hill 1970): Estimated mechanical (and thermodynamic) properties of a whole muscle preparation. This example also allows us to define some key terminology, as well as to introduce a number of concepts related to lumped-parameter modeling.

In essence, Hill used the tools of a systems physiologist/engineer to perform controlled experiments that helped him identify key phenomena that ultimately resulted in what became known as the Hill muscle model, which has been widely used (e.g., used and/or addressed in Chapters 7, 8, 10, 11, 28, 29, 30, 31, 32, 35, 36, 43) and helped him capture a Nobel prize (actually for his work that combined muscle mechanical properties with muscle thermodynamics).

2.2.1 Input–Output Perspective

If thermodynamic effects are ignored, there are two basic inputs and one output to an isolated muscle

(a MISO system). To best study its properties, one keeps one input steady while changing the other and measuring the output. As seen in Figure 1.2, one input (n_{in}) is always the neural excitation of a muscle via electrical stimulation [see, for example, Zahalak (1990) or Chapter 2 for a treatment of the muscle excitation–activation coupling process]. Because motor neuron (MN) signals affect muscle, but muscle activity does not affect the MN directly, this is a true *unidirectional* signal to a high degree of accuracy—we will see later that the nature of chemical synapses makes this so.

There are two classic types of neuromotor inputs that Hill and colleagues would provide to the muscle via this input node: (1) a brief electric pulse that caused *action potentials* (sharp changes in transmembrane potential, of about 2 ms duration) to spread by active electrical transport along the nerve axons (at speeds of up to about 100 m/s) to muscle fibers (propagating at about 1–5 m/s) that ultimately causes a muscle response (called a *twitch response* by physiologists, an *impulse response* by engineers because the time of the input is brief relative to the subsequent output response); and (2) a high stimulation frequency (e.g., 100 Hz) that resulted in a maximal (saturating) ability to generate muscle force (called *tetanus* by physiologists, a maximal *step response* by engineers).

Hill's other input was from the mechanical interface port (i.e., where the muscle couples to its environment). Notice from Figure 1.2 that there are two independent variables crossing this boundary node: the contact force f and a kinematic variable. The key candidate kinematic variables (length l, velocity v, acceleration a) are related as follows:

$$\frac{dl(t)}{dt} = v(t) = \int a(t)\, dt \qquad (1.2)$$

and v was chosen as the most convenient describing variable. The product of force and velocity

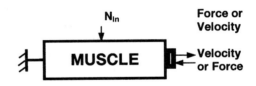

FIGURE 1.2. Muscle as a simple "system" with two inputs (one at mechanical port) and one output.

gives the instantaneous power P that is transferred between the system and environment:

$$P = f v \qquad (1.3)$$

Notice that this node is *bidirectional* (or *bicausal*): the system can influence the environment and visa-versa (i.e., *information flows both ways*). For Hill's controlled tests, one of these two variables needs to be defined as the second input, and the other becomes the measured output (heat was also measured). The following terminology is often used to characterize certain specialized input conditions across this node:

Isokinetic: velocity is a constant (specified) input, and force is the measured output (*isometric* is the special case where this prescribed velocity is zero),

Isotonic: force is a constant input, and velocity is the measured output (*unloaded* or *free* contraction is the special case where the applied force is zero).

2.2.2 Classic Testing Conditions

Given two inputs and one output, the following classic tests can then be defined (where n_{in} is neural input, v is velocity, and f is force; see also Winters 1990):

Isometric twitch: n_{in} = impulse, v_{in} = 0, f_{out} measured

Isometric tetatus: n_{in} = max step, v_{in} = 0, f_{out} measured

Free twitch: n_{in} = impulse, f_{in} = 0, v_{out} measured

Max isokinetic: n_{in} = max, v_{in} = const, f_{out} measured

Max isotonic: n_{in} = max, f_{in} = const, v_{out} measured

Quick release: n_{in} = max, f_{in} = high-to-low-step, v_{out} measured

By systematically putting in different combinations of inputs and interpreting measured outputs, Hill and many others since then have been able to learn a good deal about the properties of muscle as a tissue. [To place this approach into context, notice that here we obtain *phenomenological* data, without directly attributing observed behavior to specific *biophysical mechanisms*. The direct study of the underlying mechanisms is often called *reductionist* science; there is a natural tension (usually healthy) between

reductionist science and systems science approaches (e.g., see Chapters in Section II).

2.2.3 Essential Phenomena Underlying Hill Model Structure

One key finding of Hill and colleagues was the observation that for a given sustained level of neural excitation n, a sudden change in force (or length) would result in nearly instantaneous change in length (or force). This suggests the relationship of a spring (Hill 1950):

$$k = \frac{\Delta f}{\Delta l} \qquad (1.4)$$

where k is often called the spring constant. Because this property is nearly instantaneous (i.e., does not depend on the past history of loading but only on this sudden change), we can view it as the behavior of a lightly damped spring. Interestingly, under other situations (e.g., changes in n), muscle does appear to exhibit damping (e.g., inherent sensitivity of muscle force to muscle velocity). This suggests that a spring-like element could be conceptualized as being connected to the node on the right of the muscle "black box" of Figure 1.2, but not to the node for the neural drive n. Letting the primary contractile tissue be called the *contractile element* (CE), we have the classic Hill model for muscle, shown in Figure 1.3, with lightly damped spring-like elements both in series (SE) and in parallel (PE) with CE (Hill 1938, 1950).

In trying to understand how this model works, think of the CE as being a bit sluggish—unable to move instantaneously. We also know that for spring-like elements in:

• *Series*: forces are the same, and extensions add,
• *Parallel*: extensions are the same, and forces add.

Using the analogy of Figure 1.3B, one sees that with a sudden change in load (i.e., change to a new f_{total}), x_m moves right away but x_{ce} does not. Thus, in this case Δl in Eq. 1.4 is $\Delta(x_m - x_{ce})$. For Figure 1.3A, this is not quite true, because PE is also spring-like, and as noted above for such springs in parallel, the forces add: $f_{total} = f_{active} + f_{passive}$ (where the former is that across SE and the latter across PE).

So why do most modelers use the form shown in Figure 1.3A? First, for lengths below the mus-

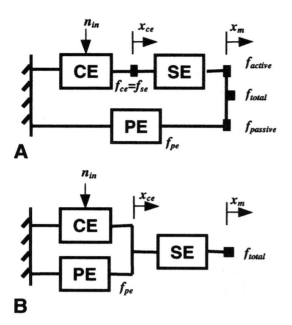

FIGURE 1.3. (A) Most common form of the Hill muscle model, with CE (representing the active contractile machinery) bridged by light-damped springs both in series (SE) and in parallel (PE). (B) Alternative form. Note that the constitutive relations for SE and PE differ for the two cases (e.g., with passive stretch where the force across CE is zero, only PE is stretched in A, yet both in B). However, because SE is usually much stiffer than PE over the primary operating range for most muscles (see Figure 1.4), it almost does not matter which form is used.

cle rest length, f_{pe} is essentially zero (see Figure 1.4B, solid line), and thus SE can be directly obtained. But also consider the case where muscle is not excited (i.e., $n = 0$ and the force across CE is zero). In Figure 1.3A, because CE and SE are in series, this implies that the force across SE is also zero (i.e., $f_{ce} = f_{se} = f_{active} = 0$). Thus as the muscle is extended, for Figure 1.3A the measured force as a result of passive stretch, $f_{passive}$, is literally that across PE. Any observed force is then simply that due to PE, which can be thought of as the passive force of the musculotendon unit that is the result of connective tissue infrastructure, such as the musculotendon sheath, the muscle fiber membranes, and the overall fluid environment within which muscle tissue lives [as pointed out by Huijing in the commentary to Chapter 7, there is in fact more of a connective tissue mesh-like infrastructure than is often appreciated]. However, for Figure 1.3B,

where SE and PE are in series, for extensions above rest length both will be stretched (although PE will likely stretch more because it is typically more compliant, i.e., less stiff). Then for a given overall extension x_m we have (in a good model) the same f_{total} for both models, but in Figure 1.3B, the PE extension equals x_{ce} rather than x_m. This is because of the confounding effect of SE still being present. Another advantage of the form of Figure 1.3A is that there are many other passive spring-like elements, structurally in parallel, that also cross joints, and these can be mathematically lumped together. Indeed, for the human system the experimental data that is available is usually already for lumped passive joint properties.

2.2.4   Constitutive Relations: SE, CE, PE

So far this chapter has focused on model structure, but not on the form of the relations and their describing parameters. Relations such as CE, SE, or PE that attempt to describe observed behavior by idealized lumped elements are often called *constitutive relations* by engineers. CE represents a more complex type of constitutive relation in that there are multiple curves—different curves for each level of activation. Indeed, there are activation-dependent force-length-velocity surfaces. We will treat the CE relation in the next section.

For SE and PE, it is normally assumed that there is a single curve (i.e., the relation of a passive spring). For both SE and PE, the axes have the units of length l and force f, and the *slope* (df/dl) is a measure of *stiffness* across the element. In both cases the slope (i.e., stiffness) increases as the element is increasingly stretched. This implies that the relation is *nonlinear*; its behavior cannot be described by a straight line, in which case a single parameter ("stiffness" or "spring constant" k) could have been used to represent its slope for any extension. Rather, the slope (df/dx) changes as a function of the value of a specified variable (force f or extension x) for at least some of the operating range.

It is often useful, however, to assume local linearity so that we can develop a feel for how springs interact. It is easy to show that for spring elements in:

- *Series*: the overall (equivalent) stiffness will be lower that the lowest, i.e. compliances (inverse of stiffness) add,

• *Parallel*: the equivalent stiffness is the sum of the individual stiffnesses.

Often constitutive relations are linearized over operating ranges. When is it sufficient to ignore or linearize SE and/or PE? Due to the Hill model structure, ignoring SE implies that we assume it to be infinitely stiff (zero compliance), while eliminating PE implies that we assume it to be infinitely compliant (zero stiffness). As seen in Figure 1.4, the peak SE strain during maximal stimulation is typically only about 2 to 6% above "slack" conditions for both muscle and tendon, while for PE the peak extension is about an order of magnitude higher. Thus, SE is usually considerably stiffer than PE, with the exception being when the muscle is stretched considerably (x_m is high) and/or the active force is very low (neural activation n is low). When n is high (and consequently f_m is high) and

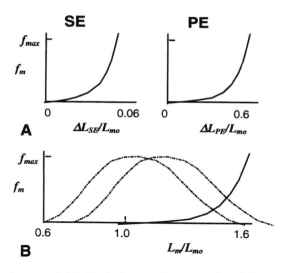

A

B

FIGURE 1.4. (A) Typical spring-like properties of the series element (SE) and parallel element (PE), showing that relative peak extension of SE is about an order of magnitude less than that for PE. In both cases, the x-axis has dimensionless *strain* units, relative to an overall muscle rest length L_{mo}. (B) The PE element is plotted with the x-axis being the *stretch ratio*, with 1.0 being the "slack" or "threshold" length above which there is a passive force. In dashed lines are several possible maximal CE force-length forces, here relative to the stretch ratio defined for PE. The curve more to the left represents a case where there is less overlap with PE ("optimum" CE length with maximum force is at about 1.0), the one to the right more overlap (optimum length at about 1.2).

$x_m - x_{ce}$ is not changing much, SE could potentially be ignored or at least linearized over the *operating range* of interest. However, how often does this happen during realistic tasks?

Often PE linearization or elimination is more justified. For most muscles the passive force tends not to be very high until the muscle is stretched about 40% above the rest length (see Figure 1.4, right), and thus both experimentalists and some modelers who have been interested in the movement midrange have often chosen to ignore the PE contribution. However, there are muscles (e.g., some neck muscles, ankle muscles, cardiac muscle) where the PE behavior is quite relevant, see also Chapter 10).

2.2.5   CE Force-Velocity Behavior

Another early key observation of Hill and others (e.g., Hill 1922; Levin and Wyman 1927; Fenn and Marsh 1935) was that the force produced by a muscle was a function of its velocity. It took the conceptual identification of a phenomenological SE element, bridging the contractile machinery, to really develop understanding of this behavior (Hill 1950; Wilkie 1950). During isotonic contractions against a load while under maximal neural drive (i.e., sustained tetanus), there was usually a period of time where the measured shortening velocity was reasonably steady. This suggests that the velocity at node x_m in Figure 1.3 was similar to that at node x_{ce}. If one additionally takes care in compensating for muscle length-sensitive effects, it is literally possible to estimate v_{ce} and f_{ce} (which is then approximately f_{total}) without the confounding influence of SE or PE, or of the well-known active CE force-length relation (e.g., see Figure 1.4B, Chapter 2, Chapter 3, Chapter 5, Chapter 7, Chapter 36). If we repeat this type of experiment for a range of isotonic loads $f_{applied} = f_{total}$ ($\sim f_{ce}$) in each case measuring a corresponding value v_m ($\sim v_{ce}$), we end up with a set of force-velocity data point pairs that can be used to obtain a CE force-velocity (CE_{fv}) constitutive equation.

In A.V. Hill's case, he focused on shortening muscle, and found that the greater the load, the lower the velocity. He fit his data with a hyperbolic function that became known as Hill's equation; this shape (for shortening velocity) is shown in Figure 1.5, and an excellent modern example is presented in Figure 10.11. This classic CE force-velocity constitutive relation was mathematically described by Hill in the form (using f and v as our variables):

$$(f + a_f f_h)(v + a_f v_{max}) = const$$
$$= a_f f_h v_{max}(1 + a_f) \quad (1.5)$$

where a_f is a dimensionless constant (Hill typically found a value of about 0.25), f_h is the "hypothetical" (as if isometric) force crossing the vertical (zero velocity) axis, and v_{max} is the velocity crossing the horizontal axis (often called the *unloaded maximum velocity*). Somewhat remarkably for his time, Hill was able to relate a_f to a thermodynamic entity, the heat of shortening. Part of the reason why Hill's equation remains popular is that the parameters are intuitive; Figure 1.5 shows how the curve changes with variation in v_{max} (Figure 1.5A) and a combination of f_h and v_{max} that is often used to represent scaling of the relation with activation (Figure 1.5B).

Through mathematical manipulation, Hill's equation can also be written (Winters and Stark 1985):

$$f_{ce} = f_h - f_b = f_h - \left[\frac{f_h + a_f}{v_{ce} + a_f v_{max}}\right] v_{ce} \quad (1.6)$$

where here one sees more explicitly that the Hill-based CE force-velocity relation can be viewed as the actual CE force f_{ce} (i.e., that "passed") being the difference between the "hypothetical" force f_h and the velocity-dependent viscous force f_b (see Figure 1.5C). If one plots the dissipated force f_d versus v_{ce}, this is mathematically the (highly nonlinear) constitutive equation of a *dashpot* element. Dashpots are inherently *energy dissipating elements*, and thus it makes sense that Hill was interested in heat loss. However, it is important to remember that we are considering a phenomenological input-output result from controlled "systems" experiments on

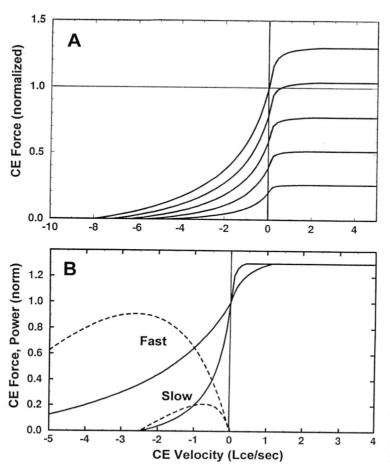

FIGURE 1.5. Hill's equation from the perspective of a motor with energy dissipation. (A) Scaling of the relation as a function of the activation, for activations of 0.2, 0.4, 0.6, 0.8, and 1.0. This assumes that the unloaded maximum velocity parameter v_{max} (which can vary substantially as a function of muscle fiber composition—see Section 3) scales linearly from −4 CE muscle lengths per second (Lce/sec) at zero activation to −8 Lce/sec at full activation. (B) Predicted CE power generation (dashed lines, shown for shortening muscle) and CE force (solid lines) as a function of CE velocity, for typical "fast" and "slow" muscle properties while under maximal activation. (Note: Remember that actual muscle power and CE power will normally differ because of the influence of SE and PE.)

whole muscle, without addressing the actual contractile mechanism and the sliding filament model of muscle (see also Chapter 2 and Chapter 8).

As with any motor, power production will be a function of ongoing conditions, and for the CE relation, the *muscle power output* P_{ce} as a function of v_{ce} (see Figure 1.5C). Notice that peak CE power is produced at about 1/3 of the peak unloading velocity; this helps explain why a bicyclist desires to change gears when the load against the foot changes (e.g., going up a hill). Of course, because v_{ce} and v_m are rarely exactly the same as a result of the presence of the spring-like SE, things are a bit trickier in reality.

Hill-type models, in various formulations, are used in many chapters in this book. CE force-velocity or force-velocity-length curves are provided in Figures 5.1, 7.5, 10.11, and 28.3. Of special note is the consideration of lengthening muscle in Chapter 10.

2.2.6 Hill's Model Versus Sliding-Filament Biophysical Models

It is important to remember that the foundation of Hill-type models is based on systematic input-output testing of whole muscle, rather than on current knowledge of the muscle contractile mechanisms at the microstructural level. Basic understanding of the microscopic level first emerged with the classic Nobel-prize winning work of A.F. Huxley and colleagues, who first documented the key observed phenomena (A.F. Huxley and Niedergerke 1954; H.E. Huxley and Hanson 1954), then put forward the basic sliding filament mechanism (Huxley 1957; Huxley and Simmons 1971) that still remains the foundation for most investigation in this area (e.g., see Zahalak 1990; Chapter 2).

However, it is also important not to reject the basic phenomenological reality: that muscle force is not a smoothed version of the EMG, but also depends on (or is sculpted by) muscle length and velocity.

3 More Terminology: Musculoskeletal Mechanics

3.1 Muscle Action

The previous section used the term "contraction" to describe muscle action. This is misleading because, over a lifespan, a muscle must lengthen and shorten

about equal amounts (joints rotate back and forth, rather than making full circles as would most DC torque motors). It is very common for muscles to be active while they are lengthening. This will happen whenever the tensile force generated by the muscle tissue is less than the force applied to the muscle. Thus the term *muscle action* is often used instead of "muscle contraction" to describe a muscle's state; but both remain in common use. Several key terms are:

- *Eccentric* action: a muscle that lengthens while activated.
- *Concentric* action: a muscle that shortens while activated.

Often these occur sequentially. For instance, an effective strategy for turning around a limb segment is to start with eccentric action (which according to the CE force-velocity relation can cause forces higher than isometric) then finish with concentric contraction. This eccentric-then-concentric scheme is common for many propulsive tasks ranging from walking to throwing, and is often referred to as the *stretch-shortening* cycle. A related but distinct (Mungiole and Winters 1990) phenomenon is *elastic bounce* (Cavagna 1970), in which a contracting muscle is stretched by a transient external load, then during rebound releases stored elastic energy. Although the relative importance of releasing stored elastic energy (primarily via SE) versus other byproducts such as the differing (and typically less) stretch of CE relative to the overall muscle remains controversial (e.g., van Ingen Schenau 1984), the reality is that the general concepts of stretch-shortening and elastic bounce are "design strategies" that are effectively utilized in many animal structures (e.g., see Alexander 1988; Alexander and Ker 1990). Chapter 13 focuses on spring-mass behavior for various legged creatures engaged in running (see also McMahon 1990).

3.2 Muscle Geometry and Skeletal Joints

Muscles cross joints, and as a consequence muscle lengths also change as a joint angle changes. Musculotendon units typically attach to a bone over a significant region, yet normally a single geometric path, or *"line of action"* (e.g., centroid of this area), is assumed. Sometimes more than one line of action is used to represent a muscle, especially fan-shaped muscles, and there exist formalized strategies for doing this (Van der Helm and Veenbaas 1991). The more *proximal* boney attachment site (i.e., the one closer to body midline) is commonly called the muscle *origin*, the more *distal* (i.e., further from midline) the *insertion*. Sometimes both are just called insertion sites. Appendix 1 describes a web-based gateway to anthropometric data that includes estimated origin and insertion sites for various human muscles.

3.2.1   Insertion Sites and Lines of Action

A challenge that always exists in utilizing anthropometric data is how to deal with muscle lines of action that often clearly cannot be represented by straight lines. Figure 1.6 assumes an idealized elbow hinge joint and an idealized biceps muscle path (or line of action) that consists of a straight line path across the elbow region when the elbow is flexed, yet when the elbow is extended, a muscle line of action that arcs around a shell. The muscle plus ten-

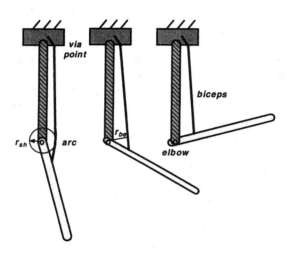

FIGURE 1.6. Conceptual diagram showing a "biceps" muscle crossing an elbow hinge joint for three elbow angles. When the arm straightens, the muscle line of action is assumed to rotate about a frictionless shell. If expanded to three dimensions, the circular arc shape at the elbow could become, for instance, spherical or cylindrical shell. We also assume here that the shoulder does not rotate; if it does, the muscle becomes biarticular (muscle crossing 2 joints). For clarity, the via point shown here slides through a point on the shoulder "base"; in reality, it glides through a groove on the humerus (represented by a "via point"), with this location moving with the upper limb.

don acts like a rope pulling on a lever around a pulley. For a muscle line of action to curve, forces must be applied to it. Because the muscle-tendon unit is covered by a sheath with very low friction, it is reasonable to assume that such forces are always applied orthogonal to the line of action (Winters and Kleweno 1993). This can be considered as traversing around a very small (frictionless) arc, and the equivalent orthogonal force thus bisects the arc. Also shown is a change in muscle path direction near the shoulder (often called a *via point*).

3.2.2   Muscle Moment (Lever) Arms

The *moment arm* represents the mechanical mapping between the muscle and joint. As might be expected at a bicausal interface, there are two mappings:

muscle force ⟷ joint moment
muscle velocity (length) ⟷ joint velocity (angle)

This muscle-joint mapping can be described mathematically by the Jacobian J_{mj} (i.e., the differential mapping between muscle length and joint angle changes). For the more general case of n_m muscles and n_j joint degrees-of-freedom (DOF) we have:

$$dx_m = J_{mj}dx_j; \qquad f_j = J_{mj}{}^T f_m \qquad (1.7)$$

where x_j are the generalized joint coordinates (usually angles) and x_m the muscle lengths, and f_m and f_j are the muscle force and the generalized joint forces (usually moments) associated with each DOF.

3.2.3   Actuator Redundancy

Normally there are more muscle actuators that joint DOFs (i.e., $n_m > n_j$). This is called *actuator redundancy*. Thus J_{mj} in Eq. 1.7 is a rectangular matrix and its inverse is not defined. If we desire to map from dx_m to dx_j, we use an approximation of the inverse called the "pseudo-inverse" $J_{mj}{}^*$; the several ways to estimate this are beyond the scope of this chapter. Chapter 32 discusses this transformation and how it relates to moment arms in more detail. Here we describe the practical reality. For the straight-line case (elbow flexed), we obtain from trigonometry (using law of cosines, then law of sines):

$$l_m = \sqrt{l_p^2 + l_p^2 - 2l_p l_d \, cos(\theta)}$$
$$r_{be} = \frac{l_p l_d}{l_m} sin(\theta) \qquad (1.8)$$

where here r_{be} is the *moment arm* of the biceps with respect to the elbow, θ is the elbow joint angle, and the lengths l_p, l_d and l_m represent the three sides of the triangle (proximal origin to joint, distal insertion to joint, and origin-insertion, respectively). For this straight-line case one could have just as easily applied the vector cross products $f_j = r \times f_m$ and $v_m = v_j \times r$, where the angular moment f_j and angular velocity v_j are directed out of the paper. For the case of greater elbow extension, where the path curves around the circular shell, we have $r_{be} = r_{sh}$, where r_{sh} is the shell radius. Also, the total muscle length is the addition of the individual lengths due to the arc (this added length is simply r_{sh} times the arc length) or the via point. The value of moment arms turns out to be very important, and often error in its estimation represents one of the key weak links in the modeling process (see also Chapter 32). This is especially true for muscles with strongly curved paths; indeed, for muscles such as the deltoid and quadriceps even the sign on the moment can be wrong if a straight line path between origin and insertion is used! Of note is that there are muscles that actually do change their sign (e.g., the sternocleidomastoid during head flexion-extension, relative to the upper cervical spine joints). Chapter 3 provides examples of measured relationships between muscle length and joint angles for human hand muscles, and relate their results to surgical planning.

3.2.4 Multiarticular Muscles

If the shoulder joint in our idealized biceps muscle of Figure 1.6 also rotates, the muscle is *biarticular* (i.e., it crosses two joints) if one considers the actual biceps in three-dimensional (3D) space, with its origin sites on the scapula and its insertion primarily on the radius bone (which also rotates axially about the forearm), it is triarticular. Such muscles spanning a number of joints are commonly termed *multiarticular* (or *polyarticular*). They are quite common, and as addressed in the book edited by Winters and Woo (1990), appear to serve many functional purposes (Crisco and Pangabi 1990; Gielen et al. 1990; Hogan, 1990; van Ingen Schenau et al. 1990). Some muscles are a mixture: the triceps (arm) and quadriceps (leg) each have one biarticular head, with the others uniarticular.

3.3 Relating Muscles: Idealized Terminology

The following terminology is often used to help characterize how muscles work together:

- *Synergists*: muscles working together to cause similar actions.
- *Antagonists*: muscles causing reciprocal actions (whatever that means!).

In anatomy textbooks certain muscles are often defined as synergists or antagonists solely based on anatomical location. Within the biomechanics community this has been criticized, and in recent years the concepts of a "synergist" or "antagonist" has gone through a state of flux, for two reasons: (1) there are many experimentally documented cases where muscles are synergists for some tasks yet antagonists for others (or even for different parts of the same task); and (2) theoretical and mechanical analysis tell us that due primarily to inertial dynamics in multisegmental systems, discerning the roles of muscles can be counterintuitive, and indeed muscles can affect motion in segmental links that they do not even cross (e.g., see Zajac and Winters 1990). We will develop this concept further in Section 3.6.

3.4 Linkage and Joint Kinematics

We start by defining the term kinematics:

- *Kinematics*: the study of motion without regard to the forces causing the motion.

3.4.1 Segmental Rigid Bodies

The previous subsection used a simple 1-DOF hinge joint to develop key terminology associated with muscle-joint attachment, and hinted at multi-DOF systems when we noted that the shoulder could also move. Here the foundations of multi-DOF mechanical linkage systems are developed by starting with the assumption that human or animal skeletal structures can be viewed as consisting of a set of *rigid bodies* (inflexible links) joined together by joints; Figure 1.7 provides an example in which there are two segments. Furthermore the mass of these rigid bodies is the sum of that of the entire segment (i.e. a lumped collection of bones,

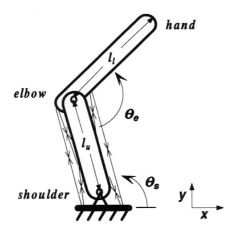

FIGURE 1.7. Idealized two-link planar linkage of the human arm (top view), idealized uni- and biarticular muscles shown as thin lines with inward-directed arrows.

muscle, etc.). Of course, there are exceptions, such as the hydroskeleton of the medicinal leech discussed in Chapter 14, in which alternative assumptions can and should be made.

3.4.2 Kinematic Chains

A *serial chain* consists of a series of links, as in Figure 1.7 or 1.8A; this is a common arrangement within the body (e.g., models of the arm or the head-neck).

An *open kinematic chain* occurs when at one end of a serial chain there is an open (unattached) distal link, such as the hand or head. The number of kinematic DOF equals the number of joint coordinates.

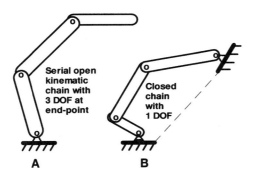

FIGURE 1.8. Examples of open (A) and closed (B) kinematic chains.

A *closed kinematic chain* occurs when the serial chain closes on itself (i.e., forms a loop). In such a case the number of DOFs is reduced; for instance, in Figure 1.8B, if we assume that the pelvis (top right joint) attaches to the bicycle seat, knee and hip are 1 DOF hinge joints, the ankle does not rotate, and the lowest joint is the bicycle crank, then the result is a *four-bar linkage* with only 1 DOF: if we know the angle of the bicycle crank, we literally can determine the orientation of the knee and hip, as well as the locations of the pedal-foot interface, the ankle (here within the rigid body) and the knee (here a joint). Notice that for closed chains there are fewer DOFs than joints that rotate. If the ankle is assumed to be able to rotate as a hinge, the result would be a five-bar mechanism with 2 independent DOFs. Other classic examples of closed kinematic chains for humans include the legs during quite standing (on two feet) and during the double support phase of walking.

3.4.3 Planar Versus Spatial Analysis

The above analysis assumes that all joints rotated about idealized hinge joints with an *axis of rotation* that is orthogonal to the page. In such a case the analysis is performed in a 2D view; this is called a *planar* analysis. The relative motion of one segment with respect to another segment, via a generalized "joint," can then be described by three terms: one rotation and two translations (often these translations are constrained).

In a more general 3D *spatial analysis, a generalized joint can consist of up to 6 terms: 3 rotations and 3 translations.* Similarly, it also takes 6 coordinates (3 translation, 3 rotation) to completely characterize the location of a rigid body in 3D space. We normally assume that joint translation is ideally *constrained* (e.g., by ligaments and bony constraints), so as not to move in that DOF; in reality it is *restrained* by such tissues to only move a small amount. For example, an idealized spherical joint (e.g., the hip) consists of 3 rotational kinematic DOFs, with the 3 translations constrained by the ball-socket arrangement.

3.4.4 Position Analysis

In motion analysis one often wants to describe the location of a point on a rigid body, for instance a link center of mass or a muscle attachment site, with respect to a point on another link. Mathematically, matrix methods are typically used to concisely describe this relation, in one of two forms:

$$p_p = R_j(p_{po} - p_j) + p_j; \qquad p_p = D_j p_{po} \quad (1.9)$$

where p_p is the point we want to know, p_{po} is its prior position, p_j is the joint location, R_j is the 3 × 3 relative *rotation matrix* (that depends on joint angles), and D_j is the 4 × 4 *displacement matrix* (that includes both rotation and translation terms, in a concise form). There are many ways to characterize the relative rotations, for instance in standard *x-y-z* Cartesian angles, Euler angles, or the screw (helical) axis, and so on; adding to the possibilities is that the order of rotation matters! Thus one research teams' definition of "Euler angles" may subtly differ from that used by another. The "best" way to characterize rotations tends to be somewhat joint-specific, and remains controversial within the movement biomechanics community despite years of effort by committees trying to establish consensus standards.

3.4.5 Velocity Analysis

As already noted, kinematics involves the study of motion (i.e., not just quantifying the configuration or position of the linkage), but also its derivatives, the first three of which are called *velocity, acceleration,* and *jerk*). For multilink systems, the relations that map joints to other locations, such as an end point, quickly get quite involved, especially for acceleration analysis, due to "cross-talk" terms (e.g., velocity products). Terminology associated with more advanced analysis include the tangential, centripetal, and Coriolis components of acceleration. The details are described in countless textbooks, and beyond our present scope.

However, it is useful to introduce the type of framework that is associated with velocity analysis. In concise mathematical terms, the velocity of a point on a link (p_p) is:

$$\dot{p}_p = W_j(p_p - p_j) + \dot{p}_j \quad (1.10)$$

where one assumes one already knows the velocity of the point at the idealized joint, and W_j is the antisymmetric (mirrored off-diagonal terms of same magnitude but opposite sign, on-diagonal terms zero) angular velocity matrix (2 × 2 for planar case, 3 × 3 for spatial case). This is essentially just a matrix representation of the classic vector cross product $v = \omega \times r$.

3.4.6 Differential Transformations and Mappings

In multi-DOF kinematics, one should be able to describe the mapping between a set of joint coordinates x_j and some global coordinates of interest, such as hand end-point coordinates x_e. In such a case, one can again describe the mapping via an appropriate differential transformation (Jacobian) matrix:

$$dx_e = J_{ej}dx_j; \qquad v_e = J_{ej}v_j \quad (1.11)$$

where this same Jacobian relation has been used to describe both differential position change (dx_e, dx_j) and velocity mapping (v_e, v_j). This is a local mapping in that the values within the matrix are sensitive to the configuration. Thus, while this relation describes a linear relationship between x_j and x_j in the local region, J_{ej} itself depends nonlinearly on the kinematic configuration. Notice also that as with J_{mj}, the Jacobian J_{ej} is in general rectangular rather than square. This shows that there is *kinematic redundancy*: more generalized joint coordinates n_q than end point coordinates n_e. Dozens of research papers are devoted to this topic, especially as related to the question of how the brain plans and controls arm reaching (e.g., see Flash 1990; Chapter 23; Chapter 27). Many argue that the brain circuitry must be involved in calculating this kinematic mapping.

Within the research community, the arm is often approximated as planar, with simple 1-DOF hinge joints for the shoulder and elbow (see Figure 1.7), in part because the Jacobian is then square, and invertible. It is instructive to show how the 2 × 2 J_{eq} is obtained for this case (see also Gielen et al. 1990): one simply determines x_e in terms of x_q:

$$x_e = \begin{bmatrix} l_u\cos(\theta_s) + l_l\cos(\theta_s + \theta_e) \\ l_u\sin(\theta_s) + l_l\sin(\theta_s + \theta_e) \end{bmatrix};$$

$$x_j = \begin{bmatrix} \theta_s \\ \theta_e \end{bmatrix} \quad (1.12)$$

where l_u and l_f are the lengths of the upper arm and forearm, respectively. We then obtain by differentiation:

$$J_{ej} = \frac{\partial x_e}{\partial x_j}$$

$$= \begin{bmatrix} -l_u\sin(\theta_s) - l_f\sin(\theta_s + \theta_e) & -l_f\sin(\theta_s + \theta_e) \\ l_u\cos(\theta_s) + l_f\cos(\theta_s + \theta_e) & l_f\cos(\theta_s + \theta_e) \end{bmatrix}$$

$$(1.13)$$

Here J_{ej} is square and invertible because there are only 2 joint DOFs, but in general J_{ej} is not square because $n_j > n_e$, and then only a "pseudoinverse" can be obtained.

3.4.7 Input–Output Kinematics

Generally speaking, the following are defined based on what we know relative to what we are after:

- *Forward kinematics*: we know the joint coordinates (angles) and link anthropometry, and want to calculate strategic positions (e.g., we could use Eq. 1.12 directly).
- *Inverse kinematics*: we know segmental positions and want to find joint angles.

Typically inverse kinematics is required when using modern 3D motion analysis systems, and in such a case one really need not actually solve the above equations, but rather use straightforward methods (numerical differentiation and trigonometry) to determine joint kinematics. A more challenging form of inverse kinematics occurs when only the endpoint coordinates are known, and one must additionally determine a viable geometric configuration. This problem is common in robotics, as well as in upper limb human movement studies.

3.5 Elastostatic Systems: Equilibrium and Stability

When a human is not moving, the problem to be solved reduces to that of: (1) solving for *static equilibrium*: sum of forces and/or moments equal zero; and (2) making sure that the equilibrium is *stable*.

3.5.1 Static Equilibrium

The classic technique for solving for static equilibrium is to isolate a system (determine a "free body diagram") and then utilize the equations for static in the form:

$$\Sigma F = 0; \qquad \Sigma M = 0 \qquad (1.14)$$

where one solves for the unknown component forces and/or moments within F and M, respectively.

In *quasistatic* analysis, one ignores velocity and acceleration terms (i.e., sets the velocities and accelerations to zero) in performing the calculations. Much of human movement is performed at moderate speed, and for all intents and purposes is essentially quasistatic (Hogan and Winters 1990). Indeed, in a study of gait (walking) in older adult females with arthritis (Fuller and Winters 1993), it was difficult to distinguish between inverse dynamic and quasistatic analysis of the same data, with the joint moment curves overlapping for all but small regions of the gait cycle.

For a *statically determinant* system, the number of *unknowns equals the number of equations*, and consequently a unique solution is possible. Often in musculoskeletal systems the number of unknowns (e.g., muscle forces, joint contact forces) exceeds the number of independent equations—the problem is then *statically indeterminant*. There are several ways to approach such problems. One approach is to make some additional assumptions (e.g., no co-contraction across a joint) in order to obtain a solution—this is equivalent to adding enough constraint equations (usually relying on heuristic common sense) such that the total number of equations equals the number of unknowns. Another option is to use static optimization, that is, use a "cost" function to essentially select one solution from among the viable solutions (see Chapter 32).

3.5.2 Stable Equilibrium

A problem with the above approaches *is that the equilibrium solution that is determined may not be stable*. In other words, if we were to perturb a certain configuration slightly from equilibrium, it might not come back, but rather drift away. This is especially a problem with inverted pendulum systems, such as the upright human head shown in Figure 1.9. This problem becomes readily apparent when using an alternative approach toward solving mechanics problems—one that is based on differentiating the total potential energy (E_{pe}) within a system with respect to the generalized joint coordinates x_j (a minimal set of coordinates that fully characterize the configuration of the system). E_{pe}, as scalar, is the sum of all elements that store potential energy, i.e.,

$$dE_{pe} = \sum_{i=1}^{n} F_i^T dx_i \qquad (1.15)$$

where in our case there are two types of energy-storing elements:

- Steady *conservative forces* (e.g., the locations of the center of masses p_{cm} and their respective values within the Earth's gravitational field), and

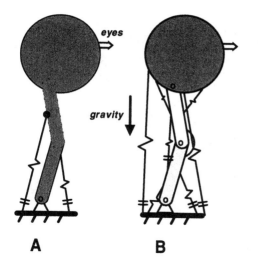

FIGURE 1.9. Inverted pendulum models, loosely based on a head-neck system. (A) Simple (single-joint) inverted pendulum with two idealized muscle springs with variable stiffness and rest length. (B) Compound (three-joint) inverted pendulum, here showing 6 uniarticular and 2 triarticular muscles.

• *Spring-like* elements (e.g., muscles, ligaments), where over a local region a change in length will cause a change in force, and thus a change in stored potential energy.

To solve for the potential energy of the steady forces due to the masses, with respect to the joint coordinates x_j, a "geometric compatibility" mapping must be done via the Jacobian J_{gj} (see also Eq. 1.11), resulting in:

$$\frac{\partial^2 E_{p-gravity}}{\partial x_j{}^2} = \frac{\partial(F_g J_{gj})}{\partial x_j} = -F_g \frac{\partial J_{gj}}{\partial x_j} \quad (1.16)$$

Muscle is locally spring-like by virtue of the common finding that, for a given level of excitation drive, its force changes when it is perturbed (e.g., see Chapters 2, 7–10). The local stiffness k_m is defined as its change in force divided by its change in length, and consequently its change in potential energy with respect to x_j is a bit more complex:

$$\frac{\partial^2 E_{p-muscle}}{\partial x_j{}^2} = \frac{\partial((k_m dx_m)J_{mj})}{\partial x_j} \frac{dx_m}{dx_j}$$
$$= J_{mj}{}^T K_m J_{mj} - F_m \frac{\partial J_{mj}}{\partial x_j} \quad (1.17)$$

As seen conceptually in Figure 1.9, for an equilibrium to be stable the following must be true:

$$\frac{\partial E_{pe}}{\partial x_j} = 0; \qquad \frac{\partial^2 E_{pe}}{\partial x_j{}^2} > 0 \quad (1.18)$$

Intuitively, this implies that with perturbation, more potential energy must be gained (e.g., via stretched springs) than lost (e.g., via falling masses), that is, there must be a potential energy "bowl" in the higher-order E_{pc}-x_j space.

It turns out that for the simple inverted pendulum elastostatic system of Figure 1.9A to be quasistatically stable, the overall joint stiffness k_j must be higher than a certain critical value, i.e. $k_j > F\,h$, where F is the weight of the head-neck system and h is the height of its center of mass with respect to the joint.

Assuming muscles must be the source of most of the joint stiffness (usually the case) and assuming a constant moment arm, and noting the relation between muscle and joint stiffness:

$$K_j = J_{mj}{}^T K_m J_{mj} \quad (1.19)$$

where the Jacobian J_{mj} is often called the moment arm matrix. For our simple example we see that k_j is hypersensitive to the muscle moment arm r_m.

Often the necessary joint stiffness goes up as the necessary joint moment goes down, and during such conditions one would expect there to be co-contraction.

Often one defines a more general "apparent joint stiffness" K_{pe} that includes the influence of both stiffness-based and conservative-force-based terms (Hogan 1990). The critical value of the "apparent stiffness" K_j is greater than zero whenever the mass is above the joint (or more generally, the conservative force has a component directed toward the joint).

$$K_j = J_{mj}{}^T K_m J_{mj} - F_m \frac{\partial J_{mj}}{\partial x_j} \quad (1.20)$$

While the mathematics can become trickier for the "apparent joint stiffness" of more realistic larger-scale systems, the basic concept is quite general. For the system to be stable, K_j must be positive definite, i.e., the eigenvalues of this symmetric matrix must all be positive:

$$K_j = \frac{\partial^2 E_{pe}}{\partial x_j{}^2} > 0 \quad (1.21)$$

3.4.3 End-Point Stiffness

Another stiffness mapping that has been of interest within the research community, especially for studies of upper arm movements, is the relation between the joint stiffness and that at the endpoint (e.g., hand) K_e. This latter stiffness can often be estimated experimentally. If one associates a n_j x n_j linear stiffness matrix K_j to represent the joint stiffnesses (off-diagonal terms are only due to multiarticular muscles), it maps to a n_q x n_q end-point stiffness matrix K_e by the relation:

$$K_j = J_{ej}^{T} K_e J_{ej} \qquad (1.22)$$

Notice that the dimensions work for matrix multiplication. For the simple planar 2-joint (shoulder-elbow) model of Figure 1.8, one can also calculate the 2×2 K_e in terms of the 2×2 K_j using:

$$K_e = J_{ej}^{-T} K_j J_{ej}^{-1} \qquad (1.23)$$

In this special case, if K_e is stable then K_j is also stable. However, it is important to recognize that for the general case where there is kinematic redundancy, it is not enough to imply system stability by considering only the end-point stiffness K_e because there can be "hidden" joint instabilities.

3.6 Dynamical Equations of Motion

Dynamics (kinetics) is the study of the actual system mechanics: the forces and the motion caused by the forces. It is an expression of Newton's laws: forces cause a mass to accelerate. Both kinematic analysis (Section 3.3) and quasistatic mechanics analysis (Section 3.4) can be viewed as subsets of dynamic analysis.

Before continuing, it is important to clear up some confusion in terminology. Within the field of movement science, a number of researchers have used the term "dynamic systems" to refer to the use of generally simple models consisting of coupled differential equations, usually used to approximate the behavior of human systems with an inherent tendency to oscillate (e.g., finger tapping). This represents a small subset of the larger field of dynamic systems. Indeed, the focus of this section, on dynamic systems consisting of a collection of interconnected rigid bodies representing biomechanical linkages, is also a subclass of the broader concept of dynamic systems.

3.6.1 Mass Moment of Inertia

For the usual assumption of skeletal rigid bodies attached via idealized joints, some additional information needed for a dynamic analysis: the mass m, center of mass vector p_{cm}, and the mass moments of inertia (with units of mass*length^2) of each rigid body. For a planar case, the mass moment of inertia I_{cm} relative to the center of mass is a single scalar quantity (one for each link), while for the spatial case I_{cm} is a 3×3 tensor matrix for each link.

3.6.2 Anthropometric Tables

Given the complexity of human and animal structures, obtaining I_{cm} can itself be challenging. Fortunately, many research groups have addressed this problem, and typically I_{cm} values for human skeletal links are estimated by using *anthropometric* tables, normally expressed either via regression equations (in terms of easily obtained metrics such as height, weight and length between key skeletal landmarks) or as 5%, 50%, and 95% percentile of normal. Until relatively recently, most of the available data was for young adult males, and "normal" meant young adult males; this is changing rapidly.

3.6.3 Forms of Equations of Motion

This subsection first briefly develops the dynamic *equations of motion* for 1 DOF systems, and then for general equations. For a 1 DOF hinge such as in Figure 1.6 or 1.9A, say with two antagonistic muscles (flexor and extensor), a single moment equation is sufficient to characterize behavior, which can be obtained by summing about either the joint or the link center of mass. Summing about the joint, we have:

$$\begin{aligned} \Sigma M_p = f_{fl} r_{fl} - f_{ex} r_{ex} &= I_j \alpha \\ &= (I_{cm} + m l_{cm}^2)\alpha \end{aligned} \qquad (1.24)$$

where the mass moment of inertia about the joint, I_j is obtained by knowing I_{cm} and then applying the parallel axis theorem of mechanics (m is the segment mass, l_{cm} the length from the joint to the center of mass), and the scalar moment arms r_{fl} and r_{ex} are already the result of the moment arm evaluation process discussed earlier (notice their opposite sign, where here one assumes the flexor moment is positive). A free-body diagram of a single limb sequent (Figure 1.10) shows the forces and moments that must be included in formulating the equations of motion.

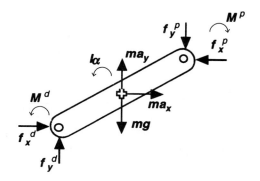

FIGURE 1.10. An isolated free body diagram for a 2D link, with joints at the distal (lower left) and proximal (upper right), and with f the force coordinates, M the joint moments, ma the mass times translational acceleration, mg the gravitational force, and $I\alpha$ the moment of inertia times the angular acceleration.

A useful form for the general linkage equations of motion is (Pandy 1990):

$$q = M(\theta)^{-1}[C(\theta)\theta^2 + G(\theta)$$
$$+ J R(\theta)f^T + T(\theta,\dot{\theta})] \quad (1.25)$$

where $M(\theta)$ is the system mass matrix, $C(\theta)$ is the vector describing the Coriolis and centrifugal effects, $G(\theta)$ is a vector containing only gravitational-based termed, J transforms joint torques into segmental torques, $R(\theta)$ is the moment arm matrix, and $T(\theta,\dot{\theta})$ is the vector of any externally applied torques.

3.6.4 Inverse Versus Forward Dynamics

In inverse dynamics problems (Figure 1.11A), the motion is known, for example, through position data from a 3D motion analysis system that is then differentiated appropriately (inverse kinematics) to estimate link translational and rotational velocities and accelerations), and generalized joint forces (typically mostly joint moments) are estimated by solving algebraic equations (i.e. solving from right to left, with everything on the right known). This is straightforward. In some case the joint moments are distributed to muscles crossing the joint, typically by heuristic assumptions or by inverse optimization techniques (see also Chapter 32).

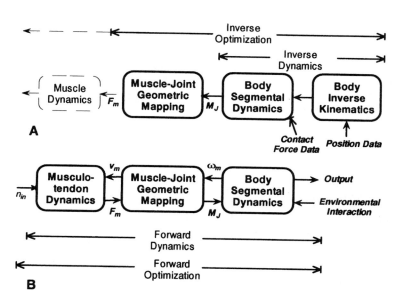

FIGURE 1.11. (A) In inverse dynamics, the goal is usually to obtain joint torques, given motion, and contact force data. The process involves solving the equations of motion in an algebraic form, and often it can be done in a distal-to-proximal fashion so as to avoid solving simultaneous equations. Distributing these to muscle forces involves using either heuristic constraints or inverse optimization. (B) For for-ward dynamics, states associated with the dynamical equations of motion and muscle dynamics must be integrated over time. Forward optimization is the "real" optimal control problem since cause–effect evolves over time, and identifying optimal inputs and/or feedback parameters so as to achieve the goals of a certain task is genuinely challenging for all but very simple models (see Chapter 32).

In *forward dynamics* (also called *direct dynamics*), the motion is not known in advance, and the equations of motion are integrated over time to determine the motion, given muscle forces or joint movements. This is considerably more challenging, and computationally intensive than inverse dynamics, for large-scale systems, often by orders of magnitude. There are three especially common methods for obtaining the terms for such equations: *Newton-Euler* (essentially the 3D application of Newton's Laws), *Lagrange* (an energy-based approach), and *Kane's Method* (Kane and Levinson 1985). The bottom line is that creating these equations can be error-prone (with any "bugs" difficult to uncover), and solving these questions can be computationally intensive. Normally researchers use canned computer packages that are dedicated to allowing users to interactively create a model, and then create and solve the equations (see Table 8.1). However, for special cases such as serial chains, recursive methods (initially developed for robotics) are available that can dramatically speed up the process.

If a task goal is specified, the process of determining a control strategy to best meet this goal is called *forward optimization*. This is the problem that the CNS must solve, since our motions must evolve within a world in which the laws of physics apply in a forward mode!

4   Neuromotor Physiology: Terminology and Function

Many textbooks and review articles are available on this topic; the purpose here is to briefly describe some of the highlights of current knowledge of neuromotor physiology, with a focus on terminology associated with neuromotor function, and on system design features.

4.1   Information Processing Via Collections of Neurons

The brain of the human and higher animals consists of billions of neurons (nerve cells). A typical neuron has a soma (cell body), a dendritic tree, and an axon (sometimes quite long). The axon ends with enlarged structures called boutons or synaptic terminals.

4.1.1   Unidirectional Synapses

For most neurons information is transferred via *chemical synapses*, across which information is transmitted unidirectionally via the release of a neurotransmitter substance from the upstream cell that then is subsequently collected by the downstream cell, a process that takes about 1 ms. Such *unidirectional connections* are the sign of a specialized system that is more focused on *information encoding and processing* and control, rather than on bicausal energetics.

4.1.2   Neuron "Summation" and Action Potentials

Generally, a neuron receives a wealth of converging information from other neurons via connections to its dendritic tree. It "synthesizes" (or "integrates") this converging information (via graded electrotonic spread) at a summing area that is normally near where the soma and axon meet (the axon hillock) by generating "all or none" *action potentials* whenever an electric threshold is reached. These action potentials, which do not attenuate as they actively propagate along the axon (which functions like an electric cable), finally pass on this information by releasing neurotransmitter at its boutons (to be received by another cell via the synapse). The speed of the active spread along an axon "cable" is a function of its size, and can reach over 100 m/s for myelinated axons with large diameters.

4.1.3   Firing Rate

The excitation level of a given neuron is essentially captured by its *firing rate* (i.e., *the number of action potentials sent down its axon per unit time*). This book is less interested in the details of this reasonably well-understood process, which is adequately described by the famous Hodgkin-Huxley equations (Hodgin and Huxley 1952). A typical neuron, for instance, might start firing at 8 Hz once its excitation "*threshold*" has been reached. It then fires at a rate that increases as the excitation drive increases (see also the "synaptic current" concept of Chapter 4), until it finally saturates at a maximal firing rate (e.g., 100 Hz).

4.1.4 Classification of Neurons

Neurons can be broadly classified into three types, depending on how they interface:

- *Sensory neurons* (SNs): are specialized cells with endings that are embedded in tissue outside of the CNS such that, in response to an incoming "stimulus" within this periphery (e.g., stretch of sensory endings), generate a graded electric potential that often becomes, via an encoding process, an action potential that then propagates along its axon.
- *Motor neurons* (MNs): represent the "final common pathway" from the CNS to the muscles, because every part of the CNS involved in the control of movement must do so by acting directly or indirectly on MNs (Liddell and Sherrington 1925). There is a large degree of convergence onto MNs; Figure 2.3 provides a schematic that shows some of the key channels.
- *Interneurons* (INs): all neurons that are neither SNs nor MNs (i.e., all neurons within the CNS). This is by far the more prevalent type of neuron, and they come in a wide variety of sizes, shapes, and properties.

This book rarely considers individual neurons, but instead focuses on their aggregate behavior, especially the orderly behavior of groups of MNs and SNs that innervate a given muscle; for example, Figure 4.1 provides an indication of effective synaptic drives by various general pathways onto MNs.

4.2 MNs: A Structural Design Perspective

4.2.1 MNs Send Commands to Muscle Fibers

In all but the most primitive organisms, mechanical output is generated by specialized *muscle cells*. We are interested in striated skeletal muscle (the other types are cardiac muscle and smooth muscle). Skeletal muscle is subdivided into parallel bundles of string-like fascicles, which are themselves bundles of even smaller string-like multinucleated cells called muscle fibers. A typical muscle cell (more commonly called a muscle fiber) has a diameter of 50 to 100 microns and a length of 2 to 6 cm. A typical muscle consists of many thousands of muscle fibers working in parallel; their individual forces thus sum to give the overall force.

4.2.2 MN-to-Muscle Mapping

A typical muscle is controlled by approximately a hundred large MNs, whose cell bodies lie in distinct clusters (*motor nuclei*) within the spinal cord or brain stem; often this set of cells is called the muscle's *motor pool*. The axon of a given MN travels out through the ventral root or a cranial nerve and through progressively smaller branches of peripheral nerves until it enters the particular muscle for which it is destined. There it branches to innervate typically 100 to 1000 muscle fibers that are scattered over a substantial part of the whole muscle; muscles in different parts of the body vary greatly in this regard. Each muscle fiber is innervated by one motor neuron in one place, typically near its midpoint. This ensemble of muscle fibers innervated by a single MN is called a *muscle unit*, and this muscle unit plus its MN is called the *motor unit*. MNs innervating one muscle are usually clustered into an elongated motor nucleus within the spinal cord that may extend over 1 to 3 segments.

The functional connection between a MN and one of its target muscle fibers is a type of chemical synapse commonly called the *motor end-plate* or *neuromuscular junction*. End-plates are usually clustered into bands that extend across some or all of the muscle. This synapse, which consists of many presynaptic boutons (each with a supply of presynaptic cholinergic vesticles), is designed so that each individual MN action potential releases sufficient transmitter to depolarize the postsynaptic membrane of the muscle fiber to its threshold, thus initiating its own action potential. The acetylcholine released from the presynaptic terminals is rapidly hydrolized by acetylcholinesterase, leaving the system ready to respond to a subsequent action potential.

4.2.3 Types of Muscle Fibers

Most mammalian muscles are composed of a mix of at least three different fiber types; indeed, muscles composed of nearly homogenous fiber type are the exception. The mechanical capabilities of each type of muscle stem from differences in the structures and metabolic properties of different types of muscle fibers. Within the muscle, all muscle fibers within a given motor unit are of the same type.

The fibers in "red meat" are predominantly *slow-twitch* (S) fibers because the force that they produce in response to an action potential rises and falls relatively slowly (e.g., with an isometric twitch response taking 100 ms or so to rise to a peak, and even longer to fall back near zero force). Muscles composed mostly of such *Type 1* fibers can work for relatively long periods of time without running down their energy stores. This fatigue-resistance (FR) results from their reliance on oxidative catabolism, whereby glucose and oxygen from the bloodstream can be used almost indefinitely to regenerate the ATP that fuels the contractile process: they are aerobic workhorses. In order to support such aerobic metabolism, these muscle fibers must have: (1) a rich extracellular matrix of blood capillaries; (2) large numbers of intracellular mitochondria with high levels of oxidative enzymes; and (3) high levels of myoglobin, a heme protein that helps absorb and store oxygen from the blood stream. While individual slow-twitch muscle fibers produce less contractile force than fast twitch fibers because they are smaller, in terms of stress (force per unit area) their force-generating capability is about the same.

The fibers within "white meat" are primarily *fast-twitch* in that their force response rises and falls more rapidly. They also tend to have a different form of myosin that possesses cross-bridges that produce force more effectively at rapid shortening velocities. Fast-twitch fibers are roughly categorized into two subtypes, depending on their metabolic processes and fatigue-resistance: (1) *fast-fatigable* (FF) or *Type 2B* fibers which rely almost exclusively on anaerobic catabolism to sustain force output and that possess relatively large stores of glycogen (which provides energy to rapidly rephosphorylate ADP as the glycogen is converted into lactic acid—this source runs out fairly quickly, and full recovery may take hours); and (2) *fast-fatigue-resistant* (FR) or *Type 2A* fibers that combine relatively fast twitch dynamics and contractile velocity with enough aerobic capability to resist fatigue for minutes to hours.

In reality this classification is a simplification; there is really more of a continuum of fiber attributes.

4.2.4 MN Size and Conduction Velocity Maps to the Number and Type of Its Muscle Fibers

MNs that control fast-twitch muscle fibers tend to innervate relatively large numbers of these fibers (e.g., 1000), and the MNs themselves have relatively large cell bodies and large-diameter axons that conduct action potential at higher speeds (e.g., 100 m/s). MNs controlling slow-twitch (Type 1) muscle fibers are smaller, slower and innervate smaller numbers of thinner muscle fibers, resulting in slower force output. As might be expected, FR MNs and muscle units tend to be intermediate in size, speed and force output.

Can fiber composition change? This question, of special interest for study of athletic performance and of functional electrical stimulation, has been asked many times, and is briefly addressed in Section 7 in discussion of tissue remodeling, as well as in Chapter 42).

4.2.5 Motor Units, MN Size, and Orderly Recruitment

In the 1960s Henneman and colleagues introduced the concept of *orderly recruitment* of motor units, often called the *size principle*, which states that motor units are recruited in order of increasing size (Henneman et al. 1965, Henneman and Mendell 1981). When only a small amount of force is required from a muscle with a mix of motor unit types, this force is provided exclusively by the small S units. As more force is required, FR and FF units are progressively recruited, normally in a remarkably precise order based on the magnitude of their force output. This serves two important purposes: (1) it minimizes the development of fatigue by using the most fatigue-resistant muscle fibers most often (holding more fatguable fibers in reserve until needed to achieve higher forces); and (2) it permits equally fine control of force at all levels of force output (e.g., using smaller motor units when only small, refined amounts of force are required).

Smaller MNs have a smaller cell membrane surface area, resulting in a higher transmembrane resistance (R_{high}). Because of Ohm's law ($E = I\,R$, where E is voltage, I is current, R is resistance), for a given synaptic current, smaller MNs produce a larger excitatory postsynaptic potential (EPSP), which reaches threshold sooner, resulting in an action potential. Larger MN may not be recruited: given a larger surface area (and thus lower overall transmembrane resistance), there may only be a subthreshold EPSP in response to the synaptic current input.

If one assumes a common drive to the motor pool (i.e., a uniform synaptic current to MNs within this pool), and assume different MN sizes within this pool, as the amount of net excitatory synaptic input to a motor nucleus increases, the individual MNs will reach threshold levels of depolarization in the order of their increasing size, with the smallest firing first—the *size principle of orderly MN recruitment*.

While the normal overall sequence of recruitment is S → FR → FF (with some overlap between types), recruitment has also been found to be orderly within the type categories. Such orderly recruitment is fairly robust, and has been seen for input drives ranging from transcortical stimulation to reflex-initiated excitation (see Binder et al. 1996 for review). However, the effect may be modified by systematic differences in the relative numbers and locations of synapses from a given source onto MNs of different unit types, usually related to dynamic task needs.

About half of the total surface area of the dedritic tree and the soma are covered by synaptic boutons, with relatively equal density on the dentrites, soma and axon hillock (Binder et al. 1996). A typical MN is contacted by about 50,000 synaptic boutons representing about 10,000 presynaptic neurons; the vast majority of these connections are made upon the dendrites, which collectively account for 93 to 99% of the total cell surface area (Binder et al. 1996), yet because many of these are farther from the axon hillock, their relative influence is smaller. MN recruitment and rate modulation depend on an integration of this barrage of converging synaptic current.

When a MN is depolarized just over its threshold for the initiation of action potentials, it tends to fire at a slow, regular rate (5–10 Hz), resulting in a partially fused train of contractions in its client muscle fibers. As its depolarization is increased by more net excitatory synaptic input, its firing rate increases. The mean level of force output increases greatly over this range of firing, saturating at a value that can be over 100 Hz (generally the smaller the MN, the lower the value). Simultaneously, other slightly larger MNs reach their thresholds for re-cruitment, adding their gradually increasing force levels as well. Because the relative timing of the individual action potentials in the various motor units is normally random and asynchronous (in a nonfatigued muscle), the various unfused contractions of all of the active motor units blend together into a smooth contraction.

It is this built-in smoothness that allows modelers to assume a lumped neuromotor drive to a lumped "macrosacromere"-tendon unit. Yet it should be remembered that the overall force depends on both the number and size of active muscle units (*recruitment*) and their individual *firing rates*. In a typical muscle, the largest MNs are not even recruited until the muscle has generated about 50% of its peak force capacity.

There are several ways to study electrical excitation and properties of MNs. Chapter 4 estimates the *effective synaptic current* (I_N) in MNs by injecting current into an identified source of synaptic input and then recording the subsequent current required to voltage-clamp the membrane at the resting potential. One can then calculate I_N at the threshold for repetitive discharge, and then once combined with the slope of the steady-state firing frequency-current relation, predict the effect of a given type of converging synaptic drive on steady-state MN firing behavior.

A key challenge in the area of functional electrical stimulation (FES) is to deal with the challenge of artificially exciting muscle without having the luxury of automatic orderly recruitment; fatigue in particular becomes an issue (see Chapter 42 and Chapter 43).

4.2.6 Orderly Recruitment in Multifunctional Muscles and the Concept of Task Groups

The concept and definition of a "motor pool" has been debated for many years. Are fan-like muscles, such as the trapezius or deltoid, best treated as one or multiple muscles? Often muscles receive MNs that exit from several spinal levels, and often certain muscles act as clear synergists for certain tasks and yet not for others. Loeb (1984) has noted that functional groupings of MNs during movements do not necessarily coincide with traditional anatomical boundaries between muscles (earlier called the motor pool) and has suggested categorization by *task group* subpopulations. An important component of the task group hypothesis is orderly recruitment of motor units within subpopulations; further research is needed in this area.

4.3   Muscle Fibers and Machinery

This relatively brief section is intended to complement Section 2 of Chapter 2, which covers the contractile mechanism in more depth. Here, the focus is on a design-oriented perspective.

4.3.1   Action Potentials Along Muscle Fibers: Slower Than for MNs but with Larger Electric Potentials

Once the postsynaptic membrane of the neuromuscular junction is depolarized to the threshold, an action potential propagates along the sarcolemma (muscle fiber cell membrane) in both directions away from the end-plate region. However, the speed along the muscle fiber is much slower that for most MNs, only 1 to 5 m/s. (Electrically, a muscle fiber is similar to a large diameter axon without a myelin sheath, and requires high transmembrane currents to propagate its action potential, giving rise to relatively large potential gradients in the extracellular fluids around the muscle cells.)

Normally, when more than minimal muscle force is required, many motor units are involved, resulting in an overlapping barrage of action potentials. This results in a complex pattern of electrical potentials (typically on the order of 100 μV) that can be recorded as an *electromyogram* (EMG) by electrodes which may be on the surface of the overlying skin or within the body. The relative timing and amplitude of these patterns provides a reasonably accurate representation of the aggregate activity of the motor neurons that innervate each muscle. Many chapters within this book make use of EMGs. It is important to recognize that the EMG is a measure of this active electrical spread (typically an integration of that from many fibers), rather than a direct measure of muscle force or activation.

4.3.2   Contractile Machinery Organization: Sarcomeres, Filaments, Cross-Bridges, and Filament Overlap

A single muscle fiber contains bundles of myofibrils, each with a regular, repeating pattern of light and dark bands. These stripes or striations change their spacing as the muscle contracts or is stretched. The longitudinal repeat in the patterns, typically defined as the length from Z-disk to Z-disk, is called the *sarcomere*; its physiological range is about 1.5 to 3.5 μm, or about 3000 sarcomeres in series for every centimeter of muscle fiber length. In models it is normally assumed that these are homogeneous building blocks, and that all are stretched equally; in reality, this is not quite true, and Chapter 6 (Huijing) discusses some of the implications.

This banding pattern is the result of partial overlap between an interspersed arrangement of thick and thin fibrillar proteins. The *thin filaments* project in both directions from thin, transverse Z-disks. *Thick filaments* are discontinuous, floating in the middle of the sarcomere; the midpoints of the adjacent thick filaments are aligned at the center of the sarcomere (M line). The main constituent of each thin filament is a pair of polymerized *actin* monomers (F actin) arranged as a helix; it also contains two other proteins, tropomyosin (a long filamentous protein lying within grooves within actin) and tropomyosin (which attach to tropomyosin at regular intervals). The thick filaments are made up of about 250 *myosin* molecules, each consisting of two identical subunits: a globular head and a 0.15 μm-long tail.

Each globular myosin head contains an ATPase that converts the chemical energy of ATP into mechanical energy, resulting in a "cocked" deformation of the myosin head. This stored mechanical energy is released when the myosin head attaches to a binding site on one of the adjacent actin filaments that has been activated by calcium (see Chapter 2). The English team that first noted this behavior were quick to use the analogy of an attached head that acts much like an oar during rowing (Huxley 1957), pulling the actin (and its load) longitudinally in a direction that increases the overlap between the two filaments, thus shortening the sarcomere (and thus muscle fiber). After an excursion of about 0.06 μm of sliding, the cross-bridge can detach, but only if ATP is present. The head is then recoiled so that it is ready to perform another "power stroke" on another actin binding site.

This process, as well as the role of calcium in the excitation-activation process, is discussed in more detail in Chapter 2, and from more of a biophysical modeling perspective in Zahalak (1990).

4.3.3 Force Generation Is a Function of Activation, Length, and Velocity of Each Muscle Fiber

If muscle force was only a function of its activation, muscle would function as a unidirectional filter. The fact that force is also a function of muscle length and velocity is of fundamental importance in this book.

We saw previously that the force-length relation is maximum in a midrange (a so-called *optimum length*), and decreases to either side. Chapter 2 discusses why this happens, in terms of actinmyosin overlap (see Figures 2.1 and 2.2). Chapter 10 provides a rationale for why at very short lengths, the thick filaments collide with Z-discs and produce a pushing force that counteracts an increasing percentage of contractile force. Finally, Chapters 5 and 6 show that this relation is a function of subtle effects such as the history of calcium activation dynamics and muscle fatigue.

As indicated in Section 2, muscles usually work on moving loads, which either may permit them to shorten (concentric action) or may force them to lengthen against their direction of contraction (eccentric action). It is well established that the faster the sarcomeres are shortening and the cross-bridges are cycling, the less force they can produce (see Chapter 2). The shortening velocity at which active force output goes to zero, v_{max}, is a function of fiber composition (see Chapter 8).

In Hill-type models, an equation such as Hill's equation (Eq. 1.5), scaled for activation (see Figure 28.4D) is used to capture the basic effect. Mathematically, this is the equation of a viscosity, with the force "lost" (viscous dissipation) being the difference between the "isometric" (zero-velocity crossing) force and the actual force (Eq. 1.6). This instantaneous "viscosity" is an ongoing function of both activation and velocity, which makes it highly nonlinear (Winters and Stark 1987). There are also many scientists involved in trying to model these basic mechanisms from first principles, and there is a parallel world of muscle modeling that develops such biophysical models (e.g., see Zahalak 1990). This represents a classic example of natural tension between "reductionist" science (focusing on understanding the underlying biophysical mechanisms) and "systems" science (focusing on capturing input–output behavior), which permeates most fields of science.

Occasionally there is true overlap between the modeling approaches; Zahalak's (1981, 1990) Distribution Moment model of muscle is one such example.

It is commonly assumed that the CE force-velocity relationship modifies force output simultaneously and independently of the force-length relationship. Abbott and Wilkie (1953) suggested this approach, and A.V. Hill spent a good deal of his later professional life carefully examining the accuracy of this assumption, coming to the conclusion that it was surprisingly good once the confounding influences of SE and PE are taken into account (Hill 1970). This means that given an activation level, we have a force-length-velocity surface that represents contractile tissue behavior; as the activation changes, the CE relation sculpts force, roughly instantaneously (see also Chapter 10). More generally, under steady conditions (e.g., during the steady part of an isotonic test), given any three of the four measures (activation, force, CE length, CE velocity), we can determine the fourth. A common approximation of this activation-dependent CE surface is to use a product relationship (e.g., see Chapter 8), where the activation signal is first scaled by the CE force-length relation to give the "hypothetical" force, and then this is used as the "zero-velocity crossing" force that defines the CE force-velocity curve; given the ongoing velocity x_{ce}, the force is then read off the CE force-velocity relation.

Of note is the assumption that the CE relation is instantaneous in no way implies that muscle does not possess intrinsic bidirectional dynamic properties, because CE must dynamically interact with SE and PE.

Several comments related to energetics are relevant. First, even though the muscle is producing little force during fast shortening, its rate of energy consumption is very high because each cross-bridge dephosphorylates one ATP molecule for each cycle of attachment and its power stroke. If an active muscle is stretched by an external load, the cross-bridges initially resist being pulled apart with a higher force than they produce isometrically for the same level of activation. Despite the higher force, the rate of energy consumption goes down because the cross-bridges do not lose their cocked state until they complete a power stroke. Cross-bridges that have been ripped away from their actin attachments tend to find another binding site on the actin and continue to contribute to the force resisting muscle stretch (only with a different "reference" length), often somewhat independent of stretch velocity. As noted previously, a muscle operating in this manner is doing what is often called *negative work*, absorbing and dissipating externally applied mechanical energy, which is generally thought to be useful in a task such as a controlled landing from a jump, and for the type of stretch-shortening behavior mentioned earlier.

4.3.4 Muscle Structural Design: Various Sizes and Shapes, with Packing Arrangements

If individual fibers of a muscle extended from bony origin to bony insertion, it would be simple to relate overall muscle length to fiber and sarcomere length, and thus compute the force for a given activation level. In many muscles, however, muscle fibers are tilted at angles (often called a *pennation angle*) relative to the overall line of action, and are attached to sheets or plates of connective tissue (*aponeuroses*) that extend over the surface and sometimes into the body of the muscle. These aponeuroses gather and transmit the force of all of the fibers that insert upon them, typically via a relatively long band of connective tissue called a *tendon*. The total force that can be generated by muscles with this pinnate architecture, per unit mass, is enhanced because the angled arrangement of shorter fibers permits a muscle with a given volume to have more fibers working in parallel.

There is a cost, of course, to this mechanical advantage: a given length change for the whole muscle will represent a larger percentage of the length of each contractile unit, and thus it more likely will function in a suboptimal position on the force-length curve.

There is also a cost in velocity, as for any mechanical transmission design: for the same amount and composition of tissue, an increase in possible force because of an increase pennation causes a roughly proportional decrease in possible velocity, such that the muscle power capability per unit mass is roughly preserved (as seen in Eq. 1.3, the product of force and velocity is power). Thus we see that pennation represents a design tradeoff. In Chapter 3, experimental results are presented that relate measured sarcomere lengths to muscle fiber length and wrist joint angles, for several human muscles.

4.4 Muscle Sensors and Their (Controversial) Role

For about 100 years, since the time of the famous physiologist Sherrington and his many colleagues (e.g., Sherrington 1910; Liddel and Sherrington 1924), there has been an evolving debate on the role of reflexes and of muscle sensors in motor control. Indeed, for much of this century the field of motor control was intimately tied to the concept of reflexes (for an interesting historical perspective, see Granit 1975 or Matthews 1981). Reflexes can be elicited by many modes of sensory information, including temperature and pressure sensors just below the skin, within joint capsules and ligaments, and within muscles and tendons.

Despite a remarkable amount of research activity, our knowledge of how sensory information from sensors such as muscle spindles and Golgi tendon organs has remained illusive (Loeb 1984). Section 3 of Chapter 2 provides an overview of reflexes and spinal neurocircuitry, and Chapter 7 and Chapter 11 discuss models of the muscle spindle; this subsection briefly provides a foundation, discussing the structural layout and basic properties of sensors within musculotendinous tissue.

4.4.1 Several Types of Sensors Are Embedded Within Musculotendon Tissue

Figure 1.12 provides an idealized view of the structural arrangement of the muscle spindles and Golgi tendon organs. All of the sensors essentially measure local tissue strain, and given that this local tissue is passive and somewhat spring-like, to a first approximation the sensors work like strain gages to (indirectly) estimate force. While the relative density of sensors within muscle tissue does vary (e.g., higher density for neck and hand muscles), the bottom line is that muscles are embedded with sensors.

For the Golgi tendon organs, because of the series arrangement, local strain would appear to be associated with muscle force, and to a reasonably good approximation this is indeed the case.

For the muscle spindles, the situation is quite involved. The traditional assumption is that the primary 1a afferents measure mostly velocity and the

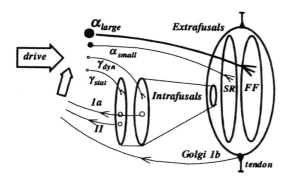

FIGURE 1.12. Simplified conceptual view of the key muscle structures: the muscle spindles are located structurally in parallel with the main (extrafusal) muscle fibers (driven by α MNs of various sizes, with idealized slow fatigue-resistant *SR* and fast fatiguing *FF* fibers shown here), while the Golgi tendon organs are structurally in series. Here idealized $\gamma_{dynamic}$ and γ_{static} drives are displayed; in reality there are multiple types of both nuclear chain and bag fibers, with chain fibers receiving γ_{static} MN drive, and nuclear bag fibers receiving both types. The two classic types of muscle spindle afferents (SNs) are the secondary (group *II*) and primary (group *1a*) sensory afferent endings, which tend to wrap around the fiber and measure local strain. While each type of intrafusal fiber may be innervated by both *II* and *1a* fibers, in general it is useful to associate secondary afferents with chain fibers and primary afferents with bag fibers. Golgi tendon organ sensors (1b SN) essentially measure local strain in tendon near muscle-tendon junctions.

secondary II afferents position (e.g., see Chapter 11). This is based primarily on various length perturbation studies in which the neural drive is constant, and the experiment involves either an isovelocity ramp (from one steady length to another) or a small sinusoidal oscillation.

For ramp-and-hold experiments, after an initial high-frequency transient, sensory action potentials are then reasonably well correlated to weighted linear sum of slightly smoothed length and velocity components. Furthermore, it is known that with increasing γ_{static} MN activation, the bias firing rate of the sensors (especially secondary) is increased, while with $\gamma_{dynamic}$ MN activity, the dynamic sensitivity is increased (especially for 1a sensors). This suggests that the γ drive just modulates the gain and a bias position (see Chapter 11), and that the process of sensing stretch and encoding this into action potentials is relatively linear.

However, it is not quite this simple—the overall response for a range of stretch is highly nonlinear even when the γ drive is constant: there is a local region of high sensitivity to initial stretch which is followed by considerably less sensitivity, and additionally there is asymmetry between lengthening and shortening. This behavior can be conceptually captured by assuming that the sensing element is in series with intrafusal muscle contractile tissue that includes the muscle apparatus mentioned earlier—cross-bridges, actin binding sites, and so on. Measuring the actual force in these small muscle spindle units has proved quite challenging. Let's assume that the spindle afferents measure stretch in what is essentially a SE-like element in series with the intrafusal contractile tissue, as in Figure 1.12, and that the "sticky" actomyosin bonds initially stretch but do not yield and then tend to follow conventional shortening and lengthening muscle behavior as for a typical extrafusal muscle). One would then anticipate an initially high force (and thus strain) across the sensing region because the cross bridges initially "hold their ground," followed by force levels that are consistent with transient behavior of a dynamic subsystem that is somewhat dominated by the CE force-velocity behavior for the intrafusal fibers. That is about what we see. Thus spindle behavior is linked to intrafusal muscle mechanics as well as transducer-encoder dynamics, and its nonlinear behavior is likely more attributed to intrafusal muscle mechanics than to sensor dynamics.

How such information might be used in a functioning animal or human remains an enigma; see Chapter 2 for a review.

5 Signal Processing of Movement Data: Key Terminology

Three types of classic data are obtained from movement science studies: *3D motion* (e.g., via markers attached to the surface of the body), *electromyograms* (EMG), and *contact forces or pressures* (e.g., via a force platform on the ground).

Typically information is measured over time and is ultimately stored in a digital computer. Each channel of measured information (or data) includes both "signal" and "noise" content. *Noise* is that aspect of the information that is not wanted, and con-

sidered spurious. It may be somewhat random fluctuation (i.e., nearly *white noise*), or it may be a
byproduct of some side effect (e.g., 60-cycle noise;
a steady drift; discretization). One normally desires
a high *signal-to-noise* ratio, and employs *filter*s to
help massage data into a more useful form. Some
filters work in real-time, others off-line.

For practical purposes, filters can be separated
into two categories: analog and digital. An *analog
filter* works on present and past data, but does not
have access to future data. Analog filters can perform many operations, but normally the purpose is
to amplify and *smooth* data, and often it is designed
to eliminate frequency content within the signal
that one expects not to be associated with the "real"
signal (e.g., 60-cycle noise). For instance, because
of the intrinsic inertia of skeletal structures, one
may not expect motion data above a certain frequency (e.g., perhaps 10 Hz), and can then employ
a *low-pass filter* to smooth the data.

A wide variety of *digital filters* are available.
Some work in real-time, others work off-line. In
the latter case "future" data can also used within
the filtering scheme. Digital filters are sensitive to
the *sampling rate*, and there are many rules of
thumb for making sure that data is collected at an
appropriate frequency (e.g., the well-known Nyquist
sampling theorem). Typical sampling rates for motion and contact force data are 50, 60, or 100 Hz,
while for EMGs, the collection rate must be considerably higher unless one first uses an analog filter to preprocess the data (e.g., rectify and smooth)
before it is sampled (Loeb and Gans 1986).

One can also distinguish filters by the general
form of the algorithm. Three classic approaches, all
in common use within biomechanics, are time-
based filtering, frequency-based filtering, and
splines. In *time-based* filtering the new data point
is a function of its old value, plus nearby past (and
often future) values. *Frequency-based* approaches
essentially map (transform) the data into the frequency domain (e.g., take a "fast Fourier transform,
or FFT), and analyze it there. In some cases the results are presented within this domain (e.g., frequency content of a signal), while in others one
then maps the filtered signal back to the time domain. *Splines* are essentially a fit of temporal data
curves by a serial sequence of fits (usually polynomial), and can be especially useful to smooth position data before it is differentiated to obtain ve-

locity or acceleration (the process of differentiation
tends to enhance the noise content).

It is important to realize that there is not one
"best" filtering approach. For instance, there are
many remarkably different approaches for reducing EMGs, as might be expected since in some investigations it might be important to carefully characterize "on" and/or "off" timing of an EMG burst
(e.g., fast tracking movements), while in others a
smoothed envelope is ideal (e.g., activities of daily
living), while in still others the frequency content
may be of primary interest (e.g., study of neuromuscular fatigue in low back muscles).

6   Neural Control Concepts

6.1   Key Concepts of Feedback Control

Our bodies are loaded with *sensors* that feed back
information to the nervous system regarding our internal state and the environment around us. Often
this information is used to affect ongoing control
strategies; this suggests a role for feedback. Most
feedback loops utilize a form of *negative feedback*.
In the classic case an "error" signal is obtained as
the difference between a "desired" (reference) signal and the actual (fed back) signal, with the goal
of the *feedback control system* being to minimize
this error (Figure 1.13). This suggests that feedback
can enhance system performance. More specifically, feedback (when used appropriately) has the
following key advantages:

- Improved transient performance.
- Less sensitivity to noise and disturbances.
- Less sensitivity to variation in internal model pa-
  rameters.

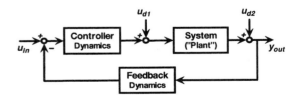

FIGURE 1.13. Classic feedback control system, designed
so that the output response y_{out} is sensitive to the primary
input u_{in}, yet relatively insensitive to disturbances u_{d1} and
u_{d2}. Lines with arrows represent signals. Parameters
within each of the three blocks are assumed not to vary.

The latter two relate to robust performance. Additionally, it is difficult to even image how learning could take place without some type of feedback of system performance. These features may seem so attractive that at first glance, the solution would seem to be to always crank up the feedback loop gains. However, as feedback gains increase, the system may become *unstable* (e.g., it might oscillate uncontrollably while diverging from a desired output). This is especially true when there are significant transmission time delays within the loop, such as for neuromotor feedback loops. Thus there is typically a *trade-off between performance and stability*, and the design of feedback control systems must be done carefully. This is why most undergraduate engineering curricula include courses in feedback control, and why scientists in many fields invest effort in learning these principles.

Tools in both the time and frequency domains exist for analyzing and designing feedback control systems, and the reader is directed to any of the many books in this area. Here we will focus on structural design of control systems and on key terminology.

6.2 Classic Structures for Adaptive Control and Parameter Estimation

In broad terms, a feedback control system consists of a model *structure* (which gives the form of the equations), *parameters* and *signals*. In a conventional feedback control system as in Figure 1.13, one designs the control system to work on signals (the lines with arrows).

In an *adaptive control system*, one uses some type of adaptation law to also work on parameters. These may be, for instance, *control parameters* such as feedback gains or *model parameters* such as stiffness. While there are many adaptive control structures, the classic adaptive control scheme consists of two "models," one of which has modifiable parameters and one of which does not. The *parallel* structure of Figure 1.14 is a very common approach: here the two models receive the same input, and through comparison of their evolving outputs, some sort of *error signal* is determined (e.g., the difference between the "desired" performance of one model and the actual performance of the other). This signal is used to adapt parameters within the adjustable model, and often ancillary signals are sent as well. This primary model will of-

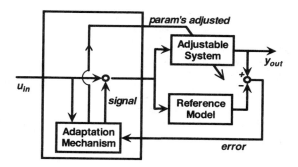

FIGURE 1.14. A classical model reference adaptive control system structure. The error between the reference model and the actual (adjustable) system is used by the adaptation mechanism to adjust certain system parameters. The "adjustable system" itself often is a model similar to that in Figure 1.13, and the parameters being adjusted may be feedback gains, local feedforward parameters, or perhaps actual plant parameters.

ten also have its own internal feedback loops that work only on signals. Such a "model reference" structure is of immense importance in engineering control systems.

An example of a neuromotor structure that has been conceived of with this form is the α-γ (extrafusal–intrafusal) "linkage." This concept, which has been packaged in various forms, assumes a "model reference" servo structure and α-γ coactivation (Granit 1981). Such coactivation is reasonably consistent with the assumption of orderly recruitment (with the γ-MNs having about the size of the smaller and first recruited α-MNs), and the concept of the spindle output as a sort of "error" signal that compares the desired performance (e.g., from the intrafusal model) and actual performance (e.g., from the actual muscle, which is subject to perturbations), has a certain appeal. The idea here is that the extrafusal system output follows the intrafusal output, with "follow-up" feedback. While there are certainly problems with such an assumed structural representation (e.g., the nature of the "reference" output (position? force? stiffness?)), the similarity of the structural layout is striking.

This concept of "on-the-fly" adjustable parameters is intriguing, yet also dangerous in that while in principle adjusting feedback gains and the like would seem advantageous, it could also lead to instability—especially for systems with significant transmission time delays. For this reason, considerable theoretical and practical efforts have been expended on designing adaptation laws that assure that parameters cannot be adjusted so as to lead to instability. A disadvantage is that in order to guarantee stability and basic robust performance, the adaptation laws often must be fairly conservative and somewhat sub-optimal.

There is an important duality between this type of structure for adaptive control, and the same kind of structure for *model parameter estimation* (Figure 1.15). In this case the actual system serves as the reference model, and the error between it and the adjustable model is used to fine tune the parameters within the latter. Typically a structure is assumed for the adjustable model that is believed to adequately capture the behavior of the reference model, if only the parameters were tuned. This classic concept has been used often in motor control modeling studies, often to identify an "inverse model" of either (e.g., see Chapter 34). An example of a structure for obtaining an inverse model, here with an "error" signal calculated at the input rather than the output, is shown in Figure 1.15.

This duality between adaptive control and parameter estimation structures has often been noted within the neuromotor literature, especially as related to structural models of the cerebellum. Here one uses a parameter adaptation scheme to create an "internal model," then uses this as part of an adaptive control scheme. It is common to then utilize some type of cascaded series-parallel arrangement. A key feature is the calculation of "error" signals by comparator elements. Indeed, virtually all control models of structures such as the cerebellum assume some type error signal. This also applies to most adaptive neural networks, especially those employing supervised learning.

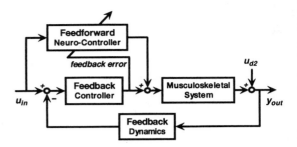

FIGURE 1.16. Adaptive neurocontrol model structure with both feedback and feedforward controllers, but structured so that the feedback error can be used to adjust the feedforward neuro-controller. In the Feedback Error Learning model of Kawato's group (e.g., Kawato et al. 1993), the error signal is used to train an adaptive artificial neural network.

It is important to recognize that connectionist neural networks are, by nature, adaptive systems since the *synaptic weights are parameters* that are adjusted. An example is shown in Figure 1.16, in which the feedforward controller is a neural network. Here the feedback error signal is used to train the network.

6.3 Feedforward Versus Feedback Control During Voluntary Movements: Issues

While all agree that sensor information is important, there has long been controversy regarding the relative importance of feedforward versus real-time feedback mechanisms. We know that feedback loop gains are variable (task-specific), and that typically they are remarkably low in comparison to most engineering control systems. Why?

In part the answer rests with intrinsic muscle mechanics. Movements of any joint crossed by a muscle will change the length and velocity of that muscle. It does not matter whether the movements were caused by the action of the muscle itself, other muscles or external forces. Because the force produced by a muscle depends on its length and velocity, such movements will produce an immediate change in the muscle's force even without any changes in its state of activation.

To better understand this important phenomenon and how it can be used by the CNS, try this experiment. Start by putting your arm out in front of

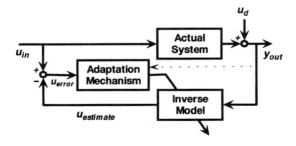

FIGURE 1.15. Classic model for model parameter estimation.

you, with your elbow bent to about 90 degrees and your eyes closed. With the help of a friend (or enemy?), perform two simple trials. In the first, which we will call the "do not resist" task, relax your arm and have your friend apply a sudden and somewhat brief transient pull at the wrist, hard enough so that the arm moves significantly. In the second, which we will call the "maximally resist" task, first "stiffen" your arm so as to try to minimize your response, and then have your friend try to apply about the same transient push at the wrist. You will likely see that the difference in the response was profound. What happened? In essence, you co-contracted muscles at either side of the elbow joint, such that even though their individual forces rose, due to their antagonistic arrangement there was no change in the net joint moment. Yet as the muscle forces went up, the intrinsic muscle stiffness to transient perturbation also went up. This initial, instantaneous response has nothing to do with reflexes (remember that there are transmission time delays), and is due to nonlinear muscle properties.

It is likely that between the two experiments, the transient elbow joint stiffness changed by a factor of 10. Furthermore, you could make this profound change in less than one-quarter of a second (i.e., the time for a isometric contraction). The initial transient is due primarily to the stiffness of the SE element, which increases with muscle force up to about 50% of maximum (e.g., Figure 1.4). During this brief time period the contractile tissue (e.g., CE) acts as a sort of "viscous ground" (Winters and Stark 1987). Using the Hill model analogy, the CE element then starts to drift, and subsequent behavior depends primarily on the dynamic interaction of CE and SE. After sufficient time for transient neural and muscle delays of likely over 50 ms, low-level reflex activity can add to the response through changes in muscle activation, although only moderately due to the rather small loop gains. Finally, more coordinated conditioned responses ("long loops," etc) can enter into the mix.

This inherent ability to vary the immediate, transient response to perturbation by modulating the initial state of a muscle is called a "preflex" in Chapter 10, or simply a change in stiffness by physiologists (Grillner 1972). Seif-Naraghi and Winters (1990) used a model and optimization to prove that during tracking tasks performed in an environment with and without pseudorandom perturbations,

when the model included nonlinear muscle properties the optimized solutions utilized significant co-activation of antagonists whenever there were unpredictable perturbations, while when muscle properties were linearized co-contraction ceased because it was not effective at stopping the effects of the perturbations.

Of course, in reality feedforward strategies that take advantage of nonlinear muscle properties and feedback loops would be expected to work together since they can often complement each other. It has often been found experimentally that reflex gains increase during periods of high muscle force or sustained co-contraction, and simulations predict that because of nonlinear properties, such variation in gains can then be better tolerated (Winters et al. 1988).

A key problem with trying to better understand the role of feedback involves our limited knowledge of the intrafusal system and muscle spindles during normal function. While treating spindles as kinematic sensors is perhaps conceptually desirable, such an assumption has considerable problems during voluntary movements. Unfortunately for the motor control community (but fortunately for living systems), animals and humans like to move. Thus perturbation studies can only take one so far in unfolding the mystery. Collecting data on spindles during natural movements is quite a challenge, and the story is far from complete. Nonetheless, there are several well-documented cases in which the traditional length-velocity sensor view breaks down completely: isometric contractions and fast voluntary shortening movements, where in both cases spindle activity is variable and often high, in direct contrast to the traditional view.

An intriguing alternative approach is Loeb's (1984) concept of an optimal transducer, in which part of the role of the γ drive is to place the sensor within a sensitive operating range (away from spindle saturation), so that it is effective at handling unpredictable perturbations. This also, however, could also get tricky during voluntary movement.

In trying to make some sense of this problem, it is useful to summarize what we know, and how it relates to other chapters in this book. Sensors, such as spindles, send corollaries up spinal tracts to various regions of the higher brain, and thus the information is used at many places other than just the local spinal level. In general, reflex gains, while normally low, are variable and can be of functional significance for tasks such as walking (see Chapter 17). Spindle information is used for so-called "long loop" (e.g., transcortical) conditioned responses as well as spinal reflexes. From whole-body perturbation experiments, such as summarized in Chapter 19, that sensor information can be used to help trigger coordinated multisegment responses. This suggests that maybe there has been too much emphasis on the concept of conventional reflex loops, as suggested in Chapter 19 and by Loeb in the commentary to Chapter 17. Under pathological conditions, such as cerebral palsy and Parkinson's disease, the excessive tone or tremor is in part the result of reflex-like phenomena, and that if the dorsal roots of selective sensory nerves of a child with cerebral palsy are cut by a neurosurgeon, the tone will typically decrease dramatically. Spindle afferents project not only to the parent muscle (both monosynaptically and via interneurons), but also (via interneurons) to both synergistic and antagonistic muscles. There are selective differences in bias levels and reflex gains between certain muscle groups, for instance extensor muscles used for posture and flexor muscles. For now, perhaps, we have to leave it at this, and enjoy the question itself.

6.4   Impedance Control: Implications of Nonlinear Muscle Properties

The force produced by a muscle has been shown to be a function of activation, length, and velocity. In many cases muscle appears spring-like, yet as suggested in the last section this property can be adjusted dramatically. A rich theory called *impedance control* has been developed that captures some of the implications of bicausal interaction (Hogan 1984, 1990). Chapter 35 addresses some of the implications for rehabilitation. This subsection briefly summarize the key features and implications of this theory, extracting heavily from Hogan (1990).

Power transmission embodies a two-way or bicausal interaction that may be characterized by mechanical impedance. Unlike information transmission, energy exchange fundamentally requires a two-way interaction. For any actuator, biological or artificial, it is important to distinguish between two aspects of its behavior. In engineering parlance they are termed the "forward-path response func-

tion" (or transfer function) and the "driving point impedance." Formally, a *mechanical impedance* is a dynamic operator that specifies the forces an object generates in response to imposed motions. Admittance is the inverse of impedance: an operator specifying a motion in response to imposed forces. For linear systems the two contain the same information, but in general they do not. Muscle-environment interaction is fundamentally two-way; because the muscle is coupled to masses that tend to impose motions, muscle is usually best regarded as providing an impedance. As seen in the previous section, intrinsic muscle impedance governs rapid interactions.

Most objects a limb contacts are *passive*; they may store energy (e.g., spring, mass) and some of that energy (potential, kinetic) may subsequently be recovered. However, the amount recovered cannot exceed the amount stored (and because of dissipation it should be less). In contrast, an actuator can (at least theoretically) supply energy indefinitely. For continuous sinusoidal inputs, to be a passive system, net power must be absorbed over each cycle; this implies that the phase angle between velocity and the force acting on the object must lie between ±90°. The necessary and sufficient condition for an object to avoid instability when coupled to any stable, passive object is simple: its driving point impedance must appear to be that of a passive system. The apparent dynamic behavior observable at any point on a musculoskeletal system (e.g., the hand) is determined by three major factors: (1) the intrinsic mechanics of muscles and skeleton; (2) neural feedback; and (3) the geometry or kinematics of the musculoskeletal system.

To get a feel for how controlling mechanical impedance could be useful, assume muscles are spring-like as in Figure 1.9 and consider this: Whereas the net moment about the joint is a weighted *difference* of muscle tensions, the net stiffness about the joint is a weighted *sum* of the contributions of each muscle. Additionally, while the moment contributed by the muscle force is proportional to the moment arm, because of the two-way interaction between force and length the joint stiffness is proportional to the square of the moment arm. Thus kinematic effects will always have a more pronounced effect on the output impedance than on the forward-path transfer function. Additionally, because impedance of a muscle increases with activation level, one means of modulating mechanical impedance is to synergistically activate both muscles; applications include hand prehension (Crago et al. 1990). Hogan (1990) provides references that document task-dependent changes in the mechanical impedence of limbs, and of how the stiffness of the joint increases with the moment of the joint.

A minimal model of nonlinear impedance modulation is the bilinear model, which can be obtained by taking a Taylor expansion of a steady-state relation of force f as a function of length l and activation u (Hogan 1990):

$$f = f(l_o,u_o) + \frac{\partial f(l_o,u_o)}{\partial l} dl + \frac{\partial f(l_o,u_o)}{\partial u} du$$

$$+ \frac{\partial^2 f(l_o,u_o)}{\partial l\, \partial u} dl\, du + ..$$

$$f_{bi-linear} \approx (k_1 + k_2 l)(c_1 + c_2 u) \qquad (1.26)$$

where the k and c are constant. This produces a fan-shaped family of force-length curves, a model that has been used extensively in exploring the consequences of the spring-like properties of muscles. Differentiating the latter with respect to l, the stiffness of this bilinear model is proportional to neural input:

$$k_{bi-linear} = \frac{\partial f_{bi-linear}}{\partial l} = k_2(c_1 + c_2 u) \quad (1.27)$$

Because the force generated at any given length is also proportional to neural input, the bilinear model predicts a proportional relation between force and stiffness, as has often been seen experimentally.

For multijoint systems, impedance also has a directional property. Multiarticular muscles can then serve to modulate impedance fields, and this ability could complement other functional roles that such muscles may play. Using transformation theory, and assuming an apparent muscle viscosity b_m, the apparent joint viscosity b_j transforms as follows (Hogan 1990):

$$b_j = J^T b_m J \qquad (1.28)$$

where $J(x_j)$ is the Jacobian matrix of partial derivatives of muscle lengths with respect to joint angles (i.e., what we have referred to as the moment arm matrix). The apparent joint stiffness transformation was given in Eq. 1.21—the second term in the equation tells us that when the force is significant, K_j is

also a function of the changes in the moment arm (see also Chapter 21). Finally, it can be shown that postural configuration modulates dynamics, as can be seen from the relation that determines the apparent inertia tensor (see Hogan 1990).

Perhaps the most important implication for multijoint control is that the mechanical impedance of the neuromuscular system coupled with the inertia of the skeletal links define the dynamics of an *attractor*. If the descending commands to MN pools are held constant, the spring-like behavior defines an equilibrium configuration. Attractor dynamics may be used to produce movement as well as to sustain posture. This idea is central to many studies of movement (Chapter 24; Chapter 26; Winters and Woo 1990, Chapters 9, 11–14, 17, 21). There is experimental evidence showing that the path between two points during a movement exhibits a measurable degree of stability, not just the endpoint (e.g., Bizzi et al. 1984; Flash 1990). This suggests that in some cases, the production of movement appears to be accomplished by a progressive movement of the neurally defined equilibrium posture, which has been termed a *virtual trajectory* (Hogan 1984, 1990). A hypothesized significance of this idea is that it implies a reduction in the computational complexity of movement generation.

7 Tissues Are Alive: Remodeling, Adaptation, Optimal Design, and Learning

In interpreting results within this book, it is important to realize that living tissues, with few exceptions (e.g., perhaps cartilage), will actively remodel based on how they are used. This holds for bones, soft connective tissues, muscles (just look how many people lift weights), blood vessels, and neural tissue. Each type of tissue has its own "criteria" for remodeling. For musculoskeletal tissues ranging from bone to muscle, the primary criteria are based on some form of mechanical loading, while for synapses within the CNS, on electrical conditions. Remodeling includes both tissue growth (e.g., muscle hypertrophy; synaptic facilitation) and decay (e.g., muscle atrophy, decreased synaptic "gain"), as summarized by the "use it our lose it" concept.

Although it is typically difficult to pinpoint the exact mechanisms that are responsible for remodeling for most tissues, what is clear is that it is a type of feedback control system, and tends to be governed primarily by local conditions. Thus, for most tissues, this constitutes a distributed, decentralized type of adaptive system. Additionally, tissue remodeling appears consistent with the concept of optimal tissue design, where an optimum design has to include a balance between task and metabolic needs of the organism.

7.1 Mechanics-Based Remodeling

Perhaps the most understood tissue remodeling process involves bone. We know that bone will atrophy if it is not adequately loaded for a period of months. This has been documented in astronauts and cosmonauts, in persons and animals kept in rigid casts for sustained periods of time, in loss of bone around internal prostheses, and in bed-ridden individuals. It's also been well documented that the ligaments of animals placed in casts will atrophy, and that animals on exercise regimes can develop stronger tendons (Woo et al. 1982). It is not uncommon for certain workers and musicians to develop calluses—thickened skin over locations under high mechanical contact. The governing principle for such phenomena is often called Wolff's Law (Fung 1993). Whether the primary driving force is tissue strain, strain rate, stress, and so on, is beyond the scope of this book. What is important is that it is a continuing, evolving process. For instance, osteoclasts are continually absorbing bone and osteoblasts laying out new bone.

Of interest here is what this principle of remodeling might tell us about the *design and properties* of tissues such as muscle and tendon. One well-accepted finding is that through athletic training, muscle can change size, and to a lesser extent, composition. It is also known that through FES training regimes, paralyzed muscles can be conditioned over time to change, to some extent, their size, strength and their fiber composition and thus resistance to fatigue (e.g., see Chapter 42).

An implication related to musculotendon structure and properties is that one might expect certain mechanical consistencies within muscle-tendon units. For instance, a tendon might be expected to remodel (grow in cross-section) if through muscle contraction, its factor of safety is consistently too low. Similarly, for a given force, SE strains might be expected to be relatively uniform across muscle, aponeurosis and tendon tissue (all include fibrous building blocks). Regarding fiber design, one might expect that the rest lengths of the collection of muscle fibers located in parallel should to some extent distribute into a functionally useful arrangement, much like the "cathedral arches" of the cancellous bone in the proximal femur (e.g., Fung 1993). Thus the fact that not all muscles have homogeneously arranged sacromeres (see Chapter 6), suggests that heterogeneous arrangements (resulting in a relatively broader CE force-length range) might be the result of an adaptive strategy that helps meet the functional needs associated with the tasks of life. A similar argument could be made that the wide variety of pennation angles that are seen for various muscles relate to optimum "packing" strategies.

Perhaps the most intriguing question is whether fundamental nonlinear muscle tissue properties are an accident of nature (a necessary evil that is the result of physiological constraints at the cellular or molecular level), or reflect a proactive "solution" to an optimum design problem. This author favors the latter interpretation (see Chapter 35), but there are certainly alternative views (e.g., see Chapter 13 or Chapter 27). One example perhaps best addresses the issue: while most muscles include both slow and fast muscle fibers in relatively equal amounts, there are clear exceptions (e.g., high number of slow muscle fibers in the soleus muscle, which is a functionally important postural muscle). Yet it is also apparent that this "heal thyself" concept has limitations; hence the need for sophisticated biomechanical studies to help guide orthopedic surgical intervention strategies (e.g., see Chapter 36).

7.2 Neurally Based Structural Remodeling

From a broad "remodeling" perspective, several features of the CNS stand out. First, there is a remarkable degree of built-in structural organization within the CNS. Areas of white matter (essentially transmission lines) and gray matter (cell bodies, synapses) are clearly distinguishable, and differ little in location or gross connectivity across a species. Thus we expect to see the same tracks (gross connections) and cell body clusters within a brain, and are able to give them names with some confidence—pyramidal track, cerebellum, superior colliculus, corpus collosum, red nucleus, etc.

Second, there is a remarkable behavioral capability in most cases for creatures to learn and adapt. Third, for some controlled conditions (as well as indirectly via clinical observations) we can often attribute learning to specific structures or entities (e.g., synapses). The process of neural tissue remodeling ("learning") is addressed within a number of chapters (e.g., Chapter 23 and Chapter 26) and occurs on multiple time scales. It can include changes in control signals (shorter time scale effects), in synapses (moderate time scale), and network structure (longer time scale).

Using the analogy of artificial neural networks (ANNs), it is known that ANNs also exhibit learning behavior (e.g., see Chapter 16, Chapter 23, Chapter 32, Chapter 34, and Chapter 35). This is typically accomplished by allowing certain "gains" ("synaptic weights") between cells to be adjusted. The algorithm behind such modulation of synaptic weights is typically based on minimizing some criteria related to "error" (e.g., difference between actual and training signal for supervised learning, or cluster "error" for unsupervised learning of self-organizing systems—see Denier van der Gon et al. 1990). This is a form of adaptive learning in which there is modulation of *parameters*, but not structure. In essence, the ANN *process* is the engineering optimization process (see also Chapter 35): Minimize a scalar (or for multicriterion optimization a finite set of scalars) by modulating a set of parameters.

Returning to neural tissue, gain changes across chemical synapses, and the creation of (or atrophy of) neurons is a remodeling process. From this context, local Hebbian reinforcement learning phenomena (Hebb 1949)—where correlation between an output signal and an input signal results in a facilitation of its synaptic gain (strength)—represents a form of local tissue remodeling.

Yet it is also possible to modulate ANN structure. The classic example of an artificial "network" with *structural* learning capabilities is the class of approaches often called genetic algorithms. Even the title suggests a biologically inspired approach.

Within engineering, genetic algorithms are considered a type of nonlinear stochastic optimization algorithm that is particularly suited for structural solving complex problems where gradient calculation is impossible or too computationally expensive.

Returning to neural tissue, there are well-documented cases were large parts of the brain appear to have structurally reorganized in response to a functional impairment. Indeed, part of the mission of therapeutic intervention for conditions such as stroke is to reorganize parts of the brain through repetitive training (Chapter 38).

References

Abbott, B.C. and Wilkie, D.R. (1953). The relation between velocity of shortening and the tension-length curve of skeletal muscle. *J. Physiol.*, 120:214–223.

Alexander, R.McN. (1988). *Elastic Mechanisms in Animal Movement.* Cambridge University Press, Cambridge, England.

Alexander, R.McN. and Ker, R.F. (1990). The architecture of leg muscles. In *Multiple Muscle Systems. Biomechanics and Movement Organization.* Winters, J.M. and Woo, S.L-Y. (eds.), Chapter 36, pp. 568–577, Springer-Verlag, New York.

Bernstein, N.A. (1967). *The Coordination and Regulation of Movement.* Pergamon Press, London.

Binder, M.D., Heckman, C.J., and Powers, R.K. (1996). The physiological control of motoneuron activity. In: *Handbook of Physiology. Section 12, Exercise: Regulation and Integration of Multiple Systems.* Rowell, L.B. and Shepherd, J.T. (eds.), American Physiological Society, New York: Oxford. pp. 3–53.

Bizzi, E., Accomero, N., Chapple, W., and Hogan, N. (1984). Posture control and trajectory formation during arm movement. *J. Neurophys.*, 4:2738–2744.

Cavagna, G.A. (1970). Elastic bounce of the body. *J. Appl. Physiol.*, 29:279–282.

Crago, P.E., Lemay, M.A., and Liu, L. (1990). External control of limb movements involving environmental interactions. In *Multiple Muscle Systems. Biomechanics and Movement Organization.* Winters, J.M. and Woo, S.L-Y. (eds.), Chapter 21, pp. 343–359, Springer-Verlag, New York.

Crisco, J.J. and Panjabi, M. (1990). Postural biomechanical stability and gross muscle architecture in the spine. In *Multiple Muscle Systems. Biomechanics and Movement Organization.* Winters, J.M. and Woo, S.L-Y. (eds.), Chapter 26, pp. 438–450, Springer-Verlag, New York.

Denier van der Gon, J.J., Coolen, A.C.C., Erkelens, C.J., and Jonker, H.J.J. (1990). Self-organizing neural mechanisms possibly responsible for movement coordination. In *Multiple Muscle Systems. Biomechanics and Movement Organization*. Winters, J.M. and Woo, S.L-Y. (eds.), Chapter 20, pp. 335–342, Springer-Verlag, New York.

Fenn, W.O. and Marsh, B.S. (1935). Muscular force at different speeds of shortening. *J. Physiol.*, 85:277–297.

Flash, T. (1990). The organization of human arm trajectory control. In *Multiple Muscle Systems. Biomechanics and Movement Organization*. Winters, J.M. and Woo, S.L-Y. (eds.), Chapter 17, pp. 282–301, Springer-Verlag, New York.

Fuller, J.J. and Winters, J.M. (1993). Joint loading during stretching exercises recommended for osteoarthritis: a biomechanical analysis. *Ann. Biomed. Eng.*, 21:277–288.

Fung, Y.C. (1993). Biomechanics. *Mechanical Properties of Living Tissues* (2nd ed). Springer-Verlag, New York.

Gielen, S., van Ingen Schenau, G.J., Tax, T., and Theeuwen, M. (1990). The activation of mono- and bi-articular muscles in multi-joint movements. In *Multiple Muscle Systems. Biomechanics and Movement Organization*. Winters, J.M. and Woo, S.L-Y. (eds.), Chapter 18, pp. 302–311, Springer-Verlag, New York.

Gottlieb, G.L., Corcos, D.M., Agarwal, G.C., and Latash, M.L. (1990). Principles underlying single-joint movement strategies. In *Multiple Muscle Systems. Biomechanics and Movement Organization*. Winters, J.M. and Woo, S.L-Y. (eds.), Chapter 14, pp. 236–250, Springer-Verlag, New York.

Granit, R. (1975). The functional role of the muscle spindles—facts and hypotheses. *Brain*, 98:531–556.

Grillner, S. (1975). Locomotion in vertebrates: Central mechanisms and reflex interaction. *Physiol. Rev.*, 55:247–304.

Hebb, D.O. (1949). *The Organization of Behavior*. John Wiley & Sons, New York.

Henneman, E. and Mendell, L.M. (1981). Functional organization of motoneuron pool and its inputs. *Handbook of Physiol. The Nervous System*, Vol. II, pt. 1, Chapter 11.

Henneman, E., Bomjen, G. and Carpenter, D.O. (1965). Functional significance of cell size in spinal motoneurons. *J. Neurophys.*, 28:560–580.

Hill, A.V. (1922). The maximum work and mechanical efficiency of human muscles, and their economical speed. *J. Physiol.*, 56:19–45.

Hill, A.V. (1938). The heat of shortening and the dynamic constants of muscle. *Proc. Roy. Soc.*, 126B:136–195.

Hill, A.V. (1950). The series elastic component of muscle. *Proc. Roy Soc.*, 137B:399–420.

Hill, A.V. (1970). *First and Last Experiments in Muscle Mechanics*. Cambridge University Press, Cambridge, England.

Hodgkin, A.L. and Huxley, A.F. (1952). A quantitative description of membrane current and its application to conduction and excitation in nerve. *J. Physiol.*, 117:500–544.

Hogan, N. (1984). An organizing principle for a class of voluntary movements. *J. Neurosci.*, 4:2745–2754.

Hogan, N. (1990). Mechanical impedance of single- and multi-articular systems. In *Multiple Muscle Systems. Biomechanics and Movement Organization*. Winters, J.M. and Woo, S.L-Y. (eds.), Chapter 9, pp. 149–164, Springer-Verlag, New York.

Hogan, N. and Winters, J.M. (1990). Principles underlying movement organization: upper limb. In *Multiple Muscle Systems. Biomechanics and Movement Organization*. Winters, J.M. and Woo, S.L-Y. (eds.), Chapter 11, pp. 182–194, Springer-Verlag, New York.

Huxley, A.F. (1957). Muscle structure and theories of contraction. *Prog. Biophys.*, 7:257–318.

Huxley, H.E. and Hanson, J. (1954). Changes in the cross-striations of muscle during contraction and stretch and their structural interpretation. *Nature*, 173:973–976.

Huxley, A.F. and Niedergerke, R. (1954). Structural changes in muscle during contraction. *Nature*, 173:971–973.

Huxley, A.F. and Simmons, R.M. (1971). Proposed mechanism of force generation in striated muscle. *Nature*, 233:533–538.

Kane, T.R. and Levinson, D.A. (1985). *Dynamics: Theory and Applications*. McGraw-Hill, New York.

Kawato, M. (1990). Computational schemes and neural network models for formation and control of multi-joint arm trajectory. In *Neural Networks for Control*. (Miller, W.T., Sutton, R.S., and Werbos, P.J., (eds.), pp. 197–228. The MIT Press, Cambridge.

Levin, A. and Wyman, J. (1927). The viscous elastic properties of muscle. *Proc. Roy Soc.*, B101:218–243.

Liddel, E.G.T. and Sherington, C.S. (1924). Reflexes in response to stretch (myotatic reflexes). *Proc. Roy. Soc. London Ser. B*, 96:212–242.

Loeb, G. (1984). The control and responses of mamalian muscle spindles during normally executed motor tasks. *Exer. Sport Rev.*, 12:157–204.

Loeb, G.E. and Gans, C. (1986). *Electromyography for Experimentalists*. University of Chicago Press, Chicago.

Loeb, G.E. and Levine, W.S. (1990). Linking musculoskeleal mechanics to sensorimotor neurophysiology. In *Multiple Muscle Systems. Biomechanics and Movement Organization*. Winters, J.M. and Woo, S.L-Y. (eds.), Chapter 10, pp. 165–181, Springer-Verlag, New York.

Matthews, P.B.C. (1981). Muscle spindles: their messages and their fusimotor supply. In *Handbook of Physiology. The Nervous System*. Kandel, E.R., (ed.), Vol. II, pt. 1, Chapter 6, pp. 189–228.

McMahon, T. (1990). Spring-like properties of muscles and reflexes in running. In *Multiple Muscle Systems. Biomechanics and Movement Organization*. Winters, J.M. and Woo, S.L-Y. (eds.), Chapter 37, pp. 578–590, Springer-Verlag, New York.

Morgan, D. (1990). Modeling of lengthening muscle: the role of inter-sacromere dynamics. In *Multiple Muscle Systems. Biomechanics and Movement Organization*. Winters, J.M. and Woo, S.L-Y. (eds.), Chapter 3, pp. 46–56, Springer-Verlag, New York.

Mungiole, M. and Winters, J.M. (1990). Overview: Influence of muscle on cyclic and propulsive movements involving the lower limb. In *Multiple Muscle Systems. Biomechanics and Movement Organization*. Winters, J.M. and Woo, S.L-Y. (eds.), Chapter 35; pp. 550–567, Springer-Verlag, New York.

Pandy, M. (1990). An analytical framework for quantifying muscular action during human movement. In *Multiple Muscle Systems. Biomechanics and Movement Organization*. Winters, J.M. and Woo, S.L-Y. (eds.), Chapter 42, pp. 653–662, Springer-Verlag, New York.

Seif-Naraghi, A.H. and Winters, J.M. (1990). Optimized strategies for scaling goal directed dynamic limb movements. In *Multiple Muscle Systems, Biomechanics and Movement Organization*. Winters, J.M. and Woo, S.L-Y. (eds.), Chapter 19, pp. 312–334, Springer-Verlag, New York.

Sherrington, C.S. (1910). Flexion-reflex of the limb, cross extension-reflex, and reflex stepping and standing. *J. Physiol.*, 40:28–121.

Van der Helm, F.C.T. and Veenbass, R. (1991). Modelling the mechancial effect of muscles with large attachment sites: application to the shoulder mechanism. *J. Biomech.*, 24: 1151–1163.

Van Ingen Schenau, G.J. (1984). An alternative view of the concept of utilization of elastic energy in human movement. *Hum Mov. Sci.*, 3:301–334.

Van Ingen Schenau, G.J. (1990). The unique action of bi-articular muscles in leg extensions. In *Multiple Muscle Systems. Biomechanics and Movement Organization*. Winters, J.M. and Woo, S.L-Y. (eds.), Chapter 41, pp. 639–652, Springer-Verlag, New York.

Wilkie, D.R. (1950). The relation betwen force and velocity in human muscle. *J. Physiol.*, 110:248–280.

Winters, J.M. (1990). Hill-based muscle models: a systems engineering perspective. In *Multiple Muscle Systems. Biomechanics and Movement Organization*. Winters, J.M. and Woo, S.L-Y. (eds.), Chapter 5, pp. 69–93, Springer-Verlag, New York.

Winters, J.M. and Kleweno, D.G. (1993). Effect of initial upper limb alignment on muscle contributions to isometric strength curves. *J. Biomech.*, 26:143–153.

Winters, J.M. and Stark, L. (1985). Analysis of fundamental movement patterns through the use of in-depth antagonistic muscle models. *IEEE Trans. Biomed. Eng.*, BME-32:826–839.

Winters, J.M. and Stark, L. (1987). Muscle Models: what is gained and what is lost by varying model complexity. *Biol. Cybern.*, 55:403–420.

Winters, J.M. and Woo, S.L-Y. (1990). *Multiple Muscle Systems. Biomechanics and Movement Organization*, Springer-Verlag, New York.

Winters, J.M., Stark, L., and Seif-Naraghi, A.H. (1988). "An analysis of the sources of muscle-joint system impedance. *J. Biomech.*, 12:1011–1025.

Woo, S.L-Y., Gomez, M.A., Woo, Y.-K., and Akeson, W.H. (1982). Mechanical properties of tendons and ligaments: II. The relationships of immobilization and exercise on tissue remodeling. *Biorheology.*, 19:397–408.

Zahalak, G.I. (1981). A distribution-moment approximation for kinetic theories of muscular contraction. *Math Biosci.*, 55:89–114.

Zahalak, G.I. (1990). Modeling muscle mechanics (and energetics). In *Multiple Muscle Systems. Biomechanics and Movement Organization*. Winters, J.M. and Woo, S.L-Y. (eds.), Chapter 1, pp. 1–23, Springer-Verlag, New York.

Zajac, F. and Winters, J.M. (1990). Modeling musculoskeletal movement systems: joint and body-segment dynamics, musculotendinous actuation, and neuromuscular control. In *Multiple Muscle Systems. Biomechanics and Movement Organization*. Winters, J.M. and Woo, S.L-Y. (eds.), Chapter 8, pp. 121–148, Springer-Verlag, New York.

Section II

2
Neural and Muscular Properties: Current Views and Controversies

Robert F. Kirsch and Richard B. Stein

1 Introduction

Much of this book is devoted to the understanding of control of posture and movement, and a number of different viewpoints will be presented. Theories regarding the control of posture and movement have arisen from analyses performed over a wide range of levels, from the microstructure of individual elements of the system to the broadest mechanical features of a whole joint or limb. This chapter will not attempt to present a unified theory regarding the control of posture and movement. A number of later chapters in this volume, however, will address these more global themes. Our role in this volume is to *briefly* present a summary of the current understanding of muscle contractile mechanisms and reflex organization on which any successful theory must be based, providing numerous references for the readers interested in more detail. The review of muscle properties will be based on the widely accepted sliding filament theory of contraction, which will then be used to discuss the mechanisms responsible for system-level properties such as length-tension, force-velocity, and muscle viscoelasticity. The review of reflex organization will emphasize first what is *possible* (mainly from classical neurophysiology in reduced animal preparations), but these properties will then be related to the functional significance of reflex effects in intact animals and human subjects. Finally, we will discuss the knowledge that is missing about basic muscle and reflex properties that has prevented the formulation of a generally accepted hypothesis regarding the control of posture and movement.

2 Muscle Properties

2.1 Basic Contractile Mechanisms

Muscle is the actuator that transforms neural signals into mechanical outputs, such as muscle force and length. This transformation is complex enough that neural control mechanisms must almost certainly take it into account. Current understanding of the basic mechanisms of muscle contraction is still well summarized by the "sliding filament" theory which was proposed Huxley in 1957 (Huxley 1957, 1974, 1980) and has undergone only minor modifications since. This theory forms the basis of explanation for many of the complex whole-muscle properties that are relevant for understanding of system-level neural control strategies (which will be briefly described below and in more detail in the accompanying perspectives), so a brief description of this hypothesis will be given below. More detailed descriptions can be found elsewhere (Stein 1980; McMahon 1984).

At the molecular level, muscle contraction (muscle force generation and the accompanying shortening) occurs because of an interaction between two muscle proteins, actin and myosin, that are arranged in a regular pattern (along with other structural and regulatory proteins) in the basic unit of muscle contraction, the sarcomere. Sarcomeres are arranged in series to form myofibrils, which are arranged in parallel to form muscle cells or "fibers." The fibers are then arranged in parallel to form the bulk of whole muscle. Within each sarcomere, actin molecules form the "thin filaments," while the myosin molecules form the "thick filaments" that have regularly spaced "heads" (cross-bridges) along their length. Different types of muscle vary somewhat in the details of the myosin and actin properties, but the basic mechanisms described below are qualitatively similar in all cases.

Contraction of skeletal muscle occurs naturally when an action potential propagating along a motoneuron axon reaches the neuromuscular junction, depolarizes vesicles in the nerve terminal and causes the release of acetylcholine. The acetylcholine diffuses across the neuromuscular junction and binds to postsynaptic receptors on the muscle, producing depolarization and giving rise to a muscle action potential. The muscle action potential then propagates along the length of the muscle fiber membrane (the sarcolemma), but it also propagates radially into the center of each of the myofibrils of the fiber through projections of the sarcolemma. These projections, termed the transverse or "T" tubules, are located along each Z-line, where the sarcomeres butt end to end. The T-tubules are closely associated with a membrane, the sarcoplasmic reticulum, which includes vesicles which contain a high concentration of calcium. Depolarization propagating along the T-tubules causes calcium to be released from the sarcoplasmic reticulum into the sarcomere. Calcium release triggers a conformational change in the regulatory protein troponin, which exposes the "active sites" on the actin molecule to which the myosin heads attach. According to the sliding filament theory, the myosin head (or cross-bridge) then hydrolyzes a molecule of ATP and rotates (or changes shape in some other manner) such that the attached thin filament is pulled toward the center of the sarcomere. This reaction converts the chemical energy of the ATP hydrolysis into mechanical work, generating force to overcome internal and external loads and causing the muscle to shorten (contract). Once the

cross-bridge has rotated, another ATP molecule is bound to the myosin head, causing it to detach from the active site on the actin molecule. The cross-bridge is now available for attachment to a new active site at a shorter length. This ratchet-like "cycling" mechanism provides for muscle shortening over a range much greater than the distance between active sites on the actin molecule. Furthermore, a large number of sarcomeres are arranged in series to provide needed muscle excursion distances, and muscle fibers are arranged in parallel in the muscle to generate needed force levels.

Figure 2.1 is a two-dimensional representation of the mechanical interactions between the myosin heads (cross-bridges) and the actin molecule. The sarcomere is actually a three-dimensional structure, however, with the thick filaments arranged in a hexagonal pattern such that each thick filament is surrounded by six actin molecules to which it can bind. The force generated by an active crossbridge (depicted as a simple rotation in Figure 2.1) is transmitted to the thin filament through an elastic (i.e., nonrigid) element. This cross-bridge compliance has been experimentally demonstrated to be an essentially purely elastic element with no additional dynamics (Ford et al. 1977) and thus is indicated as a simple spring with a spring constant K in Figure 2.1. The source of this elasticity is still controversial (Huxley 1980; Jung et al. 1992), but it could arise from the compliance of the actin and/or myosin filaments, from deformation of the bonds between the myosin head and the actin binding site, from the myosin head (Lombardi et al. 1995), or from some combination of all of these. Functionally, the elasticity of all of the attached cross-bridges in a muscle combine to produce its "spring-like" properties (see below). As also indicated in Figure 2.1, the cross-bridge can attach to the thin filament only at discrete locations (the "active

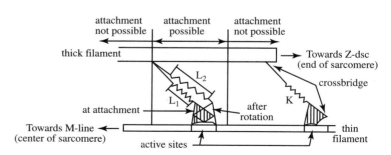

FIGURE 2.1. Schematic of actions of muscle cross-bridge.

sites"), with the likelihood of attachment decreasing rapidly with lateral distance from this optimal alignment (Huxley 1957, 1974; Stein 1980; McMahon 1984). It should be noted that although the general principles described above are widely accepted, several of the details regarding muscle contractile mechanisms (e.g., the "conformational change" that occurs in the crossbridge to produce contraction) are still controversial (Huxley 1980).

2.2 Mechanisms of System-Level Muscle Properties

The force magnitude that can be generated by a muscle has long been known to depend upon its length (Gordon et al. 1966) and its velocity of shortening or lengthening (Fenn and Marsh 1935; Hill 1938). Furthermore, muscle has been recognized to exhibit more general viscoelastic ("spring-like") properties (Hill 1938; Nichols and Houk 1976; Hoffer and Andreassen 1981a,b). These "intrinsic" muscle properties are functionally important because they can automatically and instantaneously (i.e., without the conduction delays inherent in all reflex systems) reject disturbances, and thus act to stabilize postures and movement trajectories. Because each of these muscle properties is likely to have a significant impact on how neural control strategies must be organized, the following sections will discuss these classical behaviors, as well as the mechanisms likely responsible for these system-level properties.

2.2.1 Length-Tension Relationship

The amount of force that can be generated by a particular sarcomere depends on how many active sites on the actin molecules are within binding proximity to the myosin crossbridges. The sliding filament theory hypothesizes (Gordon et al. 1966) that changes in sarcomere length (because of contrac-

tion or to imposition of external loads) result in a different amount of overlap between the regions of the thick filaments containing cross-bridges and the regions of the thin filaments containing active binding sites. Figure 2.2 shows this schematically for four different sarcomere lengths. For very long sarcomere lengths (case A in Figure 2.2), there is no overlap between these regions and no force can be generated. For shorter sarcomere lengths, however, the degree of overlap increases progressively (case B) and the available force increases in parallel. At a certain length, all of the cross-bridges are within binding proximity to an actin active site (case C), and force reaches a maximum. This maximum force will be maintained for even shorter lengths (i.e., the "plateau" region of the length-tension relationship) until the thin filaments on opposite ends of the sarcomere begin to overlap and interfere with force generation (case D). For extremely short sarcomere lengths, no force can be generated because of this interference.

The length-tension property of whole muscle arises from the combination of the length-tension properties of the multitude of sarcomeres, which make up the muscle. Although all the sarcomeres in a muscle do *not* have identical length-tension curves (see Chapters 5 and 6), they are often similar enough so that the whole-muscle length-tension curve has similar features, i.e., total force tends to increase with length up to a certain point, after which it plateaus and then declines. This relationship is functionally important because it will determine the upper limit on the isometric force a muscle can generate for different limb positions; examples of its functional importance (e.g., as related to orthopedic biomechanics) are provided in Chapters 3 and 36.

Muscles are typically assumed to act on the ascending portion of their length-tension curves (i.e., force increasing with length), with the maximum length in vivo being the peak of the length-tension

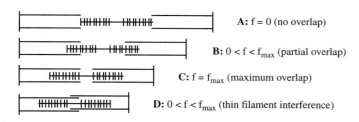

A: f = 0 (no overlap)

B: $0 < f < f_{max}$ (partial overlap)

C: $f = f_{max}$ (maximum overlap)

D: $0 < f < f_{max}$ (thin filament interference)

FIGURE 2.2. Sarcomere force magnitude depends on overlap between thin and thick filaments.

curve or the "optimal length." Operation on the ascending limb of the curve is stable in terms of the interactions between serially arranged sarcomeres, because if one sarcomere is slightly weaker than others with which it is in series, it will be lengthened while the others shorten. As its force increases with length, it will quickly reach equilibrium with the others at a slightly longer sarcomere length. If a particular muscle fiber operates on the descending limb of its length-tension curve, however, weaker sarcomeres will again be lengthened at the expense of the stronger ones, but rather than generating more force at longer lengths, they will generate less force and lengthen even more. Although this unstable situation would appear to be something to be avoided, recent evidence (see Chapter 3) indicates that at least some human muscles regularly act on the descending limb of their length-tension curve. Other mechanisms which prevent this runaway instability appear to exist. For example, it has been recently proposed (Morgan 1990, 1994) that the instability of muscle fibers acting on the descending limb of the length-tension property is prevented by the "popping" of weaker sarcomeres, which rapidly lengthen to their maximum lengths, and become essentially passive transmitters of force between the stronger sarcomeres in series with them. These longer "popped" sarcomeres thus allow the other stronger sarcomeres to operate at shorter lengths where the fiber as a whole is stable. Furthermore, muscles typically act on joints through moment arms which vary with joint angle. If the moment arm of a muscle acting on the descending limb of its length-tension curve increases more than the decrease in force capacity as the muscle lengthens, the net joint angle-joint moment relationship of the muscle acting at the joint can be stable. Joint angle-dependent changes in muscle moment arm can also be destabilizing, however, if the moment arm decreases as muscle length increases.

The mechanisms and behavior of muscle length-tension properties presented above are idealized versions of the actual properties. The shape of the length-tension relationship varies with activation level (Rack and Westbury 1969), and more recent work (Chapters 5 and 6) has indicated strong dependencies on previous activation and movement history. As emphasized by Huijing in his perspectives (Chapters 5 and 6), inhomogeneities in sar-

comere properties with a single fiber may also play a significant role in shaping whole muscle properties. However, the length-tension relationship can be useful for certain well-defined situations, but assuming that a single static relationship can describe all length-dependent force properties of muscle could lead to a misunderstanding of other important mechanisms.

2.2.2   Force-Velocity Relationship

The length-tension properties of muscle are static in nature (i.e., muscle length is experimentally fixed and force is measured, often with supramaximal activation). In most natural activities, however, the muscle is either allowed to shorten or is forced to lengthen by external loads. It has long been known that the amount of force that a muscle can produce for a fixed activation level depends upon how rapidly the muscle length is changing and in which direction (i.e., shortening or lengthening) (Fenn and Marsh 1935; Hill 1938). In contrast to the length-tension relationship (which results primarily from overlap between the thin and thick filaments), the force-velocity relationship arises primarily from cross-bridge cycling dynamics. The various chemical reactions in the cross-bridge cycle all require a finite length of time to complete, and these dynamics impose an upper limit on how rapidly the sarcomere can generate force. The primary mechanisms acting to produce force-velocity properties during lengthening and shortening are different and will be dealt with separately.

In shortening muscle, the probability that a given "active site" on an actin molecule will bind a myosin head decreases with shortening velocity because the myosin heads sliding by the active site remain within the possible attachment zone (see Figure 2.1) for increasingly shorter periods of time. Even if the cross-bridge does bind and cycle, the contribution of the cross-bridge rotation to muscle fiber force will progressively decrease with shortening velocity because more and more of this rotation is expended in merely keeping up with the shortening of the sarcomere because of the actions of other cross-bridges. Eventually, a limit will be reached where all of the cross-bridge rotation simply keeps up with the shortening motion and no force can be developed (i.e., unloaded shortening).

The classical shape of the whole muscle force-velocity relationship (Fenn and Marsh 1935; Hill 1938) for *shortening* muscle, where the maximum force is developed for zero velocity (i.e., for isometric conditions) and force drops off in a hyperbolic fashion with increasing shortening velocity, results from the combined contributions of all of the active crossbridges.

An activated muscle that develops less force than the external load imposed upon it will lengthen rather than shorten (McMahon 1984). The properties of muscle lengthened in this manner are rather different and more complex than those of shortening muscle. The most obvious difference is that lengthening muscle can actually generate more force than isometric muscle. This most likely arises from time-dependent crossbridge cycling properties, in this case the detachment dynamics. Because detachment requires a finite interval of time, the elasticity associated with an attached crossbridge (see Figure 2.1) will be extended by increasing stretch velocities to increasingly greater lengths before it can detach. Because of the elastic nature of the cross-bridges, a greater extension leads to a larger force being generated by each attached crossbridge. This effect is ultimately limited by factors similar to that for shortening muscle (i.e., the time during which the cross-bridge is within binding proximity to the actin active sites decreases with stretch velocity, so the average number of bound cross-bridges decreases). In fact, when the external load exceeds approximately 1.2 to 1.8 times the isometric maximum force (Katz 1939; Chapter 10), the muscle abruptly "yields" (i.e., the resistance to movement will not increase regardless of the stretch velocity). At these higher velocities, the decreasing probability of attachment meets and then exceeds the increased force per cross-bridge, so that force declines somewhat. At even higher velocities, force plateaus (i.e., remains constant with velocity) rather than declines further, which presumably results from a difference in cross-bridge detachment mechanisms at high lengthening velocities. For such velocities, the normal mechanism for cross-bridge detachment is probably physical disruption rather than completion of the cross-bridge cycle, and recent evidence (Lombardi and Piazzesi 1990) suggests that such cross-bridges can reattach at a much higher rate than those completing the normal cross-bridge cycle, and thus can keep up with the lengthening muscle to maintain a relatively constant force.

The standard force-velocity relationship, although it has been used successfully in a number of applications, represents only a cross-section of the dynamic properties of muscle force generation. The force-velocity relationship is typically constructed from measurements obtained from maximally activated muscle and constant loading conditions. However, the shape of this relationship for whole muscle varies with activation level and is highly dependent on previous movement history (Joyce and Rack 1969), and muscle activation and velocity will rarely be constant during normal activities. Movement history effects can be at least partially accounted for by the dynamic interaction between the velocity-dependent contractile properties and the elastic elements (presumably produced by the tendon compliance and intrinsic crossbridge stiffness) with which they are in series. This is the reason why Hill-type muscle models can approximate this effect for some conditions. However, activation and movement history-related changes in cross-bridge cycling properties themselves indicate that processes more detailed than classical length-tension and force-velocity relationships may be needed to describe the dynamics of muscle contraction for general conditions.

2.2.3 Viscoelasticity

The length-tension and force velocity properties described above are simple (and sometimes accurate) approximations to the behavior of muscle under general conditions. During many activities, however, neither muscle force nor muscle velocity are constant, and external loads will act to both lengthen and shorten the muscle. The manner in which muscles interact with these external conditions is critically important for the control of posture and movement by the nervous system, and can be described in terms of muscle *viscoelasticity*. Viscoelasticity is a system property that is used here to describe the general relationship between muscle force and displacement, and can be used to describe muscle behavior when either forces or displacements are imposed. The length-tension and force-velocity properties of a muscle are special steady-state cases of its viscoelasticity. Specific aspects of viscoelasticity can be described by stiffness (length-to-force properties), viscosity (velocity-to-force properties), and higher order effects, but the term is generally used here to include all aspects of muscle responses to mechanical disturbances, including nonlinear properties.

Viscoelastic muscle properties arise from a number of different mechanisms, including muscle contractile properties and the passive properties of noncontractile tissues within the muscle. We will concentrate upon the viscoelastic properties due to contractile mechanisms (however, see Huijing commentary on Chapter 7). These active viscoelastic properties arise both from the intrinsic elastic properties of attached cross-bridges and from the cycling dynamics of the cross-bridges. Although these "intrinsic" (to the muscle) properties are produced by the same mechanisms as force generation, they serve a functionally distinct role in the control of posture and movement. Specifically, the intrinsic viscoelastic properties of muscle provide instantaneous force responses to imposed disturbances, which is something that cannot be provided by delayed neural reflexes. Furthermore, these intrinsic responses can be quite large, at least transiently, relative to other responses and thereby play an important role in maintaining postural stability in the face of unexpected disturbances.

Muscle elasticity (i.e., the component of viscoelasticity which relates muscle forces and displacements) has received considerably more attention than other components of muscle viscoelasticity, presumably because of its simplicity and its potential for producing a steady state restoring force during maintained external disturbances. Muscle elasticity exhibits both steady-state and transient components. The steady-state component (which provides a maintained response to a maintained disturbance) arises primarily from the slope of the muscle length-tension relationship (at least for muscles acting on the ascending limb), but the stiffness magnitude provided by this mechanism is typically rather modest. A much larger elastic component is produced by the properties of the cross-bridges themselves. As described above (see Figure 2.1), force is transmitted between myosin and actin filaments through elastic elements, so any imposed displacement will be resisted by the combined stiffnesses of all of the crossbridges attached at any instant. This resistance can be substantial, and is typically much greater than the slope of the length-

tension properties. However, this stiffness component does not persist indefinitely because the extended cross-bridges will eventually cycle, detach, and reattach in a shorter (and lower force) position. This stiffness component provided by cross-bridge stiffness has been shown to be purely elastic (i.e., proportional to displacement and with no dynamics) for displacements that are small enough that cross-bridges are not forced to detach. In fact, stiffness measured in this way is typically assumed to be proportional to the number of instantaneously attached cross-bridges (Ford et al. 1977, 1981; Lombardi and Piazzesi 1990). Recent evidence, however, suggests that the proportional relationship between stiffness and the number of attached cross-bridges may not hold under all conditions (Jung et al. 1995), especially during muscle lengthening (van der Linden et al. 1996).

Many typical activities subject muscle to displacements far in excess of the range of individual cross-bridges, so the cycling properties of the cross-bridges become important. Several prominent muscle features result from these cross-bridge properties, including "yielding," "short-range stiffness," muscle force-velocity properties, and muscle viscosity. Muscle yielding was described above in terms of lengthening *velocity*, but it is defined in another context as the abrupt decrease in muscle force which occurs when an active muscle is lengthened beyond the range of individual cross-bridges and a significant fraction are disrupted simultaneously (Joyce et al. 1969; Nichols and Houk 1976; Flitney and Hirst 1978). A less abrupt form of this property is "short-range stiffness" (Rack and Westbury 1974), which describes the large resistance provided by attached crossbridges for small displacements that gives way to a lower (but nonzero) stiffness when crossbridges are required to cycle. Yielding behavior has been studied most extensively in the cat soleus muscle (Joyce et al. 1969; Nichols and Houk 1976), but it has also been reported in the human first dorsal interosseus muscle (Carter et al. 1990) and in frog muscle (Flitney and Hirst 1978). Short-range stiffness has been described for the cat soleus (Rack and Westbury 1974; Walmsley and Proske 1981) and medial gastrocnemius muscles (Walmsley and Proske 1981). The length at which yielding occurs and the excursion range of short-range stiffness both increase with increasing muscle velocity (Rack and Westbury 1974; Nichols and Houk 1976), presumably because the cross-bridges are extended to greater lengths (and thus to higher forces produced by extension of the cross-bridge elasticity) before they can detach.

Functionally, short-range stiffness has generally been assumed to contribute to postural stability, with yielding (or at least reduced stiffness) treated as an unwanted side effect that is perhaps prevented by muscle stretch reflex actions (Nichols and Houk 1976; Houk et al. 1981a,b; Carter et al. 1990). Malamud et al. (1996) reported that the magnitude of short-range stiffness is larger in the slow muscle fibers where yielding is most prominent. They hypothesized that the higher stiffness of these units is advantageous for the postural tasks normally performed by these fibers while yielding is a protective mechanism to prevent damage to the muscle that would occur if the initially high stiffness was maintained. Prominent yielding behavior has been described in only a few muscles, however, and yielding usually occurs (even in the cat soleus muscle) only for active muscle which has been previously isometric. Although muscle subjected to continuous motion (e.g., sinusoidal or stochastic) exhibits a stiffness somewhat lower than the preyield short-range stiffness observed in the same muscle, it does not exhibit either yielding or significant short-range stiffness even for perturbation amplitudes far in excess of that described for previously isometric muscle (Kirsch et al. 1994). Furthermore, preliminary work by Huyghues-Despointes et al. (1997) has demonstrated the lack of short-range stiffness in cat soleus muscle fibers subjected to an imposed shortening just prior to stretch, indicating that the mechanism for this behavior is intrinsic to the contractile mechanism itself. The functional role for these relatively well-behaved properties of nonisometric muscle is not established, but such behavior is consistent with the lower stiffnesses required during movement.

Another manifestation of the cycling properties of cross-bridges is viscosity, the term used to describe the general relationship between muscle force and velocity. The force-velocity relationship described above is the muscle viscosity measured for constant velocities. For small continuous perturbations (such as sinusoids and stochastic displacements), muscle viscosity is fairly linear (Kirsch et al. 1994). However, viscosity decreases with the frequency of movement (Kirsch et al. 1994) as expected from the force-velocity relationship and is thus nonlinear for general conditions (Gielen and Houk 1984; Lin and Rymer 1993). These nonlinear viscosity properties may be functionally advantageous, since viscosity is highest at low velocities (thus enhancing postural stability by providing greater resistance to movement), and lowest at high velocities (thus reducing resistance during rapid voluntary movements).

2.2.4 Summary

The sliding filament theory of muscle contraction is widely accepted and can be used to explain most if not all of the basic system-level properties of muscle, including its length-tension and force-velocity relationships. Furthermore, this theory can account for most observed viscoelastic properties, which characterize the way in which a muscle resists externally imposed disturbances. Viscoelastic properties are thus of fundamental importance in controlling posture and movement during interactions with the environment. These muscle properties are rather complex under general conditions, however, so they are often simplified to special cases (e.g., length-tension for fixed length and maximum activation, viscoelastic properties for fixed activation) to make consideration of the system as a whole practical. However, the effects of such simplifications (e.g., Hill-based muscle models) on the understanding of muscle behavior during movement (e.g., using large-scale musculoskeletal models) has never been adequately quantified. More complex muscle models based on crossbridge properties (e.g., Huxley 1957, 1974, 1980; Zahalak 1981, 1986) have been developed, but the computational demands of this approach have prevented it from finding application in any large-scale model to date. With ever-increasing computational power becoming available, however, arguments for using simpler models based solely on computational expediency should become less convincing. Even these cross-bridge-based models are relatively gross approximations to actual muscle properties. As discussed by Huijing in his perspectives, no current muscle model takes in account potentially important sarcomere inhomogeneities along single fibers, and the different contractile properties of different motor unit types within a muscle are not included in any model.

The challenge for muscle modeling in the future will therefore be to weigh the need for accuracy with practical concerns about complexity. More complex muscle models should be applied only where the need for such models is clearly established. Conversely, Hill-type muscle models are an almost universal feature of large-scale models of the musculoskeletal system, yet the impact of describing muscle in this simple manner on the understanding of muscle function and neural control of movement has never been adequately addressed. Some of these issues are examined in the perspective by Winters, who describes complex muscle properties that can be predicted by relatively simple muscle models and emphasizes the potential functional benefits provided by these muscle nonlinearities.

3 Neural Control of Muscle Contraction

All of the muscle properties used to maintain posture and produce movements are modified by changes in the activation levels produced by the motoneuron pool of the muscle. Roughly speaking, increases in the activation of single muscles lead to increases in force and stiffness, but the relationship between activation and the mechanical properties of muscle are dynamic and nonlinear. Furthermore, it is likely that global strategies for controlling limb posture and movement are organized at the level of the limb as a whole rather than for individual muscles, so the pattern of activation to a particular muscle may reflect these more global concerns as well as local ones. Indeed, many years of neurophysiological research has demonstrated the richness of inputs to the motoneuronal pool, which include direct actions by descending brain systems and direct monosynaptic reflex connections from some muscle receptors. However, the excitation to the motoneuronal pool of a muscle is supplied by many different types of signals (descending, intersegmental, and segmental) which are more commonly relayed through interneurons. Furthermore, some of these signals probably act to modulate transmission through interneuronal networks rather than directly exciting motoneurons. There are many different types of inputs to a mo-

toneuron pool and their organization is very complex, with the result that there is currently no consensus regarding the general principals used by the nervous system to control posture and movement. The following sections will briefly describe the complexity of the segmental reflex system, mention several important nonsegmental mechanisms, and discuss the potential function relevance of these mechanisms.

3.1 Basic Reflex Neurophysiology

The activation commands to a muscle are relayed via a set of motoneurons (typically several hundred for a typical limb muscle), which act to integrate commands from a variety of descending systems and from local spinal circuits, thus serving as the "final common pathway" to the muscles. Figure 2.3 is a schematic representation of several of the primary neural mechanisms which directly or indirectly influence the excitability of the motoneuron pool of a muscle (which is represented in Figure 2.3 by a single a motoneuron). This diagram is intended primarily to illustrate several general principals observed with these mechanisms rather than providing a comprehensive summary of all possible influences on motoneuron activation. Far more detailed reviews of these systems are available to the interested reader (e.g., Baldissera et al. 1981; Schomburg 1990; Jankowska 1992).

3.2 Descending Systems

This subsection focuses primarily on neural mechanisms intrinsic to the spinal cord, but it is important to note that a number of descending systems from the brain exert important influences, both direct and indirect, on the motoneuronal pools of muscles. Several descending systems synapse directly onto motoneurons, with both excitatory and inhibitory actions. These direct connections provide the theoretical ability for descending connections to directly control the output of the motoneuronal pool, although such connections represent a small fraction of the total input to the motoneuron pool in most muscles. More commonly, descending systems exert their actions indirectly via interneurons which then synapse upon motoneurons. The upper left portion of Figure 2.3 summarizes the effects (again, both excitatory and

FIGURE 2.3. Examples of neural systems influencing muscle contraction.

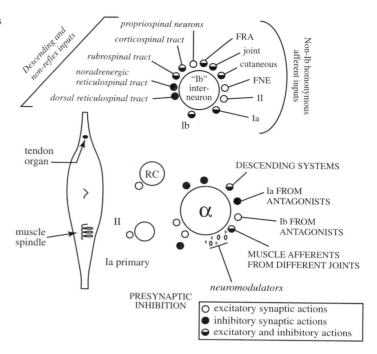

inhibitory) of descending systems on the "Ib" interneuron. This interneuron is used as an example, but the basic principle of descending systems exerting indirect effects via interneurons (including those interposed in reflex pathways) is typical (Eccles and Lundberg 1959; Lundberg 1975). These indirect actions potentially provide the ability for descending systems to selectively excite or inhibit interneurons participating in different aspects of motoneuronal output. In the example Ib interneuron shown in Figure 2.3, descending systems could selectively increase (via excitation) or decrease (via inhibition) the effectiveness of force-related reflex actions served by this interneuron.

3.3   Homonymous Muscle Receptor Reflex Systems

Sensory receptors located within muscle have long been known to give rise to neural afferents which relay information regarding the state of the muscle to the spinal cord. We will focus upon the reflex effects of these receptors on motoneurons, either by direct synapses or indirectly via interneurons. Figure 2.3 illustrates several of the more prominent reflex systems, although we again are not attempt-

ing to be comprehensive (see Baldissera et al. 1981; Schomburg 1990; Jankowska 1992).

Muscle spindles are capsular muscle receptors that are located in parallel with the fibers of the main muscle and are functionally responsive to muscle length and its derivatives. We will not discuss muscle spindle structure or receptor properties in any detail, but the interested reader is referred to an excellent review by Matthews (1981). Very briefly, muscle spindles usually give rise to two afferents, the larger Ia and smaller II. Both afferents exhibit some sensitivity to both muscle length and velocity, although the Ia afferent exhibits a more dynamic (i.e., velocity sensitive) response, while the II afferent is more akin to a static muscle length detector. Spindle responses to movement are nonlinear, and can be modified by the efferent (gamma) innervation to contractile (intrafusal) fibers within the spindle. The reflex control of muscle contraction has long focused upon the actions of spindle afferents because of their large axon diameters, the monosynaptic excitation of motoneurons by their afferents, and the large changes in motoneuron excitability produced by their activation. Furthermore, the steady-state length-related signals provided by the spindle secondary (II) afferents are theoretically capable of implementing a servo-like control of muscle length (Marsden et al. 1972).

Golgi tendon organs are capsular receptors which are located at the junction of muscle fibers and their tendon. These receptors are in series with the muscle fibers and respond in a monotonic manner with muscle force (Crago et al. 1982). The Ib afferents associated with these receptors are nearly as large as the Ia spindle primary afferent, and the pathway between a tendon organ afferent and the motoneuron pool for the muscle from which it arises contains only a single inhibitory relay interneuron. This connection is appropriate for closing a servo-like feedback pathway to regulate muscle force, although a number of attempts to demonstrate the functionality of such a feedback system have produced inconclusive results. Several studies (Houk et al. 1970; Jack and Roberts 1980; Rymer and Hasan 1980) performed in decerebrate cats failed to demonstrate a functionally large force feedback effect, although the Ib interneuron is known to be tonically inhibited by descending systems in the decerebrate preparation (Engberg et al. 1968). Subsequent work has indicated neural compensation for fatigue in human subjects (Kirsch and Rymer 1987, 1992) that might result from tendon organ-mediated force feedback, but other mechanisms probably contribute to this compensation as well. In any case, the "Ib" interneuron is really a misnomer, given the wide range of systems which synapse on it. As shown in Figure 2.3, the actions of Ib afferents are just one of several receptor systems (Lundberg et al. 1975), including cutaneous afferents, joint afferents, free nerve endings, spindle Ia and II afferents, and a number of descending systems which contribute to this "Ib" interneuron. In fact, several groups (Gossard et al. 1994; Guertin et al. 1995; McCrea et al. 1995; Pratt 1995) have recently described net positive (i.e., excitatory) reflex effects arising from tendon organ afferents during locomotor activities. Thus, the role of the tendon organ in the reflex control of muscle force has never been firmly established, and recent evidence suggests that this role may be different during movement than during posture.

A number of other systems also contribute to the reflex control of muscle. Cutaneous afferents arising from the foot and toe contact areas have significant effects on leg motoneurons (see Baldissera et al. 1981), while cutaneous afferents in the skin of the human fingers have been shown to implement a "slip compensation" scheme (Johansson et al. 1992a,b). Joint receptors have the capability of providing joint angle control, but current evidence suggests that they are primarily active at the extremes of joint excursion and thus serve a protective function (Baldissera et al. 1981). Free nerve endings exist in large numbers in muscle and are responsive to a number of different modalities, including stretch, force, pH, and temperature (Hertel et al. 1976; Mense and Stahnke 1983). These afferents have been implicated in the flexor withdrawal reflex (see Schomburg 1990 for a review), the clasp knife reflex (Rymer et al. 1979), and the control of motor unit firing rate during fatigue (Hayward et al. 1991). Motoneurons themselves send off branches which synapse upon Renshaw cells (indicated as the cell labeled "RC" in Figure 2.3), which in turn have inhibitory actions on the same muscle, thus implementing a "recurrent inhibition" system that presumably contributes to the regulation of motoneuron firing rates (Baldissera et al. 1981; Binder et al. 1996). Although these systems are not described in detail, it is important to note that a number of different sensory modalities contribute to motoneuron excitability even at a segmental level.

3.4   Intermuscle and Multijoint Reflexes

The above description focused on homonymous reflex effects (that is, reflex effects of muscle afferents on the motoneuron pool from which they arose (and those of close synergists). The reflex actions of muscle receptors are much more widely distributed, however (see Baldissera et al. 1981 for a review). The most obvious is "reciprocal inhibition," by which muscle afferents from one muscle have inverse effects on antagonistic muscles. Muscle spindle afferents from one muscle thus inhibit motoneurons to antagonist muscles and tendon organ afferents from one muscle excite antagonistic motoneurons. These effects simply accentuate the homonymous effects of the afferents and prevent the muscles from inappropriately opposing each other during normal tasks. On an even wider scope, afferent signals from one muscle have been demonstrated to effect the motoneuron output of muscles acting at different joints. For example, Ib afferents from the triceps surae muscles in the cat have effects on both knee and hip muscle motoneurons (Eccles et al. 1957), and spindle Ia afferents have been shown to influence the motoneurons of muscles distributed throughout the limb (Baldissera et al. 1981). More recently, Nichols and coworkers (Nichols 1994; Bonasera and Nichols 1994) have demonstrated (in unanesthetized decerebrate cats) that the pattern of these connections follows function: muscles predominantly involved in stereotyped activities such as locomotion (e.g., soleus and quadriceps) tend to exhibit strong force-related inhibition of other muscles in this group, while muscle which participate in more nonperiodic movements exhibited much weaker inhibitory effects. Finally, "propriospinal" neurons, which are defined as neurons that project to limb areas in the spinal gray matter while having cell bodies in the spinal cord but outside of these limb regions (Baldissera et al. 1981; Schomburg 1990), are thought to play an important role in both integrating supraspinal commands to motoneurons and in interlimb coordination. Thus, classical neurophysiology in reduced animal preparations has shown the potential for wide-ranging reflex interactions between different muscles acting at distant joints, perhaps in different limbs.

On a more functional level, evidence for feedback between elbow and shoulder muscles in intact human subjects has been presented (Lacquaniti and Soechting 1984, 1986a,b; Smeets and Erkelens 1991), and reflex effects between the two arm of human subjects has been demonstrated during bimanual tasks (Lum et al. 1992). In general, however, understanding of the reflex control of multi-muscle and multijointed systems in normally operating intact systems is rather minimal.

3.5 Presynaptic Inhibition

All of the systems described above exhibited their effects by synaptic transmission from axon terminals onto the soma or dendrites of motoneurons or associated interneurons. Other mechanisms for effecting the excitability of reflex systems have been demonstrated, however, including *presynaptic inhibition*. Presynaptic inhibition describes the inhibition of one neuron by direct actions of another neuron near the synaptic terminals of its axon. Presynaptic inhibition

is produced by and affects a wide variety of both descending and segmental systems. The functional significance of presynaptic inhibition in the segmental control of muscle action is that it provides the capability for descending systems and other afferent systems to reduce the effectiveness of a particular reflex system in a focused manner without effecting the overall excitability of a motoneuron or interneuron. This provides the potential for task-dependent modulation of different reflex actions, a topic that will be discussed below. The pattern of presynaptic inhibition is complex and significantly different for different muscle receptor afferents. For example, Ia afferents from cat flexor muscles are presynaptically inhibited primarily by homonymous Ib and Ia afferents, while Ia afferents from extensors are effected primarily by antagonistic flexor Ia and Ib inputs with smaller effects from homonymous afferents (Baldissera et al. 1981). Ib and cutaneous afferents receive much more widespread presynaptic inhibition from smaller afferents (free nerve endings, spindle II afferents, and cutaneous afferents) (Baldissera et al. 1981; Rudomin 1990, 1994). Presynaptic inhibition provided by different descending systems also appears to be organized quite differently for different muscle receptor afferents (see Rudomin 1994 for a review).

3.6 Neuromodulation

It has more recently been discovered that the certain substances (e.g., noradrenalin and serotonin) released by the terminals of descending systems into the intracellular medium can directly depolarize motoneurons (see Binder et al. 1996 for a review of the ionic basis of these effects) and interneurons interposed in reflex pathways (Miller et al. 1996). This relatively nonselective mechanism has been termed "neuromodulation." Neuromodulators have been shown to depolarize subthreshold motoneurons, thus reducing the synaptic current needed to initiate action potentials and generally enhancing the excitability of the motoneuron pool. During repetitive firing, neuromodulators can produce maintained firing ("bistable behavior") even after the synaptic input leading to the onset of activity is removed. The functional significance of neuromodulation is not well established, although it has been suggested that this mechanism is used to facilitate motor output during locomotion and postural tasks (Binder et al. 1996).

3.7 Modulation of Reflex Properties

The reflex effects of muscle receptors were once represented as fixed, automatic mechanisms, but it has been repeatedly demonstrated in intact human systems that the magnitude of reflex responses to external disturbances depends upon the task performed and the instructions given to the subject (Hammond 1956; Marsden et al. 1976; Gottlieb and Agarwal 1980; Soechting et al. 1981; Doemges and Rack 1992a,b). Specifically, the magnitude of reflex responses has been shown to change significantly and rapidly during voluntary position tracking (Gottlieb and Agarwal 1979; Dufresne et al. 1980), ballistic movements (Gottlieb and Agarwal 1978, 1980; Cooke 1980; Mortimer et al. 1981; Kirsch et al. 1993), locomotion (Yang et al. 1991), ball catching (Lacquaniti et al. 1991), and rapid imposed movements (Kirsch and Kearney 1993). In general, the magnitude of reflex responses have usually been found to increase with background muscle activation level (Matthews 1986), to be larger during postural tasks than during movement (Capaday and Stein 1986; Stein and Capaday 1988), to be strongly modulated during locomotion (Capaday and Stein 1987; Yang et al. 1991; Duysens et al. 1992), and to increase during movement as the target is approached (Bennett 1994). Several investigators (Gottlieb and Agarwal 1980; Dufresne et al. 1980; Soechting et al. 1981) have reported that changes in reflex gain often precede the onset of voluntary motor activity, suggesting that the nervous system pre-specifies a desired set of reflex "gains" appropriate for the intended task.

A number of the mechanisms potentially responsible for modulation of reflex magnitude were described above. For polysynaptic reflex systems, transmission through the interneurons in the loop can be modified by synaptic inputs from other systems. For example, reflex transmission through the "Ib" interneuron illustrated in Figure 2.3 can be impeded or facilitated by inputs from several different receptor and descending systems. Thus, descending systems could potentially modulate reflex properties to make them appropriate for a certain condition, and responses in one afferent system could potentiate or inhibit transmission through an-

other reflex pathway (e.g., joint afferents might potentiate force-related inhibition at the extremes of the range of motion of a joint to prevent damage). As described above, presynaptic inhibition can potentially be more selective in modulating feedback in even monosynaptic systems, and this again can be done either by descending systems (presumably to set up the gains according to task) or by segmental systems (Stein 1995). Locomotor effects are not shown in Figure 2.3, but the reflex response to an imposed stretch has been shown to be significantly modulated during the step cycle in the mesencephalic cat (Akazawa et al. 1981; Shefchyk et al. 1984) and in human subjects (Capaday and Stein 1986; Yang et al. 1991). As described above, locomotor rhythm has also been shown to even change the net sign of reflex action (e.g., changing Ib gain from negative to positive).

Other less sophisticated mechanisms may also contribute to reflex modulation. The most obvious of these is recruitment nonlinearity. As the recruitment of a muscle's motoneurons progresses, the size and force capacity of the recruited motoneurons increases so that the magnitude of the reflex activation produced by a fixed synaptic input increases with background activation level. This has been termed the "automatic gain principle" (Matthews 1986), and its presumed functional role is to maintain the relative contributions of intrinsic muscle properties (which also increase with motoneuron pool activation level) and reflex action. Otherwise, the reflex contribution would decrease relative to the increasing intrinsic component and a degree of neural control could be lost. More recent work (Stein et al. 1995b; Bennett et al. 1996) has emphasized that this mechanism acts to maintain a constant mechanical reflex loop gain, rather than the increase in gain proposed originally by Matthews.

In summary, it is well established that the functional effects of reflex action are significantly modulated during different tasks and for different conditions. In many cases, this modulation appears to be appropriate for attaining the goals of the task. As discussed below, an understanding of the relative contribution of reflex feedback to net muscle activation during different conditions has significant implications for our understanding of how the nervous system controls posture and movement. Because these reflex properties can clearly be modulated over a wide range during typical behaviors, an increased understanding of the rules used by the nervous system to modulate reflex effectiveness is needed. However, such general understanding has been elusive, primarily because of the sheer number of systems which contribute to net reflex excitability and the difficulties in assessing many of these mechanisms in intact systems where a normal balance between these many factors is maintained.

3.8   Functional Implications

We have attempted to present a summary of the current understanding of basic muscle properties and their control by segmental reflex systems, and have highlighted several more recently discovered aspects of each. Although this summary represents an enormous amount of basic knowledge about a number of specific entities that is interesting in its own right, the task of the current section is to identify those effects which are *functionally* important and to discuss the implications of the above findings for our understanding of the control of posture and movement.

As described above, a myriad of different systems affect the output of the motoneuron pool of a muscle, including descending systems, local spinal systems, and segmental and intersegmental reflex systems. The perspective by Binder in this volume describes one approach for determining the relative importance of different synaptic inputs to the motoneuron pool. In this approach, the effective synaptic current that can be injected by different input systems is estimated using a voltage-clamp of a motoneuron while the appropriate system is activated. This approach has indicated that descending systems appear to have surprisingly greater effects on motoneuron excitability than the tested segmental reflex pathways (which included the Ia monosynaptic pathway). Perhaps more significantly, none of these systems can provide sufficient depolarization alone to produce maintained firing of the motoneuron, indicating that many systems must contribute under normal conditions.

In intact systems, it is not possible to examine individual input systems with the same level of detail, but it is possible to evaluate the net properties of reflex action relative to the magnitude of the intrinsic viscoelastic properties of muscle. These intrinsic properties can provide an instantaneous and large resistance to external disturbances, thus enhancing postural stability. However, the steady-state intrinsic response to a steady-state disturbance is limited to the slope of the length-tension relationship, which is typically rather small. Disturbance rejection can be enhanced by reflexively-mediated increases in activation, however, and several global hypotheses (e.g., the equilibrium point hypothesis: Feldman 1966, 1986; Latash 1993) rely on reflex actions of significant magnitude.

Is the contribution of reflex pathways relative to intrinsic muscle properties large enough to support such hypotheses? A large body of information regarding reflex systems has been developed in the triceps surae muscles of the decerebrate cat. The magnitude of reflex action in these muscles has been found (e.g., Nichols and Houk 1976; Hoffer and Andreassen 1981a,b) to provide more than 50% of the net stiffness during stretches imposed upon isometrically contracting muscle (especially at low forces). Bennett et al. (1996) have recently reported that triceps surae reflex "gain" (i.e., the ratio of reflex and intrinsic force responses to sinusoidal displacements) was somewhat lower during Clonidine-induced locomotion than during tonic contractions, although reflexes still provided approximately 23% of the net force. Results from the decerebrate cat have thus provided much insight into the actions of reflex pathways, but the disrupted balance of descending inputs in this preparation lead to hyperactive reflexes (Hoffer et al. 1990; Sinkjaer and Hoffer 1990) which may overestimate reflex contributions in intact systems.

Functionally deafferented human subjects show large deficits in the control of arm movements (Sanes et al. 1985; Sainburg et al. 1993, 1995), especially when deprived of visual feedback of the limb, indicating that reflex action is an important component of the intact system for control of posture and movement. Furthermore, a number of previous studies in able-bodied human subjects have also shown significant reflex contributions at the ankle joint (Allum and Mauritz 1984; Sinkjaer et al. 1988; Toft et al. 1991; Yang et al. 1991), in the first dorsal interosseus (Carter et al. 1990; Doemges and Rack 1992a), in elbow flexor muscles (Crago et al. 1976; Kirsch and Rymer 1992), and at the wrist joint (Doemges and Rack 1992b). However, the effectiveness of reflex activity in generating a mechanical response is highly dependent on the properties of the perturbation imposed. For example, Kirsch and Rymer (1992) found that reflex contributions to human elbow joint stiffness were abolished for wide bandwidth stochastic perturbations, but significant reflex effects could be seen for more restricted perturbation bandwidths. Recently, Stein and Kearney (1995a) reported that the reflex ankle torque elicited by a rapid stretch of the human triceps surae muscles decreased in amplitude as the mean absolute velocity (rather than bandwidth *per se*) of superimposed pseudorandom perturbations was increased.

On the other hand, others (Grillner 1972; Valbo 1973; Bizzi et al. 1978) have reported much weaker reflex contributions, and the importance of reflex actions in posture and movement control has been minimized by others (Hollerbach 1982; Hogan 1985; Bizzi et al. 1992) on theoretical grounds because the delays in such systems would render any feedback system implemented by them unstable at gains sufficient to provide a significant functional improvement. As described above, the transient stiffness provided by intrinsic muscle properties can be substantial, and co-contraction of antagonistic muscles can be used to further increase net joint stiffness (and hence increase resistance to external disturbances) in an open-loop manner without increasing net joint moments. While the importance of intrinsic muscle properties in the control of posture and movement are well established, it is far from clear that these properties completely eliminate the need for a substantial reflex contribution. The studies cited above which found reflex action very weak were designed to quantify the "load compensation" actions of a presumed muscle length feedback system, a role for reflexes that few would support today. Furthermore, it is probably inappropriate to dismiss the contributions of reflex action solely because of its feedback delay, because such an argument uses linear engineering conventions to describe a highly nonlinear, multiinput, multioutput system, and there is little evidence to suggest that any of its outputs are even controlled in a servo-like manner by feedback (see Stein 1982). Indeed, it has been suggested that reflex pathways act together to maintain or regulate composite system properties such as stiffness (Nichols and Houk 1976; Houk and Rymer 1981a) or "equilibrium point" (Feldman 1966, 1986; Latash 1993) in the face of significant muscle nonlinearities such as yielding. The delayed timing of reflex action, rather than being a detriment to con-

trol of posture and movement, may in fact be precisely timed to compensate for such nonlinear intrinsic muscle properties (Houk et al. 1981b).

It thus appears that net reflex actions contribute significantly to the control of intact human posture and movement, at least under some conditions. Apart from enhancing the force response to an imposed disturbance, however, the global objective of these reflex actions remains quite controversial. A number of theories regarding the role of reflex action in the control of posture and movement have been advanced, some based on the microstructure of the spinal circuitry with little regard to functional requirements, while others have been based upon almost trivial mechanical properties bereft of underlying mechanisms. It is beyond the scope of this chapter to discuss these more global control theories in any more detail. However, several of these views will be discussed in chapters throughout this volume.

4   Future Directions

In suggesting directions for future work, it is perhaps appropriate to consider what factors currently limit our understanding of the control of posture and movement. Do we need more information about basic muscle contractile mechanisms or about the spinal connections of muscle afferents? Alternatively, are we accumulating increasing levels of fine details about particular system elements without the appropriate "big picture" to provide true insight? Will the evolution of current approaches and techniques eventually lead to an adequate understanding of the control of posture and movement, or is a radically different approach or technique required?

Some of these questions will be addressed by later chapters in this volume, which examine more global modeling and control issues. In reading these chapters, however, it should be noted that current theories and models of the control of posture and movement already neglect *most* of the vast amount of knowledge of muscle and reflex mechanisms that has accumulated over the years. Simplified descriptive muscle models proposed in the 1930s are almost universally used in large-scale models because of their reduced computational requirements relative to more recent models. However, these models predate the development of the sliding filament theory of contraction, the identification of motor unit types with different mechanical and metabolic properties, the discovery of the "size principal," and a number of other fundamental advances upon which our understanding of muscle properties is now based. Furthermore, the role of reflexes in posture and movement control is almost always discussed in terms of homonymous, servolike mechanisms, when substantial evidence indicates that intermuscular connections may be at least as important as homonymous ones and that the overall control strategies may not be accurately reflected at the single muscle or single joint level. Likewise, the use of constant-gain reflex gains in current large-scale models may significantly distort any insight into system performance, since these gains are known to be deeply modulated during different tasks. Although the inclusion of these (and other) properties in future models and theories will not guarantee full understanding of the control of posture and movement, we believe that a more realistic model will explain at least some of the inconsistencies in previous work.

The above suggestions were related to specific aspects of the neuromuscular system. On a more general level, we believe that global schemes for the control of posture and movement can be determined only from intact systems operating under reasonably natural conditions. Focused study on reduced elements of the neuromuscular system has provided (and will continue to provide) essential information about muscle and reflex mechanisms that cannot be obtained in other ways, but placing these mechanisms into a functional perspective has been less successful. Another challenge for the future will be to extract information on the normal interactions between the various elements in these complex systems without making the sweeping simplifications required today.

The authors of the five perspectives in this section have also provided suggested "Future Directions" that in many cases are consistent with those advocated above. Readers are urged to consider the more detailed suggestions that are provided in these chapters.

References

Akazawa K., Aldridge, J.W., Steeves, J.D., and Stein, R.B. (1982). Modulation of stretch reflexes during lo-comotion in the mesencephalic cat. *J. Physiol.*, 329: 553–567.

Allum, J.H.J. and Mauritz, K.-H. (1984). Compensation for intrinsic muscle stiffness by short-latency reflexes in human triceps surae muscles. *J. Neurophys.*, 52:797–818.

Baldissera, F., Hultborn, H., and Illert, M (1981). Integration in spinal neuronal systems. In *Handbook of Physiology. The Nervous System*. Brooks, V.B. (ed.) Vol. II, pp. 509–595, American Physiological Society, Bethesda, Maryland.

Bennett, D.J. (1994). Stretch reflex responses in the human elbow joint during a voluntary movement. *J. Physiol.*, 474:339–351.

Bennett, D.J., DeSerres, S.J., and Stein, R.B. (1996). Gain of the triceps surae stretch reflex in decerebrate and spinal cats during postural and locomotor activities. *J. Physiol.*, 496:837–850.

Binder, M.D., Heckman, C.J., and Powers, R.K. (1996). The physiological control of motoneuron activity. In *Handbook of Physiology. Exercise: Regulation and integration of multiple systems*. Rowell, L.B. and Sheperd, J.T. (eds.), Section 12, pp. 3–53, American Physiological Society, New York: Oxford.

Bizzi, E., Dev, P., Morasso, P., and Polit, A. (1978). Effect of load disturbances during centrally initiated movements. *J. Neurophysiol.*, 41:542–556.

Bizzi, E., Hogan, N., Mussa-Ivaldi, F.A., and Giszter, S. (1992). Does the nervous system use equilibrium-point control to guide single and multiple joint movements. *Behav. Brain Sci.*, 15:603–613.

Bonasera, S.J. and Nichols, T.R. (1994). Mechanical actions of heterogenic reflexes linking long toe flexors with ankle and knee extensors of the cat hindlimb. *J. Neurophysiol.*, 71:1096–1110.

Capaday, C. and Stein, R.B. (1986). Amplitude modulation of the soleus H-reflex in the human during walking and standing. *J. Neurosci.*, 6:1308–113.

Capaday, C. and Stein, R.B. (1987). Difference in the amplitude of the human soleus H reflex during walking and running. *J. Physiol.*, 392:513–522.

Carter, R.R., Crago, P.E., and Keith, M.W. (1990). Stiffness regulation by reflex action in the normal human hand. *J. Neurophysiol.*, 64:105–118.

Cooke, J.D. (1980). The role of stretch reflexes during active movements. *Brain Res.*, 181:493–497.

Crago, P.E., Houk, J.C., and Hasan, Z. (1976). Regulatory actions of human stretch reflex. *J. Neurophys.*, 39:925–935.

Crago, P.E., Houk, J.C., and Rymer,W.Z. (1982). Sampling of total muscle force by tendon organs. *J. Neurophysiol.*, 47:1069–1083.

Doemges, F. and Rack, P.M.H. (1992a). Changes in the stretch reflex of the human first dorsal interosseous muscle during different tasks. *J. Physiol.*, 447:563–573.

Doemges, F. and Rack, P.M.H. (1992b). Task-dependent changes in the response of human wrist joints to mechanical disturbance. *J. Physiol.*, 447:575–585.

Dufresne, J.R., Soechting, J.F., and Terzuolo, C.A. (1980). Modulation of the myotatic reflex gain in man during intentional movements. *Brain Res.*, 193:67–84.

Duysens, J., Tax, A.A., Trippel, M., and Dietz, V. (1992). Phase-dependent reversal of reflexly induced movements during human gait. *Exp. Brain Res.*, 90:404–14.

Eccles, J.C. and Lundberg, A. (1959). Supraspinal control of interneurones mediating spinal reflexes. *J. Physiol.*, 147:565–584.

Eccles, J.C., Eccles, R.M., and Lundberg, A. (1957). Synaptic actions on motoneurones caused by impulses in Golgi tendon organ afferents. *J. Physiol.*, 138:227–252.

Engberg, I., Lundberg, A., and Ryall, R.W. (1968). Reticulospinal inhibition of transmission in reflex pathways. *J. Physiol.*, 194:201–223.

Feldman, A.G. (1966). Functional tuning of the nervous system with control of movement of maintenance of a steady posture III. Mechanographic analysis of the execution by man of the simplest motor tasks. *Biophys., USSR*, 11:766–775.

Feldman, A.G. (1986). Once more on the equilibrium-point hypothesis (lambda model) for motor control. *J. Motor Behav.*, 18:17–54.

Fenn, W.O. and Marsh, B.S. (1935). Muscle force at different speeds of shortening. *J. Physiol.*, 85:277–297.

Flitney, F.W. and Hirst, D.G. (1978). Cross-bridge detachment and sarcomere "give" during stretch of active frog's muscle. *J. Physiol.*, 276:449–465.

Ford, L.E., Huxley, A.F., and Simmons, R.M. (1977). Tension responses to sudden length change in stimulated frog muscle fibres near slack length. *J. Physiol.*, 269:441–515.

Ford, L.E., Huxley, A.F., and Simmons, R.M. (1981). The relationship between stiffness and filament overlap in stimulated frog muscle fibres. *J. Physiol.*, 311:219–249.

Gielen, C.C. and Houk, J.C. (1984). Nonlinear viscosity of human wrist. *J. Neurophysiol.*, 52:553–569.

Gordon, A.M., Huxley, A.F., and Julian, F.J. (1966). The variation in isometric tension with sarcomere length in vertebrate muscle fibres. *J. Physiol.*, 184:170–192.

Gossard, J.P., Brownstone, R.M., Barajon, I., and Hultborn, H. (1994). Transmission in a locomotor-related group Ib pathway from hindlimb extensor muscles in the cat. *Exp. Brain Res.*, 98:213–228.

Gottlieb, G.L. and Agarwal, G.C. (1978). Stretch and Hoffmann reflexes during phasic voluntary contractions of the human soleus muscle. *Electroencephal. Clin. Neurophysiol.*, 44:553–561.

Gottlieb, G.L. and Agarwal, G.C. (1979). Response to

sudden torques about ankle in man: Myotatic reflex. *J. Neurophysiol.*, 42:91–106.

Gottlieb, G.L. and Agarwal, G.C. (1980). Response to sudden torques about ankle in man. III. Suppression of stretch-evoked responses during phasic contraction. *J. Neurophysiol.*, 44:233–246.

Grillner, S. (1972). The role of muscle stiffness in meeting the changin postural and locomotor requirements for force development of the ankle extensors. *Acta Physiol. Scand.*, 86:92–108.

Guertin, P., Angel, M.J., Perreault, M.C., and McCrea, D.A. (1995). Ankle extensor group I afferents excite extensors throughout the hindlimb during fictive locomotion in the cat. *J. Physiol.*, 487:197–209.

Hertel, H.L., Howaldt, B., and Mense, S. (1976). Responses of group IV and group III muscle afferents to thermal stimuli. *Brain Res.*, 113:201–205.

Hammond, P.H. (1956). The influence of prior instruction to the subject on an apparently involuntary neuromuscular response. *J. Physiol.*, 132:17–18.

Hayward, L., Wesselmann, U., and Rymer, W.Z. (1991). Effects of muscle fatigue on mechanically sensitive afferents of slow conduction velocity in the cat triceps surae. *J. Neurophysiol.*, 65:360–370.

Hill, A.V. (1938). The heat of shortening and the dynamic constants of muscle. *Proc. Roy. Soc. B.*, 126:136–1958.

Hoffer, J.A. and Andreassen, S. (1981a). Regulation of soleus muscle stiffness in premammillary cats: intrinsic and reflex components. *J. Neurophysiol.*, 45:267–285.

Hoffer, J.A. and Andreassen, S. (1981b). Limitations in the servo-regulation of soleus muscle stiffness in premammillary cats. In *Muscle Receptors and Movement.* Taylor, A. and Prochazka, A. (eds.), pp. 311–324, Macmillan, London.

Hoffer, J.A., Leonard, T.R., Cleland, C.L., and Sinkjaer, T. (1990). Segmental reflex action in normal and decerebrate cats. *J. Neurophysiol.*, 64:1611–1624.

Hogan, N. (1985). Impedance control: an approach to manipulation. Part II. Implemenation. *Trans. ASME*, 107:8–16.

Hollerbach, J.M. (1982). Computers, brains and the control of movement. *Trends Neurosci.*, 5:189–192.

Houk, J.C. and Rymer, W.Z. (1981a). Neural control of muscle length and tension. In *Handbook of Physiology. The Nervous System II.* Brooks, V.B. (ed.), pp. 257–323, Bethesda, Maryland, American Physiological Society.

Houk, J.C., Singer, J.J., and Goldman, M.R. (1970). An evaluation of length and force feedback to soleus muscles of decerebrate cats. *J. Neurophysiol.*, 33:784–811.

Houk, J.C., Crago, P.E., and Rymer, W.Z. (1981b). Function of the spindle dynamic response in stiffness regulation. A predictive mechanism provided by nonlinear feedback. In *Muscle Receptors and Movement*, Taylor, A. and Prochazka, A. (eds.), pp. 299–309, Macmillan, London.

Huyghues-Despointes, C.M.J.I., Cope, T.C., and Nichols, T.R. (1997). Intrinsic properties of cat triceps surae muscles in posture and locomotion. Soc. Neurosci. Abstr., (in press).

Huxley, A.F. (1957). Muscle structure and theories of contraction. *Prog. Biophys. Biophys. Chem.*, 7:255–318.

Huxley, A.F. (1974). Muscular contraction. *J. Physiol.*, 243:1–43.

Huxley, A.F. (1980). Reflections on muscle. Princeton University Press, Princeton, New Jersey.

Jack, J.J.B. and Roberts, R.C. (1980). Gain of the autogenetic Ib reflex pathways from the soleus muscle of the decerebrate cat. *J. Physiol.*, 305:42–43P.

Jankowska, E. (1992). Interneuronal relay in spinal pathways from proprioceptors. *Prog. in Neurobiol.*, 38: 335–378.

Johansson, R.S., Häger, C., and Riso, R. (1992). Somatosensory control of precision grip during unpredictable pulling loads. II. Changes in load force rate. *Exp. Brain Res.*, 89:192–203.

Johansson, R.S., Riso, R., Häger, C., and Bäckström, L. (1992). Somatosensory control of precision grip during unpredictable pulling loads. I. Changes in load force amplitude. *Exp. Brain Res.*, 89:181–191.

Joyce, G.C., Rack, P.M.H., and Westbury, D.R. (1969). The mechanical properties of cat soleus muscle during controlled lengthening and shortening movements. *J. Physiol.*, 204:461–474.

Jung, D.W.G., Blangé, T., de Graaf, H., and Treijtel, B.W. (1992). Cross-bridge stiffness in Ca^{2+}. Activated skinned single muscle fibers. *Pflügers Arch.*, 420:434–445.

Katz, B. (1939). The relation between force and speed in muscular contraction. *J. Physiol.*, 96:45–64.

Kirsch, R.F. and Kearney, R.E. (1993). Identification of time-varying dynamics of the human triceps surae stretch reflex. II. Rapid imposed movements. *Exp. Brain Res.*, 97:128–138.

Kirsch, R.F. and Rymer, W.Z. (1987). Neural compensation for muscular fatigue: evidence for significant force regulation in man. *J. Neurophysiol.*, 57:1893–1910.

Kirsch, R.F. and Rymer, W.Z. (1992). Neural compensation for fatigue induced changes in muscle stiffness during perturbations of elbow angle in man. *J. Neurophysiol.*, 68:449–470.

Kirsch, R.F., Boskov, D., and Rymer, W.Z. (1994). Muscle stiffness during transient and continuous movements of cat muscle: perturbation characteristics and physiological relevance. *IEEE Trans. Biomed. Eng.*, 41:758–770.

Kirsch, R.F., Kearney, R.E., and MacNeil, J.B. (1993). Identification of time-varying dynamics of the human triceps surae stretch reflex. I. Rapid isometric contractions. *Exp. Brain. Res.*, 97:115–127.

Lacquaniti, F. and Soechting, J.F. (1984). Behavior of the stretch reflex in a multi-jointed limb. *Brain Res.*, 311:161–166.

Lacquaniti, F. and Soechting, J.F. (1986a). EMG responses to load perturbations of the upper limb: effect of dynamic coupling between shoulder and elbow motion. *Exp. Brain Res.*, 61:482–496.

Lacquaniti, F. and Soechting, J.F. (1986b). Responses of mono- and bi-articular muscles to load perturbations of the human arm. *Exp. Brain Res.*, 65:135–144.

Lacquaniti, F., Borghese, N.A., and Carrozzo, M. (1991). Transient reversal of the stretch reflex in human arm muscles. *J. Neurophysiol.*, 66:939–954.

Latash, M. (1993). Control of human movement. Human Kinetics Publishers, Champaign, Illinois.

Lin, D.C. and Rymer, W.Z. (1993). Mechanical properties of cat soleus muscle elicited by sequential ramp stretches. Implications for control of muscle. *J. Neurophysiol.*, 70:997–1008.

Lombardi, V. and Piazzesi, G. (1990). The contractile response during steady lengthening of stimulated frog muscle fibers. *J. Physiol.*, 431:141–171.

Lombardi, V., Piazzesi, G., Ferenczi, M.A., Thirlwell, H., Dobbie, I., and Irving, M. (1995). Elastic distortion of myosin heads and repriming of the working stroke in muscle. *Nature*, 374:553–555.

Lum, P.S., Reinkensmeyer, D.J., Lehman, S.L., Li, P.Y., and Stark, L.W. (1992). Feedforward stabilization in a bimanual unloading task. *Exp. Brain Res.*, 89:172–180.

Lundberg, A. (1975). Control of spinal mechanisms from the brain. In *The Nervous System. The Basic Neurosciences*. Tower, D.B. (ed.), Vol. I, pp. 253–265, Raven Press, New York.

Lundberg, A., Malmgren, K., Schomburg, E.D. (1975). Convergence from Ib, cutaneous, and joint afferents in reflex pathways to motoneurons. *Brain Res.*, 87:81–84.

Malamud J.G., Godt, R.E., and Nichols, T.R. (1996). Relationship between short-range stiffness and yielding in type-identified, chemically skinned muscle fibers from the cat triceps surae muscles. *J. Neurophysiol.*, 76:2280–2289.

Marsden, C.D., Merton, P.A., and Morton, H.B. (1972). Servo action in human voluntary movement. *Nature*, 238:140–143.

Marsden, C.D., Merton, P.A., and Morton, H.B. (1976). Servo action in the human thumb. *J. Physiol.*, 257: 1–44.

Matthews, P.B.C. (1981). Muscle Spindles: their messages and their fusimotor supply. In *Handbook of Physiology. The Nervous System*. Bethesda, Maryland, Vol. II, American Physiological Society, Brooks, V.B. (ed.), pp. 189–228.

Matthews, P.B.C. (1986). Observations on the automatic compensation of reflex gain on varying the pre-existing level of motor discharge in man. *J. Physiol.*, 374:73–90.

McCrea, D.A., Shefchyk, S.J., Stephens, M.J., and Pearson, K.G. (1995). Disynaptic group I excitation of synergist ankle extensor motoneurones during fictive locomotion in the cat. *J. Physiol.*, 487:527–39.

McMahon, T.A. (1984). Muscles, reflexes, and locomotion. Princeton University Press, Princeton, New Jersey.

Mense, S. and Stahnke, M. (1983). Responses in muscle afferent fibers of slow conduction velocity to contractions and ischemia in the cat. *J. Physiol.*, 342:383–397.

Miller, J.F., Paul, K.D., Lee, R.H., Rymer, W.Z., and Heckman, C.J. (1996). Restoration of extensor excitability in the acute spinal cat by the 5-HT$_2$ agonist DOI. *J. Neurophysiol.*, 75:620–628.

Morgan, D.L. (1990). New insights into the behavior of muscle during active lengthening. *Biophys. J.*, 57:209–221.

Morgan, D.L. (1994). An explanation for residual increased tension in striated muscle after stretch during contraction. *Exp. Physiol.*, 79:831–838.

Mortimer, J.A., Webster, D.G., and Dukich, T.G. (1981). Changes in short and long latency stretch responses during the transition from posture to movement. *Brain Res.*, 229:337–351.

Nichols, T.R. (1994). A biomechanical perspective on spinal mechanisms of coordinated muscular action: an architecture principle. *Acta Anatomica*, 151:1–13.

Nichols, T.R. and Houk, J.C. (1976). Improvement in linearity and regulation of stiffness that results from actions of stretch reflex. *J. Neurophysiol.*, 29:119–142.

Pratt, C.A. (1995). Evidence of positive force feedback among hindlimb extensors in the intact standing cat. *J. Neurophysiol.*, 73:2578–2583.

Rack, P.M.H. and Westbury, D.R. (1969). The effects of length and stimulus rate on tension in the isometric cat soleus muscle. *J. Physiol.*, 204:443–460.

Rack, P.M.H. and Westbury, D.R. (1974). The short range stiffness of active mammalian muscle and its effect on mechanical properties. *J. Physiol.*, 240:331–350.

Rudomin, P. (1990). Presynaptic control of muscle spindle and tendon organ afferents in the mammalian spinal cord. In *The segmental motor system*. Binder, M.D. and Mendell, L.M. (eds.), pp. 349–380, Oxford University Press, New York.

Rudomin, P. (1994). Segmental and descending control of the synaptic effectiveness of muscle afferents. *Prog. Brain Res.*, 100:97–104.

Rymer, W.Z. and Hasan, Z. (1980). Absence of force-feedback in soleus muscle of the decerebrate cat. *Brain Res.*, 184:203–209.

Rymer, W.Z., Houk, J.C., and Crago, P.E. (1979). Mechanisms of the clasp-knife reflex studied in an animal model. *Exp. Brain Res.*, 37:93–113.

Sainburg, R.L., Poizner, H., and Ghez, C. (1993). Loss of proprioception produces deficits in interjoint coordination. *J. Neurophysiol.*, 70:2136–2147.

Sainburg, R.L., Ghilardi, M.F., Poizner, H., and Ghez, C. (1995). Control of limb dynamics in normal subjects and patients without proprioception. *J. Neurophys.*, 73:820–835.

Sanes, J.N., Mauritz, K.-H., Dalakas, M.C., and Evarts, E.V. (1985). Motor control in humans with large-fiber sensory neuropathy. *Hum. Neurobiol.*, 4:101–114.

Schomburg, E.D. (1990). Spinal sensorimotor systems and their supraspinal control. *Neurosci. Res.*, 7:265–340.

Shefchyk, S.J., Stein, R.B., and Jordan, L.M. (1984). Synaptic transmission from muscle afferents during fictive locomotion in the mesencephalic cat. *J. Neurophysiol.*, 51:986–997.

Sinkjaer, T. and Hoffer, J.A. (1990). Factors determining segmental reflex action in normal and decerebrate cats. *J. Neurophysiol.*, 64:1625–1635.

Sinkjaer, T., Toft, E., Andreassen, S., and Hornemann, B.C. (1988). Muscle stiffness in human ankle dorsiflexors: intrinsic and reflex components. *J. Neurophysiol.*, 60:1110–1121.

Smeets, J.B.J. and Erkelens, C.J. (1991). Dependence of autogenetic and heterogenetic stretch reflexes on preload activity in the human arm. *J. Physiol.*, 440:455–465.

Soechting, J.F., Dufresne, J.R., and Lacquaniti, F. (1981). Time-varying properties of myotatic response in man during some simple motor tasks. *J. Neurophysiol.*, 46:1226–1243.

Stein, R.B. (1980). Nerve and muscle. Plenum Press, New York.

Stein, R.B. (1982). What muscle variable(s) does the nervous system control in limb movements? *Behav. Brain Sci.*, 5:535–577.

Stein, R.B. (1995). Presynaptic inhibition in humans. *Prog. Neurobiol.*, 47:533–544.

Stein, R.B. and Capaday, C. (1988). The modulation of human reflexes during functional motor tasks. *Trends Neurosci.*, 11:328–332.

Stein, R.B. and Kearney, R.E. (1995a). Nonlinear behavior of muscle reflexes at the human ankle joint. *J. Neurophysiol.*, 73:65–72.

Stein, R.B., Hunter, I.W., Lafontaine, S.R., and Jones, L.A. (1995b). Analysis of short-latency reflexes in human elbow flexor muscles. *J. Neurophysiol.*, 73:1900–1911.

Toft, E., Sinkjaer, T., Andreassen, S., and Larsen, K.

(1991). Mechanical and electromyographic responses to stretch of the human ankle extensors. *J. Neurophysiol.*, 65:1402–1410.

Valbo, A.B. (1973). The significance of intramuscular receptors in load compensation during voluntary contractions in man. In *Control of posture and locomotion.* Stein, R.B., Pearson, K.G., and Smith, R.S. (eds.), pp. 211–226, Plenum, New York.

van der Linden, B.J.J.J., Koopmen, H.F.J.M., Huijing, P.A., and Grootenboer, H.J. (1996). Is high-frequency stiffness a measure for the number of attached crossbridges? *Proc. 18th Ann. Int. Conf. IEEE Eng. Med. Biol. Soc.*, 18:Section 2.8.1.

Walmsley, B. and Proske, U. (1981). Comparison of stiffness of soleus and medial gastrocnemius muscles in cats. *J. Neurophysiol.*, 46:250–259.

Yang, J.F., Stein, R.B., and James, K.B. (1991). Contribution of peripheral afferents to the activation of the soleus muscle during walking in humans. *Exp. Brain Res.*, 87:679–87.

Zahalak, G.I. (1981). A distribution-moment approximation for kinetic theories of muscular contraction. *Math. Biosci.*, 55:89–114.

Zahalak, G.I. (1986). A comparison of the mechanical behavior of the cat soleus muscle with a distribution-moment model. *J. Biomech. Eng.*, 108:131–140.

3

Intraoperative Sarcomere Length Measurements Reveal Musculoskeletal Design Principles

Richard L. Lieber and Jan Fridén

1 Summary

This chapter describes a series of experiments in which sarcomere length was measured in human upper extremity muscles in order to understand the design principles of these muscles. Such measurements have been combined with studies on cadaveric extremities to generate biomechanical models of human muscle function and to provide insights into the design of upper extremity muscles.

Intraoperative measurements of the human extensor carpi radialis brevis (ECRB) muscle during wrist joint rotation reveal that this muscle appears to be designed to operate on the descending limb of its length tension curve and generates maximum tension with the wrist fully extended. Interestingly, the synergistic extensor carpi radialis longus (ECRL) also operates on its descending limb but over a much narrower sarcomere length range. This is because of the longer fibers and smaller wrist extension moment arm of the ECRL compared to the ECRB. Sarcomere lengths measured from wrist flexors are shorter compared to the extensors. Using a combination of intraoperative measurements on the flexor carpi ulnaris (FCU) and mechanical measurements of wrist muscles, joints and tendons, the general design of the prime wrist movers emerges: both muscle groups generate maximum force with the wrist fully extended. As the wrist flexes, force decreases due to extensor *lengthening* along the descending limb of their length-tension curve and flexor *shortening* along the ascending limb of their length-tension curve. The net result is a nearly constant ratio of flexor to extensor torque over the wrist range of motion and a wrist that is most stable in full extension.

Similar measurements have been made intraoperatively during surgical tendon transfer procedures. These procedures are used to restore lost muscle function after head injury, peripheral nerve injury, and stroke. In transfers of the FCU to either the ECRL or extensor digitorum communis (EDC) muscles, the relatively short fibers of the FCU make setting sarcomere length a critical choice that determines the ultimate functionality of these transfers.

Taken together, these experiments demonstrate the elegant match among muscle, tendon, and joints acting at the wrist. Overall, the wrist torque motors appear to be designed for balance and control rather than maximum torque generating capacity.

2 Introduction

Limb movement results from mechanical interaction among skeletal muscles, tendons, and joints. The anatomy of these structures has been studied extensively at both the gross and microscopic levels (Huxley, 1974; Butler et al, 1979; Ebashi et al, 1980; Squire, 1981; Gans, 1982). Skeletal muscles are responsible for force generation during movement and have been shown to have a wide range of designs that appear to be matched to their functional tasks (Walmsley et al. 1978; Brand et al. 1981, Lieber et al. 1992, Sacks and Roy 1982). An added factor in understanding this system is that these muscles of varying design insert onto bones with a wide range of mechanical advantages yielding "torque motors" of designs that result from unique juxtaposition of muscle and joint properties (Figure 3.1).

Because relative muscle force is highly dependent on sarcomere length within the muscle, we have focused on sarcomere length measurements to provide insights into the design and function of the neuromuscular system. Prior studies of this sort in a variety of animal systems have yielded intriguing results. For example, it was demonstrated that fish fast and slow skeletal muscles operate near the peak of their power-velocity relationships at near optimal sarcomere length (Rome et al. 1988; Rome et al. 1991, 1993). During locomotion in cats, medial gastrocnemius and soleus muscles are activated in such a manner as to exploit their metabolic and force generating properties (Walmsley et al. 1978, Walmsley and Proske 1981). Finally, frog skeletal muscles appear to be designed so as to maximize either power production during hopping (Lutz and Rome 1994) or moment transfer in the biarticular musculature (Lieber and Boakes 1988, Mai and Lieber 1990). These studies suggest an impressive coordination among muscle, tendon, and joint structural properties that are suited to the functional tasks. In some of these studies, understanding the design of the system was based on a description of the relationship between sarcomere length in a particular muscle and the joint on which it operates. If muscle moment arm is relatively large compared to the number of serial sarcomeres, sarcomere length change and thus force change will be large as the joint rotates whereas the same muscle acting on a smaller moment arm will have a more modest force change during rotation (Figure 3.1).

The studies reviewed in this chapter rely on intraoperative sarcomere length measurements in human upper extremity muscles combined with studies of cadaveric extremities to yield biomechanical models of human muscle function and to provide insights into the design and function of upper extremity muscles. We developed an intraoperative laser diffraction method for measuring human muscle sarcomere length (Friden and Lieber 1994, Lieber et al. 1994). This method is noninjurious to the muscle and simply relies on the fact that laser light is diffracted by the striation pattern present in all skeletal muscles (Lieber et al. 1984). Since striation spacing is a direct manifestation of sarcomere length and sarcomere length is a good predictor of relative isometric muscle force, intraoperative sarcomere length measurements may provide insights into normal muscle and joint function.

3 Methods

3.1 Patient Populations

The subjects included in these studies were undergoing radial nerve release because of compression at the level of the supinator fascia (for combined extensor carpi radialis brevis [ECRB] and extensor carpi radialis longus [ECRL] measurements,

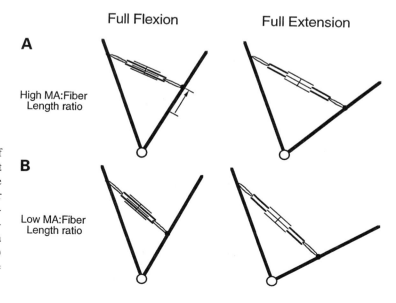

FIGURE 3.1. Schematic depiction of sarcomere length change during joint rotation for two hypothetical "torque motors." In (A), moment arm is larger relative to sarcomere number compared to (B). Thus, as the "joint" extends, sarcomere length changes a greater amount in (A) compared to (B) with a correspondingly larger change in force.

$n = 7$), surgical lengthening of the ECRB tendon for treatment of chronic lateral epicondylitis ("tennis elbow," $n = 5$), or tendon transfer for radial nerve palsy ($n = 5$). All patients received detailed instruction regarding experimental protocols. All procedures performed were approved by the Committee on the Use of Human Subjects at the University of Umeå and the University of California, San Diego.

3.2 Laser Diffraction Device

The device used was a 7-mW helium-neon laser (Melles-Griot, Model LHR-007, Irvine, California) onto which was mounted a custom component that permitted alignment between the laser beam and a small triangular prism (Melles-Griot Model 001PRA/001, Irvine, California). The laser beam was incident onto one of the short sides of the triangular prism, reflected off of the aluminum-coated surface of the other prism face, exiting 90° to the incident beam (Figure 3.2). Thus the prism could be placed

underneath a muscle bundle and transilluminated to produce a laser diffraction pattern (Lieber et al. 1994).

The diffractometer was calibrated using diffraction gratings of 2.50 μm and 3.33 μm spacings placed directly upon the prism. Diffraction order spacings from the ±2nd order were measured to the nearest 0.1 mm using dial calipers which corresponded to a sarcomere length resolution of about 0.05 μm. Repeated measurement of the same diffraction order spacing by different observers resulted in an average calibration of the 3.33 μm grating as 3.38 μm ± 0.13 μm (mean ± standard deviation).

All sarcomere lengths were calculated using the +2 to −2 diffraction order spacing. Redundant measurements of +1 to −1 and +3 to −3 were also made to ensure calculation accuracy. The 2nd order was chosen because the larger order spacing made small absolute spacing measurement errors proportionately less than for the 1st order and the intensity was usually greater than that of the 3rd order. Diffraction an-

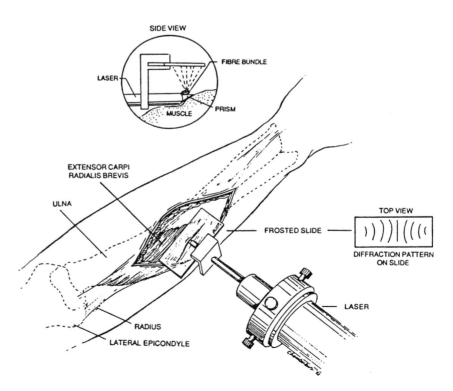

FIGURE 3.2. Device used for intraoperative sarcomere length measurements. The He-Ne laser is aligned normal to the transmitting face of the prism for optimal transmission of laser power into the muscle. Second order diffraction spacing was measured manually using calipers. Inset shows a transverse view of the illuminating prism placed beneath a muscle fiber bundle. (Used with permission from Lieber et al. 1994.)

gle (q) was calculated using the grating equation, $nl = d\sin(q)$, where l is the laser wavelength (0.632 μm), d is sarcomere length, and n is diffraction order (2nd in all cases) and assuming that the 0th order bisected the orders on either side.

3.3 Intraoperative Protocols

3.3.1 ECRB and ECRL Sarcomere Length Measurements

Immediately after administration of regional anesthesia, the distal ECRB and ECRL were exposed using a dorsoradial incision approximately 10 cm proximal to the radiocarpal joint. The overlying fascia was divided exposing the underlying muscle fibers. A small fiber bundle was isolated near the insertion site using delicate blunt dissection in a natural intramuscular fascial plane, with care not to overstretch muscle fibers. Because both muscles have characteristic appearance in this distal region, care was taken to isolate muscle fibers from the same region in all patients studied.

3.3.2 FCU Sarcomere Length Measurements

After induction of general anesthesia, the muscle was exposed through a 3-cm longitudinal incision on the

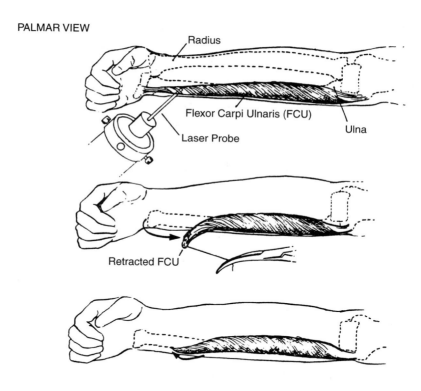

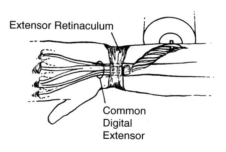

FIGURE 3.3. Intraoperative laser diffraction of the flexor carpi ulnaris muscle during transfer into the common tendons of the extensor digitorum communis. Sarcomere length is measured prior to and after tendon transfer. (Used with permission from Lieber et al. 1996.)

ulnar–palmar surface of the distal forearm. After exposure of the FCU by incision of the fascial sheath, a 5-mW He-Ne laser beam was directed into the distal portion of the FCU muscle fibers (Figure 3.3). With the elbow in 20° of flexion and the forearm supinated, FCU sarcomere length was measured with the wrist in full flexion, neutral, and full extension. The FCU was then transferred into the ECRL tendon with the wrist in a neutral position. Wrist joint angle was measured with a goniometer and was not identical in all subjects because of variations in anatomy and intraoperative positioning. After transfer, the wrist, forearm, and elbow were returned to their pre-transfer configuration and FCU sarcomere length was again measured in each of the three wrist positions. Preoperatively, the FCU was thus examined as a wrist flexor and postoperatively the FCU was examined as a wrist extensor. Digits were held in the flexed position for all measurements.

3.4   Biomechanical Simulation

To predict the muscle force and wrist extension moment generated by the selected muscles, the biomechanical model previously described (Loren et al. 1996) was implemented. This model was based on experimental measurement of prime wrist mover muscle architecture (Lieber et al. 1990) and the me-

chanical properties of each wrist tendon (Loren and Lieber 1995). Muscle properties were predicted based on architectural values obtained on cadaveric forearms (Lieber et al. 1990) yielding muscle length-force curves scaled to fiber length and physiological cross-sectional area. Briefly, the model predicts muscle moment as a function of joint angle using the known properties of the muscle and tendon and the appropriate moment arms. Sarcomere lengths predicted by the model compare favorably to those obtained intraoperatively from the extensor carpi radialis brevis and FCU muscles (cf. Loren and Lieber, 1995, Figure 7).

4   Results

4.1   ECRB Sarcomere Lengths

The average flexed wrist joint angle was approximately $-50°$ while average extension angle was approximately $+50°$. Therefore, sarcomere lengths were measured over a 100° range of wrist motion. With the wrist in full extension, sarcomere length was about 2.6 μm (Figure 3.4), which was significantly shorter (and thus would develop about 50% of the tension) than the 3.4 μm sarcomere length measured in the flexed position ($p < 0.005$). Sarcomere length with the joint in the neutral position

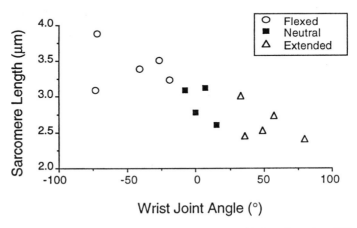

FIGURE 3.4. Sarcomere length versus wrist joint angle relationship determined for the 5 experimental subjects. Negative angles represent wrist flexion relative to neutral, while positive angles represent wrist extension. One-way ANOVA revealed a significant difference between wrist joint angles and sarcomere lengths in the three positions. (O) flexed angles, (■) neutral angles, (△) extended angles. Note that one point is missing from subject LA at a neutral joint angle. Schematic sarcomeres placed upon data are based on quantitative measurement of filament lengths from human muscle biopsies. (Used with permission from Lieber et al. 1994.)

was intermediate between these two values (3.0 μm) and significantly different than that observed in the flexed (p < 0.05) but not the extended position.

Note that these sarcomere lengths are significantly longer than those found in frog or fish muscle, due to the longer actin filaments contained in human muscle (Walker and Schrodt, 1973). Thus, quantitative electron microscopy of human ECRB muscle tissue revealed actin filament lengths within half of the I-band of 1.30 ± .027 μm while myosin filaments were 1.66 ± .027 μm long, yielding an optimal sarcomere length of 2.60 to 2.80 μm and a maximum sarcomere length for active force generation of 4.26 μm (Lieber et al. 1994) compared to an optimal sarcomere length of 2.20 μm and a maximum sarcomere length for active force generation of 3.65 μm in frog and fish muscle (Sosnicki et al. 1991; Gordon et al. 1966).

4.2   Average Slope of ECRB and ECRL Relationships

For all seven subjects studied, the slope of the sarcomere length-joint angle curve (i.e., dSL/dw) was greater for the ECRB muscle compared to the ECRL. A relatively large degree of variability in dSL/dw was observed between subjects for both the ECRB and ECRL (Figure 3.5). For example, ECRB dSL/dw

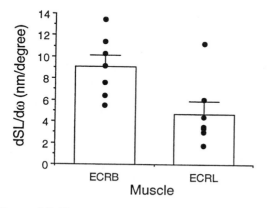

FIGURE 3.5. Slope dSL/dw of the sarcomere length-joint angle relationshp for the ECRB and ECRL muscles. Values are actually calculated as sarcomere length change per degree joint extension and are thus negative numbers since sarcomeres shorten with wrist extensions. They are plotted as positive values for convenience. (Used with permission from Lieber et al. 1997.)

ranged from −5.4 nm/° to −13.5 nm/° while those for the ECRL ranged from −1.7 nm/° to −11.2 nm/°. Across all subjects, average dSL/dw for the ECRB was −9.06 nm/° while that of the ECRL was about half of this value, or 4.69 nm/°. We computed the ratio of dSL/dw between the ECRB and ECRL for each subject and averaged these values across subjects. This average dSL/dw ratio for the intraoperative sarcomere length data was 2.45.

4.3   Surgical Transfer of FCU to ECRL

Sarcomere lengths measured after transfer into the ECRL tendon were systematically longer than those measured prior to transfer. For example, maximum sarcomere length was significantly shorter pre- as compared to posttransfer (corresponding to the wrist extended pretransfer and flexed posttransfer) by 0.7 μm (p < 0.01). The post-transfer sarcomere length of 4.82 ± .11 μm corresponded to a sarcomere length that would actually be well beyond filament overlap given the actin filament length of 2.6 μm and myosin filament length of 1.66 μm previously measured in human upper extremity muscles (Lieber et al. 1994). It was noted intraoperatively that with the wrist fully flexed, posttransfer, significant passive tension opposed wrist flexion at these very long sarcomere lengths. Overall, the sarcomere length change throughout the range of motion posttransfer was 1.63 ± .09 μm compared to only 1.32 ± .15 μm, the difference between which was not statistically significant (p > 0.1, b = 0.55).

4.4   Predicted Sarcomere Lengths

Sarcomere length pretransfer showed a relatively linear increase from ~2.5 μm to ~4.5 μm as the wrist was extended from ~60° of flexion to ~60° of extension. These length changes were well-approximated by the relationship predicted (solid line, Figure 3.6A) from muscle, tendon and joint properties (Loren et al. 1996). In fact, the coefficient of determination (r^2) between the experimental data and linear model was 0.84 demonstrating explanation of over 80% of the experimental variability by theory. After transfer, the sarcomere length predicted was curvilinear and provided a somewhat poorer fit to the data with an r^2 value of only 0.68 (Figure

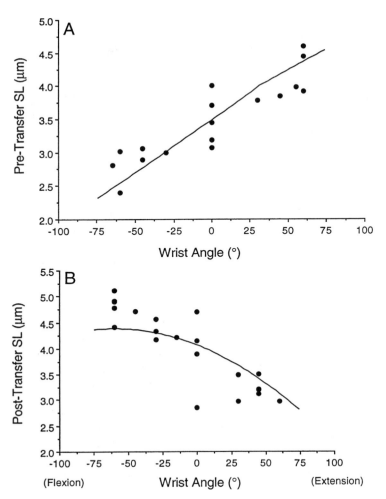

FIGURE 3.6. Relationship between wrist joint angle and sarcomere length measured intraoperatively. (A) Pretransfer. (B) Posttransfer into ECRL tendon. Solid line represents predicted sarcomere length from biomechanical model. Small fluctuations in lines are printer artifacts. (Used with permission from Lieber and Friden 1997.)

3.6B). Again, however, the basic data form was well approximated by the predicted relationship.

4.5 Model of FCU-to-ECRL Tendon Transfer

Given the good fit between experiment and theory, we further predicted all possible wrist extensor moments produced by the transferred FCU as a function of wrist angle and FCU muscle length at the time of transfer (Figure 3.7); see Chapter 2 for a review of the force-length relation of muscle. In general, joint moment increased as the wrist extended. This was largely the result of the linear increase in ECRL moment arm with extension as the extensor tendon elevated off of the radius up to the underside of the extensor retinaculum. Superimposed on this

kinematic relationship is the fact that changing FCU length at the time of surgery changes the region of the sarcomere length-tension curve over which the FCU operates. For example, when the muscle is inserted at very long lengths (250–260 mm; sarcomere length ~3.9 μm) it operated on the extreme end of the descending limb of the length-tension curve and muscle force increased with extension as the muscle shortened "up" the descending limb. Given the actual intraoperative sarcomere length measurement of 3.89 μm in this joint configuration, this is essentially the situation that describes the surgical patients in the current study. At the other extreme, when the muscle is inserted at very short lengths (200–210 mm; sarcomere length ~2.1 μm), it operated on the ascending limb and plateau of its length-tension curve but the overall joint moment was rel-

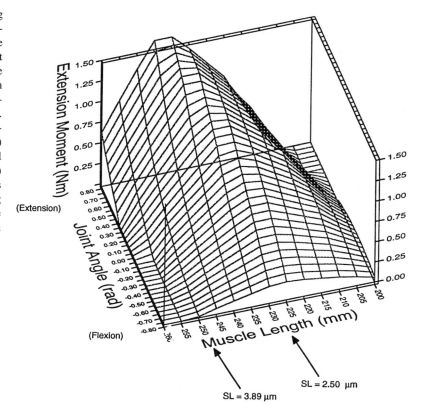

FIGURE 3.7. Relationship among predicted wrist extension moment (Nm) and wrist joint angle (radians) and length of muscle at time of transfer (mm). Muscle lengths corresponding to 3.89 μm and 2.5 μm are shown and correspond to the data in Figure 3.6. Note that a relatively large sarcomere length change (~50%) corresponds to a relatively small muscle length change (~10%) because the FCU muscle has very short fibers arranged along the muscle length (cf. Table 11.1). (Used with permission from Lieber and Friden 1997.)

atively low since, as the muscle developed maximal force, the moment arm decreased to a minimum. The largest joint moments were predicted at intermediate muscle lengths of 220 to 230 mm (sarcomere length 2.5 μm) because the muscle generated near maximum force on the plateau of the length-tension curve with the wrist extended and moment arm was nearly maximal with the wrist extended.

5   Discussion

The main result of these studies is that sarcomere length operating range varies between muscles of the upper extremity in a way that suggests a balance among muscle, tendon and joint properties.

5.1   Proposed Design of the ECRB Muscle

We have collected the majority of our data on the ECRB muscle because of its accessibility and common involvement in surgical procedures. We found

that human passive ECRB sarcomere lengths varied from 2.6 μm to 3.4 μm throughout the full range of wrist joint motion, while active sarcomere lengths ranged from 2.44 μm to 3.33 μm. Given the measured actin filament length of 1.30 μm and myosin filament length of 1.66 μm, these data suggest that the muscle operates primarily on the plateau and descending limb of its sarcomere length-tension curve (Figure 3.8). Assuming that human muscles generate force as do frog skeletal muscles, optimal sarcomere length would occur between 2.60 μm and 2.80 μm, which agrees well with the optimal sarcomere length of 2.64 μm to 2.81 μm predicted by Walker and Schrodt (1973) based on filament length measurements. These data suggest that the ECRB muscle would develop near-maximal isometric force at full wrist extension, force would remain relatively constant as the sarcomeres lengthened "over" the plateau region, and then force would decrease to about 50% maximum at full wrist flexion. This result contrasts with the generally accepted notion that skeletal muscles generate maximum forces with the joint in a neutral position. We conclude, therefore, that muscle force

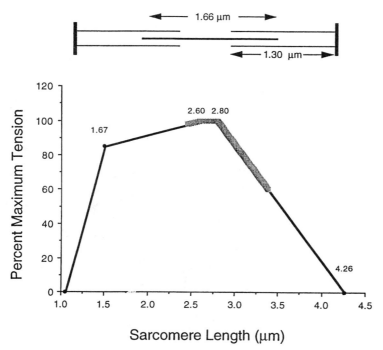

FIGURE 3.8. Hypothetical length-tension curve obtained using measured filament lengths and assuming the sliding filament mechanism (Gordon et al. 1966b). Shaded area represents sarcomere length change during wrist flexion (causing sarcomere length increase) and wrist extension (causing sarcomere length decrease). (Top) Schematic of filament lengths measured in the current study. Numbers over graph represent calculated inflection points based on filament lengths measured and a Z-disk width of 100 nm. (Used with permission from Lieber et al. 1994.)

change as a result of sarcomere length changes during joint rotation is "built-in" as part of the control in the musculoskeletal system and not simply a consequence of muscle microanatomy. Of course, the actual muscle force generated at a given angle depends not only on sarcomere length, but also on the number and firing frequency of motor units. Thus, the change in sarcomere length might be viewed as the "upper limit" for force production at a given joint angle.

5.2 Comparison Between ECRB and ECRL Muscles

Based on the measurement of sarcomere length changes of the ECRB and ECRL muscles with wrist joint rotation, we conclude that these muscles, while synergistic in terms of location and pattern of activation (Backdahl and Carlsoo, 1961; McFarland et al. 1962, Riek and Bawa 1992), have quite different anatomical designs that are predicted to produce different functional properties of the muscle-tendon-joint torque generating system.

To predict the expected ratio between dSL/dw for the two muscles, muscle lengths and number of serial sarcomeres were obtained from upper extremity architecture data (Lieber et al. 1990) while wrist

extensor moment arms can be obtained from published kinematic data (Loren et al. 1996). Sarcomere length change during wrist extension was thus modeled by integrating the moment arm equation over the range from about 40° of flexion to about 10° of extension (the average flexed and neutral positions respectively) for the ECRB:

$$r_B = \int_{\varphi=-.5}^{\varphi=.25} 16 + 8.95 \cdot \varphi + 2.84 \cdot \varphi^2 - 11.0 \cdot \varphi^3 - 12.9 \cdot \varphi^4 \qquad (3.1)$$

and for the ECRL:

$$r_L = \int_{\varphi=-.5}^{\varphi=.25} 10.5 + 9.9\varphi \qquad (3.2)$$

where wrist joint angle is in radians and negative angles refer to wrist flexion. Given the relationship between muscle excursion (s), moment arm ($r(\varphi)$) and angular rotation (f) of d$s = r$dφ, the ratio of sarcomere length changes with wrist extension for the two muscles is given by the expression:

$$\frac{dSL_R}{dSL_L} = \frac{r_B \cdot n_L}{r_L \cdot n_B} \qquad (3.3)$$

where dSL is change of sarcomere length with joint rotation (i.e., dSL/dw), n is number of serial sarcomeres, and the subscripts B and L refer to values for ECRB and ECRL, respectively. The integrated moment arms for ECRB and ECRL were 11.37 and 6.94 respectively, while the serial sarcomere numbers were 27,143 and 17,143 respectively. Inserting these values into Eq. 3.3 yields a sarcomere length change ratio of 2.57 which is close to the experimentally measured value of 2.45. This close agreement not only supports the modeling approach, it permits us to use this mechanical model to predict the joint moment of the wrist extensors with confidence.

In the context of the current study, the major factor determining the properties of the torque motor is the ratio between the moment arm and the number of sarcomeres in series within the muscle. Using ordinary muscle architectural terminology, this would be expressed as the moment arm:fiber length ratio (Zajac 1992). From a design point of view, high moment arm:fiber length ratio results in a torque motor in which large fiber length changes produce large force changes during joint rotation and, thus, this motor would vary torque output greatly as the joint rotated. This is more or less the design of the ECRB torque motor for wrist extension motions, based on its relatively short fibers and large moment arm (Figure 3.9). In contrast, the significantly longer fibers and smaller moment arm of the ECRL results in a torque motor with differ-

ent functional properties. The ECRL-based motor retains a more constant torque output with joint rotation because, for a given amount of joint rotation, sarcomere length changes less (Figure 3.4). In radial deviation, the situation is quite different. In this case, the increased ECRL radial deviation moment arm compared to the ECRB almost exactly matches their fiber length ratio (Loren et al. 1996). Thus, for radial deviation movements, the sarcomere length change per radial deviation angle for both muscles is probably nearly equivalent. The fact that the ECRL has a substantial moment arm at the elbow while the ECRB has almost none does not seem to provide insight into its long-fibered design. The functional effect of such a design would be to maintain sarcomere length relatively constant in the ECRL with simultaneous wrist flexion and elbow extension while not affecting ECRB function.

5.3   Implications for FCU-to-ECRL Tendon Transfer Surgery

The relatively complete description of the biomechanics of this tendon transfer permits a rational choice regarding the setting of muscle length at the time of transfer. This intraoperative decision represents the most important application of the use of these data. Using the data from the biomechanical

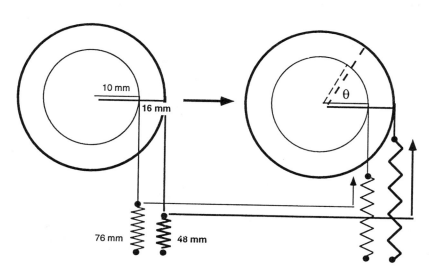

FIGURE 3.9. Schematic diagram of the interrelationship between fiber length and moment arm for the ECRB and ECRL torque motors. The ECRB (bold print, thick lines) with its shorter fibers and longer moment arm changes

sarcomere length about 2.5 times as much as the ECRL with its longer fibers and smaller moment arm. (Used with permission from Lieber et al. 1997.)

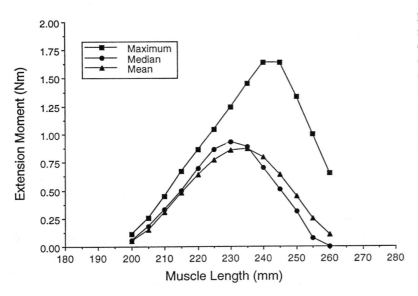

FIGURE 3.10. Wrist extension moment parameters as a function of FCU muscle length at the time of tendon transfer. Values were determined from the biomechanical model shown in Figure 3.7. (■) maximum extension moment, (●) median extension moment, (▲) mean extension moment. (Used with permission from Lieber and Friden 1997.)

model (Figure 3.7), if the clinical goal is to maximize overall joint moment, the intraoperative decision would be to insert the muscle at a length of ~245 mm (Figure 3.9, squares). From a different point of view, high force may be desired throughout the entire range of motion in which case it would be preferable to optimize median or mean muscle force (Figure 3.10, circles and triangles respectively). If this is the choice, the muscle should be sutured into the ECRL tendon at a shorter muscle length of ~230 mm. Of course, it is not yet possible to answer this question definitively since no prospective clinical tri-

als have yet been performed to quantify the outcome resulting from choice between these possibilities.

5.4 Design of the Wrist as a "Torque Motor"

Based on the relatively complete data set regarding wrist muscle and joint properties, a few general trends emerge. Wrist extensors are predicted to operate primarily on the plateau and descending limb of their sarcomere length-tension curve (Figure 3.11) with all muscles generating maximal force in full ex-

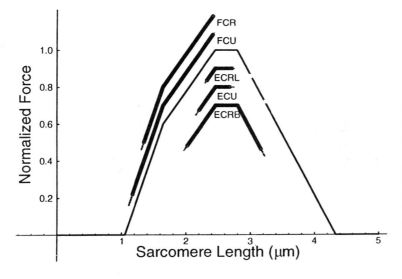

FIGURE 3.11. Operating ranges of the wrist motors on the isometric sarcomere length-tension relation. Extensors operated primarily on the plateau region while the flexors operated predominantly along the ascending and steep ascending limbs. Mean sarcomere operating ranges were determined independently of the ensemble average muscle force- and torque-joint angle relations.

tension. Only the ECRB was predicted to operate at sarcomere lengths corresponding to less than 80% P_o in the normal range of motion. Wrist flexors are predicted to operate predominantly on the ascending and steep ascending limbs of their length-tension curve with both flexors generating maximal force in full wrist extension (Figure 3.11). Note that in full flexion, it is possible for wrist flexors to generate forces that are less than 50% P_o.

Such a design presents interesting implications for the design of the wrist as a torque motor. Both flexor and extensor muscle groups generate maximum force with the wrist fully extended. As the wrist extends, maximum isometric extensor force increases due to extensor shortening up the descending limb of the length-tension curve and maximum isometric flexor force increases due to flexor lengthening up the ascending limb of the length-tension curve. This effect is superimposed upon an increasing extensor moment arm as the extensor muscles elevate off of the wrist under the extensor retinaculum and a decreasing flexor moment arm as

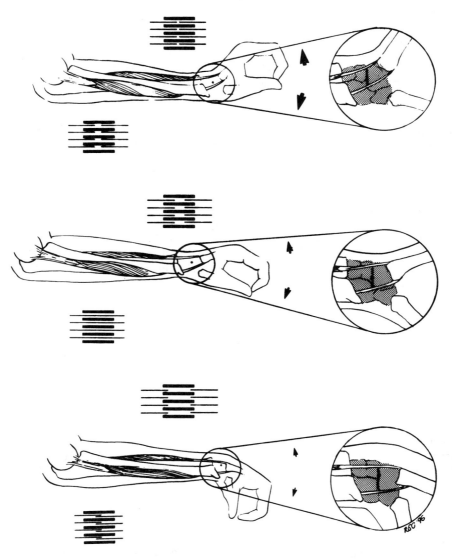

FIGURE 3.12. Schematic depiction of sarcomere length, moment arm, and moment in wrist flexors and extensors as the wrist moves throughout its range of motion.

the flexors juxtapose the wrist from the flexor reti- naculum. Combining the muscle and joint effects, extensor muscle force is amplified by an increasing extensor moment arm and flexor muscle force is at- tenuated by a decreasing flexor moment arm (Fig- ure 3.12). Interestingly, because the flexors as a group develop significantly greater force that the extensors (due to their larger physiological cross- sectional area), the net result is a nearly constant ra- tio of flexor to extensor torque over the wrist range of motion (Figure 3.13). Additionally, wrist stabil- ity increases as the wrist is moved to full extension because both flexor and extensor moments increase in a similar fashion (Figure 3.13). The take-home lesson from these results appears to be that the co- ordination between muscle and joint properties re- sults in a torque system that is balanced throughout the range of motion. This balance is achieved at the expense of maximum moment generation at the wrist and therefore, we conclude that this system is not simply designed to provide the most "bang for the buck" or operate at maximum efficiency as is often assumed in musculoskeletal models.

6   Future Directions

1.   *How is muscle fiber recruitment coordinated with design to enhance the properties of a torque mo- tor?* The data described in our studies only define the

"peak performance curve" of a musculoskeletal torque motor. Obviously, the nervous system is able to grade muscle force to a great extent simply by al- tering fiber stimulation frequency and recruitment. It will be interesting to determine whether neural strate- gies used for torque generation are, in some way, co- ordinated with the design presented here.

2.   *Is muscle fiber type between muscles a signif- icant factor that affects performance or is it sim- ply a consequence of the pattern of use in a par- ticular muscle?* Muscle fiber types vary widely between limb muscles in rodents, cats and dogs, but to a smaller extent in humans. The case is of- ten made for differences in performance between muscles being a consequence of this difference in fiber type composition. However, since muscle fiber type is significantly affected by patterns of neural activation, it could be that the neural and musculoskeletal factors dominate the relatively modest effect of fiber type transformation that may occur with a particular use pattern.

3.   *What are the mechanical and cellular signals that regulate serial sarcomere number within mus- cle fibers?* There is such consistency between mus- cles in serial sarcomere number and that precise number is determined during muscle development. It is also possible to alter serial sarcomere number by immobilization, tenotomy, or chronic stimula- tion. Because this number is a large determinant of muscle performance, it must be regulated fairly

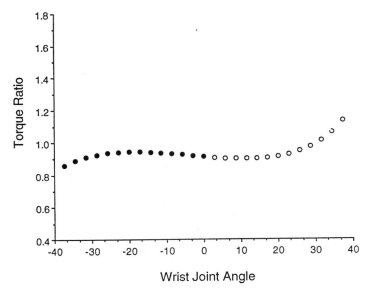

FIGURE 3.13. Ratio between extension and flexion torque by the prime movers of the wrist, calculated using the biomechanical model described. Note that, in spite of large differences in muscle forces, mo- ment arms, and sarcomere length operat- ing ranges, the moments are nicely bal- anced throughout the range of motion. The wrist is in its most stable position in full extension where the moment produced by each muscle group is maximized.

tightly. What are the signals that a muscle fiber transduces to alter its sarcomere number?

4. *Through what portion of the normal range of motion are sarcomeres actually activated? Is this consistent with their overall operating range?* The studies presented in this chapter define sarcomere length operating range throughout the entire range of motion. It is possible that muscles are only activated in a restricted range, and this could affect the interpretation of our data. It will be important to document the range over which sarcomeres are used compared to the anatomical range of motion as is presented here.

5. *How do sarcomere length operating ranges compare between upper extremity muscles, which have a relatively general pattern of use, and lower extremity muscles, whose use patterns are fairly stereotypical?* It is often not possible to simply extrapolate from one muscle group to another. The extent to which the design principles of the wrist are applicable to other joints or even to the entire lower extremity is not known.

6. *How representative are individual sarcomere length measurements of the entire muscle?* This is a technically difficult question to answer based on the large number of sarcomeres throughout a muscle. It is clear that fiber length can vary within a muscle, but it is difficult to relate this to a sarcomere length change without a detailed knowledge of the muscle kinematics during use. Some evidence of the significant sarcomere length heterogeneity that occurs within rodent muscle is presented in Chapters 5 and 6.

Acknowledgments. This work was supported by the Departments of Veteran Affairs, NIH Grant AR35192, the Swedish Medical Research Council (Project #11200) and the Medical Faculty at the Göteborg University. We acknowledge Drs. Gordon Lutz and Reid Abrams for helpful discussions and critical manuscript review. We also thank Thomas Burkholder for assistance generating the biomechanical models.

References

Bäckdahl, M. and Carlsöö, S. (1961). Distribution of activity in muscles acting on the wrist (an electromyographic study). *Acta. Morph. Neerl.-Scand.*, 4:136–144.

Brand, P.W., Beach, R.B., and Thompson, D.E. (1981). Relative tension and potential excursion of muscles in the forearm and hand. *J. Hand Surg.*, 3:209–219.

Butler, D.L., Grood, E.S., Noyes, F.R., and Zernicke, R.F. (1978). Biomechanics of ligaments and tendons. In: *Exercise and Sport Sciences Review* p. 125–182, The Franklin Institute Press.

Ebashi, S., Maruyama, K., and Endo, M. (1980). *Muscle Contraction: Its Regulatory Mechanisms.* Springer Verlag, New York.

Fridén, J., and Lieber, R.L. (1994). Physiological consequences of surgical lengthening of extensor carpi radialis brevis muscle-tendon junction for tennis elbow. *J. Hand Surg.*, 19A:269–274.

Gans, C. (1982). Fiber architecture and muscle function. In *Exercise and Sport Science Reviews*, p. 160–207, Franklin University Press, Lexington, Massachusetts.

Gordon, A.M., Huxley, A.F., and Julian, F.J. (1966). Tension development in highly stretched vertebrate muscle fibres. *J. Physiol., (London)*, 184:143–169.

Gordon, A.M., Huxley, A.F., and Julian, F.J. (1966). The variation in isometric tension with sarcomere length in vertebrate muscle fibres. *J. Physiol., (London)*, 184:170–192.

Huxley, A.F. (1974). Muscular contraction. *J. Physiol., (London)*, 243:1–43.

Lieber, R.L., and Boakes, J.L. (1988). Sarcomere length and joint kinematics during torque production in the frog hindlimb. *Am. J. Physiol.*, 254:C759–C768.

Lieber, R.L., and Fridén, J. (1997). Intraoperative measurement and biomechanical modeling of the flexor carpi ulnaris-to-extensor carpi radialis longus tendon tranfer. *J. Biomech. Eng.*, 119:386–391.

Lieber, R.L., Fazeli, B.M., and Botte, M.J. (1990). Architecture of selected wrist flexor and extensor muscles. *J. Hand Surg.*, 15:244–250.

Lieber, R.L., Jacobson, M.D., Fazeli, B.M., Abrams, R.A., and Botte, M.J. (1992). Architecture of selected muscles of the arm and forearm: anatomy and implications for tendon transfer. *J. Hand Surg.*, 17:787–798.

Lieber, R.L., Loren, G.J., and Fridén, J. (1994). In vivo measurement of human wrist extensor muscle sarcomere length changes. *J. Neurophys.*, 71:874–881.

Lieber, R.L., Ljung, B.-O., and Fridén, J. (1997). Intraoperative sarcomere measurements reveal differential musculoskeletal design of long and short wrist extensors. *J. Exp. Biol.*, 200:19–25.

Lieber, R.L., Pontén, E., and Fridén, J. (1996). Sarcomere length changes after flexor carpi ulnaris-to-extensor digitorum communis tendon transfer. *J. Hand Surg.*, 21A:612–618.

Lieber, R.L., Yeh, Y., and Baskin, R.J. (1984). Sarcomere length determination using laser diffraction. Effect of beam and fiber diameter. *Biophys. J.*, 45:1007–1016.

Loren, G.J. and Lieber, R.L. (1995). Tendon biomechanical properties enhance human wrist muscle specialization. *J. Biomech.*, 28:791–799.

Loren, G.J., Shoemaker, S.D., Burkholder, T.J., Jacobson, M.D., Fridén, J., and Lieber, R.L. (1996). Influences of human wrist motor design on joint torque. *J. Biomech.*, 29:331–342.

Lutz, G.J. and Rome, L.C. (1994). Built for jumping: the design of the frog muscular system. *Science, (Washington, D.C.)*, 263:370–372.

Mai, M.T. and Lieber, R.L. (1990). A model of semitendinosus muscle sarcomere length, knee and hip joint interaction in the frog hindlimb. *J. Biomech.*, 23:271–279.

McFarland, G.B., Krusen, U.L., and Weathersby, H.T. (1962). Kinesiology of selected muscles acting on the wrist: electromyographic study. *Arch. Phys. Med. Rehab.*, 43:165–171.

Riek, S. and Bawa, P. (1992). Recruitment of motor units in human forearm extensors. *J. Neurophys.*, 68:100–108.

Rome, L.C. and Sosnicki, A.A. (1991). Myofilament overlap in swimming carp II. Sarcomere length changes during swimming. *Am. J. Physiol.*, 163:281–295.

Rome, L.C., Funke, R.P., Alexander, R.M., Lutz, G., Aldridge, H., Scott, F., and Freadman, M. (1988). Why animals have different muscle fiber types. *Nature*, 335:824–827.

Rome, L.C., Swank, D., and Corda, D. (1993). How fish power swimming. *Science*, 261:340–343.

Sacks, R.D. and Roy, R.R. (1982). Architecture of the hindlimb muscles of cats: functional significance. *J. Morphol.*, 173:185–195.

Sosnicki, A.A., Loesser, K.E., and Rome, L.C. (1991). Myofilament overlap in swimming carp. I. Myofilament lengths of red and white muscle. *Am. J. Physiol.*, 260:C283–C288.

Squire, J. (1981). *The structural basis of muscular contraction*. Plenum Press, New York.

Walker, S.M. and Schrodt, G.R. (1973). I Segment lengths and thin filament periods in skeletal muscle fibers of the rhesus monkey and humans. *Anat. Rec.*, 178:63–82.

Walmsley, B. and Proske, U. (1981). Comparison of stiffness of soleus and medial gastrocnemius muscles in cats. *J. Neurophys.*, 46:250–259.

Walmsley, B., Hodgson, J.A., and Burke, R.E. (1978). Forces produced by medial gastrocnemius and soleus muscles during locomotion in freely moving cats. *J. Neurophys.*, 41:1203–1216.

Zajac, F.E. (1992). How musculotendon architecture and joint geometry affect the capacity of muscle to move and exert force on objects: a review with application to arm and forearm tendon transfer design. *J. Hand Surg.*, 17A:799–804.

Commentary

Wendy M. Murray and Scott L. Delp

Understanding the contribution of a muscle to the maximum isometric moment developed about a joint requires accurate estimates of the muscle's operating range on its force-length curve, its physiological cross-sectional area, and its moment arm (e.g., see Chapter 2; Chapter 7). Lieber and Fridén present a method for direct measurement of a muscle's operating range and demonstrate the type of comprehensive analysis that can be performed utilizing a musculoskeletal model and quantitative anatomical data. Our understanding of individual muscle function can be significantly enhanced by this type of analysis; however, one must also consider the limitations inherent to this method if we are to apply it appropriately in motor control research and clinical practice.

One of the limitations of the work presented by Lieber and Fridén is the invasive nature of the measurement technique. Because the sarcomere length measurements are made intraoperatively, it seems unlikely that this method will be commonly used in basic research studies. Also, it is difficult to make sarcomere length measurements in some muscles with this method. The optimal candidates for intraoperative sarcomere length measurements are superficial muscles, like the ECRB, ECRL, and FCU, which are readily accessible from a surgical incision. Deeper muscles like the pronator quadratus, for example, present significant obstacles to intraoperative measurement. Understanding the source of intersubject variability is another difficulty of intraoperative measurements. Lieber and Fridén report a relatively large degree of variability in the slope of the sarcomere length-joint angle relationship between subjects (Figure 3.5), implying differences in muscle operating ranges among individuals. However, without the capacity to measure other anatomical data in these subjects, such as optimal fiber lengths and moment arm-joint angle relations, it is difficult to determine the source of the variability.

The alternative, or supplement, to intraoperative sarcomere length measurements is to estimate the operating range of a muscle by combining a biomechanical model, such as the one Lieber and Fridén describe in Section 3.4, with measurements of muscle sarcomere length at one joint angle, obtained from cadavers (e.g., Cutts 1988; Loren et al. 1996). As demonstrated in Figure 3.11, model estimates can provide reasonable approximations of muscle operating ranges (see also Chapter 7, Section 5). The advantage of a cadaver study is that data can be obtained from deep as well as superficial muscles. In addition, anatomical studies can investigate the variability in moment arms and optimal fiber lengths between specimens and address the question of intersubject variability in muscle operating ranges. Of course, anatomical studies also have important limitations. For instance, it is unclear how well sarcomere lengths measured in cadaver specimens represent actual in vivo lengths. Also, the accuracy of the estimated operating ranges depends upon the accuracy of both the musculoskeletal model and the anatomical measurements.

Estimates of muscle function based on musculoskeletal models and anatomical data, whether obtained intraoperatively or from cadavers, should always be compared to experimental measurements of joint properties. Measurements of the moments generated when the muscles are passive and during maximum voluntary contractions provide reasonable estimates of the summed passive properties and the moment generating capacity of the muscles. Although comparison of a biomechanical model with the moments generated by a group of muscles does not test the validity of the forces estimated for individual muscles, the moment generating capacities of the individual muscles estimated by a model should sum to equal the moments generated by experimental subjects. This is the only step that is missing from the otherwise comprehensive work of Lieber and Fridén. It is important to note that wrist moments generated by ten subjects during maximum voluntary isometric contractions (Delp et al. 1996) were much larger (12.2 Nm on average for the flexors and 7.1 Nm for the extensors) than the moments estimated with the model presented by Loren et al. (1996). Also, the experimental data (Delp et al. 1996) show that moments generated during maximum voluntary contractions of the flexors and extensors of the wrist are not equal, as suggested by Figure 3.13. This discrepancy does not imply the torque profiles for the individual wrist flexors and extensors described in this chapter are incorrect. Rather, the differences in measured moments and the estimated wrist joint properties suggest that the contributions of the finger flexors and extensors, which also cross the wrist and have relatively large PCSAs and moment arms (Brand et al. 1981; Lieber et al. 1992), can make significant contributions to moments generated about the wrist (Gonzalez et al. 1997). This indicates that moment generating properties of the wrist joint are not completely defined by properties of the "prime movers" of the wrist (ECU, ECRL, ECRB, FCR, FCU).

Lieber and Fridén demonstrate the usefulness of intraoperative sarcomere length measurements and comprehensive biomechanical models in the analysis of surgical interventions such as tendon transfers. While biomechanical simulations are performed using models based on anatomical data from nonpathological muscle, tendon transfers are frequently performed in individuals with spinal cord injury or cerebral palsy who may exibit spasticity or contracture. The force-generating properties of muscle may be altered under these pathological conditions. Thus, experiments that elucidate muscle function in these pathological states must be performed to accurately represent the biomechanical consequences of surgical procedures performed on individuals with spasticity or contracture. The utility of new data regarding muscle function in pathologic states will be maximized if it is incorporated into a detailed model of muscle and joint function, such as the one that has been developed by Lieber and Fridén.

References

Brand, P.W., Beach, R.B., and Thompson, D.E. (1981). Relative tension and potential excursion of muscles in the forearm and hand. *J Hand Surg.*, 6:209–219.

Cutts, A. (1988). The range of sarcomere lengths in the muscles of the human lower limb. *J. Anat.*, 160:79–88.

Delp, S.L., Grierson, A.E., and Buchanan, T.S. (1996). Maximum isometric moments generated by the wrist muscles in flexion-extension and radial-ulnar deviation. *J. Biomech.*, 29:1371–1375.

Gonzalez, R.V., Buchanan, T.S., Delp, S.L. (1997). How muscle architecture and moment arms affect wrist flexion-extension moments. *J. Biomech.*, 30:705–712.

Lieber, R.L., Jacobson, M.D., Fazeli, B.M., Abrams, R.A., and Botte, M.J. (1992). Architecture of selected muscles of the arm and forearm: anatomy and implications for tendon transfer. *J. Hand Surg.*, 17:787–798.

Loren, G.J., Shoemaker, S.D., Burkholder, T.J., Jacobson, M.D., Fridén, J., and Lieber, R.L. (1996). Human wrist motors: biomechanical design and application to tendon transfers. *J. Biomech.*, 29:331–342.

4

Comparison of Effective Synaptic Currents Generated in Spinal Motoneurons by Activating Different Input Systems

Marc D. Binder

1 Introduction

Understanding how synaptic inputs from segmental and descending systems shape motor output from the spinal cord requires descriptions of the relative magnitudes of the synaptic currents produced by the different systems and their patterns of distribution within a motoneuron pool (Anderson and Binder 1989; Binder 1989). Although most quantitative assessments of the synaptic input to motoneurons are based on measurements of the amplitude of postsynaptic potentials (PSPs; rev. in Binder and Mendell, 1990; Binder et al. 1996), my colleagues and I have argued that the synaptic current reaching the soma (i.e., the "effective synaptic current") is a more functionally relevant measure of the magnitude of a synaptic input (Heckman and Binder, 1988, 1990, 1991b; Lindsay and Binder 1991; Binder et al. 1993, 1996, 1998; Powers et al. 1992, 1993; Binder and Powers 1995a; Westcott et al. 1995). Effective synaptic currents (I_N) can be readily measured under steady-state conditions using a modified voltage-clamp procedure (Heckman and Binder 1988; Lindsay and Binder 1991; Powers and Binder 1995a). This technique not only yields the value of I_N at the motoneuron's resting potential, but further provides an estimate of its somatic voltage dependence, somatic reversal potential, and the conductance change the synaptic input produces at the soma. From these data, one can readily calculate the effective synaptic current at the motoneuron's threshold for repetitive discharge. This value can then be combined with the slope of the steady-state frequency-current (f/I) relation to predict the effect of the synaptic input on

motoneuron firing behavior (Heckman and Binder, 1990; Powers et al. 1992; Binder et al. 1993; Binder et al. 1995a,b; Binder et al. 1998).

2 Distribution of Effective Synaptic Currents from Identified Input Systems

Thus far, we have studied the distribution of effective synaptic currents for six different input systems to cat lumbar motoneurons. Results from these studies are briefly reviewed below and summarized in Figure 4.1. The magnitudes of these currents can be calibrated by the minimal currents required to generate repetitive firing in these motoneurons, ranging from about 3.5 nA in the lowest-threshold cells to 40 nA in the highest-threshold cells (Heckman and Binder 1991a).

2.1 Monosynaptic Ia Input

The contacts between Ia afferent fibers and motoneurons constitutes the most extensively studied of all synaptic input systems (reviewed in Munson 1990). The observed covariance between Ia EPSPs and motoneuron input resistance (R_N) has generally been assumed to result from an approximately constant synaptic current applied to cells of varying R_N values (reviewed in Heckman and Binder 1990), in keeping with the original *size principle* of recruitment (Henneman et al. 1965). Although the amplitudes of steady-state Ia EPSPs are highly correlated with R_N, measurements of the underlying effective synaptic currents indicate that this co-

FIGURE 4.1. Graphical representation of the magnitude and distribution of the effective synaptic currents (I_N) from six different input systems measured in cat lumbar motoneurons at rest. Abcissa is the input resistance (R_N) of the motoneurons. The dark, stippled band represents I_N from homonymous Ia afferent fibers (Heckman and Binder 1988); the striped band represents I_N from Ia-inhibitory interneurons (Heckman and Binder, 1991a); the black band represents the I_N from Renshaw interneurons (Lindsay and Binder 1991); the thick lines outline the I_N from contralateral rubrospinal neurons (Powers et al. 1993); the light stippled band represents I_N from ipsilateral Deiter's nucleus (Westcott et al. 1995); and the striped line represents I_N from the contralateral pyramidal tract (Binder et al. 1998).

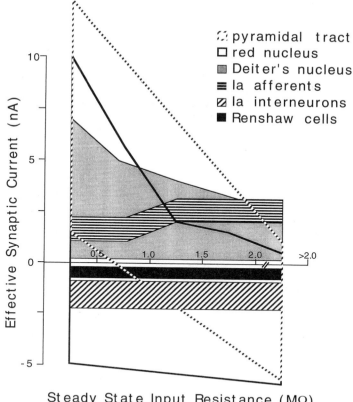

variance results from systematic variance in both I_N and R_N. Within the cat medial gastrocnemius motoneuron pool, the effective synaptic currents (I_N) generated by homonymous Ia afferents display a wide range of values that covary with R_N and other motoneuron properties: I_N is about twice as large on average in motoneurons with high R_N values as in those with low R_N values (Heckman and Binder 1988). Adding activation of the heteronymous Ia afferent fibers from the synergist lateral gastrocnemius and soleus muscles approximately doubles the magnitude of I_N (mean value MG Ia afferents : 2 nA; mean value triceps surae Ia afferents: 4.2 nA; Westcott 1993).

2.2 Reciprocal Ia Inhibition

Comparisons of reciprocal Ia IPSPs with homonymous Ia EPSPs generated in cat hindlimb motoneurons demonstrate a strong correlation between these two synaptic input systems (Burke et al. 1976). As is the case for Ia EPSPs, the amplitudes

of composite reciprocal Ia IPSPs vary systematically in type-identified motoneurons: Ia IPSPs in type S are larger than those in type FR, which in turn are larger than those in type FF motoneurons (Burke et al. 1976; Dum and Kennedy 1980). However, there is a steeper gradient of synaptic strength across the motoneuron pool for Ia excitation than for Ia inhibition (Stein and Bertoldi 1981), and thus, one would expect that differences in the effective synaptic currents underlying Ia inhibition within a motoneuron pool would be considerably less than those found for the excitatory Ia input. This appears to be the case, as we found that the amplitudes of the inhibitory effective synaptic currents generated in antagonist motoneurons by the activation of MG Ia afferent fibers were not correlated with the intrinsic motoneuron properties or with putative motor unit type (Heckman and Binder 1991b). The average value of the steady-state Ia inhibitory synaptic current in the putative type F motoneurons was 1.60 + 0.64 nA, whereas that for putative type S motoneurons was 1.65 + 0.71 nA.

Because the synaptic boutons of Ia inhibitory interneurons lie predominantly on the soma of motoneurons (R.E. Fyffe, personal communication), as expected, they generate substantial changes in motoneuron input resistance. As was the case for the effective synaptic currents, there was no systematic relationship between the steady-state change in conductance and the electrical properties of the motoneuron (Heckman and Binder 1991b).

2.3   Recurrent Inhibition

The role of recurrent inhibition in motor control remains uncertain. One long-held hypothesis is that recurrent inhibition acts to suppress the activity of low threshold motoneurons when higher threshold motoneurons are recruited (Eccles et al. 1961; Granit and Renkin 1961; Friedman et al. 1981). Support for this hypothesis came from the demonstration that recurrent IPSP amplitude varies systematically with motor unit type in the order FF<FR<S (Friedman et al. 1981) and the anatomical finding that the number of end branches and "swellings"(presumed synaptic boutons) of recurrent axon collaterals of type F motoneurons are greater than those of type S motoneurons (Cullheim and Kellerth 1978). An alternative hypothesis has been offered by Hultborn and colleagues (1979), who proposed that recurrent inhibition (and its modulation by descending systems) provides a variable-gain regulator of the overall output of a motoneuron pool. This hypothesis is predicated on a uniform distribution of recurrent inhibition within a motoneuron pool.

The distribution of effective synaptic currents underlying recurrent inhibition (RC I_N) in triceps surae motoneurons does appear to be uniform (Lindsay and Binder 1991). We found that RC I_N was entirely independent of all intrinsic motoneuron properties measured, although the amplitudes of the steady-state recurrent IPSPs were correlated with motoneuron input resistance (R_N), rheobase (I_{rh}) and afterhyperpolarization half decay time ($AHP_{t1/2}$), as has been previously reported for transient recurrent IPSPs (Friedman et al. 1981; Hultborn et al. 1988). The mean RC I_N measured at rest was only 0.4 nA, considerably smaller than comparable values derived for other input systems (cf. Figure 4.1). Because the Renshaw input was activated in this study

by stimulating the synergist motor axons at 100 Hz, which produces steady-state IPSPs that are about 50% maximum value (Lindsay, Heckman and Binder, unpublished data), the maximal RC I_N at rest is still likely to be <1 nA on average.

Motoneurons are more likely to receive recurrent inhibition while they are firing repetitively than when they are quiescent, and several hypotheses of the role of recurrent inhibition in motor control emphasize the possible effects of recurrent inhibition on firing frequency (e.g., Hultborn et al. 1979). The mean RC I_N calculated at threshold was still only 1.25 nA, and the distribution within the pool was no different from that measured at resting potential (Lindsay and Binder 1991). These results suggest that the effect of recurrent inhibition on the firing frequencies of motoneurons are uniform and quite modest. Using an average value for the slope of the motoneuron firing rate-injected current relationship (*f/I* curve) of 1.5 imp/s/nA (Kernell 1979), the average change in firing frequency produced by maximal recurrent inhibition (2.5 nA) should be less than 4 impulses/s. Because the slope of the *f/I* curve does not appear to covary with other motor unit properties (Kernell, 1979), the change in firing frequency should not vary systematically within a motoneuron pool. The maximum effective synaptic current calculated at threshold (4.6 nA) would only decrease firing frequency by about 7 impulses/s, a value similar to that observed in the experiments of Granit and Renkin (1961) in which whole ventral roots were stimulated to activate the Renshaw cells.

2.4   Rubrospinal Input

The rubrospinal system is one of several oligosynaptic pathways in the cat that generate qualitatively different synaptic potentials within hindlimb motoneuron pools (Burke et al. 1970; see also Hongo et al. 1969; Endo et al. 1975). Both excitatory and inhibitory last-order interneurons are activated by the contralateral red nucleus, leading to short-latency EPSPs and IPSPs, respectively. Low-threshold motoneurons receive predominantly inhibitory input, whereas the high-threshold motoneurons receive predominantly excitatory input.

We have measured the effective synaptic currents produced by stimulating the hindlimb projection area of the contralateral magnocellular red nucleus in cat triceps surae motoneurons (Powers et al. 1993). At resting potential, the distribution of effective synaptic currents from the red nucleus was qualitatively similar to the distribution of synaptic potentials: 86% of the putative type F motoneurons received a net depolarizing effective synaptic current from the red nucleus stimulation, whereas only 38% of the putative type S units did so. However, at threshold the distribution was markedly altered. Inhibition continued to predominate in the type S cells, but among the type F cells, half received net excitatory effective synaptic currents and half received net inhibitory effective synaptic currents. Other surprising features of these data are the enormous range of effective synaptic currents observed in different cells and the fact that they are often three times larger than comparable inputs from segmental pathways (Figure 4.1). Activation of red nucleus synaptic input reduced motoneuron input resistance by 40%, on average (Powers et al. 1993) The effect on input resistance was most pronounced in those motoneurons that received hyperpolarizing effective synaptic currents.

The data on effective synaptic currents indicate that the red nucleus input may provide a powerful source of synaptic drive to some high threshold motoneurons, while concurrently inhibiting low threshold cells. Thus, this input system can potentially alter the gain of the input–output function of the motoneuron pool, change the hierarchy of recruitment thresholds within the pool, and mediate rate limiting of discharge in low-threshold motoneurons (Burke et al. 1970; Burke 1981; Heckman and Binder, 1990, 1991a, 1993a,b; Binder et al. 1993; Powers et al. 1993; Heckman, 1994).

2.5 Lateral Vestibulospinal (Deiter's Nucleus) Input

In cats, the lateral vestibulospinal or Deiter's nucleus (DN) neurons project primarily to neck and forelimb motoneurons, but a small percentage descend to the lumbosacral area (Fukushima et al. 1979; Shinoda et al. 1988a,b). Mono- and disynaptic DN EPSPs have been recorded in cat triceps motoneurons (Grillner et al. 1970). Although all of monosynaptic EPSPs were all quite small (<2.2 mV), they did display a wide range of amplitudes,

that were not correlated with the duration of the AHP. Similarly, the amplitudes of monosynaptic EPSPs produced by stimulation of DN axons within the ipsilateral ventral funiculus were not related to motor unit type (Burke et al. 1976).

Based on the dependence of PSP amplitude on R_N, the data on DN synaptic potentials suggest that the underlying effective synaptic currents are inversely related to R_N. Our measurements of the effective synaptic currents produced by DN stimulation in triceps surae motoneurons of the cat demonstrated that this is indeed the case (Westcott et al. 1995). The DN effective synaptic currents were primarily small and depolarizing in both medial gastrocnemius (MG) and lateral gastrocnemius-soleus (LGS) motoneurons (mean = 2.5 nA). The DN input tended to be larger in putative type F motoneurons. The amplitudes of the steady-state DN synaptic potentials were similar to those previously reported for transient PSPs (Grillner et al. 1970; Burke et al. 1976). The DN PSPs showed no correlation to rheobase, resting membrane potential or either steady state or maximal input resistance of the cell. The DN synaptic input caused no significant change in motoneuron input resistance.

Comparison of the Deiter's nucleus input to that of the red nucleus shows that the range of effective synaptic currents are similar (Figure 4.1). Also in both descending systems, there are larger depolarizing currents to the putative type F motoneurons. However, the mean depolarizing red nucleus I_N was twice as large as that from DN, and the difference in distribution of red nucleus I_N between putative type F and S motoneurons was more substantial.

2.6 Corticospinal Input

Endo and colleagues (1975) demonstrated that stimulation of the contralateral motor cortex in cats gererates predominantly excitatory PSPs in medial gastrocnemius motoneurons, whereas the PSPs evoked in soleus motoneurons are predominantly inhibitory. When they stimulated the contralateral red nucleus in the same motoneurons, the pattern of PSPs they observed was equivalent in almost all cases.

We have recently studied the effective synaptic currents produced by stimulating the contralateral pyramidal tract in cat triceps surae motoneurons

(Binder et al. 1998). The magnitudes and distri-bution of the effective synaptic currents were quite similar to those observed earlier for the red nucleus (Powers et al. 1993). Measured at rest, the effective synaptic currents from the pyramidal tract were depolarizing in all but one motoneuron (a putative type S cell). However, the currents in the putative type F motoneurons were more than three times larger, on average, than those in the putatative type S motoneurons (Figure 4.1). At threshold, the distribution was markedly altered. All but one type S cell received a hyperpolarizing current from the pyramidal tract, but among the type F cells, half received net excitatory effective synaptic currents and half received net inhibitory effective synaptic currents.

3 Utility and Limitations of Effective Synaptic Current Measurements

There is a large body of experimental work supporting the utility of current measurements in predicting the effects of synaptic inputs on motoneuron discharge rate. Experiments in which injected current has been combined with steady-state synaptic currents demonstrate that synaptic and injected currents are usually entirely equivalent with respect to their effects on repetitive firing within the primary range (e.g., Schwindt and Calvin 1973a). Moreover, during motoneuron discharge, input currents normally sum algebraically (Granit et al. 1966; Schwindt and Calvin 1973a,b). Even synaptic inputs leading to large changes in motoneuron input resistance simply shift the f/I relation along the current axis, without producing a change in slope (Kernell 1970; Schwindt and Calvin 1973a,b), indicating that synaptic inputs add a constant amount of depolarizing or hyperpolarizing current regardless of the background firing rate (see, however, Kernell 1966 and Shapovalov 1972, in which some synaptic inputs alter the slope of the f/I relation, indicating nonequivalence of synaptic and injected current).

Based on these results, Schwindt and Calvin (1973a) postulated that one could infer the effective, steady-state current delivered by any synaptic input from the slope of the motoneuron's f/I relation and the change in discharge produced by activating the synaptic input. This inference leads to a simple, quantitative expression for steady-state motoneuron behavior: the change in motoneuron discharge is equal to the product of the net effective synaptic current and the slope of the frequency-current curve in the primary range ($\Delta F = I_N * f/I$; Powers et al. 1992).

The validity of this expression has been tested by measuring each of these three quantities (ΔF, I_N, and f/I) in the same motoneurons (Powers et al. 1992; Powers and Binder 1995a). The actual increases or decreses in firing rates produced in cat motoneurons by the steady-state activation of a synaptic input are generally in excellent agreement with the predicted changes in firing rate based on the product of effective synaptic current and the slope of the f/I relation (Binder and Powers 1995; Powers and Binder 1995a; Binder et al. 1996). The correspondence between the observed and predicted changes in motoneuron discharge rate validates the measurement of effective synaptic current as a quantitative index of synaptic efficacy. This expression also simplifies quantitative analysis of neural circuitry because the effective synaptic current already subsumes all of the factors governing current delivery from the dendrites to the soma, and therefore does not require any detailed information about the electrotonic architecture of the postsynaptic cell or information about the precise location of the presynaptic boutons.

The major limitations to the usefullness of effective synaptic currents are first that they are based on steady-state measurements and second, that they are derived from recordings made in anesthetized animals. Marked non-linearities in synaptic transduction accompany the introduction of time-varying inputs (Powers and Binder 1995b, 1996; Poliakov et al. 1996a,b). Thus, the simple relationship betweeen effective synaptic current and motoneuron discharge rate described above fails to predict the responses of motoneurons to more transient synaptic inputs (Powers and Binder 1995b; 1996). Moreover, the presence of active dendritic conductances mediating inward currents and their neuromodulation will no doubt alter the delivery of synaptic current to soma under different behavioral states (rev. in Binder et al. 1996). Whether or not the relative magnitudes and distribution patterns of effective synaptic currents from different input systems will persist under altered states is simply not known.

4   Summary

Although the organization of synaptic inputs to motoneurons and the intrinsic properties of motoneurons are among the most thoroughly studied aspects of the neural control of movement, it is not yet possible to account for all of the observations on motoneuron and muscle behavior in terms of the underlying physiology. However, the problem can be simplified by describing the transformation of synaptic inputs into motor output in three stages: (1) the delivery of synaptic current from the synaptic boutons to the motoneuron soma, where it can be measured as I_N; (2) the transduction of that synaptic current into motoneuron firing rate; and (3) the transformation motoneuron firing rate to muscle unit force. Even though many of the biophysical mechanisms underlying these three processes are not completely understood, the relevant quantities and relations can be empirically determined, and thus, it is already possible to generate useful models of the relation between synaptic input and motor output under certain restricted conditions (e.g. Heckman and Binder 1991a, 1993a,b; Fuglevand et al. 1993; Heckman 1994).

5   Future Directions

We are presently working to extend our experimental and modeling efforts on several fronts. First, we are attempting to formulate a general description of the input–output transform of motoneurons using the white-noise systems identification procedure (Poliakov et al. 1996). This will permit us to expand our simulations of motoneuron pool behavior beyond simple steady-state conditions (Heckman and Binder 1991a; 1993a,b; Heckman 1994). Likewise, the application of an improved single electrode voltage-clamp technique will allow us to measure the effective synaptic currents underlying transient synaptic inputs. (Lee and Heckman 1996). Second, we are now studying how the inputs from different synaptic systems are integrated when they are activated concurrently (Binder and Powers 1999; Powers and Binder

2000). Finally, a more complete description of the dynamic properties of motor units is emerging (e.g. Heckman et al. 1992), such that it will soon be feasible to model the relationship between synaptic input and motor output over a much wider set of conditions than have been examined to date.

Acknowledgments. Nearly all of the material presented here has appeared previously in refereed papers and reviews coauthored by my colleagues Randall K. Powers, C.J. Heckman, Farrel R. Robinson, Amy D. Lindsay, Sarah L. Westcott, and Mark A. Konodi. Work in my laboratory is supported by Grants NS 26840, NS 31592, NS 31925 from the National Institutes of Health.

References

Anderson, M.E. and Binder, M.D. (1989). Spinal and Supraspinal Control of Movement and Posture. In *Textbook of Physiology, 21st ed.* Patton, H.D., Fuchs, A.F., Hille, B., Scher, A.M., and Steiner, R. (eds.), Chapter 26, pp. 563–581, W.B. Saunders, Philadelphia.

Binder, M.D. (1989). Functional organization of motoneuron pools. In *Textbook of Physiology, 21st ed.* Patton, H.D., Fuchs, A.F., Hille, B., Scher, A.M., and Steiner, R. (eds.), Chapter 25, pp. 549–562, W.B. Saunders, Philadelphia.

Binder, M.D. and Mendell, L.M. (1990). *The Segmental Motor System.* Oxford University Press, New York.

Binder, M.D. and Powers, R.K. (1995). Effective synaptic currents generated in cat spinal motoneurons by activating descending and peripheral afferent fibers. In *Alpha and Gamma Motor Systems.* Taylor, A., Gladden, M., and Durbaba, R. (eds.), pp. 14–22, Plenum Press, New York.

Binder, M.D. and Powers, R.K. (1999). Synaptic integration in spinal motoneurons. *J. Physiol.* (Paris), 93:71–79.

Binder, M.D., Heckman, C.J., and Powers, R.K. (1993). How different afferent inputs control motoneuron discharge and the output of the motoneuron pool. *Curr. Opin. Neurobiol.*, 3(6):1028–1034.

Binder, M.D., Heckman, C.J., and Powers, R.K. (1996). The physiological control of motoneuron activity. In *Handbook of Physiology. Section 12, Exercise: Regulation and Integration of Multiple Systems.* Rowell, L.B. and Shepherd, J.T. (eds.), pp. 3–53, American Physiological Society, New York, Oxford.

Binder, M.D., Robinson, F.R., and Powers, R.K. (1998). Distribution of effective synaptic currents in cat triceps surae motoneurons. VI. Contralateral pyramidal tract. *J. Neurophysiol.*, 80:241–248.

Burke, R.E., Jankowska, E., and ten Bruggencate, G. (1970). A comparison of peripheral and rubrospinal input to slow and fast twitch motor units of triceps surae. *J. Physiol. (London)*, 207:709–732.

Burke, R.E., Rymer, W.Z., and Walsh, J.V. (1976). Relative strength of synaptic input from short-latency pathways to motor units of defined type in cat medial gastrocnemius. *J. Neurophysiol.*, 39:447–58.

Cullheim, S. and Kellerth, J. (1978). A morphological study of the axons and recurrent axon collaterals of cat a-motoneurones supplying different functional types of muscle unit. *J. Physiol. (London)*, 281:301–313.

Dum, R.P. and Kennedy, T.T. (1980). Synaptic organization of defined motor unit types in cat tibialis anterior. *J. Neurophysiol.* 43:1631–1644.

Eccles, J.C., Eccles, R.M., Iggo, A., and Ito, M. Distribution of recurrent inhibition among motoneurons. *J. Physiol. (London)*, 159:479–499, 1961.

Endo, K., Araki, T., and Kawai, Y. (1975). Contra- and ipsilateral cortical and rubral effects on fast and slow spinal motoneurons of the cat. *Brain Res.* 88:91–98.

Friedman, W.A., Sypert, G.W., Munson, J.B., and Fleshman, J.W. (1981). Recurrent inhibition in type-identified motoneurons. *J. Neurophysiol.*, 46:1349–59.

Fuglevand, A.J., Winter, D.A., and Patla, A.E. (1993). Models of recruitment and rate coding organization in motor-unit pools. *J. Neurophysiol.*, 70:2470–2488.

Fukushima, K., Peterson, B.W., and Wilson, V.J. (1979). Vestibulospinal, reticulospinal and interstitiospinal pathways in the cat. *Prog. Brain Res.*, 50:121–136.

Granit, R., and Renkin, B. (1961). Net depolarization and discharge rate of motoneurones, as measured by recurrent inhibition. *J. Physiol. (London)*, 158:461–475.

Granit, R., Kernell, D., and Lamarre, Y. (1966). Algebraical summation in synaptic activation of motoneurones firing within the 'primary range' to injected currents. *J. Physiol. (London)*, 187:379–99.

Grillner, S., Hongo, T., and Lund, S. (1970). The vestibulospinal tract. Effects on alpha-motoneurones in the lumbosacral spinal cord in the cat. *Exp. Brain Res.*, 10:94–120.

Heckman, C.J. (1994). Computer simulations of the effects of different synaptic input systems on the steady-state input-output structure of the motoneuron pool. *J. Neurophysiol.*, 71:1717–1739.

Heckman, C.J. and Binder, M.D. (1988). Analysis of effective synaptic currents generated by homonymous Ia afferent fibers in motoneurons of the cat. *J. Neurophysiol.*, 60:1946–1966.

Heckman, C.J. and Binder, M.D. (1990). Neural mechanisms underlying the orderly recruitment of motoneurons. In *The Segmental Motor System.* Binder,

M.D. and Mendell, L.M. (eds.), pp. 182–204, Oxford University Press, New York.

Heckman, C.J. and Binder, M.D. (1991a). Computer simulation of the steady-state input-output function of the cat medial gastrocnemius motoneuron pool. *J. Neurophysiol.*, 65:952–967.

Heckman, C.J. and Binder, M.D. (1991b). Analysis of Ia-inhibitory synaptic input to cat spinal motoneurons evoked by vibration of antagonist muscles. *J. Neurophysiol.*, 66:1888–1893.

Heckman, C.J. and Binder, M.D. (1993a). Computer simulations of motoneuron firing rate modulation. *J. Neurophysiol.*, 69:1005–1008.

Heckman, C.J. and Binder, M.D. (1993b). Computer simulations of the effects of different synaptic input systems on motor unit recruitment. *J. Neurophysiol.*, 70:1827–1840.

Heckman, C.J., Weytjens, J.L.F., and Loeb, G.E. (1993). Effect of velocity and mechanical history on the forces of motor units in the cat medial gastrocnemius muscle. *J. Neurophysiol.*, 68:1503–1115.

Henneman, E., Somjen, G., and Carpenter, D.O. (1965). Excitability and inhibitability of motoneurons of different sizes. *J. Neurophysiol.*, 28:599–620.

Hongo, T., Jankowska, E., and Lundberg, A. (1969). The rubrospinal tract. I. Effects on alpha-motoneurones innervating hindlimb muscles in cats. *Exp. Brain Res.*, 7:344–64.

Hultborn, H., Katz, R., and Mackel, R. (1988). Distribution of recurrent inhibition within a motor nucleus. II. Amount of recurrent inhibition in motoneurons to fast and slow units. *Acta Physiologica Scandinavia*, 134:363–374.

Hultborn, H., Lindstrom, S., and Wigstrom, H. (1979). On the function of recurrent inhibition in the spinal cord. *Exp. Brain Res.*, 37:399–403.

Kernell, D. (1966). The repetitive discharge of motoneurones. In *Muscular Afferents and Motor Control. Nobel Symp. I.* Granit, R. (eds.), pp. 351–362, Almqvist and Wiksell, Stockholm.

Kernell, D. (1970). Synaptic conductance changes and the repetitive impulse discharge of spinal motoneurones. *Brain Res.*, 15:291–294.

Kernell, D. (1979). Rhythmic properties of motoneurones innervating muscle fibres of different speed in m. gastrocnemius medialis of the cat. *Brain Res.*, 160:159–62.

Lee, R.H. and Heckman, C.J. (1996). Influence of voltage-sensitive conductances on bistable firing and effective synaptic current in cat spinal motoneurons in vivo. *J. Neurophysiol.*, 76:2107–2110.

Lindsay, A.D. and Binder, M.D. (1991). Distribution of effective synaptic currents underlying recurrent inhibition in cat triceps surae motoneurons. *J. Neurophysiol.*, 65:168–177.

Munson, J.B. (1990). Synaptic inputs to type-identified motor units. In *The Segmental Motor System*. Binder, M.D. and Mendell, L.M. (eds.), pp. 291–307, New York: Oxford University Press.

Poliakov, A.V., Powers, R.K., and Binder, M.D. (1996). Dynamic responses of motoneurons to current transients studied with the white noise method. *Soc. Neurosci. Abstr.*, 22:1845.

Poliakov, A.V., Powers, R.K., Sawczuk, A., and Binder, M.D. (1996). Effects of background noise on the response of motoneurons to excitatory current transients. *J. Physiol. (London)*, 495:143–157.

Powers, R.K. and Binder, M.D. (1995a). Effective synaptic current and motoneuron firing rate modulation. *J. Neurophysiol.*, 74:793–810.

Powers, R.K. and Binder, M.D. (1995b). Quantitative analysis of motoneuron firing rate modulation in response to simulated synaptic inputs. In *Alpha and Gamma Motor Systems*. Taylor, A., Gladden, M., and Durbaba, R. (eds.), pp. 42–44, Plenum Press, New York.

Powers, R.K. and Binder, M.D. (1996). Experimental evaluation of input -output models of motoneuron discharge. *J. Neurophysiol.*, 75:367–379.

Powers, R.K. and Binder, M.D. (2000). Summation of effective synaptic currents and firing rate modulation in cat spinal motoneurons produced by concurrent stimulation of different synaptic input systems. *J. Neurophysiol.* (in press).

Powers, R.K., Robinson, F.R., Konodi, M.A., and Binder, M.D. (1992). Effective synaptic current can be estimated from measurements of neuronal discharge. *J. Neurophysiol.*, 68:964–968.

Powers, R.K., Robinson, F.R., Konodi, M.A., and Binder, M.D. (1993). Distribution of rubrospinal synaptic input to cat triceps surae motoneurons. *J. Neurophysiol.*, 70:1460–1468.

Schwindt, P.C. and Calvin, W.H. (1973a). Equivalence of synaptic and injected current in determining the membrane potential trajectory during motoneuron rhythmic firing. *Brain Res.*, 59:389–94.

Schwindt, P.C. and Calvin, W.H. (1973b). Nature of conductances underlying rhythmic firing in cat spinal motoneurons. *J. Neurophysiol.*, 36:955–973.

Shapovalov, A.I. (1972). Extrapyramidal monosynaptic and disynaptic control of mammalian alpha-motoneurons. *Brain Res.*, 40:105–115.

Shinoda, Y., Ohgaki, T., Futami, T., and Sugiuchi, Y. (1988). Structural basis for three-dimensional coding in the vestibulospinal reflex. *Ann. New York Acad. Sci.*, 545:216–227.

Shinoda, Y., Ohgaki, T., Futami, T., and Sugiuchi, Y. (1988b). Vestibular projections to the spinal cord: the morphology of single vestibulospinal axons. *Prog. Brain Res.*, 76:17–27.

Stein, R.B. and Bertoldi, R. (1989). The size principle: a synthesis of neurophysiological data. In *Progress in Clinical Neurophysiology. Motor Unit Types, Recruitment, and Plasticity in Health and Disease*. Desmedt, J.E. (ed.), pp. 85–96, Basel, Karger.

Westcott, S.L. (1993). Comparison of vestibulospinal synaptic input and Ia afferent synaptic input in cat triceps surae motoneurons. PhD dissertation, University of Washington.

Westcott, S.L., Powers, R.K., Robinson, F.R., and Binder, M.D. (1995). Distribution of vestibulospinal input to cat triceps surae motoneurons. *Exp. Brain Res.*, 107:1–8.

Commentary: Nonlinear Interactions Between Multiple Synaptic Inputs

Thomas M. Hamm and Mitchell G. Maltenfort

Measurement of effective synaptic current provides a useful description of the distribution of synaptic input to a motoneuron pool. Studies by Binder and his colleagues using this technique have made an important contribution toward understanding the contribution of individual synaptic systems to the organization of activity within a motoneuron pool. However, several issues must be considered when attempting to apply these results to the problem of how two or more synaptic inputs combine to influence motoneuron activity. This problem is critical to analyzing the control of motoneurons when multiple synaptic inputs are active.

The demonstrations of linear summation of injected and synaptic current cited by Binder cannot be taken as evidence that pairs of synaptic currents will also sum linearly. While recurrent inhibition and stretch-evoked excitation sum linearly in motoneurons (Granit and Renkin 1961), Ia monosynaptic excitation and Ia reciprocal inhibition do not (Burke et al. 1971). An inhibitory conductance proximal to an excitatory conductance on the dendrite may reduce the excitation reaching the motoneuron soma by a greater amount than if the effective inhibitory synaptic current were simply subtracted from the excitatory synaptic current. The magnitude of such interactions might not be readily assessed from the current and somatic conductance change determined for the inputs individually, even if their electrotonic distance from the soma is known. Recent studies have demonstrated that synaptic inputs to motoneurons may be unevenly distributed to branches in the dendritic arborization (Rose et al. 1995; Burke and Glenn 1996). If such localization of synaptic inputs occurs as a rule, then overlapping pairs of inputs may have significant nonlinear interactions, while nonoverlapping pairs may have none.

When considering synaptic inputs to motoneurons mediated by interneuronal pathways, the extent of interaction within the interneuronal pools must also be taken into account. Extensive convergence of different pathways to motoneurons may occur within a set of interneurons (e.g., Harrison and Jankowska 1985), providing ample opportunity for occlusion or facilitation when different pathways act in concert. In addition, studies from Jankowska's laboratory have demonstrated mutually inhibitory projections within two different functional groups of interneurons (Brink et al. 1983; Edgley and Jankowska 1987). Indirect evidence also suggests the presence of mutually inhibitory connections between Renshaw cells with projections to synergistic motoneuron pools (McCurdy and Hamm 1994). If subsets of inhibitory interneurons are mutually inhibitory, then simultaneous activation of both subsets will produce less total inhibition on the postsynaptic motoneuron than the sum of the inhibitions produced by each set individually.

We also wish to remark on comparison of supraspinal and segmental currents. Stimulation of the pyramidal tract or magnocellular red nucleus seems likely to have activated a larger part of these descending pathways than were activated in the Ia monosynaptic or recurrent inhibitory pathways by stimulation of individual nerves. Activity in the afferents of several muscles or in the activity of several motor nuclei would be expected in most motor activities. Thus the value given for Ia effective synaptic current from triceps surae is more comparable to the supraspinal estimates than is the value for homonymous Ia current.

Binder and his colleagues have pinpointed the essential component of synaptic input that determines the activity of the motoneuron, the synaptic current that reaches the soma and the initial segment. Estimation of the effects of combined inputs requires consideration of the mechanisms of dendritic integration, and of interactions between separate synaptic inputs in interneuronal pathways in order to determine the total effective current.

References

Brink, E., Jankowska, E., McCrea, D.A., and Skoog, B. (1983). Inhibitory interactions between interneurones in reflex pathways form group Ia and group Ib afferents in the cat. *J. Physiol.*, (London), 343:361–373.

Burke, R.E. and Glenn, L.L. (1996). Horseradish peroxidase study of the spatial and electrotonic distribution of group Ia synapses on type-identified ankle extensor motoneurons in the cat. *J. Neurophysiol.*, 372:465–485.

Burke, R.E., Fedina, L., and Lundberg, A. (1971). Spatial synaptic distribution of recurrent and group Ia inhibitory systems in cat spinal motoneurones. *J. Physiol.*, (London), 214:305–326.

Edgley, S.A. and Jankowska, E. (1987). An interneuronal relay for group I and II muscle afferents in the midlumbar segments of the cat spinal cord. *J. Physiol.*, (London), 389:647–674.

Harrison, P.J. and Jankowska, E. (1985). Organization of input to the interneurones mediating group I nonreciprocal inhibition of motoneurones in the cat. *J. Physiol.*, (London), 361:403–418.

McCurdy, M.L. and Hamm, T.M. (1994). Spatial and temporal features of recurrent facilitation among motoneurons innervating synergistic muscles of the cat. *J. Neurophysiol.* 72:227–234.

Rose, P.K., Jones, T., Nirula, R., and Corneil, T. (1995). Innervation of motoneurons based on dendritic orientation. *J. Neurophysiol.*, 73:1319–1322.

5
Length, Shortening Velocity, Activation, and Fatigue Are Not Independent Factors Determining Muscle Force Exerted

Peter A. Huijing

1 Introduction

In muscle and movement modeling it is almost invariably assumed that force actually exerted by any given muscle is determined by several independent factors:

1. Actual length.
2. Actual velocity of length change.
3. Degree of activation.

In mathematical form this can be expressed as:

$$F_m\ (l_m,\ dl_m/dt,\ u,\ t) = q(u,t)\ K(l_m)\ W(dl_m/dt \quad (5.1)$$

where l_m indicates muscle length and $q(u,t)$ the active state function, which is dependent on active state u and time t. In most applications of muscle modeling, if submaximal activation levels are considered at all, this is achieved by linear scaling of length-force characteristics assumed for maximal activation (i.e., if the muscle is activated at 50% of its maximal level, 50% of the maximal force will be exerted) (Figure 5.1).

An important assumption incorporated in such a model is the equivalence of isometric length-force curves and those of zero velocity of force velocity curves.

In Figure 5.1 (as in most modeling), linear scaling is applied at all muscle lengths (i.e., no muscle length effect) so that the length-velocity-normalized force curves are identical.

It is the purpose of this chapter to consider if such a simplification of reality will allow reasonable estimates of actual muscle length-force characteristics. The main reason for this approach is the work of Rack and Westbury (1969), which indi-cated substantial changes of the muscle length-force characteristic as a function of the stimulation frequency of the nerve.

Additional experimental evidence has been accumulating in the last two decades that indicates that length-force characteristics are not a fixed property of muscle, but rather they should be considered the product of a substantial number of interacting factors.

2 Methods

In these experiments the sciatic nerve was stimulated at supramaximal currents to obtain fully recruited muscles. All branches innervating muscles other than the gastrocnemius medialis muscle (GM) were cut. The experiments were performed at many muscle lengths to be able to construct a length-force curve for each particular condition. It should be noted in comparison to the in-vivo situation, a major drawback of this experiment is the synchronization of motor units by the stimulation.

2.1 Stimulation Frequency Experiment

The frequency of stimulation was altered as described below. Frequencies used were 100, 50, 40, 30, and 15 Hz, respectively. At lower frequencies the tetanus was not completely fused anymore but peak forces were used in the analysis.

Two types of experiments were performed.

1. *Constant stimulation frequency* (CSF) tetanic contractions: muscle force was measured across

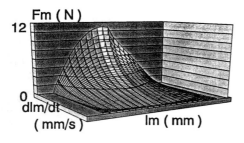

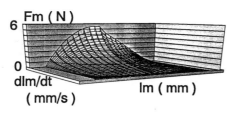

FIGURE 5.1. Modeled rat GM length-shortening velocity-force curves for maximal (top) and 50% activation (bottom) obtained by linear scaling of maximal activation muscle properties.

time for a fixed stimulation frequency. Data from such contractions were used to assess changes in the length-force curves across time, and the time course of force was also used to compare to that obtained with the force generated by the "decreasing stimulation frequency" paradigm at identical times.

2. *Decreasing stimulation frequency* (DSF) tetanic contractions: During each contraction a stimulation frequency staircase (decreasing order) was imposed. For details of experimental methods see (Roszek et al. 1994; Roszek and Huijing 1997).

2.2 Effects of Previous Shortening Experiments

After an intial isometric contraction at the starting length, the muscle was allowed to shorten isokinetically under control of an ergometer until the target length was reached. A second isometric phase then ensued during which the muscle was allowed to redevelop isometric force at the target length. The experiments were performed at many starting lengths and target lengths, so that initial as well as post-shortening isometric length-force curves could be constructed. For details of experimental methods see Meijer et al. (1995a, b, 1997).

3 Results

3.1 Interaction Between Level of Activation and Length-Force Characteristics

3.1.1 Constant Stimulation Frequency

Level of activation (Stephenson and Wendt 1981) or its extracellular equivalent, firing rate (Rack and Westbury 1969; Roszek et al. 1994; Roszek and Huijing 1997), is a very important determinant of actual isometric length-force characteristics. Only very few modellers have attempted to take such effects into account. Hatze (1981) should be credited for first applying changes of shape of the frequency force curve for a limited range of muscle lengths in his model on the basis of results of Rack and Westbury (1969). It should be noted, however, that changing shape factors of the curve takes into account changes of optimum but not of active slack length and therefore overestimates the length range of the ascending limb of the length-force curve. Figure 5.2 shows length-stimulation frequency-isometric force characteristics. Note that optimum length as well as active slack length (defined as the smallest length at which muscle can exert active force on its environment) are encountered at substantially greater lengths for lower stimulation frequencies. Also note the sizable changes in shape of the frequency-force curve as evident at planes orthogonal to the length axis. If these effects are neglected and linear scaling is used, sizable errors will be made in the estimation of muscle force.

It should be pointed out that if a phase delay of sufficient magnitude is found between varying muscle force and consequent length changes due to the viscoelastic properties of the tendon and aponeurosis, the length-force characteristics will no longer be what their name indicates but may also contain some velocity effects (see also Huijing 1996). This may be true for the experimental data shown here. If it is, it is also likely to be the case for in vivo conditions.

3.1.2 Decreasing Stimulation Frequency

Changes in firing rate not only affect instantaneous properties, but such changes also produce short-term history dependence, that is, potentiation (Sweeney et al. 1993). Results are shown in Fig-

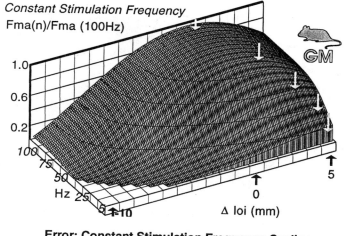

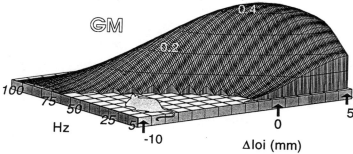

FIGURE 5.2. Rat GM length-stimulation frequency-force characteristics for conditions of constant stimulation frequency) and error of estimate of force when applying linear scaling of length-force properties of maximally activated muscle instead of CSF properties. Force and difference in force were normalized for isometric optimum force during 100 Hz stimulation. Length plotted is muscle tendon complex length. (Top) The length-frequency-force curve is plotted by fitting a sigmoid curve (e.g., Roszek et al. 1994) to frequency force data for many lengths and using this curve for nonlinear adaptation the 100 Hz length-force curve. Note shifts of optimum and active slack lengths. (Bottom) Results of subtracting linear scaling results from CSF results. Note that the error is very much dependent on muscle length but reaches sizable magnitudes (up to maximally +45% of 100 Hz stimulation optimum force). Also note that relatively small negative errors ($<-8\%$ Fmao (100Hz)) at low lengths and frequencies were not plotted.

ure 5.3 for the particular decreasing stimulation frequency staircase protocol described in the methods (see also Roszek and Huijing 1997).

Compared to the condition of maximal activation (100 Hz), optimum length shifts also to higher length at lower frequencies, but the changes in active slack length are relatively small. Compared to CSF conditions, shifts of optimum and active slack lengths to lower lengths are found: optimum length does not shift as much as in CSF conditions and the change of active slack length is very small.

It is clear (e.g., Figure 5.3, lower panel) that if

this factor is not taken into account, as is the case in linear scaling of force, substantial errors (up to 50% for rat muscle) will be introduced at most muscle lengths if decreasing frequencies from high levels are modeled as CSF properties.

3.2 Effects of Sustained Isometric Contraction at Constant Stimulation Frequency

It is well established that during sustained contraction, fatigue (e.g., Allen et al. 1992) will play a ma-

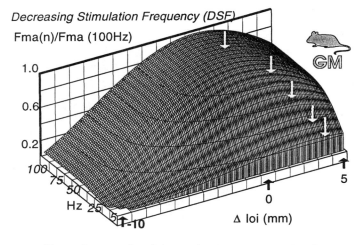

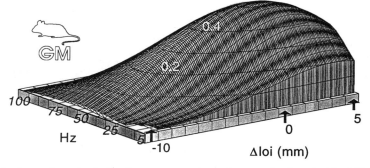

FIGURE 5.3. Effects of a decreasing stimulation frequency staircase protocol on muscle length-force characteristics (top) and error of estimate of force when applying constant stimulation frequency length-force properties of submaximally activated muscle (bottom). Force and difference in force were normalized for isometric optimum force during 100 Hz stimulation. Length plotted is muscle tendon complex length. (Top) Note shifts of optimum length to higher lengths at lower stimulation frequencies and much smaller shifts of active slack length. (Bottom) Note that substantial errors (up to 50% depending on length) would be made if linear scaling of length-force properties would be used in modelling the experimental DSF protocol.

jor role in determining how much force is exerted by the muscle. It is common practice to define such fatigue operationally as the decrease in force seen in time during a contraction. Recently, Huijing, and Baan (1995) reported the effects of fatigue on isometric length-force characteristics during tetanic contraction at almost maximal activation (100 Hz). Small shifts of optimum length but not of active slack length were hypothesized by these authors to be related to intracellular inhomogeneities of calcium concentration (Allen et al. 1992). In effect, these results indicate that during sustained maximal contraction, isometric muscle length-force characteristics are changing as a function of time.

It is less well established that very complicated interactions between fatigue (Allen et al. 1992) and potentiation (Roszek et al. 1994) are observed during sustained submaximal isometric contractions. Initially, as in the case of fatigue, potentiation was defined operationally, but more recently it has been described in terms of effects of phosphorylation of the myosin light chain (Sweeney et al. 1993).

It has been argued (Huijing and Roszek 1996; Huijing 1997) that the mechanisms of fatigue may even lead to an increase of force at higher muscle lengths (Figure 5.4). Crucial to these arguments is that the maximal length that a muscle can produce any force is considered to be a fixed property of muscle regardless of its degree of activation and potentiation. In work on skinned fibers (e.g., Stephenson and Wendt 1981), evidence for this assumption was found, but this property has not been studied in intact fibers. As a consequence, net changes in force generated during an isometric contraction, even under conditions of constant firing rate, can *not* be adequately assigned to be the consequence of of fatigue (Huijing and Baan 1995, Huijing and Roszek 1996). Similarly, the effects of MLC phosphorylation (Figure 5.4) may lead to a decrease of force (Huijing and Baan 1995; Huijing and Roszek 1996; Huijing 1997). The top two panels indicate the effects of two mechanisms thought to be active during fatiguing sustained isometric contractions. Note that in both cases fatigue shifts the optimum length to longer muscle lengths, but that there are differences in the effects on active slack length. If these changes are executed with maximal length of active force exertion being constant, the pre- and postcurves will have to cross,

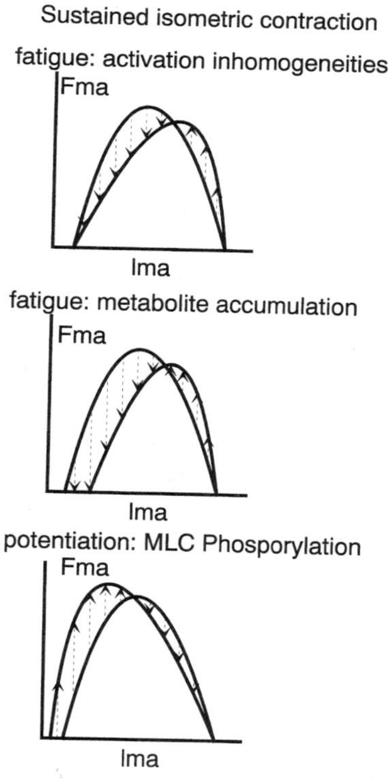

leading to increased force at higher lengths. The magnitude of any change in force is determined by the relative magnitude of optimum length shift and force decrease at optimum length.

The bottom panel indicates similar effects for myosin light chain phosphorylation related potentiation. In this case, the magnitude of the force decrease is governed by the relative magnitude of the shift of optimum length to lower lengths and the increase of force at optimum length. Note that the direction of change in force does not agree for all lengths with operational definitions used commonly for fatigue and potentiation.

It should be noted that in vivo the central nervous system responds to sustained isometric contractions by lowering firing rates (Bigland Ritchie et al. 1986). In such cases the effects of potentiation such as described for decreasing stimulation frequency protocols should be taken into account as well.

3.3 Effects of Previous Shortening

It has been known for quite some time (e.g., Edman et al. 1993) that previous shortening leads to what is called a deficit in isometric force. After shortening, the redeveloped isometric force does not attain values that are as high as that of a comparable isometric contraction at the target length. These effects may result primarily from inhomogeneities in the lengths of sarcomeres in series within one fiber (Edman et al. 1993), but some differences have also been encountered for conditions in which no changes of such distribution could be shown (Sugi and Tsuchiya 1988).

Until recently very little attention was paid to possible changes in length-force curves because of this mechanism. Meijer et al. (1995a,b, 1997) showed that isometric length-force characteristics are influenced by previous shortening (Figure 5.5). Note that the force deficit is very much length dependent: At rather short lengths the differences with the fully isometric curve are much smaller than at lengths over optimum length. Note also that the force deficit is also dependent on the amount the muscle shortened.

The upper panel shows muscle relative tendon complex length (Δl_{oi}) as a function of time. Note that isometric contraction at different lengths are followed by isokinetic contraction at identical shortening velocity to the target length. The contraction that was isometric at the target length is kept isometric

Sustained isometric contraction

fatigue: activation inhomogeneities

fatigue: metabolite accumulation

potentiation: MLC Phosporylation

FIGURE 5.4. Schematic representation of the principle of effects of fatigue and potentiation on muscle length-force characteristics.

and the results used as a reference for comparison. The middle panel shows muscle force as a function of time. Note that at the target length larger previous shortening causes a larger force deficit at the target length shown as an example. The lower panel shows length-force curves obtained for isometric contractions with equal history of previous shortening by performing the above described experiment at many target lengths. Active muscle length (l_{ma}) was determined photographically.

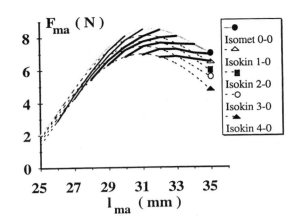

FIGURE 5.6. Isometric length-force characteristics for a shortening muscle.

The same authors (Meijer et al. 1997) realized that the isometric length-force curve of a muscle would not be equivalent to following the curve for one shortening condition but should be constructed from points of curves of different shortening history (Figure 5.6). Note the absence of negative slopes in the curves that take effects of previous shortening into account even at high lengths.

The curves of the lower panel of Figure 5.5 is plotted in this figure as a reference. The length-force curve for fully isometric contractions is shown in gray and the length-force curves after different amounts of shortening as dotted lines. The thick solid lines connect points with appropriate previous shortening history. These isometric length-force curves take effects of previous shortening into account.

Because short-term movement history has been found to be an important determinant of length-force properties, an unexplored consequence of this phenomenon is that muscle force-velocity characteristics as classically determined in isotonic or isokinetic contractions can be expected to be contaminated by length-history and or velocity-history effects in a way that does not reflect in vivo characteristics.

4 Conclusions and Discussion

The major conclusion is that muscle length-velocity-force characteristics are not unique properties of muscle. This conclusion is based on experimental results obtained under more controlled conditions,

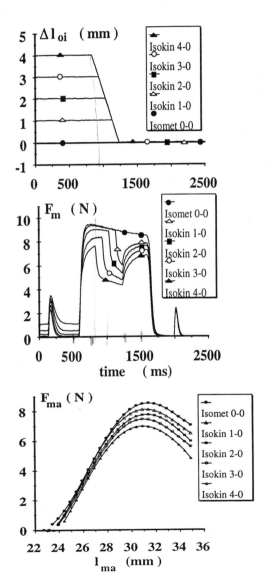

FIGURE 5.5. Effects of previous shortening on isometric length-force characteristics.

which are also rather artificial conditions compared to those encountered during in vivo movement.

The lack of the uniqueness of length-force-velocity characteristics is caused by the fact the determinants of force, length, velocity of shortening and degree of activation are not independent factors. Models of muscle force should follow a description in similar to that shown in equation (2):

$$F_m\ (l_m, dl_m/dt,\ dl^2{}_m/d^2t,\ u,\ du/dt,\ du^2/d^2t,\ t) =$$
$$q\ (u, du/dt, du^2/d^2t, t)\ K\ (l_m, dl_m/dt, du^2/d^2t,\ t)\quad (3.2)$$

where l_m indicates muscle length and q the active state function. The second derivative of variables with respect to time is used here in a rather unconventional way to indicate history effects.

The effects of interaction among these factors have been found to be of a large enough magnitude for the strict but artificial conditions imposed in animal experimentation, that they are expected to play major roles also during in vivo movement. This poses the following possibly unsolvable dilemma for biomechanists active in modeling muscle and movement: Either effects such as the ones presented here are incorporated into realistic models, that then become so complex that they can not be handled and most likely can not be understood anymore, or substantial deviations from reality will be built into the models, preventing accurate prediction of variables important for muscle function.

A major lesson to be drawn from these arguments is that muscle models should be built for specific well defined purposes. These purposes can not be other than being heuristic (i.e., using the model to generate idea's about important aspects of muscle function).

Modelers should be much more explicit about the explanations and justification of their choice of including or not including certain parameters in their model. If this is done, the drawbacks of a certain approach will be openly seen and it may help people to realize that there is no model that even comes close to reality. A major topic for modelers will then be to search for differences between modeling and experimental outcomes rather than working towards agreement by manipulation of parameters. In my experience it is finding differences that usually leads to increased understanding.

Such an approach will require also a changed attitude for journal referees and funding agencies, but it will lead to better science and probably also better chances for application.

Acknowledgments. The author gratefully acknowledges Boris Roszek, Kenneth Meijer, and Peter Bosch for their experimental contributions as well as continuing discussions on issues raised in this chapter.

References

Allen, D.G., Westerblad, H., Lee, J.A., and Lannergren, J. (1992). Role of excitation contraction coupling in muscle fatigue. *Sports Med.*, 13:116–126.

Bigland-Ritchie, B., Dawson, N., Johansson, R., and Lippold, O. (1986). Reflex origin for the slowing of motoneurone firing rates in fatigue of human voluntary constractions. *J. Physiol.*, 379:451–459.

Edman, K.A.P., Caputo, C., and Lou, F. (1993). Depression of tetanic force induced by loaded shortening of frog muscle fibers. *J. Physiol.*, 466:535–552.

Hakkinen, K., Keskinen, K.L., Komi, P.V. (eds.). (1995). Proc. XVth Congress of the International Society of Biomechanics, University of Jyvaskyla, Finland.

Hatze, H. (1981). Myocybernetic control models of skeletal muscle. University of South Africa Press, Pretoria.

Huijing, P.A. (1995). Parameter interdependence and succes of skeletal muscle modelling. *Hum. Mov. Sci.*, 14:443–486.

Huijing, P.A. (1996). Important experimental factors for skeletal muscle modelling. *Eur. J. Morph.*, 34:47–54.

Huijing, P.A. (1997). Muscular properties in experiments, modelling and function. *J. E.MG. Kinesiol*, (in press).

Huijing, P.A. and Baan, G.C. (1995). Fatigability and length-force characteristics of rat gastrocnemius muscle during sustained tetanic contraction. In (4) Hakkinen et al., pp. 408–409.

Huijing, P.A. and Roszek, B. (1996). Length-force charactereristics of rat muscle during isometric contraction are determined by independent and cumulative effects of fatigue and potentiation. In *Proc. 18th Conference of IEEE Engineering in Medicine and Biology Society*, Boom, H.B.K., and Rutten, W.L.C. (eds.), pp. 215–216. IEEE Publishing Service, New York.

Meijer, K., Grootenboer, H.J., Koopman, H.F.J.M., and Huijing, P.A. (1995b). The isometric length-force relationship during concentric contractions of the rat medial gastrocnemius muscle. In *Integrated Biomedical Engineering for Restoration of Human Function*. Feijen, J. (ed.), pp. 73–76, Institute for Biomedical Engineering, Universiteit Twente, Enschede.

Meijer, K., Grootenboer, H.J., Koopman, H.F.J.M., and Huijing, P.A. (1997). Length-force curves during and after isometric contractions differ from the fully isometric length-force curve in rat muscle. *J. Appl. Biomech.*, (in press).

Meijer, K., Koopman, H.F.J.M., Grootenboer, H.J., and Huijing, P.A. (1995a). The effect of shortening history on the length-force relationship of the muscle. In (4) Hakkinen et al., pp. 618–619.

Rack, P.M.H. and Westbury, D.R. (1969). The effects of length and stimulus rate on tension in the isometric cat soleus muscle. *J. Physiol.*, 204:443–460.

Roszek, B. and Huijing, P.A. (1997). Stimulation frequency history alters length-force characteristics of fully recruited rat muscle. *J. EMG Kinesiol.*, (in press).

Roszek, B., Baan, G.C., and Huijing, P.A. (1994). Decreasing stimulation frequency dependent length-force characteristics of rat muscle. *J. Appl. Physiol.*, 77: 2115–2124.

Stephenson, D.G. and Wendt, D.A. (1981). Effects of sarcomere length on the force—pCa relation in fast- and slow-twitch skinned muscle fibres from the rat. *J. Physiol.*, 333:637–653.

Sugi, H. and Tsuchiya, T. (1988). Stiffness changes during enhancement and deficit of isometric force by slow length changes in frog skeletal muscle fibers. *J. Physiol.*, 407:215–229.

Sweeney, H.L., Bowman, B.F., and Stull, J. (1993). Myosin light chain phosphorylation in vertebrate striated muscle: regulation and function. *Am. J. Physiol.*, 264:C1085–C1095.

Commentary:
What Is the Use of Models That Are Not Even True?

Steve L. Lehman

In Salman Rushdie's children's novel, *Haroun and the Sea of Stories* (1990), the young hero asks his father, a storyteller, "What's the use of stories that aren't even true?" The answer gets Haroun into a lot of trouble, but the consequent adventure is wonderful, and much is learned. The same question may be asked about the models we commonly use to represent muscle mechanics. Peter Huijing explores two aspects of this question in Chapters 5 and 6.

Huijing takes on two fundamental assumptions often included in muscle models: That the effects of activation and length on force are independent of each other, and independent of the history of length and activation (Chapter 5); That the mechanics of maximally activated muscle are the same as the mechanics of a single fiber (single sarcomere) (Chapter 6). In both chapters, he produces evidence from his own laboratory, working with whole muscle stimulated in situ, that the assumption is false.

Chapter 5 gives evidence that the dependencies of force on length and activation are not separable. The experiments in which force is measured at different lengths and stimulus frequencies show that there is not a single length/tension curve, scaled by activation, but that the length at which the activated muscle first produces force, the length at which it produces most force, and the shape of the relationship between length and force all depend on stimulus frequency. Is force of an isometric muscle completely determined by its present length and activation? The experiments with decreasing stimulus frequency show that past stimulation can be important. The experiments showing effects of previous shortening demonstrate that the force is also dependent on past values of length. In short, the length, activation and time dependencies are not separable.

As the chapter notes, the facts that length-tension curves change with stimulus frequency, and that force-frequency curves are a function of length were shown by Rack and Westbury (1969), and a length dependence of activation has also been exhibited in skinned fibers (Stephenson and Williams 1982). Likewise, the decrement in force because of previous shortening has been demonstrated in whole muscle and in fibers. By controlling for segment inhomogeneity, Edman et al. (1993) convincingly demonstrated that the decrement due to shortening in fibers is caused by sarcomere inhomogeneity. The length dependence of this effect noted here may also be attributable to sarcomere inhomogeneity, as longer fibers should be less homogeneous (Julian and Morgan 1979).

All the results in Chapter 5 have their analogs (and possibly causes) in single fibers. Perhaps the fiber is a good model for whole muscle in situ. In contrast, Chapter 6 lists evidence that there may be no representative fiber. The direct evidence is that the length-tension relationship for the whole muscle is not the same as the length-tension relationship for selected motor units. Indirect evidence supports this: at optimal muscle length, selected fibers do not have optimum mean sarcomere length. There is substantial individual variation in the length-tension relationships of individual muscles. A finite element model shows that, even if fibers start out uniform in length, the uniformity is likely to go away once the fiber is activated. So if muscle is modeled as a single fiber, what is the sarcomere length in the fiber? If, as is more common, muscle is modeled as a single sarcomere, what is the length of the sarcomere? The more serious possibility of inhomogeneity in sarcomere length and contractile properties along each fiber is not mentioned.

There are two different ways in which models may be "unrealistic." First, something may be left out. The effects in Chapter 5 are arguably examples. In this case, a sensitivity analysis will show whether the missing effect is important, and if it is, the model may be amended. If the model is not precise enough for prediction, it may still be useful for "heuristic purposes." But a model may be unrealistic in a much more troublesome way: its structure may be contradicted by data. The lack of a representative fiber in Chapter 6 may be an example. The lack of homogeneity among sarcomeres within a fiber may be another. The separation of the "series elasticity" from the "contractile component" in the Hill model (1938) is a classical, and still topical, example, contradicted long ago by both physiological (Jewell and Wilkie 1958) and structural (Huxley 1957) data. Sensitivity analysis will not reveal errors in model structure, and the lack of an effect cannot be fixed by amendment. If such a model is used heuristically, the difference between

model output and data may lead to new discoveries, but may as easily lead to rediscovery that the model structure is not right. At worst, the faulty model will be fit to the data, anyway. So what is the use of models that are not even true? Models that are "unrealistic" because they leave something out can still be heuristically useful. Thinking about them may get us into lots of trouble, but the adventure is wonderful, and much is likely to be learned. But models that are "untrue," because they are structurally incorrect can get us into an altogether more serious kind of trouble.

References

Edman, K.A.P., Caputo, C., and Lou, F. (1993). Depression of tetanic force induced by loaded shortening of frog muscle fibers. *J. Physiol.*, 466:535–552

Hill, A.V. (1938). The heat of shortening and the dynamic constants of muscle. *Proc. Roy. Soc. B.*, 126: 136–195.

Jewell, B.R. and Wilkie, D.R. (1958). An analysis of the mechanical components in frog's striated muscle. *J. Physiol.*, 143:515–540.

Julian, F.J. and Morgan, D.L. (1979). The effect on tension of non-uniform distribution of length changes applied to frog muscle fibres. *J. Physiol.*, 293:379–392.

Rack, P.M.H. and Westbury, D.R. (1969). The effects of length and stimulus rate on tension in the isometric cat soleus muscle. *J. Physiol.*, 204:443–460.

Rushdie, S. (1990). Haroun and the sea of stories, Granta, New York.

Stephenson, D.G. and Williams, D.A. (1982). Effects of sarcomere length on the force-pCa relation in fast- and slow-twitch skinned muscle fibres from the rat. *J. Physiol.*, 333:637–653.

6
Modeling of Homogeneous Muscle: Is It Realistic to Consider Skeletal Muscle as a Lumped Sarcomere or Fiber?

Peter A. Huijing

1 Introduction

Much scientific effort has been devoted toward modeling skeletal muscle and movement. This effort started in the 17th century and continues at an enormous volume because of the development of easily accessible computational facilities. Most frequently, muscle is modeled as homogeneous with respect to muscle geometry and fiber properties (e.g., Benninghoff and Rollhäuser 1952; Gans and Bock 1965; Hatze 1981; Hawkins and Hull 1991; Huijing and Woittiez 1984, 1985; Woittiez et al. 1984; Bobbert et al. 1986; Bobbert and van Zandwijk 1996; Otten 1988; Zajak 1989; Zajac and Winters 1990; Heslinga and Huijing 1993; Scott and Winter 1991).

Only fairly recently some modelers interested in morphological details of active muscle have tried to study aspects of functional consequences of inhomogeneous muscle (e.g. Otten 1988; van Leeuewen and Spoor 1996), or to apply finite element models to skeletal muscle (van den Donkelaar et al. 1996; van der Linden et al. 1995a,b; van Kan et al. 1996).

The question to be addressed in this chapter is whether it is realistic to model muscles in a homogeneous way *even when only the single condition of maximal activation is considered.*

The problem of assumed homogeneity does not only play a role in modeling but also in the interpretation of experimental results. For human muscles it has been found that joint angle-moment characteristics of human skeletal muscle are hard to explain on the basis of fiber lengths or the number of sarcomeres in series within fibers (e.g., Huijing et al. 1986, 1987; Herzog and ter Keurs 1988).

Recently Lieber, Friden, and coworkers (Lieber and Boakes 1988; Lieber et al. 1992; Lieber and Brown, 1992) extended their previous experimental animal work dealing with the relationship of joint angle and sarcomere length to architectural studies of human muscles (Lieber et al. 1990), as well as to invasive experiments on humans (Lieber et al. 1992). This must be considered a major contribution to the field of human movement sciences because surprisingly little is known about the length range of the length-force curve actually used by humans in vivo. Knowledge about the joint angle-sarcomere length is very important to the modeling of muscle, because many patients are hampered severely by limitations of joint excursion, which may be caused by the faulty regulation of number of sarcomeres in series in a fiber or by other factors determining this relationship. It is therefore very important to be aware of the factors that should be taken into account in interpreting joint angle-sarcomere length data.

2 Methods

2.1 Mechanical Muscle Models

Van der Linden et al. (1995a,b) presented a finite element model of the model rat gastrocnemius containing a muscle element and a tendinous element. The size and shape of an element is defined by the positions of a number of nodes. The behavior of an element is based on the displacement of these nodes. The displacement of points within an element is based on the displacements of the nodes at the boundary of the element. Properties of a muscle element are based on the displacement of six nodes, with quadratic interpolation functions in the fiber direction (h) and linear functions in the radial direction perpendicular (x) to the fibers. A tendinous element can be used for the modelling of tendinous structures such as the tendon, as well as the aponeurosis. Because of the linear tendinous properties assumed, a two dimensional linear beam element of the ANSYS program library was used as a model. Fixed end fiber length-force characteristics as obtained by Zuurbier et al. (1995) were used rather than sliding filament theory inspired estimates of sarcomere length-force characteristics, thereby taking into account at least some possible effects of inhomogeneities of lengths of sarcomeres in series within one fiber.

2.2   Experiments

Anesthesia was induced in the experimental animals by intraperitoneal injection of sodium-pentobarbitone (initial dose 8.0 mg/100 g body mass). Supplementary doses were given as necessary. In these experiments, the nerve innervating the target muscle was stimulated at supramaximal currents to obtain full recruitment. All nerve branches innervating muscles other than the target muscle were cut. The experiments were performed at many muscle lengths to be able to construct a length-force curve for each particular condition. It should be noted in comparison to the in-vivo situation, a major drawback of this experiment is the synchronization of motor units by the stimulation.

2.1.1   Experiments on Medial Gastrocnemius Muscle

After removing the skin and the superficially located biceps femoris muscle, the medial head of m. gastrocnemius (GM) was exposed and isolated from the lateral head by cutting tissues at the expense of the lateral head. During contraction, photographic images of the muscle belly were taken. The number of sarcomeres in fibers teased from different locations within the muscle were determined. The experimental methods were similar to those described by Huijing et al. (1994).

2.2.2   Experiments on Semimembranosus Muscle

The lateral head of this muscle, which may be considered functionally as a parallel-fibered muscle (i.e., angular and elastic contributions to muscular length change are neglible), was studied. Estimates were made of its proximal and distal fiber length, as well as of the number of sarcomeres in series. These experiments were described in detail by Willems and Huijing (1994). Distribution of fiber mean sarcomere length was quantified for each individual muscle by calculating the coefficient of variance of mean sarcomere length (i.e., fiber length / # sarcomeres in series) at optimum muscle length.

3   Results

Do muscle optimum length and fiber mean sarcomere optimum length have to coincide? For a pennate muscle such as GM, the situation is somewhat more complicated because secondary distributions have to be taken into account as well. It has been estimated previously (Heslinga and Huijing 1990) that in a homogeneously modeled GM fiber, optimum fiber length should occur at a muscle length approximately 0.12 mm below muscle optimum length because of interactions between decreasing angular effects on muscle force at increasing muscle lengths and increasing fiber force as a function of length in this length range.

3.1   Finite-Element Modeling and GM Experimental Results

Results presented by Van der Linden et al. (1995a,b) will be discussed here. Figure 6.1 shows an important feature of the model. Initially, the muscle is modeled in the passive state with homogeneous properties just as in the geometric modeling. In this initial state, the muscle fibers are assumed to be arranged linearly. As the muscle is activated, a distribution of fiber length, and thus of fiber mean sarcomere length, develops. Such a distribution has been referred to as a secondary distribution of fiber mean sarcomere length (Huijing 1995, 1996). Also, curved fibers and aponeuroses develop, with the amount of curvature dependent

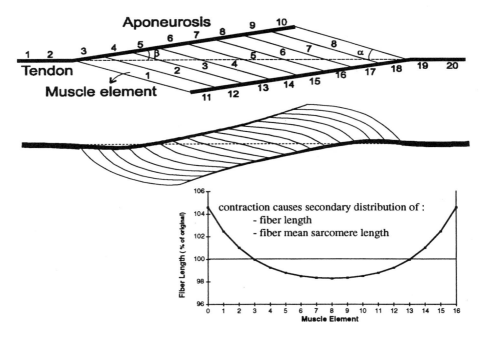

FIGURE 6.1. Finite-element model of rat gastrocnemius medialis muscle in passive and active condition. (Top) The muscle modeled homogeneously, with uniform fiber length and angles as well as aponeurosis length and angles. Muscle elements (representing bundles of muscle fibers) are indicated (through 8) in addition to tendon and aponeurosis elements (through 20). (Middle) Muscle geometry after isometric contraction. Note the varying curvature in aponeuroses and muscle fiber bundles that developed after activation of the model muscle. (Bottom) Distribution of fiber length with respect to the initial (homodeneous) condition.

on the location within the muscle. It was concluded that the mechanical interaction between the elements of the model makes the continued presence of homogeneous properties within the muscle impossible.

The top panel of Figure 6.1 shows the muscle modeled homogeneously, with uniform fiber length and angles as well as aponeurosis length and angles. Muscle elements (representing bundles of muscle fibers) are indicated (1 through 8) in addition to tendon and aponeurosis elements (1 through 20). The middle panel shows muscle geometry after isometric contraction. Note the varying curvature in aponeuroses and muscle fiber bundles that developed after activation of the model muscle. The bottom panel shows the distribution of fiber length with respect to the initial (homogeneous) condition.

Despite this added feature, van der Linden et al. (1995a,b) concluded that muscle length-force characteristics were not adequately described by the model because substantial differences with actual length-force data remained. It was concluded that

factors of major importance must not have been included in the models used.

3.2 Primary Inhomogeneities of Fiber Mean Sarcomere Length

3.2.1 Indirect Evidence

Circumstantial evidence regarding the existence of a distribution of fiber mean sarcomere length is available for rat muscle (Grottel et al. 1990; Heslinga and Huijing 1990, 1993; Huijing et al., 1994; Ettema and Huijing, 1994; Willems and Huijing 1994). The majority of this indirect evidence is based on the finding that the mean sarcomere length of certain groups of fibers at muscle optimum length is not equal to sarcomere optimum length as expected from a simple homogeneous muscle. For example, GM distal fibres were found experimentally (Huijing et al. 1994) to reach optimum sarcomere length at muscle lengths 1.27 mm over optimum length. This amounts to more than 10% of the length range between optimum and active slack lengths. Models have been made to assess the effects on length-force characteristics of inhomogeneous sarcomere length in different GM fibres (Ettema and Huijing 1994) and for the EDL muscle (Bobbert et al. 1986).

3.2.2 Direct Experimental Evidence

Particularly for cat muscle, direct evidence has been accumulating in the last two decades that, at any muscle length, sarcomere length may be distributed (Lewis et al. 1972; Stephens et al. 1978). This has been done by determining length-force characteristics of whole muscle and of single motor units stimulated through their isolated ventral roots. In some cat muscles (flexor digitorum longus muscle), the distribution was shown to be systematic for motor unit size but for others this could not be shown. It should be noted that if the distribution is regulated by motor unit size, muscle length-force characteristics would be very much dependent on the degree of recruitment of motor units.

More recently, direct evidence for distribution of length-force characteristics of groups of motor units was also reported for the rat GM (de Ruiter, 1995; de Ruiter et al. 1996). This study also reported that larger motor units tended to have their optimum length at longer muscle lengths (e.g., Figure 6.2).

This indicates that to obtain total fiber force for a parallel fibered muscle, fiber length-force curves should not be added "in phase with respect to fiber- or sarcomere length," but rather that the addition should take place with the curves shifted a certain amount with respect to each other in relation to muscle length (see also Huijing 1996). In such a muscle at muscle optimum length the grand mean of the sarcomere lengths should be equal to sarcomere optimum length. In a pennate muscle the situation is complicated by length dependent angular effects (e.g., Huijing 1995, 1996).

3.3 Individual Variation

It was noted that individual SMl muscles showed considerable differences in the length range between optimum length and active slack length, and thus in the ascending limb of the muscle length-force curve (Figure 6.3). Surprisingly, the mean number of sarcomeres in series showed no correlation ($r = -0.14$) with these individual length ranges. In contrast, the coefficient of variation of fiber mean sarcomere length at optimum muscle

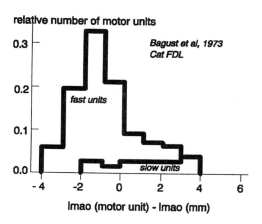

FIGURE 6.2. Distribution of motor unit optimum length with respect to muscle length reported for cat flexor digitorum longus muscle. Note that the optimum lengths of individual motor units were dispersed with respect to the whole muscle optimum length over a sizable length range of 8 mm. Also note that the distribution was systematic for fast and slow units, and thus presumably for motor unit size.

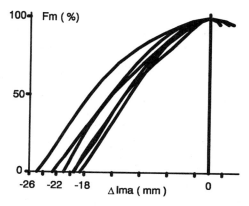

FIGURE 6.3. Rat semimembranosus muscle individual length normalized force curves for muscles for animals of comparable body mass (approximately 300 g). Note that the length range of force exertion (between optimum and active slack lengths) showed individual differences of up to 8 mm (i.e., approximately 30% of the largest length range found). These differences could not be explained by different number of sarcomeres in series within muscle fibers but correlate highly with an estimate of fiber mean sarcomere length distribution.

length correlated significantly ($r = 0.88$) with the length range.

As some of these parallel-fibered muscles showed no evidence of distributed sarcomere length, it is concluded that most of this variation of sarcomere length was most likely not the result of secondary effects, but rather was caused by a primary distribution of sarcomere length. This concept was reviewed in detail by Huijing (1995, 1996).

4   Discussion

The question addressed in this chapter is whether it is realistic to consider a muscle as one giant sarcomere. The results summarized above indicate that the answer to such a question has to be unequivocally no. Whether modelers and experimenters should continue to report results that are based, implicitly or explicitly, on assumptions related to homogeneity can not be answered as unequivocally. Either the effects such as the ones presented here are incorporated into realistic models, or substantial deviations from reality will be built into the models, preventing accurate prediction of variables important for muscle function. In the examples described in this chapter, it is very unlikely we will be able to accurately obtain the parameters needed for the more complex distributed muscle model. However, as discussed in Chapter 5, how deviations of models from reality are dealt with very much depends on the purpose of the studies involved. It is clear that if muscle is modeled as lumped sarcomere or lumped fiber, success can be obtained for heuristic purposes. An example can even be found within the material presented in this chapter: Most of the indirect evidence for distribution of fiber mean sarcomere length could only be interpreted because experimental data was compared to expectations based on very simple models of rat muscles. However it should not be expected that such models will yield realistic descriptions even of muscle functioning under isometric conditions with maximal activation. The possibility of very high individual variation in sarcomere lengths make sensible clinical or other applications, for which accurate prediction of muscular capabilities is necessary, highly unlikely.

Similar arguments seem valid for the results of other experiments. The data collected by Lieber and Fridén about the sarcomere length range used in vivo (see Chapter 2) are of prime importance for dealing with the widespread but false concept that during movement sarcomeres only change length within a narrow range near optimum length. It is also very useful in generating testable hypothesis concerning many aspects of human movement. However, one should be aware of the limitations of invasive human experimentation in sampling a sufficiently large sample of sarcomere lengths to be representative.

In interpreting experimental results, it is essential that the limitations and assumptions of a particular model be openly acknowledged. This may make the main purpose of an article harder to accomplish: The sale of idea. However it will not decrease the value of such ideas and it will help the generation of new ideas.

Acknowledgments. The author gratefully acknowledges Bart Koopman, Henk Grootenboer, Kenneth Meijer, and Bart van der Linden at the University of Twente for their continuing discussions on issues raised in this chapter.

References

Bagust, J., Knott, S., Lewis, D.M., Luck, J.C., and Westerman, R.A. (1973). Isometric contractions of motor units in a fast twitch muscle of the cat. *J. Physiol.*, 231:87–104.

Benninghoff, A. and Rollhäuser, H. (1952). Zur inneren Mechaniek des gefiederte muskels. *Pflüger's Arch. Ges. Physiol.*, 254:527–548.

Bobbert, M.F. and van Zantwijk, J.P. (1994). Dependence of human maximal jump height on moment arms of the biarticular m. gastrocnemius; a simulation study. *Hum. Mov. Sci.*, 13:696–716.

Bobbert, M.F., Ettema, G.J., and Huijing, P.A. (1990). Force length relationship of a muscle tendon complex: experimental results and model calculations. *Eur. J. Appl. Physiol.*, 61:323–329.

Bobbert, M.F., Huijing, P.A., and van Ingen Schenau, G.J. (1986). A Model of human triceps surae muscle-tendon complex applied to jumping. *J. Biomech.*, 19: 887–898.

de Ruiter, C.J. (1996). Physiological properties of skeletal muscle units vary with intra-muscular location of their fibres. Doctoral dissertation, Vrije Universiteit Amsterdam.

de Ruiter, C.J., de Haan, A., and Sargeant, A.J. (1995). Physiological characteristics of two extreme muscle compartments in gastrocnemius medialis of the anaesthetized rat. *Acta Physiol. Scand.*, 153:313–324.

Ettema, G.J.C. and Huijing, P.A. (1994). Effects of distribution of fiber length on active length-force characteristics of rat gastrocnemius medialis. *Anat. Rec.*, 239:414–420.

Gans, C. and Bock, W.J. (1965). The functional significance of muscle architecture: a functional analysis. *Erg. Anat. Entw. Ges.*, 38:115–142.

Hakkinen, K., Keskinen, K.L., and Komi, P.V. (eds.), (1995). Proc. XVth congress of the International Society of Biomechanics, University of Jyvaskyla, Finland.

Hatze, H. (1981). Myocybernetic control models of skeletal muscle. University of South Africa Press, Prudery.

Hawkins, D.A. and Hull, M.L. (1991). A computer simulation of muscle mechanics. *Comput. Biol. Med.*, 21:369–382.

Herzog, W. and ter Keurs, H.E.D.J. (1988). A method for determination of force-length relation. Pflügers Arch. 411:637–641.

Heslinga, J.W. and Huijing, P.A. (1990). Effects of growth on architecture and functional characteristics of adult rat gastrocnemius muscle. *J. Morph.*, 206:119–132.

Heslinga J.W. and Huijing, P.A. (1993). Muscle length-force characteristics in relation to muscle architecture: a bilateral study of gastrocnemius medialis muscles of unilaterally immobilized rats. *Eur. J. Appl. Physiol.*, 66:289–298.

Hill, A.V. (1970). First and last experiments in muscle mechanics. Cambridge University Press, Cambridge.

Huijing, P.A. (1995). Parameter interdependence and success of skeletal muscle modelling. *Hum. Mov. Sci.*, 14:443–486.

Huijing, P.A. (1996). Important experimental factors for skeletal muscle modelling: non-linear changes of muscle length-force characteristics as a function of degree of activity. *Eur. J. Morph.*, 34:47–54.

Huijing, P.A. and Woittiez, R.D. (1984). The effect of architecture on skeletal muscle performance: a simple planimetric model. *Neth. J. Zool.*, 34:21–32.

Huijing, P.A. and Woittiez, R.D. (1985). Notes on planimetric and three-dimensional muscle models. *Neth. J. Zool.*, 35:521–525.

Huijing, P.A., Greuell, A.E., Wajon, M.H., and Woittiez, R.D. (1987). An analysis of human isometric voluntary plantar flexion as a function of knee and ankle angle. In *Biomechanics: Basic and Applied Research*, Bergman, G., Kolbel, R., Rohlman, R.A. (eds.), pp. 662–667, Nijhoff, Dordrecht.

Huijing, P.A., Nieberg, S.M., van de Veen, E.A., and Ettema, G.J.C. (1994). A comparison of rat extensor digitorum longus and gastrocnemius medialis muscle architecture and length-force characteristics. *Acta Anat.*, 149:111–120.

Huijing, P.A., Wajon, M.H., Greuell, A.E., and Woittiez, R.D. (1986). Muscle excitation during voluntary maximal plantar flexion: a mechanical and electromyographic analysis. In *Proc. VIIIth Conf. IEEE Eng. Med. Biol. Soc.*, Kondraske, G.V. and Robinson, C.J. (eds.), Vol I, pp. 637–639, IEEE Publishing Service, Piscataway.

Lieber, R.L. and Boakes, J.L. (1988). Sarcomere length and joint kinematics during torque production in frog hindlimb. *Am. J. Physiol.*, 254:C759–C768.

Lieber, R.L. and Brown, C.G. (1992). Sarcomere length-joint angle relationship of seven frog hindlimb muscles. *Acta Anat.*, 145:289–295

Lieber, R.L., Fazeli, B.M., and Botte, M.J. (1990). Architecture of selected wrist flexor and extensor muscles. *J. Hand Surg.*, 15:244–250.

Lieber, R.L., Loren, G.J., and Friden, J. (1994). In vivo measurement of human wrist extensor muscle sarcomere length changes. *J. Neurophysiol.*, 71:874–881.

Lieber, R.L., Raab, R., Kashin, S., and Edgerton, V.R. (1992). Short communication sarcomere length changes during fish swimming. *J. Exp. Biol.*, 169:251–254.

Lewis, D.M., Luck, J.C., and Knott, S. (1972). A comparison of isometric contractions of the whole muscle with those of motor units in a fast twitch muscle of the cat. *Exp. Neur.*, 37:68–85.

Otten, E. (1988). Concepts and models of functional architecture in skeletal muscle. *Exerc. Sport Sci. Rev.*, 16:89–139.

Scott, S.H. and Winter, D.A. (1991). A comparison of three muscle pennation assumptions and their effect on isometric and isotonic force. *J. Biomech.*, 24:163–167.

Stephens, J.A., Reinking, R.M., and Stuart, D.G. (1978). The motor units of cat medial gastrocnemius: Electrical and mechanical properties as a function of muscle length. *J. Morph.*, 146:495–512.

van den Donkelaar, C.C., Drost, M.R., van Mameren, H., Tuinenburg, C.F., Janssen, J.D., and Huson, H.A. (1996). Three dimensional reconstruction of the rat triceps surae muscle and finite element mesh generation of the gastrocnemius medialis muscle. *Eur. J. Morph.*, 34:31–37.

van der Linden, B.J.J.J., Meijer, K., Huijing, P.A., Koopman, F.F.J.M., and Grootenboer, H.J. (1995a). Finite element model of anisotropic muscle: effects of curvature on fibre length distribution. In (10) Hakkinen et al., pp. 956–957.

van der Linden, B.J.J.J., Huijing, P.A., Meijer, K., Koopman, H.F.J.M., and Grootenboer, H.J. (1995b). In *Integrated biomedical engineering for restoration of human function*, Feijen, J. (ed.), pp. 53–57. Institute for Biomedical Engineering, Universiteit Twente, Enschede.

van Kan, W.J., Huyghe, J.M., Jansen, J.D., and Huson, A. (1996). A 3D finite element model of blood perfused rat gastrocnemius muscle. *Eur. J. Morph.*, 34: 19–24.

van Leeuwen, J.L. and Spoor, C.W. (1996). A two dimensional model for the prediction of muscle shape and intramuscular pressure. *Eur. J. Morph.*, 34:25–30.

Willems, M.E.T. and Huijing, P.A. (1994). Heterogeneity of mean sarcomere length in different fibres: effects on length range of active force production in rat muscle. *Eur. J. Appl. Physiol.*, 68:489–496.

Woittiez, R.D., Huijing, P.A., Boom, H.B.K., and Rozendal, R.H. (1984). A three dimensional muscle model: a quantified relation between form and function of skeletal muscles. *J. Morphol.*, 182:95–113.

Zajac, F.E. (1989). Muscle and tendon: properties, models scaling and application to biomechanics and motor control. *CRC Crit. Rev. Biomed. Eng.*, 17:359–411.

Zajac, F.E. and Winters, J.M. (1990). Modeling musculoskeletal movement systems: Joint and body segmental dynamics; musculoskeletal actuation, and neuromuscular control. In *Multiple muscle systems*, Winters, J.M. and Woo, S.L-Y. (eds.), pp. 121–164.

Zuurbier, C.J., Heslinga, J.W., Lee-de Groot, M.B.E., and van der Laarse, W.J. (1995). Mean sarcomere length-force relationship of rat muscle fibre bundles. *J. Biomech.*, 28:83–87.

Commentary:
The Role of Distributed Properties in Muscle Mechanics

Michael P. Slawnych

The basic premise of Peter Huijing's article is that the characteristics of individual sarcomeres can significantly influence the mechanical behavior of whole muscle systems. I wholeheartedly agree. As Huijing states, there is now ample evidence that whole muscle force-length relations are influenced by sarcomere length heterogeneity, and, as a result, muscle optimum length does not necessarily correspond to sarcomere optimum length. However, it

should be noted that the force-length relation is only one aspect of muscle behavior, and at that only provides a "static" picture of how muscles operate. One of the more interesting lines of current research concerns how inter-sarcomeric behavior contributes to the dynamic characteristics of muscle response. At least at the level of the single muscle fiber, such research has been going on for some time. Perhaps the best known example of inter-sarcomere dynamics is the "creep" phase associated with fiber-isometric contractions performed at long sarcomere lengths. This late phase in which the tension slowly increases has been attributed to internal fiber motion in which the end regions shorten while the rest of the fiber lengthens (Huxley and Peachey 1961). Another, more recent example is that of the increased tension observed when the muscle is stretched during activation. Specifically, for a muscle fiber operating at a length corresponding to the descending limb of the length-tension curve, stretch during active contraction results in a final force level that is higher than that observed when the muscle is isometrically activated at the final, stretched length. It has been proposed that this is due to non-uniform sarcomere behavior in which the weaker sarcomeres "pop" to longer lengths in order to match the force produced by the stronger sarcomeres (Morgan 1994).

It is important to note that both of the aforementioned phenomenon occur only when the sarcomeres are operating on the descending limb of the length-tension curve. It has long been recognized that a series connection of sarcomeres whose tensions decrease with increasing sarcomere length can potentially lead to instability (Hill 1953). A widely employed method of dealing with this instability is to keep the sarcomere or segment length of a specific region constant using a clamping technique.* Using such a method, Bagni and colleagues (Bagni, Cecchi, et al. 1988) were able to completely abolish tension creep from their records. Obviously, actual in vivo muscle systems do not do not employ a clamping technique per se. However, there may be a form of indirect clamping taking place in the form of lateral interactions between structures at the

---

*Such a method was employed by Gordon and colleagues in the derivation of their now classic tension-length relation (Gordon, Huxley et al. 1966).

various muscle levels. Evidence supporting such interaction has recently been reported by Friedman and Goldman (1996). Specifically, in characterizing the mechanical properties of small groups of myofibrils, these investigators observed in some instances that the myofibrillar preparations contained a myofibril that was not attached to the transducer wires, and that these "decoupled" myofibrils noticeably restrained the myo-fibrillar shortening. Ultimately, one must ask the question of to what degree does intersarcomeric behavior contribute to the dynamic behavior of muscle?

Given that muscle structure is hierarchical in nature, one must also ask to what degree distributed properties in higher order structures contribute to muscle behavior. While sarcomeres can be thought of as the basic contractile unit at the microscopic level, the equivalent structure at the macroscopic level would be the motor unit. As with sarcomeres, motor units also have distributed properties. Huijing cites the example of there being a positive correlation between the size of the motor unit and the optimum muscle length at which peak force is observed. It has long been known that the sizes of individual motor units (in terms of the numbers of their constituent muscle fibers) vary considerably within any given muscle. The size of a motor unit is well correlated with many of its properties such as the twitch tension, contraction time, fatigueability, and excitability (and hence its recruitment during reflex and voluntary contractions). This collection of properties have been referred to as the *size principle,* which is one of the major organizing principles of muscle physiology. However, much work remains to be done in terms of quantifying the contractile characteristics of individual motor units and understanding how these characteristics sum to produce the overall muscle response.

Over the course of the last several decades, much progress has been made in terms of understanding the detailed structure of muscle and relating it to muscle function. The field of muscle mechanics has progressed to the state in which it is now possible to examine systems in which the contractile machinery consists of a single actin filament in association with a single crossbridge (Finer, Simmons,

et al. 1994). While significant effort is being directed toward a comprehensive understanding of muscle function at the crossbridge level, it is important to be able to relate this back to what is happening at the levels of the single fiber, single motor unit and whole muscle. That is, the accurate characterization of the mechanical properties of whole muscle systems requires the quantification of the *distributed* properties of their constituent muscle fibers and motor units in the form of

1. Quantifying the operating ranges of sarcomeres during active contraction,
2. Quantifying the degree of sarcomere heterogeneity in whole muscle systems,
3. Quantifying the numbers of motor units in nerve–muscle systems, and
4. Quantifying the contribution of individual motor units to force generation and/or movement in the form of their relative sizes and recruitment characteristics.

Once these properties have been characterized, the challenge will then be to incorporate them into a unified muscle model.

References

Bagni, M.A., Cecchi, G., et al. (1988). Plateau and descending limb of the sarcomere length-tension relation in short length-clamped segments of frog muscle fibres. *J. Physiol.,* 401:581–595.

Finer, J., Simmons, R., et al. (1994). Single myosin molecule mechanics: piconewton forces and nanometre steps (see comments). *Nature,* 368:113–119.

Friedman, A. and Goldman Y. (1996). Mechanical characterization of skeletal muscle myofibrils. *Biophys. J.,* 71:2774–2785.

Gordon, A.M., Huxley, A.F., et al. (1966). The variation in isometric tension with sarcomere length in vertebrate muscle fibres. *J. Physiol.,* 184:170–192.

Hill, A.V. (1953). The mechanics of active muscle. *Proc. Roy. Soc. B.,* 141:104–117.

Huxley, A. and Peachey, L. (1961). The maximum length for contraction in vertebrate striated muscle. *J. Physiol.,* 156:150–165.

Morgan, D.L. (1994). An explanation for residual increased tension in striated muscle after stretch during contraction. *Exper. Physiol.,* 79(5):831–838.

7
Subtle Nonlinear Neuromuscular Properties Are Consistent with Teleological Design Principles

Jack M. Winters

1 Introduction

The other chapters in this section address certain subtle details of neuromuscular properties, such as interaction between properties (e.g., activation and contractile force-length) and effects of fatigue (see Chapters 5 and 6, Huijing), convergence of information on motoneurons (see Chapter 2), and muscle force-length operating range variability (see Chapter 3). In general, such details may be interpreted as adding complexity to neuromusculoskeletal (NMS) models, and there is thus a natural resistance on the part of modellers to include every single observed behavior. Yet this opens the door to a criticism of models as not being sufficiently accurate, and therefore inadequate; this was a recurring theme during several of the discussions at the recent EFC on Biomechanics and Neural Control of Movement. It needs to be addressed.

In line with the theme of Section II, this chapter addresses the implications of a range of observed nonlinear neuromuscular properties (not the models that try to capture these properties), only from the perspective of a modeler with extensive experience in using simulation as a tool to gain insight into multiple muscle systems. The primary purpose is to suggest that much of the information obtained through recent experiments is consistent with teleological expectations, and furthermore is remarkably synergistic with the subtle changes that are needed to improve our models. The secondary purpose is to suggest that barriers to utilizing synthesized neuromusculoskeletal models are falling, with much to be gained from utilizing models that intimately intertwine biomechanics and neurocir-

cuitry. It is time to go beyond the two conventional extremes: biomechanics-centered neuro*musculoskeletal* approaches (e.g., where musculoskeletal models utilize a neurocontrol "final common pathway" input to muscle; Figure 7.1A) and neuroscience-centered *neuro*musculoskeletal approaches (e.g., use of simplified biomechanics and overconstrained experiments), while keeping the best of each. Remember also that connectionist neural networks (CNNs) typically thrive on nonlinearity. We will build on insights from the fourth-order nonlinear "muscle-reflex" actuator of Figure 7.1B developed elsewhere (Winters 1995a,b).

2 Underlying Principles

Underlying this perspective are two presuppositions that here will be assumed to be fundamental principles.

2.1 Profoundly Strong Biocybernetic Drive

We assume here that NMS structure and function is remarkably consistent with what would be expected based on a thoughtful consideration of neuromechanically based teleological design principles, driven by strong biocybernetic tendencies that span three time scales:

1. *Evolutionary* (very long, multilifespan time scale): Neural and musculoskeletal structures have coevolved toward inherently bidirectional "optimal" design solutions that take advantage of non-

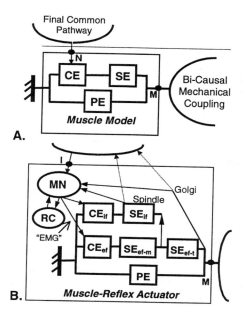

FIGURE 7.1. Two approaches for a muscle actuator. (A) A more classic view using a Hill-type muscle model, captured by a second or third-order nonlinear muscle model. (B) The fourth-order muscle-reflex actuator, with Hill-type models for both extrafusal (ef) and intrafusal (if) muscle (see subscripts on elements, where SE, -m is muscle and -t is tendon). MN is motoneuron, RC is Rhenshaw cell, CE is contractile element, SE is series element, PE is parallel element, **N** and **I** are unidirectional neural input ports, **M** is a bidirectional port.

linearities within the system. The optimum design "criteria" involves satisfactory completion of that collection of key tasks most necessary for survival and pleasure of the organism, the "system" consists of a collection of intertwined adaptive subsystems, the "algorithm" is related in part to Darwinian theory, and the "training set" is the laws of physics and composition of the environment (e.g., predator–prey). Alexander's thought-provoking writings on animal mechanics (e.g., Alexander 1988; Alexander and Ker 1990) capture the concept as it relates biomechanics to teleological function—muscles and tendons have sizes, lengths, and shapes that appear "designed" to address teleological needs, with "solutions" that balance competing criteria (e.g., force generation versus segmental mass; tendons for elastic energy vs tendons for control). The level of neuromechanical design intimacy also cannot be underestimated—for instance, consider the intertwined nature of the information coupling between

motoneuronal properties and muscle fiber composition during early formation of animal structure and function (Close 1969).

2. *Tissue Remodeling* (medium-length time scale, e.g., weeks and months): Tissues are alive and capable of a remarkable degree of remodeling that is primarily a function of the recent history of mechanical loading, coupled with the level of support within the local biochemical and bioenergetic environment. Careers in fields, such as physical therapy—indeed the very concept of rehabilitation, are based on this principle. Tissue remodeling also dominates research in areas such as orthopedic biomaterials (e.g., "Wolff's law" [(Wolff 1884) and Roux's (1985) principles of functional adapation and maximum–minimum design (see Fung 1993)]. The key consequence is so obvious that it is easily forgotten by researchers: that observed tissue behavior should be treated as locally near-optimal (if one knew the tissue utilization history). This principle applies to any tissue with an adequate blood supply, including neural tissue. Within this context, skill development may be considered as "remodeling" of neural tissue (which here includes chemical synapses); for instance, Sperry (1959), a pioneering psychobiologist, refers to "disuse atrophy."

3. *Neurocontrol and Adaptive Learning* (short time scale, e.g., seconds): Neuromotor structures in larger-scale organisms are designed for fast "on-the-fly" learning of many novel tasks, including postural and tracking tasks. It is well documented for humans and larger mammals that for simple tasks (such as target tracking) humans and mammals will quickly adapt their performance when faced with a new goal (e.g., new target locations) or a novel environmental conditions (e.g., an added helmet mass), usually within a window of about 4 to 8 trials (e.g., Gauthier et al. 1986; see also Chapter 26).

Thus, information on tissue properties should be interpreted in light of this profound drive, on multiple time scales, toward tissue optimality. Note that (1) and (2) relate to optimal design; (3) to optimal control.

2.2   Use Biomodels That Do Not Break Down

The second principle of the test of a good biomodel of nature is that it must be satisfactory for a broad range of experiments, that is, it never fundamentally breaks down (Winters 1990, 1995a). This is often inconsistent with two approaches that are traditionally taught to science and engineering students: (1) linearization of systems over operating ranges, and (2) the concept of using the simplest possible models and experiments to study a problem. So be it.

The motivation for linearization is strong, with its foundation in linear differential equations and matrix methods, and its implications related to signal superposition and scalability. Yet observation of natural phenomena—from geophysical to biological—suggests that nature thrives on designs that are inherently nonlinear and is governed by strategies that may be transparent to linear systems analysis (Winters and Stark 1987). For instance, when using linearized Hill-type models (e.g., as sometimes used in Chapters 11 and 29), antagonistic cocontraction is not a viable feedforward strategy for varying muscle/joint/contact stiffness and viscosity because the slopes for the contractile element (CE) force-length (CE_{fl}), CE force-velocity (CEfl), series element (SE), and parallel element (PE) relations are not a function of the ongoing muscle force (Winters and Stark 1987; Winters et al. 1988). Reality cannot always be packaged by mathematical convenience.

The motivation for using the simplest possible models and experiments is also strong, and certainly often worthwhile (e.g., study of running by McMahon 1990); yet it is abused. Indeed, a case could be made that with the exception of eye movements, the hundreds of studies of single-joint movement (including some by this author) have yielded limited knowledge of principles that might govern realistic movements. Such studies do not bring to the table key fundamental challenges that seem to permeate movements performed outside of research laboratories. "Dynamic systems" input-output behavioral models that seek noncausal explanations without regards to dedicated mechanisms or tissue structures (e.g., see Beek et al. 1995) are also of limited value—essentially mathematical descriptions of attractor dynamics. Thus as a community we have become adept at avoiding classic problems that have been on the table since Bernstein (1967): planning, selection, and control of action systems that include vast "redundancy" (kinematic, actuator) but also have embedded neuromotor "synergy" circuitry.

3 Muscle-Reflex Actuator Model Framework

3.1 Minimal Model for Muscle Mechanics

Reviews of Hill-type muscle models can be found many places (e.g., see Chapter 8; Winters 1990, 1995b; Zajac 1989). One key finding from simpler movement simulations, using an antagonistic muscle-joint model (Winters and Stark 1985), is that it is easy to show through systematic use of simulations for a various classes of movements for which experimental data exists (e.g., unloaded point-to-point tracking, isometric, isotonic, and external perturbation responses), that the minimal representation for muscle is the Hill model structure with nonlinear CE_{fl}, CE_{fv}, SE, and PE properties and different kinetics for activation and deactivation (Winters and Stark 1987). For these simulations there was no task-specific adjustment of model parameters; this is a guiding principle for all modeling studies by this author, and reflects the concept that since "tissue is tissue," any need for task-specific parameter adjustment suggests a weak mechanical model!

Consider the popular second-order model structure of the form, often used for studying impedance control (Hogan 1990) or systems identification (see review in Chapter 9):

$$J(\theta)\frac{d^2(\theta)t}{dt^2} + B(n,\theta)\frac{d\theta(t)}{dt} + K(n,\theta)\theta(t) = n(t) + m(t) \quad (7.1)$$

where J is generalized mass (due to a limb segment), B is viscosity, K is stiffness (due to muscles), θ is position, n is a controllable neuro-input (with units of a moment), and m is an environmental (applied) moment. Notice that, unlike in Figure 7.1A, where ports **N** and **M** are structurally distinct, in Eq. 1 the signals n and m simply add. Furthermore, in interpreting the "stiffness" K we have a fundamental problem: Is it closest to SE, CE_{lt} or PE? Often it is interpreted as CE_{fl} (e.g., Chapter 11; Hogan 1990), yet for most perturbation and frequency response experiments, over the time/frequency range under experimental consideration it is often most closely related to SE (e.g., Cannon and Zahalak 1982; detailed review in Winters 1990).

For external perturbation studies under a limited operation range and steady neural conditions (n = constant and B and K may be a function of n), Eq. 1 may be useful. But what about for a voluntary movement task? This model breaks down at a profoundly fundamental level (Winters and Stark 1987). Even task-specific tuning, as developed in Chapter 9, cannot prevent such breakdown because the assumed structure cannot capture, for instance, dynamic interplay between CE and SE.

To summarize: If one's purpose is to reasonably replicate reality for a range of movement tasks, there is simply no case to be made for using anything simpler than a nonlinear Hill-type model. An added advantage of starting with a muscle model of at least this level of sophistication is that task-specific sensitivity analysis (Winters and Stark 1985; see also Chapter 10; Chapter 43) or task-specific linearization (Winters and Seif-Naraghi 1988; see also Chapter 11) can be used to justify its simplification when only a small task set is of interest.

3.2 Problem of Postural Drift

Now consider a basic problem that occurs for any type of musculoskeletal model that assumes a physiologically realistic CE force-length (CE_{fl}) property: open-loop muscle models tend to lack adequate steady-state (quasistatic) stiffness to capture spring-like muscle-reflex behavior that is experimentally seen (e.g., Mussa-Ivaldi et al. 1985) and clearly needed (e.g., Winters and Van der Helm 1994) for studies of quasisteady posture under normal (submaximal) levels of contraction (see also Winters 1995a,b; Chapter 10). The overall model has a tendency to slowly drift, because of the low intrinsic stiffness of the CE force-length property, which in dimensionless tissue terms cannot be higher than about 2 Mpa (2 $F_{max}/(L_{rest-length}*$ $A_{pysiol-cross-sec})$, which is an order of magnitude below the SE element stiffness (which to a roboticist is already low). This helps explain why musculoskeletal models have been utilized mostly for dynamic tasks (tracking, propulsion, etc).

One possible solution is to employ cross-bridge models, because it is well established that there is a small local stiffness when stable bonds are stretching and not broken (Hill, 1978)—the local observed "short range" stiffness would then be that of the overall SE. For Hill-type models, this can be

approximated by a "sticky" discontinuity (dry friction) at the zero velocity region of the CE force-velocity (CE_{fv}) property (suggestion by Peter Rack to this author in 1986); although the high slope and slope discontinuity that most advanced models use at zero velocity (e.g., see Chapter 10 for experimental justification) essentially accomplishes a similar effect.

3.3 Partial Solution: Employ Muscle-Reflex Structure

In the opinion of this author the above explanation is not always enough to provide a necessary level of functional stiffness at low activation in an intact system (see data of Hoffer and Andreasson 1981), especially for certain inverted pendulum systems (Winters and Van der Helm 1994; see also Chapter 32).

A reasonable solution—especially for larger-scale systems—is to add homonymous reflex activity (Winters 1995a,b). The key biosensors that are embedded in viscoelastic musculotendonous tissue are muscle spindles and Golgi tendon organs (see Figure 7.2), which are discussed elsewhere (see Loeb 1984; see also Chapter 2). Historically, the former have been assumed, based mostly on controlled ramp-and-stretch external perturbation experiments, to transduce muscle length and ve-

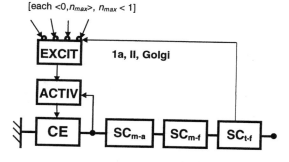

effective synaptic currents
[each $<0, n_{max}>$, $n_{max} < 1$]

FIGURE 7.2. Hill-based muscle model, for either EF or IF, emphasizing interactions to either side of the contractile element (CE). Thick lines between elements represent bicausal mechanical interaction. SC = series compliance, subscript m is muscle, t is tendon, -a represents activation-dependence, -f force-dependence, 1a and II are primary and secondary sensors for IF muscle, Golgi for EF muscle.

locity, and to some extent acceleration and a product of length times velocity to a low power (see Houk and Rymer 1981); Chapter 11) also make this assumption. But here we employ our presupposition, which requires reasonably satisfactory behavior for any task. We then see that the assumption of a muscle spindle as a kinematic (e.g., position-velocity) sensor fails for voluntary tasks, such as isometric contraction, where variable (and reasonably high) levels of spindle activity are clearly seen (e.g., reviewed in Loeb 1984) that are simply not compatible with the classic kinematocentric length-velocity assumption for spindles. This is a remarkably basic finding that makes physical sense when one conceptualizes the anatomical location of the spindle and assumes α–γ motoneuron coactivation (Figure 7.1B) as the default—the sensory apparatus is essentially a strain gauge across a series compliance element (as in Figure 7.2), more tied to SE_{IF} stretch and thus IF force than to absolute (or even relative) muscle length. Thus we need to go back to structural anatomy, and to IF muscle mechanics—a concept that is not new (Matthews 1981, Schaafsma et al. 1991). Figure 7.2 is intended to be applicable to both EF and IF muscle.

Of note is that this structural framework is teleologically consistent with advances in "model reference" adaptive control and CNNs, where dimensionless "error"-like signals play a key role. Nature appears to find a winning structure that only recently has come into use by control engineers. Of note is that, unlike in Van der Helm and Rozendaal (Chapter 11), we choose not to associate units to these sensory signals—CNNs need not always associate units with "error" signals or invariant characteristics.

3.4   Neuroparameter Identification

The critical concern with the proposed type of framework is that unlike musculoskeletal parameters, for which reasonable approaches are available for neuroparameter estimation, the addition of internal reflex infrastructure and recurrent inhibition (via Renshaw cells) adds new parameters that lack adequate methods for parameter estimation. This explains why leading movement biomechanists have been hesitant to make this jump. Yet here we utilize insights from the collection of studies that are discussed by Binder (Chapter 4). What Binder

tells us, in essence, is that there are experimentally documentable limits to the synaptic strengths of the various pathways that converge onto motoneurons. This provides guidance. For instance, the upper limits on steady-state drives for spindle, Golgi, and Rhenshaw inhibition are all well under 10% of maximum, which is consistent with that used in Winters (1995a,b), based on an empirical review of available experimental data combined with simulation experience. Thus Figure 7.2 calls the input to the motoneuron "effective synaptic current" (see Chapter 4) rather than "excitation," and each driving current input has a finite maximum strength that can be expressed as a percentage of maximum. In essence the suggested muscle-reflex model of Fig. 7.1B utilizes conservative feedback as a default for the actuator (cf. the implicit hyperconservative assumption of our usual open-loop models). What interesting is that even conservative feedback provides the magnitudes of stiffness increment in the low-to-moderate force regions that are needed to replicate results such as Hoffer and Andreasson (1981), as well as the types of "invariant" shifts such as those identified by Feldman (1986)! Why? Reasons will emerge below.

3.5   What Is Gained with a Muscle-Reflex Actuator

Notice in Figure 7.1B that the EMG becomes an internal signal, rather than an estimate of the neurocontrol input to muscle; this may appear painful to some. Yet this broader interpretation of actuation as an integrated muscle-reflex process is not without its benefits, especially for postural neuroscience. For instance, Feldman et al. (1990) advocate this interpretation of the EMG, and indeed the muscle-reflex concept is at the foundation of theories such as stiffness regulation (Houk 1979) and the λ equilibrium-point model (e.g., Feldman 1986). Theoretical requirements for spring-like actuator passivity are also satisfied because the new input at node I is not a function of feedback, which pulls in the rich science of impedance control (Hogan 1990). The concept of treating intrinsic muscle properties as "preflexes" (Chapter 10) becomes more formalized, and the concept of neuromotor "primitives" (e.g., Chapter 25) also benefits from this broader view of the actuation process.

4 Teleological Interpretation of Nonlinear Subtleties

We now systematically consider a range of muscle nonlinear properties (not how they are modeled), pointing out 18 cases in which this author suggests they are advantageous. To help make up for a historical injustice, we will, whenever appropriate, focus on IF muscle. We utilize idealized curves to make our key points (see Winters 1995a,b) for examples of constitutive relations and parameter values for the various elements).

4.1 Great Excitation/ Activating Dynamics

1. *Sigmoidal soft saturation.* It is well established that there is soft saturation in the isometric firing rate to force relation (e.g., Rack and Westbury 1969 and Mannard and Stein 1973). This implies greater midrange sensitivity and gradual end-range sensitivity; the reemergence of CNNs during the early 1980s was in part due to use of nonlinear sigmoidal mapping between key signals. It can be easily modeled as a static, unidirectional mapping.

2. *General activation–deactivation temporal dynamics.* Considerable infrastructure is invested in activating and deactivating muscle (e.g., sacroplasmic reticulum (SR) and t-tubules). Activation is a fast process, and interestingly is faster for muscles with fast muscle fibers. Indeed, from sensitivity analysis studies it is clear that this process—represented by an activation "time constant" parameter on the order of 10 to 20 ms—has low-to-moderate sensitivity, as if it was "designed" to be just fast enough not to be ratelimiting. Deactivation, which includes an active ion pumping process, is slower to develop, and indeed simulation experiences and sensitivity analysis show that it can be rate limiting. A reasonable model of this dynamic process is classic Michaelis Menten kinetics (see Zahalak 1990), and this has built-in stabilizing phenomena related to gradual saturation (Winters 1995b): faster deactivation when from higher activation levels, but slower once lower activation is approached (and there is less calcium to pump into SR).

3. *Jump starting.* There are also nonlinear phenomena within both the motoneuron integration and activation processes that provide excitation–activation jump-start capabilities. For instance, it is well documented that there is "catch-like" enhancement of firing rate and force at the motoneuronal level (modeled in Hannaford and Winters 1990), and enhanced force response to action potentials in close succession such as doublets. In such cases the firing frequencies and subsequent force are disproportionately greater that the input current when "fast contraction" is desired. Furthermore, it is well established that in some cases the more conventional orderly recruitment strategy can be bypassed.

4. *Activation dynamics are dependent on mechanics!* Zahalak (1990) distinguished between two approaches for modeling activation dynamics: (1) loose coupling, in which activation (i.e., binding/unbinding of calcium and troponin) is assumed to be a function of the excitation drive and not of the state of binding of actin and myosin; and (2) tight coupling in which these processes are interdependent. From a Hill-model perspective, tight coupling implies that the state of activation depends on the state of the muscle, that is, the length (and perhaps velocity or force) of the CE. Several activation-dependent effects related to CE_{fl} relation scaling are documented in Huijing (Chapter 5) and discussed below, because one way to approximately capture this effect is to scale CE_{fl} differently. When put together, the excitation-activation-deactivation process seems wonderfully tuned to teleological needs!

4.2 The CE_{flv}: Designed First for Stability/Control, Then for Energetics/Power

We assume that the CE_{flv} property is instantaneous, that is, a nonlinear force-length-velocity surface exists that is specified by the ongoing activation, and changes instantaneously with this activation (Hill 1970; see Chapters 1, 8, 10, 11). There are two reasonably orthogonal nonlinear relationships, which for our design-oriented purposes will be considered separately: CE force-length (CE_{fl}) and CE force-velocity (CE_{fv}).

5. *CE_{fl} scaling with activation.* Huijing (Chapter 5) tells us that the optimum length of the CE_{fl} relation scales to longer lengths with lower activations—a phenomena shown in Rack and Westbury

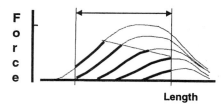

Length

FIGURE 7.3. Scaling of CE force-length constitutive relation as a function of activation, assuming that both the optimum length and the active slack length scale to the left with increasing activation. Arrows define a typical in-situ operating range, and the thick lines the type of region hypothesized to dominate for everyday movements.

(1969) and Willems (1994) but generally not included in most Hill-based muscle models (which usually employ multiplicative scaling). Such a model is shown graphically in Figure 7.3. It turns out that for the EF-IF muscle reflex model discussed here, this is a critical result! Simulations show that without such shifting of the intrafusal CE_{fl} curve to the right (Figure 7.4) for submaximal activation, its capability to predict quasistatic data such as that of Feldman (1986) or Hoffer and Andresson (1981) vanishes, and a model with clearly unstable region emerges (Winters 1995b)! This helps explain why Schaafsma et al. (1991), in their mechanical model of the muscle spindle, assumed a positive slope operating range (without really explaining their rationale). This is so important that this author has suggested, on teleological and simulation grounds, that the shifting is probably more pronounced in IF muscle (and/or the relative optimum length for IF is longer than EF). This shifting is easy to incorporate into Hill-type models (Winters 1995a,b); yet better experimental data is needed to refine this modulation.

6. *CE_{fl} shift with fatigue.* Huijing (Chapter 5) also provides results indicating that with fatigue, the CE_{fl} optimum length moves to a longer length (and the active slack length changes little). While fatigue is, operationally, known to be have many sources (ranging from neural factors to muscle energetics), this activation-dependent CE_{fl} shifting appears favorable, because if anything, it enhances the operating range with a positive slope; here nature appears to provide nonlinear behavior that at least changes in a direction that enhances muscle stability.

7. *CE_{fl} spread and sarcometric heterogeneity.* There is mounting evidence of muscle tissue architectural heterogeneity in some muscles, both in parallel and series (e.g., see Chapters 5 and 6). While this creates an additional concern for modelers, from a teleological perspective it is a useful result in that it allows a design tradeoff between peak force and relative spread. As sarcomeric inhomogeneity increases, the relative spread of the CE_{fl} curve increases, and the force tends to have a lower peak but more of a plateau region. The existence of a range of behavior—some muscles more heterogeneous than others—suggests that such variety within the CE_{fl} relation is not an unfortunate limitation of nature, but a proactive design feature. Additionally, because of actuator redundancy across joints, a similar effect happens if muscles crossing a given joint have different optimum lengths within the joint range of motion.

8. *CE_{fl} operating range design.* The CE_{fl} property clearly has its foundation in biological constraints of the cross-bridge apparatus (actin–myosin overlap), with force falling off to either side of "optimum" rest length (e.g., see Chapters 2, 3, 5, 6, 10). This suggests that it is a suboptimal design, but as also noted in Chapter 10, perhaps such a view is misplaced. Consider again the assumption that the first and foremost consideration in CE design is postural stability and robust control, with force and power generation more secondary (albeit important) concerns.

Figure 7.4 displays three classic types of torque-angle strength curves, with the distinction being the location of joint angle that corresponds to the op-

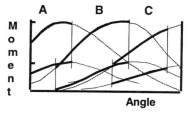

A   B   C

FIGURE 7.4. Three moment-angle "strength" curves, scaled relative to an overall physiological range of motion, then overplotted (A: descending; B: bell-shaped; C: ascending). Below the strength curve is a representative submaximal curve. Also shown as thick regions are idealized "primary" operating ranges, for maximal activation (top) and submaximal activation (bottom).

timum rest length of muscle fibers, with respect to its physiological range. It is suggested that all three design strategies can make sense when placed within a functional context, and that this allows considerable flexibility that virtually always follows teleological expectations. Case B is employed by many muscles, including heart muscle. They function on the ascending limb for most of their operating range, often peaking near the end of the physiological range. This provides built-in spring-like stabilization over the primary range, while retaining significant force and power production capabilities. However, there are Case C muscles, such as the wrist flexors discussed by Lieber (Chapter 3), in which there is a clear positive slope for the whole region, with the theoretical optimum length sometimes even located outside of the physiological range. This can also be an effective strategy for muscles where robust control is a greater priority than power generation or efficiency, for instance the inherent length shifting of the CE_{fl} curve with activation (Figure 7.4) seems often to be very useful (Feldman et al. 1990). Two classic examples of Case A are the elbow and knee extensors in the human, in which the optimum length during maximal stimulation is about 30 to 40 deg from full extension (i.e., the torque-angle relation has a negative slope for most of the operating range). Yet this also makes teleological sense, especially for the knee, since the primary operating range during key tasks (e.g., standing, locomotion) is between 0 to 40 deg for most activities of life, where the slope is positive. Furthermore, these muscles provide significant power production capabilities. Another Case A muscle is identified by Lieber (Chapter 3), in which wrist extensors are on the descending limb— stronger in extension. Lieber's explanation—that the flexor-extensor ratio balance is preserved— makes sense, especially given the importance of contraction (for impedance modulation) for the wrist subsystem. Also, most manipulation and lifting tasks in life are performed with a slightly extended wrist.

Finally, notice for Figure 7.4 that if we assume the type of activation-dependent curve shift as in Figure 7.3, all three examples display a positive slope over the primary submaximal range.

9. *Mechanics and nonlinear spindle temporal response.* Figure 7.5 assumes that the spindle sensing element is part of a viscoelastic SE element in

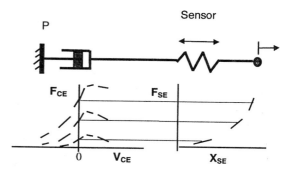

FIGURE 7.5. Conceptual behavior of the CE_{fv} relation in key regions, with a dashpot (representing CE_{fv} behavior) serving as a "soft ground" for a series spring. Slopes for slow (by F_{CE} axis), medium and fast regions are identified for each of shortening and lengthening. F is force, V_{CE} is CE velocity, X_{SE} is spring extension.

series with the CE element for the IF. What happens during a moderate-speed ramp stretch at point P (unfortunately the test of choice for experiments)? First we note that for low velocities and moderate or high ongoing activation levels the CE_{fv} relation initially functions as a sort of "viscous ground" for the SE relation (Winters and Stark 1987). This is because of the relatively high CE_{fv} slope near zero velocity (e.g., Scott and Loeb 1995; Figure 10.11): for Hill-type models the CE_{fv} slope literally describes a viscosity, with the viscous force being the difference between the CE_{fv} force and the "hypothetical" zero-velocity force (Winters and Stark 1985). This CE_{fv}-SE tandem, rather than CE_{fl}, is at the heart of the "preflex" concept discussed in Brown and Loeb, and the "short range stiffness" described by Rack and Westbury (1974): the CE serves as a temporary "ground," with the SE stiffness expressed. For the IF muscle and spindle this helps explain the inherently nonlinear response to stretch (Matthews 1981; Schaafsma et al. 1991; Winters 1995a,b): it is straightforward to show via simulation that even with a linear spindle sensing element embedded within the spindle tissue, the sensor behavior displays the classically observed nonlinearities of initial hypersensitivity to stretch followed by lower sensitivity (e.g., Winters 1995b). While only a temporary "first line of defense" since cross-bridge breaking and recycling, this "sticky" low-velocity CE_{fv} transient behavior represents a key advantageous use of intrinsic nonlinearity—what was necessary, nature provides.

10. *Variety in CE force-velocity design.* Most actuators—manufactured or biological—have a force-velocity constitutive property, and posses a velocity region where power (the product of force (or torque) and velocity) is most efficiently supplied; muscle is no exception. What is unusual about skeletal muscle is actuator design variety, in which nature has provided a range of muscle fiber speeds (from about 2 fiber lengths/sec for slow fibers to about 10 fiber lengths/sec for fast fibers), and a packing strategy (fiber pennation within a whole muscle; fiber lengths) that essentially trades off force and velocity. While rarely noted, it is actually remarkable that most mammalian muscles possess a relatively equal distribution of fast and slow muscle fibers—one might have expected that optimum design solutions would be all fast muscle fibers (or all slow fibers). Since exceptions do exist (e.g., the soleus, eye muscles) that make teleological sense, it is suggested that this seems to be a purposeful design.

Given that it is well documented that fast muscles can generate considerably more power (e.g., Faulkner et al. 1984), why slow muscles? Part of the answer lies in metabolic cost: for submaximal activation (which dominates the living process for most organisms); some faster fibers (with less well-developed metabolic infrastructure) are rarely recruited, thus trading off added mass with the occasional need for a propulsive contraction. However, the primary answer may rest with robust control: the low-region force-velocity slope ("viscosity") is higher for a slow-fibered muscles, which allows less sensitivity to perturbation. Furthermore, the fact that optimum power occurs at lower velocities may be advantageous for muscles not as needed for propulsive tasks.

11. *Selective actuator invisibility via CE$_{fv}$ (and SC).* Notice that at the lower right of Figure 7.5, the CE "viscosity" in a deactivating, lengthening muscle is very low (low slope). Thus a lengthening antagonist muscle tends to become "mechanically invisible" (drifts) at an appropriate time; this would not be the case for a linearized muscle, in which all of the force-velocity slopes would be constant, independent of both activation and velocity (Winters and Stark 1987). If the latter had been the case, life as we know it would not exist—"spastic"

resistance would be the norm. Furthermore, we will see in the next section that deactivation-dependent SC behavior further complements this effect since the SC compliance is high (stiffness low) as activation and force lower.

4.3 Series Compliance (SC) and a Bicausal Reality

Muscle tissue, even when activated strongly, exhibits a significant degree of compliance, relative to how we design most manufactured actuators (Hannaford and Winters 1990). Because of the series nature of the property, if it was more rigid it would functionally disappear; hence our reason for calling it a SC property for the series element SE. From a standard engineering perspective, skeletal muscle compliance represents a horrible, substandard design: muscle impedance is so low that the system has trouble maintaining a desired position! Furthermore, this compliance is nonlinear, being lower for muscles with low activation and force. Yet from a teleological perspective, a different story emerges.

12. *Activation dependence of SC has advantages.* We can conceptually break the SC property into two history-dependent force-extension relations in series, in which the extensions add: one a function of activation (i.e., attached bonds), the other that of a classic nonlinear passive viscoelastic soft tissue (i.e., lightly damped spring). Both make teleological sense. Figure 7.6 shows that when activation leads force because of increased muscle excitation drive, the SE extends less than might be expected because there are a greater number of attached bonds, while during deactivation, the tissue may be temporarily quite compliant (Winters 1995c). While for most everyday movements, activation-dependency of the overall SE appears to be a mi-

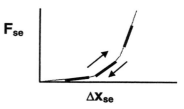

FIGURE 7.6. Nonlinear and history-dependent nature of the SC force-extension relation.

nor effect, at least the direction of the "hysteresis" assists with force development and release during voluntary movement tasks. That passive damping is light also makes sense—otherwise storage of strain energy in musculotendinous structures, which we know to be functionally important (e.g., Cavagna 1970; Alexander and Ker 1990; McMahon 1990), would be inefficient.

13. *Nonlinearity allows impedance modulation.* Because of the nonlinear nature of the SE relation (Figure 7.6), with stiffness a function of activation and force, through co-contraction of antagonistic muscles joint transient impedances can be modulated (e.g., Winters et al. 1988). For a "typical" musculotendinous actuator, this effect continues up to about 50% of maximum force. The details of this effect and its implications for EF muscle have been discussed many places (e.g., Hogan and Winters 1990), and in clinical settings this intrinsic ability to modulate stiffness is taken as gospel. Yet linearize muscle properties, and it vanishes (as does the ability to modulate viscosity)!

While the theory of impedance control (Hogan 1984, 1990) has to date mostly been applied to study contact tasks, it should be extended to apply to impedance matching and impedance control between critical substructures within the body. It is, for instance, well established that muscles involved in postural maintenance are often co-contracting before or in unison with prime movers for self movement (e.g., Bouisset and Zatara 1990) or in preparation for body landing (Chapter 20). Once again, nature appears to have proactively designed in nonlinearities that make teleological sense.

14. *Nonlinearity enhances spindle sensitivity.* Focusing on the same property as above, we now consider the effects of stiffness changes on the IF sensory transduction process, assuming that the spindle sensors are embedded in nonlinear passive IF tissue (SC property). Because of more stretch per incremental change in force in the low-moderate SC force region (see Figure 7.6), it has been seen in simulations (Winters 1995a,b) that the spindles are predicted to be more sensitive at lower forces—this is useful because this is the primary physiological region for postural regulation studies, where significant spindle activity is needed to replicate experimental data (e.g., Hoffer and Andressson 1981; Feldman 1986). This is likely further enhanced since there tends to be a higher sensor sensitivity

within a midoperating range of the sensory transducer, with soft saturation to either side, as is common for biological transducers (see Loeb 1984).

15. *Length-independent compliance modulation.* The SE element (with the SC property) can be conceptualized as being placed between the CE length (which shifts) and overall muscle length, and thus is not "grounded." Thus its compliance is not directly sensitive to absolute muscle length, but rather to muscle force and activation (which of course are length sensitive). Many transient perturbation studies have documented how local stiffness increases with force (e.g., Joyce et al. 1974; Cannon and Zahalak 1982; Lacquaniti et al. 1982), and its been shown that this measured perturbation stiffness is indeed a way to estimate the SC element (e.g., Zahalak 1983; Winters 1990). This suggests, in consistency with our knowledge of tremor (Rack 1984), that transient (short-range) compliance is reasonably length independent over a significant midrange, which appears to be important for neuromotor coordination; again, an excellent design strategy. Note again that short-range stiffness is associated with SE, not CE_{fl}.

16. *SC is distributed through many structures.* The relation for series springs is straightforward—forces are the same and extensions add, with the relative extensions being a function of the local tissue strain times its length. No tissue within the structure is rigid, nor is any structure exceptionally compliant—tendon, aponeurosis, z-disc, and filaments (including cross-bridges) all strain under physiological stresses. There is no reason for passive tissues within the system to display large strain variation—it would be remodeled away and thus is not ecologically sustainable. Experimentally reported local strains below 1% or above 10% should always be suspect—either the testing protocol loads the tissue nonphysiologically or there is measurement error (these are difficult experiments!). And variation in strain between passive tissues such as tendon and aponeurosis should be very small (see Scott and Loeb 1995 for supporting evidence). With regards to the muscle filaments, while a peak strain that is somewhat than that for tendon might be expected, there is no teleological rationale for nature to invest in the infrastructure and mass required to fabricate truly rigid structures. This is true despite the intimate 3D nature of the myosin bridge sites—neither kinetic phenomena such as calcium

diffusion nor mechanical phenomena such as cross-bridge coupling appear hypersensitive to mild filamentary extensions (e.g., 1–4% strains). Even deep muscles with very little tendon will possess significant series compliance. To summarize: let the arguments cease—both the evidence and teleological expectations suggest fairly uniform strains among the sources of overall SC properties.

17. *Multisegment coordination.* There are many reasons for multisegment coordination, ranging from taking advantage of kinematic redundancy (e.g., see Chapter 35) to applying forces in arbitrary directions (Gielen et al. 1990). But because of intrinsic series compliance and contractile impedance, multisegment coordination is the rule and not the exception, at least any multiarticular mammals. The reason is that joints cannot become rigid (i.e., possess near-infinite impedance) upon demand, and thus mechanical interactions (spring-like, inertial) within musculoskeletal substructures are inherent. Professionals do their best to minimize this reality by designing exercise equipment that "isolates" joints and constrains unwanted motion through high contact impedance, and they design experiments in which subjects wear special braces that prevent unwanted rotation at joints. Yet as so eloquently developed by Alexander and colleagues in their studies on animal mechanics (e.g., Alexander 1988), SC behavior has many uses, from energy storage-release ("stretch-shortening") for creatures ranging from insects to kangaroos to humans, to impedance modulation during manipulation tasks. The co-dependency among neuromotor control structures, muscles, and body segments starts with the nonlinear SC.

18. *Mechanical transient impedance and feedback gains.* As developed in Chapter 2, higher feedback gains often occur with higher activation. Models with nonlinear CE and SE predict that as coactivation increases, the threshold loop gain that is stable also increases (Winters et al. 1988); linearize CE and SC, and this useful effect, which complements open-loop "preflex" impedance modulation (item 13), disappears (cf discussion in Chapter 29).

4.4 Closure

In all, Section 4 has 18 key points, few of which should be controversial except for details (or teleo-

logical interpretation, if one does not agree with the presuppositions of Section 2). Others were omitted for reasons of space! Several well-documented, significant nonlinear phenomena were not addressed because teleological explanations are not yet fully clear to this author: some subtle ultratransient stiffness-force effects (reviewed in McMahon 1984; Winters 1990), "yielding" during ramp stretches in a few muscles like the soleus (Joyce et al. 1969), and force "enhancement" (e.g., greater-than-expected isometric force after stretch, Edman 1978). The latter two both can be modeled via adding an "attachment" state variable to Hill-type models (see Winters 1990). Perhaps these are biophysical "quirks;" but perhaps not—for instance, this author would not object to force enhancement for certain tasks.

5 Future Directions

This perspective has focused on nonlinear properties. Although a muscle-reflex framework was briefly developed, this is not a paper on modeling. It is interesting that while nearly all 18 of the identified phenomena can be incorporated into Hill-type model structures, all disappear either when the Hill structure is reduced or its key constitutive relations are linearized. Many involved intertwined properties, as discussed by Huijing (Chapter 6), that cannot be perfectly captured by Hill models. Yet from a systems engineering perspective, why not continue letting muscle be a big sarcomere with lumped properties! The process of having multiple variables modulate constitutive relations is intrinsically insightful, and forms a useful "bridge to the 21 century" between experimental biologists and systems engineers. However, sensitivity analysis tools must be used more often.

Time will tell whether a muscle-reflex actuator represents a more fruitful approach. For fast and propulsive movements, this new framework is not very important (but does not hurt) because reflex activity only mildly sculptures the feedforward drive. However, for studies of posture and more everyday movements, and especially for larger-scale models, it may prove especially useful to neuroscientists, for two reasons. First, it provides experimental neuroscientists with a rather dependable and robust "black box" actuator model that produces realistic spring-like behavior during lower activations, and provides a more robust estimate of the λ_{static} spindle signal (Winters 1995a,b). Second, because it is designed to be reasonably task-independent with regards to capturing nonlinear properties, accepting converging neuromotor signals and handling low-level interaction, it would seem to be ideal for computational neuroscientists studying biologically-inspired control via CNNs. Modular "expert" CNNs (e.g., see Chapter 34) and neuromotor primitives (e.g., see Chapter 25) can take advantage of "redundancy" and nonlinearity, yet need a dependable model of mechanics. Such synthesized neuromechanical models, with nonlinear muscle properties as described here and in Chapter 10 (Brown and Loeb), are suggested to have a bright future.

References

Alexander, R.McN. (1988). *Elastic Mechanisms in Animal Movement.* Cambridge University Press, Cambridge.

Alexander, R.McN. and Ker, R.F. (1990). The architecture of leg muscles. In *Multiple Muscle Systems: Biomech. and Movem. Organiz.*, Winters, J.M. and Woo, S.Y. (eds.), Chapter 36, pp. 568–577, Springer-Verlag, New York.

Beek, P.J., Peper, C.E., and Stegeman, D.F. (1995). Dynamical models of movement coordination, *Hum. Mov. Sci.*, 14:573–608.

Bernstein, N. (1935, 1967). *The Co-ordination and Regulation of Movements.* Pergamon, New York.

Cannon, S.C. and Zahalak, G.I. (1982). The mechanical behavior of active human skeletal muscle in small oscillations. *J. Biomech.*, 15:111.

Cavagna, G.A. (1970). Elastic bounce in the body. *J. Appl. Physiol.*, 29:279–282.

Close, R. (1969). Dynamic properties of fast and slow skeletal muscles of the rat after nerve cross-union. *J. Physiol.*, 204:331–346.

Edman, K.A.P., Elzinga, G., and Noble, M.I.M. (1978). Enhancement of mechanical performance by stretch during tetanic contractions of vertebrate skeletal muscle fibres. *J. Physiol.*, 280:139–155.

Faulkner, J.A., Claflin, D.R., and McCully, K.K. (1986). Power output of fast and slow fibers from human skeletal muscles. In *Human Muscle Power.* Human Kinetics, Champaign.

Feldman, A.G. (1986). Once more on the equilibrium-point hypothesis (λ model) for motor control. *J. Motor Behav.*, 18:17–54.

Feldman, A.G., Adamovich, S.V., Ostry, D.J., and Flanagan, J.R. (1990). The origin of electromyograms—Explanations based on the equilibrium point hypothesis. In *Multiple Muscle Systems.* Winters, J.M. and Woo, S.L-Y. (eds.), Chapter 12, pp. 195–213, Springer-Verlag, New York.

Fung, Y.C. (1993). *Biomechanics: Mechanical Properties of Living Tissues.* 2nd ed., Springer-Verlag, New York.

Gauthier, G.M, Martin, B.J., and Stark, L. (1986). Adapted head and eye movement responses to added-head inertia. *Aviat. Space Environ. Med.*, 57:336–342.

Hannaford, B. and Winters, J.M. (1990). Actuator properties and movement control: biological and technological models. In *Multiple Muscle Systems.* Winters, J.M. and Woo, S.L-Y. (eds.), Ch. 7, pp. 101–120, Springer-Verlag, New York.

Henneman, E., Somjen, G., and Carpenter, D. (1965). Excitability and inhibitability of motoneurons of different sizes. *J. Neurobiol.*, 28:599–620.

Hill, A.V. (1970). *First and Last Experiments in Muscle Mechanics.* Cambridge University Press, Cambridge.

Hoffer, J.A. and Andreasson, S. (1981). Regulation of soleus muscle stiffness in premammillary cats: intrinsic and reflex components. *J. Neurophys.*, 45:267–285.

Hogan, N. (1990). Mechanical impedance of single- and multi-articular systems. In *Multiple Muscle Systems.* Winters, J.M. and Woo, S.Y. (eds.), Chapter 9, pp. 149–164, Springer-Verlag, New York.

Houk, J.C. (1979). Regulation of stiffness by skeletomotor reflexes. *Ann. Rev. Physiol.*, 41:99–114.

Houk, J.C. and Rymer, W.Z. (1981). Neural control of muscle length and tension. *Handbook of Physiol. The Nervous System II*, Chapter 8, pp. 257–323.

Lacquaniti, F.F., Licata, R., and Soetching, J.F. (1982). The mechanical behavior of the human forearm in response to transient perturbations. *Biol. Cybern.*, 44:35–46.

Loeb, G.E. (1984). The control and response of mammalian muscle spindles during normally executed motor tasks. *Exerc. Sport Rev.*, 12:157–204.

Mannard, A. and Stein, R.B. (1973). Determination of the frequency response of isometric soleus muscle in the cat using random nerve stimulation. *J. Physiol.*, 229:275–296.

Matthews, P.B.C. (1981). Muscle Spindles: their messages and their fusimotor supply. In *Handbook of Physiology.* The *Nervous System II*, Brooks, V.B. (ed.), pp. 189–228, Bethesda, Maryland, American Physiological Society.

McMahon, T.A. (1984). *Muscles, Reflexes and Locomotion.* Princeton University Press, Princeton.

McMahon, T.A. (1990). Spring-like properties of muscle and reflexes in running. In *Multiple Muscle Systems.* Winters, J.M. and Woo, S.Y. (eds.), Chapter 5, pp. 69–93, Springer-Verlag, New York.

Rack, P.M.H. and Westbury, D.R. (1969). The effects of length and stimulus rate on tension in isometric cat soleus muscle. *J. Physiol.*, 204:443–460.

Rack, P.M.H. and Westbury, D.R. (1974). The short range stiffness of active mammalian muscle and its effect on mechanical properties. *J. Physiol.*, 240:331–350.

Roux, W. (1895). *Gasmmelte Abhandlungen uber Entwicklungs-mechanik der Organismen V. I & II. Engelmann, Leipzing.*

Seif-Naraghi, A.H. and Winters, J.M. (1989). Effect of task-specific linearization on musculoskeletal system control strategies. *Biomech. Symp.*, ASME, AMD-98:347–350, San Diego.

Sperry, R.W. (1959). The growth of nerve circuits. *Sci. Amer.*, November, pp. 1–9.

Winters, J.M. (1990). Hill-based muscle models: a systems engineering perspective. In *Multiple Muscle System.* Winters, J.M. and Woo, S.Y. (eds.), Chapter 5, pp. 69–93, Springer-Verlag, New York.

Winters, J.M. (1995a). How detailed should muscle models be to understand multi-joint movement coordination? *Hum. Mov. Sci.*, 14:401–442.

Winters, J.M. (1995b). An improved muscle-reflex actuator for use in large-scale neuromusculoskeletal models. *Ann. Biomed. Eng.*, 23:359–374.

Winters, J.M. (1995c). Concepts in neuro-muscular modelling, In *3–D Analysis of Human Movement.* Allard et al. (eds.), Chapter 12, pp. 257–292, Human Kinetics.

Winters, J.M. and Stark, L. (1985). Analysis of fundamental movement patterns through the use of in-depth antagonistic muscle models. *IEEE Trans. Biomed. Eng.*, BME-32:826–839.

Winters, J.M. and Stark, L. (1987). Muscle models: what is gained and what is lost by varying model complexity. *Biol. Cybern.*, 55:403–420.

Winters, J.M. and Van der Helm, F.C.T. (1994). A field-based musculoskeletal framework for studying human posture and manipulation in 3-D, pp. 410–415, *Proc. Symp. on Modeling Control of Biomed. Sys.*, IFAC, Galveston.

Winters, J.M., Stark, L., and Seif-Naraghi, A.H. (1988). An analysis of the sources of muscle-joint system impedance. *J. Biomech.*, 12:1011–1025.

Wolff, J. (1884). Das Gesetz der transformation der inneren architektur der knochen bei pathologischen veranderungen der ausseren knochenform. *Sitz, Ber. Preuss. Akad. Wiss. 22. Sitzg., phys-math. Kl.*

Zahalak, G.I. (1982). The dynamics of active human skeletal muscle in vivo. In *Mechanics of Skeletal and Cardiac Muscle.* Phillips, C.A. and Petrofsky, J.S. (eds.), Springfield, Charles C Thomas.

Zahalak, G.I. (1990). Modeling muscle mechanics (and energetics). In *Multiple Muscle Systems* Winters, J.M. and Woo, S.Y. (eds.), Chapter 1, pp. 1–23, Springer-Verlag, New York.

Zajac, F. (1989). Muscle and tendon: properties, models, scaling and application to biomechancis and motor control. *CRC Crit. Rev. Biomed. Eng.*, 17:359–415.

Commentary:
Analysis of Nonlinear Neuromuscular Properties— Teleology or Ideology?

Robert E. Kearney and Michael P. Slawnych

This chapter reviews a extensive array of nonlinear properties of the neuromuscular system and argues that they are "remarkably consistent with what would be expected based on a thoughtful consideration of neuromechanically based teleological design principles." If this hypothesis could be substantiated it would provide a solid theoretical basis for the development of a field which has for the most part been very empirical in nature. However, the teleological analysis is dependent upon two underlying assumptions:

1. The neuromuscular control system is in some sense optimal.
2. Muscle can be considered as "a big sarcomere with lumped properties."

Unfortunately, we believe both assumptions to be flawed.

Consider optimality first. This is an extremely attractive assumption because it provides a unifying conceptual approach to the study of neuromuscular systems which is readily formulated in quantitative terms. However, to be useful we must know what is being optimized and whether the optimum has been reached. We know neither.

Certainly a variety of optimal criteria can be formulated based on maximizing the system output (e.g., force, speed or power generation), minimizing operating costs (e.g., energy consumption, stress, etc.), or control considerations (e.g., stability, robustness). However, it is by no means clear which, if any, of these criteria is optimized by the neuromuscular system. Certainly the wide differences in the organization and structure of muscles within a single organism (e.g., eye muscles versus soleus), among individuals of the same species (e.g., marathon runners versus weight lifters), and between species (e.g., cheetahs versus elephants) indicates that there are a wide range of viable solutions. Are each of these a solution to the same optimal problem or do they represent solutions to different problems? The latter seems more likely and suggests that neuromuscular system finds solutions to some weighted combination of optimality criteria with the particular balance of weights depend upon the specialization of individual or organism. Moreover even if a well-defined optimality criteria existed and were known, questions would arise as to the current state of the system: has it reached an optimum? Is it progressing toward the optimum? Or is it caught in some local minimum?

The proposition that the neuromuscular system is in some sense optimized is attractive—but it must be regarded as a hypothesis and tested as such. Unfortunately, evidence supporting the hypothesis of optimization in the neuromuscular system criteria is as yet limited. Thus, for example, none of the many attempts to use optimization to apportion the forces generated by the different muscles acting about an overdetermined joint have provided a convincing general solution to this apparently "easy" problem. Similarly, attempts to use optimality to predict the trajectory of arm movements have not met with universal acceptance.

Thus, given the wide range of constraints and possible solutions it seems dangerous to assign a teleological purpose to a property of the system purely on the basis of how it might be useful in meeting a particular design requirement. It would seem wise to consider what effects it might have on other requirements before drawing such conclusions. Thus, for example, while a low intrinsic viscosity for muscle might be desirable in order to minimize energy dissipation as argued in the target article, it could be regarded equally as undesirable from a control viewpoint by reducing stability and facilitating oscillations.

A second underlying assumption of the analysis is that it is reasonable to represent the overall mechanical behavior of muscle as an equivalent sarcomere modeled with a Hill type model. Certainly, such an approach offers a number of advantages, including a close connection with basic macroscopic muscle properties such as the force-length and force-velocity relations and also numerical tractability. However, it must be remembered that Hill-type models are phenomenological in nature, and have little connection to the underlying physiological mechanisms of muscle contraction. Thus, for example, the division of forces between the parallel and contractile elements and the division of extensions between the contractile and series elements are completely arbitrary. For this reason, attempts to elucidate the exact nature of mechanical transduction in muscle have focused on models at the cross-bridge level. While more realistic, these cross-bridge level models rapidly become numerically intractable and so are unusable in simulations of neuromuscular systems. The question to be addressed is whether a model based on general whole muscle characteristics is adequate for such purposes? Winters argues that it is. However, recent studies of intersarcomere dynamics suggest otherwise, as there it is now evidence that interactions between sarcomeres significantly influence the dynamic response of muscle. For example, eccentric contractions have been shown to lead to increased sarcomere heterogeneity which plays a significant role in behavior such as stretch-enhanced activation (e.g., Morgan 1994). It is not obvious that simple lumped parameter models can account for such behavior. (Indeed, we suggest that many of the other nonlinear properties listed in Winters' article cannot be well predicted by his lumped muscle model without modulating its parameters, even though he states otherwise). Perhaps a more useful approach would be to use a distributed model with multiple Hill-type elements.

In summary, the analysis of this article is based on two assumptions that we find open to question. As such it is more ideological (defined as visionary speculation) than teleological (defined as the use of design, purpose or utility as an explanation of any natural phenomenon) in nature and might better be regarded as a guide for future research rather than a summary of current understanding.

References

Morgan, D.L. (1994). An explanation for residual increased tension in striated muscle after stretch during contraction, 1994. *Exp. Physiol.*, 79:831–838.

Commentary:
Remarks Regarding the Paradigm of Study of Locomotor Apparatus and Neuromuscular Control of Movement

Peter A. Huijing

1   Introduction

In Chapter 36, Delp and his coauthors describe a number of successful applications of classical musculoskeletal models in orthopaedics and rehabilitation. Winters (1997) draws attention toward a neuromechanical design intimacy of muscle, which should be considered for models to be successful.

Below we will consider a classical view first and then present an alternative view which is in an early stage of development (see also Huijing and Winters, 1997) and takes into account a neuromyomysial intimacy, in which connective tissue plays a central role, in addition to the sizable roles for muscle and neural tissues.

A Classical View

Classically the human and vertebrate locomotor apparatus is viewed as a rather stiff skeleton (in modelling of movement usually represented as a set of rigid bodies, connected by joints with one degree of freedom) and muscle tendon complexes which are attached to the skeleton. The musculoskeletal arrangement has to be controlled to be able to perform movement and maintain postures.

Connective Tissue and Its Role

In such a view of the apparatus, connective tissue other than tendons only plays a minor role at best and is neglected in most modelling studies.

Huijing and Winters (1998) showed with a linguistic analysis that different aspects of connective tissue function are high lighted. Most of the focus is on binding, sometimes even with a connotation of "supporting." Even though this is not considered in "connective tissue", the idea of a supporting function is present. This can be illustrated by a quote from Rowe (1981): "It is well established that skeletal muscles are made up of muscle fibers held together in various configurations by an extracellular connective tissue frame work."

Even if one considers a cross-section of muscle it usually does not become immediately apparent that connective tissue is a very important structure within a muscle. One reason for that is the visual impact of muscle fibers and their content, which in most preparations prevents the connective tissue from being seen very well. At best, attention is only drawn to the fact that muscle is subdivided into bundles by connective tissue (see Bloom and Fawcet 1969, Figure 11.9).

2   An Alternative View

To develop an alternative view let us start by describing a different view of the structure of the locomotor apparatus. For practical purposes, during locomotion the skeleton may be simplified as described above. However for the connective tissue a quite different model is proposed: Connected to the bony skeleton is a very complex and intricate network of connective tissue which is interconnected and extends from connective tissue in joints to that of the general fascia of the body, intramuscular septa as found in limbs and the connective tissue of muscles. The muscular connective tissue consists of a fascia around the muscle called epimysium (or perimysium externum in the Nomina Anatomica), a fascia around bundles of muscle fibers called perimysium (or perimysium internum), and last but not least a fascia surrounding each muscle fiber called endomysium. The interface between the muscle fiber's sarcolemma and the endomysium is formed by the basement membrane, which is always found throughout the body on the border between connective tissue and other tissues. The arrangement of connective tissue is referred to as a stroma, a word whose etymology also indicates the concept of support. Usually these words do not, for most people, create an image that can be easily applied in the process of contemplation of the functions. Therefore we suggest that the reader should look at the scanning electron micrographs of muscular connective tissue published by Trotter and Purslow (1992). It is clear that the intramuscular connective tissue forms a set of tubular cavities or tunnels in which the muscle fibers can be placed. It should be noted that all elements of connective tissue within the muscle are interconnected (Bloom and Fawcet 1969, Figure 11.9). Once such an image is created, it is less difficult to extend it to a whole limb.

Only very rarely it is argued explicitly that the connective tissue, particularly that related to muscle, is the carrier of not only vessels of circulation, but also of nerves and its specialized endings mechanoreceptors on the afferent side and motor end plates on the efferent side. Good examples of the role of connective tissue in relation to the location of mechanoreceptors is provide in the work of van der Wal (1988) and Strassmann et al. (1990). From this work it becomes clear that connective tissue in a limb is arranged in such a way as to allow interaction between several morphological entities.

The New Role of Muscle

In this new concept the role of muscle will be quite different from that of the classical view. Force is not only exerted onto the tendon but also on the connective sheets. In effects muscle will act as the interface between the central and peripheral nervous system and the complex of bony skeleton and connective tissue. This new way of looking at the role of muscle does not mean that muscular properties will be less important for movement than in the classical way of looking. It is evident that the functioning of the whole system will be very much determined by the properties of the interface. Therefore all that is indicated in other chapters of this section about the non-linear properties of muscle and their importance has not become irrelevant.

It is very likely that as this concept develops and is tested further for validity, revisions will have to be made in our ideas concerning muscular identity, units of control and many more concepts that are very familiar to us now.

Experimental Evidence Consistent with the New Role

This evidence can be distinguished according to three levels of organization:

1. In a small bundle of muscle fibers, Street (1983) showed that an active fiber with one end un-

attached will be kept at length by surrounding passive fibers. It will not go to its active slack length but will generate external force.

2. For whole muscle, work (Huijing, Baan, and Rebel 1997; Huijing 1999) on rat extensor digitorum longus (EDL) muscles indicates force transmission between parallel parts of that muscle inserting on digits of toes II through V. If the morphology of this muscle is scrutinized, it becomes clear that proximally there is one aponeurosis on which fibers from all parts insert. Therefore proximally the muscle looks very much like a simple unipennate muscle. Distally, however, four heads can be distinguished with individual tendons for each head. In the classical view one would expect that if one of these tendons is cut, the force on the proximal tendon would decrease by an amount related to the quantity of muscle disconnected from its tendon. In contrast it was found experimentally that approximately 55% of the muscle (i.e., head II through IV) could be removed with only a loss of 10% of optimum force. This is interpreted to mean that force is transmitted from the heads disconnected from their tendons to heads still connected to their tendons. An intact connective tissue between heads proved to be essential for this transfer of force. It should be noted that the idea of intramuscular lateral transfer of force by shearing is not a completely new idea, but no direct experimental evidence has been provided and application has been mostly limited to muscles which have fibers that do not course from tendon to tendon but end in the middle of the bundle (Tidball 1983; Purslow and Duance 1990; Trotter and Purslow 1992; Heron and Richmond 1993; Trotter 1993; Trotter, Richmond, and Purslow 1995).

3. For intermuscular interaction evidence is presented by Riewald and Delp (1997). They studied four patients in whom the rectus femoris muscle distal tendon was transplanted to the flexor side of the knee without disturbing the connective tissue more proximally. After recovery, intramuscular electrical stimulation of this muscle did not cause flexion, but still extended the knee.

It is concluded that sufficient evidence is found to indicate further exploration of the concepts described. If the idea holds it will have far reaching consequences for myology, biomechanics, and modeling of human movement.

References

Bloom, W. and Fawcett, D.W. (1968). Textbook of histology. Saunders, Philadelphia.

Delp, S.L., Arnold, A.S., and Piazza, S.J. (1997). Clinical applications of musculoskeletal models in orthopaedics and rehabilitation. This volume.

Heron, M.I. and Richmond, F.J.R. (1993). In series fiber architecture in long human muscles. *J. Morph.*, 216: 35–45.

Huijing, P.A. (1999). Muscle as a collagen fiber reinforced composite material: Force transmission in muscle and whole limbs. *J. Biomech.*, 32:329–345.

Huijing, P.A. and Winters, J.M. (1998). Toward a new paradigm of locomotor apparatus and neuromuscular control. In *Models in human movement sciences.* Bosch, P., Boschker, M.S.J., van Lenthe, H., Post, A.A., Pijpers, J.R., Roeleveld, K., Steenbergen, J., and Tanck, E. (eds.). Free University Press, Amsterdam, pp. 45–50.

Huijing, P.A., Baan, G.C., and Rebel, G. (1998). Nonmyotendinous force transmission in the extensor digitorum longus muscle of the rat. *J. Exp. Biol.*, 201:683–691.

Purslow, P. and Duance, V.C. (1990). Structure and function of intramuscular connective tissue. In *Connective tissue matrix*, Hukins, D.L. (ed.), Part 2. pp. 127–166, CRC Press, Boca Raton.

Riewald, S.A. and Delp, S.L. (1997). The action of the rectus femoris muscle following distal tendon transfer: does it generate a knee flexion moment? Dev. Med. Child Neurol., 39:99–105.

Rowe, R.D.W. (1981). Morphology of perimysial and endomysial connective tissue in skeletal muscle. *Tissue Cell*, 13:681–690.

Strassmann, T., van der Wal, J.C., Halata, Z., and Drukker, J. (1990). Functional topography and ultrastructure of periarticular mechanoreceptors in the lateral elbow region of the rat. *Acta Anat.*, 138:1–14.

Street, S.F. (1983). Lateral transmission of tension in frog myofibers: a myofibrilar network and transverse cytoskeletal connections are possible transmitters. *J. Cell. Physiol.*, 114:346–364.

Tidball, J.G. (1983). The geometry of actin filament-membrane associations can modify adhesive strength of the myotendinous jumction. *Cell. Motil.*, 3:439–447.

Trotter, J.A. (1993). Functional morphology of force transmission in skeletal muscle. *Acta Anat.*, 146:205–222.

Trotter, J.A. and Purslow, P.P. (1992). Functional morphology of the endomysium in series fibred muscles. *J. Morph.*, 212:109–122.

Trotter, J.A., Richmond, F.J.R., and Purslow, P.P. (1995). Functional morphology and motor control of series-fibered muscles. *Exerc. Sport Sci. Rev.*, 23:167–213.

van der Wal, J.C. (1988). The organization of the substrate of proprioception in the elbow region of the rat. Doctoral Dissertation, University of Limburg, Maastricht, The Netherlands.

Section III

8
Creating Neuromusculoskeletal Models

Patrick E. Crago

1 The Modeling Process

Neuromusculoskeletal models play an important role in developing an understanding of motor control by permitting quantitative tests of concepts, designing crucial experiments, and estimating otherwise unmeasurable quantities. Models also play a role in designing methods to optimize motor performance, treat movement disorders, and assist or restore motor function in cases of disability.

A mathematical model is simply a quantitative description between inputs and outputs, with or without an explicit relationship to a system's structure. Models of the real world are never exact. Errors in models of physiological systems arise frequently from the two practical modeling necessities of leaving out many factors that either are of secondary importance to the problem being studied or are simply unknown, and of describing spatially distributed systems by lumped systems.

The process of creating a neuro-musculoskeletal model depends on the intentions of the model user, but there are some common steps.

1.1 Choosing Model Complexity

The first step is to choose the model scope, including the behavior to be described, and the inputs, outputs (dependent and independent variables), and operating conditions. Ideally, one could envision having one general model that could be used for all applications. This is not a practical goal for several reasons. Generality increases the mathematical complexity of the model, which in turn increases the computation time. A model that is too general will also have more parameters and require more inputs, increasing the difficulty of obtaining input data and parameter estimates for the desired application.

1.2 Choosing Model Structure

The second step in the modeling process is to choose the form of the model. One of the primary decisions that must be made is whether to use a black-box model, or a structurally based model (also referred to as a-priori modeling in Chapter 9, or reductionist modeling in Chapter 10). Both black-box and structural models have advantages and both have a place in modeling the neural control of movement. Black-box models generally use mathematical formulations to relate inputs to outputs without knowing (or assuming) anything about how the system operates internally. The parameters and internal variables in a black-box model may have no relationship to identifiable physical characteristics of the system. In fact the relationships may be described nonparametrically (e.g., graphically), without even assuming anything about the model order. Applications of black-box modeling to the neuromuscular system are described in Chapter 9.

In a structural model, elements of the system are isolated on the basis of known or assumed system structure. Models of individual elements are combined into multielement systems to describe the overall behavior. The advantages of a structural model are that knowledge of internal structure and operation can be incorporated, and internal variables and parameters can sometimes be directly related to the physical properties and experimental measurements.

It should be pointed out that most structural models are built up from black box-models of the elements.

A possible disadvantage of structural models is that the individual element models can introduce many parameters that are impossible to estimate from input–output data of the combined system. On a positive note however, models of the subsystems can be identified from reductionist experiments where the behavior of the subsystem is isolated, and subsequently used in larger systems.

1.3   Parameter Estimation

The third step, parameter estimation, is the process of fitting a model to input–output data. In many cases, the fit is overdetermined and inexact, so one must find the set of parameter values that gives the best fit according to some mathematical criterion (objective function), such as the least squared error between the experimental and model predicted output, or the percentage of the total variance accounted for by the model. Often, nonlinear optimization techniques must be used, since most neuromusculoskeletal models are nonlinear (in the parameter space). With nonlinear models, there might be many local minima, and thus the search might not yield the global minimum. A practical way to address this difficulty is to start the optimization process with widely ranging initial guesses of parameter values and see if the solution comes to the same final parameter values, or finds a different set of parameter values with a lower valued objective function. If the locally minimum objective functions are nearly the same value, then the parameter values are uniquely determined.

Because most models are nonlinear, it is important to choose the set of inputs for parameter estimation carefully so that it is diverse enough to contain data adequate to permit estimation of all parameters. It is important to test the data set and parameter estimation process with data obtained by simulating the system with the expected parameter values. At the very least, the data and parameter estimation process should reproduce the known parameters.

1.4   Sensitivity Analysis

The fourth step is to test model sensitivity to parameter values. Models essentially never reproduce data exactly, and there is a temptation to increase model complexity (number of terms, number of parameters) to reduce modeling errors. For example, a Fourier series expansion can approximate periodic functions to arbitrary accuracy by including a greater number of harmonics and scaling coefficients in the series. Each added term decreases the approximation error of the lower harmonic terms. At what point does one stop increasing model complexity? One of the tools for answering this question is sensitivity analysis.

Sensitivity analysis consists of calculating the change in the system output (or the change in the parameter estimation objective function), produced by a change in parameter value(s). The sensitivity is often normalized so that sensitivities to parameters with different units can be compared. The normalized sensitivity is given by (Lehman and Stark 1982):

$$S_P = \left[\left. \frac{\Delta Y}{Y} \middle/ \frac{\Delta P}{P} \right] \cdot 100\% \right. \qquad (8.1)$$

where S_P is the normalized sensitivity to changes in parameter P and Y is the output. Sensitivity can be calculated analytically for models with explicit analytical forms. However, for models based upon numerical methods with no analytical form, sensitivity must be estimated from a series of simulations with different parameter values.

Sensitivity analysis is one of the main applications of a model, because it allows one to assess the relative importance of different physiological systems or properties to particular behaviors. For example, by choosing a measure of postural stability as the output variable Y, and by incrementally changing a muscle's length-tension slope parameter, one could calculate the importance of the slope to postural stability.

If the sensitivity is high, then the output depends strongly on the parameter value, and it is extremely important to obtain accurate estimates of the parameter value to preserve the overall accuracy of the model. If the sensitivity is very low, then the output is not strongly influenced by the parameter and this portion of the model could probably be removed with little consequence. In nonlinear models, the sensitivity is not constant, but varies with operating conditions (inputs) and parameter values. Thus, it is important to test sensitivity under the full range of operating conditions expected or of interest.

Sensitivity is also important for parameter estimation from input–output data when output error is the objective function. If the sensitivity is low, then errors in the parameter will only lead to small output errors, and the parameter values may not be well determined.

1.5 Model Validation

The fifth step is to verify the model's ability to predict behavior, rather than just fit data. The distinction is that the model should be able to predict behavior that was not used to estimate the parameter values. This step tests the scope of applicability of a model. A model that can only reproduce the data used to generate the model is not very useful, since the data set itself is a better model. Ideally one wants a model based on a small set of data that is able to predict behavior under a wide range of conditions. Failure to predict other known data could indicate several problems. One is that important independent variables have been omitted or have not been controlled (i.e., an error in defining model scope, Step 1). A second possibility is that the form of the model is incorrect (Step 2). In this case, the model is a curve fit that is not applicable outside the range of the original data. This type of error is most common in black-box models because they do not reflect the underlying physiological structure. A third possibility is that the parameters have not been estimated well, because of a poorly chosen set of input conditions for the parameter estimation data (Step 3).

All neuromusculoskeletal models (such as those presented in other chapters of this book) should be evaluated with respect to the above considerations. The remainder of this chapter will expand on the first two steps, to give a common background with respect to the better known elements of neuromusculoskeletal models.

2 Structural and Functional Divisions of the Neuromusculoskeletal System

Organisms move about in the world to achieve their goals, but their movement is not independent of the world. The nervous system plans and regulates movements, but cannot do so without taking into account the mechanical properties of the muscles, limbs, and loads. Thus, the structural diagram of the sensorimotor control system shown in Figure 8.1 includes not only the nervous system and the muscles, but also the entire skeletal system, the soft tissue interfaces between the skeleton and the environment, and the environment itself. Each of the elements of the system is physically identifiable and can be modeled independently, but there are significant interactions between elements that make it difficult to understand natural movement if the elements are considered independently. Thus, models of sensorimotor control must account explicitly for interactions when appropriate.

One of the goals of neuromusculoskeletal modeling is to identify and quantify the roles of the different divisions of the movement control system. The four structures in Figure 8.1 serve unique roles. The central nervous system is responsible for generating the commands to muscles via the motor efferents (including commands to muscle spindle sensory receptors). At least in part, these commands are feedforward (open-loop), because functional movements can be made even in the absence

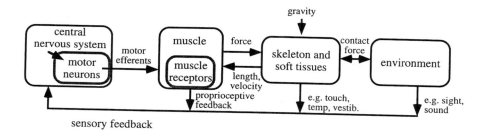

FIGURE 8.1. Structural diagram of movement control system. Each of the blocks in the diagram represents a physically identifiable division of the system.

of feedback from sensory receptors. To do this, the nervous system must be able to plan movements and generate patterned motor outputs. Feedback from sensory receptors (e.g., in the muscles and skin, the vestibular system, or even exteroceptors such as the eyes and ears) play a role in movement planning, but also provide mechanisms for control and regulation of preplanned movements.

Muscles produce the active controlled forces that are required to make movement possible. However, muscles should not be viewed as simple force generators, because the force is a function of not only the neural input, but also the mechanical load. In a sense, a muscle is a neurally controlled viscoelastic mechanical element, which has been described as an impedance, or dynamic stiffness.

The skeleton and soft tissues transform the mechanical properties of the muscles into mechanical properties of limbs, and couple the body to the environment. Muscle forces are transmitted through the skeleton and the soft tissues to objects in the environment. Positions and lengths change in order to maintain force equilibrium. Therefore, the environment is just as much a factor in determining muscle lengths and limb posture as the efferent command from the nervous system.

3   Putting the Parts Together to Simulate Movements

The goal of many model studies is to predict the movements that would be produced by a given set of neural inputs. This is sometimes required when trying to solve the inverse problem of deducing what neural inputs are required to produce a given movement. With numerical models, it is often impossible to directly calculate the inputs from the desired outputs. Instead, the optimal inputs may be have to be calculated by iteratively altering the inputs until the desired movement is obtained, a process of optimization.

The process of calculating the output given a set of inputs is called forward simulation. It generally involves solving the differential equations of motion by numerical integration. The equations of motion are based on Newton's second law of motion relating acceleration (a) to the resultant force (F) applied to a mass (m), i.e., $F = ma$. By defining the resultant force and mass (which may be time

and position dependent), the acceleration can be calculated and integrated once to calculate the velocity and again to calculate the position. In spite of the simplicity of Newton's law, the equations of motion for a multijoint system are complicated due to the nonlinearities of many of the subsystems, and the interactions between subsystems. It is therefore prudent to carefully define the scope of a model at the outset.

3.1   Specifying the System

Choosing the model scope (Step 1) determines the number of degrees of freedom (df) of the system and thus the number of equations of motion. The number of degrees of freedom is the number of *independent* kinematic variables needed to completely specify the kinematics of the system. The number of df for the system being modeled may be less than the total number of df at the joints, because of kinematic constraints imposed by the environment. Consider the example shown in Figure 8.2 of picking up a book from a shelf. To begin with, assume that the location (X_S, Y_S) of the shoulder center of rotation is fixed. Then if the lower edge of the book is in contact with the shelf (constraining Y_B, as shown in the drawing), only two degrees of freedom are needed to specify all of the angles (θ_S, θ_E, θ_W) and the book position on the shelf (X_B). The system is a closed kinematic chain. If the arm is moving freely in space, the number of

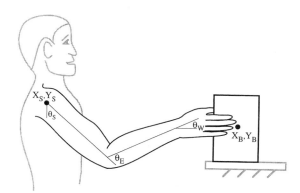

FIGURE 8.2. An example multijoint task with environmental interaction: picking up a book from a shelf. In this example, the system has been isolated to three rigid body segments, (1) the upper arm, (2) the forearm, and (3) the hand, with and without the book.

df is three (the same as the number of joint *df*). In this situation, the system being modeled is a three-link open kinematic chain, and the book location can be calculated from knowledge of the joint angles and segment lengths. Thus, the equations of motion required to model the task of picking up the book from the shelf change with the different phases of the task.

The sequence of system configurations required to model the task of picking up the book is as follows. When the hand is moving to the book, three equations of motion are needed. After the hand has grasped the book, but the book is still being supported by the shelf, only two equations of motion are needed, and the mass properties of the book must be added to those of the hand. Once the book is off the shelf, three equations of motion are needed again, but the parameters are changed because of adding the book to the system. Thus system structure and system parameters change when contact is made/broken with objects in the environment. Such changes occur routinely in normal behavior, such as during foot-to-floor contact during locomotion. In simulations, care must be taken to match positions and forces at the time of making and breaking contact, and change the equations of motion appropriately.

One equation of motion is needed for each degree of freedom, but the choice of coordinate systems for writing the equations of motion is not unique. When the book is on the shelf in the example above, any two of the four degrees of freedom can be chosen as the coordinate system, such as X_B & θ_S or θ_S and θ_W. Because translational and angular coordinates can be mixed, the coordinates are referred to as *generalized coordinates*.

What if the shoulder location is not fixed during the task? Two effects are added. First, translational movement of the shoulder produces inertial interaction moments at all of the joints. If the shoulder translation trajectory is known, then these can be calculated. In general, however, moments and forces produced by muscles in the arm and by the environment will result in forces and moments at the shoulder that will in turn affect the shoulder trajectory (i.e., there will be interaction as discussed below). Modeling the interaction requires information about the mechanical properties at the shoulder, rather than just the trajectory, and makes the equations of motion much more complicated. Thus,

there is a strong desire on the part of modelers to restrict the scope of simulation models, and also to restrict the scope of experiments in order to reduce the complexity of modeling and interpretation.

3.2 Subsystems and Subsystem Interactions

The different subsystems required for forward simulation, and the interactions between the different subsystems are shown explicitly in a block diagram of a system for solving the equations of motion numerically (Figure 8.3). Models for the components and subsystems are discussed below, but it is advantageous to see the interactions first to appreciate the overall system structure.

In Figure 8.3, the equations of motion are assumed to be in joint rotation coordinates, hence the angular accelerations are calculated from the net joint moments.

$$\ddot{\theta} = I^{-1}[M_{muscle} + M_{gravity} + M_{coupling} \\ + M_{passive} + M_{external}] \quad (8.2)$$

Only the muscle moments are functions of the neural inputs to the system, but all of the moments and the inertia are functions of the system outputs, $\dot{\theta}$ and θ, which are calculated by numerically integrating the accelerations.

$$\dot{\theta} = \int \ddot{\theta} dt \text{ and } \theta = \int \dot{\theta} dt \quad (8.3)$$

The muscle moments are the products of the muscle forces and the moment arms. The moment arms depend directly on joint angle, whereas the muscle force depends on joint angle and angular velocity via the (inverse) kinematic relationship between joint angle and muscle-tendon-unit length. Muscle force also depends on the neural input (α) to the muscle-tendon-units, which in turn is altered by feedback (Ia,II,Ib) from muscle spindle and tendon organ receptors. Because of the series tendon compliance of muscle-tendon-units, the length (as well as length derivative) inputs to muscle spindles must be calculated from the contractile element lengths of the muscles rather than the muscle-tendon-unit lengths. Thus, the outputs of muscle spindles depend not only on their neural inputs (γ,β), but also on the contractile element lengths and forces produced by the muscles, which depend on joint angles and velocities. The gravitational moments depend on the posture of the limb in the

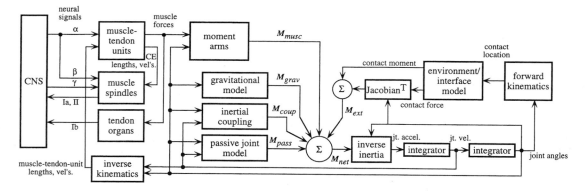

FIGURE 8.3. Block diagram for forward simulation of movements produced by a neuromusculoskeletal system interacting with the environment. All signal lines repre-sent vectors, rather than scalars, because it is assumed there are multiple muscles and multiple degrees of free-dom.

gravitational field, and thus depend on joint angles, but not velocities.

Multisegment limbs have two inherent sources of interaction that are the result of the distribution of mass among the rigid body segments (Zajac and Gordon 1989). First, the inertia affecting any joint depends on other joint angles in the chain. For ex-ample, the moment of inertia at the hip increases with knee extension, because knee extension moves the mass of the lower leg further away from the hip and inertia is proportional to the square of the dis-tance of the mass from the center of rotation. There-fore, the inertia matrix I is posture (i.e., joint an-gle) dependent.

Second, the mechanical coupling between seg-ments makes the inverse inertia matrix nondiago-nal. This means that a moment at one joint can po-tentially induce accelerations at any or all other joints (Zajac and Gordon 1989). Classification of a muscle as a flexor or extensor of a joint can not therefore be based on whether or not the muscle spans that joint. For example, the soleus can act as a knee extensor even though it only spans the an-kle joint.

Inertial coupling moments as a result of cen-tripetal and Coriolis effects are an additional source of interaction in the system. The moments depend on the joint angle configuration, as well as on joint velocities.

The contributions of passive structures show up in two places. The muscle-tendon-unit forces in-clude those components of the passive force that depend on muscle length. Other passive moments,

due to joint-based structures such as ligaments and joint capsules (see below), are modeled separately because they do not depend on muscle length. The separation allows explicit modeling of the coupling across joints due to passive forces in multiarticular muscles.

External joint moments can be produced by both contact forces and contact moments. The location of contact is calculated by a forward kinematic re-lationship (joint angles to Cartesian coordinates). The interaction between the limb and the environ-ment is modeled separately, with the contact loca-tion as the input, and contact moment and contact force as the outputs (i.e., as a dynamic stiffness). The contact force must be transformed by the in-verse Jacobian matrix, J^T, to calculate the resultant joint moments. The Jacobian matrix is joint angle dependent.

$$M_{ext} = M_{contact} + J^T F_{contact} \qquad (8.4)$$

4 Neuromusculoskeletal Component and Subsystem Models

4.1 Body Segments

Analysis and modeling of multisegment movement is almost always based on the simplifying as-sumption that the limb segments behave as rigid bodies connected by joints with defined axes of translation/rotation, as assumed in Figure 8.2 above. Realistic limb models include multiple seg-ments, and joints can have more than one degree

TABLE 8.1. Commercial software packages suitable for multisegment dynamic modeling.

Name	Company (WebPage)	Comments	Reference
ADAMS	Mechanical Dynamics Inc., Ann Arbor, MI, USA http://www.adams.com	Generates equations for generic systems, dynamic simulation, graphical display	Lemay and Crago 1996
SD/FAST	Symbolic Dynamics, Inc. Mountain View, CA, USA http://symdyn.com	Generates equations for generic systems	
SIMM	Musculographics Inc. Evanston, IL, USA http://www.musculographics.com	Written for musculoskeletal kinematic modeling, graphical display; requires SD/FAST, Dynamics Pipelines for dynamic simulation	Delp and Loan 1995
TSi Dynamics (a Mathematica package)	Techno-Sciences, Inc., Lanham, MD, USA info@technosci.com	Generates equations for generic systems; requires Mathematica, links to MATLAB	
Working Model	Knowledge Revolution, San Mateo, CA, USA http://www/krev.com	Simulates generic systems; graphical programming and animation interface	

of freedom. Each segment is characterized by the relative locations of successive joint centers of rotation, the mass, the location of the center of mass, and the mass moment of inertia about each rotational degree of freedom. Because inertial properties are not measured easily, they are often estimated with regression equations derived from studies over a range of subject sizes and conditions (e.g., McConville et al. 1980; Hinrichs 1985; Zatsiorski et al. 1990). Care must be taken in the application of regression equations to subjects that do not belong to the same population as the subjects in the database, e.g. in cases of deformity due to disease or injury.

A multilink chain of rigid bodies exhibits complex dynamic properties because of centripetal and Coriolis interaction torques and gravitational torques. The derivation of the equations describing these torques is tedious for all except the simplest systems. However, writing the equations is amenable to automation, and there are numerous computer programs that will generate the equations of motion for dynamic mechanical systems. Some equation generators are tightly linked to programs that perform the numerical integration and graphically display variables of interest. Some also have graphical programming interfaces that allow simple specification of links and joints, which can be used to build up complex systems that can be visualized easily. Some allow contact between objects to be

made and/or broken during simulation. A listing of several commercial programming tools is given in Table 8.1. Most of the programs are extremely powerful, reducing significantly the programming burden required to implement neuromusculoskeletal models. However, none are complete, and it is mandatory that any programming system allow linking to custom written routines to implement custom subsystems, so that each of the subsystems shown in Figure 8.3 can be included. Some noncommercial packages are described in Appendices 2 through 5.

4.2 Muscle

4.2.1 Basic Model Structure

Muscle models typically contain three elements (Figure 8.4). The parallel elastic element models the force generated under passive conditions. The contractile element produces the active force, which is transmitted through the series elastic element to the point of attachment. The combination of the contractile and series elastic elements is referred to as the muscle-tendon-unit, to emphasize the fact that the force produced by the contractile element can not be determined without taking the series elastic element into account (Zajac 1989). The inputs to the model are the neural input, and the muscle-tendon-unit length. One output is the total force,

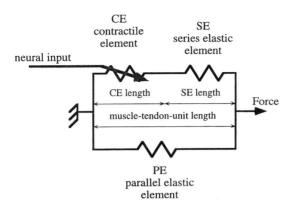

FIGURE 8.4. Muscle model structure.

which is the sum of the active and passive forces. A second output, the contractile element length, must be available if muscle spindles are to be modeled in parallel with the contractile element (see Section 4.5.1).

The series elastic and parallel elastic elements are generally modeled as nonlinear springs, with no damping. The series elastic element represents the muscle tendon, as well as some aspects of the contractile element. The compliance of the series elastic element is highest in muscles with long thin tendons (e.g., the long finger muscles), and lowest in muscles that attach directly to bone (e.g., hip extensors). The parallel elastic element is generally modeled as producing no force below the slack length, with the force rising exponentially at greater lengths.

$$F_p = c_0(e^{c_1(l-l)} - c_2) \qquad (8.5)$$

where F_p is the passive force, c_i are constants, l is the length, and l_s is the slack length. Generally in muscle models, F_p is assumed to be zero for $l < l_s$, and l_s is assumed to be l_0 (the length at which the contractile element produces maximal force, the optimal length). In this case, $c_2 = 1$.

Exponential models are used frequently for the series elastic element, although straight-line (constant stiffness) models are also used. Note that in exponential force-extension models such as this, the incremental (tangential) stiffness is proportional to force.

Two types of contractile element model have been developed. One is empirical, building on the work of Hill (1938). The second is structural, and

is based on cross-bridge kinetics as formulated by Huxley (1957).

4.2.2 Hill-Based Contractile Elements

The detailed formulation and parameterization of Hill-based contractile element models have been described extensively in recent years (e.g., Zajac 1989; Winters 1990) and will not be repeated here. It is widely used by the research community, as evidenced by its use in other chapters of this book. It is arguably the least complex model of neurally activated muscle that should be used in forward simulation of neurally controlled movement.

In Hill-based contractile element models, the normalized force is the product of three independent experimentally measured factors describing the length-tension property ($LT(l_m)$), the force-velocity property ($FV(\dot{l}_m)$), and the dynamics of activation by neural inputs ($A(u)$). (Each of the three factors is also normalized).

$$\frac{F}{F_o} = A(u) \cdot LT(l_m) \cdot FV(\dot{l}_m) \qquad (8.6)$$

Together, the length-tension and force-velocity factors are referred to as contractile dynamics. They describe the modulation of force by the contractile element length and velocity at a fixed activation level. The length-tension factor describes the steady state relationship between contractile element length and isometric force under tetanic stimulation conditions. It is normalized to the force at the length producing maximal force. The force-velocity factor describes the force developed by the muscle during constant velocity movement at the length producing maximal isometric force. It is also measured during tetanic stimulation and is normalized in the same way as the length-tension factor.

The activation dynamics factor models the dynamic process of force modulation when the neural input is varied, including the processes of excitation-contraction coupling. The neural input ($0 \leq u(t) \leq 1$) describes both motor unit recruitment and rate modulation with a single scalar. EMG, measured experimentally, often serves as an indirect measure of the neural input. The output, activation, scales the isometric force, which is further modulated by the length-tension and force-velocity factors. The dynamics are usually described by a low pass filter which can be linear or nonlinear, but generally is of low order (usually first order).

The popularity of Hill-based models is based on its structure.

1. The multiplicative structure captures the major nonlinearity of muscle, which is the scaling of static and dynamic properties by each of the three factors.
2. The three individual factors, and their parameters, describe familiar and intuitive concepts that have been well verified and which are observable in simple experiments.
3. The parameters for the force-velocity and length-tension properties can be related to the sarcomere structure, allowing them to be scaled on the basis of anatomical, rather than physiological, studies (assuming muscle acts as one giant sarcomere).
4. The single scalar neural input is analogous to the EMG, which is measurable experimentally.
5. The model is computationally simple.
6. Although the model certainly does not capture all known behavior of muscle, there are various empirical fixes to describe particular properties (e.g., catch) if they are deemed important in certain situations (see Chapter 7).

Models should be chosen on the basis of their validity for particular applications, rather than their popularity, ease of use, or mathematical sophistication. (These issues are not necessarily independent, but models sometimes take on a life of their own.) Hill-based models are suitable *mathematically* for simulating many arbitrary conditions, but they may not be valid *physiologically*.

Hill-based models have not been validated widely under conditions where the neural input and the length/velocity inputs are changing simultaneously. Validity under these conditions depends on the independence of the three factors and this is not proven (e.g., see Chapter 5). Furthermore, the length-tension and force-velocity properties are measured under conditions of tetanic stimulation, which may not be realistic physiologically.

Perhaps the most undervalidated aspect of the Hill model is the dependence of contractile dynamics on the neural input. The shapes of the length-tension and force-velocity relationships depend strongly on firing rate (Crago 1991; Joyce and Rack 1969; Rack and Westbury 1969). During vol-

untary contractions, motor unit firing rate varies with effort and recruitment level, but the firing rate of freshly recruited motor units is generally well below tetanic levels (e.g., Enoka 1995). Thus, the assumption of tetanic activation is likely to hold only during maximal effort contractions, which is the condition where joint moment-angle curves are well fit by the modeled musculoskeletal geometry and length-tension relationships (e.g., Gonzalez, Buchanan, and Delp 1997). Without validation, the shapes should not be assumed to be invariant at submaximal contractions.

Hill-based models do not predict all the effects of mechanical inputs, which can themselves alter muscle mechanical properties and perhaps affect activation. Increasing the velocity spectrum or amplitude of random length perturbations reduces mean muscle force and stiffness (Kirsch, Boskov, and Rymer 1994; Shue, Crago, and Chizeck 1995). Similarly, prior movement history affects the shape of the force-velocity property (Joyce, Rack, and Westbury 1969). Some of these effects can be interpreted as an effect of movement on activation dynamics, since movement alters the number of attached cross-bridges, and perhaps the rate of cross-bridge turnover.

4.2.3 Cross-Bridge-Based Contractile Elements

Muscle models based on the structure and chemistry of muscle [Huxley 1957] offer an attractive alternative to the inherently black-box Hill-based models. This approach has been advanced significantly by Zahalak, who has developed a structural model describing excitation-contraction coupling and contraction dynamics (Zahalak and Ma 1990). The model is made computationally tractable by assuming a Gaussian distribution for the actin-myosin bond lengths, rather than calculating the distribution directly. Since stiffness, force and work are functions of the mathematical moments of the bond-length distribution, rather than the distribution itself, the macroscopic properties of the muscle can be computed by numerically solving ordinary differential equations, rather than partial differential equations. For a muscle in series with an elastic element, the model is fifth order. The model does a good job of describing the dynamic dependence of force on stimulus frequency, as well as force transients due to varying length inputs.

The largest component of a muscle that can be described by Zahalak model is one motor unit, since the firing pattern is the neural input to the model. Thus, although the basic model is only fifth order, the overall order of a complete muscle model would be five times the number of motor units. Such an approach might be reasonable for a detailed model of muscle combining a motoneuron pool with the peripheral biomechanics, but it is rather cumbersome for simulating multiple muscle systems.

4.3   Joint Mechanics

The muscle tendon path relative to the joint center of rotation determines the moment arm of the muscle at the joint, which in turn determines the relationship between force and moment, and the relationship between muscle length and joint angle. In one respect, the relationships are straightforward, because the moment is proportional to the moment arm, r, and the change in muscle-tendon-unit length is similarly proportional to the change in joint angle.

$$M(\theta) = r(\theta)F(\theta)$$
$$\partial l_m = r(\theta)\partial\theta \qquad (8.7)$$

(Note the sign convention: length is assumed to increase with θ; muscle force and joint moment are positive in opposite directions from muscle length and joint angle.)

These two equations show up independently in Figure 8.3 as "moment arms" and "inverse kinematics," respectively. The significance of these equations is made stronger (and interesting) by the fact that many muscle moment arms vary with joint angle (e.g., Visser et al. 1990; Murray et al. 1995), and the force produced by muscle also varies with joint angle because of the length-tension property and the inverse kinematic relationship between length change and muscle length. Thus, the moment arm scales force to moment in proportion to the moment arm, but scales muscle stiffness to joint stiffness in a different and nonlinear way. The relationship between muscle and joint elastic stiffness (K_m, K_j) can be derived from the two above equations (Lin and Crago 1994; McIntyre et al. 1996).

$$K_j = \frac{\partial M_m}{\partial\theta} = r\frac{\partial F}{\partial\theta} + \frac{\partial r}{\partial\theta}F \qquad (8.8)$$

Note that $\dfrac{\partial F}{\partial\theta} = \dfrac{\partial F}{\partial l}\dfrac{\partial l}{\partial\theta} = K_m r$.

Therefore,

$$K_j = r^2 K_m + \frac{\partial r}{\partial\theta}F. \qquad (8.9)$$

Thus, muscle makes two contributions to joint stiffness. First, muscle stiffness contributes directly to joint stiffness in proportion to the square of the instantaneous moment arm (the first term above). Second, muscle force contributes to joint stiffness in proportion to the slope of the moment arm dependence on joint angle (the second term above). As a result, if moment arm increases with joint angle, then moment will increase even if the muscle force stays constant. It is easy to see that if the moment arm decreases with joint angle, then the second term can be negative and actually decrease or reverse the sign of the total joint stiffness. This is the source of the geometric contribution to joint stiffness discussed in Chapter 11. Note that K_m is also negative on the descending limb of the length-tension curve, giving a second source for negative stiffness. The condition for positive joint stiffness from muscle and joint geometry is obtained by rearranging the previous equation to collect the muscular and geometrical terms.

$$\frac{K_m}{F} > -\frac{1}{r^2}\frac{\partial r}{\partial\theta} \qquad (8.10)$$

This formulation highlights the importance of the muscle stiffness to force ratio in overcoming the destabilizing effects of a negative moment arm-angle slope. Negative slopes can be compensated if the intrinsic muscle stiffness to force ratio is high enough. Note that increasing cocontraction can not compensate for this source of instability, since cocontraction increases muscle stiffness and muscle force proportionally.

The possibility of negative joint stiffnesses (with signs defined as above) has both mathematical and physiological significance. It is impossible to maintain a fixed posture in the presence of disturbances if the net joint stiffness (including *all* contributions, not just those given above) is negative. Any perturbation, including computational inaccuracies, will drive the system away from postural equilibrium. As a consequence, the ability of the model to maintain a fixed posture will depend on the accuracy and completeness of the model. If the model does not accurately describe the properties of stiffness that are required for postural stability (e.g., muscle short-range stiffness), then the model may be unstable even when the real system would be stable. In addition, if the model does not include some of the subsystems in Figure 8.3 (e.g., reflex pathways), then postural stability will not be achievable under all conditions.

4.4   Passive Viscoelastic Properties

The net passive viscoelastic joint moment includes all moments not due to active contraction of muscle, segment inertias, coupled inertias, or gravity. The sources of passive viscoelastic joint moments include (1) the passive properties of muscle mentioned above, (2) ligaments that reduce the degrees of freedom at a joint by restraining movement in certain directions, (3) the joint capsule, (4) blood vessels, nerves and fascia that cross the joint, (5) skin crossing the joint, and (6) joint surface friction. Synovial joints have low friction, so the contributions of friction are small, but could be significant in certain pathologies such as arthritis. Because the contributing sources are so diverse and difficult to measure or predict, joint based passive moment models are empirical curve fits.

The passive moment M_p has both elastic and viscous components. The elastic component contributes an exponentially rising moment magnitude at the extremes of the range of motion, and the viscous component gives rise to hysteresis. A general formulation for a joint based visco-elastic model for the passive moment M_p is given by the following equation

$$
\begin{aligned}
M_p &= c_0(e^{c_1(\theta-\theta_s)} - e^{c_2(\theta-\theta_s)}) + c_3(\theta - \theta_s) \\
&\quad + G(\theta)D(\dot{\theta}) \\
G(\theta) &= sign(\dot{\theta})(g_0 + g_1(\theta - \theta_s) + g_2(\theta - \theta_s)^2) \\
D(\dot{\theta}) &= \dot{\theta}^n
\end{aligned}
$$

$$(8.11)$$

where the c_i, g_i coefficients, and the slack angle θ_s are constants determined by curve fitting. The first two terms of the passive moment describe the elastic moment, and the last term, $G(\theta)D(\dot{\theta})$, describes the hysteresis. This nonlinear model of damping (Esteki and Mansour 1996) assumes that the hysteresis effects are separable into angle dependent $G(\theta)$ and angular velocity dependent $D(\dot{\theta})$ factors.

The exponent n is typically 0.1 (i.e., much less than one). The angle dependence of damping is fit to a second order polynomial. In some studies, linear damping is assumed by setting $G(\theta)$ to a constant and $n = 1$.

The importance of passive moments in the total moment balance is often overlooked, based on the relatively small passive moment, compared to the maximum active moment, in the middle of the joint range of motion. However, the region of interest in movement simulation is much larger than this. For instance, in the lower extremity during quiet standing, only the ankle is near the center of the range of motion, while the knee and hip are near the extremes of extension where passive moments are high. In addition, the level of muscle activation is quite low in the absence of postural perturbations. Passive moments are even more important in cases of muscle weakness or paralysis, where they can provide function (e.g., passive tenodesis grasp in tetraplegia) or can create deformity. In neuroprostheses, passive moments are sometimes the only significant balance to the moment produced by a stimulated muscle (e.g., Chapter 44).

As pointed out above, the total passive moment includes contributions from the parallel elastic element of muscle. Ideally, this contribution should be included in the muscle model, rather than the passive joint model, because many muscles are multiarticular and thus contribute passive moments at more than one joint. The practical limitation to this is that it is difficult to separate experimentally the contributions of muscles from the contributions of other structures. In the absence of separation, the joint passive moments will be in error during multijoint movement.

4.5   Neural Components

Muscle spindles and tendon organs are major sources of afferent feedback, and are responsible in part for augmenting muscle stiffness in the stretch reflex (Houk 1979; Matthews 1981). Our understanding of the role of feedback in movement control has been developed almost exclusively by meticulous experimentation (Nichols and Houk 1976; Prochazka et al. 1979). The tools of mathematical simulation have been underutilized and have been of limited value, partly because of the lack of general, robust mathematical models of afferent behavior under a wide range of conditions.

4.5.1 Spindles

A general model of muscle spindles must include all the neural inputs (gamma dynamic (γ_d), gamma static (γ_s) and skeletofusimotor fibers (β)), mechanical inputs (spindle length and its time derivatives), and neural outputs (Ia and II afferents).

Black-box spindle models relating inputs and outputs without regard to the internal structure of the spindle were developed either to describe the behavior of the muscle spindle in the linear range (using very small stretch magnitude) (e.g., Poppele & Bowman 1970) or to reduce the complexity of the whole system when the muscle spindle is a component of a large system (e.g., Gossett et al. 1994). The difficulty with these models is the limited range of applicability. The sensitivity of the muscle spindle decreases significantly and becomes nonlinear for muscle length changes greater than a few hundred microns (Hasan and Houk 1975). Thus for linear transfer function models, different transfer functions must be used for different length ranges (Chen and Poppele 1978).

A simplified structural model developed by Hasan (1983) separated the mechanical filtering process from the later transducing and filtering processes. The model successfully explained many phenomena of the Ia afferent when gamma efferent levels were fixed, but the range of operation was still limited and parameters had to be changed with different gamma efferent levels. Schaafsma et al. (1991) took a major step by including submodels for individual intrafusal muscle fibers and sensory regions. The model described spindle properties under many conditions (passive, dynamic and static γ stimulation), but only modeled the Ia afferent and did not provide a mechanism for recovery from thixotropic effects. Thus, neither of these models can be used for a wide range of general movement conditions.

Schaafsma's structural approach was expanded by Lin (1996) to include all three types of intrafusal fiber (bag 1, bag 2, and chain) as well as both types of neural input and both afferent fibers (as shown in Figure 8.5). The polar regions of the intrafusal muscle fibers are modeled as Hill-type contractile elements, with series and parallel elastic elements. The sensory region is modeled as a linear elastic element that is stretched in proportion to the force exerted by the fiber. Each afferent arises from branches to two fibers (e.g., the Ia afferent has branches to sensory regions on both the bag 1 and bag 2 fibers). In the model the force in each sensory region is filtered (linearly) to model the transduction and encoding processes, producing a discharge rate in each branch. The discharge rate of the afferent is the higher of the discharge rates in the two branches (the function of the comparator in the model), mimicking the process of action potential competition that has been observed experi-

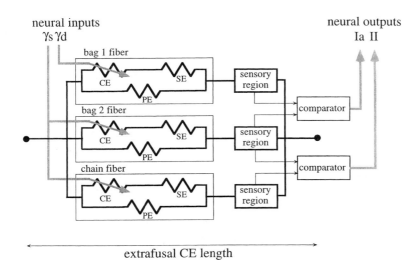

FIGURE 8.5. Structural model for a muscle spindle. The inputs to the model are γ_d, γ_s, and extrafusal contractile element length. The outputs are the discharge rates in the Ia and II afferents. Conduction delays for the afferent fibers are not shown.

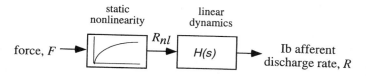

FIGURE 8.6. Force feedback model for a population of tendon organs.

mentally. Afferent conduction delays must also be included to account for the limited action potential conduction velocity to the central nervous system.

Structural spindle models are not used extensively in neuro-musculoskeletal simulation. Instead, spindles are often modeled as ideal (linear) length and velocity sensors (e.g., He et al. 1991), or even as joint angle and angular velocity sensors (e.g., Iqbal et al. 1993). Although simple models are expedient and may be justifiable for small perturbations around an operating point, they are unrealistic models of individual spindles, which display prominent nonlinearities, especially for large length ranges. What is needed is a model of the feedback signals provided by a population of muscle spindles, including the processing that goes on in the central nervous system. The properties of feedback from a population of afferents could have quite different properties than the individual afferents, depending on how the information is actually combined and processed.

4.5.2 Tendon Organs

Tendon organs sample the active muscle force. During physiologically graded contractions, the thresholds of individual tendon organs are low and the time-averaged discharge rate increases linearly with force (Crago et al. 1982). Furthermore, transfer function models have been developed that describe well the dynamic behavior of individual receptors (Houk 1971; Anderson 1974). The distribution of thresholds and slopes has been documented for the cat soleus stretch reflex (Crago et al. 1982). Thus, the information was available for Lin (1996) to develop a model of the population response in the stretch reflex.

The population model assumes that the outputs of all tendon organs in a muscle are spatially averaged to produce an equivalent single feedback signal. The idea that the nervous system performs such averaging is speculative.

The force feedback signal was modeled as a force sensor with a static nonlinearity preceding

linear dynamics (see Figure 8.6). The static nonlinearity accounts for the distribution of tendon organ thresholds and slopes and is the curve fit of a logarithmic function to the population response (Crago et al. 1982; Figure 8).

$$R_{nl} = G_r * \ln(F/F_n + 1) \qquad (8.12)$$

where R_{nl} is the output rate of the static nonlinearity stage, G_r (60 impulses/s) and F_n (4 N) are constants and F is the extrafusal muscle force. The linear dynamics transfer function H(s) was derived from the experimental data of Houk and Simon (1967)

$$\frac{R_f(s)}{R_{nl}(s)} = H(s) = \frac{40(1.7s^2 + 2.58s + 0.4)}{s^2 + 2.2s + 0.4} \qquad (8.13)$$

where the output rate of the population of tendon organs is R_f.

5 Summary and Future Directions

The topics presented above show that there has been considerable effort and success in identifying principles and modeling subsystems that are required to simulate neuromusculoskeletal systems. Most of the modeling success has occurred in the musculoskeletal area, rather than in the neural area, and this is where progress is needed most urgently.

Although feasible, it is still tedious to develop models of peripheral musculoskeletal systems from scratch. The availability of commercial modeling and simulation software makes the task easier, but the wealth of detail that must be incorporated into a model makes it attractive to develop and deploy standard models that can be validated and enhanced by independent teams of researchers.

Current models of the central nervous system's role in motor control are primarily conceptual. Even the gross structure of the system is up to debate. However, the availability of biomechanical simulation models will facilitate the understanding of central nervous system function. The main contributions of simulation will be to help understand the complex interactions and nonlinearities that are involved in movement control, and to provide quantitative estimates of the separate contributions to complex motor behavior.

References

Anderson, J.H. (1974). Dynamic characteristics of Golgi tendon organs. *Brain Res.*, 67:531–537.

Chen, W.J. and Poppele, R.E. (1978). Small-signal analysis of response of mammalian muscle spindles with fusimotor stimulation and a comparison with large-signal responses. *J. Neurophysiol.*, 41:15–27.

Crago, P.E. (1992). Muscle input-output model: the static dependence of force on length, recruitment and firing period, *IEEE Trans. Biomed. Eng.*, 39:871–874.

Crago, P.E., Houk, J.C., and Rymer, W.Z. (1982). Sampling of total muscle force by tendon organs. *J. Neurophysiol.*, 47:1069–1083.

Delp, S.L. and Loan, J.P. (1995). A graphics-based software system to develop and analyze models of musculoskeletal structures, *Comput. Biol. Med.*, 25:21–34.

Enoka, R.M. (1995). Morphological features and activation patterns of motor units, *J. Clin. Neurophysiol.*, 12:538–559.

Gonzalez, R.V., Buchanan, T.S., and Delp, S.L. (1997). How muscle architecture and moment arms affect wrist flexion-extension moments, *J. Biomech.*, 30:705–716.

Gossett, J.H., Clymer, B.D., and Hemami, H. (1994). Long and short delay feedback on one-link nonlinear forearm with coactivation. *IEEE Trans. Sys. Man Cybern.* 24:1317–1327.

Hasan, Z. (1983). A model of spindle afferent response to muscle stretch. *J. Neurophysiol.*, 49:989–100.

Hasan, Z. and Houk, J.C. (1975). Transition in sensitivity of spindle receptors that occurs when muscle is stretched more than a fraction of a millimeter. *J. Neurophysiol.*, 38:673–689.

He, J., Levine, W.S., and Loeb, G.E. (1991). Feedback gains for correcting small perturbations to standing posture, *IEEE Trans. Auto. Contr.*, 36:322–332.

Hill, A.V. (1938). The heat of shortening and the dynamic constants of muscle. *Proc. Roy. Soc.*, 126B: 136–195.

Hinrichs, R.N. (1985). Regression equations to predict segmental moments of inertia from anthropometric measurements: an extension of the data of Chandler et al. (1975). *J. Biomech.*, 8:621–624.

Houk J.C. (1971). A viscoelastic interaction which produces one component of adaptation in responses of Golgi tendon organs. *J. Neurophysiol.*, 34:1482–1493.

Houk, J.C. (1979). Regulation of stiffness by skeletomotor reflexes. *Ann. Rev. Physiol.*, 41:99–114.

Houk, J.C. and Simon, W. (1967). Responses of Golgi tendon organs to forces applied to muscle tendon. *J. Neurophysiol.*, 30:1466–1481.

Huxley, A.F. (1957). Muscle structure and theories of contraction, *Prog. Biophys. Biophys. Chem.*, 7:257–318.

Iqbal, K., Hemami, H., and Simon, S. (1993). Stability and control of a frontal four-link biped system, *IEEE Trans. Biomed. Eng.*, 40:1007–1018.

Joyce, G.C. and Rack, P.M.H. (1969). Isotonic lengthening and shortening movement in cat soleus muscle, *J. Physiol.*, 204:475–491.

Joyce, G.C., Rack, P.M.H., and Westbury (1969). The mechanical properties of cat soleus during controlled lengthening and shortening movements, *J. Physiol.*, 204:461–474.

Kirsch, R.F., Boskov, D., and Rymer, W.Z. (1994). Muscle stiffness during transient and continuous movements of cat muscle: perturbation characteristics and physiological relevance. *IEEE Trans. BME*, 41:758–770.

Lehman, S.L. and Stark, L. (1982). Three algorithms for interpreting models consisting of ordinary differential equations: sensitivity coefficients, sensitivity functions, global optimization. *Math Biosci.*, 62:107–122.

Lemay, M.A. and Crago, P.E. (1996). A dynamic model for simulating movements of the elbow, forearm, and wrist. *J. Biomech.*, 29:1319–1330.

Lin, C.-C. (1996). Neuromechanical interaction in the stretch reflex: a structural simulation model. Ph.D. Dissertation, Case Western Reserve University, Cleveland, Ohio.

Lin, C-C. and Crago, P.E. (1994). Biomechanical factors determining postural stability at the elbow joint, *16th Ann. Int. Conf. IEEE Eng. Med. Biol. Soc.*, Baltimore, November.

Matthews, P.B.C. (1981). Evolving views on the internal operation and functional role of the muscle spindle. *J. Physiol. Lond.*, 320:1–30.

McConville, J.T., Churchill, T.D., Kaleps, I., Clauser, C.E., and Cuzzi, J. (1980). *Anthropometric Relationships of Body and Body Segment Moments of Inertia*. National Technical Information Service, Springfield, Virginia.

McIntyre, J., Mussa-Ivaldi, F.A., and Bizzi, E. (1996). The control of stable postures in the multijoint arm. *Exp. Brain Res.*, 110:248–264.

Murray, W.M., Delp, S.L., and Buchanan, T.S. (1995). Variation of muscle moment arms with elbow and forearm position. *J. Biomech.* 28:513–525.

Nichols, T.R. and Houk, J.C. (1976). Improvement in linearity and regulation of stiffness that results from actions of stretch reflex. *J. Neurophysiol.*, 39:119–142.

Poppele, R.E. and Bowman, R.J. (1970). Quantitative description of linear behavior of mammalian muscle spindles. *J. Neurophysiol.*, 33:59–72.

Prochazka, A., Stephens, J.A., and Wand, P. (1979). Muscle spindle discharge in normal and obstructed movements. *J. Physiol. Lond.*, 287:57–66.

Rack, P.M.H. and Westbury, D.R. (1969). The effects of length and stimulus rate on tension in the isometric cat soleus muscle. *J. Physiol.*, 204:443–460.

Schaafsma, A., Otten, E., and van Willigen, J.D. (1991). A muscle spindle model for primary afferent firing based on a simulation of intrafusal mechanical events. *J. Neurophysiol.*, 65:1297–1312.

Shue, G., Crago, P.E., and Chizeck, H.J. (1995). Muscle-joint models incorporating activation dynamics, torque-angle and torque-velocity properties, *IEEE Trans. Biomed. Eng.*, 42:212–223.

Visser, J.J., Hoogkamer, J.E., Bobbert, M.F., and Huijing, P.A. (1990). Length and moment arm of human leg muscles as a function of knee and hip-joint angles. *Eur. J. Appl. Physiol.*, 61:453–460.

Winters, J.M. (1990). Hill-based muscle models: a systems engineering perspective. In *Multiple Muscle Systems: Biomechanics and Movement Organization*, Winters, J.M. and Woo, S.L-Y. (eds.), Chapter 5, pp. 69–93, Springer-Verlag, New York.

Zahalak, G.I. and Ma, S.-P. (1990). Muscle activation and contraction: constitutive relations based directly on cross-bridge kinetics, *J. Biomech. Eng.*, 112:52–62.

Zajac, F.E. (1989). Muscle and tendon: properties, models, scaling, and application to biomechanics and motor control. *CRC Crit. Rev. Biomed. Eng.*, 17:359–411.

Zajac, F.E. and Gordon, M.E. (1989). Determining muscle's force and action in multi-articular movement. *Exerc. Sport Sci. Rev.*, 17:187–230.

Zajac, F.E. and Winters, J.M. (1990). Modeling musculoskeletal movement systems: joint and body segmental dynamics, musculoskeletal actuation, and neuromuscular control. In *Multiple Muscle Systems: Biomechanics and Movement Organization*. Winters, J.M. and Woo, S.L-Y. (eds.), Chapter 8, pp. 121–148, Springer-Verlag, New York.

Zatsiorsky, V.M., Seluyanov, V.N., and Chugunova, L.G. (1990). Methods of determining mass-inertial characteristics of human body segments. In *Contemporary Problems of Biomechanics*, Chernyi, G.G. and Regirer, S.A. (eds.), pp. 272–291, CRC Press, Boca Raton, Florida.

9
System Identification and Neuromuscular Modeling

Robert E. Kearney and Robert F. Kirsch

1 Introduction

"System identification" is a term that describes mathematical techniques used to infer the properties of an unknown system from measurements of the system inputs and outputs. Typically, the inputs to the system are controlled by the experimenter, although the only real requirements are that all the inputs and outputs are known and measurable and that the inputs sufficiently excite the system. System identification techniques have been applied to a wide range of problems, this chapter focuses on applications related to the neuromuscular system (i.e., muscle, joint, and limb mechanics) as well as the neural signals that control posture and movement.

There are two general approaches to system identification: "black-box" and "a priori modeling." In the "black-box" approach, the system is regarded as performing a purely mathematical transformation between the inputs and the outputs, and there need not be any direct relationship between the model and the underlying physical processes giving rise to the system properties. On the other hand, only very general assumptions, such as linearity and model order, need be made. Although we will concentrate on system identification techniques that characterize *dynamic* systems, many different types of analyses can be classified as "black-box" system identification, ranging in complexity from simple linear regression to functional expansions of dynamic nonlinear systems in terms of Wiener or Volterra kernels.

In contrast, the "a priori modeling" approach starts with the development of a mathematical model, based on first principles (e.g., Newton's laws) or other prior knowledge. The system identification problem is that of determining the "correct" model and then estimating its parameters. Most "neuromuscular modeling" is actually system identification of this type, although this is rarely recognized. Thus, the aim is to describe the behavior of muscles, joints, or limbs based on the anatomical structure of the limb, of the sensory receptors in the muscles, joint, and skin, and of the neural control circuitry. Anatomical features of the limb typically included are the masses and moments of inertia of each of the limb segments, points of origin and insertion of each of the muscles of the limb, and contractile models of each of the individual muscles. The responsiveness of muscle spindles, tendon organs, and other receptors, as well as the reflex properties of these receptor signals, may also be included in the model. The individual elements are characterized either experimentally or (more typically) using model structures and parameter values obtained from the literature. The models of the individual elements may be based upon the detailed structure of the element, or they may be empirical in nature. Optimization procedures may be used to adjust unknown parameters (e.g., reflex feedback "gains") to best match variations in the system outputs.

Both approaches to system identification have been used to characterize the dynamic properties of human limbs during posture and movement. The results are then used to deduce underlying neural control strategies or to determine optimal artificial control strategies utilizing rehabilitation interventions such as functional neuromuscular stimulation.

Although black-box system identification and more structurally based modeling approaches overlap in scope to some degree, they have different strengths and weaknesses and tend to be applied at different levels of analysis. We therefore believe that these two approaches are complementary in nature rather than being mutually exclusive. The following sections will describe the basic ideas behind the "black-box" system identification approach (hereafter referred to simply as "system identification"), briefly describe the range of currently available system identification procedures, describe how system identification techniques can contribute to the understanding and modeling of neuromuscular systems, and describe important applications of system identification techniques that are ready for future study.

2   A System Identification Primer

2.1   What Is System Identification?

A general definition of system identification is the process of building mathematical models of systems based on measurements of their input(s) and output(s) (Ljung 1987). The general identification problem can be posed as illustrated in Figure 9.1. The system is comprised of everything within the dotted box, and consists of two parts. (1) A stochastic part driven by a white noise process, $w_2(t)$, which is not available to the experimenter; this element of the model represents measurement errors, noise, and so on. (2) A deterministic part

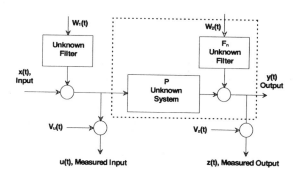

FIGURE 9.1. Schematic diagram of the system identification context. (Redrawn from Verhaegen and DeWilde 1992.)

(i.e., the system to be characterized) driven by the sum of the controlled input, $x(t)$, and a filtered version of an inaccessible white noise process, $w_1(t)$ (which represents measurement and other noise). The signals available for analysis are noise corrupted versions of the input, $u(t)$ and output, $z(t)$. Given this structure, the objective can be to identify the:

- *Deterministic Model*: Find the relationship between $u(t)$ and $z(t)$, assuming that the noise signal $w_1(t)$ is zero. This is the approach used when the objective is to understand the underlying function of the system.
- *Noise Model*: here the problem is to determine, F_n, the relation between the output and noise model $w_2(t)$ given observations of the output. The input signal is usually assumed to be zero. This type of identification is used when the inputs cannot be measured; chaos theory and nonlinear dynamics are examples of this type of identification.
- *Complete Model*: Here the problem is to identify both the stochastic, F_n, and the deterministic, P, models in order to generate accurate predictions of the observed outputs for use in design of model-based control systems.

For the purposes of this chapter, we will restrict our discussion to the identification of deterministic models. A number of mathematical techniques exist for relating system inputs to outputs, ranging from simple regression (for *static* linear and nonlinear systems), to spectral- and correlation-based methods (for *dynamic* linear systems), to Volterra and Wiener kernel functional expansions (for dynamic nonlinear systems). The one feature that all of these techniques (several of which will be described below in more detail) have in common is that they relate the inputs to the outputs using arbitrary basis functions rather than anything related to underlying anatomy and physiology of the system. For example, linear frequency-domain approaches (which relate input to output via a gain and phase or "Bode" plot) express system properties in terms of weighted sinusoids of different frequencies. Linear time-domain methods typically express the output in terms of a weighted sum of current and past inputs. Nonlinear methods use more complex functions to describe system behavior, but all of these functions are chosen for their mathematical properties or their relation to the input and output signals rather than for any relationship to the underlying physical processes of the system.

2.2 Parametric Versus Nonparametric System Identification

System identification can be performed using a number of different approaches. We have divided these approaches into two types, "parametric" and "nonparametric." Parametric approaches adopt a relatively specific model of the system; in some cases this may be based on a priori knowledge, but in other cases the model may be purely mathematical and not based on the underlying physical structure of the system. On the other hand, nonparametric approaches make only very general assumptions regarding system structure (e.g., linearity, stationarity). Nonparametric approaches can reveal potentially unknown system properties and mechanisms, while parametric approaches (while drawing on a priori knowledge) have already decided upon structure and determine the best parameters *for that model* which fit the measured data. The following sections will describe examples of these two classes of system identification techniques, with somewhat more emphasis on the nonparametric approaches.

Parametric models are defined by analytic expressions with relatively "few" parameters. For linear systems, parametric identification methods are usually discrete time and typically involve formulating a regression problem relating the current output to the weighted linear combination of current and past input and/or output samples; the identification procedures determine the weighting coefficients which "best" (i.e., via least-squares or some other error criteria) relate the input signal to the output signal over the time interval of interest. Usually, the order of the system (i.e., the number of past input samples which must be included to adequately predict the current output) must be defined in advance. If the model order is not known, the identification must be performed several times and a model-order criteria such as the Akaike criteria (Akaike 1974) or the minimum descriptor length (Rissanen 1978) must then be used to select the appropriate order (Ljung 1987). Some recent "subspace" algorithms (Verhaegen and DeWiles 1992) estimate both the model order and its parameters directly from the experimental data.

For nonlinear systems, parametric system identification again typically involves prediction of the current output from current and past inputs, but in this case the inputs are combined in a nonlinear fashion (e.g., polynomial functions, products, etc.) and the weighting factors are estimated using a nonlinear minimization technique. The Nonlinear Autoregressive Moving Average with eXogenous input (NARMAX) model (Chen and Billings 1989), which can represent a wide range of nonlinear systems, is perhaps the best known model of this class. As with linear parametric identification, model structure and order must be determined in advance of applying the identification algorithm.

A major advantage of parametric methods is that they are flexible enough to easily incorporate highly nonlinear and nonstationary behavior. Thus, they can incorporate (indeed, they depend on) a priori knowledge about the system (such as model order and nonlinearity type), and they can often provide an accurate characterization of a system with a relatively small number of parameters. The major disadvantage of the approach is that the model structure must be selected correctly a priori; large errors will result if the model structure is incorrect. Determining the variance of parameter estimates, especially for nonlinear systems, is therefore difficult (Ljung 1989). Another limitation is that many of the methods are formulated in discrete time, and converting the results to continuous time for interpretation is not straightforward (Kukreja et al. 1996).

Nonparametric models are described by blackbox relations having "many" parameters. These may take a variety of forms. Linear and quasilinear systems (e.g., muscle and joint stiffness properties during stochastic displacements of posture) may be described using frequency response curves (Bendat and Piersol 1986) or impulse response functions (Hunter and Kearney 1983b; Westwick and Kearney 1996a). Time-varying, linear identification methods may be used to approximate nonlinear systems (e.g., changes in muscle or joint stiffness during movement) by linearizing the system about a reference trajectory defined by a sequence of operating points (MacNeil et al. 1992). Some nonlinear systems [e.g., reflex electromyographic (EMG) responses] can be modeled using a combination of linear dynamic and static nonlinear blocks (Hunter and Korenberg 1986; Korenberg and Hunter 1986). Simple models of this type include the Wiener (or LN—a linear dynamic system followed by a static nonlinearity) system, the Hammerstein (or NL) system, and the sandwich (or LNL) model. More general *dynamic* nonlinear systems may be described by functional expansions such as the Volterra and Wiener series (Wiener 1958; Volterra 1959; Marmarelis and Marmarelis 1978). Although these functional expansion methods have the theoretical potential to describe systems of arbitrary complexity, they have been used in only a few applications due to their large computational requirements, need for white inputs, and sensitivity to noise (Westwick and Kearney 1996b). Recently, however, it has been shown that such systems may also be modeled as the sum of a finite number of parallel cascade LNL (Palm 1979) or Wiener (Korenberg 1991) systems, which can be estimated efficiently and then converted into the corresponding functional expansion form if desired. The Wiener–Bose model, in which the outputs of orthogonal linear filters are transformed by a multiinput static nonlinearity shows promise in this regard as well (Boyd and Chua 1985; Marmarelis 1989a).

The major advantage of nonparametric methods is that they require few a priori assumptions, such as linearity, stationarity, the length of an impulse response, or the range of frequencies used for frequency responses. Furthermore, these assumptions can usually be verified experimentally. The nonparametric approaches can also provide results directly in continuous time, and the relationship between parameter variance and input selection is relatively straightforward (Kearney and Hunter 1990). Of particular interest to the understanding of underlying physical structure, nonparametric representations can provide insight into system order, structure (low frequency gain, natural frequency, reflex delays, and other general behavior).

A major disadvantage of nonparametric approaches is that they may require many more "parameters" than the parametric approaches, especially if the basis function is not well matched to the actual system structure. (This limitation is particularly important in nonlinear models.) Furthermore, methods for dealing with the types of nonlinear systems prevalent in the neuromuscular system have been quite limited until the recent development of the parallel cascade methods (Westwick and Kearney 1996b).

2.3 Design of Appropriate Inputs

The ability to identify a system or to validate a model is determined to a large part by the properties of the input signal. For linear systems, the requirements for using nonparametric methods to identify a system are well known: the input signal must be persistently exciting and must excite all modes of the system. In the frequency domain, this means that the input signal must have power at all frequencies where the system has a significant response. Conversely, the identification will only be valid for the range of frequencies which are identified; by definition, a linear system will have an output bandwidth which is the same or smaller than the input bandwidth (which is defined by the experimentally-determined sampling rate and record length). Parametric methods can estimate system dynamics using less rich inputs (i.e., more restricted range of input amplitudes and frequencies), provided the model structure is known accurately (e.g., Ljung 1989).

For nonlinear systems, the design of an appropriate input is much more difficult since inputs at a particular frequency may produce output power not only at that frequency but at other frequencies as well. In general, the output bandwidth will be greater than the input bandwidth (Marmarelis and Marmarelis 1978). Furthermore, system behavior in some cases (e.g., spindle afferent responses to vibration) may be so nonlinear that certain input frequencies may distort or obliterate relevant system behavior at other frequencies (Kirsch and Rymer 1992; Kirsch et al. 1994). Thus, determining an appropriate input bandwidth and sampling rate for identifying a general dynamic nonlinear system is not straightforward. In addition, the amplitude structure of the input signal will have a strong influence on what can and cannot be identified. For example, the range of validity of a Hammerstein (NL) system model is directly determined by the amplitude range of the input because the input is directly imposed on the static nonlinear subsystem. However, the effects are less clear with Wiener structures or parallel cascades, where the imposed input is first filtered (and thus reduced in amplitude) before being nonlinearly transformed

(Westwick 1996). Thus, while it is evident that the choice of the input signal is very important for non-linear identification, there are as yet few guidelines as to how to do this effectively (but see Kearney and Hunter 1990 for a more detailed discussion of these issues).

Many different types of input signals can be used for system identification (Kearney and Hunter 1990), including pulses, steps, ramps, sinusoids, pseudorandom binary sequences, white noise, and others. The choice of an appropriate input type, as well as the specific properties of an individual input (e.g., amplitude, rate of change, bandwidth, etc.), is very application dependent. General guidelines were described above, and a much more detailed discussion is given in Kearney and Hunter (1990).

2.4   Effects of Noise

Measurement and other noise components have an important influence on the validity of model estimates. For linear systems, noise at the output will generate only a random error that will not bias estimates of system properties, provided that appropriate estimators are used (Bendat and Piersol 1986). Moreover, the magnitude of this random error can in principle be reduced to an arbitrary level by increasing the length of the data record (Bendat and Piersol 1986). In contrast, input noise generates a bias error which cannot be eliminated by increasing the record length, although its effects can sometimes be accounted for by using specialized experimental and analysis methods to estimate the input noise and correct for its effects (Bendat and Piersol 1986). Little is known as yet about the effects of noise in nonlinear identification. The same general principles can be expected to apply but the effects of noise are likely to be considerably more important than in linear systems.

2.5   Feedback

The presence of feedback significantly complicates the task of separately identifying the properties of the "open loop" and "feedback" elements of the system, and many neuromuscular systems operate in the presence of reflex or other neural feedback pathways. The "forward path" (i.e., open loop) properties of a linear system operating with feed-back can be identified provided that it is possible to insert a disturbance signal within the loop. If this cannot be accomplished, attempts to identify the system from input-output measurements will only yield estimates of the overall "closed loop" transfer function (Ljung 1989). If input noise is present, the resulting model may represent the dynamics of the forward path, the feedback pathway, or some arbitrary combination of the two (Kearney and Hunter 1990). The identification of nonlinear feedback systems is more difficult in general, although for some specific structures the identification may actually be easier than for linear systems (Marmarelis 1989b).

3   Why Use System Identification in Neuromuscular Studies?

System identification techniques have been applied to the study of the neuromuscular system in several different but overlapping ways. The following sections will describe how system identification has been used by itself to obtain basic knowledge about the neuromuscular system, to advance the artificial control of neuromuscular systems, to aid the development of models of individual elements in large-scale models, and to characterize the overall properties of complex systems.

3.1   Fundamental Knowledge

To date, one of the primary uses of system identification techniques has been to characterize discrete properties of neuromuscular systems (e.g., muscle and joint stiffness properties, reflex EMG properties). In some cases, such models are intended for use as components in a larger scale model of system behavior (see below), but many of these studies were intended primarily to gather basic information about the properties of the system. In general, the objective of this use of system identification is to predict the responses of the system over a wide range of operating points with high accuracy and temporal resolution. The individual values in the input-output description (e.g., frequency response or impulse response values) themselves do not relate directly to underlying mechanisms. However, the way the parameters vary across frequencies or across experimental conditions often reveals important information about the underlying mechanisms that can subsequently be experimentally tested more directly and/or modeled with a parametric description more related to mechanism. The following sections will describe applications of different types of system identification procedures specifically designed to increase basic understanding of the neuromuscular system.

3.1.1 Quasistatic Linear Applications

Applications of system identification techniques to the neuromuscular system have focused predominantly on relatively simple systems tested under conditions for which their behavior is linear about a constant operating point (e.g., constant mean forces and joint positions, small perturbation amplitudes). For example, accurate estimates of dynamic mechanical properties, in terms of linear stiffness or compliance impulse responses or frequency responses, have been identified during quasistatic, postural conditions for many joints including the ankle (Agarwal and Gottlieb 1977; Hunter and Kearney 1982a; Kearney and Hunter 1982; Evans et al. 1983), wrist (Lakie et al. 1984), elbow (Joyce et al. 1974; Zahalak and Heyman 1979; Kirsch and Rymer 1992), jaw (Cooker et al. 1980), knee (Robinson et al. 1994; Tai and Robinson 1995), and neck (Viviani and Berthoz 1975; Bizzi et al. 1976). Typically, a second-order stiffness model of the form:

$$TQ(t) = I(l)\frac{d^2q(t)}{dt^2} + B(l)\frac{dq(t)}{dt} + K(l)q(t) \quad (9.1)$$

has been found to accurately predict the joint torque, $TQ(t)$, elicited by small perturbations of joint position, $q(t)$. The model is quasilinear, because the inertial (I), viscous (B), and elastic (K) parameters vary significantly with operating point (Kearney and Hunter 1990). This approach has been used to demonstrate that the stiffness of the human ankle joint decreases with displacement amplitude (Kearney and Hunter 1982), increases with muscle activation (Hunter and Kearney 1982a; Weiss et al. 1988), increases near the extremes of mean joint position (Weiss et al. 1986a,b), and does not change significantly during fatigue (Hunter and Kearney 1983). These same system identification techniques have also been used to characterize the properties

of the reflexive electromyographic (EMG) signal evoked in the human triceps surae (Kearney and Hunter 1983) and tibialis anterior (Kearney and Hunter 1984) muscles by imposed stochastic perturbations. The nonparametric impulse responses that resulted quantified the reflex delay in these muscles, the duration of the reflex EMG response, and its modulation with muscle activation level and perturbation amplitude. Others have used similar techniques to investigate the cross-bridge cycling dynamics of isolated muscle fibers (Barden 1981; Kawai and Schachat 1984; Calancie and Stein 1987) and the receptor dynamics of muscle spindles (Poppele 1981; Kröller et al. 1985).

Quasistatic characterizations of the type described above have yielded a wealth of important information about the neuromuscular system, even though specific "parameters" (e.g., individual frequency response values) do not directly relate to underlying mechanisms. Insight into system properties obtained from nonparametric input-output relationships usually requires experience and intuition, since conclusions are typically drawn from the overall structure of the relationship and often from across different experimental conditions. In other words, the manner in which different mechanisms affect an input-output relationship must be recognized. This is easier for a linear system, where engineering approaches of this type (e.g., Bode plots) are common, but certain properties of nonlinear systems can also be recognized in Wiener or Volterra Kernel descriptions (Korenberg and Hunter 1990). There are a number of examples of how important system properties can be recognized in input–output relationships. The low frequency behavior of joint stiffness indicates its steady-state stiffness (important for postural stability) and its higher frequency behavior is related to the mass of the moving limb segment. In isolated muscle fiber work, the time constants associated with various states in the crossbridge cycle can be visualized as peaks and valleys in the muscle stiffness frequency response (Kawai and Schachat 1984; Calancie and Stein 1987); the variation of these frequency response features can thus be used to differentiate between fast and slow muscle fibers and to examine the effects of temperature and other variables on muscle function.

In addition to the input–output relationship itself, simple descriptive models have been widely used to summarize the behavior of quasistatic linear systems. In particular, second order models have been used extensively (as described above) to describe the stiffness dynamics of human joints. The use of this simple model has been criticized because its parameters (with the exception of the joint inertia) arise from a mixture of different mechanisms and, even worse, vary over a wide range with both internal and external variables. Nonetheless, the second order model description has proven useful because of the functional relevance of the elastic and viscous estimates obtained: higher elastic stiffness indicates that external disturbances will be more vigorously rejected (i.e., higher postural stability), while higher viscous stiffness indicates smoother and more damped responses. Thus, we feel that the results of simple linear system identification, as well as those from simple models of the identified properties, have contributed significantly to understanding a range of neuromuscular systems.

3.1.2 Time-Varying Applications

Many systems can be accurately characterized by the quasistatic linear approaches described above only for fixed operating points. For more general conditions, this simple description is usually not valid; for example, the variation of joint stiffness with operating point demonstrates that the underling behavior is nonlinear. Nonlinear behavior in neuromuscular systems arises from two primary sources: inherently nonlinear responses to an input (e.g., muscle contractile properties vary with length) and nonlinear interactions among multiple inputs (e.g., muscle stiffness varies strongly with activation level). Although mathematical methods exist for characterizing such nonlinear dynamic systems (see below), the application of these techniques to the study of many neuromuscular systems is often not feasible. In particular, some of the inputs may not be accessible for measurement (e.g., muscle activation).

As an intermediate step, systems whose behavior is influenced by inaccessible inputs or which are nonlinear during specific tasks can be studied using "time-varying" identification methods. These approaches linearize the system about a sequence of operating points rather than about a fixed operating point, and thus provide a separate linear description at each time sample throughout the task. Soechting and coworkers (Soechting et al. 1981) pioneered a nonparametric approach which used a novel "shifted-input" method to track changes in

reflex EMG properties associated with going from a "do not resist" to a "resist" paradigm. An improved ensemble identification technique was used to characterize human ankle joint stiffness (MacNeil et al. 1992; Kirsch and Kearney in press) and triceps surae EMG (Kirsch and Kearney 1993; Kirsch et al. 1993) properties during rapid changes in isometric contraction level and during a large imposed stretch. Several others have used parametric (usually second-order) time-varying models. Bennett et al. (1992) examined the stiffness of the human elbow joint during unconstrained cyclic movements using an ensemble technique and small force perturbations provided by an air-jet apparatus. He subsequently (Bennett 1993) superimposed sinusoidal position perturbations upon a nominal joint trajectory to study reflex modulation of joint stiffness during rapid elbow joint movements. Lacquaniti et al. (1993) examined joint and endpoint stiffness during a multijoint ball catching task with an ensemble time-varying identification technique. Recently, Gomi (1996) used a high-performance manipulandum to estimate inertial, viscous and elastic parameters of the shoulder and elbow throughout point-to-point arm movements

The primary advantage of the time-varying identification approaches is that functional modulation of important quantities such as joint stiffness and reflex EMG can be characterized with high accuracy and temporal resolution. The primary disadvantage of the approach is that the resulting time-varying description is valid only for the particular trajectory tested and is not easily generalized. For example, modulation of joint stiffness during changes from one activation level to another will not necessarily generalize to other combinations of initial and final activation level, or to different contraction speeds. Interpretation of results has generally followed that used for quasistatic conditions, that is, from features of the nonparametric relationships (e.g., low frequency gain) or from the parameters of simple models. As described above, second-order models have been used widely for time-varying identification, primarily because of the simplicity of this parametric approach relative to more general nonparametric approaches. However, the two nonparametric, time-varying ankle stiffness studies to date (MacNeil et al. 1992; Kirsch and Kearney in press) have found that the second-order model performed rather poorly during rapid (although physiological) changes in system properties. This underscores the need to demonstrate that the assumed model structure is indeed adequate for the tested condition (i.e., to estimate and report goodness of fit measures in addition to parameter variations).

3.1.3 Nonlinear Applications

The ultimate goal of any system identification approach is to characterize the unknown system over its entire range of operation with a single set of parameters. Existing general nonlinear techniques (Marmarelis and Marmarelis 1978), including Volterra (1959) and Wiener (1958) kernels, are theoretically capable of providing such a characterization, yet their application to neuromuscular problems has been extremely limited. Kearney and Hunter (1988) described the application of Wiener kernel analysis to human triceps surae reflex EMG properties, and a number of studies have examined the properties of individual neurons or sensory receptors (Marmarelis 1989; Sakai 1992). Some nonlinear systems can be described by a combination of linear dynamic and static nonlinear subsystems that can be identified by relatively simple iterative procedures (Hunter and Korenberg 1986; Korenberg and Hunter 1986). This approach has been used to characterize muscle stiffness (Hunter 1986) and human reflex EMG properties (Kearney and Hunter 1988). Apart from this relatively small number of applications, however, nonlinear techniques have been largely ignored in the study of the neuromuscular system.

The reasons for this limited application of general nonlinear identification techniques to the neuromuscular system are both experimental and analytical in nature. Standard dynamic nonlinear identification requires nonrealizable white noise inputs and very long data records that are not possible in many neuromuscular systems due to fatigue, deterioration of the preparation, and other nonstationary properties (Marmarelis and Marmarelis 1978). Furthermore, conventional implementations of the identification algorithms have enormous computational requirements and are very sensitive to measurement noise (Westwick and Kearney 1996b). Finally, the nonlinear methods return multiple kernels which have proven difficult to interpret in terms of mechanisms except in a relatively few cases. Recent advances in nonlinear system identification theory, however, may result in the

more general application of these techniques (see "Future directions").

3.1.4 Multiinput Applications

Most postural tasks and movements involve coordinated activity between muscles acting at and across several joints; inertial properties further couple the mechanics of different joints. System identification has been applied with somewhat less frequency to these more complex systems. Several studies (Mussa-Ivaldi et al. 1985; Lacquaniti et al. 1993; Tsuji et al. 1995; Gomi et al. 1996; McIntyre et al. 1996) have used various types of system identification procedures to characterize the multijoint stiffness of the human arm, and several additional studies have examined intermuscle reflex EMG properties (Soechting et al 1981; Lacquaniti et al. 1991). All of the studies of limb stiffness, however, have used rather simple "system identification" procedures which either completely ignore dynamic properties or which assume a simple model structure a priori.

More general applications of system identification to such large-scale systems have been limited to some extent by the typical desire for the model structure to correspond with the underlying "physical structures." The use of system identification techniques in conjunction with such structurally based models will be discussed in more detail in the following section. Perhaps more importantly, applications of system identification to more complex systems has been limited by the lack of appropriate multiinput, multioutput system identification procedures, as well as experimental devices capable of providing the required multidimensional input perturbations. However, Tsuji et al. (1995) recently extended the earlier work of Mussa-Ivaldi et al. (1985) to include dynamic properties, and Gomi et al. (1996) have examined the time variations in human arm endpoint-stiffness during movement. Other work (Acosta and Kirsch 1996a,b; Kirsch et al. 1996) has begun to address deficiencies in multijoint identification procedures by developing a powerful, wide bandwidth manipulator capable of imposing adequate perturbations onto the human arm and implementing multiinput, multi-output (MIMO) identification procedures capable of characterizing linear multiinput systems (such as the multijoint stiffness of the human arm) where the inputs are not completely independent. Thus, although the application of multiinput identification procedures has been sparse, recent developments indicate that such applications will grow significantly in the coming years (see "Future directions").

3.1.5 Control

External systems are increasingly being used to control portions of the neuromuscular system which have been damaged by injury or disease. One particularly relevant example of such external control is the use of functional neuromuscular stimulation to activate paralyzed muscle and thus to restore movement to disabled individuals (Crago et al. 1990; Peckham and Keith 1992; see also Section 9). In such applications, there is usually a need for input–output descriptions of system elements (e.g., muscle force generation) that are efficient (so that real-time control algorithms can utilize the model), accurate, and formulated in discrete rather than continuous time (again because the controllers are typically discrete-time in nature). Although discrete approximations to more mechanism-based models [e.g., Shue et al. 1995] have been utilized, there usually is no need for the model to have any mechanistic underpinning. Thus, parametric approaches such as time-series models can also be used (Chizeck et al. 1991). Nonlinearities and time-varying behavior can be dealt with conveniently through the use of techniques such as linearization and adaptive control, which again substitute purely mathematical descriptions with good predictive value for a more general description of the complex underlying mechanisms.

3.2 Assisting the Modeling of Large-Scale Systems

A very active area in biomechanics and motor control (as witnessed by the various chapters devoted to this topic in this volume, e.g., Section 7) is the development of large-scale models of the human body, in particular of the upper and lower extremities. Such large-scale models typically combine models of the individual elements (e.g., moments of inertia of different limb segments, muscle contractile properties) into an overall model whose objective is to provide insight into the neural (or artificial) control of the limb and/or to predict how changes in the system (e.g., surgical procedures

such as muscle tendon transfer, functional neuromuscular stimulation) might improve performance. Although the potential impact of these modeling efforts is immense, their utility to date has been rather modest. One reason for this has been the difficulty in dealing with the complexity of these models. To reasonably reflect the actual system, many different elements must be included, models must be formulated for each element, and values obtained for the model parameters. To reduce model complexity to a manageable level, simple models of some of the elements are typically adopted. For example, the Hill model of muscle contractile properties is almost universally used in musculoskeletal models of the limbs. Even so, the models are usually left with large numbers of unknown parameters which must be determined in some way. Modelers are not always experimenters as well, so the standard approach to parameterizing these complex models has been to use generic values from the literature, perhaps with some scaling procedure to adjust for individual size differences. Parameters for which no reasonable estimate is available, or perhaps parameters of importance for a particular hypothesis, may be estimated using optimization techniques which determine the parameters values which most closely generate some desirable output.

We believe that system identification can make several important contributions to the development of these models, but that the use of these powerful techniques has been extremely underutilized. System identification has the ability to model individual elements more accurately than the grossly simplified models often employed, and model parameters can be determined by fitting directly to experimentally obtained input–output relationships rather than relying on an assortment of literature-based values. Furthermore, system identification can be used to validate overall model structure and indicate the conditions under which the model should be tested to obtain accurate parameter values. These potential contributions will be described in more detail in the following sections.

3.2.1 Models of Individual Elements and Parameterization

Large-scale models are typically constructed by combining models of the individual components of the system, see Chapter 8. For example, a neuromusculoskeletal model of a human limb may contain models of skeletal geometry, muscle attachment sites, muscle contractile properties, sensory receptor properties, neural processing schemes, and so on. The usual objective for a large-scale model is to use it in simulation to provide increased insight into the behavior of the system *as a whole*; thus, the primary requirements for the individual element models are accuracy and reasonable simplicity. In some cases, a "black-box" model of individual system elements could be used directly in the large-scale model. In many cases, however, these black-box models do not have a sufficient range of validity to be used directly. As described above, however, these input–output descriptions often provide enough insight so that a simpler model can be developed to replace the more cumbersome black-box description. Such a parametric model can be obtained from one or several input–output descriptions obtained under different conditions, and the model can be based on a knowledge of underlying mechanisms (e.g., the Huxley (1957) or Zahalak (1981) muscle models) or they can be basically curve fits (e.g., a series of second order systems, Hill-based muscle models, etc.). The result in either case is that fitting a model to an input–output description arising from system identification would provide model parameters that would otherwise be specified generically or left to optimization.

Even for relatively straightforward components of a large-scale model, the use of system identification in combination with modeling has several advantages. As described above, system identification theory is well developed in several key areas relevant to modeling. The effects of measurement noise on the system identification results can be anticipated and the properties of the input perturbation required to optimally characterize the dynamic properties of the system can usually be determined. Furthermore, the use of system identification to estimate model parameters may in many cases be simpler and more accurate than more conventional approaches to parameterization. For example, model-based controllers for multijoint robots rely upon accurate models of the kinematic and inertial parameters of the robotic arm segments; these parameters could presumably be determined readily from the dimensions and material properties of the links. However, it has been suggested (An et al.

1988) that accurate values can be obtained more easily using system identification procedures to perform "kinematic calibrations" (measuring joint angles and endpoint Cartesian positions and finding the unknown link lengths by solving the kinematic equations) and "dynamic calibrations" (measuring joint angles, velocities, accelerations, and moments and solving for the unknown link masses and moments of inertia). These same requirements will obviously exist in the development of human limb models, but the determination of inertial properties from limb dimensions is even more approximate. Similarly, the use of system identification techniques to estimate other parameters may also be valuable in efforts to personalize musculoskeletal models for individual subjects.

The use of system identification to parameterize different model components is clearly not a solution to all problems. For example, it will be impossible to use system identification to obtain models of all of the individual muscles in a limb model because neither the individual inputs (activation and muscle length) nor the output (muscle force) are typically available for measurement. However, we believe that this use of system identification has significant unrealized potential in the modeling of individual components of large-scale neuromuscular models.

3.2.2 Model Validation

Large-scale neuromuscular models are typically developed using data from the literature to specify both the structure of individual model components and the parameter values. Simulations are then performed using the model to investigate some theory or to replicate some classical experiment. However, it is highly unusual for the model to be validated over a range of conditions. Often, very rough qualitative agreement with experimental data is taken as acceptable (e.g., a lower extremity model can be made to "walk" without falling), even if the manner in which this is achieved (muscle forces, joint angle trajectories) exhibits large errors. Furthermore, there is generally no attempt to demonstrate that the test conditions used to validate these models are appropriate, and the conditions examined tend to lack the characteristics defined by system identification theory as being necessary to identify a model.

As with model parameterization, system identification cannot provide solutions to all these problems. However, system identification can contribute to the validation of complex large-scale models in at least two significant ways. First, the theories of input design from system identification discussed above could be used determine optimal test conditions for evaluating complex neuromuscular models and for demonstrating that the model structure is indeed capable of being validated from the available experimental data. Second, complex neuromuscular models could be required to predict input–output relationships obtained using system identification. Single input–output relationships are typically valid over only small operating point ranges, but systems tend to be tested over a range of conditions that span the functional range, and the characterized properties (e.g., stiffness) are usually functionally important. Although large-scale models are intended to be valid over a much larger range, they should *at least* be capable of predicting experimentally obtained input–output relationships for important conditions. The needed input–output data will become increasingly more detailed and complete as the use of system identification on the same complex systems which are typically described by large-scale, structurally-based models (e.g., the stiffness properties of the human arm) becomes more widespread.

Again, although we believe that system identification has an role to play in the validation of complex neuromuscular models, many important aspects of neuromuscular function cannot be conveniently described by experimentally measured input–output relationships. Furthermore, other important tools such as sensitivity analysis (also currently under-utilized) are also critical for model development and validation. However, system identification provides an avenue for at least validating certain functionally important features of system performance (such as joint and limb stiffness) for which experimental data already exists.

4   Future Directions

The preceding sections have described a number of different system identification techniques and how these techniques could be used, in principle, to aid the development and validation of complex neuro-

muscular models. This final section will describe several new applications of system identification techniques that we believe will eventually have a significant impact both on fundamental understanding of the neuromuscular system and on the development of accurate large-scale neuro-musculoskeletal models.

4.1   Validation and Parameterization of Complex Neuromusculoskeletal Models

The previous section described a number a ways in which system identification could be used to aid the development of complex models of human limbs and their neural control. Although these methods have rarely been used in the past, we believe that pressures to increase accuracy, to incorporate dynamic properties, and to personalize models for individual subjects will lead to the increased use of experimental data and system identification procedures to model individual elements within the overall model and to obtain parameters for the models. We also believe that the complex models should be required, as a minimum, to predict the input–output relationships obtained for specific experimental conditions using system identification. New and/or improved techniques for characterizing nonlinear, time-varying, and multiinput systems can all contribute to the improvement of the large-scale models.

4.2   Advanced Nonlinear Identification Techniques

As described above, nonlinear system identification techniques have been used sporadically to describe neuromuscular systems in the past. We expect this to change in the near future because recent developments have made available nonlinear identification methods which are robust in the presence of noise, work with nonwhite inputs, and yield models with as few parameters as possible (Korenberg 1991; Westwick and Kearney 1996b; Marmarelis 1993). Obvious applications of these nonlinear techniques will be to reexamine systems previous characterized by linear techniques for a range of fixed operating points or using linear time-varying approaches, but in this case determining a *single* nonlinear dynamic description that is valid for all relevant conditions. Given the general nature of these techniques, however, we expect their applicability to expand to a wider range of neuro-muscular problems.

4.3 Multijoint Stiffness Dynamics

As described above, the use of multi-input, multi-output descriptions of systems such as the stiffness properties of the human arm is already expanding (Mussa-Ivaldi et. al. 1985, Tsuji et al. 1995, Gomi et al. 1996, Acosta and Kirsch 1996a,b; Kirsch et al. 1996), but we believe that these applications will both expand and improve to include full dynamic properties (i.e., more than just steady-state stiffness), time-varying properties, and nonlinear descriptions. Such applications will enhance the understanding of dynamic properties (e.g., limb inertia and muscle-based damping) in the control of whole-limb posture and movement, and will allow the characterization of intermuscle and interjoint reflex action in intact human subjects.

4.4 Mechanical Effects of Reflex Action

The role of reflex pathways in the control of posture and movement has been a controversial topic for more than 30 years, and there still is no general consensus. Reflex pathways are often included in neuromusculoskeletal models, but the critical "gain" factor associated with each of these pathways is usually not measured directly, but rather is deduced from optimization procedures. Recently, a technique for extracting the reflex contribution to intact human ankle joint stiffness has been developed (Kearney et al. 1996). This or similar techniques can be applied to a variety of human joint or whole-limb systems to provide critical information regarding the mechanical impact of reflex action in intact human systems under a range of relevant conditions. This will allow a more informed decision regarding the role of reflex action to be made.

Acknowledgments. Much of the work of REK on quasistatic, time-varying, and nonlinear identification has been supported by the Canadian MRC and NSERC. Recent multijoint system identification development has been supported by a Whitaker Foundation grant to RFK. The authors would like to thank Eric Perreault for reviewing an earlier draft of this chapter.

References

Acosta, A.M. and Kirsch, R.F. (1996a). Endpoint stiffness estimation for assessing arm stability. *1st Ann. Conf. Int. Funct. Elec. Stim. Soc.*, 1:60.

Acosta, A.M. and Kirsch, R.F. (1996b). A planar manipulator for the study of multi-joint human arm posture and movement control. *9th Eng. Found. Conf. Biomech. Neur. Cont. Mov.*, 9:1–2.

Agarwal, G.C. and Gottlieb, G.L. (1977). Compliance of the human ankle joint. *Trans. ASME*, 99:166–170.

Akaike, H. (1994). A new look at the statistical model identification. *IEEE Trans. Auto. Cont.*, 19:716–23.

An, C.H., Atkeson, C.G., and Hollerbach, J.M. (1988). *Model-Based Control of a Robot Manipulator.* The MIT Press, Cambridge, Massachusetts.

Barden, J.A. (1981). Estimate of rate constants of muscle crossbridge turnover based on dynamic mechanical measurements. *Physiol. Chem. and Phys.*, 13:211–219.

Bendat, J.S. and Piersol, A.G. (1986). *Random Data: Analysis and Measurement Procedures (Second edition).* Wiley-Interscience, New York.

Bennett, D.J. (1993). Torques generated at the human elbow joint in response to constant position errors imposed during voluntary movements. *Exp. Brain Res.*, 95:488–498.

Bennett, D.J., Hollerbach, J.M., Xu, Y., and Hunter, I.W. (1992). Time-varying stiffness of human elbow joint during cyclic voluntary movement. *Exp. Brain Res.*, 88:433–442.

Bizzi, E., Polit, A., and Morasso, P. (1976). Mechanisms underlying achievement of final head position. *J. Neurophysiol.*, 39:435–444.

Boyd, S. and Chua, L.O. (1985). Fading memory and the problem of approximating nonlinear operators with Volterra series. *IEEE Trans. Cir. Sys.*, CAS-32:1150–1161.

Calancie, B. and Stein, R.B. (1987). Measurement of rate constants for the contractile cycle of intact mammalian muscle fibers. *Biophys. J.*, 51:149–159.

Chen, S. and Billings, S.A. (1989). Representation of non-linear systems: the NARMAX model. *Int. J. Cont.*, 49:1013–1032.

Chizeck, H.J., Lan, N., Streeter-Palmieri, L., and Crago, P.E. (1991). Feedback control of electrically stimulated muscle using simultaneous pulse width and stimulus period modulation. *IEEE Trans. Biomed. Eng.*, 38:1224–1234.

Cooker, H.S., Larson, C.R., and Luschei, E.S. (1980). Evidence that the human jaw stretch reflex increases the resistance of the mandible to small displacements. *J. Physiol.*, 308:61–78.

Crago, P., Lemay, M., and Liu, L. (1990). External control of limb movements involving environmental interactions. In *Multiple Muscle Systems: Biomechanics and Movement Organization*. Winters, J. and Woo, S.-Y. (eds.), pp. 343–359. Springer-Verlag, New York.

Evans, C.M., Fellows, S.J., Rack, P.M.H., Ross, H.F., and Walters, D.K.W. (1983). Response of the normal human ankle joint to imposed sinusoidal movements. *J. Physiol.*, 344:483–502

Gomi, H. and Kawato, M. (1996). Equilibrium-point control hypothesis examined by measured arm stiffness during multi-joint movement. *Science*, 272:117–120.

Hunter, I.W. (1986). Experimental comparison of Wiener and Hammerstein cascade models of frog muscle fiber mechanics. *Biophys. J.*, 49:81a.

Hunter, I.W. and Kearney, R.E. (1982a). Dynamics of human ankle stiffness: variation with mean ankle torque. *J. Biomech.*, 15:747–752.

Hunter, I.W. and Kearney, R.E. (1982b). Two-sided linear filter identification. *Med. Biol. Eng. Comput.*, 21:203–209.

Hunter, I.W. and Kearney, R.E. (1983). Invariance of ankle dynamic stiffness during fatiguing muscle contractions. *J. Biomech.*, 16:985–991.

Huxley, A.F. (1957). Muscle structure and theories of contraction. *Prog. Biophys. Biophys. Chem.*, 7:255–318.

Joyce, G.C., Rack, P.M.H., and Ross, H.F. (1974). The forces generated at the human elbow joint in response to imposed sinusoidal movements of the forearm. *J. Physiol.*, 240:351–374.

Kawai, M. and Schachat, F. H. (1984). Differences in the transient response of fast and slow skeletal muscle fibers. *Biophys. J.*, 45:1145–1151.

Kearney, R.E. and Hunter, I.W. (1982). Dynamics of human ankle stiffness: variation with displacement amplitude. *J. Biomech.*, 15:753–756.

Kearney, R.E. and Hunter, I.W. (1983). System identification of human triceps surae stretch reflex dynamics. *Exp. Brain Res.*, 51:117–127.

Kearney, R.E. and Hunter, I.W. (1984). System identification of human stretch reflex dynamics: tibialis anterior. *Exp. Brain Res.*, 56:117–127.

Kearney, R.E. and Hunter, I.W. (1988). Nonlinear identification of stretch reflex dynamics. *Ann. Biomed. Eng.*, 16:79–94.

Kearney, R.E. and Hunter, I.W. (1990). System identification of human joint dynamics. *Crit. Rev. Biomed. Eng.*, 18:55–87.

Kearney, R.E., Stein, R.B., and Parameswaran, L. (1996). Identification of intrinsic and reflex contributions to human ankle stiffness dynamics. *IEEE Trans. Biomed. Eng.*, (in Press).

Kirsch, R. and Kearney, R. (1993). Identification of time-varying dynamics of the human triceps surae stretch reflex: II. Rapid imposed movement. *Exp. Brain Res.*, 97:128–138.

Kirsch, R. and Rymer, W. (1992). Neural compensation for fatigue induced changes in muscle stiffness during perturbations of elbow angle in man. *J. Neurophysiol.*, 68:449–470.

Kirsch, R., Boskov, D., and Rymer, W. (1994). Muscle stiffness during transient and continuous movements of cat muscle: perturbation characteristics and physiological relevance. *IEEE Trans. Biomed. Eng.*, 41:700–758.

Kirsch, R., Kearney, R., and MacNeil, J. (1993). Identification of time-varying dynamics of the human triceps surae stretch reflex: I. Rapid isometric contraction. *Exp. Brain Res.*, 97:115–127.

Kirsch, R.F., Perreault, E.J., and Acosta, A.M. (1996) Identification of multi-input dynamic systems: limb stiffness dynamics. *9th Eng. Found. Conf. Biomech. Neur. Cont. Mov.*, 9:46–47.

Kirsch, R.E. and Kearney, R.E. (1996). Identification of time-varying stiffness dynamics of the human ankle during an imposed movement. *Exp. Brain Res.*, (in Press).

Korenberg, M. and Hunter, I. (1990). The identification of nonlinear biological systems: Wiener kernel approaches. *Ann. Biomed. Eng.*, 18:629–54.

Korenberg, M.J. (1991). Parallel cascade identification and kernel estimation for nonlinear systems. *Ann. Biomed. Eng.*, 19:429–455.

Korenberg, M.J. and Hunter, I.W. (1986). The identification of nonlinear biological systems: LNL cascade models. *Biol. Cybern.*, 55:125–134.

Kröller, J., Grüsser, O.-J., and Weiss, L.-R. (1985). The response of primary muscle spindle endings to random muscle stretch: a quantitative analysis. *Exp. Brain Res.*, 61:1–10.

Kukreja, S.L., Kearney, R.E., and L., G.H. (1996). Estimation of continuous-time models from sampled data via the bilinear transform. *Proc. 18th Annu. Int. Conf. IEEE Eng. Med. Biol. Soc.*, Amsterdam, The Netherlands.

Lacquaniti, F., Borghese, N.A., and Carrozzo, M. (1991) Transient reversal of the stretch reflex in human arm muscles. *J. Neurophysiol.*, 66:939–954.

Lacquaniti, F., Carrozzo, M., and Borghese, N.A. (1993). Time-varying mechanical behavior of multi-jointed arm in man. *J. Neurophysiol.*, 69:1443–1464.

Lakie, M., Walsh, E.G., and Wright, G.W. (1984). Resonance at the wrist demonstrated by the use of a torque motor an instrumental analysis of muscle tone in man. *J. Physiol.*, 353:265–285.

Ljung, L. (1987). *System Identification for the User*. Englewood Cliffs. Prentice-Hall, New Jersey.

MacNeil, J.B., Kearney, R.E., and Hunter, I.W. (1992). Identification of time-varying biological systems from ensemble data. *IEEE Trans. Biomed. Eng.*, 39:1213–1225.

Marmarelis, P.Z. and Marmarelis, V.A. (1978). *Analysis of Physiological Systems: The White Noise Approach.* Plenum Press, New York.

Marmarelis, V.Z. (1989a). Signal transformation and coding in neural systems. *IEEE Trans. Biomed. Eng.*, 36:15–24.

Marmarelis, V.Z. (1989b). Volterra-Wiener analysis of a class of nonlinear feedback systems and application to sensory biosystems. In *Advanced Methods of Physiological Systems Modeling*. Marmarelis, V.Z. (ed.), pp. 302, Plenum Press, New York.

Marmarelis, V.Z. (1993). Identification of nonlinear biological systems using laguerre expansions of kernels. *Ann.' Biomed. Eng.*, 21:573–589.

McIntyre, J., Mussa-Ivaldi, F., and Bizzi, E. (1996). The control of stable arm postures in the multi-joint arm. *Exp. Brain Res.*, 110:248–264.

Mussa-Ivaldi, F.A., Hogan, N., and Bizzi, E. (1985). Neural, mechanical and geometric factors subserving arm position in humans. *J. Neurosci.*, 5:2732–2743.

Palm, G. (1979). On representation and approximation of nonlinear systems. *Biol. Cybern.*, 34:49–52

Peckham, P.H. and Keith, W. (1992). Motor prostheses for restoration of upper extremity function. In *Neural Prostheses: Replacing Motor Function After Disease or Injury*. Stein, R.B., Peckham, P.H., and Popovic, D.P. (eds.), Oxford University Press, New York.

Poppele, R.E. (1981). An analysis of muscle spindle behavior using randomly applied stretches. *Neuroscience*, 6:1157–1165.

Rissanen, J. (1978). Modelling by shortest data description. *Automatica*, 14:465–471.

Robinson, C.J., Flaherty, B., Fehr, L., Agarwal, G.C., Harris, G.F., and Gottlieb, G.L. (1994). Biomechanical and reflex responses to joint perturbations during electrical stimulation of muscle: instrumentation and measurement techniques. *Med. Biol. Eng. Comput.*, 32:261–272.

Sakai, H.M. (1992). White-noise analysis in neurophysiology. *Physiol. Rev.*, 72:491–505.

Shue, G., Crago, P.E., and Chizeck, H.J. (1995). Muscle-joint models incorporating activation dynamics, moment-angle, and moment-velocity properties. *IEEE Trans. Biomed. Eng.*, 42:212–223.

Soechting, J., Dufresne, J., and Lacquaniti, F. (1981). Time-varying properties of myotatic response in man during some simple motor tasks. *J. Neurophysiol.*, 46:1226–1243.

Stein, R.B., Rolf, R., and Calancie, B. (1986). Improved methods for studying the mechanical properties of biological systems with random length changes. *Med. Biol. Eng. Comput.*, 24:292–300.

Tai, C. and Robinson, C.J. (1995). Variation of human knee stiffness with angular perturbation intensity. *Proc. 17th Annu. Int. Conf. IEEE Eng. Med. Biol. Soc.*, Montreal, Canada.

Tsuji, T., Morasso, P., Goto, K., and Ito, K. (1995). Hand impedance characteristics during maintained posture. *Biol. Cybern.*, 2:475–485.

Verhaegen, M. and DeWilde, P. (1992). Subspace model identification part 1. The output-error state-space model identification class of algorithms. *Int. J. Cont.*, 56:1187–1210.

Viviani, P. and Berthoz, A. (1975). Dynamics of the head-neck system in response to small perturbations: analysis and modeling in the frequency domain. *Biol. Cybern.*, 19:19–37.

Volterra, V. (1959). Theory of functionals and of integral and integro-differential equations. Dover, New York.

Weiss, P.L., Hunter, I.W., and Kearney, R. (1988). Human ankle joint stiffness over the full range of muscle activation levels. *J. Biomech.*, 21:539–544.

Weiss, P.L., Kearney, R.E., and Hunter, I.W. (1986a). Position dependence of ankle joint dynamics: passive mechanics. *J. Biomech.*, 19:727–735.

Weiss, P.L., Kearney, R.E., and Hunter, I.W. (1986b). Position dependence of ankle joint dynamics: active mechanics. *J. Biomech.*, 19:737–751.

Westwick, D.T. (1996). Methods for the identification of multiple-input nonlinear systems. Ph.D. Thesis. *Depart. Elec. Eng. Biomed. Eng.*, McGill University, Montreal.

Westwick, D.T. and Kearney, R.E. (1996a). Identification of physiological systems: a robust method for nonparametric impulse response estimation. *Med. Biol. Eng. Comput.*, (in press).

Westwick, D.T. and Kearney, R. E. (1996b). Robust Nonlinear System Identification Using Band-Limited Inputs. *Ann. Biomed. Eng.* (in press).

Wiener, N. (1958). *Nonlinear Problems in Random Theory.* John Wiley & Sons, New York.

Zahalak, G.I. (1981). A distribution-moment approximation for kinetic theories of muscular contraction. *Math. Biosci.*, 55:89–114.

Zahalak, G.I. and Heyman, S.J. (1979). A quantitative evaluation of the frequency-response characteristics of active human skeletal muscle in vivo. *Trans. ASME*, 101:28–37.

10
A Reductionist Approach to Creating and Using Neuromusculoskeletal Models

Ian E. Brown and Gerald E. Loeb

1 Introduction

There are many possible approaches for developing models of physical systems. At one extreme exists the black-box model in which only the inputs and outputs of the system are considered important aspects of the model (see Chapter 9). Alternatively one can divide a system into separate components and model each component separately (see Chapter 8). The most obvious difference between these approaches is that there is more information in the latter model than just the inputs and outputs of the system. This latter approach is loosely termed reductionism—the form of the model for the system is 'reduced' into smaller components, each of which should have some testable relationship to a corresponding physical structure.

For the beginner modeler who is exploring various avenues of model development, the key question is which approach to use? The short answer is that the optimal approach will depend upon the desired use of the model. If all that is of concern are the inputs and outputs of the system, then the black-box approach (which is often easiest) is probably the most logical. On the other hand, if there is a desire to understand the relative importance of various internal components or to associate various emergent properties of the whole system with one or more of the internal components, then the reductionist approach has an obvious advantage. Furthermore, models often need to be extrapolated to predict outputs under conditions that lie outside those for which black-box data are available; reductionist models may be more robust in such applications.

There are two other potential advantages to using reductionist models. First, it is easier to design experiments for and to model small, simple components from a reduced system, than it is for large, complex systems. Second, reductionist models can often be produced more easily than black-box models when the goal is to create models of many similar systems. For example, consider two systems that are similar but not identical. The black-box approach would have to be fully validated twice, once for each system. Conversely, the reductionist approach would only need to revise the models for those components that are not identical.

Because this book is about neuromusculoskeletal (NMS) systems, the rest of this chapter will focus upon modeling the NMS system. We believe that the reductionist approach is useful for modeling this system for the reasons supplied above; the model that is presented here is in the form of a reductionist model. To demonstrate the advantages of this approach, published examples of observations that were understood or explained with the help of a reductionist model will be given. The chapter will then end with a detailed example of a previously unpublished study that could only be undertaken with the use of a reductionist model.

2 Reduction to Practice

The neuromusculoskeletal (NMS) system is particularly well suited for reductionist modeling. A simple examination of the NMS apparatus reveals several distinct types of components that occur more often than once. Although the individual components within various NMS systems may be similar, the relative numbers and proportions of these components may vary. Black-box modeling would require independent models for each and every system, whereas a well-designed reductionist model could reuse components from one model to the next.

A key issue in reductionist models is how much to reduce the system. For example, there are many useful approaches for reducing muscle: one can break it into tendon and fascicle; tendon, aponeurosis, and fascicle, or even a population of collagen fibers for each of tendon and aponeurosis and a population of cross-bridges, z-lines and myofilaments for each of the fibers within each fascicle. To resolve the issue of how much to reduce a system requires a precise goal (i.e., different studies with different goals will produce or use different degrees of reduction).

The model that we use and are continuing to build is based on the following goal: to be able to correctly predict for any NMS system operating under physiological conditions the resultant kinematics and kinetics for a given neural input and to be able to associate various emergent properties of the NMS system with specific anatomical structures. This is a rather lofty goal that is pursued by many laboratories, but by defining it for ourselves, the appropriate structure for our model becomes clearer.

Given the previously stated goal for our NMS model, it is obvious that we must use a reductionist approach for our design. By choosing components that are common to various NMS systems and modeling these components individually, every time that we are interested in a new NMS system, we are not required to start from the beginning. The question remains: how far do we keep reducing? The other consideration that helps define the answer to this question is the complexity of the model. If reducing a model offers no further insight (in the context of the stated goal) but increases the complexity, then it is not a useful step to take. Similarly, if reducing a given model further provides a marginal improvement of accuracy but a substantial increase in complexity, then it, too, is not a useful step to make.

As an example, the tendon is described as one component and is not modeled as a population of collagen fibers. A single component describing tendon as a nonlinear spring is far simpler then a population of collagen fibers, each of which is has its own complex properties. Introducing the population of collagen fibers does not produce any insight into how the whole system behaves under various conditions. On the other hand, researchers interested in understanding a nonlinear behaviour known to exist in the tendon (e.g., rupture) may want to model the tendon as a population of collagen fibers.

We divide our NMS model into three obvious subsystems: neural, muscular and skeletal (see Chapter 8 for a similar division into subsystems, and description of subsystem properties). Each of these large subsystems is then subdivided into useful smaller components. We provide examples from published studies to help demonstrate the usefulness of dividing the system as we have done.

2.1 The Skeletal System

The skeletal system refers to the skeleton itself. Physically this includes the bones and the joints connecting the bones. The mass associated with muscles, skin and other soft tissues has traditionally been lumped with the skeletal segments when analyzing NMS kinetics, although this may cause computational instabilities (He et al. 1991 and below) or even inaccuracies (Cavagna 1970). Because the forces produced by the muscular system act upon the skeletal system to produce torques around the various joints, the NMS model typically includes sufficient information about the position of the muscles and tendons to permit computation of muscle moment arms.

The components that we have chosen to use are linked segments and moment arms (see Figure 10.1). Mechanical segments are not necessarily identical to individual bones. For example, the shank (which includes two bones: the tibia and the fibula) could be considered a single segment. To be able to model external rotation of a single segment shank (which realistically entails a joint between the tibia and fibula) a rotational degree of freedom can be assigned to the knee itself without introducing significant error into the kinematics of muscles attaching on those bones. Segments have a length, a mass (and associated inertia) and are linked to other segments by joints with restricted

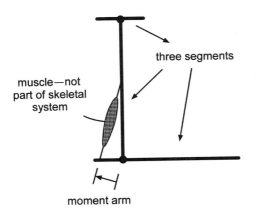

FIGURE 10.1. A sample musculoskeletal model of the arm to demonstrate the model components of the skeletal system. There are three segments and one moment arm in this example skeletal system.

degrees of freedom and ranges of motion. The choice of linkage usually reflects the purpose of the model as will be demonstrated in example A1. Conversely, moment arms are usually determined from the actual anatomy of the origin, insertion and tendon paths of muscles. As will be seen in example A2, however, it may be important, more efficient and more accurate to infer moment arms from kinematics instead of gross anatomy.

2.1.1 Example A1: Using an Appropriate Number of Segments

This example (from Zajac and Gordon 1989) compares two models of the body. The first model uses two segments to describe the body while the second model uses three segments. The question being asked was: what effect does activating soleus muscle (usually classified as an ankle extensor) have on the skeletal system?

If the body is modeled as two segments (Figure 10.2A), one for the foot and the other for the rest of the body, the equation relating angular acceleration around the ankle and net torque is:

$$\ddot{\theta} = \left(\frac{1}{\tilde{I}}\right) T^{NET} \qquad (10.1)$$

Clearly, when soleus is activated in this model (increasing the net torque), it extends the ankle joint.

If the body is modeled as three segments (Figure 10.2B), foot, shank, and rest of body, then the

equations become more complex (for a complete derivation, see Zajac and Gordon, [1989]):

$$\ddot{\theta}_1 = \frac{1}{\beta}\left\{\left[\frac{1}{\tilde{I}_1}\right]T_1^{NET} \right. $$
$$\left. - \left[\frac{\tilde{I}_{CS}\cos(\theta_1 - \theta_2)}{\tilde{I}_1\tilde{I}_2}\right]T_2^{NET}\right\} \qquad (10.2)$$

$$\ddot{\theta}_2 = \frac{1}{\beta}\left\{\left[\frac{1}{\tilde{I}_2}\right]T_2^{NET} \right. $$
$$\left. - \left[\frac{\tilde{I}_{CS}\cos(\theta_1 - \theta_2)}{\tilde{I}_1\tilde{I}_2}\right]T_1^{NET}\right\} \qquad (10.3)$$

where

$$\beta = [\tilde{I}_1\tilde{I}_2 - \tilde{I}_{CS}^2\cos^2(\theta_1 - \theta_2)]/(\tilde{I}_1\tilde{I}) > 0$$
$$\text{for all } (\theta_1 - \theta_2)$$

Although these equations are not trivial, they clearly show that both of the angular accelerations (knee and ankle) are dependent upon both of the net torques (about the knee and ankle). In particular, if soleus is activated to produce a torque around the ankle, this will cause extension in the ankle AND extension in the knee. The significance of these results is that a muscle that was typically clas-

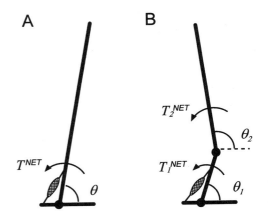

FIGURE 10.2. (A) A two-segment model of the body including the soleus muscle—one segment for the foot and the other segment for body. (B) A three-segment model of the body including the soleus muscle—one segment for the foot, one for the shank and one for the rest of the body. In both cases, each segment has a mass and associated inertia I. The net torques (T) are calculated from the torque due to soleus activation and the torque due to gravity. I_1 and I_2 refer to the inertias of the shank and body, respectively.

sified as an ankle extensor (because it only crosses the ankle joint), was shown to be able to produce knee extension in a particular circumstance. Only by incorporating an appropriate number of segments in the model was it possible to understand the system properly and consider more fully the possible implications of activating a particular muscle.

2.1.2 Example A2: Changing Moment Arms

When considering the role of a muscle, physiologists look at the muscle's moment arm and electromyogram (EMG) pattern during normal usage. Often the complexities of moment arm are not well understood or well modeled. The following example (from Young et al. 1993) shows how looking at muscle moment arms in three dimensions versus two dimensions can reveal large, previously unknown effects. The study by Young et al. (1993) also demonstrated that it is important to examine the full range of moment arms that a muscle might experience at different joint angles.

Eleven muscles that cross the ankle joint of the cat were examined by measuring length changes in response to angular motion of the ankle joint in all three planes. Most of these muscles have been classified according to their action in the parasagittal plane alone; the moment arms in the other planes were examined to see if this classification was appropriate. The results demonstrated that some of the muscles previously classified as ankle flexors/extensors had larger moment arms in the adduction/abduction axis (see Figure 10.3). The study

went further to look at joint angle effects on moment arm and discovered that some moment arms changed dramatically with joint angle (see Figure 10.3, abduction/adduction moment arm). These results are significant because they alter our understanding of what these muscles can actually do. Such "details" have been discovered to correlate with the natural patterns of recruitment of the muscles (Abraham and Loeb 1985; Loeb 1993) and with their patterns of spinal reflex connectivity (Bonasera and Nichols 1996).

Muscles that have a larger moment arm in nonparasagittal planes may be used for ankle flexion/extension, but cannot do so alone as their activation will also produce moments in other axes. Their primary usage may, in fact, not be as a flexor/extensor, but rather as an adductor/abductor. Furthermore, if a muscle has a very small moment arm when the ankle is in a neutral position, then the muscle is not particularly useful in moving the ankle out of neutral position. However, if the moment arm becomes larger as the joint angle moves away from the neutral position (as occurs for many of these ankle muscles), then the muscle can become very useful for returning the joint to neutral position (see Figure 10.4). Thus a muscle classified as an adductor may only be able to adduct so as long as the ankle is abducted. It may not be able to adduct the joint when it is in the neutral position or partially adducted. Only by examining the mo-

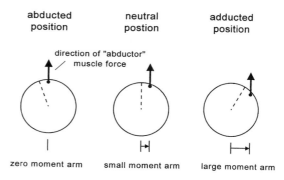

FIGURE 10.4. A schematic caudal view of the ankle with an "abductor" muscle. As the abduction/adduction angle changes (indicated by dashed lines), the moment arm produced by the muscle changes. While in the adducted position, the muscle has a large moment arm and would be a strong abductor. But as the angle changes towards abduction, the ability of the muscle to abduct decreases until the moment arm becomes zero.

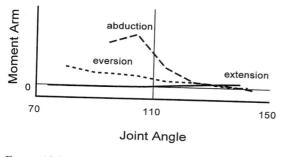

FIGURE 10.3. Moment arm versus joint angle relationship for peronius brevius. (Data originally published in Young et al. 1993.) The moment arms for the three different axes are plotted on the same figure for various joint angles (extension/flexion.)

ment arms thoroughly and individually can the potential usefulness of various muscles for certain tasks be understood.

2.2 The Muscular System

The muscular system includes individual muscles composed of tendon, aponeurosis and fascicles (bundles of fibers) as shown in Figure 10.5A. The major division into components for our model is along these anatomical lines as connective tissue and contractile tissue.

Connective tissue includes both the tendon and aponeurosis. These two tissues have nonlinear spring-like properties that are similar to each other (Scott and Loeb 1995). Because their properties are nearly identical, the two anatomically separate tissues are grouped together into a single component

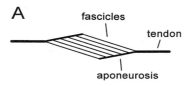

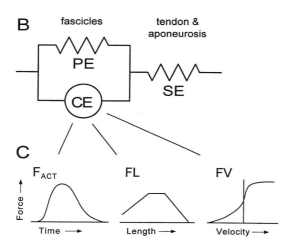

FIGURE 10.5. (A) Schematic figure of a typical muscle including the tendon, aponeurosis and fascicles. (B) Model components representing the muscle. The tendon and aponeurosis are combined to form the series elasticity (SE), while the fascicles are composed of the parallel elasticity (PE) and the contractile element (CE). (C) The CE can be subdivided into three components: activation (F_{ACT}), the force-length relationship (FL) and the force-velocity relationship (FV).

(series elastic element SE in Figure 10.5B). The fascicles are classified as contractile tissue. It is generally assumed for simplicity that fibers within a muscle are identical and act together so that a single component (the fascicles) may be used to represent them. However, there are often different populations of fiber types within a muscle. In these cases, separate fascicles with different properties representing the populations of fibers may be used in parallel with each other. The fascicles in a real muscle may actually consist of shorter muscle fibers in series as well as in parallel (Loeb et al. 1987); this may be important for issues of mechanical stability (Loeb and Richmond 1994) but can be ignored for most normal behaviors by using the physiological cross-sectional area of the muscle instead of the morphometry of individual muscle fibers.

The fascicles themselves can be broken into two smaller mechanical subcomponents (Figure 10.5B)—the passive elastic element (PE) and the active contractile element (CE). In the passive state (no neural activation) fascicles behave much like a nonlinear spring. This property can be modeled and included as a passive component of the fascicles. When fascicles are activated neurally, they produce (active) force in parallel with and adding to the passive force.

The active component can be divided even further into three smaller subcomponents (Figure 10.5C; see also Chapters 7, 8, 11, 28, 43). The force-length (FL) component arises because of the change in actin/myosin overlap as the length of fascicles (and hence sarcomeres) changes. As the overlap between the myofilaments changes, the number of cross-bridge sites available for force generation changes (see Chapter 2). The force-velocity (FV) component is thought to arise because of cross-bridge dynamics (Huxley 1957). As muscles change length, cross-bridges complete their cycles, detach and reattach. The dynamics of attachment and detachment thus effect the shape of the FV relationship (see review in Chapter 2). The activation component (F_{ACT}) is associated with calcium kinetics. F_{ACT} represents the percentage of myosin binding sites on the thin filament that are available for cross-bridge formation, which depends upon sarcoplasmic calcium concentration. F_{ACT} thus consists of a function of time that relates motoneuron activity to calcium release and re-uptake by the sarcoplasmic reticulum as well as sarcoplasmic calcium concentrations to myosin binding site availability.

The active component is divided in such a manner because historically it was observed that active force was affected by length, velocity and activation. The simplest approach to understanding these phenomena is thus to assume that they are independent and examine them one by one (i.e., hold two of them constant while varying one and measuring force). It has been through subsequent experiments that the underlying determinants for each of these factors were understood.

At this point, we should remind the readers of our goal: to be able to correctly predict for any NMS system under physiological conditions the resultant kinematics and kinetics for a given neural input and to be able to associate various emergent properties with anatomical structures. We are not interested in the biochemistry or molecular energetics involved in muscular contraction. Although those are both fascinating subjects, they require different models than the one presented here. Developing and testing theories about molecular energetics also may lead the experimenter to apply conditions of activation or kinematics that are unphysiological, such as tetanic stimulation or small, abrupt length changes. Models that account for the effects observed under such conditions may be excessively complex or unreliable when extrapolated to physiological conditions of muscle work. The following examples will address each of the levels of division that are used in our model

2.2.1 Example B1: Separating the Contractile Tissue from the Connective Tissue

As stated earlier, reducing a system increases the number of components and so tends to increase the complexity of the resultant model. However, not reducing a system sufficiently can result in difficulties in modeling the system accurately. Whether to separate the aponeurosis from the fascicles has been a key problem in recent years (Huijing and Ettema 1989). To address this issue Scott et al. (1996) recorded fascicle length and aponeurosis length during muscular activation of cat soleus to compare the effects of including the aponeurosis with separating it out.

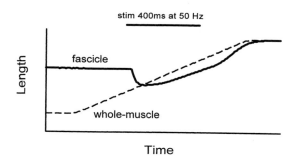

FIGURE 10.6. Simultaneous length records for fascicle and whole-muscle length during a whole-muscle stretch. When the muscle as maximally activated (indicated by solid bar above length records), the fascicles shorten even though the whole-muscle continues lengthening. (Reprinted with kind permission of Kluwer Publishers from Fig. 2a of Scott et al. 1996; © 1996 Chapman & Hall.)

A sample trial from Scott et al's (1996) study is shown in Figure 10.6 during which the length of the whole musculotendon and the length of the fascicles were recorded. While the whole musculotendon was stretched at a constant rate, when the muscle was activated the fascicles initially shortened. Because of the steep nature of the FV relationship, a model that did not separate the fascicles from the connective tissue would incorrectly predict the forces produced by the muscle during the initial stage of activation. That model would assume that the active component was stretching when in reality it was shortening, resulting in a large error for predicted force.

A second potential problem that develops if fascicle and connective tissue are not separated occurs when a model is intended to be generic. Garies et al. (1992) collected a series of FL curves for a variety of cat muscles (Figure 10.7) and used fascicle PLUS aponeurosis as the relevant length for scaling the active component. As can be seen clearly in their results, the FL curves are different for each muscle (similar results have been reported for rat muscles by Woitteiz et al. (1983)). For those of us trying to create a generic model that can describe all muscles, the conclusion from the these studies and that of Scott et al. (1996) is that the system has not been reduced enough if the aponeurosis is included with the fascicles—the fascicles and aponeuroses must be separated. In fact, the complexity of the overall system need not be increased by this reduction because

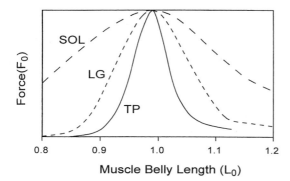

FIGURE 10.7. FL curves from three cat muscles—soleus (SOL), lateral gastrocnemius (LG) and tibialis posterior (TP). Force is normalized to maximal isometric force (F_0) and the muscle belly length at which that occurs (L_0). Muscle belly length is fascicle length PLUS aponeurosis length. (Data originally published in Baries et al. 1992.)

the aponeurosis tends to have mechanical properties that are similar to the tendon of the muscle (Scott and Loeb 1995); the two probably can be lumped into a single element in most muscles.

2.2.2 Example B2: Separating the Active and Passive Components of the Contractile Component

The most obvious reason for separating the active and passive components of the contractile component is that the active component scales with activation (because it is active) and the passive component does not. Unfortunately, in traditional models only part of the passive component is usually removed. The active and passive FL curves for a typical muscle are shown in Figure 10.8 as suggested by Brown et al. (1996a). PE2 is the well-known passive elasticity that can be observed simply by stretching a muscle and recording tension. PE1 is harder to observe, as it can only be seen when there is active force to counteract its action as a compression spring within the sarcomeres. PE1 exists because the myosin filament is stiff and resists compression. Normal passive shortening of a muscle to very short lengths does not reveal this phenomenon because the muscle buckles. Only by activating the muscle does PE1 become obvious.

The fact that myosin compression affects the FL curve was first suggested by Gordon et al.

(1966) as an explanation for the steep portion of the ascending limb of the FL curve, but only recently has there been evidence to support it. This same evidence reveals that treating PE1 as it should be (i.e., passive) results in a simpler model. Scott et al. (1996) recorded FV curves from soleus muscle at various lengths (during maximal activation). Given our presumption of FL, FV, and F_{ACT} independence, FV curves scaled to the isometric force should all produce congruent curves. FV data (recorded at different lengths during shortening) scaled to isometric WITHOUT accounting for PE1 is shown in Figure 10.9A. The same data scaled to isometric WITH accounting for PE1 is shown in Figure 10.9B, in which the shortening half of the FV curve is indeed independent of length. A model that omitted element PE1 would actually need a considerably more complex FV component with a length-dependent term to account for these data.

A second reason for separating the active and passive FL curves is that they do not scale similarly between muscles. The active FL curve is normally scaled by L_0. Brown et al. (1996b) compared passive FL curves (PE2) from various parallel-fibered, strap-like muscles in the cat hind limb and demonstrated that different muscles have different

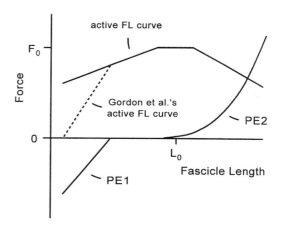

FIGURE 10.8. The shapes of the active and passive FL curves are shown here. Note the difference between Gordon et al.'s (1966) FL curve with the steep portion (dashed line) and our version (Brown et al. 1996a). The difference between the two curves is our recognition that PE1 exists as a separate component. PE1 resists compression and so produces 'negative' force. PE2 represents the well-recognized, spring-like properties the passive muscles exhibited when stretched.

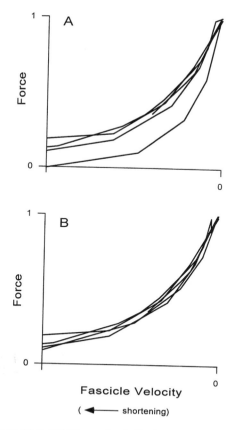

FIGURE 10.9. (A) Shortening half of FV curves from one muscle collected at different lengths without PE1 accounted for. (B) Shortening half of FV curves from one muscle collected at different lengths with PE1 accounted for—note the difference in congruity between A and B. (Data originally published in Scott et al. 1996.)

2.2.3 Example B3: Independence Versus Interdependence of the FL and FV Relationships

As stated earlier, a common assumption among muscle models is the independence of the FL and FV relationships. This obviously makes the model easier to design as the two factors can be studied and modeled separately. But how realistic an assumption is it? Scott et al. (1996) designed an experiment to answer this question.

The cat soleus muscle was activated maximally over a range of velocities to produce a FV curve at a particular length. This paradigm was then repeated on the same muscle for other lengths and the resulting FV curves compared. As was shown earlier in example B2 (Figure 10.9), when the passive elements of the muscle are properly accounted for the shortening half of the FV relationship appears to be independent of length. However, the story is quite different in the lengthening half. Figure 10.11 shows the entire FV curve scaled to isometric (with the passive elements properly accounted for). Although the curves for the shortening half are congruent, the lengthening curves are not.

These results indicated that the common assumption about FL and FV independence is not entirely correct. Fortunately, these data can be described reasonably accurately by adding a length dependence to the equation describing the lengthening half of the FV curve (Brown et al. 1996a) without changing the overall form of the model. An interesting result of

passive FL curves. The range of passive FL curves from five different muscles are shown in Figure 10.10 with all forces normalized to physiological cross-sectional area and all lengths normalized to L_0. Although there is some consistency between muscles of one type from different animals, there is a significant degree of variability between different muscles (e.g., caudofemoralis and sartorius). Treating the passive and active FL curves similarly is thus not appropriate for a generic model. Although not shown here, the conclusion of Brown et al.'s (1996b) study was that the passive FL curves (PE2) should be normalized to L_{MAX} (maximal in situ length of the muscle) and not L_0, consistent with the suggestion that much of PE2 arises from extrasarcomeric connective tissue rather than the myofilaments themselves.

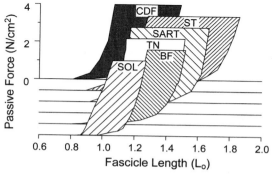

FIGURE 10.10. A large range of passive FL curves is shown from six different cat muscles. Note the large differences between muscles. N varies from 5–9 specimens for each muscle. (Reprinted with kind permission of Wiley-Liss, Inc., a subsidiary of John Wiley & Sons, Inc. from Fig. 3a of Brown et al. 1996b; © 1996 Wiley-Liss, Inc.)

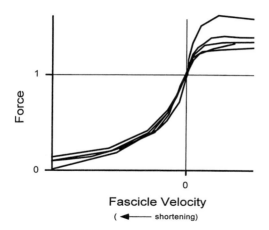

FIGURE 10.11. FV curves from one muscle collected at different fascicle lengths (from $0.67\ L_0$ to $1.0\ L_0$). Note that the shortening halves appear congruent, while there are some large differences on the lengthening half. (Data originally published in Scott et al. 1996.)

reducing the fascicle's active component into separate FL and FV components is that the observed extra length dependence upon force during active lengthening must be due to some as yet unknown property of cross-bridge dynamics—it is not simply a change in the filament overlap.

2.3   The Nervous System

The nervous system includes all neural circuitry, including both the central nervous system and the peripheral nervous system. In our models, the neural circuitry is divided into three components: Planner, Controller, and Regulator. These divisions are not based upon anatomy as was the case for both the skeletal and muscular systems, but instead are conceptual, relating to the computational properties of the circuitry and the available input–output signals at various levels of the nervous system.

We conceptualize our three divisions as follows. Planner makes the strategic decision about what is to be done (much as the general of an army does so). Controller identifies the appropriate tactics (much like a colonel or major). Regulator interprets the commands from the Controller and implements the proscribed tactics, taking local conditions into account (as a platoon commander would in the army). To provide some anatomical landmarks for the model, we might assign the premotor cortex and extrapyramidal structures to be the Planner, motor

cortex as the Controller, and spinal cord as the Regulator. Models of sensorimotor control based on such divisions are under development (Loeb et al. 1990; Loeb and Brown 1995).

3   Implications for Control

So far, we have considered some of the reductionist principles that could be used to guide the development of any model. We will now build a specific model according to these principles in order to answer two specific questions about the NMS, questions that could be addressed successfully only by such a reductionist model.

3.1   Objective

The objective of the following study (presented in preliminary form by Brown et al. 1995) was to answer the following two questions:

1. Are the "complex" intrinsic properties of muscle important in responding to perturbations?
2. How does co-activation of an "overcomplete" set of muscles modify these responses?

In the first question, the reference to "complex" intrinsic properties refers to the FL and FV curves. In large systems with many muscles, simplifying muscle properties is a very good way to simplify the overall system, but leads potentially to a loss in accuracy. How do systems containing muscles with real muscle properties compare to systems containing muscles without real muscle properties?

The second question makes a reference to "overcomplete" muscles. Overcompleteness, or redundancy, refers to the observation that for any given joint, there are usually many more muscles that cross that joint than are required to achieve independent control of each degree of freedom of the joint (e.g., a one degree-of-freedom hinge joint requires two and only two muscles operating as an antagonist pair). It also refers to the fact that there are biarticular muscles (muscles that cross two joints) that have actions that appear to be redundant with respect to existing mono-articular muscles that already cross the same joints.

3.2   The Model

We chose to examine a system's response to perturbations as our representative motor task. The model is shown pictorially in Figure 10.12A and is composed of the components listed below. Also shown in Figure 10.12A is part of a sample simulation. The task requires holding a weight in the hand (simulating a gun) with the forearm horizontal and the upper arm vertical. A perturbation (gun reaction force) is applied to the hand-held gun (circle) at the beginning of the simulation; the direction of which depends on the direction in which the gun is pointing (horizontal, 45° up or 45° down). The arm moves in response to the perturbation and the position of the hand is plotted every 10 ms. The arm position after 50 ms is shown to give a better idea graphically of what is happening. In this particular example, there is only one active muscle. The other five passive muscles are indicated by lines.

An important aspect of this model is that it is a musculoskeletal model. There is no nervous system attached. All muscle activations are held constant throughout various simulations meaning that this is a reflex-free model. The importance of this point will be made clear further on.

3.2.1 Model Components

• Three-segment robotic arm.
• Each segment is 45cm × 5cm, 2kg.
• Six massless actuators (muscles) representing the

mono- and biarticular muscles found in real musculoskeletal systems (Figure 10.12B).

• Muscle lengths at 90°; joint angles defined as 90% optimal ($0.9 \, L_0$).
• Optimal isometric muscle force (F_o) of 1000N for each muscle.
• A 1kg hand-held "gun" (shown as a circle).
• Gun reaction force of 2000N lasting 10ms (equivalent to 50g bullet fired at 400 m/s).
• Second version of model includes tendons.
• Fascicle length:tendon length = 1:2.
• Moment arm decreased to one third of no-tendon version to maintain relative fascicle range of motion.
• Optimal isometric muscle force increased to 3000 N for each muscle to compensate for reduced moment arm.
• Muscle mass included (0.5 kg for monoarticulars and 1.0 kg for biarticulars; needed for mechanical stability when a velocity-sensitive actuator operates in series with a spring; He et al. 1991].

3.3 The Simulation

The details of the various simulations are listed below, with the results of the simulations shown in Figure 10.13. The response of the arm to the force perturbation was tracked for 100 ms. This time interval was chosen because it represents the time during

A

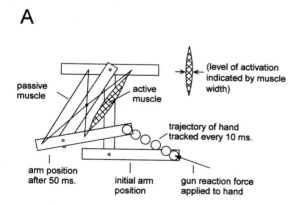

B

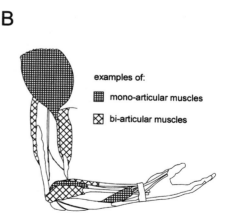

FIGURE 10.12. (A) The model is composed of three segments (two joints) and six muscles. Two of the muscles cross two joints (biarticulars) and four cross one joint (monoarticulars). Note how passive muscles are indicated with lines and active muscles with a thick muscle belly. The arms starts in the initial position with joint angles at 90°. The gun is fired which produces a gun re-

action force and a kickback on the hand (circle). Only the hand is plotted as the trajectory is tracked every 10 ms. This particular simulation ends at 50 ms with the complete final arm position shown. (B) This figure shows the human arm with various mono- and biarticular muscles indicated to demonstrate their existence.

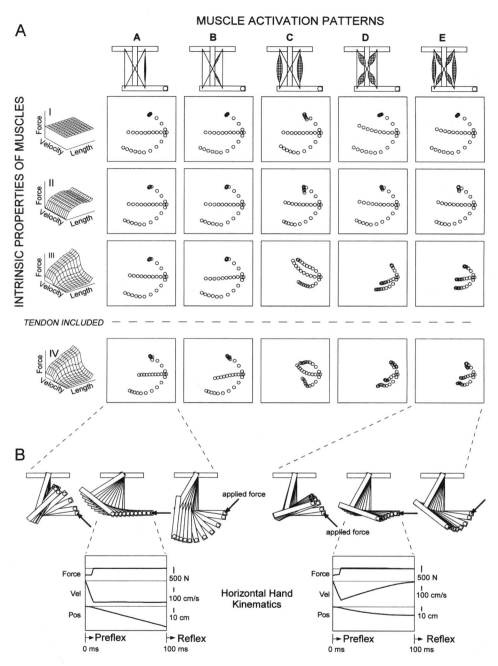

FIGURE 10.13. (A) This figure shows the results from sixty simulations. For each set of intrinsic muscle properties (I-IV) and activation patterns (A-E), three trajectories are plotted for the three different perturbation forces. To simplify the figure only the hand trajectory was plotted. The complete arm trajectories of six simulations (three from IV-A and three from IV-E are shown in part B to help clarify the figure. Each simulation lasted 100 ms with the hand plotted every 10 ms. (B) Six of the simulations from part A are shown with complete arm trajectories. The horizontal hand kinematics are shown for the simulations with the horizontal perturbation forces.

which one would expect an essentially reflex-free response (reflex loops + activation delays in human muscle ~ 100 ms). Three parameters were varied during various simulations: activation levels, intrinsic muscle properties and direction of perturbation.

All five activation patterns (A–E) used in the simulations were equivalent in the absence of a perturbation; the arm would remain stationary with enough net muscle torque to counteract gravity. Two of the activation patterns were equivalent 'minimal' activations: one for the biarticular flexor and one for the monoarticular flexors. The other three activation patterns used coactivation of various antagonist pairs. Each of these coactivation patterns activated the same total volume of muscle tissue.

The four sets of intrinsic muscle properties (I-IV) range from muscles with no length or velocity dependence and no tendons, to a realistic muscle with FL and FV properties plus a tendon. The one set that is not shown here but was also examined is the FV only set (no tendon). The results of that simulation were almost identical to that which had both FL and FV (no tendon).

Each of the above five activation patterns and four sets of muscle properties was examined in two ways. The first was to perturb the system with one of three different perturbations (equal magnitude, different direction) and track the response. The resulting hand trajectory is shown in Figure 10.13A with some detailed examples in Figure 10.13B. The second way in which each of these patterns and property sets was examined was by calculating the mechanical impedance at the initial position.

Mechanical impedance has three components: stiffness, viscosity, and inertia. We only calculated the stiffness and viscosity because for the majority of the simulations, there were no changes to the inertia (the lone exception being the no-tendon versus tendon version of the model). For a given stiffness measurement, the arm was displaced 1 cm from initial position in 24 evenly spaced directions. The resulting restoring forces were measured and plotted, creating the ellipses shown in Figure 10.14 (after the technique of Mussa-Ivaldi et al. (1985). Viscosity was measured in a similar manner by moving the arm through the initial position at 10 cm/s and recording the restoring force.

An important detail about the impedance measurements made on this model is that they are 100% reflex free. Although impedance tests on humans are sometimes attempted on a short enough time scale to avoid reflexes, they are often conducted over a longer time scale that results in reflex dependent impedances. Both types of impedance measurements have their virtues, but because they are different, conclusions based upon one or the other should reflect the differences inherent in the measurement technique.

3.3.1 Simulation Details

- Simulated on Working Model 3.0 with an integration time-step of 2.5 ms.
- Applied a constant level of muscle activation to counteract gravity.
- Applied a perturbation to the hand in the form of a gun reaction force.
- Tracked the resulting pre-reflex trajectory for 100 ms at 10 ms intervals.
- Five patterns of muscle activation (% activation indicated by width of muscles):
  1. Minimum biarticular (~5% activation).
  2. Minimum monoarticular (~5% activation).
  3. Biarticular coactivation (~70% activation).
  4. Monoarticular coactivation (~70% activation).
  5. Mono- and biarticular coactivation (~35% activation).
- Four sets of intrinsic properties of muscle (shown as force-length-velocity surfaces, equations and parameter values from Brown et al. 1996a):
  1. Constant force.
  2. Force-length relationship (active and passive).
  3. Force-length and force-velocity relationship (active and passive).
  4. Force-length and force-velocity relationship (active and passive)—tendon included.
- Three directions of gun reaction forces:
  1. 45° up.
  2. horizontal.
  3. 45° down.

3.4 Results and Discussion

3.4.1 Are the "Complete" Intrinsic Properties of Muscle Important in Responding to Perturbations?

The answer to this question can be seen clearly in Figure 10.13A. If we start with row I (flat intrinsic muscle properties), we see that as the level of

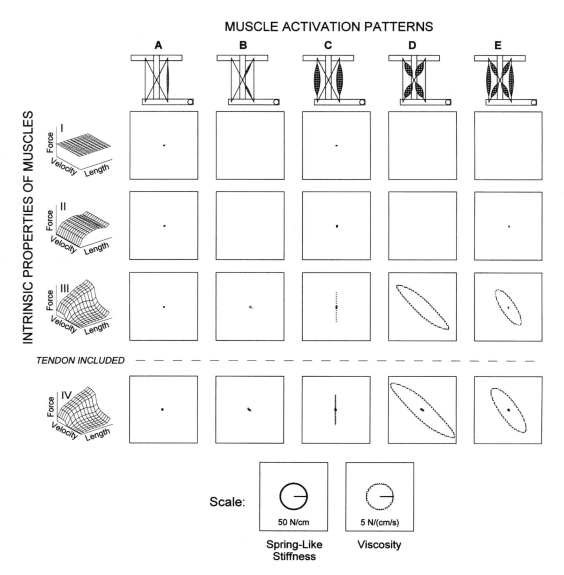

FIGURE 10.14. This figure shows the results of the imped-ance calculations for the four sets of intrinsic muscle prop-erties (I-IV) and five patterns of activation (A-E). Both stiff-ness and viscosity ellipses are shown. For a given set of parameters, if no ellipse is shown, that is because there was an unstable (negative) impedance. The scaling factors were chosen based on the expected extreme range of displace-ments and velocities (estimated as a 1:10 ratio) so that the resulting ellipses indicated the relative force response ex-pected from each component of impedance. Although not visible in this figure, the stiffness ellipses in rows III and IV had orientations parallel to those of the viscosity ellipses.

activation changes, there is no observable change in the arm's response to the perturbation. A com-parison with row II (FL properties) shows that the responses are almost identical to those in row I sug-gesting that the FL properties of muscle provide al-most no resistance to brief perturbations. We ini-tially found this result surprising given the large

amount of attention focused on the FL properties of muscle in various theories of motor control (e.g., equilibrium point hypothesis [Bizzi et al. 1992; Feldman and Levin 1995]). However, a close look at the FL curve reveals that the slopes in the re-gions where muscles tend to operate are not very steep.

If we look further in Figure 10.13A to row III (FL and FV properties) we finally see a change in the arm's response. As the level of activation increases above minimal, we see the arm slow and stop itself (compare the horizontal hand kinematics for IV-A and IV-E shown in Figure 10.13B). The significance of these results is that they occur in the absence of a nervous system. Row IV demonstrates that the addition of tendons and muscle masses does not significantly alter the results.

We can summarize our response to question (1) by saying that the FV properties of muscle provide a strong restoring force to perturbations, while the FL properties offer little resistance. These findings are supported by the impedance calculations (Figure 10.14) that show large viscosity impedances for rows III and IV under activation patterns C, D, and E. Note that the magnitude of the simulated impedances is of a similar magnitude to those estimated from human arm experiments (Tsuji et al. 1995). Probably the most significant aspect of these intrinsic responses is that they occur with zero time delay. Although a large perturbation such as the one in this model would likely elicit reflexes in an arm, the intrinsic properties of muscle can provide a useful response before those reflexes could effect any significant response. Because these intrinsic responses occur prereflex, we have coined the term "preflex" to describe them.

To avoid confusion, we will specifically define a preflex as: *the zero-delay, intrinsic response of a neuromusculoskeletal system to a perturbation.* Preflexes occur because of the intrinsic properties of muscle, therefore they are programmable, high-gain, and occur with zero time delay. Each and every simulation shown in Figure 10.13A has a preflex, only some of which appear to be useful for the goal of stabilizing the hand in this task (III-C, D, E and IV-C, D, E). Preflexes are not the same as reflexes nor are they a subset of reflexes.

Both preflexes and reflexes are under the control of the CNS, but via different mechanisms. The preflexes depend on the CNS's selection of a particular pattern of muscle activation to perform the nominal task from the infinitely many possible combinations created by the "overcompleteness" of the available muscles. To decide what constitutes a useful preflex, there has to be some set of expected perturbations. This set may be the Null set, in which case the gunman of our model would

probably choose activation pattern A or B to conserve energy. Alternatively, if the gunman in our model wanted to fire a gun horizontally while holding it underneath a table, he would set up activation pattern D or E instead of C so that his arm would always be deflected downward in response to the firing of his gun.

Reflexes may also be programmed to deal with expected perturbations by adjusting the bias on the various interneurons that control the gain between afferent input and motor output and the bias on the unrecruited motoneurons themselves. It is common to ascribe learned responses to such reflexes, particularly when they occur over a time-course where this is feasible, but motor psychologists are beginning to appreciate the importance of preflexes under such circumstances (Almeida et al. 1995).

3.4.2   How Does Coactivation of "Overcomplete" Muscles Modify These Responses?

This question too is answered clearly in Figure 10.13A and 10.13B. Simply, the muscles appear overcomplete only when the task is underspecified. In the absence of a perturbation, all of the muscle patterns are identical in function—they hold the arm stationary, counteracting the effects of gravity. Based upon this observation, one might suggest that the muscular system is thus 'over-complete' or redundant. However, when a perturbation is applied, there are a variety of responses from the various activation patterns. Depending upon the expected set of perturbations and the desired preflex, the actual choice of activation pattern becomes well defined. In order to define a task completely, tasks must be specified in terms of the performance criteria AND the expected set of perturbations.

4   Future Directions

The model as it has been presented is still incomplete. The largest problem is the one that was least talked about in this chapter—modeling the nervous system. The lack of adequate models reflects not any lack of research in this area, but rather demonstrates the complexities found in the nervous system. Although there have been many models in the past of one portion of the nervous system or another, we feel that future directions must account for the distinct properties and potential contributions of the various components of both the nervous system and the neuromusculoskeletal apparatus. The challenge is to create a model that is complex enough to replicate the interactive structure of the NMS system in such a way that the individual components are still recognizable and can be related to their structure and function as studied in isolation.

The model of muscle used in this study is still far from adequate to account for the full range of skeletal muscle properties that will affect the performance of NMS systems under at least some physiological conditions. This model was based on as complete a dataset as has yet been collected for the purpose of creating a model (Scott et al. 1996). Yet the dataset upon which the model was based did not describe submaximal activation, nor did it describe certain output properties that depend on activation history (e.g. yielding [Joyce, Rack, and Westbury 1969]). The muscle used in Scott et al.'s study (1996) was composed of 100% slow-twitch muscle fibers and, as yet, there is no good data-set on fast-twitch muscle fibers. The model also has no way to handle posttetanic potentiation, which is known to occur in fast-twitch muscle fibers. Experiments are under way to collect these data so that models can be built to describe NMS systems more realistically. Until this is done, our models remain incomplete in that they describe slow-twitch muscle during maximal activation—a poor representation of most muscles and most natural activities.

Acknowledgments. This work was supported the Medical Research Council of Canada. The authors would like to thank Tiina Liinamaa for helpful comments regarding the manuscript.

References

Abraham, L.D. and Loeb, G.E. (1985). The distal hindlimb musculature of the cat. Patterns of normal use. *Exp. Brain Res.* 58:580–593.

Almeida, G.L., Hong, D., Corcos, D., and Gottlieb, G.L. (1995). Organizing principles for voluntary movement: extending single-joint rules. *J. Neurophysiol.*, 74:1374–1381.

Bizzi, E, Hogan, N., Mussa-Ivaldi, F.A., and Giszter, S. (1992). Does the nervous systgem use equilibrium-point control to guide single and multiple join movements? *Behav. Brain. Sci.*, 15:603–613.

Bonasera, S.J. and Nichols, T.R. (1996). Mechanical actions of heterogenic reflexes among ankle stabilizers and their interactions with plantarflexors of the cat hindlimb. *J. Neurosci.*, 75:2050–2055.

Brown, I.E., Scott, S.H., and Loeb, G.E. (1995). "Preflexes"—programmable, high-gain, zero-delay intrinsic responses of perturbed musculoskeletal systems. *Soc. Neurosci. Abstr.*, 21:1433 (Abstract).

Brown, I.E., Scott, S.H., and Loeb, G.E. (1996a). Mechanics of feline soleus: II. Design and validation of a mathematical model. *J. Muscle Res. Cell Motil.*, 17: 219–232.

Brown, I.E., Liinamaa, T.L., and Loeb, G.E. (1996b). Relationships between range of motion, L_o and passive force in five strap-like muscles of the feline hindlimb. *J. Morphol.*, 230:69–77.

Cavagna G.A. (1970). Elastic bounce of the body. *J. Appl. Physiol.*, 29:279–282.

Feldman, A.G. and Levin, M.F. (1995). Positional frames of reference in motor control. The origin and use. *Behav. Brain. Sci.*, 18:723–806.

Garies, H., Solomonow, M., Baratta, R., Best, R., and D'Ambrosia, R. (1992). The isometric length-force models of nine different skeletal muscles. *J. Biomech.*, 25:903–916.

Gordon, A.M., Huxley, A.F., and Julian, F.J. (1966). The variation in isometric tension with sarcomere length in vertebral muscle fibres. *J. Physiol.*, 184:170–192.

He, J., Levine, W.S., and Loeb, G.E. (1991). Feedback gains for correcting small perturbations to standing posture. *IEEE Trans. Auto. Cont.*, 36:322–332.

Huijing, P.A. and Ettema, G.T. (1989). Length-force characteristics of aponeurosis in passive muscles and during isometric and slow dynamic contractions of rat gastrocnemius muscle. *Acta. Morphol. Neerl. Scand.*, 26:51–62.

Huxley, A.F. (1957). Muscle structure and theories of contraction. *Prog. Biophys. Mol. Biol.*, 7:255–318.

Joyce, G.S., Rack, P.M.H, and Westbury, D.R. (1969). Mechanical properties of cat soleus muscle during controlled lengthening and shortening movements. *J. Physiol.*, 204:461–474.

Loeb, G.E. (1993). The distal hindlimb musculature of the cat: I. Interanimal variability of locomotor activity and cutaneous reflexes. *Exp. Brain Res.*, 96:125–140.

Loeb, G.E. and Richmond, F.J.R. (1994). Architectural features of multiarticular muscles. *Hum. Mov. Sci.*, 13:545–556.

Loeb, G.E. and Brown, I.E. (1995). Realistic neural control for real musculoskeletal tasks. *Soc. Neurosci. Abstr.*, 21:1433 (Abstr).

Loeb, G.E., Levine, W.S., and He, J. (1990). Understanding sensorimotor feedback through optimal control. Cold Spring Harbor Symposia on Quantitative Biology, 55:791–803.

Loeb, G.E., Pratt, C.A., Chanaud, C.M., and Richmond, F.J.R. (1987). Distribution and innervation of short, interdigitated muscle fibers in parallel-fibered muscles of the cat hindlimb. *J. Morphol.*, 191:1–15.

Mussa-Ivaldi, F.A., Hogan, N., and Bizzi, E. (1985). Neural, mechanical, and geometric factors subserving arm posture in humans. *J. Neurosci.*, 5:2732–2743.

Scott, S.H. and Loeb, G.E. (1995). The mechanical properties of the aponeurosis and tendon of the cat soleus muscle during whole-muscle isometric contractions. *J. Morphol.*, 224:73–86.

Scott, S.H., Brown, I.E., and Loeb, G.E. (1996). Mechanics of feline soleus: I. Effect of fascicle length and velocity on force output. *J. Muscle Res. Cell Motil.*, 17:205–218.

Tsuji, T., Morasso, P.G., Goto, K., and Ito, K. (1995). Human hand impedance characteristics during maintained posture. *Biol. Cybern.*, 72:475–485.

Woittiez, R.D., Huijing, P.A., and Rosendal, R.H. (1983). Influence of muscle architecture on the length-force diagram of mammalian muscle. *Pflugers Arch.*, 399:275–279.

Young, R.P., Scott, S.H., and Loeb, G.E. (1993). The distal hindlimb musculature of the cat: II. Multiaxis moment arms of the ankle joint. *Exp. Brain Res.*, 96:141–151.

Zajac, F.E. and Gordon, M.E. (1989). Determining muscle's force and action in multi-articular movement. *Exerc. Sport Sci. Rev.*, 17:187–230.

11
Musculoskeletal Systems with Intrinsic and Proprioceptive Feedback

Frans C.T. van der Helm and Leonard A. Rozendaal

1 Introduction

The Central Nervous System is unique in its capacity to control a wide variety of tasks, ranging from standing, walking, and jumping to fine motor tasks, such as grasping and manipulating. Typically, the actions of a controller require knowledge about the system to be controlled. It is likely that the CNS takes advantage of, or at least takes into account, nonlinear dynamic features of the musculoskeletal system resulting from multiple degree-of-freedom joints, ligaments, muscles, but also kinematic and actuator redundancy.

Much research in biomechanics has focused on the individual components of the musculoskeletal system. Although this has resulted in a cumulated knowledge about more and more microscopic properties, the scope of this approach has its limitations. In this perspective it will be argued that an integrative approach is necessary, not to know more about the components, but to assess the importance of each component in relation to the complete system.

As an example, much research effort has been spent in the development of muscle models (Hatze 1976; Zahalak 1981; Winters and Stark 1985a; Otten 1987; Zajac 1989). In these models, the activation dynamics and contraction dynamics are represented in detail. However, all these muscle models are open-loop models, transferring neural input into force. There have been some attempts to combine the feedback of muscle spindles with the muscle properties (Hasan 1983; Gielen and Houk 1987; Schaafsma et al. 1992; Otten et al. 1994; Winters 1995). However, these models are applied on single muscles and do not study the interaction with dynamic inertial properties of the limb.

State-of-the-art musculoskeletal modeling employs open-loop muscle models (i.e., without proprioceptive feedback) in combination with an inertial system. The stability of the optimized solutions has never been a subject of discussion. Whenever open-loop muscle models are applied in a musculoskeletal model, the stability of the whole model depends completely on the intrinsic viscoelastic properties of the muscles, resulting from the cross-bridge stiffness (if represented), the force-length and force-velocity relationships, in relation to the passive dynamic properties such as segment inertia and joint viscoelasticity. Inverse-dynamic simulations can result in unstable solutions, and forward-dynamic simulations will tend to be borderline stable systems, meaning that small perturbations can not be compensated because no additional 'effort' will be spent on stability. Ergo, open-loop muscle models will typically *underestimate* the effort needed for stabilization of the limbs, since presumably human beings will always keep a certain safety region from unstable positions. In open-loop musculoskeletal systems, co-contraction is the only means for increasing the impedance (i.e., the stiffness and viscosity) of the system.

However, experiments on reflexive muscle actuators and intact limbs revealed that the emerging viscoelastic behavior of muscles is the result of the proprioceptive feedback of muscle spindles and Golgi Tendon organs. Hoffer and Andreassen (1981) showed a large increase in (quasistatic) stiffness at small activation levels, because of feedback. The results of Feldman (1966) and Mussa-Ivaldi et al. (1985) can only be explained by the stiffness resulting from length feedback, since intrinsic muscle viscoelasticity is much smaller.

The goal of this chapter is to outline a musculoskeletal model containing muscle dynamics and inertial properties in combination with muscle spindle and Golgi Tendon Organ feedback loops. The specific role of force, velocity and position feedback will be discussed, as well as the optimal gain of these feedback loops. The effect of co-contraction in combination with proprioceptive feedback is analyzed. Finally, a framework for implementing proprioceptive feedback including estimation of optimal feedback gains in a large-scale musculoskeletal model is described.

2 General Description of the Musculoskeletal Model Including Feedback

In Figure 11.1 a general model of a musculoskeletal model including feedback is presented. The principles of such a model will be demonstrated for a single degree-of-freedom (DOF) system, whereas ex-

tension of the approach to large-scale system will be discussed in the last paragraph. The input of the model is the supraspinal neural input to the closed-loop actuator (a muscle *inside* its proprioceptive feedback loops). The output of the model is the joint angle of a limb. The neural input to the model contains a set-point signal for the feedback loops, in combination with a feedforward signal from an internal model containing the inverse dynamics of the system. This concept will be further described in Chapter 29.

In the straight path one can discern a muscle block, moment arm and segmental inertia. The muscle block contains the muscle dynamics, transferring the neural input signal (α-activation) to muscle force. The moment arm r transfers muscle force into muscle moment. In the inertia block muscle moment is transferred into accelerations, subsequently twice integrated to obtain the actual position. Proprioceptive feedback can be divided into force feedback in the Golgi tendon organ, and length and velocity feedback (represented by a gain K_l and K_v, respectively, and a time-delay τ_d) in the muscle spindle. The moment arm r transfers joint angle to muscle length (and joint angular velocity to muscle velocity). The negative moment arm is

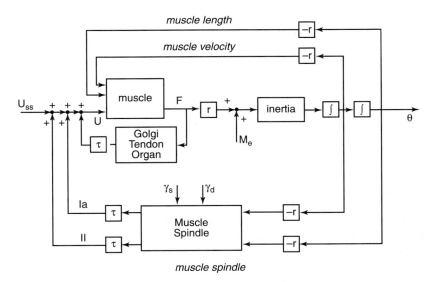

FIGURE 11.1. A general scheme of a musculoskeletal system and its proprioceptive feedback loops. Input u_{ss} is a supraspinal neural signal, the output is joint angle θ. The moment arm r transfers muscle force into moment, and joint angle into muscle length. τ represents the time delay because of neural signal transport and processing.

Joint angle and angular velocity are fed back through the intrinsic muscle properties (force-length and force-velocity relation) and through the muscle spindle, resulting in Ia and II muscle spindle afferents. The Golgi tendon organ is sensitive to muscle force.

used here: If force-to-moment is positive, the resulting angular motion results in muscle shortening, which is by convention negative.

For an analysis of the role of proprioceptive feedback some simplifications of the model are introduced, without affecting the most important dynamic features. For the feedback loops only low-level spinal reflexes ($\sim$ short-latency reflexes) are considered, generally assumed to be mono- or bi-synaptic. The dynamic effects of the GTO are presumably negligible, so the force feedback can be represented by a gain k_f and a time-delay τ_d. Muscle spindles contain nuclear bag and nuclear chain fibers, each with a sensory organ and intrafusal muscle fibers. Input to the muscle spindle are the γ_s and γ_d motorneuron activation of the (static) nuclear chain and (dynamic) nuclear bag, respectively, and length and velocity of the extrafusal muscle fiber. Outputs of the muscle spindle are the Ia-afferent signals, containing length and contraction velocity information, and II-afferent signals containing only length information. Some remarks on the functioning of the muscle spindle are important:

- The output of the muscle spindle (Ia or II) is the result of the combined input signals (i.e., neural input and mechanical input). As such, the muscle spindle is not an absolute length or velocity sensor: Length or velocity are likely to be deduced from the γ input and afferent output.
- Hence, the γ input can not specify any reference length. Whenever a γ input is present, the intrafusal muscle fiber will exert force, and the sensory part will be strained. Any servo-type motor control scheme in which γ denotes the reference length and the α motor neuron will follow the γ is therefore not very likely.
- Ia-afferents contain the combined length and velocity signal. The relative contribution of length and velocity depends on the relation between γ_d and γ_s. It will be shown in this chapter why the combination of a length and velocity signal is in fact very useful.
- Despite the complex structure of the muscle spindle, it essentially provides length and velocity information to the CNS. It has been shown that intrafusal muscle fibers are slow twitch fibers. Fast dynamic behavior would in fact distort the quality of this information, and would in any case not be useful in regard of the time-delays for the

neural transport and processing times, which dominate the dynamic behavior. Therefore, simplification of the muscle spindle to a gain (i.e., no dynamic behavior), and a time delay seems to be justified.

- The gain of the muscle spindle feedback loop can be affected by supraspinal signals inhibiting or exciting interneurons, or by γ-activation. These effects can be combined in one gain, which can be adapted to the circumstances.

The problem in the control scheme presented is to find reasonable values for the feedback gains. Though the control scheme may contain highly nonlinear elements, for sake of analysis it can be linearized in any state. Using available analysis tools for linear systems, the feedback gains can be optimized.

For the muscle model the third-order model proposed by Winters and Stark (1985a) is used. It contains excitation dynamics (from hypothetical motor control signal to neural signals), activation dynamics (from neural signal to "active state," representing the calcium uptake/release dynamics), and contraction dynamics (i.e., a force-velocity relation, in combination with fiber force-length relation and series-elastic force-length relation). Without loss of essential dynamic features the model can be linearized and simplified. The excitation dynamics and activation dynamics can be represented by a first-order linear model. The dynamic contribution of the series-elastic element only becomes somewhat important for frequencies above 4 Hz, and can reasonably be neglected. Hence, contractile element (CE) length and velocity are directly related to joint angle and angular velocity. By linearizing the force-length and force-velocity relation about any current working point, the intrinsic muscle stiffness (df/dl) and muscle viscosity (df/dv) are obtained. This is related to joint stiffness ($dM/d\theta$) and joint viscosity ($dM/d\dot\theta$) by: $M = r.f$

$$\frac{dM}{d\theta} = \frac{dr}{d\theta}f + r\frac{df}{d\theta} = \frac{dr}{d\theta}f + r\frac{df}{dl}\frac{dl}{d\theta}$$

$$= \frac{dr}{d\theta}f + r^2\frac{df}{dl} = K_{\theta f} + K_{\theta s} \qquad (11.1)$$

$$\frac{dM}{d\dot\theta} = r\frac{df}{d\dot\theta} = r\frac{df}{dv}\frac{dv}{d\dot\theta} = r^2\frac{df}{dv} = B_{\theta v} \qquad (11.2)$$

where $K_{\theta f} = \dfrac{dr}{d\theta}f$, $K_{\theta s} = r^2\dfrac{df}{dl}$, $r = \dfrac{dl}{d\theta}$, M is the

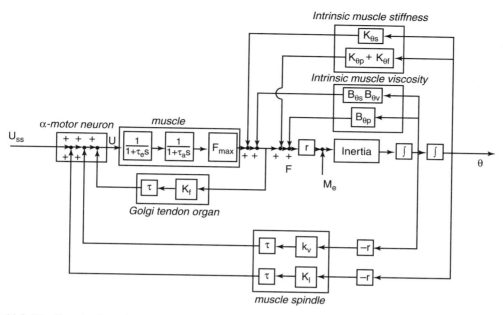

FIGURE 11.2. The linearized version of Figure 11.1. The active and passive joint stiffness and viscosity are attributed to the skeletal part. Proprioceptive feedback pathways are represented by a gain and a time delay.

joint moment, r is the moment arm, f is muscle force, l is muscle length, v is muscle velocity, θ is joint angle and $\dot{\theta}$ is joint angular velocity. $K_{\theta f}$ and $K_{\theta s}$ are the *muscle force* contribution (depending on f) and *muscle stiffness* contribution (depending on df/dl) to *joint* stiffness, respectively. Similarly, $B_{\theta v}$ is the *muscle viscosity* contribution (depending on df/dv) to *joint* viscosity. The impedance of the skeletal system can be described in the Laplace domain by:

$$M = Js^2\theta +$$
$$(B_{\theta v} + B_{\theta p})s\theta + (K_{\theta s} + K_{\theta f} + K_{\theta p})\theta \quad (11.3)$$

where $K_{\theta p}$ and $B_{\theta p}$ are the passive joint stiffness and viscosity, respectively. The importance of the distinction between the geometrical and force contribution of the joint stiffness and viscosity will be shown below. In Eq. (11.3) it is shown that the linearized force-length and force-velocity relation can be regarded as part of the skeletal system, affecting the joint stiffness and viscosity. Hence, the activation dynamics can be described by the transfer function H_{ad}

$$H_{ad}(s) = H_{exc}(s).H_{act}(s).F_{max} \quad (11.4)$$

where

$$H_{exc}(s) = \frac{1}{1 + \tau_e.s}$$

$$H_{act}(s) = \frac{1}{1 + \tau_a.s}$$

in which τ_e is the excitation time-constant, and τ_a is the activation time constant, in the linearized case the average between the activation and deactivation time-constant. F_{max} is the maximal isometric muscle force, depending on the physiological cross-sectional area of the muscle.

In Figure 11.2 the linearized version of Figure 11.1 is shown, with the active and passive joint stiffness and viscosity contributions attributed to the skeletal part, and the activation dynamics according to Eq. (11.4). Simulations will show the role of force, velocity and length feedback for the behavior of the musculoskeletal system. A more extensive description of the results can be found in Rozendaal (1997). Values for the variables in Figure 11.2 are given in Table 11.1.

3   Role of Force Feedback

In Figure 11.3A the block diagram with the activation dynamics and the force feedback is separated. The input is the neural input and the output is the muscle force. The muscle force is fed back by a gain and a time-delay to the neural input. From

TABLE 11.1. Parameters used in the musculoskeletal model.

Parameter	Symbol	Value	Unit
Sensory time delay	τ_d	0.035	s
Muscle excitation time const.	τ_e	0.040	s
Muscle activation time const.	τ_a	0.030	s
Maximum muscle force	F_{max}	1000	N
Muscle moment arm	r	0.04	m
Arm inertia	θ	0.25	kgm^2
Inherent arm viscosity	B_θ	0.2	Nms/rad
Inherent arm stiffness	K_θ	0.0	Nm/rad

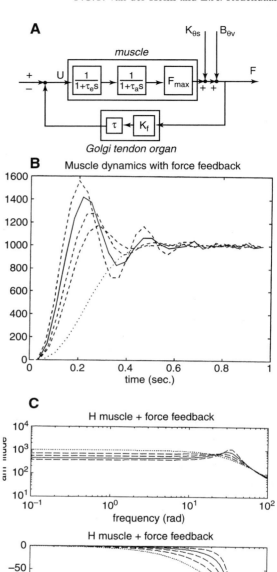

FIGURE 11.3. The effect of force feedback on the muscle activation dynamics. A: Block diagram of the force feedback. The muscle activation dynamics are represented by an excitation and activation block, and the maximal isometric force. Force feedback is represented as a gain and a time-delay. B: Step response simulations with different gain values in the time-domain. ..... : NO force feedback; ____ "optimal" force feedback (k_f*Fmax = 1.27); - - - intermediate values. C: Similar to B, presented in the frequency domain.

Figure 11.3A it can be seen immediately that the summing point must have a negative sign, indicating negative feedback. Otherwise, positive feedback would result immediately in a unstable system: An increase in muscle force would result in an increase in muscle activation, which would result in an increase in muscle force, etc. If the force feedback is added, the transfer function Had_ff of the closed-loop system is

$$H_{ad\_ff}(s) = \frac{H_{exc}(s).H_{act}(s).F_{max}}{1 + H_{exc}(s).H_{act}(s).F_{max}.kf.e^{-\tau d.s}} \quad (11.5)$$

where $e^{-\tau_d s}$ is the time-delay of the force-feedback loop resulting from neural transport and processing, and k_f is the gain of the force-feedback. In Figure 11.3B and 11.3C, the transfer function H_{ad_ff} is shown for several values of k_f in the time-domain and frequency domain, respectively. Force feedback is very effective in increasing the frequency bandwidth of the muscle activation dynamics, or in other words, the muscle can react faster. However, if the gain is too high, the system will start oscillating, thereby decreasing the effectiveness of the feedback path. Another important feature of the force feedback gain k_f is that it is not affected by the dynamics of the skeletal system. It depends only on the activation dynamics and F_{max}, which are all fixed constants. Rozendaal (1997) showed that the maximal loop gain k_f*F_{max} is approximately 1.27 (for a ms time delay). However, for a suitable performance the loop gain should be much smaller. If

$$kf = \frac{1.27}{F_{max}} \quad (11.6)$$

the gain at the resonance frequency is twice the static gain, being a compromise between a fast response and oscillating behavior. Rozendaal (1997) showed that the bandwidth of the actuator is increased by a factor 2.7, from 18 rad/s to 50 rad/s (from 2.9 to 8 Hz). It is concluded that force feedback is very important whenever muscle dynamics are described, and may not be neglected.

4 Intrinsic Muscle Viscoelasticity Versus Length and Velocity Feedback

In Figure 11.2 two feedback pathways are described for the muscle length and velocity. One describing the intrinsic muscle stiffness and viscosity, and one describing the proprioceptive feedback including muscle spindles. The intrinsic muscle viscoelasticity feedback path does not have a time-delay present: the contribution is a direct function of the muscle force exerted by the force-length and force-velocity relation. By increasing the force through co-contraction, the system will become stiffer and more viscous (for nonlinear version of the model). For a lumped shoulder-elbow model but using realistic values derived from a detailed shoulder-elbow model (Van der Helm et al. 1992), Rozendaal (1997) estimated that the maximal stiffness of the shoulder is way below experimental values as presented by Mussa-Ivaldi et al. (1985), or derived from Feldman (1966). This means that the (quasistatic) stiffness is dominated by proprioceptive feedback. A similar conclusion can be drawn from the experiments of Hoffer and Andreassen (1981) on decerebrated cats, where the intrinsic and proprioceptive contributions were of the same order.

In one of the previous paragraphs a distinction has been made between the force, geometric and passive contribution to joint stiffness. If we focus on the summing point of the various stiffness contributions (Figure 11.4), it is shown why this distinction is important. If the muscle force increases because of the muscle stiffness or viscosity, this force increase is reduced by the force feedback through the Golgi Tendon organ! Therefore, the force contribution to stiffness becomes

$$K_{\theta f\_ff}(s) = \frac{K_{\theta f}}{1 + H_{exc}(s).H_{act}(s).F_{max}.k_f.e^{-\tau d.s}} \quad (11.7)$$

and a similar relation is found for the viscosity contribution $B_{\theta v}$. Obviously, the geometrical and passive contributions do not change, since no increase of muscle force is involved. The geometrical contribution, depending on $dr/d\theta * F$, is almost always negative (i.e., destabilizing). Increasing muscle force increases the destabilizing geometrical contribution. In fact, Rozendaal (1997) showed that for the shoulder the force contribution to stiffness (reduced by force feedback!) is almost canceled out by the geometrical contribution. This means that co-contraction does not add directly to the joint stiffness!

Muscle fiber length and contraction velocity are sensed by the muscle spindle. Though the spindle is a highly non-linear processor of the length and velocity information, one can assume that the CNS is capable of deriving the original length and velocity. The muscle spindle dynamics *must* be an order of magnitude slower than the extrafusal muscle

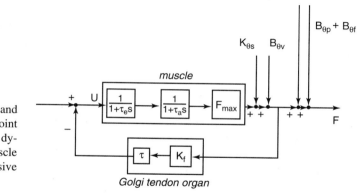

FIGURE 11.4. The muscle stiffness ($K_{\theta s}$) and muscle viscosity ($B_{\theta v}$) contribution to joint stiffness and viscosity are filtered by the dynamics of the force feedback loop. The muscle force contributions ($K_{\theta f} + B_{\theta f}$) and the passive contributions ($K_{\theta p} + B_{\theta p}$) are not filtered.

fiber dynamics, otherwise the sensor information would be truly distorted. In addition, for the feedback loop dynamics the muscle spindle dynamics can be neglected in regard to the time delay. Therefore, for our model it is acceptable to model the muscle spindle and neural pathways by a simple gain and a time delay. The most important features of the length and velocity feedback can be shown, and the optimal feedback gains can be estimated.

The gains of the proprioceptive feedback loops are limited by stability requirements that are well known in control engineering: For frequencies where the phase lag is 180°, the total loop gain must be considerably below 1. Figure 11.2 shows that the total loop gain includes the muscle dynamics, but also the skeletal dynamics and moment arms. Therefore, the length and velocity gain of the muscle spindle depends to a large extent on the specified musculoskeletal system and its current position. This is a very important consideration whenever experiments on single muscles or other species (cats!!: Hoffer and Andreassen 1981) are evaluated.

If the muscle dynamics and time delays were neglected, the length and velocity feedback gains, k_l and k_v respectively, would directly result in a certain linear impedance (mass-damping-stiffness) behavior of the musculoskeletal system. Here, the desired impedance is defined as the transfer function $H^d_{com}(s)$ between external perturbations Me and joint angle:

$$H^d_{com}(s) = \frac{\theta}{Me} = \frac{1}{Js^2 + Bs + K} \quad (11.8)$$

However, in the presence of muscle dynamics and time-delays, the resulting impedance can only approximate a true second-order (passive mass-spring-damper) system:

$$H_{com}(s) = \frac{\theta(s)}{Me(s)}$$

$$= \frac{1}{\substack{Js^2 + (B + k_v r^2 H_{ad\_ff}(s))s \\ + (K + k_l r^2 H_{ad\_ff}(s))}} \quad (11.9)$$

However, the optimal length and velocity feedback gains, k_l and k_v respectively, can be estimated by minimizing the difference between the actual impedance and desired linear impedance with the following optimization criterion:

$$J = \int_{\omega=0.1}^{100} (\log(|H^d com(j\omega)|) \\ - \log(|Hcom(j\omega)|)^4 d \log(\omega) \\ + \omega \int_{\omega=0.1}^{100} \max(0,(\angle H^d com(j\omega) \\ - \angle Hcom(j\omega))^4 d \log(\omega) \quad (11.10)$$

i.e., weighting the differences between $H^d_{com}(s)$ and $H_{com}(s)$ in the Bode-plot format.

5   Simulation Results

The model as shown in Figure 11.2 has two inputs (supra-spinal neural input signal u_{ss} and perturbing moment M_e) and one output (joint angle θ). The transfer function from u_{ss} to θ denotes the tracking behavior, i.e. how fast can the system track an input signal. The tracking behavior and the origin of the signal u_0 is discussed in Chapter 29. Here, the transfer function of M_e to joint angle θ is treated, being the compliance (= inverse of the impedance) of the system. As a typical example, the optimized behavior will be shown for the case

$$J_\theta^d = m_. = 0.25 \text{ kgm}^2$$
$$B_\theta^d = 3 \text{ Nms/rad}$$
$$K_\theta^d = 25 \text{ Nm/rad}$$

which is a reasonably demanding task for the system. It is assumed that no active or passive joint stiffness is present and only a small passive joint viscosity ($B_{\theta p} = 0.2$) (i.e., a relaxed arm) as is the case in most experimental studies. Figure 11.5 shows results for length feedback only, both length and velocity feedback, and length, velocity and force feedback. Because the open loop system itself, containing the muscle activation dynamics, the inertia of the limb and the time delay, has a phase-lag of 180 degrees at 0.5 Hz, length feedback only can carry a very small gain factor, no more that $k_l = 0.001$. Obviously, this is not enough for good servocontrol. Just length feedback in the presence of time-delays is not a feasible option in musculoskeletal systems. The addition of velocity feedback and the accompanying phase lead results in a stable system. However, there are still limitations on the bandwidth of the velocity feedback path, and hence also on the bandwidth of the length feedback. The result is shown in Figure 11.5A and 11.5B (time-domain and frequency-domain respectively). The desired stiffness can not be obtained by the

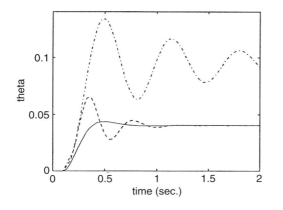

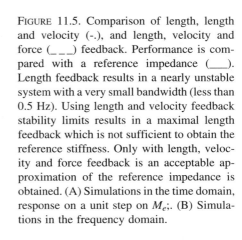

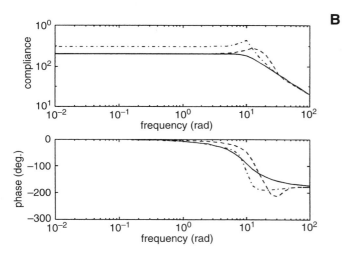

FIGURE 11.5. Comparison of length, length and velocity (-.), and length, velocity and force (_ _ _) feedback. Performance is compared with a reference impedance (___). Length feedback results in a nearly unstable system with a very small bandwidth (less than 0.5 Hz). Using length and velocity feedback stability limits results in a maximal length feedback which is not sufficient to obtain the reference stiffness. Only with length, velocity and force feedback is an acceptable approximation of the reference impedance is obtained. (A) Simulations in the time domain, response on a unit step on M_e;. (B) Simulations in the frequency domain.

length and velocity feedback. Only the combination of length, velocity and force feedback results in a system that can approximate the desired impedance. However, in Figure 11.5B it can be seen that because of the dynamics of the feedback path (time-delays, muscle activation dynamics), the stiffness and viscosity are not instantaneous properties of the system. However, it is concluded that force feedback increases the muscle activation dynamics, whereas the length and velocity feedback can approximate the desired stiffness and viscosity. The intrinsic muscle stiffness and viscosity due to the force-length and force-velocity relation hardly contribute to the joint impedance, due to the filtering of the force feedback path. The increase of especially the intrinsic muscle viscosity is functional in the respect that the length and velocity

feedback gains can increase further, and a larger range of impedance levels can be obtained.

6   Physiological Comparisons

At the present, there are hardly any experiments described in literature to back up the model and modeling assumptions presented here. Nonetheless, some observations are in surprisingly good agreement with the model results. Let us summarize first the strategy in the present optimization study. A model structure has been chosen which is closely related to the known anatomy and physiology (Figure 11.1). The muscle model and skeletal model are not very detailed, though reasonably good for the present purpose. Parameter values for a shoulder system are lumped from a very detailed large-scale shoulder model. Subsequently, the model is linearized, assuming that in the close neighborhood of the working point the effect of nonlinear properties are small. This is reasonably the case for posture tasks, though less true during motions. Subsequently, the gains in the proprioceptive feedback loops are optimized in order to achieve a desired impedance, without any a priori restrictions on the feedback gains (except for the obvious requirement that the system should be stable). How can the resulting feedback gains be valued?

The force feedback gain k_f is always negative. This is in agreement with the observation that there is an inhibitory interneuron present in the projection of the Golgi tendon organ to the α motor neuron (bisynaptic reflex). From Figure 11.3A it is easily shown that a positive feedback gain would result in oscillations. The force feedback gain k_f only depends on prefixed constants, being the muscle dynamics, F_{max} and the time delay. No modulation of k_f is necessary, and no modulation has been shown in literature.

The length and velocity feedback gains k_l and k_v can vary over quite a large range, in order to adapt to the environment. However, their mutual relation remains between quite fixed boundaries: $0.06 < k_v/k_l < 0.14$ for most impedance levels (Rozendaal 1997). Using an average value of $k_v/k_l = 0.1$ (same as used in Winters and Stark 1985b), the transfer function of the muscle spindle itself can be deduced. The results are shown in Figure 11.6, together with experimental results of Chen and Poppele (1977). Given the simplifications of the present model, the results agree very well!

Another source of experimental evidence can be found in perturbation experiments of the arm. Since it is necessary to perturb the intact system, an experimental problem arises in measuring BOTH inputs (u_0 and M_e) to the system. Especially, u_0, being a supraspinal signal, can not be accessed. Therefore, the well-known experimental circumstance "do not intervene" (Feldman 1966; Mussa-Ivaldi et al. 1985) has been developed. To what extent subjects are capable of refraining from "intervention" can not be assessed. Without proprioceptive feedback and only intrinsic muscle impedance present, the results are way below experimental results. Using proprioceptive feedback the results come reasonably close to experimental re-

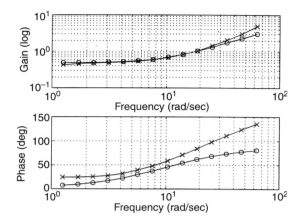

FIGURE 11.6. Transfer function of a muscle spindle in the frequency domain measured by Chen and Poppele (1977), (xx) compared with the result of optimization of k_l and k_v, resulting in an average relation $k_v/k_l = 0.1$ (oo).

sults, though they are still at the low side. We conclude that the present approach of implementing proprioceptive feedback loops and estimating the gains is viable.

7 Application in a Large-Scale Musculoskeletal Model

Using the analysis above, it is feasible to implement proprioceptive feedback loops and the accompanying gains in a large-scale musculoskeletal model (Rozendaal 1997). The procedure starts with a linearization of the model to make a linear state-space description of the muscles, skeletal system and the interaction between both. Time delays can be incorporated using a second-order Padé approximation, adding two states per time delay. This results in:

$$
\begin{bmatrix} \dot{exc} \\ \dot{act} \\ \dot{Vce} \\ \dot{p}_1 \\ \dot{p}_2 \\ \dot{\theta} \\ \ddot{\theta} \end{bmatrix}
=
\begin{bmatrix} A_{11} & A_{12} & A_{13} \\ A_{21} & A_{22} & A_{23} \\ A_{31} & 0 & A_{33} \end{bmatrix}
\cdot
\begin{bmatrix} exc \\ act \\ lce \\ p_1 \\ p_2 \\ \theta \\ \dot{\theta} \end{bmatrix}
$$

$$+ \; B \cdot u_{ss} \qquad (11.11)$$

where (exc, act, l_{ce}) are the muscle states, (p_1, p_2) are the states of the Padé approximation, and (θ, θ') are the states associated with the skeletal system. The state space matrix A is a function of k_l, k_v and k_f. For a stable system it is required that the eigenvalues of A are smaller/equal to zero. A priori assumptions are that the loop gain $k_f.F_{max}$ is fixed, and the relation $k_v/k_l = 0.1$. The optimization criterion weighs the speed of the response with the oscillating behavior (Rozendaal 1997). Whenever the optimal gains are assessed for a sequence of positions (the linearization allows for transient states θ and θ'), a stable forward dynamic simulation is warranted.

8 Conclusions

In principle, there are two strategies for increasing the joint stiffness: by co-contraction and by proprioceptive reflexes. Co-contraction increases the intrinsic muscle viscoelasticity. The intrinsic muscle viscoelasticity is an almost instantaneous process and is therefore effective for the whole frequency range. On the other hand, for frequencies where proprioceptive feedback is not effective, the impedance is dominated by the mass effects, not by the stiffness effect and only slightly by viscous effects. The major disadvantage of cocontraction is that it costs energy. In addition, in the present simulations it is shown that the effect of co-contraction is decreased by the force feedback through Golgi tendon organs, and the net effect of cocontraction on joint stiffness is negligible (at least for the shoulder region). Co-contraction is only useful as it increases the joint viscosity as well, thereby allowing for higher feedback gains for the length and velocity feedback.

The use of proprioceptive reflexes does not directly cost energy. Through the time delay it has a limited frequency bandwidth, and is therefore only effective for low frequencies (~posture tasks). However, a reasonable extent of impedance levels can be achieved by the adaptation of length and velocity feedback gains.

In this chapter some guidelines are given for the estimation of force, velocity and length feedback gains. The force feedback gain only depends on the muscle activation dynamics and maximal isometric force. The ratio between velocity and length feedback gains is approximately 1 to 10. These guidelines can be used for estimation of the feedback gains in large-scale musculoskeletal systems.

9 Future Directions

The simulations in this chapter do show that the intrinsic muscle properties through co-contraction and proprioceptive feedback can complement each other. Also a methodology is outlined for estimating the feedback gains. This definitely shows that only very little is known about the feedback properties in combination with the dynamics of a musculoskeletal system. Much research should be devoted into this area which will show the importance of stability for muscle co-ordination, but will also give new insights in the research of muscle spindles and Golgi tendon organs.

Experimental evidence is very difficult to obtain. For analyzing feedback systems, a good approach is to perturb the system and measure its response. A good example of this approach can be found in Kirsch et al. (1993). Robot manipulators with high performance are needed for imposing perturbations while providing different task conditions as well. Then, the impedance of the arm can be measured, as well as the contribution of the length, velocity and force feedback (Brouwn 2000).

References

Brouwn, G.G. (2000). Postural control of the human arm. PhD thesis, Delft University of Technology, Delft, The Netherlands.

Chen, W.J. and Poppele, R.E. (1978). Small-signal analysis of response of mammalian muscle spindles with fusimotor stimulation and a comparison with large-signal responses. *J. Neurophysiol.*, 41:15–26.

Feldman, A.G. (1966). Functional tuning of the nervous system with control of movement or maintenance of a steady posture: 2. Controllable parameters of the muscle. *Biophysics*, 11:565–578.

Gielen, C.C.A.M. and Houk, J.C. (1987). A model of the motor servo: incorporating nonlinear spindle receptor and muscle mechanical properties. *Biol. Cybern.*, 57:217–231.

Hasan, Z. (1983). A model of spindle afferent response to muscle stretch. *J. Neurophysiol.*, 49:989–1006.

Hatze, H. (1976). The complete optimization of a human motion. *Math. Biosc.*, 28:99–135.

Hoffer, J.A. and Andreasson, S. (1981). Regulation of soleus muscle stiffness in premammillary cats: are flexive and reflex components. *J. Neurophysiol.,* 45:267–285.

Kirsch, R.F., Kearney, R.E. and MacNeil, J.B. (1993). Identification of time-varying dynamics of the human triceps surae stretch reflex: 1. Rapid isometric contraction. *Exp. Brain Res.,* 97:115–127.

Mussa-Ivaldi, F.A., Hogan, N., Bizzi, E. (1985). Neural and geometric factors subserving arm posture. *J. Neurosci.,* 5:2732–2743.

Otten, B. (1988). Concepts and models of functional architecture in skeletal muscle. *Exerc. Sport Sci. Rev.,* 16:89–137.

Otten, E., Scheepstra, K.A. and Hulliger, M. (1994). An integrated model of the mammalian muscle spindle.

Rozendaal, L.A. (1997). *Stability of the Shoulder: Intrinsic Muscle Properties and Reflexive Control.* PhD thesis, Delft University of Technology, Delft, The Netherlands.

Schaafsma, A., Otten, B., and Van Willigen, J.D. (1991). A muscle spindle model for primary afferent firing based on a simulation of intrafusal mechanical events. *J. Neurophysiol.,* 65:1297–1312.

Van der Helm, F.C.T., Veeger, H.E.J., Pronk, G.M., Van der Woude, L.H.V. and Rozendal, R.H. (1992). Geometry parameters for musculoskeletal modelling of the shoulder mechanism. *J. Biomech.,* 25:129–144.

Winters, J.M. (1995). An improved muscle-reflex actuator for use in large-scale neuromusculoskeletal models. *Ann. Biomed. Eng.,* 23:359–374.

Winters, J.M. and Stark, L. (1985a). Analysis of fundamental human movement patterns through the use of in-depth antagonistic muscle models. *IEEE Trans. Biomed. Eng.,* 32:826–839.

Winters, J.M. and Stark, L. (1985b). Task-specific second-order movement models are encompassed by a global eighth-order nonlinear musculo-Skeletal model. *Proc. IEEE Eng. Med. Biol.,* CH2253-3/85, pp. 1111–1115. Chicago.

Zahalak, G.I. (1981). A distribution-moment approximation for kinetic theories of muscular contraction. *Math. Biosc.,* 55:89–114.

Zajac, F.E. (1989). Muscle and tendon: properties, models, scaling and application to biomechanics and motor control. *CRC Crit. Rev. Biomed. Eng.,* 17:359–419.

Section IV

12
Neuromechanical Interaction in Cyclic Movements

James J. Abbas and Robert J. Full

1 Introduction

Cyclic behaviors, such as breathing, chewing, and locomotion, serve our basic needs for respiration, nutrition, and transportation. They are performed on a daily basis in a wide variety of situations and often for extended periods of time. Therefore, the mechanisms by which they are generated must be reliable, versatile, durable, and efficient. In some cases, these specific needs have resulted in the evolutionary development of specialized neural, muscular, and/or skeletal structures.

This chapter describes some of these structures that are specialized for cyclic movements and explain (or speculate on) how they fit the needs of the overall system. The focus is on locomotor systems, with a particular emphasis on the interaction between the neural and mechanical (muscular and skeletal) systems. The approach is decidedly integrative, drawing from studies on a wide variety of locomotor behaviors (walking, crawling, swimming, flying) in a wide variety of animals (bipedal, polypedal, aquatic, winged).

Locomotion can be viewed as a multiobjective control problem. While the primary concern is "get me over there," several other objectives might come into play. First, the animal might try to avoid any physical harm that might occur during the process of moving. This could include escaping from a predator (which would put severe time constraints on "get me over there"), avoiding obstacles, and maintaining balance. Second, the animal might try to keep energy expenditure low during locomotion. Achieving an absolute minimum of energy expenditure may not be a primary concern,

but keeping expenditures low is definitely advantageous. Finally, the animal might also be concerned about conveying a specific appearance during locomotion such as authority, ferocity or attractiveness.

1.1 Interactions Determine Movement Pattern

The neural control system that achieves these multiple, complex objectives has often been viewed as having a hierarchical structure (Ghez 1985). A purely hierarchical system would be organized in such a way that the high level goal of "get me over there" is systematically broken down to subtasks and parceled out to lower and lower levels of the organizational structure. For example, the high level command for locomotion would result in the coordinated activity of neural subsystems for controlling posture, balance, forward progression, and so on. Each subsystem would further parcel out its task to lower-level subsystems; one such low-level subsystem may be charged with the task of controlling hip flexion toward the end of swing phase. Although this view may at times be convenient, it is clear that it does not adequately describe the structure of neural control systems. In a modified version of the hierarchical structure, the chain of command would be less clearly defined and specific goals of each subsystem may be more global in nature. Here, coordination is achieved through the interaction of various components, rather than through the successive delegation of tasks from higher to lower level structures. Ghez (1985) describes this as a parallel structure that complements the heirarchy; Cohen (1992) goes further to describe the overall structure as a "heterarchy." Thus, even at a general block diagram level of description, there is ambiguity and limitations to our understanding of the organization of neural control systems for locomotion.

This issue of hierarchy in the nervous system can also be used as a framework in which to view the neuromechanical system. Muscles, skeleton, and environment are often studied separately from the nervous system and are viewed as mechanical components that constitute the system to be controlled. Motor commands from the lowest level in the hierarchy of the nervous system (motoneurons) are specified in such a way that the resulting movement/posture of the skeletal system meets the higher level objectives (i.e., the movement pattern is directed from above). In a more integrated view, the movement pattern is a result of the interactions among the neural and mechanical components (Chapter 14; see Raibert and Hodgins 1992; Chiel and Beer 1993; McGeer 1993; Chapter 13). Here, the mechanical system is no longer just an object with properties to be reckoned with, it is one of many system components whose properties influence the overall system behavior. This complex interaction of the neural and mechanical dynamic systems has been described as a "self-organizing" process that results in the formation of movement patterns (Schoner and Kelso 1988).

These concepts regarding neural organization and neuromechanical interactions are by no means new ideas. It has long been accepted that neural control systems are not entirely hierarchical and that movement patterns are the result of interaction between neural and biomechanical systems. However, motor control studies have primarily focused on a specific neural or biomechanical component for two reasons: (1) there is still a lack of detailed understanding of the various components and (2) many of the experimental techniques that are available necessitate the use of reduced preparations of neural system components or musculoskeletal system components in which the interactions are removed. Note that there are exceptions in which clever experimental techniques have been used to study the interactions in an intact or semi-intact preparation, but these are the exceptions. In many cases, this lack of suitable experimental techniques has often been successfully overcome through the use of mathematical mod-

els and computer simulations (e.g., Loeb and Levine 1990; Chapter 15).

This chapter briefly describes the neural and biomechanical components and then focuses on the interactions among the components, as revealed by both experimental and modeling studies. Throughout this chapter, and more generally in the study of neural control of locomotion, there are several concepts that recur. One is the notion that the neural and biomechanical components have evolved together and that they develop and adapt together in a given individual animal. A second is that there is conservation across species (i.e., over a wide range of sizes, body structures, and modes of locomotion we see similarities in the neural and biomechanical components as well as in the manner in which the components are organized). A third is that an approach to studying motor control that utilizes a combination of biological and engineering techniques will lead to greater insight than a less integrated approach.

2   Features of Biomechanical Systems for Locomotion

Many chapters in this book include material on muscle biomechanics, soft tissue biomechanics, and skeletal biomechanics, especially in Sections II, III, and VII. This section focuses on specific issues that are important for understanding cyclic movements in general, and locomotion in particular.

Biological systems use a variety of modes of locomotion, including bipedal locomotion, polypedal locomotion, flying, and swimming. Many animals are capable of utilizing several modes of locomotion, such as birds that can fly, walk, and even swim. Obviously, in a given animal, certain biomechanical constraints must be met in order to utilize a given form of locomotion. For example, appendages capable of providing lift are required for flight; appendages capable of supporting body weight are required for legged locomotion; and some means of propulsion in water is required for swimming. The mechanical structures used for the various forms of locomotion include endoskeletons, exoskeletons, and hydrostatic skeletons (a fluid-filled tube surrounded by muscle).

Chapters in this section include descriptions of studies on locomotion of the leech (hydrostatic skeleton) (Chapter 14), legged locomotion of the stick insect (exoskeleton) (Chapter 16), swimming of the lamprey (aquatic vertebrate) (Chapter 15), walking in humans and cats (legged vertebrates) (Chapter 18; Chapter 17), as well as studies on a wide variety of legged animals (Chapter 13).

2.1 Muscle and Skeletal Dynamics

In segmented skeletal systems, movements that involve just a single segment can exhibit complex dynamic properties because of passive joint stiffness and damping as well as the inertial properties of the limb (Audu and Davy 1985). Over limited ranges of movements, passive joint properties are often assumed to be linear, but characterization over the normal range of movement has often demonstrated that nonlinear properties can be functionally significant (Chapter 7). Musculotendon actuators exhibit complex length-, velocity-, and activity-dependent behavior (Chapter 2; McMahon 1984; Zajac 1989; Ettema and Huijing 1990; Hof 1990; Mungiole and Winters 1990; Zajac and Winters 1990) that are functionally significant. The physical properties of tendons vary widely for different muscles within or across animals and are clearly specialized for critical tasks. A good example of the important role of tendon properties is the ability of the plantar flexor tendon to store elastic energy (Alexander and Vernon 1975; Biewener and Baudinette 1995).

Locomotion, however, typically involves movements of several segments of two or more limbs and therefore the situation is much more complex than the mechanics of single segment movements (Winter 1987, 1990; Hinrichs 1990; Yamaguchi 1990; Zernicke et al. 1991). Here, one must also consider the effects of intersegmental coupling (e.g., a swinging thigh exerts forces and moments that tend to accelerate the shank) as well as interlimb coupling (e.g., a swinging right leg exerts forces and moments that tend to accelerate the left leg). Thus, in systems with segmented skeletons, it is clear that movements such as locomotion must involve the coordinated activity of a set of muscles and that the mechanics of the musculoskeletal system will greatly influence the pattern of neural activity that drives the movement.

A remarkable feature of locomotor systems, however, is the degree to which running, hopping and trotting of complex multi-segment mechanical

systems can be explained by relatively simple spring-mass type models (Chapter 13). For legged locomotor systems, it appears as though the complex details of the various system components are integrated in order to result in an overall spring-mass type behavior while simultaneously providing the ability to perform a range of locomotor and nonlocomotor activities.

2.2 Locomotion as a Multistate Process

The locomotor cycle has often been described as having multiple phases, or 'states' (Winter 1987; Inman et al. 1994). That is, at any point in time the body may be in one of several states and the process of locomotion involves a repeated set of transitions among these states. For example, we can use a very broadly defined set of states (see Figure 12.1) to define bipedal locomotion: right double-support (both feet on ground with right leg forward), left swing, left double-support, right swing. A more detailed description might include such states as midstance, initial contact, terminal swing, and so on. A full description of behavior would include some nonlocomotor states (standing, sitting, falling, etc.) as well as states for different forms of locomotion (walking, running, hopping, crawling, etc). Note that as speed gradually increases from a slow walk

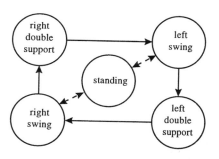

FIGURE 12.1. Simple multistate model of biped locomotion. Each circle represents a possible state in the system. Individual states may be differentiated by different biomechanical configurations or by different control goals. In this diagram, normal gait would involve a clockwise progression of the system from one state to another around the perimeter following the solid lines. Cyclic activity may be interrupted by entering the "standing" state in the center of the diagram. A more complete multistate model of this sort may include a further breakdown of the locomotor states listed here, as well as additional states for other cyclic and noncyclic tasks.

to a fast run one would see changes in the state sequence, changes in the time between state transitions as well as modifications within a given state (e.g., double-support during a fast walk would not just be a fast version of the kinetic and kinematic trajectories of double-support during a slow walk (Winter 1987)).

One important feature of the mechanics of locomotor systems is that there are state-dependent effects that are often significant. That is, in many instances, the result of a neural action may strongly depend on the biomechanical configuration of the body. For example, activation of the gastrocnemius muscle towards the end of stance phase will result in forward propulsion of the body (and knee extension) while activation of the same muscle at the beginning of stance phase will result in backward propulsion of the body (and knee flexion).

A second important feature of the mechanics of locomotor systems is that although movement patterns are stereotyped, there is a wide range of variability within a given movement pattern for a given individual. That is, the kinematic and kinetic trajectories for a given state may vary widely from one cycle to another—there is not a unique solution to the problem of getting from one state to another. The sources of this variability may be environmental (e.g., obstacles, interactions with other movement patterns such as lifting an arm), mechanical (e.g., muscle fatigue or response variability) or neural (intrinsic variability in motor pattern output).

In summary then, locomotor systems include a wide range of biomechanical structures that can utilize different modes of locomotion in a variety of movement patterns. The range of possible configurations for a biomechanical system can be viewed as having a finite number of possibilities, called "states." Locomotion can be represented as a cyclic movement of the biomechanical system through a subset of all possible states and the details of movement within a given state may vary from one locomotor cycle to another.

3   Neural Systems

Experimental evidence in a wide variety of animals has indicated the presence of pattern generating neural circuits that drive cyclic movements (Grillner 1981; Cohen et al. 1988). The general structure

of the neural system is shown in Figure 12.2. Typically, the "pattern generator" receives inputs from other neural centers and from the periphery that modulate its outputs. The neural mechanisms used to generate the oscillatory patterns may include cellular properties of specific neurons (voltage and time-dependent membrane channels) as well as network properties (patterns of interconnections among neurons in the pattern generator circuit). The general structure shown in Figure 12.2 as well as the neural mechanisms used to generate oscillatory patterns are similar in a wide variety of animals; the variety appears to be in the details of how

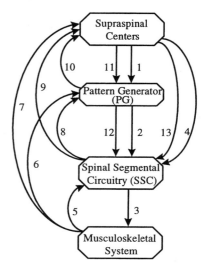

FIGURE 12.2. Interactions among components of the neural system and the musculoskeletal system. Numbered pathways are described as follows:
Motor commands:
  1. descending inputs to PG
  2. PG motor commands
  3. motoneuron outputs
  4. descending supraspinal motor commands
Feedback pathways:
  5. local spinal reflex pathways
  6. reflex pathways to PG
  7. supraspinal reflex pathways
  8. ascending signals from SSC to PG
  9. ascending signals from SSC to supraspinal centers
  10. ascending PG outputs
Modulatory pathways:
  11. supraspinal modulation of PG reflexes
  12. PG modulation of spinal reflexes
  13. supraspinal modulation of spinal reflexes

the basic elements are utilized to achieve specific motor output patterns. The following sections describe the organization of pattern generating circuits and how their behavior is influenced by modulatory inputs from other neural centers and from the periphery.

3.1 Pattern Generator

A "pattern generator" (or "central pattern generator") is a group of neurons that is capable of generating oscillatory outputs in the absence of phasic inputs. A classic and well-studied example of a vertebrate pattern generator is the spinal circuitry of the lamprey (Chapter 15; Cohen and Wallen 1980; Grillner et al. 1991). In this eel-like vertebrate, an *isolated* piece of the spinal cord (consisting of as little as a few spinal segments) can generate stereotyped patterns of activity that are similar to the motor patterns exhibited during locomotion in the intact animal (Cohen and Wallen 1980). This "fictive locomotion" preparation clearly demonstrates existence of pattern generating circuitry in the spinal cord and has been an important tool in characterizing the neural mechanisms that generate the oscillatory behavior. Similar pattern generator circuits have been studied in several invertebrate [e.g., Tritonia (Getting and Dekin 1985), Lymnaea stagnalis (Elliot and Benjamin 1985), crayfish (Mulloney et al. 1993)] and vertebrate preparations [e.g., rat (Cazalets et al. 1995; Smith et al. 1988), cat (Pearson and Rossignol 1991)]. In many of these animals, there may not be conclusive evidence of a pattern generator in the truest sense (i.e., "in the absence of any phasic inputs"), but many pattern generator properties, structures and mechanisms have been identified. Several recent studies in humans have suggested the existence of a spinal pattern generator for locomotion (Calancie et al. 1994; Illis 1995; Gerasimenko et al. 1996), although its properties have not yet been characterized.

Neural circuits utilize a variety of mechanisms in order to generate oscillatory output patterns. The oscillatory behavior of some neural circuits is driven by the intrinsic oscillatory membrane properties of an individual neuron in the circuit. These cells, often called "pacemaker" neurons, are sometimes capable of generating oscillatory outputs when isolated from all other cells in the network. Pacemaker properties of isolated neurons are the result of membrane channels with voltage- and time-dependent ionic conductances (Epstein and Marder 1990). The oscillatory behavior of other neural circuits is driven by the pattern of synaptic connections among the various neurons in the network, none of which has specialized membrane pacemaker properties. This type of circuit is termed a "network oscillator." Examples of synaptic connectivity that are often components of a network oscillator are mutual inhibition and recurrent inhibition (see Figure 12.3). Several computer modeling studies have clearly demonstrated that pure forms of pacemakers and network oscillators are capable of generating oscillatory patterns (Brodin et al. 1991; Ekeberg et al. 1991; Buchanan 1992; Jung et al. 1996). However, most neural circuits appear to utilize a combination of pacemaker and network properties to generate oscillatory patterns (Getting and Dekin 1985; Grillner et al. 1991; Pearson 1993; Rossignol and Dubuc 1994).

3.2 Modulatory Inputs to the Pattern Generator

The pattern generator produces a cyclic set of neural trajectories, but often it is capable of generating many different patterns (Harris-Warrick and Marder 1991; Katz and Frost 1996). The variety in such patterns may be in such characteristics as the relative timing, frequency, duty cycle, and/or amplitude or it may be that the PG circuit is capable of generating patterns that are qualitatively very different (e.g., switching from a pattern suitable for running to one suitable for hopping.) This variety in output patterns can be generated by one or more of several different mechanisms. Although there may be mechanisms that are entirely internal to the pattern generator, we will focus on mechanisms that involve modulatory inputs from neurons other than the core pattern generating circuitry.

One of the more important mechanisms for modulating pattern generator activity is that of "tonic drive" (i.e., a steady nonphasic input that results in modifications to the output pattern). An example of this type of modulatory tonic drive is the excitation of spinal PG circuit from supraspinal centers in the reticular formation of the brainstem, as shown in pathway #11 in Figure 12.2. In several animals, it has been shown that increased levels of tonic activity in the reticular formation leads to in-

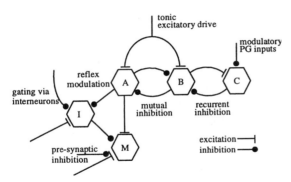

FIGURE 12.3. Neural connectivity for pattern generation and reflex integration. This simplified schematic illustrates several types of connections that may be used by neural circuits to generate oscillatory patterns and to incorporate afferent information. In this schematic, neurons A, B, and C belong to the pattern generator, while neurons I and M are part of the spinal segmental circuitry. The primary oscillatory circuit is formed by neurons A and B, which are in *mutual inhibition* and therefore will oscillate out-of-phase with each other. *Tonic excitatory drive* to PG neurons A and B may be required for oscillation and may be used to modulate frequency. Neuron B excites C which in turn inhibits B, thus demonstrating *recurrent inhibition*. This mechanism may be used in pattern generating circuitry to influence duty cycle as well as frequency. *Modulatory PG inputs*, possibly from supraspinal centers or reflex pathways, may alter the PG output via inhibition (or excitation). In the simplified schematic, they are shown as acting on neuron C, but they may act on any or all of the PG neurons. Motoneurons (M) receive motor commands from the PG and supraspinal centers as well as from afferents. The reflex actions may be directly onto the motoneurons and may be mediated by interneurons (I). The input labeled *"gating via interneurons"* demonstrates an inhibitory input that could act to shut down the reflex pathways by preventing neuron I from firing. The source of the gating signal may be from afferents, supraspinal centers or the PG. *Reflex modulation* is similar to gating, except that the effect may be graded, rather than on or off. In the schematic, reflex modulation is indicated as a cyclic modulation of interneuron mediated reflexes. *Pre-synaptic inhibition* is a mechanism where the inhibitory effect is on the pre-synaptic terminal, rather than the dendrite or cell body. This mechanism may provide a way to selectively deactivating specific inputs to the post-synaptic neuron.

creased frequency of oscillation of the spinal pattern generator (Armstrong 1988). The tonic drive signals may influence PG activity by directly depolarizing the membrane of PG neurons, thereby

bringing them closer to threshold and reducing the time required for the pacemaker or network properties to drive them above threshold. Another possible scenario is that the tonic drive may influence the PG indirectly through the action of neuromodulators. Here, the tonic drive signals may cause the release of neuromodulators (e.g., neuropeptides, calcium ions) which in turn modify the membrane properties of some or all of the PG neurons. These types of changes in membrane properties can produce subtle changes in the PG output pattern such as modulation of frequency, or they can produce drastic changes in the configuration of the network that would result in a functionally distinct output pattern (Harris-Warrick 1988; Harris-Warrick and Marder 1991; Skinner et al. 1994; Brodfuehrer et al. 1995; Katz and Frost 1996).

Another important mechanism for modulating PG activity is via phasic inputs that may produce transient or long-lasting effects. These phasic inputs to the PG are often derived from afferent signals (as described in the next section and illustrated as pathway #6 in Figure 12.2), but they also may originate from within the nervous system. For example, descending phasic inputs (Barnes 1984; Armstrong 1988) may be responsible for initiating or terminating oscillatory activity (via pathway #11 in Figure 12.2). Inputs from supraspinal centers also play an important role in interlimb and intersegmental coordination (Armstrong 1986; Drew 1991; Rossignol et al. 1993). Studies in humans have shown that voluntary noncyclic motor actions performed during locomotion, such as raising an arm or stepping over an obstacle, may result in modifications to motor pattern output that may last for one or more cycles. (Hirschfeld and Forssberg 1991; Patla et al. 1991). While these specific studies could not distinguish between effects on the pattern generator and effects that bypassed the pattern generator, the extensive coordination that was exhibited suggest a pattern generator-mediated mechanism.

4   Interactions in Neuromechanical Systems

Neural signals from motoneurons cause muscle fibers to contract; as muscle fibers contract they generate forces that act on the skeletal structure to result in posture and movement. This "forward" path, although fairly straightforward at a general level of description, is exquisitely complex at a detailed level of analysis. Ionic concentrations in the sarcoplasmic reticulum, heterogeneity in sarcomere lengths (Chapter 3; Morgan 1990), distribution of fibers of a given motor unit, muscle length-tension and force-velocity properties, muscle moment arm dependence on joint angle (Zajac 1989), and the passive mechanics of tendons, ligaments, joint capsule, and skeletal segments (Audu and Davy 1986; Alexander and Ker 1990; Alexander 1993) can all have profound effects on the resulting posture and movement. Several chapters in this book describe many of the more important features of this "forward" path, from neural outputs to movement. This section focuses primarily on the feedback pathway (i.e., how do the mechanical components of the system influence the pattern of signals generated by the neural control system?). Inputs from the mechanical system (periphery) are transmitted to the neural control system via proprioceptive and exteroceptive afferent pathways. Note that these pathways may involve direct transduction of mechanical quantities (e.g., touch sensors in the feet) or may follow a less direct pathway (e.g., visual feedback of limb position).

4.1   Influence on Multiple Time Scales

Signals from the periphery can influence neural output patterns on several different time scales. First, the two systems have coevolved to meet the demands of various environments. Skeletal structure, passive musculotendon properties, active muscle contractile properties, motoneuron structure, pattern generator network architecture, neuromodulator properties, and so on, have simultaneously evolved to result in locomotion that is adequate for a given animal. Obviously, the mechanical properties of the body have influenced the evolution of the neural control system.

Over the course of an animal's life, the neural and mechanical systems codevelop. The neural control system in a given individual develops to meet the requirements of the mechanical system as it develops (and vice versa, but probably to a lesser extent). The combined processes of evolution and development result in the detailed structure of the neural control system.

Throughout the duration of a cyclic activity, such as locomotion, signals from the mechanical system are used to adapt motor patterns. For example, neural activity patterns may be adapted during the course of a walk to account for changing loads, changing slopes or muscle fatigue.

Within the locomotor cycle, sensory inputs are used to detect discrete events that may trigger a transition from one state to another. For example, activation of flexor muscle spindle afferents can facilitate the initiation of swing phase. These inputs may be used to detect "ordinary" events that would be a part of the normal cycle, to detect "extraordinary" events, such as a stumble or slip, or to initiate or terminate locomotion. These inputs affect the PG activity patterns and may, in some cases, be an essential component of pattern generator function.

Finally, sensory inputs are used within specific phases, or states, of the locomotor cycle to influence the form of the pattern being generated. For example, stretch reflex inputs may influence the activity of agonist and antagonist muscles during one phase of the gait cycle, but may not have any effect during other phases. These modulated reflexes, which appear to be widely used in neural control systems, can be described in terms of multi-state models as state-dependent reflexes. The mechanism for such modulation may involve inputs to the pattern generator or may only involve inputs to the local spinal circuits that process pattern generator signals.

The sections that follow describe some specific examples of interactions between neural and mechanical systems.

4.2   Modulation of PG Activity via Afferent Pathways

In normal operation, the pattern generator activity is continuously modulated (Katz and Harris-Warrick 1990; Baev et al. 1991) by input signals from afferents that may be either periodic or non-periodic (see pathway #6 in Figure 12.2). Periodic inputs to the PG can result in *entrainment* of the PG oscillator by the periodic stimulus. An oscillator is said to be entrained by a stimulus when the frequency of the oscillator adjusts to match that of the stimulus. A good example of entrainment of the PG oscillator is demonstrated with the semi-intact preparation of the lamprey where an imposed mechanical oscillation of the tail entrains the spinal oscillator (McClellan and Sigvardt 1988; McClelland and Jang 1993). A similar, but more complicated example, is demonstrated by the spinal cat walking on a treadmill (Forssberg and Grillner 1973). Here, the movement of the treadmill provides a mechanical stimulus by passively moving the limbs into extension. This mechanical stimulus generates neural activity to generate active flexion and therefore results in stepping movements. The speed of the treadmill influences the frequency of passive limb movement, which entrains the neural oscillator. A third example of entrainment is that of a human walking to the beat of a metronome (Chapter 18).

Phasic inputs to the PG that are nonperiodic may also result in modifications to the pattern of PG activity. For example, a bump on the leg may result in modifications the PG pattern for the present cycle, but have no effect on subsequent cycles. It is also possible, however, for a phasic input to trigger a transition from one oscillatory pattern to another oscillatory pattern (e.g., walk to run) or it may result in cessation of all oscillatory activity (e.g., walk to stand). A third possibility is that the effect of the input may be transient in that the system rapidly returns to its original pattern, but that the pattern is phase-shifted with respect to where it would have been without the stimulus (see Figure 12.4). This type response is termed a *phase-resetting* reflex, and appears to be used often in locomotor (Hiebert et al. 1996) and other cyclic activities.

Another important issue to consider regarding transient inputs to the PG is that of *phase-dependence*. That is, the response of the PG to an input may depend upon the phase at which the signal arrives. For example, activation of the flexor muscle afferents during stance phase results in a shortened stance phase and a resetting of the locomotor rhythm, but activation during swing phase does not (Hiebert et al. 1996). In the multistate model of locomotion described above, this phase-dependent reflex would be described as a state-dependent effect, but phase-dependence may also be used to describe changes in the response within a specified state.

It should be noted that the mechanisms that re-

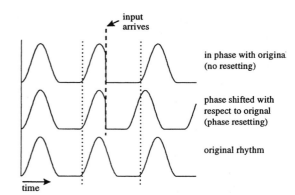

FIGURE 12.4. Effects of phasic inputs to the pattern generator. The traces above illustrate two possible effects of perturbations on the pattern generator output. The bottom trace is the "unperturbed" rhythm. The middle trace demonstrates a reflex in which the phase has been "reset" (i.e., after one short cycle, the phase is shifted with respect to its original rhythm). The top trace demonstrates a reflex without resetting (i.e., the oscillator returns to its original pattern and phase after the perturbation).

sult in entrainment, phase-resetting, and phase dependence are most likely related to the cellular and network properties that are responsible for the dynamic oscillatory behavior of the PG. Voltage-dependent and time-dependent membrane channels and network configurations have all been demonstrated to be capable of altering the effect of inputs on PG activity (Brodin et al. 1991; Traven et al. 1993; Rossignol and Dubuc 1994; Abbas 1996).

4.3 Processing PG Outputs with Afferent Signals

Signals from the pattern generator must eventually reach motoneurons, but it many animals these signals are substantially processed along the way. In more complex vertebrates, for example, the signals pass through the complex network of motoneurons, primary afferent neurons, Renshaw cells and other interneurons, and so on. (Baldissera et al. 1981; Burke 1990). There is considerable evidence to demonstrate that these cells do not belong to the core pattern generating circuitry and that they receive oscillatory inputs during locomotion (Pratt and Jordan 1987), presumably emanating from the pattern generator. The local circuits formed by these spinal neurons are some of the earliest identified components of neuromotor control systems and are probably the most well understood components of the mammalian central nervous system. The monosynaptic, disynaptic and polysynaptic reflex pathways mediated by these neurons (pathway #5 in Figure 12.2) have been very well studied because they are relatively accessible: a quantifiable motor response can be readily evoked by a repeatable stimulus.

The most widely studied reflex pathway is the muscle stretch reflex in which signals from muscle spindles travel via the Ia-afferent pathway to cause monosynaptic activation of the homonymous alpha motoneurons and disynaptic inhibition of the antagonist motoneurons (Carew 1985). An interesting aspect of this reflex loop is that the muscle spindles are innervated by gamma motoneurons. An early view of role of gamma system (the gamma-loop length-servo hypothesis) stated that movement is generated by activation of gamma motoneurons, thus reflexly activating the alpha motoneurons, resulting muscle contraction (Carew 1985). According to this view, the motor control system works much like a classic feedback control system with specification of a setpoint that is achieved via a feedback loop. While it is now appears that this mode of operation is not as dominant as was once believed, it clear that these reflex pathways exist and that they are active during goal-directed movements as well as during locomotion. Therefore, the signal that is sent to a given muscle during the locomotor cycle includes a component from the local afferent pathways as well as a component from the pattern generator.

One issue that arises, then, is to determine the manner in which the two signals are combined—is it just simple addition of the feedforward component from the PG to the feedback component from afferents? Experimental and computer simulation studies have indicated that motoneurons alone are capable of performing complex integration of signals from multiple sources (Burke 1990; Segev et al. 1990). The mechanisms that mediate the interactions between the two components may be at the network level [e.g., presynaptic inhibition (Segev 1990) as shown in Figure 12.3.] or at the cellular level (e.g., involving second messenger systems).

An important aspect of this afferent component is that the feedback gains can be highly modulated depending upon several factors such as attention, movement pattern selection, movement speed and phase within the movement cycle (see pathway #12 in Figure 12.2). This concept of "sensorimotor gain control" (Prochazka 1989) appears to be widely used in the nervous system for a variety of movements, and it has been widely observed during locomotion as well as other cyclic activities (Sillar and Roberts 1988; Koerber and Mendell 1991; Murphy and Martin 1993). First, the reflex gain can be increased at times when feedback might be particularly useful and it can be learned (Evatt et al. 1989; Wolpaw and Lee 1989; Abbruzzese et al. 1991). This type of modulation, which is linked to such higher cortical functions as anticipation, is an example of reflex modulation mediated by higher centers in the nervous system. Second, the reflex gain can be modulated depending upon the type of activity. It has been shown that the human soleus H-reflex gains are lower during running than during walking (Capaday and Stein 1987), while gains during walking are lower than stance (Capaday and Stein 1986). Third, studies on walking at different speeds have demonstrated that lower reflex gains are utilized at higher movement speeds (Zill and Moran 1981). Animals with a slower locomotor cycle appear to utilize higher reflex gains than animals with faster locomotor cycles to the extent to which the reflex mechanisms may dominate and actually be a critical component of the pattern generating circuitry (Chapter 16; Nothof and Bassler 1990). Lastly, studies on cats (Duenas et al. 1990) and humans (Stein and Capaday, 1988; Edamura et al. 1991) have demonstrated that reflex gains are actively modulated throughout the gait cycle. This modulation has been observed in reflex gains from muscle afferents (Dietz et al. 1985), cutaneous afferents (Forssberg et al. 1975; Yang and Stein 1990) and proprioceptors (Hasan and Stuart 1988). The source of the modulatory inputs to the spinal reflex circuits may be directly from the pattern generator, from cyclic afferent inputs, or from oscillatory supraspinal centers. Many aspects of reflex modulation are thoroughly reviewed in (Prochazka 1989); activity-dependent and phase-dependent modulation are discussed in Chapter 17.

Thus, to summarize the most important aspects of these spinal reflex pathways: (1) they are often active during locomotion, (2) they operate to supplement and interact with the signals from the pattern generator, (3) their strength can be modulated through the course of the locomotor cycle, and (4) the relative importance of the feedforward path (from the pattern generator) and the feedback path (via local afferent pathways) may depend upon the speed of the movement being generated as well as other factors.

5   Implications for Biology and Engineering

5.1   Understanding Cyclic Movement Generation

This view of the locomotor control system is centered around a pattern generator and includes supraspinal centers, spinal segmental circuitry, and musculoskeletal mechanical components. The motor pattern results from the interaction of the dynamics of the neural system components with the dynamics of the mechanical system components. Several examples of different forms of interactions amongst the various neural and mechanical components are described.

In its purest form, a pattern generator can produce oscillatory signals in the absence of phasic inputs. On the other hand, in the absence of pure pattern generator circuitry, oscillatory patterns may result from a repeating sequence of reflex-driven movements. Although biological examples of each of these extremes may exist, it may be best to view these as two ends of a spectrum, with most animals using a combination of intrinsic pattern generator and reflex mechanisms. The pattern generator mechanism may be more prominent in animals with fast locomotor cycles, while the reflex mechanisms may be more prominent in animals with slower locomotor cycles. Similarly, in a given animal, the relative importance of the two mechanisms may depend upon walking speed and/or environmental conditions.

The examples in this chapter and the perspectives that follow have drawn from a wide variety of studies on a wide variety of animals. The work cited ranges from studies that utilized intracellular recordings in isolated invertebrate ganglia to studies that utilized kinematic and kinetic analysis of humans. The future of the study of locomotor control systems will undoubtedly include many studies on isolated cells to characterize cellular mechanisms as well as studies on behaving humans to characterize various movement patterns and these studies will continue to provide important contributions to our understanding of the various aspects of the motor control system. The challenge is, as it has always been, to explain the behavioral data in terms of the mechanisms identified in the reduced preparations. An understanding of the interactions between the neural and mechanical systems is an integral part of this challenge.

To understand the neuromechanical interactions in locomotor systems, two approaches may prove to be particularly useful. First, there is a need to develop and exploit experimental paradigms to investigate the interactions between the various components. Isolated muscle preparations and fictive locomotor preparations, while extremely useful for understanding components of the motor control system, can provide only limited information regarding the interactions amongst the components. Technological innovations such as those that allow for neural recordings, muscle length changes and muscle force measurements in behaving animals and experimental paradigms, such as those that use behaving animals to investigate neural connectivity patterns, may provide even more important contributions to the field in the future. These data can also provide the neural activation and muscle strain conditions actually seen by the musculotendon system during rhythmic behavior. More controlled experimental paradigms, such as musculo-tendon work-loop analysis (Josepshon 1985), which take advantage of these relevant conditions, providing an avenue for integration of isolated preparations with whole animal behavior. The second approach that may be critical is to utilize mathematical models of combined neural and mechanical systems (Loeb and Levine 1990). This approach draws from a long history of development of neural (Koch and Segev 1989; Schwartz 1990; Calabrese and DeSchutter 1992; Cohen et al. 1992; Selverston 1993; Bower 1996) and musculoskeletal models (Zajac 1989; Zajac and Winters 1990) and exploits recent advances in computing technology. Here, the use of mathematical models may be particularly useful because experimental analysis and modeling efforts have led to the development of (arguably limited) models of neural and mechanical components, but detailed experimental investigations of the interactions are not currently possible. Several recent examples of studies that have utilized models of combined neural and mechanical systems can be found in this book. (Chapter 15) as well as in the literature (Beer and Chiel 1993; Ekeberg 1993; McFadyen et al. 1994; Taga 1995; Winters 1995; Blum and Leung 1996; Hatsopoulos 1996).

5.2 Designing Improved Movement Control Systems

In many fields of engineering there is a growing interest in the area of biomimicry; that is, how can the principles derived from our understanding of biological systems be utilized to design better engineering systems. Advanced engineering designs of novel materials, lubricants, and image processing systems have successfully incorporated ideas inspired by biological systems into artificial systems. Several research groups involved in the development of locomotion systems have borrowed from biological systems in the design of either the mechanical or control systems component.

Perhaps the most active area in this regard is the development of robots that locomote and perform useful functions. For example, biological foundations have been used to design intelligent systems for motor control that achieve a degree of autonomy (Brooks 1989; Beer et al. 1992). Many of these new engineering systems are targeted at replacing and/or supplementing human operators with the development of robot systems that can move, manipulate objects, gather information, make decisions, and act on them. These systems will require that the robots have intelligence and motor control capabilities that approach those of animal cognitive and motor control systems.

A second important engineering application is the design of controllers for use in systems that electrically stimulate paralyzed muscles to restore function to people with neurological disorders (see Section IX, for descriptions of Functional Neuromuscular Stimulation (FNS) systems). Biologically inspired systems for controlling cyclic movements have been developed for intended use in restoring locomotor function (Abbas and Chizeck 1995; Abbas 1996; Chapter 46). In addition, effective utilization of the ability of the

spinal cord to modulate reflexes has been proposed as a means of enhancing FES system function (Fung and Barbeau 1994).

The view of locomotor systems presented in this chapter includes several key features that may be important in the design of engineering systems for cyclic movement. To design a system that will generate cyclic movements such as locomotion, the mechanical properties of the system should be well suited for the "regular" movement pattern(s) and the controller should be capable of exploiting those properties. A control system that utilizes such ideas as a pattern generator, phase-resetting reflexes, reflex modulation, and so on, may provide advantages in terms of mechanical and/or computational efficiency and the capability of generating a variety of movement patterns at different speeds. While it may be possible to incorporate these features using standard control system structures, mathematical models of neural processing systems may be an effective and efficient way to achieve these objectives (Beer 1990; Abbas 1996).

References

Abbas, J.J. (1996). Using neural models in the design of a movement control system. *Computational Neurosci.*, Bower, J.M. (ed.), pp. 305–310, Academic Press, New York.

Abbas, J.J. and Chizeck, H.J. (1995). Neural network control of functional neuromuscular stimulation systems. *IEEE Trans. BME*, 42:1117–1127.

Abbruzzese, M., Reni, L., and Favale, E. (1991). Changes in central delay of soleus H-reflex after facilitatory or inhibitory conditioning in humans. *J. Neurophys.*, 65:1598–1605.

Alexander, R.M. (1993). Optimization of structure and movement of the legs of animals. *J. Biomech.*, 26:1–6.

Alexander, R.M. and Ker, R.F. (1990). The architecture of leg muscles. *Multiple Muscle Systems: Biomechanics and Movement Organization*. Winters, J.M. and Woo, S.L.-Y. (eds.), pp. 568–577, Springer-Verlag, New York.

Alexander, R.M. and Vernon, A. (1975). Mechanics of hopping by kangaroos (Macropodidae). *J. Zool. Lond.*, 177:265–303.

Armstrong, D.M. (1986). Supraspinal contributions to the initiation and control of locomotion in the cat. *Prog. Neurobiol.*, 26:273–361.

Armstrong, D.M. (1988). The supraspinal control of mammalian locomotion. *J. Physiol.*, 405:1–37.

Audu, M.L. and Davy, D.T. (1985). The influence of

muscle model complexity in musculoskeletal motion modeling. *J. Biomech. Eng.*, 107:147–157.

Baev, K.V., Esipenko, V.B., and Shimansky, Y.P. (1991). Afferent control of central pattern generators: experimental analysis of locomotion in the decerebrate cat. *Neuroscience*, 43:237–247.

Baldissera, F., Hultborn, H., and Illert, M. (1981). Integration in spinal neuronal systems. *Handbook of Physiology, Sect. 1: The Nervous System II, Motor Control*. Brooks, V.B. (ed.), pp. 509–595, American Physiological Society. Waverly Press, Bethesda, Maryland.

Barnes, C.D. (1984). *Brainstem Control of Spinal Cord Function*. Academic Press, New York.

Beer, R.D. (1990). *Intelligence as Adaptive Behavior: An Experiment in Computational Neuroethology*. Academic Press, Boston.

Beer, R.D. and Chiel, H.C. (1993). Simulations of cockroach locomotion and escape. *Biological Neural Networks in Invertebrate Neuroethology and Robotics*. Beer, R.D., Ritzmann, R.E., and McKenna, T. (eds.), pp. 267–286, Academic Press, New York.

Beer, R.D., Chiel, H.J., Quinn, R.D., Espenschied, K.S., and Larsson, P. (1992). A distributed neural network architecture for hexapod robot locomotion. *Neural Computation*, 4:356–365.

Biewener, A.A. and Baudinette, R.V. (1995). In vivo muscle force and elastic energy storage during steady-speed hopping of tamar wallabies (Macropus eugenii). *J. Exp. Biol.*, 198:1829–1841.

Blum, E.K. and Leung, P.K. (1996). Modeling and simulation of human walking: a neuro-musculo-skeletal model. *Computational Neurosci.* Bower, J.M. (ed.), pp. 323–327. Academic Press, New York.

Bower, J.M. (ed.). (1996). *Computational Neurosci.*, Academic Press, New York.

Brodfuehrer, P.D., Debski, E.A., O'Gara, B.A., and Friesen, W.O. (1995). Neuronal control of leech swimming. *J. Neurobiol.*, 27:403–418.

Brodin, L., Traven, H., Lansner, A., Wallen, P., and Grillner, S. (1991). Computer simulations of N-methl-D-aspartate receptor-induced membrane properties in a neuron model. *J. Neurophys.*, 66:473–484.

Brooks, R.A. (1989). A robot that walks: emergent behaviors from a carefully evolved network. *Biological Neural Networks in Invertebrate Neuroethology and Robotics*. Brooks, R.A. (ed.), pp. 355–363, The MIT Press, Cambridge, Massachusetts.

Buchanan, J.T. (1992). Neural network simulations of coupled locomotor oscillators in the lamprey spinal cord. *Biol. Cybern.*, 66:367–374.

Burke, R.E. (1990). Spinal cord: ventral horn. *The Synaptic Organization of the Brain*. Shepherd, G.M. (ed.),

pp. 88–132, 3rd edn. Oxford University Press, New York.

Calabrese, R. and DeSchutter, E. (1992). Motor-pattern-generating networks in invertebrates: modeling our way toward understanding. *TINS*, 15:439–444.

Calancie, B., Needham-Shropshire, B., Jacobs, P., Willer, K., Zych, G., and Green, B.A. (1994). Involuntary stepping after chronic spinal cord injury. Evidence for a central rhythm generator for locomotion in man. *Brain*, 117:1143–1159.

Capaday, C. and Stein, R.B. (1986). Amplitude modulation of the soleus H-reflex in the human during walking and standing. *J. Neurosci.*, 6:1308–1313.

Capaday, C. and Stein, R.B. (1987). Difference in the amplitude of the human soleus H-reflex during walking and running. *J. Physiol.*, 392:513–522.

Carew, T.J. (1985). The control of reflex action. *Principles of Neural Science*. Kandel, E.R. and Schwartz, J.H. (eds.), pp. 457–468, Elsevier Publishers, New York.

Cazalets, Jean-R., Borde, M., and Clarac, F. (1995). Localization and organization of the central pattern generator for hindlimb locomotion in newborn rat. *Neuroscience*, 15:4943–4951.

Chiel, H.J. and Beer, R.D. (1993). Neural and peripheral dynamics as determinants of patterned motor behavior. *The Neurobiology of Neural Networks*. Gardner, D. (ed.), pp. 137–164, The MIT Press, Cambridge, Massachusetts.

Cohen, A., Ermentrout, B., Kiemel, T., Kopell, N., Sigvardt, K., and Williams, T. (1992). Modelling of intersegmental coordination in the lamprey central pattern generator for locomotion. *TINS*, 15:434–438.

Cohen, A.H. (1992). The role of heterarchical control in the evolution of central pattern generators. *Brain, Behav. Evol.*, 40:112–124.

Cohen, A.H. and Wallen, P. (1980). The neuronal correlate of locomotion in fish. 'fictive swimming' induced in an in vitro preparation of the lamprey spinal cord. *Exp. Brain. Res.*, 41:11–18.

Cohen, A.H., Rossignol, S., and Grillner, S. (eds.). (1988). *Neural Control of Rhythmic Movements in Vertebrates*. John Wiley & Sons, New York.

Dietz, B., Quintern, J., and Berger, W. (1985). Afferent control of human stance and gait: evidence for blocking of group I afferents during gait. *Exp. Brain Res.*, 61:153–163.

Drew, T. (1991). Functional organization within the medullary reticular formation of the intact unanesthetized cat III. Microstimulation during locomotion. *J. Neurophys.*, 66:919–938.

Duenas, S.H., Loeb, G.E., and Marks, W.B. (1990). Monosynaptic and dorsal root reflexes during locomotion in normal and thalamic cats. *J. Neurophys.*, 63:1467–1476.

Edamura, M., Yang, J.F., and Stein, R.B. (1991). Factors that determine the magnitude and time course of human H-reflexes in locomotion. *J. Neurosci.*, 11:420–427.

Ekeberg, O. (1993). A combined neuronal and mechanical model of fish swimming. *Biol. Cybern.*, 69:363–374.

Ekeberg, O., Wallen, P., Lansner, A., Traven, H., Brodin, L., and Grillner, S. (1991). A computer based model for realistic simulations of neural networks. *Biol. Cybern.*, 65:81–90.

Elliot, C.J.H. and Benjamin, P.R. (1985). Interactions of pattern-generating interneurons controlling feeding in lymnaea stagnalis. *J. Neurophys.*, 54:1396–1421.

Epstein, I.R. and Marder, E. (1990). Multiple modes of a conditional neural oscillator. *Biol. Cybern.*, 63:25–34.

Ettema, G.J.C. and Huijing, P.A. (1990). Architecture and elastic properties of the series elastic element of muscle-tendon complex. *Multiple Muscle Systems: Biomechanics and Movement Organization*. Winters, J.M. and Woo, S.L.-Y. (eds.), pp. 57–68, Springer-Verlag, New York.

Evatt, M.L., Wolf, S.L., and Segal, R.L. (1989). Modification of human spinal stretch reflexes: preliminary studies. *Neur. Lett.*, 105:350–355.

Forssberg, H. and Grillner, S. (1973). The locomotion of the acute spinal cat injected with clonidine i.v. *Brain Res. Bull.*, 50:184–186.

Forssberg, H., Grillner, S., and Rossignol, S. (1975). Phase dependent reflex reversal during walking in chronic spinal cats. *Brain Res.*, 85:103–107.

Fung, J. and Barbeau, H. (1994). Effects of conditioning cutaneomuscular stimulation on the soleus H-reflex in normal and spastic paretic subjects during walking and standing. *J. Neurophys.*, 72:2090–2104.

Gerasimenko, Y., McKay, W.B., Pollo, F.E., and Dimitrijevic, M.R. (1996). Stepping movements in paraplegic patients induced by epidural spinal cord stimulation. *Soc. Neurosci. Abstracts*, 22:1372.

Getting, P.A. and Dekin, M.S. (1985). Tritonia swimming: a model system for integration within rhythmic motor systems. *Model Neural Networks and Behavior*. Selverston, A.I. (ed.), pp. 3–20, Plenum Press.

Ghez, C. (1985). Introduction to the motor systems. *Principles of Neural Science*. Kandel, E.R. and Schwartz, J.H. (eds.), pp. 427–442, 2nd edn. Elsevier, New York.

Grillner, S. (1981). Control of locomotion in bipeds, tetrapods and fish. *Handbook of Physiology, Sect. 1: The Nervous System II, Motor Control*. Brooks, V.B. (ed.), pp. 1179–1236, American Physiological Society, Waverly Press, Bethesda, Maryland.

Grillner, S., Wallen, P., Brodin, L., and Lansner, A. (1991). Neuronal network generating locomotor behavior in Lamprey: circuitry, transmitters, membrane properties, and simulation. *Annu. Rev. Neurosci.*, 14:169–199.

Harris-Warrick, R.M. (1988). Chemical Modulation of Central Pattern Generators. *Neural Control of Rhythmic Movements in Vertebrates*. Cohen, A.H., Rossignol, S., and Grillner, S. (eds.), pp. 285–331, John Wiley & Sons, New York.

Harris-Warrick, R.M. and Marder, E. (1991). Modulation of Neural Networks for Behavior. *Annu. Rev. Neurosci.*, 14:39–57.

Hasan, Z. and Stuart, D.G. (1988). Animal solutions to problems of movement control: the role of proprioceptors. *Annu. Rev. Neurosci.*, 11:199–223.

Hatsopoulos, N.G. (1996). Coupling the neural and physical dynamics in rhythmic movements. *Neural Computation* 8:567–581.

Hiebert, G.W., Whelan, P.J., Prochazka, A., and Pearsons, K.G. (1996). Contribution of hind limb flexor muscle afferents to the timing of phase transitions in the cat step cycle. *J. Neurophys.*, 75:1126–1137.

Hinrichs, R. (1990). Whole body movement: coordination of arms and legs in walking and running. *Multiple Muscle Systems: Biomechanics and Movement Organization*. Winters, J.M. and Woo, S.L.-Y. (eds.), pp. 694–716, Springer-Verlag, New York.

Hirschfeld, H. and Forssberg, H. (1991). Phase-dependent modulations of anticipatory postural activity during human locomotion. *J. Neurophys.*, 66:12–18.

Hof, A.L. (1990). Effects of muscle elasticity in walking and running. *Multiple Muscle Systems: Biomechanics and Movement Organization*. Winters, J.M. and Woo, S.L.-Y. (eds.), pp. 591–607, Springer-Verlag, New York.

Illis, L.S. (1995). Is there a central pattern generator in man? *Paraplegia*, 33:239–240.

Inman, V.T., Ralston, H.J., and Todd, F. (1994). Human locomotion. *Human Walking*. Rose, J. and Gamble, J.G. (eds.), pp. 1–22, 2nd ed. Williams & Wilkins, Baltimore.

Josephson, R.K. (1985). Mechanical power output from striated muscle during cyclic contraction. *J. Exp. Biol.*, 114:493–512.

Jung, R., Kiemel, T., and Cohen, A.H. (1996). Dynamic behavior of a neural network model of locomotor control in the lamprey. *J. Neurophys.*, 75:1074–1086.

Katz, P.S. and Forst, W.N. (1996). Intrinsic neuromodulation: altering neuronal circuits from within. *TINS*, 19:54–61.

Katz, P.S. and Harris-Warrick, R.M. (1990). Neuromodulation of the crab ploric central pattern generator by serotonergic/cholinergic proprioceptive afferents. *J. Neurosci.*, 10:1495–1512.

Koch, C. and Segev, I., Editors, (1989). *Methods in Neuronal Modeling: From Synapses to Networks.* The MIT Press, Cambridge, Massachusetts.

Koerber, J.R. and Mendell, L.M. (1991). Modulation of synaptic transmission as Ia-afferent connections on motorneurons during high-frequency afferent stimulation: dependence on motor task. *J. Neurophys.,* 65:1313–1320.

Loeb, G. and Levine, W. (1990). Linking musculoskeletal mechanics to sensorimotor neurophysiology. *Multiple Muscle Systems: Biomechanics and Movement Organization.* Winters, J.M. and Woo, S.L.-Y. (eds.), pp. 165–181, Springer-Verlag, New York.

McClellan, A.D. and Jang, W. (1993). Mechanosensory inputs to the central pattern generators for locomotion in the lamprey spinal cord: resetting, entrainment and computer modeling. *J. Neurophys.,* 706:2442–2454.

McClellan, A.D. and Sigvardt, K.A. (1988). Features of entrainment of spinal pattern generators for locomotor activity in the lamprey spinal cord. *J. Neurosci.,* 8:133–145.

McFadyen, B.J., Winter, D.A., and Allard, F. (1994). Simulated control of unilateral, anticipatory locomotor adjustments during obstructed gait. *Biol. Cybern.,* 72:151–160.

McGeer, T. (1993). Dynamics and control of bipedal locomotion. *J. Theor. Biol.,* 163:277–314.

McMahon, T.A. (1984). *Muscles, Reflexes, and Locomotion.* Princeton University Press, Princeton, New Jersey.

Morgan, D. (1990). Modeling of lengthening muscle: the role of inter-sarcomere dynamics. *Multiple Muscle Systems: Biomechanics and Movement Organization.* Winter, J.M. and Woo, S.L.-Y. (eds.), pp. 45–56, Springer-Verlag, New York.

Mulloney, B., Murchison, D., and Chrachri, A. (1993). Modular organization of pattern-generating circuits in a segmental motor system: the swimmerets of crayfish. *Sem. Neurosci.,* 5:49–57.

Mungiole, M. and Winters, J.M. (1990). Overview: influence of muscle on cyclic and propulsive movements involving the lower limb. *Multiple Muscle Systems: Biomechanics and Movement Organization.* Winters, J.M. and Woo, S.L.-Y. (eds.), pp. 550–567, Springer-Verlag, New York.

Murphy, P.R. and Martin, H.A. (1993). Fusimotor discharge patterns during rhythmic movements. *TINS,* 16:273–278.

Nothof, U. and Bassler, U. (1990). The network producing the "active reaction" of stick insects is a functional element of different pattern generating systems. *Biol. Cybern.,* 62:453–462.

Patla, A.E., Prentice, S.D., Robinson, C., and Neufeld, J. (1991). Visual control of locomotion: strategies for changing direction and for going over obstacles. *J. Exp. Psych.,* 17:603–634.

Pearson, K.G. (1993). Common principles of motor control in vertebrates and invertebrates. *Annu. Rev. Neurosci.,* 16:265–297.

Pearson, K.G. and Rossignol (1991). Fictive motor patterns in chronic spinal cats. *J. Neurophys.,* 66:1–14.

Pratt, C.A. and Jordan, L.M. (1987). Ia inhibitory interneurons and renshaw cells as contributors to the spinal mechanisms of fictive locomotion. *J. Neurophys.,* 57:56–71.

Prochazka, A. (1989). Sensorimotor gain control: a basic strategy of motor systems? *Prog. Neurobiol.,* 33:287–307.

Raibert, M.H. and Hodgkins, J.K. (1993). Legged robots. *Biological Neural Networks in Invertebrate Neuroethology and Robotics.* Beer, R.D., Ritzmann, R.E., and McKenna, T. (eds.), pp. 319–353, Academic Press, New York.

Rossignol, S. and Dubuc, R. (1994). Spinal pattern generation. *Curr. Opin. Neurbiol.,* 4:894–902.

Rossignol, S., Saltiel, P., Perreault, M.C., Drew, T., Pearson, K., and Belanger, M. (1993). Intralimb and interlimb coordination in the cat during real and fictive rhythmic motor programs. *Sem. Neurosci.,* 5:67–75.

Schoner, G. and Kelso, J.A.S. (1988). Dynamic pattern generation in behavioral and neural systems. *Science,* 239:1513–1520.

Schwartz, E.L. (ed.). (1990). *Computational Neuroscience.* The MIT Press, Cambridge, Massachusetts.

Segev, I. (1990). Computer study of presynaptic inhibition controlling the spread of action potentials into axonal terminals. *J. Neurophys.,* 63:987–998.

Segev, I., Fleshman, J.W., and Burke, R.E. (1990). Computer simulation of group Ia EPSPs using morphologically realistic models of cat alpha-motoneurons. *J. Neurophys.,* 64:648–660.

Selverston, A. (1993). Modeling of neural circuits: what have we learned? *Annu. Rev. Neurosci.,* 16:531–546.

Sillar, K.T. and Roberts, A. (1988). A neuronal mechanism for sensory gating during locomotion in a vertebrate. *Nature,* 331:262–265.

Skinner, F., Kopell, N., and Mardner, E. (1994). Mechanics for oscillation and frequency control in reciprocally inhibitory model neural networks. *J. Comp. Neurosci.,* 1:69–87.

Smith, J.C., Feldman, J.L., and Schmidt, B.J. (1988). Neural mechanisms generating locomotion studied in mammalian brain stem-spinal cord in-vitro. *FASEB J.,* 2:2283–2288.

Stein, R.B. and Capaday, C. (1988). The modulation of human reflexes during functional motor tasks. *TINS,* 11:328–332.

Taga, G. (1995). A model of the neuro-musculo-skeletal

system for human locomotion I: emergence of basic gait. *Biol. Cybern.*, 73:97–111.

Traven, H.G.C., Brodin, L., Lansner, A., Ekeberg, O., Wallen, P., and Grillner, S. (1993). Computer simulations of NMDA and non-NMDA receptor-mediated synaptic drive: sensory and supraspinal modulation of neurons and small networks. *J. Neurophys.*, 70:695–709.

Winter, D.A. (1987). *The Biomechanics and Motor Control of Human Gait.* University of Waterloo Press, Waterloo, Canada.

Winter, D.A. (1990). *Biomechanics and Motor Control of Human Movement.* John Wiley & Sons, New York.

Winters, J.M. (1995). An improved muscle-reflex actuator for use in large-scale neuromusculoskeletal models. *Ann. Biomed. Eng.*, 23:359–374.

Wolf, H. and Laurent, G. (1994). Rhythmic modulation of the responsiveness of locust sensory local interneurons by walking pattern generating networks. *J. Neurophys.*, 71:111–118.

Wolpaw, J.R. and Lee, R.L. (1989). Memory traces in primate spinal cord produced by operant conditioning of H-reflex. *J. Neurophys.*, 61:563–572.

Yamaguchi, G.T. (1990). Performing whole-body simulations of gait with 3-D, dynamic musculoskeletal models. *Multiple Muscle Systems: Biomechanics and Movement Organization.* Winters, J.M. and Woo, S.L.-Y. (eds.), pp. 663–679, Springer-Verlag, New York.

Yang, J.F. and Stein, R.B. (1990). Phase-dependent reflex reversal in human leg muscles during walking. *J. Neurophys.*, 63:1109–1117.

Zajac, F. and Winters, J. (1990). Modeling musculoskeletal movement systems: joint and body segmental dynamics, musculoskeletal actuation, and neuromuscular control. *Multiple Muscle Systems: Biomechanics and Movement Organization.* Winters, J.M. and Woo, S.L.-Y. (eds.), pp. 121–148, Springer-Verlag, New York.

Zajac, F.E. (1989). Muscle and tendon: properties, models, scaling, and application to biomechanics and motor control. *CRC Crit. Rev. Biomed. Eng.*, 17:359–411.

Zernicke, R.F., Schneider, K., and Buford, J.A. (1991). Intersegmental dynamics during gait. Patla, A.E. (eds.), pp. 187–202, Elsevier Science Publishers, New York.

Zill, S.N. and Moran, D.T. (1981). The exoskeleton and insect proprioception: III Activity of tibial campaniform sensilla during walking in the american cockroach. Periplaneta americana. *J. Exp. Biol.*, 94:57–75.

13
Musculoskeletal Dynamics in Rhythmic Systems: A Comparative Approach to Legged Locomotion

Robert J. Full and Claire T. Farley

1 Introduction

1.1 Advantage of Rhythmic Systems

The challenge to integrate neural control with musculoskeletal dynamics has and will continue to benefit from the study of rhythmic systems. Rhythmic systems offer at least two major advantages over systems that are episodic, discontinuous or ballistic. First, experiments on individual organisms which manipulate a single variable—*direct experiments*—can be more conclusive because differences in continuous, patterned outputs are often easier to discern (Figure 13.1). Second, rhythmic systems are ubiquitous in nature. Fliers, swimmers and runners cycle their bodies and appendages at frequencies that range from less than 1 cycle per second in the largest animals such as whales to as high as 1000 Hz in flying insects. The extraordinary pervasiveness of rhythmic systems allows us to conduct *natural experiments* on choice animals by using the comparative method (Figure 13.1). By choice animals, we mean the ones most amenable to a particular experimental procedure. The giant squid axon and the robust gastrocnemius muscle of frogs are notable examples. August Krogh said it best in 1929 at the 13th International Congress of Physiology in Boston: "For many problems there is an animal on which it can be most conveniently studied" (Krebs 1975). By the comparative method or natural experiment, we mean the process of inferring function by comparing two or preferably more species that differ in the variable of interest due to evolutionary history as opposed to an experimenter's procedure.

1.2 Advantage of the Comparative Method

The advantage of natural experiments is substantial. Comparing systems that have evolved over millions of years can results in enormous differences in variables of interest. Organismal diversity can enable discovery. Comparing systems which differ naturally can avoid the disruption in function to a finely integrated system that can result from direct experimental perturbations pushed too far in search of a significant difference. For example, the metabolic cost of locomotion often varies by much less than ten-fold when speed, stride frequency, inclines or added loads are altered in individuals, whereas cost naturally differs by over five orders of magnitude when all legged animals are compared. Large variation in dependent variables found in natural systems permits isolation and investigation of processes of interest in nearly an ideal setting—one of exaggerated function in a normally operating system. Large differences in function are associated with differences in body mass, environmental extremes, and lifestyles. Fortunately, variation in dependent variables shows remarkably general patterns and correlations which can be used to infer function and predict performance in animals not yet studied. Equally important, however, are those systems that demonstrate spectacular performance and deviate from the general pattern. Characterization of these specialized systems can allow extrapolation to other systems in which the properties of interest are not present in the extreme, but in which the principles of function are the same.

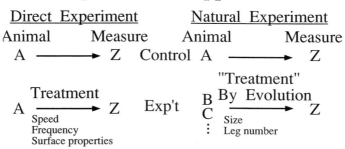

FIGURE 13.1. Comparative method. Direct experiments with imposed treatments and tight controls are most effective at establishing cause and effect. Natural experiments compare species that differ by the "treatment" variable because of evolution. With a comparative approach, large differences can be measured among species in a naturally operating system unperturbed by the investigator. Natural experiments may be less controlled because variables other than those of interest may also differ among species.

For example, hopping red kangaroos can increase speed without an increase in metabolic energy cost (Dawson and Taylor 1973). Further examination provided evidence of elastic strain energy storage in the tendons of kangaroo leg muscles (Alexander and Vernon 1975; Alexander 1988; Biewener and Baudinette 1995). It is reasonable to conclude, at least in large vertebrates such as humans, that tendons serve a similar role, albeit to a lesser extent than in specialized, bipedal hoppers.

By using a phylogenetic approach with three or more species, the effects of history can be examined. Although powerful, the comparative method is best used in conjunction with a knowledge of evolutionary history or phylogeny (Huey 1987; Garland and Adolph 1994). The reason is simple. Natural experiments can be viewed as imperfect because they may lack an appropriate control. Seldom do the species being compared differ only by the variable of interest. The ideal comparison—very closely related species possessing a large difference in the process being studied—is rare. Fortunately, recently developed techniques in phylogenetic analysis (Felsenstein 1985; Garland et al. 1992) offer a tool to remove the effects of history or use them to hint at present function. If the process of interest has severe functional/structural constraints or complete adaptation has taken place, then the potentially confounding effects of historical differences may be of little consequence. If, however, functional constraint and adaptation have been less than completely dominant, then the most parsimonious assumption is that the process should operate as it did in the ancestor.

Adopting the comparative method has made it possible to use a three pronged approach toward integrating neural control with musculoskeletal dynamics. The *forward dynamics* approach has emphasized circuit breaking in animals such as arthropods (see Chapter 16), lamprey (Chapter 15), and annelids (see Chapter 14). Fundamental neural control principles, which include central pattern generation and parallel, distributed control, have been derived from animals that exhibit rhythmic behavior. Musculoskeletal *system identification* techniques are finally allowing the characterization of muscle function under stimulation and strain conditions actually found during rhythmic behavior. One such approach is termed musculoskeletal work-loop analyses (Josephson 1985). Lastly, the mechanical behavior of the complete rhythmic system can be characterized first by using an *inverse dynamics* approach. For each approach, we advocate both direct experimental manipulations in close coordination with modeling which employs extensive sensitivity analyses.

The present perspective focuses on an inverse dynamics approach for several reasons. First, it argues for a broader view of the plant—the musculoskeletal system. Second, it illustrates the utility of the comparative approach. Third, it demonstrates how a simple model of the musculoskeletal system can be used as a starting point to limit the possible ways in which legs function.

1.3 Control Offered by the Musculoskeletal System

The rigid characterization of the controller as the central and sensory neural networks and the plant as the musculoskeletal system is artificial and counterproductive for several reasons (Zajac and Gordon 1989). First, the information flow among neural and musculoskeletal units is closed loop, not unidirectional. Joint and muscle sensors feed back information about limb and body position, velocity and force. In a way, feedback makes the musculoskeletal system act as the controller and the neural system function as the plant. Second, feedback can occur within the musculoskeletal system itself where joint position and velocity can determine musculoskeletal dynamics (i.e., preflexes of Brown and Loeb in Chapter 10). Moreover, joint angle can affect joint geometry by altering moment arms which, in turn, affect joint dynamics through torque development.

The third and most important reason for not characterizing the musculoskeletal system as simply a plant is the increasing evidence that the mechanical system as a whole exerts a form of control. Raibert and Hodgins (1993) stated it this way, "Many researchers in neural motor control think of the nervous system as a source of commands that are issued to the body as direct orders. We believe that the mechanical system has a mind of its own, governed by the physical structure and laws of physics. Rather than issuing direct commands, the nervous system can only make suggestions which are reconciled with the physics of the system and task at hand." In this vein, Mochon and McMahon (1980, 1981) showed the importance of a passive leg swing in ballistic walking. Playter and Raibert (1994) demonstrated how movable arms attached by springs can stabilize rotation by passive dynamics. McGeer (1990a, 1990b) showed that completely passive, somewhat anthropomorphic 2D mechanisms, including kneed models, could walk stably down shallow slopes. Ruina and his students have built similar mechanisms and studied related 2D (Garcia et al. 1998) and 3D (Coleman and Ruina 1998) passive-dynamic models. These walking machines have no sensors or actuators and lack any computer control. For years, Raibert has devised spectacular hopping, running and trotting robots with only very simple control systems (Raibert and Hodgins 1993). Perhaps it would be easier to identify active neural control if we adopted the view that the musculoskeletal or mechanical system functions also as a control system—a more passive, yet dynamic, control system. In passive dynamic control, the control algorithms are simply embedded in the form of the machine or animal itself. Control results from the properties of the parts and their morphological arrangement. Musculoskeletal units, leg segments, and legs do much of the computations on their own by using segment mass, length, inertia, elasticity, and dampening as "primitives." Nevertheless, passive dynamic control lacks the plasticity of active neural control, since suites of integrated structures which have evolved over millions of years take longer to modify. Passive dynamic controllers can respond, however, immediately (i.e., zero order) and effectively to a variety of perturbations.

2 The Spring-Mass Model of Running

The purpose of this short perpective is to illustrate how a simple, passive dynamic model can be an effective starting point in attacking complex, indeterminant biological systems. We refer the reader to benchmark publications and insightful reviews of the history, development and characterization of the spring-mass model of running available (Cavagna et al. 1976, 1977, 1988; Heglund et al. 1982; McMahon 1985, 1990; Alexander 1988, 1992; Blickhan, 1989; Thompson and Raibert 1989; McMahon and Cheng 1990).

Terrestrial locomotion is accomplished by legged animals that could hardly be more diverse. Legged locomotors differ in leg number (from forty-four on a centipede to two on a human), body form (tall in bipedal birds, round in crabs, long in reptiles), skeletal type (exo- versus endoskeletons), body mass (from less than 1 g to hundreds of kg), and thermal strategy (warm- versus cold-blooded). With this degree of diversity, one reasonable assumption is that each taxa would have a unique set of kinematic and kinetic relationships which power their bodies forward. Abundant diversity does not suggest the possibility that simple rules for locomotion of legged animals would emerge when comparisons are made.

2.1 Comparable Ground Reaction Force Patterns

Surprisingly similar ground reaction forces are exerted by animals that do differ in leg number, body mass, body form and type of skeleton (Figure 13.2). Two-, four-, six- and eight-legged animals can produce similar force patterns during locomotion. They all can bounce as they run using two alternating sets of legs. One human leg works like two legs of a trotting dog, two legs of a trotting lizard, three legs of an insect and four legs of a crab.

Cavagna et al. (1977) provided the first evidence that the musculoskeletal system behaves like a sin-

gle linear spring in running, hopping and trotting gaits in mammals. These data led to the development of a spring-mass model, consisting of a single linear massless "leg spring" and a mass (Figure 13.3). The "leg spring" represents the spring-like characteristics of the overall integrated musculoskeletal system during locomotion. The mass is equivalent to the mass of the animal. This spring-mass model has been shown to describe and predict the mechanics of running gaits remarkably well (Ito et al. 1983; Alexander 1988; Blickhan 1989; Thompson and Raibert 1989; McGeer 1990a; McMahon 1990; McMahon and Cheng 1990; Farley et al. 1991, 1993; He et al. 1991; Blickhan and

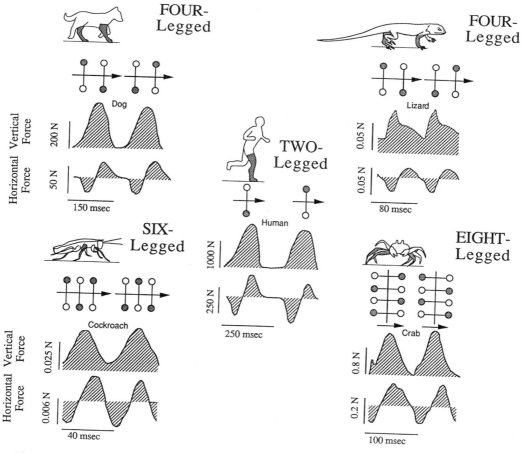

FIGURE 13.2. Comparable ground reaction forces in diverse animals. Each legged animals generates a similar pattern of forces as they bounce along like a pogo stick or spring-mass system. In each case two sets of propulsors operate as leg springs to slow down and then speed up the animal. The leg spring is composed of one leg in a human, two legs in a trotting dog, three legs in a cockroach and four legs in a trotting crab. Closed circles in gait diagrams represent legs contacting the ground. (Modified from Full 1989; data on lizard is from Farley and Ko, 1997.)

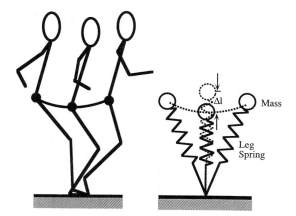

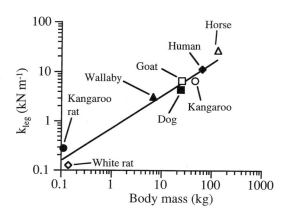

FIGURE 13.3. Trotting, hopping, and running animals are modeled as simple spring-mass systems bouncing along the ground. The model consists of a single linear leg spring and a point mass, equivalent to the mass of the animal. This figure depicts the model at the beginning of the stance phase (left-most position), at the middle of the stance phase (leg spring is oriented vertically) and at the end of the stance phase (right-most position). The arc shows the movement of the mass during the stance phase. The leg spring is maximally compressed at the middle of the stance phase. The dashed spring-mass model shows the length of the uncompressed leg spring. Thus, the difference between the length of the dashed leg spring and the maximally compressed leg spring represents the maximum compression of the leg spring (Δl). (Modified from Farley et al. 1993.)

FIGURE 13.4. The stiffness of the leg spring (k_{leg}), calculated from the ratio $F/\Delta l$, increases with body mass ($k_{leg} \propto M^{0.67}$). This comparison was made at moderate trotting, running, or hopping speeds where the duty factor (0.4) and the Froude number (1.5) were similar in all of the animals. (Modified from Farley et al. 1993.)

Full 1993; Farley and Gonzalez 1996). The stiffness of the leg spring is defined as the ratio of the ground reaction force to the compression of the leg spring at the instant at midstance when the leg is maximally compressed (Figure 13.3). If F is the ground reaction force and Δl is the compression of the leg spring, the leg stiffness (k) can be calculated from Equation 13.1.

$$k = F \, \Delta l^{-1} \qquad (13.1)$$

2.2   Leg Spring Stiffness Scales with Body Size

The spring-mass model for bouncing gaits describes the mechanics of locomotion in animals ranging in body size from a cockroach (0.001 kg) to a horse (135 kg). Force platform studies of locomotion have revealed that larger animals have stiffer leg springs (Figure 13.4). For example, a trotting horse has a leg spring stiffness 100-fold

higher than a trotting white rat. Comparison of mammals with a 1000-fold range of body masses has revealed that the stiffness of the leg spring increases in proportion to $M^{0.67}$ (M = body mass). As a result, the resonant period of vibration of the spring-mass system increases in larger animals ($M^{0.19}$), exactly paralleling the observed increase in the time of foot-ground contact with increasing body mass (Farley et al. 1993).

2.3   Similarity in Relative Individual Leg Spring Stiffness

The number of legs used to support the weight of the body during the ground contact phase of bouncing gaits varies from one to four among the legged animals studied thus far (Blickhan and Full 1993). A relative leg stiffness of all limbs in contact with the surface can be calculated by dividing the relative force of the spring (normalized for body weight) by the relative compression (normalized by leg length), and we can calculate the relative stiffness of the leg spring (k_{rel}) as

$$k_{rel} = \frac{\dfrac{F}{mg}}{\dfrac{\Delta l}{l}} \qquad (13.2)$$

where m is body mass, g acceleration due to gravity and l is the "hip" height. Six-legged trotters (in-

TABLE 13.1. Legged animals as spring-mass systems.

Gait	Legs used per step	Steps per cycle	Relative force	Relative compression	Relative stiffness (sum of legs used per step)	Relative force per leg	Relative stiffness per leg
Trotters (dog)	2	2	F/mg	1/2 Δl/l	2k	1/2 F/mg	$k_{rel_{leg}}$
Trotters (cockroach)	3	2	F/mg	1/3 Δl/l	3k	1/3 F/mg	$k_{rel_{leg}}$
Runners (human)	1	2	F/mg	Δl/l	k	F/mg	$k_{rel_{leg}}$
Hoppers (kangaroo)	2	1	2F/mg	Δl/l	2k	F/mg	$k_{rel_{leg}}$

Each column labelled relative represents the relative magnitude of a dimensionless ratio. For example, the relative force of hoppers (2F/mg) is twice that of trotters and runners (F/mg). Blickhan and Full (1993).

sects using 3 legs on the ground at once) compress their virtual leg spring by only a small amount (1/3) relative to runners and hoppers (Table 13.1). Because the relative force is the same as in runners, the whole body stiffness of the insect's virtual leg spring is 3-fold greater than for runners. Since the stiffness of the virtual leg spring is determined by the number of legs that hit the ground, the relative individual leg stiffness can be estimated by dividing the relative whole body stiffness of the insect's virtual leg spring by the number of legs (e.g., for insect 3k / 3). Given this reasoning, the relative leg stiffness is surprisingly similar in trotters, runners and hoppers using 1 to 3 legs per step (Figure 13.5). Relative force is about 10-fold greater than relative compression in 6-legged trotters (cockroaches), 4-legged trotters (dogs, horses), 2-legged runners (humans, birds) and 2-legged hoppers (kangaroos; Table 13.1).

2.4 Constant Leg Spring Stiffness Versus Speed

Studies have shown that the stiffness of the leg spring remains nearly the same at all forward speeds in a variety of running, hopping, and trotting mammals including humans (Figure 13.6A) (He et al. 1991; Farley et al. 1993). As animals run faster, the body's spring system is adjusted to bounce off the ground more quickly by increasing the angle swept by the leg spring during the ground contact phase rather than by increasing the stiffness of the leg spring (Figure 13.6B). By increasing the angle swept by the leg spring at higher speeds, the vertical excursion of the center of mass (Δy) is reduced without changing the leg stiffness (Figure 13.7). This method of adjusting the spring-mass system for different speeds is the same in all of the mammals studied to date (He et

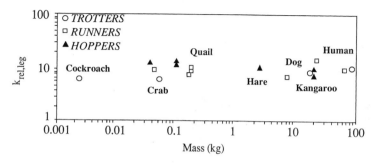

FIGURE 13.5. Relative individual leg stiffness as a function of body mass. Relative leg stiffness was determined by normalizing peak vertical ground reaction force by body weight and virtual leg spring compression by hip height. To determine relative individual leg stiffness, relative leg stiffness was simply divided by the number of legs used in a step. See relative magnitudes in Table 13.1. (Blickhan and Full 1993.)

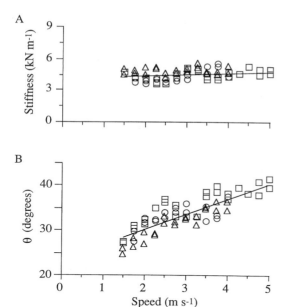

FIGURE 13.6. Leg spring stiffness and landing angle as a function of speed for trotting dogs. (A) The stiffness of the leg spring is independent of speed (Farley et al. 1993). (B) Half the angle swept by the leg spring (θ; Figure 13.7) increases at higher speeds. This adjustment to the geometry of the spring-mass system reduces the vertical displacement of the center of mass during the stance phase at higher speeds. Different symbols are used for each of the trotting dogs in the study ($n = 3$). A similar pattern has been observed for trotting goats and horses, hopping kangaroos, and running humans. (He et al. 1991; Farley et al. 1993.)

al. 1991; Farley et al. 1993). Interestingly, it is also the method used to change the forward speed of running robots with spring-based legs (Raibert and Hodgins 1993).

2.5 Leg Spring Stiffness Is Adjustable to Bouncing Frequency

Although the stiffness of the leg spring remains the same at all locomotor speeds in mammals, experimental evidence shows that it is possible for humans to alter the stiffness of their leg spring. The first evidence comes from a study of humans hopping in place, an ideal experimental system for examining leg stiffness adjustment because it follows the same basic mechanics and spring-mass model as forward running yet has simpler kinematics. When humans hop in place at a range of frequencies, the stiffness of the leg can be more than doubled to accommodate increases in hopping frequency or to accommodate increases in hopping height at a given frequency (Figure 13.8) (Farley et al. 1991). Similarly, when humans run at a given speed using a range of stride frequencies, they more than double their leg stiffness between the lowest and highest possible stride frequencies (Figure 13.9) (Farley and Gonzalez 1996). A similar range of possible leg stiffnesses was observed in an earlier study that showed that when humans bounce vertically on a compliant board, the stiffness of the leg spring can change by as much as two-fold in response to changes in knee angle (Greene and McMahon 1979).

2.6 Leg Spring Stiffness Is Adjustable to Surface Stiffness

Almost all of the research to date on the mechanics and control of legged locomotion has focused on walking and running on hard and smooth laboratory floors. However, when legged animals run in the natural world, they encounter a variety of surfaces that have a range of mechanical proper-

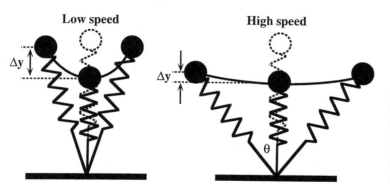

FIGURE 13.7. At higher speeds, the angle swept by the leg spring during the stance phase increases, and as a result, the vertical displacement of the center of mass (Δy) decreases. In both examples, the displacement of the leg spring (Δl) is the same.

ties. Clearly, the properties of the surface will have substantial effects on the mechanics of locomotion, if the mechanical behavior of the legs is not adjusted to accommodate changes in surface properties. For example, when lead weights are dropped onto a variety of surfaces, the impact force varies by much as 400% depending on the properties of the surface (Bodmer and Herzog 1980). This observation clearly indicates that if the mechanical properties of the legs are not adjusted to accommodate changes in surface properties, the mechanics of locomotion will be greatly affected.

Research has examined the mechanical behavior of the human leg during hopping in place on surfaces with a range of stiffnesses (Ferris and Farley 1997). The findings show that the stiffness of the leg is increased by as much as 3-fold to accommodate decreases in surface stiffness (Figure 13.10A) (Ferris and Farley 1997). The increased stiffness of the leg during hopping on surfaces of lower stiffness is reflected in reduced leg compression (Figure 13.10B). As a result of the in-

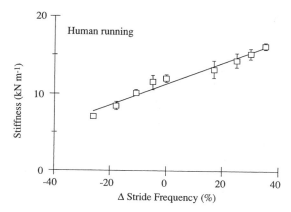

FIGURE 13.9. When humans run at a moderate speed, they can use stride frequencies other than their preferred stride frequency by changing leg spring stiffness. "Δ Stride Frequency" is the percentage difference between the stride frequency used and the preferred stride frequency for that speed (speed = 2.5 m/s; preferred stride frequency = 1.3 Hz). The error bars are the standard errors of the means. (From Farley and Gonzalez 1996.)

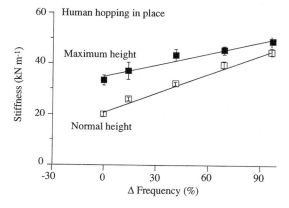

FIGURE 13.8. When humans hop in place, they can more than double the stiffness of the leg spring to increase hopping frequency. In addition, they increase the stiffness of the leg spring to hop higher at a given frequency. In both cases, the higher leg spring stiffness allows the hopper to have a shorter ground contact time. The open symbols represent data in which the subjects were simply instructed to hop at a given frequency set by a metronome; the closed symbols represent data in which the subjects were instructed to hop as high as possible at a given frequency set by a metronome. "Δ Frequency" represents the percentage difference between the hopping frequency and the preferred frequency of 2.2 Hz. The error bars are the standard errors of the means. (Modified from Farley et al. 1991.)

creased leg stiffness on more compliant surfaces, the total stiffness of the series combination of the leg spring and the surface is independent of surface stiffness at the preferred hopping frequency (Figure 13.10A). This allows the peak ground reaction force, the ground contact time, and the vertical displacement of the center of mass to be nearly the same over a 1000-fold range in surface stiffness. Based on these recent studies of the interaction of leg stiffness and surface stiffness during human hopping in place, we hypothesize that the stiffness of the leg is increased to accommodate running on compliant surfaces. Although they did not examine submaximal sustainable speeds, McMahon and Greene (1979) did show that track stiffness influences the maximum speed that runners can attain.

3 Future Directions

3.1 Increase Awareness of the Comparative Approach

Greater interaction between comparative biomechanists and those more interested in the human condition would benefit both groups. Realizing that the comparative method can be more than simply the study of one species as a model animal for humans

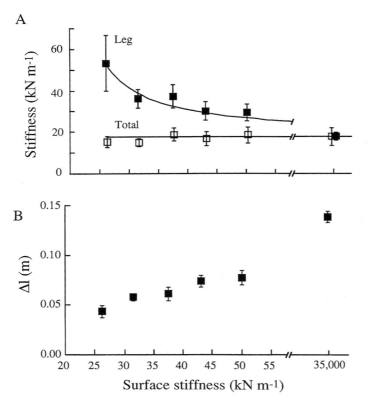

FIGURE 13.10. Leg spring stiffness and displacement of the leg spring as a function of surface stiffness. (A) The leg spring stiffness (k_{leg}) is higher during hopping on less stiff surfaces (closed symbols). However, the stiffness of the series combination of the surface and the leg spring (k_{tot}) is independent of surface stiffness (open symbols). The error bars are the standard errors of the means. (Modified from Ferris and Farley 1997) (B) When humans hop on surfaces with a range of stiffnesses, the displacement of the leg spring (Δl) is lower on less stiff surfaces.

could open new doors toward the transfer of novel concepts and hypotheses among comparative biologists, researchers on humans and biomedical engineers. Species very different from ourselves and often more amenable to experimental manipulation can provide new insights, so long as we understand if or how the differences among species alter our conclusions (Full 1997). We suggest greater use of phylogenetic analysis in cases where evolutionary history is the prime reason why structures function the way they do. If no clear selection or constraint on function is apparent, then instead of pursuing unfounded dynamic optimization studies, perhaps we should look at function in closely related species, our sister taxa.

3.2   The Explicit Goal Should Be Integration in an Effort to Understand Mechanism

Raibert and Hodgins (1993) state, "It is ironic that while workers in neural motor control tend to minimize the importance of the mechanical characteristics of an animal's body, few workers in biome-

chanics seem very interested in the role of the nervous system. We think that the nervous system and the mechanical system should be designed to work together, sharing responsibility for the behavior that emerges." Certainly, part of the reason for the division is that each group asks different questions. Neurobiologists want to know how control systems can function. Muscle physiologists wish to characterize the capability of muscles. Biomechanists desire to define the performance of the mechanical system. Specialists must continue to specialize, but each group could benefit from the common goal of integration and the complete understanding of the system as a whole. Whether by direct experimentation or optimization and sensitivity analyses using large scale models, we all benefit from restricting the set of possible input or output functions. A trainable artificial neural network model may be extraordinarily valuable to a biomechanist as a controller for one behavior, but wholly inadequate for another behavior. Characterization of the capabilities and limitations of actual neural networks could be invaluable toward explaining observed behaviors. We must remember that complex biological systems with all their redundancies have evolved to be "just good enough" at a whole variety of tasks (subsystem criteria of Chapter 35). In addition, however, we must not be fooled into thinking that evolution results in optimal design of the whole system or the sub-systems. Animals are constrained by their evolutionary history, by how their ancestors looked and functioned in the past. If we are to understand the unique design solution of one species, then we must move to an integrative approach using a variety of species where each sub-system can limit the set of possible solutions for another (see Chapter 14 and Chapter 16).

3.3 Limit the Number of Possible Solutions: The Leg-Spring Example

We could imagine a seemingly infinite number of possible solutions to explain the locomotion of many-legged animals, especially given the tremendous diversity present. The field simply had no starting point to begin to understand how legs function, let alone how multiple muscles produced leg movement. Fortunately, a single solution gives us a place to begin an effort at integration with lower levels of organization. The concept of the "leg spring" allowed biomechanists to characterize the spring-like properties of the overall musculoskeletal system during locomotion. This characterization lead to the surprising discovery of leg spring constancy, despite differences in leg number and speed of locomotion. These findings, in turn, sparked investigation into how the system could be adjusted during changes in frequency and surface stiffness. Before these studies, it was not possible to predict leg stiffness from joint stiffness, since it wasn't known whether leg spring stiffness even changed or instead if the geometry of leg operation (i.e., landing angle, θ) was altered while leg spring stiffness remained constant.

Knowledge of how the whole system behaves like a spring limits the range of potential solutions for the behavior of the individual joints and their interactions. Joint stiffness is complex and depends on the number of muscles (agonists and antagonists) that are active and the stiffness of the individual muscles acting about the joint (Feldman 1966, 1980; Smith 1981; Abend et al. 1982; Akazawa et al. 1983; Bizzi et al. 1984, Carter et al. 1993; Nielsen et al. 1994). The control of muscle stiffness is equally elaborate and depends on a variety of factors including activation, reflexes, muscle force, muscle length, and architecture (Rack and Westbury 1969, 1974; Nichols and Houk 1976; Agarwal and Gottlieb 1977; Gottlieb and Agarwal 1978, 1980, 1988; Akazawa et al. 1983; Hoffer and Andreassen 1981; Rack et al. 1983; Nichols 1987; Sinkjaer et al. 1988; Weiss et al. 1988; Gollhofer et al. 1992). Given the complexity of stiffness control in a single muscle or joint, it was nearly impossible to use a forward dynamics approach that begins at the level of individual muscle stiffness control and attempts to explain the mechanics of locomotion. In such a complex system, it was useful to begin by using an inverse dynamics approach that characterizes the overall behavior of the musculoskeletal system. In this case, we learned that the whole musculoskeletal system behaves like a spring-mass system during locomotion, and have uncovered some of the rules underlying the link between the stiffness of the complete musculoskeletal system and the mechanics of locomotion. Thus, the broad understanding of the overall behavior of the leg during locomotion provided by the spring-mass model can guide future work examining the integration of the actions of multiple joints and muscles whose function must result in spring-like behavior.

3.4 Practical Applications

Finally, the ability to represent the complex musculoskeletal system as a single spring will have practical applications. The properties of the leg spring give information about the mechanical interactions between the human musculoskeletal system and the ground during running. This information is valuable to engineers who are designing equipment for use during human locomotion. For example, knowledge of the stiffness of the overall musculoskeletal system during locomotion has aided in the design of running tracks for maximizing sprint speed (McMahon and Greene 1979) and the design of running shoes. In the future, it will undoubtedly aid in the design of surfaces that minimize overuse injuries during aerobic exercise, the design of spring-based prostheses for running that allow normal running mechanics, and the design of legs for running robots.

Acknowledgment. Supported by ONR Grant N00014-92-J-1250 (R.J.F.) and NIH R29 AR44008 (C.T.F.)

References

Abend, W., Bizzi, E., and Morasso, P. (1982). Human arm trajectory formation. *Brain*, 105:331–348.

Agarwal, G.C. and Gottlieb, G.L. (1977). Oscillation of the human ankle joint in response to applied sinusoidal torque on the foot. *J. Physiol.*, 268:151–176.

Akazawa, K., Milner, T.E., and Stein, R.B. (1983). Modulation of reflex EMG and stiffness in response to stretch of human finger muscle. *J. Neurophysiol.*, 49:16–27.

Alexander, R. McN. (1988). *Elastic Mechanisms in Animal Movement.* Cambridge University Press, Cambridge.

Alexander, R. McN. (1992). A model of bipedal locomotion on compliant legs. *Phil. Trans. Roy. Soc. B.*, 338:189–198.

Alexander, R. McN. and Vernon, A. (1975). Mechanics of hopping by kangaroos (Macropodidae). *J. Zool. Lond.*, 177:265–303.

Biewener A.A. and Baudinette R.V. (1995). *In vivo* muscle force and elastic energy storage during steady-speed hopping of tammar wallabies (*Macropus eugenii*). *J. Exp. Biol.*, 198:1829–1841.

Bizzi, E. Accornero, N., Chapple, W., and Hogan, N. (1984). Posture control and trajectory formation during arm movement. *J. Neurosci.*, 4:2738–2744.

Blickhan, R. (1989). The spring-mass model for running and hopping. *J. Biomech.*, 22:1217–1227.

Blickhan, R. and Full, R.J. (1993). Similarity in multi-legged locomotion: bouncing like a monopode. *J. Comp. Physiol. A.*, 173:509–517.

Bodmer, H., and Herzog, M. (1980). *Materialeigenschaften von sportboeden und schuhen.* Deplomarbeit in Biomechanik, ETH, Zurich.

Carter, R.R. Crago, P.E., and Gorman, P.H. (1993). Nonlinear stretch reflex interaction during cocontraction. *J. Neurophysiol.*, 69:943–952.

Cavagna, G.A., Franzetti, P., Heglund, N.C., and Willems, P. (1988). The determinants of step frequency in running, trotting, and hopping in man and other vertebrates. *J. Physiol. Lond.*, 399:81–92.

Cavagna, G.A., Heglund, N.C., and Taylor, C.R. (1977). Mechanical work in terrestrial locomotion: Two basic mechanisms for minimizing energy expenditure. *Am. J. Physiol.*, 233(5):R243–R261.

Cavagna, G. A., Thys, H., and Zamboni, R. (1976). The sources of external work in level walking and running. *J. Physiol.*, 262:639–657.

Coleman, M. and Ruina, A. (1998). An uncontrolled walking toy that cannot stand still. Physical Review Letters April, 80(16):3658–3661.

Dawson, T.J. and Taylor, C.R. (1973). Energeic cost of locomotion in kangaroos. Nature, 246:313–314.

Farley, C.T., Blickhan, R., Saito, J., and Taylor, C.R. (1991). Hopping frequency in humans: a test of how springs set stride frequency in bouncing gaits. *J. App. Physiol.*, 71(6):2127–2132.

Farley, C.T. and Gonzalez, O. (1996). Leg stiffness and stride frequency in human running. *J. Biomech.*, 29:181–186.

Farley, C.T. and Ko, T.C. External mechanical power output in lizard locomotion. *Exp. Biol.*, 200:2177–2188.

Farley, C.T., Glasheen, J., and McMahon, T.A. (1993). Running springs: speed and animal size. *J. Exp. Biol.*, 185:71–86.

Feldman, A.G. (1966). Functional tuning of the nervous system during control of movement or mainatenance of a steady posture. II. Controllable parameters of the muscles. *Biophysics*, 11:498–508.

Feldman, A.G. (1980). Superposition of motor programs. I. Rythmic forearm movements in man. *Neuroscience*, 5:81–90.

Felsenstein, J. 1985. Phylogenies and the comparative method. *Am. Nat.*, 125:1–15.

Ferris, D.P. and Farley, C.T. (1997). Interaction of leg stiffness and surface stiffness during human hopping. *J. Appl. Physiol.*, 82:15–22.

Full, R.J. (1989). Mechanics and energetics of terrestrial locomotion: from bipeds to polypeds. In *Energy Transformation in Cells and Animals.* Wieser, W. and Gnaiger, E. (eds.), pp.175–182, Thieme, Stuttgart.

Full, R.J. (1997). Invertebrate locomotor systems. In *The Handbook of Comparative Physiology.* Dantzler, W. (ed.), pp. 853–930. Oxford University Press.

Garcia, M., Chatterjee, A., Ruina, A., and Coleman, M. (1998). 'The simplest walking model: stability, complexity, and scaling'. *ASME J. Biomechan. Eng.*, 120:281–288.

Garland, T.J. and Adolph, S.C. (1994). Why not to do two-species comparative studies: limitations on inferring adaptation. *Physiol. Zool.*, 67(4):797–828.

Garland, T., Jr., Harvey, P.H., and Ives, A.R. (1992). Procedures for the analysis of comparative data using phylogenetically independent contrasts. *Sys. Biol.*, 41:8–32.

Gollhofer, A., Strojnik, V., Rapp, W., and Schweizer, L. (1992). Behaviour of triceps surae muscle-tendon complex in different jump conditions. *Eur. J. Appl. Physiol.*, 64:283–291.

Gottlieb, G.L., and Agarwal, G.C. (1978). Dependence of human ankle compliance on joint angle. *J. Biomech.*, 11:177–181.

Gottlieb, G.L. and Agarwal, G.C. (1980). Response to sudden torques about the ankle in man: III. Suppression of stretch-evoked responses during phasic contraction. *J. Neurophysiol.*, 44:233–246.

Gottlieb, G.L. and Agarwal, G.C. (1988). Compliance of single joints: elastic and plastic characteristics. *J. Neurophysiol.*, 59:937–951.

Greene, P.R. and McMahon, T.A. (1979). Reflex stiffness of man's anti-gravity muscles during kneebends while carrying extra weights. *J. Biomech.*, 12:881–891.

He, J., Kram, R., and McMahon, T.A. (1991). Mechanics of running under simulated reduced gravity. *J. Appl. Physiol.*, 71:863–870.

Heglund, N.C., Cavagna, G.A., and Taylor, C.R. (1982). Energetics and mechanics of terrestrial locomotion. III. Energy changes of the centre of mass as a function of speed and body size in birds and mammals. *J. Exp. Biol.*, 97:41–56.

Hoffer, J.A. and Andreassen, S. (1981). Regulation of soleus muscle stiffness in premammillary cats: intrinsic and reflex components. *J. Neurophysiol.*, 45:267–285.

Huey, R.B. (1987). Phylogeny, history and the comparative method. In *New Directions in Ecological Physiology*. Pp. 76–97. Cambridge University Press, Cambridge.

Ito, A., Komi, P.V., Sjodin, B., Bosco, C. and Karlsson, J. (1983). Mechanical efficiency of positive work in running at different speeds. *Med. Sci. Sports Exerc.*, 15:299–308.

Josephson, R.K. (1985). Mechanical power output from striated muscle during cyclic contraction. *J. Exp. Biol.*, 114:493–512.

Krebs, H.A. (1975). The August Krogh Principle: "For many problems there is an animal on which it can be most conveniently studied". *J. Exp. Zool.*, 194:221–226.

McGeer, T. (1990a). Passive bipedal running. *Proc. R. Soc. Lond.*, B240:107–134.

McGeer, T. (1990b). Passive dynamic walking. *Int. J. Robot. Res.*, 9(2):62–82.

McMahon, T.A. (1985). The role of compliance in mammalian running gaits. *J. Exp. Biol.*, 115:263–282.

McMahon, T.A. (1990). Spring-like properties of muscles and reflexes in running. In: *Multiple Muscle Systems*. Winters, J.M. and Woo, S.L.-Y. (eds.), pp. 578–590, Springer, New York.

McMahon, T.A. and Cheng, G.C. (1990). The mechanics of running: how does stiffness couple with speed? *J. Biomech.*, 23(Suppl. 1), 65–78.

McMahon, T.A. and Greene, P.R. (1979). The influence of track compliance on running. *J. Biomech.*, 12:893–904.

Mochon, S. and McMahon, T.A. (1980). Ballistic walking. *J. Biomech.*, 13:49–57.

Mochon, S. and McMahon, T.A. (1981). Ballistic walking: an improved model. *Math. Biosci.*, 52:241–260.

Nichols, T.R. (1987). The regulation of muscle stiffness: implications for the control of limb stiffness. *Med. Sport. Sci.*, 26:36–47.

Nichols, T.R. and Houk, J.C. (1976). Improvement of linearity and regulations of stiffness that results from actions of stretch reflex. *J. Neurophysiol.*, 39:119–142.

Nielsen, J., Sinkjaer, T., Toft, E., and Kagamihara, Y. (1994). Segmental reflexes and ankle joint stiffness during co-contraction of antagonistic ankle muscles in man. *Exp. Brain Res.*, 102:350–358.

Playter, R.R. and Raibert, M.H. (1994). Passively stable layout somersaults. Eighth Yale workshop on adaptive and learning systems, June 13–15, pp. 66–71, New Haven, Connecticut.

Rack, P.M.H. and Westbury, D.R. (1969). The effects of length and stimulus rate on tension in the isometric cat soleus muscle. *J. Physiol.*, 204:443–460.

Rack, P.M.H. and Westbury, D.R. (1974). The short range stiffness of active mammalian muscle and its effect on mechanical properties. *J. Physiol.*, 240:331–350.

Rack, P.M.H., Ross, H.F., Thilmann, A.F., and Walters, D.K.W. (1983). Reflex responses at the human ankle: the importance of tendon compliance. *J. Physiol.*, 344:503–524.

Raibert, M.H. and Hodgins, J.K. (1993). Legged robots. In *Biological Neural Networks in Invertebrate Neuroethology and Robotics*. Beer, R.D., Ritzmann, R.E., and McKenna, T. (eds.), pp. 319–354, Academic Press, San Diego.

Sinkjaer, T., Toft, E., Andreassen, S., and Hornemann, B.C. (1988). Muscle stiffness in human ankle dorsiflexors: intrinsic and reflex components. *J. Neurophys.*, 60:1110–1121.

Smith, A.M. (1981). The co-activation of antagonist muscles. *Can J. Physiol. Pharmacol.*, 59:733–747.

Thompson, C. and Raibert, M. (1989). Passive dynamic running. In *International Symposium of Experimental Robotics* Hayward, V. and Khatib, O. (eds.), pp. 74–83, Springer-Verlag, New York.

Weiss, P.L., Kearney, R.E., and Hunter, I.W. (1988). Human ankle joint stiffness over the full range of muscle activation levels *J. Biomech.*, 21:539–544.

Zajac, F.E. and Gordon, M.E. (1989). Determining muscle's force and action in multi-articulate movement. *Exerc. Sport Sci. Rev.*, 17:187–230.

Commentary: Cyclic Movements and Adaptive Tissues

Jack M. Winters

The chapters in Section IV of this book focus on rhythmic movements in animals. Such cyclic movements are important in life, especially with regard to moving around within one's environment. By choosing to look at such movements via smaller animals, we are able to study more closely the neuromechanical behavior underlying movements. While some chapters within this section have focused more on the neural components and others more on mechanical aspects (e.g., this chapter), a recurring theme is that neural and mechanical subsystems have *co-evolved* (in both structure and properties) to meet the demands of various environments, and *furthermore co-developed* over the course of an animal's life. Chapter 12, Section 4 states that, if anything, the neural control system develops primarily to meet the requirements of the mechanical system as it develops, rather than visa-versa. I agree. For instance, neural pattern generators (PGs) often must adapt to the natural resonant frequencies of the musculoskeletal system, and modulatory tonic drives can often be effective in initiating cyclic movement because of such subservient PGs that are complemented by sensory input that detect discrete mechanical events that may trigger transitions between states. This clearly makes the phylogenetic approach advocated by Full and Farley intriguing, and the authors successfully articulate the advantages of utilizing the comparative method and focusing on natural experiments. However, it is suggested that there are also limitations to such approaches, and this will be our focus here. Two are addressed.

1. *Approaches based on observation of macrobehavior, interpreted using simple macromodels, have limitations.* In Figures 13.2 to 13.10, the authors' use the concept of leg springs, and spring-mass systems in general, to provide a fascinating classification scheme for legged locomotion behavior. But rather than being awestruck by the data in these figures, a worthwhile question is: Could the results have been otherwise? Why should the general similarity of ground reaction forces between diverse animals (relative to their weight) be expected, given that all of these legged creatures locomote while subject to the laws of physics (e.g., inertial dynamics, the same gravitational field)? Indeed, after examining Figure 13.2, I was more surprised by the magnitude of the *differences* in shape of the ground reaction forces than by the similarities. And what about the remarkable plots of relative leg stiffness vs body mass (Figure 13.4)? I would suggest an alternative, *rule-based* interpretation: animals don't like to be overly squashed during propulsion. Using a fuzzy rule analogy, if the variable $\Delta l/l$ is too large, say over about 50%, the degree of membership in the fuzzy set "squashed" is pretty high. Being squashed is bad, and creatures might not like to be squashed for several reasons: mechanical disadvantage at joints, a less stable visual field, societal humiliation. Given that F/mg per leg in ground contact should be over 1.0 but not by too much, it seems to me rather unsurprising that the relative individual

leg stiffness is reasonably independent of mass (Figure 13.5)—especially on a logarithmic scale. What about speed variation within a given animal? The story is clean: stiffness does not change, while the angle swept by the legs grows with speed. The trouble I have with the global spring-mass macromodel is that while it effectively tells us *what* happens, it does not really tell us *why or how* this happens. I prefer relating function (and behavior) to specific tissue structure and properties. There has a track record here, for instance in the evolving writings of Alexander. Based on biomechanical inspection of the different musculoskeletal *structures*, Alexander (1968) sheds light on why the horse is better at running fast, yet the armadillo better at borrowing holes. Based on a consideration of the function of human leg vs arm muscles (i.e., propulsion vs manipulation), Alexander and Ker (1990; see also Alexander 1988) were able to show why the factor of safety of key tendons (related to the tendon cross sectional area relative to the muscle physiological cross-sectional area) differs.

2. *Adaptive tissues and optimum design.* Full and Farley state that "we must not be fooled into thinking that evolution results in optimal design of the whole system or the sub-systems" after noting that "complex biological systems with all their redundancies have evolved to be 'just good enough' at a whole variety of tasks (subsystem criteria, Chapter 35, Winters). . . ." While I agree in spirit with the latter statement, the authors' come painfully close to denying a role for optimization as an adaptive process that can lead to change in structure and tissue. With this I would have to disagree. As noted in Chapter 7, adaptive processes can happen on roughly three time scales: (1) very long (evolutionary, multigenerational effects); (2) moderate length (adaptive tissue/behavioral effects, occuring during animal lifespan); and (3) on-the-fly learning (the "iterations" or "epochs" of an optimal neurocontrol problem). The first two of these especially apply here.

In engineering terms, the first of these relates primarily to changes in structure, and to optimal design. A rapidly emerging tool for dealing with such problems are genetic algorithms (Holland 1975; Davis 1991), which are literally categorized as a type of stochastic search optimization algorithm (e.g., Michaleqicz 1992; Lin and Lee 1996, Chapter 35)! They are especially adept at using the optimization *process* to address *structural* optimization, with the emerging "solution" being near-optimum (and never static). As the criteria for success changes (e.g., a natural predator starts to disappear; the climate slowly changes), so will this solution—it evolves, using at minimum mathematical descriptions of the three classic operations of reproduction, crossover and mutation. In engineering practice one often uses hybrid approaches, with the core genetic algorithm performing global search to get close to a solution ("substantial convergence"), with a local optimization procedure (usually gradient-based) refining the solution. Of note is that these algorithms can still get caught in local minima, and perhaps this helps explain the wonderful collection of solutions (creatures) that exist.

In engineering terms, the second relates to tissue adaptation, a generalization of the well developed "use it or lose it" principle. Let's develop this further. Consider the animal as a wonderful structural assembly of tissues: connective, skeletal, muscular, neural, and so on. With remarkably few exceptions (e.g., certain types of cartilage), tissues (with adequate blood supply) are known to adapt in response to how they are used—hence the reason for weight rooms and therapists. This implies that within limits, tissues can change their properties (e.g., size, responsiveness to "stimulation"). In Chapter 35, I make a case is that this represents a local optimization process, where change is governed by tissue adaptation laws. As in a recent commentary by Huijing and Winters (1998), here we do not distinguish between the adaptive properties of "mechanical" tissues such as ligaments from "electrical" tissues such as neurons, and view muscle as a electrochemicomechanical transducer between these subsystems. Some tissues happen to be most sensitive to mechanical inputs, others electrical, still others chemical—heart muscle (a tissue responsible for a cyclic movement) provides an exquisite example of a tissue that is very sensitive to all three. Thus calluses form on well-used skin, bones and muscles decay when underused for a while, and skills develop with practice.

Some would argue the following: So why are we all not able to just adapt to have near-optimum athletic skills like Michael Jordan?). In other words, is not the inherent variety within a species a case against optimization? My answer is still no. Genetic (optimum search) algorithms require "mutations" to work, and variety adds robustness to a species (an important performance subcriteria). Optimization is a cybernetic process toward a biosolution defined by criteria, subject to constraints. One constraint is that structural and tissue changes must occur relatively slowly (but then, we are free to choose our relevant time scales). And criteria change; for instance, a short five generations ago shooting balls through a 10 ft high hoop was not high on the priority list. And even now sometimes a leisurely life is fun. When such proactive local processes prove beneficial to the creature, this adaptive process may gradually affect the ongoing evolution of the species.

In conclusion, we return to our start, on synthesized neuromechanical systems. Perhaps effective leg springs exist because the structural assembly of tissues is tuned for this task. But the more intriguing question might be: Why do the tissues have the shapes and properties that they do? If these biosystems are so clearly suboptimal, why could we not just interchange parts between some of the species shown in the figures—say a femur here, a knee flexor there, and so on. My guess is that we would then really experience suboptimal design, and Figures 13.2 to 13.10 would change dramatically. What governs why structures and tissues are what they are, if not an evolving optimization process?

References

Alexander, R.McN. (1968). *Animal mechanics.* University of Washington Press, Seattle.

Alexander, R.McN. (1988). *Elastic Mechanisms in Animal Movement.* Cambridge University Press, Cambridge, England.

Alexander, R.McN. and Ker, R.F. (1990). The architecture of leg muscles. In *Multiple Muscle Systems: Biomechanics and Movement Organization*, Winters, J.M. and Woo, S.Y. (eds), Chapter 36, pp. 568–577, Springer-Verlag, New York.

Davis, L. (ed.). (1991). *Handbook of Genetic Algorithms.* Van Nostrand Reinhold, New York.

Holland, J.H. (ed.). (1975). *Adaptation in Neural and Artificial Systems.* University Press of Michigan, Ann Arbor.

Michalewicz, Z. (1992). *Genetic Algorithms + Data Structures = Evolution Programs.* Springer-Verlag, New York.

Huijing, P.A. and Winters, J.M. (1998). Toward a new paradigm of locomotor apparatus and neuromuscular control of movement? In *Models in human movement sciences.* Post, A.A., Bosch, J.R.P.P., and Boschker, M.S.J. (eds.), pp. 45–50. Amsterdam, Instituut voor Fundamentele en Klinische Bewegingswetenschappen.

Lin, C.-T. and Lee, C.S.G. (1996). *Neural Fuzzy Systems: A Neuro-Fuzzy Synergism to Intelligent Systems.* Prentice Hall, Upper Saddle River, New Jersey.

14
Biomechanics of Hydroskeletons: Studies of Crawling in the Medicinal Leech

William B. Kristan, Jr., Richard Skalak, Richard J.A. Wilson,
Boguslaw A. Skierczynski, James A. Murray, F. James Eisenhart,
and Timothy W. Cacciatore

1 Introduction

How well can patterns of motor neuronal activity account for behavior? How much of the behavior is produced by "central patterns" and how much depends on the mechanics of the system? To what extent do motor patterns reflect mechanical necessities imposed on the nervous system by the properties of muscles and their arrangements? To approach these kinds of questions, we have started to develop a biomechanical model of the leech (Skalak et al. 1996), based on the structure of the leech body, the mechanical properties of their muscles, and the transformation of neural signals into activation of the muscles. These properties are the standard ones needed for any biomechanical description of a neuromuscular system, but this effort is complicated by the fact that leeches, like many other invertebrate animals, have no rigid skeleton—either internally or externally—with which to produce movements. Leech muscles must stiffen all or part of its body that serves as a skeleton. This kind of skeleton, because it ultimately depends upon building up a force on an enclosed incompressible fluid, has been variously called a *hydrostatic skeleton, hydraulic skeleton,* or *hydroskeleton.*

The nervous system of the medicinal leech has proven to be a valuable tool for determining the neuronal basis of several behaviors: swimming, shortening, local bending, and crawling. Typically, these behaviors are first characterized kinematically, determining the detailed changes in body shape that constitute these behaviors. Then the animal is cut open and motor activity is recorded from dissected nerves, to show the pattern of motor activity responsible for each of the behaviors. The identification of the neurons involved in the behavior and the connectivity pattern among the neurons have been done mainly in these and even more isolated nervous system preparations.

2 Moving Without a Rigid Skeleton: The Hydroskeleton

A variety of animals—worms, octopi, jellyfish—make all their movements using a hydroskeleton. In addition, most animals move some body parts using a hydroskeleton. Consider the movements produced by our tongues or our intestines, by the trunk of an elephant, or by the tentacles of a squid or a snail. In none of these organs and organisms is there a permanently rigid skeleton. Given the large diversity in size, behavior, and evolutionary history of these different hydroskeletons, are there general principals to be gleaned? Because of the remarkable array of different animal forms that use hydroskeletons, as well as the different ways they have been used—for locomotion, for capturing prey, as a prehensile tool, to move food through the body, and so on—one might expect a variety of mechanisms. Fortunately, there do seem to be a limited number of ways that evolution has found to produce these useful devices.

2.1 General Structure of Hydroskeletons: The Pressure Vessel Model

Hydroskeletons can have a variety of shapes, from the nearly globular bodies of jellyfish to the highly elongated forms of worms and tentacles. In all, the

FIGURE 14.1. Types of muscle arrangements found in elongated, constant-volume hydroskeletons. (A) The general form of such hydro-skeletons is either a *muscular hydrostat* (left), in which muscles and other tissues form a solid structure, or muscles in the body wall surround a *fluid-filled cavity* (right). The muscles in either type of hydroskeleton are classified by their arrangement relative to the long axis: parallel (B), perpendicular (C), or oblique (D). Some typical subtypes of each of the arrangements are shown: (B) Parallel muscles are often in bands located either centrally or peripherally. (C) Perpendicular muscles may form circular bands around the structure, or transverse or radial bands through the structure. (D) Oblique muscles may form helical bands or diagonal bands.

A. Elongated body form

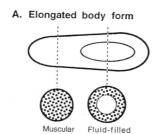

Muscular   Fluid-filled
hydrostat   cavity

B. Parallel muscles

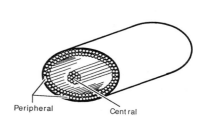

Peripheral   Central

C. Perpendicular muscles

Circular   Transverse   Radial

D. Oblique muscles

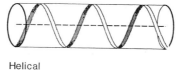 

Helical   Diagonal

muscles are arranged to build up pressure among the muscle fibers or in a fluid-filled chamber enclosed by the muscles (Figure 14.1A). Hydrostats without a separate enclosed fluid volume have been termed "muscular hydrostats" (Kier and Smith 1985). Some hydroskeletons are closed, so that the volume of the structure remains constant during movements, whereas others are open, taking up and expelling fluid, to produce such behaviors as jet propulsion in squids. Because the body of the medicinal leech is an elongated structure of constant volume, only this type of hydroskeleton will be considered.

Hydroskeletons with constant volume have been modeled as pressure vessels (Chapman 1950; Chiel et al. 1992; Skierczynski et al. 1996), which relates the forces produced by contractions of different muscle types in terms of the geometry of the vessel. To use this model requires describing the shape of the animal geometrically, as well as determin-

ing the spatial extent of the pressure changes caused by contraction of a set of muscles. For instance, the bodies of round worms (such as soil nematodes and *Ascaris*, an intestinal parasite) constitute a single vessel so that the pressure caused by contraction of any muscle in the body wall exerts an equal pressure to all parts of the body wall. Other worms (such as earthworms) localize their internal pressure by having segments separated by septa (Chapman 1950, 1958). Even better localization of pressure is provided by muscular hydrostats (as used in tongues and tentacles in a variety of animals, and in trunks by elephants), where the pressure built up by contraction of any muscle influences only the muscles and other tissues nearby (Kier and Smith 1985; Chiel et al. 1992) because the incompressible fluid is primarily contained within cells.

For localized movements, muscular hydrostats are more efficient than fluid-filled cavities, because a local contraction does not dissipate energy as it would in a fluid-filled hydrostat. Fluid-filled vessels can produce localized movements—a bend, for instance, or a local shortening—only by increasing the pressure in the whole structure. This is clearly inefficient, unless there is a lessening of total pressure by well-coordinated relaxation of other muscles (see below for a discussion of muscle arrangement). Conversely, fluid vessels more easily maintain a uniform pressure throughout a structure. For stiffening a whole structure—as an elephant does, for instance, before it lifts a heavy weight with its trunk (Wilson et al. 1991)—the muscular hydrostat is at a disadvantage because all the contributing muscles must be contracted to a comparable extent to have a uniform stiffening of the structure.

Having made the distinction between muscular hydrostats and fluid-filled vessels, it is worth noting that the distinctions are not so clear in many cases. For instance, even in what appear to be clear examples of fluid-filled vessels such as round worms or our own intestines, the pressures within the muscle layers are likely to be greater than the pressures within the cavity. In our own studies of the medicinal leech, we have found that the pressures generated by muscles in one segment are confined to that segment during certain behaviors (Wilson et al. 1996b), despite the fact that leeches do not have frank septa between the segments such as are found in their annelid cousins, the earthworms. Because medicinal leeches can consume, at one feeding, a volume of blood that is nine times their body weight (Lent et al. 1988), and then store the blood for many months before all of it is digested, it may well be that these leeches are more like fluid-filled hydrostats just after a meal and become muscular hydrostats as the blood is digested over several months.

2.2 Geometrical Arrangement of Muscles

Because hydroskeletons have no joints to bend or rigid skeletons to resist compression, how is movement actually produced? As is true for hard skeletons, hydroskeletons depend upon contractions of reciprocally opposed muscles to produce coordinated movements. The reciprocity is accomplished by increases in pressure caused by one set of muscles producing elongation of the antagonistic

muscles, and vice versa. Which muscles are antagonistic depends upon the geometry of the hydroskeletal structure and the arrangement of the muscles. It is not surprising, with the great diversity of form and function among hydroskeletal elements, that the arrangements of the muscles are also quite different in detail. In general form, however, there are just three arrangements of muscles: parallel to the long axis, perpendicular to the long axis, and oblique to both the parallel and perpendicular axes (Figure 14.1B–14.1D). Each of the muscles of one class is antagonistic to those in the other classes. The functions of the different muscle groups, too, have some simplifying features: parallel fibers usually shorten or bend the hydroskeleton, perpendicular muscles lengthen it, and oblique muscles twist or bend it.

2.2.1 Muscles Parallel to the Long Axis

Muscles parallel to the long axis are termed *longitudinal muscles*. They are usually distributed uniformly about the center of the cross-section of the hydroskeleton, so that contraction of all the longitudinal muscles at once produces shortening (Figure 14.1B). Essentially all hydroskeletons have longitudinal muscles, and use them to produce shortening. In most cases, the longitudinal muscles on one side can also be activated separately, thereby producing a bend to that side of the hydroskeleton. (This assumes that the hydroskeleton is being held rigid; if the pressure in the hydroskeleton were very low—i.e., if it were very flaccid—contraction of longitudinal muscles on one side would shorten the structure with very little bending.) In some structures, as in the base of the "arms" (the eight short tentacles around the mouth) of the squid, the longitudinal muscles run exclusively down the center of the hydroskeleton, so that they can cause only shortening, not bending.

One interesting consequence for the longitudinal arrangement of muscles in a hydroskeleton is that there are no levers to magnify length changes in these muscles: all the shortening must be accomplished by changes in length of the muscle fibers. For instance, for the leech to show a 3-fold change in length in its various behaviors, the longitudinal muscle fibers must be able to exert force over this whole range of length (Wilson et al. 1996a). Leech muscles do not have a fixed sarcomere length (Lanzavecchia et al. 1985), and have mechanical properties similar to those of smooth muscles, including slow contraction and stress relaxation (Fung 1993).

2.2.2 Muscles Perpendicular to the Long Axis

Muscles perpendicular to the long axis of a hydroskeleton are always antagonistic to longitudinal muscles. They are arranged in such a way that their contraction causes a decrease in the cross-sectional area, either by decreasing the circumference or by distorting the cross-sectional shape. The most commonly described muscles of this sort are *circular* (or *circumferential*) muscles, which form a continuous ring around the hydroskeleton; they are usually found near the outside edge of the hydroskeleton (Figure 14.1C). They are usually assumed to function as a unit, like a sphincter, causing the hydroskeleton to get narrower and longer. Individual leech motor neurons innervate non-overlapping quadrants of circular muscles in each segment (Stuart 1970), so circular muscles could potentially be used to produce more localized movements, probably bending, but to date, they have been found to be activated only as a unit, to produce elongation movements during crawling steps (Baader and Kristan 1992; Eisenhart et al. 1995).

A second arrangement of muscles perpendicular to the long axis of the hydroskeleton are *transverse* muscles, which run from one side of the hydroskeleton to the other (Figure 14.1B). Transverse muscles can be arranged in two sheets, as is found in squid tentacles (Kier 1982), thereby functioning like circular muscles, to cause elongation of the hydroskeleton. Alternatively, there may be a single sheet of transverse fibers, which would serve to flatten (and elongate) the hydroskeleton. Leeches have such a single sheet of muscles, which insert on the dorsal and ventral surfaces of the body wall in each segment. Contraction of these muscles produces a dorsoventral flattening of the body, which leeches use during swimming to increase the area of the leading surface of the body that pushes against the fluid as they produce the undulations that constitute swimming (Kristan et al. 1974). A third kind of transverse muscle are *radial* muscles, that span from the center of the cross-section of the hydroskeleton to its outer surface. Such diverse structures as the delicate tentacles of snails and the robust trunks of elephants use radial muscles to elongate themselves.

Despite there being no joints to magnify the size of muscle contractions, there can be a mechanical advantage given by the shape of the structure. For instance, in a very elongated structure, even small contractions of the circular muscles produce large and fast elongations of the structure (Chapman 1950).

2.2.3 Muscles Oblique to the Long Axis

In addition to muscles that run either longitudinal or perpendicular to the long axis of the hydroskeleton, there are muscles that cut through both planes; as a class, these are called *oblique muscles* (Figure 14.1D). One common arrangement of oblique muscles is a helical one, usually running in separate left- and right-handed helices around the outside of the hydroskeleton. Contraction of either one alone would probably produce torsion movements (Kier 1982; Alexander 1987). Contracting both at the same time produces shortening if the hydroskeleton is elongated (e.g., if the circular muscles are contracted), and elongation if the hydroskeleton is in a shortened state (e.g., if the longitudinal muscles are contracted). Hence, co-contraction of these muscles alone would probably produce stiffness at a length intermediate between fully shortened and fully relaxed (Mann 1962). A second kind of oblique muscle is seen near the tail end of a leech, in which muscles pass through the body cavity insert on the ventral surface of one segment and on the dorsal of a more posterior segment; these muscles are probably used to produce bending of the whole animal as it attaches to the substrate with its posterior sucker (Wilson et al. 1996a).

2.3 Biomechanical Models of Hydroskeletons

There have been only a few previous efforts at incorporating mechanical and neuronal properties into biomechanical models of hydroskeletons. In all cases, both the geometric arrangements of the muscles and the biomechanics of the muscles have been simplified. For instance, a model of the protrusion and retraction of the tongue by lizards (Chiel et al. 1992) uses just two intrinsic muscles: longitudinal muscles running down the center of a cylinder surrounded by a layer of circular muscles. (This simulates one lateral half of the tongue.) The muscles are activated by motor patterns obtained from lizard tongue muscles (Smith 1984), but the biomechanics used were those for cat soleus muscles (Rack and Westbury 1969). Despite its simplicity and the mosaic nature of the data, this model reproduced many of the detailed features of moving reptilian tongues.

Wadepuhl and Beyn (1989) have modeled a segmented worm using data from leech muscles to simulate the qualitative predictions of Chapman (1950). They used simplified properties of leech muscles and greatly simplified the geometry of the leech body, so that the mathematical description of the hydroskeleton was more tractable. Jordan (1994) measured leech muscular properties, both active and passive, and incorporated them into a model of leech swimming. The mechanical aspects of the Jordan model, however, are much less realistic than the Wadepuhl-Beyn model. It models the leech body as a series of rigid rods running down the middle of each segment, with hinges that are bent up and down by longitudinal muscles arranged dorsally and ventrally. Such a model is valuable for studying some specific questions relating to swimming efficiency, but it does not model the hydroskeleton, so it cannot be used to simulate other behaviors, such as crawling or shortening.

3   Specific Example: Crawling Behavior in the Medicinal Leech

We have simulated the hydroskeleton of the medicinal leech (Skierczynski et al. 1996), based upon the anatomy, arrangement, and mechanical proper-

ties of its muscles (Wilson et al. 1996a), and the activation of the muscles by the motor neurons (Mason and Kristan 1992). We have provided the simulation with motor neuron firing patterns that produce the movements seen during crawling. The parameters of the simulation were modified to optimize the shapes seen throughout a crawling step, and the pressures calculated from the simulation were compared to pressures measured in intact leeches (Wilson et al. 1996b). Although we are still at early stages in the modeling, and we have not yet measured the dynamic properties of leech muscles, we have been able to use our simulation to develop the important features of the motor neuronal firing pattern that underlies crawling.

3.1   Functional Anatomy of the Leech

A leech is an annelid worm, composed of 21 midbody segments plus a head and a tail; the head and tail each consisting of several segments compressed into a small space with specialized structures (Figure 14.2A). The most obvious specializations are the suckers that are used for attaching the front and the rear ends to the substrate. The body is a closed sac, with a muscular body wall surrounding a central body cavity that contains the viscera. The muscles responsible for whole body movements are largely contained within the body wall (Figure 14.2B): longitudinal muscles, circular muscles, and helical oblique muscles. A fourth set, the dorsoventral muscles, are perpendicular to the long axis and produce a dorsoventral flattening of the leech's body. To study the neuronal basis of leech behaviors, we have used three kinds of preparations (Figure 14.2C): (1) *intact animals*, with markers sewn into the skin to measure changes in the shape of

---

FIGURE 14.2. The functional anatomy of the medicinal leech. (A) An outline of the body indicating some of the external features (e.g., suckers, eyes, and annuli) as well as the location of the 21 midbody ganglia and the brains, one anterior and another posterior. (B) A drawing of the major muscle layers present in all body segments. In particular, the circular and longitudinal muscles produce the elongations and contractions used to produce crawling steps. The more simplified drawing below iindicates the geometric features incorporated into the biomechanical model. (From Wilson et al. 1996a) (C) The preparations used to perform experiments on behaving leeches. (a) In-

tact animals, with threads sewn into the skin to mark segments, are used to measure the changes in shape of the segments during different behaviors. (b) Semi-intact animals, with several segmental ganglia exposed from intracellular and extracellular recording, are used to correlate neuronal activity to the behavior and locate neurons involved in the behaviors. (c) Isolated nerve cords are used to determine the connectivity pattern among the neurons, using multiple intracellular and intracellular recordings. The exposed portions of the nervous systems in (b) and (c) produce activity patterns that would produce many leech behaviors, at least qualitatively.

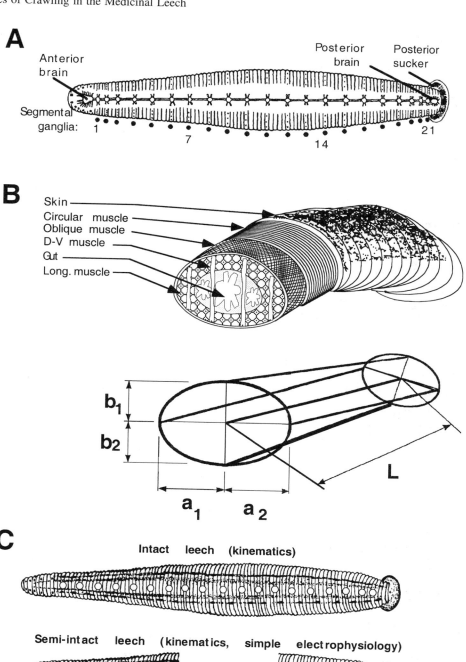

A

Anterior brain

Posterior brain

Posterior sucker

Segmental ganglia: 1

7

14

21

B

Skin

Circular muscle

Oblique muscle

D-V muscle

Gut

Long. muscle

b_1

b_2

a_1

a_2

L

C

Intact leech (kinematics)

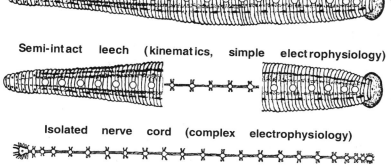

Semi-intact leech (kinematics, simple electrophysiology)

Isolated nerve cord (complex electrophysiology)

body segments as the animals move around; (2) *semi-intact animals*, in which part of the animal's nerve cord is exposed for recording motor neuronal activity as the remaining, intact part of the body behaves; and (3) *isolated nerve cords*, which can be induced to produce patterns of motor neuronal activity that are clearly remnants of particular behaviors, such as swimming, crawling, local bending, and shortening.

3.2   Muscle Properties

We have determined the passive properties of pieces of body wall, before and after anesthetizing the muscles (Wilson et al. 1996a). These experiments showed that leech muscles have passive properties similar to smooth muscle, in that they show *stress relaxation*: they resist stretch very strongly at first, but this resistance falls off to a relatively low, maintained level (Figure 14.3A). The rate at which the muscles reach their maintained tension level was approximated by two exponential time constants. A model of these passive properties (Figure 14.3B), using two nonlinear dashpots in parallel to simulate the stress relaxation and a nonlinear spring to simulate the maintained level of tension, produced a good approximation of the passive properties, except at the longer lengths of stretch.

When the longitudinal muscle is stretched out, the amount of isometric tension generated depends on the length at which the muscle is held (Figure 14.3C): at about 75% of the longest length ever attained during any leech behavior, the longitudinal muscles achieve their maximal force for any given firing rate of the motor neurons; at maximal or minimal length, the muscles generate nearly zero force (Wilson et al. 1995). The general form of this active length-tension curve is not unusual, except that the very large range of lengths over which leech muscle exerts force is, again, more like a smooth muscle than a skeletal muscle (Fung 1993). The rate of rise and maximal force of contractions produced by muscles, held at a given length, are determined by the firing rates of the motor neurons (Figure 14.3D).

From preliminary experiments, the dynamics of longitudinal muscles stimulated electrically show the hyperbolic relationship between contraction rate and initial length that is described by the Hill equation (Fung 1993). For simulating fast behaviors, such as swimming (which consists of repeating undulations up to 2.5 Hz), we will need to generate more detailed data. To date, we have limited our modeling to slower behavior, like crawling, in which quasisteady state values for muscle forces provide reasonable estimates of the movements.

3.3   Components of the Crawling Behavior

Leech crawling is a cyclic act consisting of elongation and contraction phases punctuated by sucker attachments and releases. Although variable in nature, simple stereotyped crawling steps achieved while leeches crawl on a smooth, dampened surface has three major features (Figure 14.4A). For convenience, we consider that a step begins when the leech releases the attachment of its front sucker while it is at minimal length. An elongation wave begins, caused by contraction of the circular muscles at the most anterior segments. The wave passes from segment to segment until the back end elongates, at which point the front sucker attaches, terminating the elongation phase of the step. Almost immediately, the front end starts to contract, by relaxing its circular muscles and contracting its longitudinal muscles. This begins a wave of contraction that, too, sweeps from front to back. As the contraction wave passes through the middle of the body (the actual location varies), the rear sucker releases and is pulled forward by the contraction wave. At minimal length, the rear sucker reattaches. The cycle is completed with a short pause before the next release of the front sucker.

From the kinematics of the movements, obtained by measuring the changes in distance between markers sewn into the leech's body at known segmental locations (Stern-Tomlinson et al. 1986; and Cacciatore et al. 2000), we have determined the detailed movement patterns. A neuronal model (Cacciatore et al. 2000), constrained by a variety of such behavioral observations on both intact and partially dissected leeches (Baader and Kristan 1995), produced motor activity patterns such as those shown in Figure 14.4B, providing our best estimate of the underlying firing pattern of the circular and longitudinal motor neurons during crawling. This motor neuronal firing pattern was used to adjust the biomechanical parameters so that they produce a crawling cycle that reproduces smooth, normal-looking crawling steps.

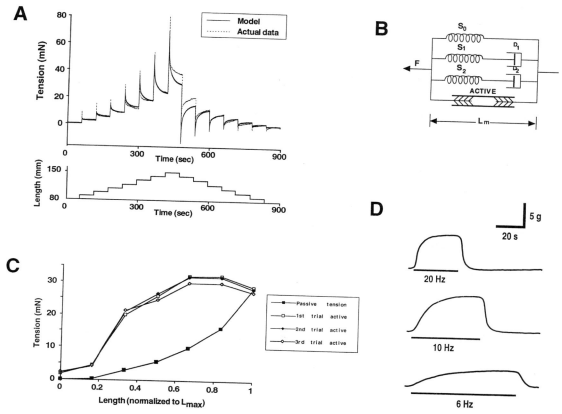

FIGURE 14.3. Properties of leech muscles and motor neurons used to generate the biomechanical model. (A) Passive tension (top graph) generated by a piece of leech body wall stretched in the longitudinal axis to different lengths (bottom graph). The solid lines are the measurements obtained from leech muscles and the dotted lines are the simulated results from the model shown in B. The response to stress was complex, consisting of a time-dependent *stress relaxation* and a length-dependent maintained force. (From Wilson et al. 1996a.) (B) The model used to simulate the passive properties of leech longitudinal muscles. The time constants of the two spring-dash pot combinations were 1.8 and 18 seconds to produce the simulated data in part A. (From Wilson et al. 1996a.) (C) The active (open symbols) and passive (closed symbols) length-tension curves for longitudinal muscles. The passive curve was generated from data as in part A. The active curve was generated by stimulating an L cell at a constant, high rate after the longitudinal muscle was pulled out to different lengths. The three curves were the maximal forces measured at three different times after the muscle was stretched, during times of high, middle, and low levels of stress relaxation. The similarity of the three curves show that the magnitude of the active contractions depended upon length, and not upon the degree of stress relaxation. (From Wilson et al. 1995.) (D) Isometric forces generated by leech longitudinal muscle from a single segment in response to firing a single motor neuron, the L cell, at three different frequencies for different durations. This motor neuron produced the strongest and fastest muscle contractions. All muscle activations were simulated as fractions of the responses to this motor neuron. (From Mason and Kristan 1982.)

In addition, we have directly recorded the motor neuron firing patterns from crawling leeches. To do this, however, it was necessary to open the leech in several segments and provide the suckers with a moving substrate while the middle of the animal was held down to permit recording the activity of motor neurons (Baader and Kristan 1992). Recently (Eisenhart et al. 1995), it has been possible to record the crawling motor activity from a semi-intact preparation using a simpler procedure (data used for Figure 14.4C), as well as from an isolated nerve cord (data used for Figure 14.4D). In both

FIGURE 14.4. Leech crawling: kinematics and motor patterns. (A) One complete crawling step. Fully contracted leeches, with both suckers attached (top drawing) begin a step by releasing their front sucker and elongating by contracting the circular muscles in the most anterior end. A wave of elongation passes from the front to the back until the animal is fully elongated, at which point the front sucker reattaches. With little delay, the front end contracts, and a wave of contraction passes backward. As the wave reaches approximately the middle of the animal, the rear sucker releases, reattaching when the animal fully shortens. After a delay, the cycle begins anew. (From Baader and Kristan, 1995.) (B) The assumed activity patterns in the circular and longitudinal motor neurons during two crawling steps, based on kinematic data from intact leeches. (From Stern-Tomlinson et al. 1986.) (C) The motor neuron activity patterns, averaged from data recorded in semi-intact preparations. (From FJ Eisenhart, unpublished data.) (D) The motor neuron activity patterns, averaged from data recorded in isolated nerve cords. (From Eisenhart and Cacciatore, unpublished data.)

A. Crawling behavior.

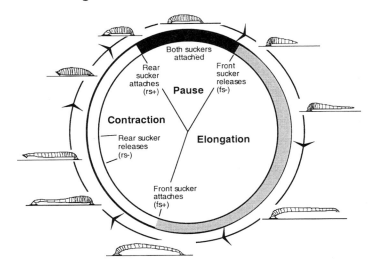

B. Co-ordination pattern: intact animals.

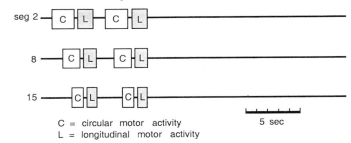

C = circular motor activity
L = longitudinal motor activity

5 sec

C. Co-ordination pattern: semi-intact animals.

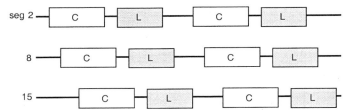

D. Co-ordination pattern: isolated nerve cords.

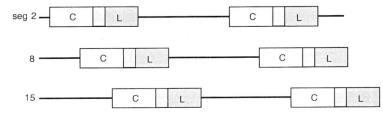

semi-intact and isolated nerve cord preparations, we recorded the activity from a segment that has been denervated, thereby eliminating sensory feedback from the segment in which the motor neurons are located, and from several to either side. In the semi-intact preparation, however, both the front and back ends of the animal are intact and are making crawling movements, so that sensory feedback from these distant sites was a possibility.

3.4   A Model of the Leech Hydroskeleton

The simulation of the leech body consists of 16 segments. (The anterior four segments and head are so small that they are lumped into a single simulated segment; likewise, the tail plus the last 4 segments are lumped into one segment.) Each segment is an ovoid cylinder, with longitudinal muscles connected near the outside edge of the cylinder and circular muscles wrapped around the outside (Figure 14.2B). The ovoids were characterized by four semiradii (a, a_2, b, and b_2), which divided the segment into four quadrants. The circular and longitudinal muscles in each quadrant can be activated separately, corresponding to the major innervation fields of the motor neurons to these muscles (Stuart 1970). Flattener muscles connect the top and bottom of the segmental body wall, although these muscles were not activated during the simulated crawling steps. Simulated crawling was driven by each of the motor neuronal activity patterns of Figure 14.4. For each increment of time, the following calculations were made:

A.   The *active forces* generated by muscles in each segment. The firing rates of the motor neurons determined these forces in two ways:
   a.   The maximal steady-state force that each muscle could generate: this is a function of the length of the muscle, defined by the active length-tension curve (Figure 14.3C).
   b.   The activation of each muscle: the ratio of the actual force being generated to the maximal steady-state force, from data shown in Figure 14.3D. The shape of the activation curve was defined by three terms: the time constant of contraction (t_c), the time constant of relaxation (t_r), and the firing rate of the motor neurons (W).

B.   The *passive forces* generated by all muscles, both active and inactive, was calculated. It is a function of the muscle's length, as defined by the passive length-tension curve (Figure 14.3C).
C.   The *segmental potential energy* was calculated, as the integrated active and passive forces generated in that segment. Each segment was regarded as an individual hydrostat, so that the longitudinal muscles in one segment directly antagonized the circular muscles in only that segment. (There are interactions between segments, however, because the longitudinal muscles of adjacent segments connect the same septum.)
D.   The *total potential energy* was calculated and the body shape was then estimated, using sequential quadratic programming, an iterative procedure that minimized the total potential energy. This calculation depended upon the assumptions that at a given activation the muscles behave like springs, and that each segment has a fixed volume.

These calculations were repeated for an average of 120 time intervals per step and the successive body shapes were video-taped and viewed as an animated sequence. Using the motor neuronal bursts generated by the model of the kinematic data, the simulated leech takes full-length steps that are smooth and look like a crawling leech (Figure 14.5A). In addition, the pressures generated within the central cavity during each step were similar to a real leech in two ways: the magnitudes of the pressure are 5 to 30 cmH$_2$0; and there are two pressure peaks, one at the maximal extension and the other at maximal contraction. The relative sizes and shapes of the peaks are not a good match, however, presumably because of details of the geometry or mechanics. Because we have not taken into account the time dependence of the contractions at different firing rates of the motor neurons (the "Hill relationships") for any muscles, our simulations to date are only quasidynamic. The fact that the match to measured parameters of shape and internal pressure are good is, therefore, encouraging.

When the data from semi-intact animals (Figure 14.4C) was used to drive the model (Figure 14.5B), the simulated steps were normal in some respects: the steps were of the appropriate length and the elongation waves moved back through the animal's body smoothly. There were, however, some anom-

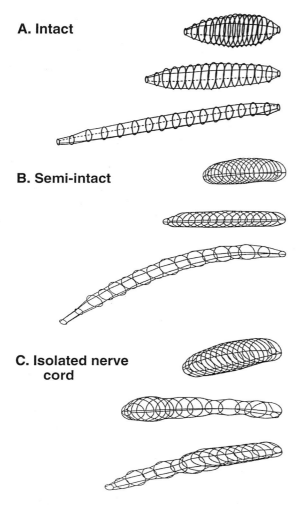

A. Intact

B. Semi-intact

C. Isolated nerve cord

bursts: the timing between activation of antagonistic muscles—particularly the delay from longitudinal muscle activation to circular muscle activation—were so long that the contractions tended to reach maximal force and fall back to minimum during each half of the step cycle, rather than maintaining intermediate forces (and thereby some stiffness) during the pauses between motor bursts.

When the data from isolated nerve cords were used (Figure 14.4D), the simulated steps were even more abnormal (Figure 14.5C): parts of the body had bulges as the contraction waves moved to the rear; the maximal elongation of the body was less than seen in other simulations, so that the progress made with each step was much less; and the maximal pressures measured during each step were much greater than in the other conditions. Both the increased pressure and the decreased step length make this pattern very much less efficient—measured as forward progress for a given amount of force generated—than the steps produced with the other two patterns. These problems resulted from two anomalies in the timing between activation of the circular and longitudinal muscles. First, the longitudinal bursts overlapped with the ends of the circular bursts. These overlaps produced the local bulges in the body as well as the very high pressures seen during these co-contractions. Second, the delays between longitudinal and circular muscle activation times were abnormally long, so that the simulated animal went slack during each step.

FIGURE 14.5. Simulated leech crawling steps. (A) The body forms of a simulated crawling step at three different times during the elongation phase: at full contraction, in the middle of extension, and at nearly full extension. Activity patterns were those of Figure 14.4B. (From Skierczynski et al. 1996.) (B) Body forms from similar times during elongation in a simulation driven by motor activity obtained from semi-intact preparations (Figure 14.4C). (C) Body forms from similar times during elongation in a simulation driven by motor activity obtained from isolated nerve cord preparations (Figure 14.4D).

4   What We Have Learned

In general, hydroskeletons do not have specialized muscles that produce the stiffness; instead, stiffness is produced by the same muscles that produce body movements. In the leech, a major source of this stiffness comes from the interplay of forces generated by antagonistic muscles. For crawling, at least, this overlap is not truly "co-contraction" of the muscles, but rather overlap in the contraction of one muscle with the relaxation of its antagonist. The timing between the contractions, therefore, becomes crucial, as evidenced by the progressive deterioration of the effectiveness of crawling in Figure 14.5, as the timing of the longitudinal and circular muscle activation times became more abnormal. In fact, we found that it is very difficult

alies: there was some variation in length as the elongations and contractions approached their maximum, and the pressures inside the simulated body cavity had somewhat larger amplitudes than those seen in the animal. These differences resulted from changes in the timing between the motor neuronal to maintain a constant pressure in the simulation during any whole-body movement by using co-contractions of these antagonistic muscles. One possibility that we are investigating, both physiologically and computationally, is that helical oblique muscles maintain a low level of stiffness while the animal is at rest, and that co-contractions of other muscles contribute to stiffness only during sustained body movements.

The data in Figures 14.4 and 14.5 have also made clear to us the importance of sensory feedback in controlling transitions between antagonistic muscles. The physiological studies have shown that these transition intervals are not built into the central pattern generator, implying a need for sensory feedback. The exposed ganglia in a semi-intact preparation produce much more functional motor neuronal spike bursts (Figure 14.5B) than does the isolated nerve cord (Figure 14.5C). These results suggest that sensory feedback from intact segments, even very distant ones (as in the semi-intact preparation), help to coordinate the timing of bursts in a given segment. It is possible that the major function of local sensory input is to slow down the rate at which elongation and contraction waves pass through a given segment.

5 Future Directions

Our model can certainly be modified to simulate other hydroskeletons, with other kinds of muscle types arranged in other orientations. Our studies have convinced us, however, that the more realistic the model, the more useful will be the insights. One model of leech swimming (Jordan 1994), for instance, assumes a rigid skeleton that is hinged segmentally. Such a model is more like a fish or a lamprey (Ekeberg 1993; Bowtell and Williams 1994; Williams et al. 1995; Chapter 15). This rigid model is a useful one for asking questions about the coupling between the animal and the fluid through which the animal moves while undulating, but it will not be useful to study other leech behaviors, such as crawling, and would not serve well to test the effects of varying the parameters of the motor burst, or testing the importance of the biomechanics of leech muscles. Hence, if the model were to be used to simulate the movements of an octopus' tentacle, for instance, it should use neu-romuscular and biomechanical properties measured in tentacular nerves and muscles. Hopefully, our experience with simulating leech behaviors will serve to show which parameters would be most usefully measured.

Acknowledgments. This research was supported by an NSF research Grant (IBN 11432) to WBK and RS, as well as an NIMH Research Grant (MH 43396) to WBK. We appreciate the enthusiastic and painstaking efforts of several dedicated undergraduates who helped to gather the data used to generate the model, including Sidney Blackwood, Jennifer Meyer, and Alvin Ting. In addition, we would like to dedicate this paper to the memory of Richard Skalak, who was an ideal collaborator and colleague. We miss his enthusiasm, his insights, and his friendship.

References

Baader, A. and Kristan, W.B. Jr. (1992). Monitoring neuronal activity during discrete behaviors: a crawling, swimming, and shortening device for tethered leeches. *J. Neurosci. Meth.*, 43:215–223.

Baader, A.P. and Kristan, W.B. Jr. (1995). Parallel pathways coordinate crawling in the medicinal leech, *Hirudo medicinalis. J. Comp. Physiol.*, 176:715–726.

Bowtell, G. and Williams, T.L. (1994). Anguilliform body dynamics: a continuum model for the interaction between muscle activation and body curvature. *J. Math. Biol.*, 32:83–91.

Cacciatore, T.W., Rozenshteyn, R., and Kristan, W.B., Jr. (2000). Kinematics and modeling of leech crawling: evidence for an oscillatory behavior produced by propagating waves of excitation. *J. Neurosci.* (in Press).

Chapman, G. (1950). Of the movement of worms. *J. Exp. Biol.*, 27:29–39.

Chapman, G. (1958). The hydrostatic skeleton in the invertebrates. *Biol. Rev.*, 33:338–371.

Chapman, G. (1975). Versatility of hydraulic systems. *J. Exp. Zool.*, 194:249–270.

Chiel, H.J., Crago, P., Mansour, J.M., and Hathi, K. (1992). Biomechanics of a muscular hydrostat: a model of lapping by a reptilian tongue. *Biol. Cybern.*, 67:403–415.

Eisenhart, F.J., Cacciatore, T.W., Kristan, W.B. Jr., and Wessel, R. (1995). Crawling in the medicinal leech is produced by a central pattern generator. *Soc. Neurosci. Abstr.*, 21:149.

Ekeberg, O. (1993). A combined neuronal and mechanical model of fish swimming. *Biol. Cybern.*, 96:363–374.

Fung, Y.C. (1993). *Biomechanics: Mechanical Properties of Living Tissues*, 2nd ed. Springer-Verlag, New York.

Jordan, C.E. (1994). Mechanics and dynamics of swimming in the medicinal leech, *Hirudo medicinalis*. Ph.D. dissertation, University of Washington, Seattle.

Kier, W.M. (1982). The functional morphology of the musculature of squid (*Loliginidae*) arms and tentacles. *J. Morph.*, 172:179–192.

Kier, W.M. and Smith, W.K. (1985). Tongues, tentacles and trunks: the biomechanics of movements in muscular-hydrostats. *Zool. J. Linn. Soc.*, 83:307–324.

Kristan, W.B. Jr., Ort, C.A., and Stent, G.S. (1974). Neuronal control of swimming in the medicinal leech. III. Impulse patterns of motor neurons. *J. Comp. Physiol.*, 94:155–176.

Lanzavecchia, G., DeEguileor, M., and Valvassori, R. (1985). Superelongation in helical muscles of leeches. *J. Musc. Res. Cell. Motil.*, 6:569–584.

Lent, C.M., Fliegner, K.H., Freedman, E., and Dickinson, M.H. (1988). Ingestive behavior and physiology of the medicinal leech. *J. Exp. Biol.*, 137:513–527.

Mann, K.H. (1962). *Leeches (Hirudinea): Their Structure, Physiology, Ecology and Embryology*. Pergamon Press, London.

Marshall, D.J., Hidgson, A.N., and Trueman, E.R. (1989). The muscular hydrostat of a limpet tentacle. *J. Moll. Stud.*, 55:421–422.

Mason, A. and Kristan, W.B. Jr. (1982). Neuronal excitation, inhibition and modulation of leech longitudinal muscle. *J. Comp. Physiol.*, 146:527–536.

Rack, P.M.H. and Westbury, D.R. (1969). The effect of stimulus rate on tension in the isometric cat soleus muscle. *J. Physiol.*, 204:443–460.

Skierczynski, B.A., Wilson, R.J.A., Kristan, W.B. Jr, and Skalak, R. (1996). A model of the hydrostatic skeleton of the leech. *J. Theoret. Biol.*, 181:329–342.

Smith, K.K. (1984). The use of the tongue and hyoid apparatus during feeding in lizards (*Ctenosaurra similis* and *Tupinambis nigropunctatus*). *J. Zool. Lond.*, 202:115–143.

Smith, K.K. and Kier, W.M. (1989). Trunks, tongues, and tentacles: Moving with skeletons of muscle. *Am. Sci.*, 77:29–35.

Stern-Tomlinson, W., Nusbaum, M.P., Perez, L.E., and Kristan, W.B. Jr. (1986). A kinematic study of crawling behavior in the leech, *Hirudo medicinalis*. *J. Comp. Physiol.*, 158:593–603.

Stuart, A.E. (1970). Physiological and morphological properties of motoneurones in the central nervous system of the leech. *J. Physiol.*, 209:627–646.

Wadepuhl, M. and Beyn, W.-J. (1989). Computer simu-

lation of the hydrostatic skeleton: the physical equivalent, mathematics and application to worm-like forms. *J. Theoret. Biol.*, 136:379–402.

Wainwright, P.C. and Bennett, A.F. (1992). The mechanism of tongue projection in chameleons I. Electromyographic tests of functional hypothesis. *J. Exp. Biol.*, 168:1–21.

Wainwright, P.C. and Bennett, A.F. (1992). The mechanism of tongue projection in chameleons II. Role of shape change in a muscular hydrostat. *J. Exp. Biol.*, 168:23–40.

Williams, T.L., Bowtell, G., Carling, J.C., Sigvardt, K.A., and Curtin, N.A. (1995). Interactions between muscle activation, body curvature and the water in the swimming lamprey. In *Biological Fluid Dynamics*, Ellington, C.P. and Pedley, T.J. (eds.). SEB Symposium, 49:49–59. Company of Biologists, London.

Wilson, J.F., Mahajan, U., Wainwright, S.A., and Croner, L.J. (1991). A continuum model of elephant trunks. *J. Biomech. Eng.*, 113:79–84.

Wilson, R.J.A., Skierczynski, B.A., Meyer, J.K., Blackwood, S., Skalak, R., and Kristan, W.B. Jr. (1995). Predicting overt behavior from the pattern of motor neuron activity: a biomechanical model of a crawling leech. *2nd J. Symp. Neural Comput.*: 176–190.

Wilson, R.J.A., Skierczynski, B.A., Meyer, J.K., Skalak, R., and Kristan, W.B. Jr. (1996a). Mapping motor neuron activity to overt behavior in the leech. I. Passive biomechanical properties of the body wall. *J. Comp. Physiol.*, 178:637–654.

Wilson, R.J.A., Skierczynski, B.A., Blackwood, S., Skalak, R., and Kristan, W.B. Jr. (1996b). Mapping motor neurone activity to overt behaviour in the leech. Internal pressures produced during locomotion. *J. Exp. Biol.*, 199:1415–1428.

Commentary: Biomechanical Studies Clarify Pattern Generator Circuits

Hillel J. Chiel and Randall D. Beer

We are extremely enthusiastic about the work described in this chapter. The results of Kristan et al. illustrate the value of focusing on the detailed biomechanics of the body in order to understand neural activity. A striking example is provided by the data that they present, which suggests that a "crawling motor program" generated by an isolated nervous

system is not terribly effective in actually controlling crawling. This result is very reminiscent of
work in other pattern generators (e.g., locust flight;
Aplysia feeding (Pearson et al. 1983; Morton and
Chiel 1993)) in which sensory feedback plays an
essential role in shaping a central motor pattern to
make it behaviorally useful (Pearson 1985). It is
clear that an essential way of evaluating a "motor
pattern" is to simulate the body, and to test the pattern's effects on the body's behavior. This may provide a much more meaningful metric for defining
the significance of perturbations of a motor pattern.

   We have a few small comments on their excellent general review of hydroskeletons:

1. They cite Chapman's conclusions, based on
kinematic arguments, that "in a very elongated
structure, even small contractions of the circular
muscles produce large and fast elongations of the
structure" (Chapman 1950). This conclusion is
based on kinematic considerations, which should
be balanced against the kinetics of the structure. In
an elongated hydrostat, circumferential muscles
will lose their mechanical advantage, and passive
forces in longitudinal muscles will antagonize elongation movements, so that further elongation will
depend critically on the load encountered by the
structure (Chiel et al. 1992).
2. It is also worth noting that circumferential, radial, or transverse muscles when contracted simultaneously all cause similar effects that are scaled
versions of one another; the key issue is whether
the added muscle mass is exploited for additional
flexibility in movement (Chiel et al. 1992).
3. It is interesting that timing and overlap of activation are essential for appropriate crawling
movements, as they are for generating appropriate
tongue kinematics in the model of reptilian lapping,
even though the details of the models differ greatly
(Chiel et al. 1992).

   The leech preparation may offer a superb opportunity for clarifying the neural control of hydrostatic structures. This problem is attracting increasing attention. For example, a recent study of
reaching in octopus found a stereotyped velocity
profile of the bending tip of the tentacle during
reaching, suggesting that stereotyped patterns of
muscle activation may simplify the difficult control problem posed by the many degrees of free-

dom present in a hydrostatic structure (Gutfreund
et al. 1996). Some of these questions may be directly addressable in the leech.

   The current review does not address the more
rapid movements that leeches perform during
swimming. It will be interesting to see the extent
to which the hydrodynamics of water, inertial properties of the body, and active muscle properties may
begin to play significant roles in this behavior. Under these circumstances, the low pass filtering properties of muscles must be overcome to activate
them effectively at higher frequencies, the inertial
properties of the animal will be more important at
higher accelerations, and viscous drag will increase
at higher velocities. In addition, the hydrodynamic
profile of the animal will be significantly different
shortly after a blood meal, when it is greatly dis-
tended, versus its hydrodynamic profile after it has
fully digested its meal.

   The results summarized in this chapter provide
strong support for a new view of adaptive behavior, in which that behavior is not merely the product of the commands of the nervous system, slavishly followed by the body. Rather, the nervous
system evokes certain features of the dynamics of
the body, and this gives rise to behavior. Moreover,
the embedding of the body within its environment
may also play important roles (Chiel and Beer
1993; Beer 1995). A striking recent example has
been provided by studies of swimming in the lamprey, which suggest that sensory feedback is essential for the normal swimming pattern, and that
in the absence of the hydrodynamics of water, the
lamprey cannot generate a normal traveling wave
(Bowtell and Williams 1994; Williams et al. 1995).

References

Beer, R.D. (1995). A dynamical systems perspective on
   agent-environment interaction. *Artif. Intell.*, 72:173–
   215.
Bowtell, G. and Williams, T.L. (1994). Anguilliform
   body dynamics: a continuum model for the interaction
   between muscle activation and body curvature.
   J. Math. Biol., 32:83–91.
Chapman, G. (1950). Of the movement of worms. *J. Exp.
   Biol.*, 27:29–39.
Chiel, H.J. and Beer, R.D. (1993). Neural and peripheral
   dynamics as determinants of patterned motor behavior. *The Neurobiology of Neural Networks*, pp.
   137–164, The MIT Press. Cambridge, Massachusetts.

Chiel, H.J., Crago, P., Mansour, J.M., and Hathi, K. (1992). Biomechanics of a muscular hydrostat: a model of lapping by a reptilian tongue. *Biol. Cyber.*, 67:403–415.

Gutfreund, Y., Flash, T., Yarom, Y., Fiorito, G, Segev, I., and Hochner, B. (1996). Organization of octopus arms movements: a model system for studying the control of flexible arms. *J. Neurosci.*, 16:7297–7303.

Morton, D.W. and Chiel, H.J. (1993). The timing of activity in motor neurons that produce radula movements distinguishes ingestion from rejection in Aplysia. *J. Comp. Physiol. A*, 173:519–536.

Pearson, K.G. (1985). Are there central pattern generators for walking and flight in insects? In *Feedback and Motor Control in Invertebrates and Vertebrates*, pp. 307–315. Croom Helm, London.

Pearson, K.G., Reye, D.N., and Robertson, R.M. (1983). Phase-dependent influences of wing stretch receptors on flight rhythm in the locust. *J. Neurophysiol.*, 49: 1168–1181.

Williams, T.L., Bowtell, G., Carling, J.C., Sigvardt, K.A., and Curtin, N.A. (1995). Interactions between muscle activation, body curvature and the water in the swimming lamprey. *Symp. Soc. Exp. Biol.*, 49:49–59.

15
Simulation of the Spinal Circuits Controlling Swimming Movements in Fish

Örjan Ekeberg

1 Introduction

Limb movements in humans and other higher vertebrates are generated through a complex interplay between the mechanical apparatus and the central nervous system. Understanding of this neuromechanical system entails more than just a knowledge of its constituent parts. It is necessary to find ways of combining our knowledge of the mechanical system with that of the controlling neuronal circuitry. Both parts by themselves are far from trivial and computer simulation has emerged as an indispensable tool in the task of understanding how they cooperate.

To make computer simulation possible, it is necessary to come up with complete mathematical models of both systems as well as of their interaction. Here, "complete" does not imply that the models have to include every known experimental fact but merely that they should incorporate the behavior of all parts of the system in a sufficiently realistic fashion.

The relevant properties of muscles and skeletal mechanics are comparatively well known. In fact, even rather simplistic models may often be sufficient when used as components in combined neuromechanical simulations. Of course, further understanding of muscle and skeletal mechanics may be useful in refining the simulations but very sophisticated models do not, at least not in the initial stages, seem necessary.

When it comes to the neuronal control circuitry, the situation is less favorable. This system can be described in very different ways depending on the purpose of the model. There are basically two approaches that can be taken. Either some sort of black-box controller is used, or a model is constructed from what is known about the actual circuitry.

In black-box models one is trying to capture the most important properties of the input–output mapping without worrying much about how things are actually accomplished in the underlying neuronal circuitry. One serious drawback with this approach is that it is hard to gradually extend and refine such a model to incorporate more detailed facts about neurons and their circuitry. From this point of view, the black-box approach may turn out to be a dead end if the goal is to eventually understand how the neuronal control system works in terms of neurons and neuronal networks.

Assembling a model from neuronal components is not, however, a trivial task. A large number of models are available that describe various neuronal mechanisms in high detail, but there are very few attempts to synthesize these into complete models of the entire neuronal control system. While models of muscle and skeletal mechanics account for many details in a realistic fashion, the lack of a correspondingly realistic model of the spinal control circuitry is gradually becoming a serious limiting factor in the further understanding of the neuromechanical system.

2 Animal Models

Data on the spinal circuitry in humans is not yet sufficiently detailed and complete to allow simulation studies based on such data alone. To proceed it is necessary to utilize what is known from animal preparations in order to come up with reasonable hypothetical circuit structures. Detailed information about the spinal circuitry involved in motor control is primarily available from studies in lower vertebrates. The conservative nature of evolution justifies such use of lower vertebrates as simplified models of corresponding systems in higher vertebrates (Cohen 1988; Grillner 1996).

The neuronal generation of vertebrate locomotion has been extensively studied in the lamprey. Superficially the swimming movements of lampreys are very different from limb movements such as walking and reaching in humans. However, this does not necessarily mean that the underlying control circuitry is fundamentally different. In fact, many features found in higher vertebrates have also been identified in the lamprey, albeit in simpler form. This includes the general anatomical organization as well as minute details such as transmitter substances and ion channels. It is therefore reasonable to assume that studies of the controlling neuronal circuitry in the lamprey, in particular those in the spinal cord, are highly relevant also in the process of assembling models of human limb movements.

2.1   Lamprey Preparation

The lamprey is unique in that its central nervous system can be isolated, by cutting the nerve fibers, excised from the body, and kept alive in a nourishing bath for hours or even days. The central nervous system of the lamprey is therefore experimentally more accessible than in most other vertebrates. This has made it possible to conduct a number of unique experiments to describe various neurons and neural circuits in high detail. In particular, the spinal generator for swimming rhythmicity has been the subject of a long series of studies (Grillner et al. 1995).

2.2   Lamprey Swimming

The lamprey swims by producing lateral body waves propagating from head to tail with increasing amplitude (Figure 15.1A). As the waves travel backwards the sides of the body are pressing against the water giving a thrust that propels the lamprey forward. The frequency of these waves varies between 0.2Hz and 10Hz depending on the

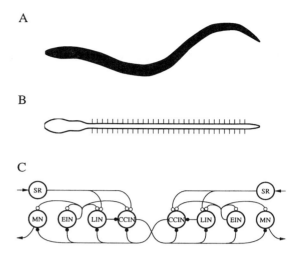

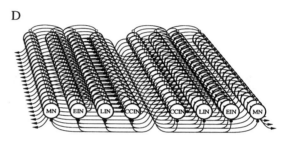

FIGURE 15.1. (A) A swimming lamprey as seen from above. (B) Schematic of the isolated CNS preparation. (C) Local spinal connectivity of the rhythm generator. (D) Full spinal network comprised of many identical local circuits.

speed of swimming. Slightly more than one full wave is normally present along the body, regardless of the swimming speed. Although the lamprey has fins, these do not seem to play a major role in this process.

These undulatory waves are caused by corresponding waves of contraction in laterally located muscles. The muscle contractions are in turn driven by waves of motoneuronal activity alternating between the two sides and propagating down the spinal cord.

In an isolated lamprey CNS, rhythmic motoneuronal activity can still be detected in the spinal cord, especially if some excitatory amino acid, like glutamate, is added to the bath. These activity patterns are remarkably similar to EMG recordings in the intact swimming animal. Since no real movement occurs, this has been termed fictive swimming. The lack of sensory feedback in this preparation tells us that it is not a prerequisite for the production of the basic locomotor rhythmicity. It should be noted, however, that sensory information, from stretch receptors and vestibular apparatus, may be important for adjusting locomotion, particularly in a changing environment.

2.3   Neuronal Control Circuitry

As in all vertebrates, the neural control of locomotion is distributed over many parts of the central nervous system. Local circuits in the spinal cord generate the basic rhythmic activity while coupling between such "local oscillators" coordinates the motor patterns. Brainstem neurons relay commands to the spinal circuitry, thus providing high level control of locomotion.

The isolated spinal cord preparation has made it possible to identify the neurons involved in the production of swimming patterns. Further, since intracellular recording techniques can be utilized this also allows for examination of a whole range of physiological as well as pharmacological details. Apart from the motoneurons (MN), there are basically three types of interneurons involved: excitatory interneurons (EIN), lateral inhibitory interneurons (LIN), and contralateral inhibitory interneurons (CCIN). Intraspinal stretch receptors (SR) have also been found that presumably participate in the generation of the locomotor rhythm.

The use of paired intracellular recording techniques has made it possible to map out the local synaptic connectivity among these neurons (Figure 15.1C). This local connectivity scheme seems to be repeated more or less identically along the spinal cord. It is generally agreed that the local connectivity alone is sufficient to account for the generation of alternating rhythmic output. This is consistent with the observation that even small pieces of isolated spinal cord are capable of generating rhythmic output, regardless of where the piece was originally located in the cord.

During a burst on one side of the cord, the EINs excite the ipsilateral neurons, thus helping to sustain the burst. At the same time the CCINs inhibit contralateral activity. The LINs presumably play a role in terminating this stable state by eventually inhibiting the ipsilateral activity, thereby allowing the contralateral side to take over.

Whereas the operation of the local network is pretty well understood, the mechanisms which coordinate the oscillations along the cord are less clear. There does not seem to be any single localized pacemaker on the spinal level. Based on this, the whole spinal network is sometimes described in mathematical terms as a chain of discrete local oscillators. However, it is even more correct to view the network as continuous sets of neurons evenly spread out along the cord without any distinct segmental boundaries (Figure 15.1D). Simulations have shown that connections restricted to a local neighborhood are sufficient to obtain traveling waves similar to those observed in the real animal. Still, it is known that long range connections also influence the coordination.

The neurons in the local spinal circuitry receive synaptic input from several sources. In addition to the brainstem projections and connections from nearby segments, they also receive input from sensory neurons. Of special interest are the stretch receptor neurons (SR) located laterally within the spinal cord. These neurons are sensitive to lateral curvature of the body and are therefore active in phase with the swimming movement. The stretch receptors synapse into the local network in a way which excites the stretched side and inhibits the other. However, it is still unclear how this actually contributes to swimming.

3   Methods

3.1   Realistic Neuronal Simulation

The combination of a well-described isolated neuronal network and an output with well-defined objectives and characteristics provides an ideal situation for computer modeling. Indeed, models at different levels of abstraction have been used in describing the lamprey spinal locomotor circuitry, from abstract nonlinear oscillators (Cohen et al. 1992), to interconnected model neurons comprised of multiple compartments, and a Hodgkin–Huxley representation of the most relevant ion-channels (Wallén et al. 1992).

In most of the more detailed simulations we have been using a five-compartment model of each neuron to account for some of the morphology. Explicitly calculating currents through different kinds of ion channels makes it possible to directly include physiological data all the way down to the channel level. Simulating the interaction between hundreds or even thousands of such realistic model neurons is a substantial computational task. To be able to perform these, a special purpose software package, "SWIM," was developed (Ekeberg, Hammarlund, Levin, and Lansner 1994). This simulator has evolved into a general purpose network simulator and is being used also in studies of other neuronal systems.

The five-compartment model can be regarded as being of intermediate complexity, incorporating many electrophysiological features without the high detail sometimes used in single cell simulations (Churchland and Sejnowski 1992). An alternate modeling approach is to utilize more simplifying neuronal models which may be able to reproduce the same behavior. Such models generally involve fewer parameters to tune and are easier to understand conceptually. On the other hand, since the parameters do not have a direct correlate in the real system it becomes harder to supply them directly from experimental data.

We have partially avoided this problem by using a two-stage approach. The parameters of our simplifying nonspiking models were set from functional properties that could be measured by simulating the activity of the five-compartment model cell under various conditions. Thus, we were able to circumvent the problem of arbitrarily setting abstract parameters in a simplified model directly. The simplifying models have been particularly useful in lamprey simulations where the activity in the entire spinal cord is mimicked.

3.2   Neuromechanical Simulation

To study if and how the output of the neuronal network model could produce swimming behavior, it was extended to include a planar, two-dimensional, model of the mechanical system. Muscle properties, body mechanics, hydrodynamics, and sensory feedback were all incorporated using simplified, yet reasonably realistic models (Ekeberg 1993).

Muscles are regarded as purely linear actuators with both elastic and viscous properties with the spring constants controlled by the output from the motoneurons. The body is modeled as a chain of links where pairs of antagonistic muscles produce torques around the joints. Inertial properties of the

links are set to correspond roughly to a 30 cm long lamprey. Counteracting forces from the surrounding water are included via a static-drag model, which implies that the effects from water movements around the body are not included. Finally, sensory feedback is provided by stretch receptors which, in the model, give an output proportional to the local curvature of the body.

In a separate series of studies, a hypothetical extended network was assembled and simulated to assess its plausibility as a model for three-dimensional steering (see below). The planar mechanical model was then extended to incorporate movements in three dimensions. This included extending the body model into a chain of links connected via spherical joints with four muscles symmetrically arranged around each joint.

4   Results

4.1   Burst Terminating Mechanisms

Several mechanisms are required for neurons to generate rhythmic bursts. There has to be an initial source of activation, mechanisms that sustain activity during the burst, and something which makes the burst terminate. In the lamprey, tonic background excitation, primarily from the brainstem, provides a basis for both initiating and sustaining bursts. Furthermore, self-excitation among the EINs provides a positive feedback which tends to sustain the bursts.

Burst termination turns out to be a more intricate matter. The original view was that LIN neurons are responsible for suppressing ipsilateral activity with some delay, and thus terminating the bursts. Detailed simulations have revealed that the situation is more complex. In fact, several other mechanisms may contribute: accumulation of intracellular calcium, intrinsic pacemaker properties of individual neurons, phasic brainstem input, sensory feedback, and possibly others.

Even with the currently available experimental data, in combination with detailed simulation studies, it is still hard to know to what extent the different mechanisms are of importance. It seems likely that they all play some role, possibly under different circumstances.

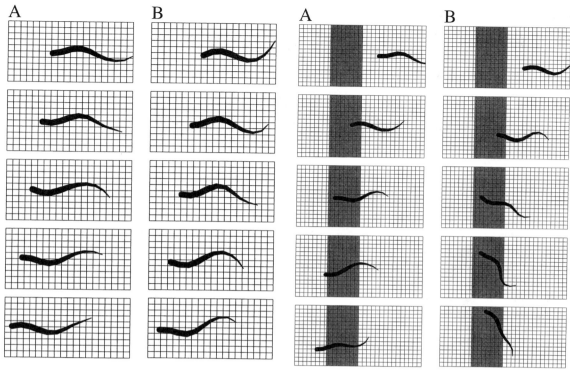

FIGURE 15.2. A neuromechanical simulation with (A) and without (B) sensory feedback.

FIGURE 15.3. A neuromechanical simulation with (A) and without (B) sensory feedback. The dark area indicates a region where the water is moving contrary to the swimming direction.

4.2   Role of Sensory Feedback

The role of the stretch receptors, which provide sensory feedback during swimming, is best analyzed in a neuro-mechanical model where the feedback loop has been closed. A somewhat surprising result is that adding or removing the sensory feedback in such simulations has very little effect on the resulting movement (Figure 15.2).

This indicates that the sensory feedback might only come into play when the movement does not proceed as predicted. Indeed, in simulations where the body is subject to natural perturbations, removing feedback from the mechano-receptors can be detrimental to the swimming performance (Figure 15.3).

4.3   Repertoire of Swimming Patterns

Neuromechanical simulation reveals that the currently known network is, in fact, capable of generating a whole repertoire of swimming patterns. Varying the overall level of brainstem input results in swimming at different speeds. By selectively setting the tonic input to different groups of spinal network neurons, several other behaviors can be evoked. For example, the wavelength and amplitude of the body waves can be regulated, enabling swimming through tight passages. The direction of the waves can be reversed, resulting in backwards swimming. Realistic lateral turns can also be produced, simply by giving asymmetric input.

Whereas lateral turns can be explained without any additions to the spinal network model, pitch and roll maneuvers require some idea about the dorsoventral dimension. It is known that separate motoneuron pools drive dorsally and ventrally located muscles. Little is known, however, about a corresponding subdivision of the underlying interneuronal circuitry. Some experimental results indicate that such a subdivision exists. We have put forward a "crossed-oscillators" hypothesis in which we assume partly separate dorsal and ventral circuits, and a preferred crossed connectivity between

FIGURE 15.4. Proposed extension of the neural control circuitry to handle pitch turns and rolls.

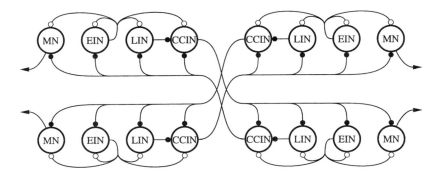

the two sides (Figure 15.4). The local network can then be regarded as two diagonally arranged coupled oscillators that normally run in synchrony. Tonic input given specifically to one of the two oscillators can, however, change the phase relation between them.

Neuromechanical simulations of this system show that it is capable of generating motor patterns that result in realistic pitch turns and rolls (Figure 15.5). Increasing the tonic input to the dorsal parts

of the network results in more emphasized dorsal motoneuronal bursts which will bend the body to give an upwards pitch. Similarly, extra input to the ventral parts gives a downwards pitch.

Perhaps more unexpected is the finding that a diagonal stimulation pattern produces a roll by affecting the timing of the dorsal and ventral bursts differently. Increasing the dorsal input on the left side and the ventral input on the right makes the dorsal muscles contract before the ventral ones when ac-

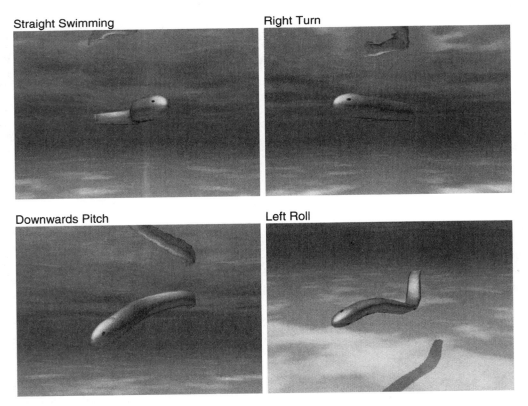

FIGURE 15.5. Example from a 3D neuromechanical simulation in which different kinds of steering commands are given through simulated brainstem input.

tivity switches over to the left side of the cord. On the other hand, the ventral muscles contract before the dorsal ones when switching back to the right side again. Mechanically, this gives rise to a spiral shape of the body which makes it roll as it pushes against the water. This spiral shape, mainly seen as a slight rotation of the tail, is consistent with behavioral studies of swimming lampreys (Ullén, Deliagina, Orlovsky, and Grillner 1995).

5 Discussion

Computer simulation is a useful tool in the study of neuronal circuitry. Computer models are well suited for synthesizing diverse experimental data into a holistic framework. Running simulations then serves to verify whether or not current pieces of information are sufficient to explain all observed behaviors. If not, they can help in directing the search for critical pieces of missing information.

A related but somewhat different use of simulation is in trying out new ideas about possible mechanisms influencing the behavior of the system. Because the system under study is so complex, it may be very hard to conceive the consequences of some hypothesis. Simulation can aid in this process by emphasizing where important, yet unforeseen secondary effects may arise. Even if the computer model is dramatically simplified when compared to the "real thing," it is often much more complete and fine grained than what is possible to envision in ones own mind. Used in this way, simulations may become a tool for reasoning about the system under study rather than for purely mimicking it.

This perspective has tried to convey how simulations of the lamprey spinal circuitry have increased our understanding of the complex interplay between different mechanisms in the system. Finding neuronal mechanisms that are capable of producing rhythmic output is not hard, but deducing which mechanisms are most influential under different working conditions is. This turns out to be important in the understanding of spinal motor control since input from the brainstem and above, which affects specific mechanisms, may or may not have a significant impact depending on whether that particular mechanism happens to be critical or not.

From the in-vitro experiments it was known that the overall excitation given to the system sets its operating speed. The simulations have complemented this by showing that the same network is also capable of generating a whole set of other maneuvers, such as turning, narrow and backwards swimming. This is achieved by giving extra excitatory input to selected parts of the spinal circuitry. Thus, such input constitutes a set of "knobs" by which different inherent capabilities of the spinal pattern generation can be activated. The neuronal correlate of these knobs is presumably input from brainstem neurons acting as specific command neurons.

Interestingly, neurons have been found in the lamprey brainstem that respond when the body is tilted at specific angles (Orlovsky, Deliagina, and Wallén 1992). These neuronal responses, originating from the vestibular system, may in fact be commands in the process of initiating a counteracting steering action. Future neuro-mechanical simulations will undoubtedly include these pathways as they become unraveled, from the vestibular organs all the way down to the spinal circuits.

In the lamprey it seems as if the spinal cord circuitry carries a set of motor patterns which can be activated and reshaped through brainstem input. Maneuvers such as steering can be thought of as reshaped propulsion patterns. From an evolutionary viewpoint, this makes sense. It seems reasonable that more sophisticated motor behaviors have evolved as successive additions to already existing circuitry rather than as completely new subsystems.

The lamprey is probably very similar to the early ancestors of all vertebrates and we should not be surprised to find parts of the lamprey's circuitry interspersed in the human control system. This raises the question of to what extent different aspects of human limb control may be explained in terms of such evolutionary building blocks. Walking and running, being rhythmic motor acts, are presumably generated by elaborated versions of the lamprey swimming pattern generator. Furthermore, it may be fruitful to view nonrhythmic motor acts, such as reaching and grasping, as reshaped rhythmic movements reusing parts of the locomotor circuitry.

Electrical stimulation of descending tracts is known to generate stereotypic limb movements that can be statically described in terms of force fields. Such stimulation may be thought of as a crude version of brainstem input which activates the spinal pattern generators. However, the pure force-field map does not capture the important dynamic properties of elicited motor patterns. Extending the force-field idea with dynamic properties may be best described in terms of partially activated rhythm generators.

6 Future Directions

We are only at the beginning of the exciting process of uncovering the neuronal circuitry involved in motor control. Understanding the spinal circuitry and how it is controlled from the brainstem is a natural first step since the rest of the brain depends on these systems in expressing movements.

As more details of the actual neuronal circuitry become known, we are likely to be faced with the insight that what, from an engineering point of view, seems as naturally separate tasks does not necessarily have to be handled by separate neuronal subsystems. Experiments carefully designed to highlight one particular aspect of the control system can often be understood in terms of simple control strategies or optimization criteria. However, such descriptions may say more about the experimental setup than about the controller itself. A neuronal controller is likely to be simultaneously involved in a number of related tasks. In order to understand its operation in more detail it will become necessary to take large parts of the neural control system into account simultaneously. Techniques for handling such holistic models, in particular computer simulation, will undoubtedly become increasingly important in the future.

Acknowledgments. I am grateful to Tom Wadden for his valuable comments on the manuscript. This work was supported by the Swedish National Board for Industrial and Technical Development (NUTEK) proj. no. 93-3147, the Swedish Natural Science Research Council (NFR) proj. no. B-TV-3531-100 and the Swedish Research Council for Engineering Sciences (TFR) proj. no. 221-97-755.

References

Churchland, P.S. and Sejnowski, T.J. (1992). *The Computational Brain*, The MIT Press, Cambridge, Massachusetts.

Cohen, A.H. (1988). Evolution of the vertebrate central pattern generator for locomotion. In *Neural Control of Rythmic Movements in Vertebrates*. Cohen, A.H., Rossignol, S., and Grillner, S. (eds.), pp. 129–166. John Wiley & Sons, New York.

Cohen, A.H., Ermentrout, B., Kiemel, T., Kopell, N., Sigvardt, K., and Williams, T.L. (1992). Modelling of intersegmental coordination in the lamprey central pattern generator for locomotion. *Trends Neurosci.*, 15:434–438.

Ekeberg, Ö. (1993). A combined neuronal and mechanical model of fish swimming. *Biol. Cybern.*, 69:363–374.

Ekeberg, Ö., Hammarlund, P., Levin, B., and Lansner, A. (1994). SWIM—A simulation environment for realistic neural network modeling. In *Neural Network Simulation Environments*. Skrzypek, J. (ed.). Kluwer, Hingham, Massachusetts.

Grillner, S. (1996). Neural networks for vertebrate locomotion. *Sci. Am.*, 48–53.

Grillner, S., Deliagina, T., Ekeberg, Ö., El Manira, A., Hill, R.H., Lansner, A., Orlovsky, G.N., and Wallén, P. (1995). Neural networks that co-ordinate locomotion and body orientation in lamprey. *Trends Neurosci.*, 18:270–279.

Orlovsky, G., Deliagina, T., and Wallén, P., (1992). Vestibular control of swimming in lamprey: 1. Responses of reticulospinal neurons to roll and pitch. *Exp. Brain Res.*, 90:479–488.

Ullén, F., Deliagina, T., Orlovsky, G.N., and Grillner, S., (1995). Spatial orientation in the lamprey. I. Control of pitch and roll, *J. Exp. Biol.*, 198:665–673.

Wallén, P., Ekeberg, Ö., Lansner, A., Brodin, L., Tråvén, H., and Grillner, S. (1992). A computer-based model for realistic simulations of neural networks. II: The segmental network generating locomotor rhythmicity in the lamprey. *J. Neurophysiol.*, 68:1939–1950.

Commentary: Computer-Simulated Models Complement Experimental Investigations of Neuromotor Control in a Simple Vertebrate

Ranu Jung

This chapter promotes the use of computer simulations as a tool to understand complex neuromotor control systems in terms of their component neural circuitry and interaction with the environment. The work presented examines locomotor control in the lamprey, a simple vertebrate using computer simulations of the spinal circuitry. The lamprey locomotor system is indeed very amenable to investigations using both the experimental and computer simulation paradigms, as indicated by the author. This commentary supports the importance of using computer simulations, with a discussion of a few instances in which computer simulation models have complemented experimental studies in the lamprey.

Most models of the lamprey central pattern generator have described it as a chain of coupled unit pattern generators; the unit pattern generators have been modeled using either a phase oscillator structure or as a neural network structure. The phase oscillator models have been used primarily to investigate intersegmental coordination (Cohen et al. 1982; 1992; Kopell 1988; Sigvardt and Williams 1991; Sigvardt 1993); the neural network models have been used to examine the roles of various neuronal and network properties in intra and intersegmental coordination and other aspects of pattern generation (Buchanan 1992; Wallen et al. 1992; Williams 1992; Jung et al. 1996; Chapter 15). The architecture of these neural network models has been constrained by the conceptual model of Buchanan and Grillner (1987) to contain three classes of neurons: EIN, LIN, and CCIN (see descriptions in the chapter). The neurons have been modeled as single compartments with a passive membrane and synpatic inputs (Buchanan 1992; Williams 1992; Jung et al. 1996) or, as by the author and his colleagues, with multiple compartments and detailed biophysical properties (Wallen et al. 1992; Grillner et al. 1993). These models have been used to address issues such as burst generation, coordination along the spinal cord, interaction between the brain and the spinal cord, and neuromechanical interaction in the intact animal. In the following paragraphs examples are presented where models have complemented experimental work for each of these issues.

The neural network models have led to a better understanding of the pattern generating mechanisms and the dynamical behavior of the network. Ekeberg describes several interpretations derived from computer simulation studies. Different views are presented for the mechanisms for generation and termination of the locomotor burst. As the author suggests, simulations can be used to examine the effects of various mechanisms on complex behavior. For example, while conceptually it has been suggested that tonic excitation of the EIN causes the initiation of the burst, and that self-excitation among the EIN sustains the burst, this cannot be verified experimentally. Computer simulation studies have demonstrated that these active mechanisms in the EIN neurons are capable of generating the observed behavior. However, in our work (Jung et al. 1996), we have used bifurcation analysis to show that, with a suitable set of model parameters, a network based only on the LIN and CIN neurons is also capable of generating the locomotor rhythm. Thus, an important caution to be observed while utilizing models is the dependence of the model behavior on parameter values. Details in the model are a double-edged sword. While detailed multicompartment models allow the modeler to utilize experimentally obtained parameters, all parameters are usually not available and even simple models with reduced sets of parameters exhibit complex dependence on parameter values.

Several investigators have used mathematical models and computer simulations to investigate rostrocaudal coordination in the lamprey, the mechanisms for which still remain controversial. During forward swimming in the lamprey, irrespective of swim frequency, there is a constant phase delay of about 1% per body segment in the rostrocaudal direction. Two different views, based on experimental and computational studies, have been discussed by Sigvardt (1993) and Grillner et al. (1993) to explain this constant phase delay. Using a coupled oscillator chain as an analogy for the spinal central pattern generator, the former view suggests that the delay arises because of asymmetric dominance between the ascending and descending coupling amongst oscillators with the same intrinsic frequency. The latter view however suggests, that the delay arises when the rostral most segment receives a higher excitability resulting in a higher intrinsic frequency thereby becoming the leading segment in a symmetrically connected chain of oscillators. The controversy arose primarily from interpretations made from analysis of mathematical models. These analyses led to experimental investigations which in turn have led to further developments and utilization of the mathematical models (Williams and Sigvardt 1994).

Models are also an important tool in the investigation of the role of the brain in the control of locomotion. As described, the brain is a source of descending tonic drive to regulate frequency and a component to relay sensory input. However, the brain itself is a dynamical system that interacts in a complex manner with the spinal cord circuitry such that the brain and the spinal cord form a loop. Experimental evidence suggests that such a loop exists in the lamprey and can profoundly affect the locomotor behavior (Kasicki et al. 1989; Vinay and Grillner 1993; Cohen et al. 1996). While the experimental data are difficult to interpret, our analysis of a mathematical model representing a unit pattern generator interacting with representative neurons in the brain has suggested that the balance between the feedforward input from the brain to the spinal cord and the intrinsic feedback from the spinal cord to the brain could be responsible for some of the puzzling behavior observed experimentally (Jung et al. 1996).

Lastly, work described in the chapter provides an example of how models are being used to understand the interaction of neural and mechanical system components. Complex neural models were simplified and combined with simple mechanical models in order to investigate complicated behaviors of the whole animal. This approach of reducing the complex models to a simpler model that captures the essential behavior exhibited by the complex model is a good one. Using a more manageable set of parameters, the model could be used to examine the roles of the neural control mechanisms and the biomechanical system in generating the different locomotor behaviors observed in the intact animal. Using this model, the author presents a crossed-oscillator hypothesis for dorsal and ventral muscle control of the roll behavior observed in the intact animal. This is an intriguing hypothesis that has been derived from computer simulation studies and we can expect that it will lead to experimental studies for verification. I agree with the author. It is important to examine the behavior of the whole animal as a dynamical neuromechanical system. The role of computational models in such studies will be increasingly important.

References

Buchanan, J.T. and Grillner, S. (1987). Newly identified 'glutamate interneurons' and their role in locomotion in the lamprey spinal cord. *Science*, 236:312–314.

Buchanan, J.T. (1992). Neural network simulations of coupled locomotor oscillators in the lamprey spinal cord. *Biol. Cybern.*, 66:367–374.

Cohen, A.H., Holmes, P.J., and Rand, R.H. (1982). The nature of coupling between segmental oscillators of the lamprey spinal generator for locomotion: a Mathematical model. *J. Math. Biol.*, 13:345–369.

Cohen, A.H., Ermentrout, G.B., Kiemel, T., Kopell, N., Sigvardt, K.A., and Williams, T.L. (1992). Modelling of intersegmental coordination in the lamprey central pattern generator for locomotion. *Trends Neurosci.*, 15(11):434–438.

Cohen, A.H., Guan, L., Harris, J., Jung, R., and Kiemel, T. (1996). Interaction between the caudal brainstem and the lamprey central pattern generator for locomotion. *Neurosci.*, 74(4):1161–1173.

Grillner, S., Matsushima, T., Wadden, T., Tegnér, J., El-Manira, A., and Wallén, P. (1993). The neurophysiological bases of undulatory locomotion in vertebrates. *Sem. Neurosci.*, 5:17–27.

Jung, R., Kiemel, T., and Cohen, A.H. (1996) Dynamical behavior of a neural network model of locomotor control in the lamprey. *J. Neurophysiol.*, 75(3):1074–1086.

Kasicki, S., Grillner, S., Ohta, Y., Dubuc, R., and Brodin, L. (1989). Phasic modulation of reticulospinal neurones during fictive locomotion and other types of spinal activity in lamprey. *Brain Res.*, 484:203–216.

Kopell, N. (1988). Toward a theory of modeling central pattern generators. In *Neural Control of Rhythmic Movements in Vertebrates*. Cohen, A.H., Rossignol, S., and Grillner, S. (eds.), pp. 369–413. John Wiley & Sons, New York.

Sigvardt, K.A. (1993). Intersegmental coordination in the lamprey central pattern generator for locomotion. *Sem. Neurosci.*, 5:3–15.

Sigvardt, K.A. and Williams, T.L. (1991). Models of central pattern generators as oscillators: the lamprey locomotor CPG. *Sem. Neurosci.*, 4:37–46.

Vinay, L. and Grillner, S. (1993). The spino-reticulospinal loop can slow down the NMDA-activated spinal locomotor network in lamprey. *NeuroReport*, 4:609–612.

Wallén, P., Ekeberg, Ö., Lansner, A., Brodin, L., Tråvén H., and Grillner, S. (1992). A computer-based model for realistic simulations of neural networks. II. The segmental network generating locomotor rhythmicity in the lamprey. *J. Neurophysiol.*, 68(6):1939–1950.

Williams, T.L. (1992). Phase coupling by synaptic spread in chains of coupled oscillators. *Science*, 258:662–665.

Williams, T.L. and Sigvardt, K.A. (1994). Intersegmental phase lags in the lamprey spinal cord: experimental confirmation of the existence of a boundary region. *J. Comp. Neurosci.*, 1:61–67.

16

A Simple Neural Network for the Control of a Six-Legged Walking System

Holk Cruse, Christian Bartling, Jeffrey Dean, Thomas Kindermann,
Josef Schmitz, and Michael Schumm

1 Introduction

All behaving systems—whether living or artificial—can be placed along a continuum spanning three types. One type comprises strictly sensory or data-driven, feedforward systems, such as simple reflex machines. The extant input determines the output. Although the internal state of such systems might be changed by learning, the system at any moment can be regarded as having a static world model. The other two types of systems have a form of world model that can be used to make predictions concerning future states of the world. Two levels may be distinguished. One is the type of system possessing an internal world model which is "manipulable" in the sense that the internal world model can be cut off from the sensory input and the motor output and then used to mentally "play" with different possibilities. Such a model could be used for thinking, for reflecting and—when connected to a value system—for decision making. A system with such a model is often called a cognitive system. Intermediate between these two extremes, namely systems with a manipulable world model and feedforward systems with a quasi-static world model, there is a third type—systems with nonmanipulable world models that nevertheless can make predictions and therefore do not slavishly rely on the extant sensory input. Important biological examples of such models are the form of central pattern generators more appropriately called central oscillators, elements that can produce rhythmic changes in activity in the absence of sensory or other external influences. Formed by evolution or design, a proper central oscillator provides a pre-

diction of future situations which can be employed to control actions and also to control and integrate sensory input. A well-known example is the circadian clock. It controls behavior (e.g., waking up in the morning), even if the sensory input is not appropriate. This is advantageous in a predictable world, but can lead to problems when the physical situation changes, as anyone suffering from jet lag can attest. Corresponding situations can be found in the control of rhythmic movement. In swimming and flying, the environment is mainly predictable and therefore a central oscillator is advantageous. In contrast, when walking on a uneven, irregular surface the situation may vary from one moment to the next and, if a central oscillator is in charge, the system has the additional problem of mediating between the possibly divergent actions commanded by the central oscillator and by the signals from the periphery. Therefore, the present chapter tests to what extent it is possible to control the quasirhythmic movement of a six-legged, 18-joint walking system without relying on central oscillators. It describes a decentralized control system in which hard physical interactions play an essential role and shows that this system generates adaptable motor rhythms. Furthermore, it illustrates how the loop through the world can be exploited to simplify computation drastically and to solve otherwise intractable computational problems. The results are based on experimental investigation of the walking behavior of the stick insect. Although walking, the behavior we discuss here, is sometimes regarded as simple and uninteresting, it involves a very strong and complex interaction with the physical environment. Thus, walking provides a particularly good behavior for examining mechanisms for integrating autonomous activity with multimodal information from multiple sources including proprioceptors and exteroceptors. Walking is therefore of interest not only in itself, but also with respect to other, more complicated behaviors where these two types of activity must be integrated.

Behavioral and physiological studies indicate that the flexibility evident in the walking system of the stick insect arises in part because control is distributed among several autonomous centers. The decentralization means that several functional problems which have to be solved can be addressed separately. One problem concerns the way the movement of the individual leg is controlled. The second refers to the coordination between legs. From biological experiments the following general answers can be given. First, each leg has its own control system which generates rhythmic step movements (for review, see Graham 1985). The behavior of this control system corresponds to that of a relaxation oscillator in which the change of state, the transition between stance and swing, is determined by thresholds based on leg position. Second, several coordinating mechanisms couple the movement of the legs to produce a proper gait (for review, see Cruse 1990). These results are briefly summarized in Sections 2 and 3 (see also Cruse et al. 1995). Section 4 explains the architecture of the swing net and Section 5 gives an extremely simple solution for the stance net.

2 Control of the Step Rhythm of the Individual Leg

The step cycle of the walking leg can be divided into two functional states. During stance, the leg is on the ground, supports the body and, in the forward walking animal, moves backwards with respect to the body. During swing, the leg is lifted off the ground and moved in the direction of walking to where it can begin a new stance. The anterior transition point (i.e., the transition from swing to stance in the forward walking animal) is called the anterior extreme position (AEP) and the posterior transition point is called the posterior extreme position (PEP). Differences in the constraints acting during the two states and in the conditions for

their termination suggest that the leg controller consists of separate control networks: a swing network controlling the swing movement, and a stance network controlling the movement of the leg during stance. (This separation may only be justified on a logical level and may not reflect a morphological separation in the nervous system). The transition between swing and stance is controlled by a selector network. The swing network and the stance network are always active, but the selector network determines which of the two nets controls the motor output. To match experimental results which failed to demonstrate a robust central pattern generator producing strong intrinsic rhythms (Bässler and Wegner 1983), this selection is done on the basis of sensory input. The selector net consists of a two-layer, feedforward net with positive feedback connections in the second layer. These positive-feedback connections serve to stabilize the ongoing activity, namely stance or swing.

3 Coordination Between Legs

In all, six different coupling mechanisms have been found in behavioral experiments with the stick insect. These are summarized in Figure 16.1. One mechanism serves to correct errors in leg placement (no. 5); another has to do with distributing propulsive force among the legs (no. 6). These will not

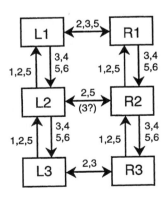

FIGURE 16.1. Summary of the coordination mechanisms operating between the legs of a stick insect. The leg controllers are labelled R and L for right and left legs and numbered from 1 to 3 for front, middle, and hind legs. The different mechanisms (1 to 6) are explained in the text.

be considered here. The other four are used in the present model. The beginning of a swing movement, and therefore the end-point of a stance movement (PEP), is modulated by three mechanisms arising from ipsilateral legs: (1) a rostrally directed inhibition during the swing movement of the next caudal leg, (2) a rostrally directed excitation when the next caudal leg begins active retraction, and (3) a caudally directed influence depending upon the position of the next rostral leg. Influences (2) and (3) are also active between contralateral legs. The end of the swing movement in the animal is modulated by a single, caudally directed influence (4) depending on the position of the next rostral leg. This mechanism is responsible for the targeting behavior—the placement of the tarsus at the end of a swing close to the tarsus of the adjacent rostral leg.

These interleg influences are mediated in two parallel ways. The first pathway comprises the direct neural connections between the step pattern generators. The second pathway arises from the mechanical coupling among the legs. That is, the activity of one step pattern generator influences the movements and loading of all legs and therefore influences the activity of their step pattern generators via sensory pathways. This combination of mechanisms adds redundancy and robustness to the control system of the stick insect (Dean and Cruse 1995).

4 Control of the Swing Movement

The task of finding a network that produces a swing movement seems to be easier than finding a network to control the stance movement because a leg in swing is mechanically uncoupled from the environment and therefore, due to its small mass, essentially uncoupled from the movement of the other legs. The geometry of the leg is shown in Figure 16.2A. The coxa-trochanter and femur-tibia joints, the two distal joints, are simple hinge joints with one degree of freedom corresponding to elevation and to extension of the tarsus, respectively. The subcoxal joint is more complex, but most of its movement is in a rostrocaudal direction around the nearly vertical axis. The additional degree of freedom allowing changes in the alignment of this axis is little used in normal walking, so the leg can be

considered as a manipulator with three degrees of freedom for movement in three dimensions. Thus, the control network must have at least three output channels, one for each leg joint. A simple, two-layer feedforward net with three output units and six input units can produce movements which closely resemble the swing movements observed in walking stick insects. In the simulation, the three outputs of this net, interpreted as the angular velocities $d\alpha/dt$, $d\beta/dt$, and $d\gamma/dt$, are fed into an integrator (not shown in Figure 16.2B), which corresponds to the leg itself in the animal, to obtain the joint angles. The actual angles are measured and fed back into the net.

This network with only 8 or 9 nonzero weights (for details see Cruse et al. 1996), which most probably represents the simplest possible network, is able to generalize over a considerable range of untrained situations, a further advantage of the network approach. Furthermore, the swing net is remarkably tolerant with respect to external disturbances. The learned trajectories represent a kind of attractor to which the disturbed trajectory returns. This compensation for disturbances occurs because the system does not compute explicit trajectories, but simply exploits the physical properties of the world. The properties of this swing net can be described by the 3D vector field in which the vectors show the movement produced by the swing net at each tarsus position in the workspace of the leg. Figure 16.3 shows the planar projections of one parasagittal section (a) and one horizontal section (b) through the work space. The complete fields are similar to those shown in Chapter 25.

This ability to compensate for external disturbances permits a simple extension of the swing net to simulate an avoidance behavior observed in insects. When a leg strikes an obstacle during its swing, it initially attempts to avoid it by retracting and elevating briefly and then renewing its swing forward from this new position. In the augmented swing net, an additional input similar to a tactile or force sensor signals such mechanical disturbances (Figure 16.2B, m.d.). This unit is connected by fixed weights to the three motor units in such a way as to produce the brief retraction and elevation seen in the avoidance reflex.

In the model, the targeting influence reaches the leg controller as the input to the swing net (Figure 16.2B). A simple feedforward net with three hid-

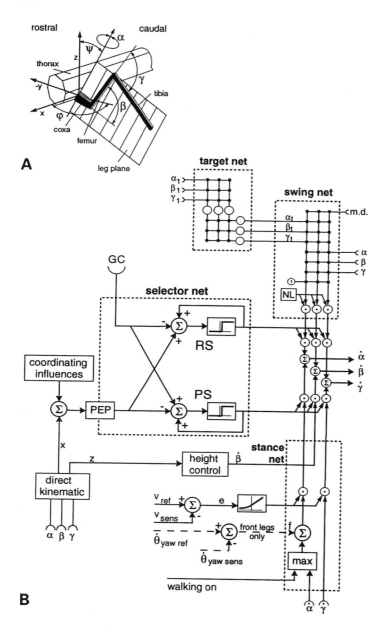

A

B

FIGURE 16.2. Summary of leg geometry and control networks for the model. (A) Schematic model of a stick insect leg showing the arrangement of the joints and their axes of rotation. (B) The leg controller consists of three parts: the swing net, the stance net, and the selector net which determines whether the swing or the stance net can control the motor output, i.e., the velocity of the three joints α, β, and γ. The selector net contains four units: the PEP unit signalling posterior extreme position, the GC unit signalling ground contact, the RS unit controlling the return stroke (swing movement), and the PS unit controlling the power stroke (stance movement). The target net transforms information on the configuration of the anterior, target leg, α_1, β_1, and γ_1, into angular values for the next caudal leg which place the two tarsi close together. These desired final values (α_t, β_t, γ_t) and the current values, α, β, and γ, of the leg angles are input to the swing net together with a bias input (1) and a sensory input (m.d.) which is activated by an obstruction blocking the swing and thereby initiates an avoidance movement. A nonlinear influence (NL) modulates the velocity profile. For details see Cruse et al. (1996).

den units and logistic activation functions (Figure 16.2B, "target net") directly associates desired final joint angles for the swing to current joint angles of a rostral leg such that the tarsi of the two legs are at the same position. There is no explicit calculation of either tarsus position. Physiological recordings from local and intersegmental interneurons (Brunn and Dean 1994) support the hypothesis that a similar approximate algorithm is implemented in the nervous system of the stick insect.

Because a network solution for controlling leg movement during stance seemed quite complicated at first sight, initial simulations used the inverse kinematics to find joint configurations to move the tarsus along a straight line parallel to the long axis of the body (Section 4 presents alternative solutions for controlling stance). This model shows a proper coordination of the legs for walks at different speeds on a horizontal plane (Cruse et al. 1995). Steps of ipsilateral legs are organized in triplets

A

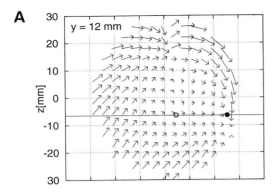

B

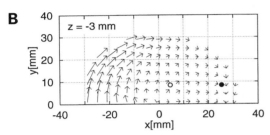

FIGURE 16.3. Vector field representing the movement of the tarsus of a left front leg produced by the swing net. (A) Projection of a parasagittal section ($y = 12$ mm). (B) Projection of a horizontal section slightly below the leg insertion ($z = -3$ mm). Left is posterior, right anterior. The average posterior extreme position (start of swing movement) and of the average anterior extreme position (end of swing movement) are shown by an open circle and by a closed circle, respectively.

forming "metachronal waves," which proceed from back to front, whereas steps of the contralateral legs on each segment step approximately in alternation. With increasing walking speed, the typical change in coordination from the tetrapod to a tripod-like gait is found. For slow and medium velocities the walking pattern corresponds to the tetrapod gait with four or more legs on the ground at any time and diagonal pairs of legs stepping approximately together; for higher velocities the gait approaches the tripod pattern with front and rear legs on each side stepping together with the contralateral middle leg. The coordination pattern is very stable. For example, when the movement of one leg is interrupted briefly during the power stroke, the normal coordination is regained immediately at the end of the perturbation. Note that the temporal sequence of the activities of the legs is implicitly determined by the connections between the leg controllers.

4 Control of the Stance Movement and Coordination of Supporting Legs

The control solution proposed above for the stance movement is feasible for straight walking on a flat surface. In more natural situations, the task of controlling the stance movements of all the legs on the ground poses several major problems. It is not enough simply to specify a movement for each leg on its own: the mechanical coupling through the substrate means that efficient locomotion requires coordinated movement of all the joints of all the legs in contact with the substrate, that is, a total of 18 joints when all legs of an insect are on the ground. However, the number and combination of mechanically coupled joints varies from one moment to the next, depending on which legs are lifted. The task is quite nonlinear, particularly when the rotational axes of the joints are not orthogonal, as is often the case in insects, particularly for the basal leg joint. A further complication occurs when the animal negotiates a curve, which requires the different legs to move at different speeds.

In machines, these problems can be solved using traditional, though computationally costly, methods, which consider the ground reaction forces of all legs in stance and seek to optimize some additional criteria, such as minimizing the tension or compression exerted by the legs on the substrate. Due to the nature of the mechanical interactions and inherent in the search for a globally optimal control strategy, such algorithms require a single, central controller; they do not lend themselves to distributed processing. This makes real-time control difficult, even in the still simple case of walking on a rigid substrate.

Further complexities arise in more complex, natural walking situations, making solution difficult even with high computational power. These occur, for example, when an animal or a machine walks on a slippery surface or on a compliant substrate, such as the leaves and twigs encountered by stick insects. Any flexibility in the suspension of the joints further increases the degrees of freedom that must be considered and the complexity of the computation. Further problems for an exact, analytical solution occur when the length of leg segments changes during growth or their shape changes through injury. In such cases, knowledge of the geometrical situation is incomplete, making an explicit calculation difficult, if not impossible.

Despite the evident complexity of these tasks, they are mastered even by insects with their "simple" nervous systems. Therefore, there has to be a solution that is fast enough that on-line computation is possible even for slow neuronal systems. How can this be done? Several authors (e.g., Brooks 1991) have pointed out that some relevant parameters do not need to be explicitly calculated by the nervous system because they are already available in the interaction with the environment. This means that, instead of an abstract calculation, the system can directly exploit the dynamics of the interaction and thereby avoid a slow, computationally exact algorithm. To solve the particular problem at hand, we propose to replace a central controller with distributed control in the form of local positive feedback (Cruse et al. 1996). Compared to earlier versions (Cruse et al. 1995), this change permits the stance net to be radically simplified. The positive feedback occurs at the level of single joints: the position signal of each is fed back to control the motor output of the same joint (Figure 16.2B, stance net). How does this system work? Let us assume that any one of the joints is moved actively. Then, because of the mechanical connections, all other joints begin to move passively, but in exactly the proper way. Thus, the movement direction and speed of each joint does not have to be computed because this information is already provided by the physics. The positive feedback then transforms this passive movement into an active movement.

There are, however, several problems to be solved. The first is that positive feedback using the raw position signal would lead to unpredictable changes in movement speed, not the nearly constant walking speed which is usually desired. This problem can be solved by introducing a kind of band-pass filter into the feedback loop. The effect is to make the feedback proportional to the angular velocity of joint movement, not the angular position. In the simulation, this is done by feeding back a signal proportional to the angular change over the preceding time interval.

The second problem is that using positive feedback for all three leg joints leads to unpredictable changes in body height, even in a computer simula-

tion neglecting gravity. Normally, body height of the stick insect is controlled by a distributed system in which each leg acts like an independent, proportional controller. However, maintaining a given height via negative feedback appears at odds with the proposed local positive feedback for forward movement. How can both functions be fulfilled at the same time? To solve this problem we assume that during walking positive feedback is provided for the α joints and the γ joints, (Figure 16.2B, stance net), but not for the β joints. The β joint is the major determinant of the separation between leg insertion and substrate, which determines body height.

A third problem inherent in using positive feedback is the following. Let us assume that a stationary insect is pulled backward by gravity or by a brief tug from an experimentor. With positive feedback control as described, the insect should then continue to walk backwards even after the initial pull ends. This has never been observed. Therefore, we assume that a supervisory system exists which is not only responsible for switching on and off the entire walking system, but also specifies walking direction (normally forward). This influence is represented by applying a small, positive input value (Figure 16.2B, walking on) which replaces the sensory signal if it is larger than the latter (the box "max" in Figure 16.2B, stance net).

To permit the system to control straight walking and to negotiate curves, a supervisory system was introduced which, in a simple way, simulates optomotor mechanisms for course stabilisation that are well known from insects and have also been applied in robotics. This supervisory system uses information on the rate of yaw ($\dot{\theta}_{\text{yaw sens}}$, Figure 16.2B, stance net), such as visual movement detectors might provide. It is based on negative feedback of the deviation between the desired turning rate and the actual change in heading over the last time step. The error signal controls additional impulses to the α joints of the front legs which have magnitudes proportional to the deviation and opposite signs for the right and left sides. With this addition and $\dot{\theta}_{\text{yaw ref}}$ set to zero, the system moves straight (Figure 16.4A) with small, side-to-side oscillations in heading such as can be observed in walking insects. To simulate curve walking (Figure 16.4B), the reference value is given a small positive or negative bias to determine curvature direction and magnitude.

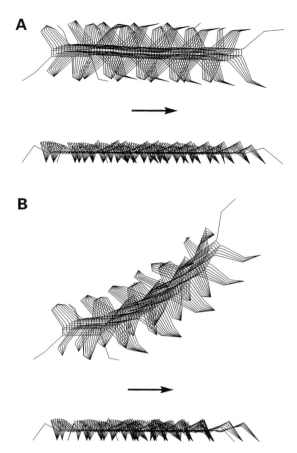

A

B

FIGURE 16.4. Simulated walk by the basic six-legged system with negative feedback applied to all six β joints and positive feedback to all α and γ joints as shown in Figure 16.2B. Movement direction is from left to right (arrow). Leg positions are illustrated only during stance and only for every second time interval in the simulation. Each leg makes about five steps. Upper part: top view, lower part: side view. (A) Straight walking ($\theta_{\text{yaw ref}} = 0$). (B) curved walking ($\theta_{\text{yaw ref}} \neq 0$).

Finally, we have to address the question of how walking speed is determined in such a positive feedback controller. Again, we assume a central value which represents the desired walking speed v_{ref}. This is compared with the actual speed, which could be measured by visual inputs or by monitoring leg movement. This error signal is subject to a nonlinear transformation and then multiplied with the channels providing the positive feedback for all α and γ joints of all six legs (Figure 16.2B, stance net).

Unexpectedly, the following interesting behavior was observed. By "brute force," for example by clamping three legs to the ground, the system can be made to fall. Although this can lead to extremely disordered arrangements of the six legs, the system was always able to stand up and resume proper walking without any help. This means that the simple solution proposed here also eliminates the need for a special supervisory system to rearrange leg positions after such an emergency.

5   Discussion

To produce an active behavior (i.e., a time-varying motor output), a system needs recurrent connections at some level. A pacemaker neuron incorporates these recurrent influences within a single neuron, one which can produce simple rhythms on its own. More interesting are networks—with or without pacemakers—where the recurrent connections involve nonlinear interactions among two or more units. In such systems, recurrent connections may occur in two ways. They may occur as internal connections, as in the various types of artificial recurrent networks (e.g., Hopfield nets or Elman nets). Alternatively, if a simple system with a feedforward structure is situated in a real or simulated environment, information flow via the environment serves as an "external" recurrent connection because any action by the system changes its sensory input, thereby closing the loop. We have described a system containing recurrent connections of both types which can be used to control hexapod walking. No central oscillators are used. One subsystem, the swing net, consists of an extremely simple feedforward net which exploits the recurrent information via the sensorimotor system as the leg's behavior is influenced by the environment to generate swing movements in time. The system does not compute explicit trajectories; instead, it computes changes in time based on the actual and the desired final joint angles.

More difficult problems have to be solved by the stance system, the second subsystem of the leg controller. During stance, the movements of many joints (9–18) have to be controlled in order that each leg contributes efficiently to support and propulsion and that the legs do not work at cross purposes during walking. This task is already inherently nonlinear, but several factors make it still more difficult. First, the number and combination of legs on the ground varies during stepping. Second, each leg has to move at a different speed during curved walking. Third, important effects are unpredictable: for example, when walking on soft ground, the speed of the individual leg changes in an unforseeable way. Finally, even the geometry of the system may change due to non-rigid suspension of the joints, or due to growth or injury. Whereas the predictable effects are solvable, although they make a centralized controller to mediate the cooperation seem unavoidable at first sight, the unpredictable effects make a computational solution still more difficult if not impossible.

Solving these complex tasks represents quite a high level of "motor intelligence." However, this does not require a complicated, or a centralized control system. This simplification is possible because the physical properties of the system and its interaction with the world are exploited to replace an abstract, explicit computation. Thus, "the world" is used as "its own best model" (Brooks 1991). Because of its extremely decentralized organization and the simple type of calculation required, this method permits very fast computation. Some central commands from a superior level are still required. These are necessary to determine the beginning and end of walking as well as its speed and direction. However, these commands do not have to be precisely adjusted to the particular configurations and states of all the legs; an approximate specification is sufficient for the positive feedback control. For example, turning is possible without having to send a corresponding command to every leg; appropriate commands to both front legs are sufficient.

Two applications of positive feedback have been discussed. The use of positive feedback in the context of motor control has been considered previously with respect to recurrent circuits within the brain (Houk et al. 1993). The interpretation advanced is that the positive feedback within motor and premotor centers serves to sustain ongoing activity. Similar results have been described for the swimming system of the leech (Kristan et al. 1992). This basically corresponds to the simple circuit used in the selector net (Figure 16.2B) to sustain power stroke or return stroke activity in the walking leg. Qualitatively, the effect is the same: ongoing patterns of movement are sustained through the recurrent connections. Sensory pathways are not part of the positive feedback although sensory inputs may effect changes from one pattern to another. An uncontrolled increase of the output signal is prevented by the Heaviside activation function which permits only a 0 or 1 output (see Figure 16.2, selector net). In the stance net presented here, in contrast, the positive feedback loop includes elements outside the nervous system (i.e., it includes the sensorimotor loop through the world), and it modulates the pattern itself by affecting relative levels of activity in different joints. Therefore, graded output values are possible and desired. The band-pass filter in the positive feedback pathway prevents uncontrolled increases in activation.

One result of this and related work is that the controlling of rhythmic movement of a complex system is possible without relying on central oscillators. Such a central controller would even deteriorate some of our solutions which rely on the physical properties of the system. Our results are sensible for slow walkers, where slow neuronal conduction speeds do not play a critical role. For fast walkers, which cannot rely on temporally adequate sensory input, central oscillators might be the better solution. On the other hand, Full (Chapter 13) shows that, in a fast walker, the mechanical, in particular the dynamical properties of the system may well be exploited to substantially decrease the control effort. This may also be true in slow walking systems where the dynamics are important such as in larger animals or in human walking in a predictable environment.

Two major questions deserve special attention in the future. One is to test whether these hypotheses which to date have only been investigated in kinematic models, are also applicable to dynamic systems. To this end, both dynamic simulations and the implementation on a real robot are planned. The second important question is to test whether, and if so, in which way these hypotheses are realized in the biological system; this requires electrophysiological investigation.

Acknowledgments. This work was supported by the Körber Foundation, by BMFT (grant no. 01 IN 104 B/1), and DFG (grant no. Cr 58/g-1,2).

References

Bässler, U. and Wegner, U. (1983). Motor output of the denervated thoracic ventral nerve cord in the stick insect Carausius morosus. *J. Exp. Biol.*, 105:127–145.

Brooks, R.A. (1991). Intelligence without reason. *IJCAI-91*, Sydney, Australia, 569–595.

Brunn, D. and Dean, J. (1994). Intersegmental and local interneurones in the metathorax of the stick insect, Carausius morosus. *J. Neurophysiol.*, 72:1208–1219.

Cruse, H. (1990). What mechanisms coordinate leg movement in walking arthropods? *Trends Neurosci.*, 13:15–21.

Cruse, H., Bartling, C., Brunn, D.E., Dean, J., Dreifert, M., Kindermann, T., and Schmitz, J. (1995). Walking: a complex behavior controlled by simple systems. *Adaptive Behav.*, 3:385–418.

Cruse, H., Bartling, C., Dean, J., Kindermann, T., Schmitz, J., Schumm, M., and Wagner, H. (1996). Coordination in a six-legged walking system: simple solutions to complex problems by exploitation of physical properties. In *From animals to animats 4*. Maes, P., Mataric, M.J., Meyer, J.-A., Pollack, J., and Wilson, S.W. (eds.), pp. 84–93. The MIT Press, Cambridge Massachusetts.

Dean, J. and Cruse, H. (1995). Motor pattern generation. In *The Handbook of Brain Theory and Neural Networks*. Arbib, M., (ed.), pp. 600–605. Bradford Book, The MIT Press.

Graham, D. (1985). Pattern and control of walking in insects. *Adv. Insect Physiol.*, 18:31–140.

Holst, E.V. (1943). Über relative Koordination bei Arthropoden. Pflügers Archiv., 246:847–865.

Houk, J.C., Keifer, J., and Barto, A.G. (1993). Distributed motor commands in the limb premotor network. *Trends Neurosci.*, 16:27–33.

Kristan, W.B., Jr., Lockery, S.R., Wittenberg, G., and Brody, D. (1992). Making behavioral choices with interneurones in a distributed system. In *Neurobiology of Motor Programme Selection*. Kien, J., McCrohan, C.R., and Winlow, W., (eds.), pp. 170–200. Pergamon Press.

Commentary:
Are Decentralized or Central Control Systems Implied in the Locomotion?

Marc Jamon and François Clarac

This chapter presents an interesting perspective within this section devoted to the neuromechanical interaction in rhythmic systems. The problem discussed provides a demonstration that decentralized control systems can manage adaptive locomotor behaviour by means of a simple network structure incorporating relaxation oscillators driven by hard physical interaction with the system geometry. This is a good example of the idea developed in Chapter 13, which stated that "the musculoskeletal or mechanical system functions also as a control system, more passive, but dynamic control system." This chapter deals with the solution of two levels of control problems, namely, the control of individual leg movements, and the coordination between the legs.

Using a comparative method that Full and Farley appealed to in Section 1 of 13, we will comment these two points referring to our own studies on another arthropod, the crayfish Procambarus clarkii. We analyzed the kinematics of various joints in the leg motion of crayfish during free locomotion. By means of a statistical analysis based on conjugate cross-correlation functions (which allowed the analysis of sustained and periodic sensorimotor controls without involving the use of a time origin) we demonstrated the existence of true interjoint synergy between joints in the leg's kinematics (Jamon and Clarac, 1997). The global oscillation of the leg implies therefore a synchronized movement of the various joints to solve the biomechanical constraints, and is supposedly regulated by local interactions based on proprioceptive reflexes. This process of the leg movement is consistent with the functionalities of the stick insect neural network: the positive feedback loop solution of the control of movement in individual legs can be represented as an assistance reflex, and suggests that the system uses the constraint of the inter-joint geometry to produce an active synergy between joints.

Similarly, the hypothesis that a decentralized control system having hard physical interactions with the external world drives interleg coupling also provides a heuristic approach of locomotion in other arthropods, as can be shown in another example. We showed, in a previous study (Jamon and Clarac 1995), that freely moving crayfish are able to switch spontaneously between two different locomotor patterns whose main characteristic is the stabilization of the fourth legs' pair in an in-phase or alternated coupling. One explanation of this phenomenon suggests that the motor system is organized peripherally around steady states, either in-phase or in alternation, depending on dynamic equilibrium, with the interleg coupling being attracted either to the in-phase or to the alternated pattern, depending on minute changes in the physical constraints applied to the mechanical system. This hypothesis is consistent with the fact that a decentralized control system is prevalent on a centrally commanded locomotor output.

These examples of crayfish locomotion suggest that the neuromechanical interaction of decentralized control centers applies to various species, living in different medium and therefore with different constraints. We cannot, nevertheless, exclude the fact that central pattern generators (CPG) have been found to exist in several invertebrate and vertebrate species in vitro preparations. In the crayfish, for instance, the isolated ganglia fire with a basal rhythmic organization reminiscent of the successive promotor-levator then remotor–depressor synchronous activation implied in the forward walking (Chrachri and Clarac 1990). On the other hand, an inter-appendicular pattern has been shown to persist in isolated neural chain preparations in the form of ipsilateral metachronal waves. The combination of behavioural and electrophysiological experimental data shows then that locomotion is not the result of decentralized or central mechanisms but of both. The question arises now as to why some species would utilize CPGs while other would utilize decentralised control centres. Apparently, organisms that require complex locomotion control (because they move on uneven substrates), or experience reduced internal constraints (because they walk slowly, as does the stick insect), or experience released postural constraints (due to underwater locomotion, such as in the case of crayfish) rely on peripheral organization, incorporating world-system interaction. On the other hand, in the organisms submitted to strong internal constraints, like fast walkers (as described by other authors in this section), the central command takes coordination in hands with a higher stereotypy and less adaptivity.

References

Chrachri A. and Clarac F. (1990). Fictive locomotion in the fourth thoracic ganglion of the crayfish, Procambarus Clarkii. *J. Neurosci.*, 10:707–719.
Jamon M. and Clarac F. (1995). Locomotor patterns in freely moving crayfish, Procambarus clarkii. *J. Exp. Biol.*, 163:187–208.
Jamon M. and Clarac F. (1997). Variability of leg kinematics in free walking crayfish, Procambarus clarkii, and related inter-joint coodination. *J. Exp. Biol.*, 200:1201–1213.

Commentary:
Neural Control and Biomechanics in the Locomotion of Insects and Robots

Randall D. Beer and Hillel J. Chiel

We find ourselves very much in agreement with the general perspective of Cruse et al., which emphasizes the value of studying insect locomotion, the importance of the body and environment in generating behavior, as well as the importance of modeling and robotics applications. In particular, he and his colleagues have characterized the network of coordinating mechanisms operating between the legs of the stick insect that appear to be capable of generating robust gait patterns without the use of central oscillators. A continuous range of statically stable gaits, robustness to lesions, and the ability to withstand perturbations such as restraining a single leg all emerge from closing the feedback loop for the locomotion control through the world, that is, by activating these influences as a function of the position of the legs relative to the anterior extreme position (AEP) or posterior extreme position (PEP). Their experimental and modeling results have had a major impact on many researchers, and have directly influenced our own work on insect walking and robotics (Espenschied et al. 1993).

The work of Cruse and his colleagues has emphasized the importance of distributed control for generating robust solutions to locomotion, which would be very difficult or impossible for more centralized controllers. An important implication of this observation, however, is that attempts to create clean functional decompositions of a behavior as a basis for designing controllers (e.g., assuming dedicated and distinctive neural networks for stance and swing phases in locomotion) may be undermined by the distributed nature of the neural networks that generate these behaviors. Both experimental evidence and modeling work suggests that dedicated circuits are relatively rare, and that nervous systems utilize a limited number of neural elements to generate a wide variety of behaviors by either reorganizing a network, or by distributing function among many neurons (Morton and Chiel 1994). For example, many different stable patterned movements are generated by the same set of neurons in the stomatogastric nervous system by reorganizing the network (Harris-Warrick et al. 1992). Another example comes from our studies of model pattern generators for locomotion, generated using genetic algorithms, in which robust locomotion results from neural networks whose elements cannot be cleanly decomposed into either swing or stance subnetworks (Beer and Gallagher 1992). However, it is clear that distributed control has great advantages for robustness and efficiency, and the concept of distributed control is likely to be relevant for designing experiments to analyze neural circuitry.

The chapter emphasizes the importance of utilizing both leg kinematics and the properties of the substrate to generate rapid, flexible locomotion. More generally, different lines of evidence strongly suggest that continuous, ongoing interactions between the nervous system and the sensory feedback it receives are essential for the generation of adaptive behavior. For example, we and our collaborators recently described a stick-insect-like robot, each of whose legs had three active and one passive degree of freedom. The controller for this robot utilized mechanisms based on those observed in stick insects, generalized for the three-dimensional work space of the leg, and incorporated local reflexes that have been observed in a variety of insects, including stepping reflexes, elevator reflexes, and searching reflexes. The interactions between the local reflexes and the distributed mechanisms generating coordinated gait patterns were essential for allowing the robot to effectively traverse extremely irregular terrain (Espenschied et al. 1996). In addition, in modeling studies of evolved pattern generators for locomotion, we have found that both the oscillatory dynamics of the central controller and sensory feedback play crucial roles in generating locomotion (Beer and Gallagher 1992). The picture that emerges from these results is that many

nervous systems may have centralized oscillatory tendencies, but sensory inputs to them have a major shaping effect. Of course, in a particular system, one may observe that sensory inputs and reflexes are dominant, or that central oscillatory control is dominant; but it is likely that most systems utilize a continuous mix of central and sensory mechanisms for maximum robustness and flexibility.

The chapter points out that the neural network can exploit leg kinematics to generate appropriate movements of legs. It is also essential to deal with the dynamics of the leg and the body in understanding and controlling locomotion, as has been emphasized by Full and Farley (Chapter 13). In the robot that we previously described, we found that incorporating dynamics was essential for simplifying the control of locomotion. Equilibrium point control was used to simplify the control of the legs. Passive and active compliances allowed this robot to conform to extremely irregular terrain (Espenschied et al. 1996). More generally, these results and those of other investigators suggest the importance of the coupling between the nervous system and the body, and between the body and the environment, in generating adaptive behavior (Chiel and Beer 1993; Beer 1995).

Studies of insect walking, of the sort described in this chapter, have begun to influence the design of autonomous legged robots. Donner used distributed control for a large hydraulically activated hexapod, and the resulting controller was robust to lesions (Donner 1987). Brooks and his students have built several hexapod robots controlled by augmented finite state machines (Brooks 1989; Ferrell 1995). Studies of the dynamics of walking in insects and vertebrates have served as an important inspiration for the robots built by Raibert and colleagues that are capable of hopping, jumping, and running (Raibert 1986). Our own previous collaborative work has utilized cockroach-based, stick-insect based, or evolved controllers for generating a continuous range of statically stable gaits in a hexapod robot capable of straight-line locomotion that is extremely robust to lesions of the central control, and other perturbations (Beer et al. 1992; Espenschied et al. 1993; Gallagher et al. 1996). As mentioned above, we have also constructed a more complex hexapod robot whose body and control were stick-insect-like and incorporated local leg reflexes that allowed it to traverse extremely irregular terrain. Pfeiffer and coworkers have also built a stick-insect-like robot whose control is based on the mechanisms observed in stick insects (Pfeiffer et al. 1994). It is likely that continued interactions between researchers studying the neurobiology and biomechanics of animal behavior, and roboticists constructing autonomous robots, will lead to even more exciting developments in the years to come (Beer et al. 1993).

References

Beer, R.D. (1995). A dynamical systems perspective on agent-environment interaction. *Artif. Intell.*, 72:173–215.

Beer, R.D. and Gallagher, J.C. (1992). Evolving dynamical neural networks for adaptive behavior. *Adaptive Behav.*, 1:91–122.

Beer, R.D., Chiel, H.J., Quinn, R.D., Espenschied, K., and Larsson, P. (1992). A distributed neural network architecture for hexapod robot locomotion. *Neural Computation*, 4:356–365.

Beer, R.D., Ritzmann, R.E., and McKenna, T., (eds.) (1993). *Biological Neural Networks in Invertebrate Neuroethology and Robotics*. Academic Press, Inc., San Diego.

Brooks, R.A. (1989). A robot that walks: emergent behaviors from a carefully evolved network. *Neural Comput.*, 1(2):253–262.

Chiel, H.J. and Beer, R.D. (1993). Neural and peripheral dynamics as determinants of patterned motor behavior. *The Neurobiology of Neural Networks*. Gardner, D. (ed.), pp. 137–164. The MIT Press, Cambridge, Massachusetts.

Donner, M. (1987). *Real-time Control of Walking*. Birkhauser Boston, Inc., Cambridge, Massachusetts.

Espenschied, K.S., Quinn, R.D., Chiel, H.J., and Beer, R.D. (1993). Leg coordination mechanisms in stick insect applied to hexapod robot locomotion. *Adaptive Behav.*, 1:455–468.

Espenschied, K.S., Quinn, R.D., Chiel, H.J., and Beer, R.D. (1996). Biologically-based distributed control and local reflexes improve rough terrain locomotion in a hexapod robot. *Robot. Auto. Sys.*, 18:59–64.

Ferrell, C.L. (1995). Global behavior via cooperative local control. *Auto. Robots*, 2:105–125.

Gallagher, J.G., Beer, R.D., Espenschied, K.S., and Quinn, R.D. (1996). Application of evolved locomotion controllers to a hexapod robot. *Robot. Auto. Sys.*, 19:95–103.

Harris-Warrick, R.M., Marder, E., Selverston, A.I., and Moulins, M. (eds.). (1992). *Dynamic Biological Networks: The Stomatogastric Nervous System. Computational Neurosci.*, The MIT Press, Cambridge, Massachusetts.

Morton, D.W. and Chiel, H.J. (1994). Neural architectures for adaptive behavior. *Trends Neurosci.*, 17:413–420.

Pfeiffer, F., Eltze, J., and Weidemann, H.J. (1994). The TUM walking machine. *Proc. of the Fifth International Symposium on Robotics and Manufacturing*, Maui, Hawaii.

Raibert, M.H. (1986). *Legged Robots That Balance*. The MIT Press, Cambridge, Massachusetts.

17
Neuromechanical Function of Reflexes During Locomotion

E. Paul Zehr and Richard B. Stein

1 Summary

Reflex responses arising from various afferent sources modify rhythmic behavior in many preparations. Functional interpretations have often been based solely on neural responses and mechanical information has been lacking. When mechanical and neural factors have been studied in concert, reflexes have shown strong functional neuromechanical effects. Muscle reflexes seem particularly important for body weight support and step cycle timing, while cutaneous reflexes function in stumble correction and placing reactions during swing phase of gait in cats and humans. We present a framework for conceptualizing reflex function during locomotion.

2 Introduction

Peripheral afferent inputs can alter ongoing rhythmic motor patterns, such as walking, in many animals (Sherrington 1906). Studies have often focused on the details of the neural control signal sent to the involved muscles as measured by electromyography (EMG) (Rossignol 1996). Different afferent inputs have been studied to determine and evaluate their effects on walking. However, interpretation of the functional significance of these reflex responses has often been done in the absence of adequate mechanical data to corroborate the implied function based upon electroneurographic or EMG analysis (reviewed in Stein 1991). The need to understand and delineate the function of reflex control spans a range from determining which peripheral inputs are required to improve models and neural networks to determining which inputs might be useful in functionally altering abnormal reflex patterns in disease and after trauma in humans. This chapter addresses the issue of reflex function by focusing upon those studies in cats and humans where mechanical and neural data have been collected together during locomotion and functional conclusions have been based upon both. Further the focus here is restricted to the functional effects of cutaneous and muscle reflexes.

3 Stretch Reflexes During Walking

Reflex responses arising from muscular afferents, particularly the velocity-sensitive Group Ia afferents from the muscle spindles, have been much studied and their functional role investigated in walking animal preparations. It was shown that during selective peripheral nerve block of fusimotor axons EMG activity of the plantarflexor muscle group triceps surae was reduced by ~50% (Severin 1970). This indicates that muscle spindles (and therefore stretch reflexes) contribute significantly to muscle activation. The functional relevance (i.e., in terms of force production) of the stretch reflex has remained enigmatic for quite some time, however (Stein et al. 1995). Akazawa et al. (1982) studied stretch reflexes during locomotion in the mesencephalic cat preparation and observed that stretch and H-reflexes were deeply modulated throughout the step cycle such that they were large during stance and small during swing phases. It was concluded that stretch reflexes assist in load compensation during gait, particularly during the extension or stance phase.

Because of methodological difficulties, stretch reflex responses have not been measured in intact, freely walking animals. Further, until quite recently there was no quantifiable term relating to the magnitude of the reflex force produced by the stretch reflex. Bennett et al. (1996) studied stretch reflex gain in the triceps surae muscle group in decerebrate and spinal cats during locomotor-like activities. A sample of their data is given in Figure 17.1. They observed that low frequency cyclic movements of the ankle during locomotion in the spinal cat produced up to 25% of the stretch-related force modulation.

In line with the observations of Akazawa et al. (1982), Capaday and Stein (1986) observed that the soleus H-reflex was high during stance and low during swing phases in human gait. This indicated that this pathway should be capable of contributing functionally to force production during the stance phase. Later, Yang et al. (1991a) used a pneumatic actuator to mechanically dorsiflex the foot and thus stretch soleus muscle during stance. Illustrated in Figure 17.2 is a sample of their data

showing the stretch reflex response in soleus and tibialis anterior (TA) muscles after the perturbation applied during early stance.

They calculated that reflexes due to a velocity-sensitive component (i.e., muscle spindle primaries) during stance accounted for an estimated 30 to 60% of soleus activation. However, with this device it was not possible to quantify the force produced by the soleus stretch reflex throughout the step cycle. Anderson and Sinkjaer (1995) made use of Bowden wire technology to construct an actuator capable of stretching the human ankle joint throughout all phases of the step cycle. This device has been used to study and compare short-latency stretch reflex modulation in soleus muscle during walking and responses obtained during standing (Sinkjaer et al. 1996a). They showed that the amplitude of the short-latency reflex had a maximum during stance, approached zero in the stance to swing transition, and reached a value of ~50% maximum during late swing. Additionally, a non-reflex component was also maximal during stance. At increased walking speeds, the late swing phase stretch reflexes approximated the maximal stance phase values. Sinkjaer et al. (1996a) concluded, as did Yang et al. (1991a), that muscle afferent input contributes to a significant activation of the ankle

Length (5 Hz stretch)

1 mm

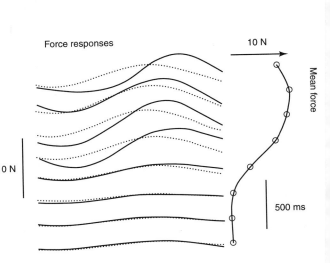

Force responses

10 N

Mean force

10 N

500 ms

FIGURE 17.1. Reflex force responses with 5 Hz sinusoidal stretch of triceps surae muscle throughout the step cycle (from stance to swing, top to bottom) in the decerebrate cat. Dotted lines are force responses (intrinsic stiffness) with cut dorsal roots. Note the large stretch reflex force (solid lines) particularly during the first 4 phases of the step cycle. (Adapted from Bennett et al. 1996.)

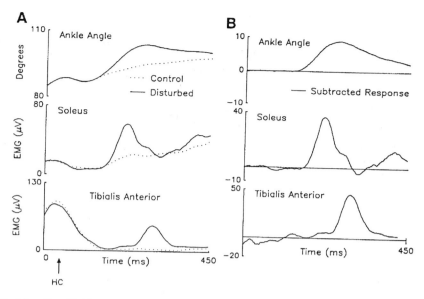

FIGURE 17.2. Stretch reflex EMG responses (A, ensemble activity; B, reflex response) in soleus muscle for one subject after mechanical dorsiflexion in stance phase during treadmill walking. Note the large soleus EMG peak at short latency after ankle stretch. (From Yang et al. 1991a.)

extensors during stance. Unfortunately, it was not possible to partition out nonreflex and reflex induced torques in these experiments. As an estimate of the reflex forces that can be elicited, Stein and Kearney (1995) showed that stretch reflexes generated by random perturbations at the human ankle joint during isometric contraction could generate reflex torques larger than 20% of maximum voluntary contraction. Using a new method in high decerebrate, walking cats Stein et al. (2000) calculated that about 30% of the force generated in ankle muscles during the stance phase is caused by phasic muscle reflexes.

Studying dysfunctional motor control in pathological states may contribute to understanding normal function. For example, Yang et al. (1991b) showed that H-reflex modulation was impaired in spastic paretic subjects during treadmill walking and suggested that abnormal control of afferent input may contribute to the clonic behavior observed during gait in these subjects. Further, Sinkjaer et al. (1995) showed a similar dysfunctional reflex modulation in spastic multiple sclerosis patients. Sinkjaer et al. (1996) have also shown impaired stretch reflex modulation during gait in these patients. However, in addition to the stretch reflex impairment, they concluded that non-reflex torque contributed substantially to the spasticity. Thus, impaired reflex input can cause severe dysfunction during human walking.

4 Load Receptor Reflexes and Step Cycle Timing

Afferent input from group Ib afferents (Golgi tendon organs) has been shown to reverse from a disynaptic inhibition to a positive force feedback disynaptic and oligosynaptic excitation during locomotion in the decerebrate cat (Pearson and Collins 1993). Pratt (1995) has shown that positive force feedback elicited by activation of group Ib afferents generates strong reflexive changes in force and joint angles in the intact standing cat. These studies suggest that extensor muscle group Ib afferent inputs contribute strongly to postural responses and stance phase weight support.

Proprioceptive inputs from muscle group I (in particular group Ib) afferents are important in regulation of the step cycle especially during phase transitions (Pearson 1995). In the cat, load-sensitive afferents in the extensor muscles have been shown to modify the stepping rhythm. Whelan et al. (1995) showed that electrical stimulation of group I afferents in extensor

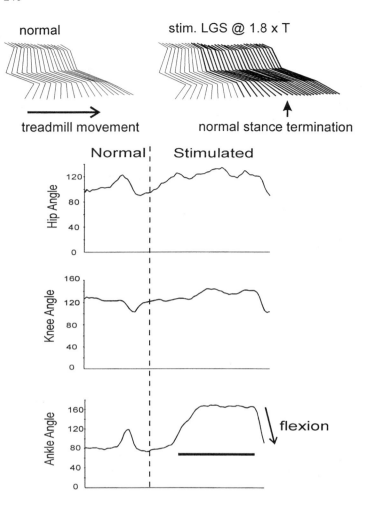

FIGURE 17.3. (Top) Stimulation of extensor (LGS) nerve during spontaneous stepping in a premamillary decerebrate cat prolongs stance phase of the step cycle. (Bottom) Hip, knee, and ankle joint angle traces during normal step cycle (left) and during stimulation (right). Note that the stance phase is prolonged for the duration of stimulation (horizontal bar). (Modified from Whelan 1996.)

nerves of the hindlimb significantly prolonged the stance phase in decerebrate cats walking on a treadmill. A sample of this data is given in Figure 17.3. The threshold for this effect (greater than 1.3 times the threshold for activation of the largest afferents) implicates the group Ib afferents. This supports the assertion that the end of stance phase is signaled by a reduction of force in the extensor muscles (Duysens and Pearson, 1980). Grillner and Rossignol (1978) suggested that another condition regulating step cycle timing is the signal from stretch-sensitive afferents in the flexor muscles which are activated at end stance (the hip flexors being stretched at this time). Hiebert et al. (1996) have shown that group I and group II afferents are crucial for forming the afferent signal for the termination of stance phase and the onset of swing in the cat.

There has been a dearth of information regarding the functional effects of load receptors in human subjects. However, Stephens and Yang (1999) demonstrated that the addition or subtraction of additional load during human gait significantly increased extensor muscle activity without affecting stance phase duration. In comparison to the reduced cat preparations, though, the overall effect was smaller in human walking, perhaps reflecting intrinsic differences in the preparations studied. Interestingly, human infants subjected to a similar loading paradigm responded in much the same way as decerebrate cats. There was a large effect on cycle timing (Yang et al. 1998). Taken together with the cat work described above, data of this type suggest that reflex activity can strongly modify and regulate step cycle timing in many preparations.

5 Cutaneous Reflexes During Walking

While powerful effects on the locomotor rhythm had been observed earlier (Sherrington 1906, 1939), it was not until the work of Forssberg (1979) that a deliberate attempt to evaluate the functional role of cutaneous reflex responses by measuring both neural responses and kinematics was undertaken. These experiments involved both electrical and mechanical stimulation of the dorsal surface of the paw in the cat distal hindlimb during locomotion. A coordinated reflex forming a functionally-relevant "stumbling corrective response" was revealed (Forssberg 1979). This response consisted of a sequential activation of the hindlimb musculature to allow the perturbed swing limb to continue past the encountered obstacle and maintain stability of ongoing locomotion. Similar responses were also observed by Wand et al. (1980) and Prochazka et al. (1978) who, in a series of experiments, systematically revealed that the origin of the corrective response lay in the cutaneous afferents arising from the paw dorsum. A similar neuromechanical linkage was demonstrated in the cat forelimb by Drew and Rossignol (1987). Buford and Smith (1993) showed that reflex responses elicited by mechanical and electrical stimulation during both forward and backward walking have relevant neuromechanical correlates. As is shown in Figure 17.4, a corrective response which was suitable for maintaining ongoing locomotion was elicited by dorsal surface stimulation during forward walking and past ventral stimulation during backward walking. In both instances the swing limb is moved over and past the mechanical or electrically simulated perturbation.

From these results it appears clear that cutaneous afferents do have strong reflex responses which serve to functionally modify ongoing quadrupedal locomotion, particularly in the swing limb. What has remained unclear, though, is to what extent the results obtained in the experiments on the cat apply to the bipedal human in whom balance and postural concerns are quite different.

Cutaneous reflexes, particularly those arising from stimulation of the sural nerve (innervating the lateral border of the foot) and the tibial nerve (innervating the ventral foot surface), have been stud-

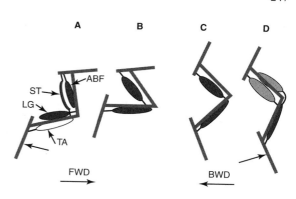

FIGURE 17.4. Schematic of responses to obstructions applied during swing phase for forward (FWD, dorsal surface stimulation) and backward (BWD, ventral surface stimulation) walking in cats. Note that coordinated muscle activation at knee and ankle lead to swing limb flexion over the obstacles. (From Buford and Smith 1993.)

ied in human subjects and have shown phase-dependent modulation of discrete reflex responses occurring at restricted latencies (Yang and Stein 1990; for review see Duysens and Tax 1994). While many functional interpretations of the EMG responses in humans have been put forth, mechanical data to support these ideas has been lacking. Movement kinetics have been quantified in these studies and limited kinematic data are available only in a few papers (Duysens et al. 1990, 1992). A significant correlation was demonstrated between a phasic middle-latency excitatory response in tibialis anterior (TA; ankle dorsiflexor) muscle and increased ankle joint dorsiflexion by Duysens et al. (1992). However the function of this biomechanical association was not directly addressed. Further, the experiments in the cat described above, particularly the results of Buford and Smith (1993) would indicate that the issue of the local sign of cutaneous reflexes deserves further attention. That is, the anatomical location of the nerve seems crucial in determining the functional nature of any reflex responses (e.g., Gassel and Ott 1970).

Experiments were conducted in which different cutaneous nerves in the human lower leg were electrically stimulated at nonnoxious intensities during treadmill walking (Zehr et al. 1997, 1998a). The nerves were tibial, sural, and superficial peroneal (SP) which innervate the plantar, lateral, and dorsal foot surfaces, respectively. Reflex EMG responses from the upper (biceps femoris, BF, knee flexor, and vastus lateralis, VL, knee extensor) and lower leg (Sol, medial, MG, and lateral gastrocnemius, LG, plantarflexors, and TA, dorsiflexor) were measured. Changes in ankle and knee joint trajectory were also recorded along with the neural responses. To provide a quantified EMG index with which to compare to the kinematic data, a recently developed EMG technique which represents the net reflex effect of stimulation (ACRE$_{150}$, Average Cumulative Reflex EMG at 150 ms post-stimulation; Komiyama et al. 1995, Zehr et al. 1995) was employed. Linear regression was calculated between ACRE$_{150}$ and the mechanical changes throughout the step cycle. This analysis revealed statistically significant correlation throughout the step cycle. This neuromechanical correlation was interpreted as being functionally relevant to correct ongoing limb trajectory after perturbation during the swing phase of locomotion.

During swing the net effect could be either TA inhibition (SP and tibial at early, mid and late swing) or excitation (sural throughout swing and tibial at stance to swing transition) and the effect on ankle trajectory increased plantarflexion or dorsiflexion depending upon which nerve was stimulated (Zehr et al. 1997, 1998a). Mixed responses were observed after sural nerve stimulation. Illustrated in Figure 17.5 is a schematic summary of the functional outcome of tibial nerve stimulation at the stance to swing transition (left, the foot is withdrawn from the simulated contact), and late swing (right, the ankle is plantarflexed in a placing reaction), and of SP n. stimulation during early swing (middle, the ankle is plantarflexed and the knee flexed to overcome the obstacle). Excitation in BF muscle correlated to increased knee flexion was ob-

served during early swing after SP stimulation (Zehr et al. 1997). The corrrelations were taken to indicate functionally relevant linkages to correct limb trajectory. Thus both phase-dependance and site-dependance, in which the location of the stimulus determined the neuro-mechanical effect, was seen during human gait, just as observed by Buford and Smith (1993) in the cat after electrical and mechanical stimulation. In the human subjects there were negligible kinematic effects during stance in the same experiments. These data indicate that cutaneous reflexes do have meaningful and mechanically relevant functional effects which stabilize human gait after perturbation and prevent stumbling or tripping.

Eng et al. (1994) studied strategies for recovering from a trip while subjects walked on a special obstacle walkway. It was shown that subjects used a lowering or an elevating strategy involving both ipsilateral and contralateral limbs and particularly the VL and BF muscles, dependent upon when in the swing phase the obstacle was encountered occurred. Some of their results at the ankle were similar to the electrical stimulation experiments described above and provide further evidence for the powerful effects of reflexes during swing phase. Further, Schillings et al. (1996) performed experiments wherein mechanical perturbations were provided to elicit stumbling reactions in human subjects walking on a treadmill. During perturbations in early swing phase, Schillings et al. (1996) showed BF excitation and knee flexion as described above after SP n. electrical stimulation. These data indicated that during treadmill walking subjects can utilize similar strategies as during walking overground. Lastly, it should be noted that cutaneous reflexes do serve an important protective role

FIGURE 17.5. Illustration of the functional outcomes of electrical stimulation of the plantar foot surface (Tibial n.) at the stance to swing transition (left) and late swing (right) and after foot dorsum (SP n.) stimulation during early swing in treadmill walking. (From Zehr et al 1997.) * indicates area of stimulation and → indicates direction of response.

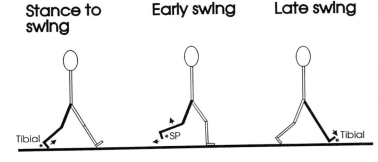

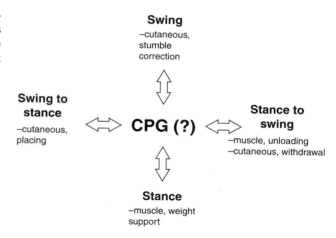

FIGURE 17.6. Schematic illustration of the proposed functional roles of muscle and nonnoxious cutaneous reflexes during stance, swing, and the transitional swing to stance and stance to swing phases of locomotion.

in response to noxious stimuli at all phases of the step cycle (Bélanger and Patla 1984, 1987).

Finally, as with the monosynaptic H-reflex, polysynaptic cutaneous reflex modulation has also been shown to be impaired in spinal-cord injured patients. Jones and Yang (1994) stimulated the tibial nerve during walking in these patients and found that, although some modulation did exist, the responses seemed to be predominantly excitatory. More recently, Zehr et al. (1998b) found predominately suppressive responses in stroke patients after SP n stimulation. Kinematic changes were observed in the stimulated limb but the neuromechanical linkage between TA reflexes and ankle movement was altered. The implications for rehabilitation require further study.

6 Future Directions

Future studies on the function of reflex responses in mammals should continue to address the mechanical outcomes and functional relevance of the afferents or mechanisms under study. At present we would broadly summarize the available data to indicate that the functional roles of cutaneous and muscular afferents are to primarily modify swing phase kinematics and stance phase kinetics, respectively. We have graphically illustrated this in Figure 17.6.

This schema does not directly address the issue of the nature of the modulation during certain phases of the step cycle, but rather presents it in a broad, conceptual framework. Viewing the data

within this scheme indicates quite clearly the value of further investigations into these responses and how these data and principles should be integrated into improved models. Further, a broader understanding of peripheral reflex mechanisms makes it more feasible that rehabilitation techniques will be usefully applied to functionally modify abnormal reflexes in disease states or after trauma.

In conclusion, many questions remain to be answered, particularly in human subjects. While understanding of the neuro-mechanical effects of cutaneous reflexes has progressed (e.g. flexion–extension movements), more complex mechanical effects need to be investigated. How do cutaneous and muscular afferent reflexes interact to affect gait mechanics? To what extent is neuro-mechanical reflex linkage altered after trauma or during disease? How can this effectively be applied to gait rehabilitation and retraining? Answering these questions remains a challenge for future studies.

Acknowledgments. This work was supported by a Medical Research Council of Canada operating grant to RBS. EPZ was supported by studentships from the Natural Sciences and Engineering Research Council of Canada and the Alberta Heritage Foundation for Medical Research.

References

Akazawa, K., Aldridge, J.W., Steeves, J.D., and Stein, R.B. (1982). Modulation of stretch reflexes during locomotion in the mesencephalic cat. *J. Physiol.*, 329: 553–567.

Andersen, J.B. and Sinkjaer, T. (1995). An actuator system for investigating electrophysiological and biomechanical features around the human ankle joint during gait. *IEEE Trans. Rehab. Eng.*, 3:299–306.

Bélanger, M. and Patla, A.E. (1984). Corrective responses to perturbation applied during walking in humans. *Neurosci. Lett.*, 49:291–295.

Bélanger, M. and Patla, A.E. (1987). Phase-dependent compensatory responses to perturbations applied during walking in humans. *J. Motor Behav.*, 19:434–453.

Bennett, D.J., DeSerres, S.J., and Stein, R.B. (1996). Gain of the triceps surae stretch reflex in decerebrate and spinal cats during postural and locomotor activities. *J. Physiol.*, 496:837–850.

Buford, J.A. and Smith, J.L. (1993). Adaptive control for backward quadrupedal walking: III. Stumbling corrective reactions and cutaneous reflex sensitivity. *J. Neurophysiol.*, 70:1102–1114.

Capaday, C. and Stein, R.B. (1986). Amplitude modulation of the soleus H reflex in the human during walking. *J. Neurosci.*, 6:1308–1313.

Drew, T. and Rossignol, S. (1987). A kinematic and electromyographic study of cutaneous reflexes evoked from the forelimb of unrestrained walking cats. *J. Neurophysiol.*, 57:1160–1184.

Duysens, J. and Pearson, K.G. (1980). Inhibition of flexor burst generation by loading ankle extensor muscles in walking cats. *Brain Res.*, 187:321–332.

Duysens, J. and Tax, A.A.M. (1994). Interlimb reflexes during gait in cats and humans. In *Interlimb Coordination: Neural, Dynamical, and Cognitive Constraints*. Swinnen, S.P., Heuer, H., Massion, J., and Casaer, P. (eds.), pp. 97–126. Academic, San Diego, California.

Duysens, J., Tax, A.A.M., Trippel, M. and Dietz, V. (1992). Phase-dependent reversal of reflexly induced movements during human gait. *Exp. Brain Res.*, 90:404–414.

Duysens, J., Trippel, M., Horstmann, G.A., and Dietz, V. (1990). Gating and reversal of reflexes in ankle muscles during human walking. *Exp. Brain Res.*, 82:351–358.

Eng, J.J., Winter, D.A., and Patla, A.E. (1994). Strategies for recovery from a trip in early and late swing during human walking. *Exp. Brain Res.*, 102:339–349.

Forssberg, H. (1979). Stumbling corrective reaction: a phase-dependent compensatory reaction during locomotion. *J. Neurophysiol.*, 42:936–953.

Gassel, M., and Ott, K.H. (1970). Local sign and late effects on motoneuron excitability of cutaneous stimulation in man. *Brain*, 93:95–106.

Grillner, S. and Rossignol, S. (1978). On the initiation of the swing phase of locomotion in chronic spinal cats. *Brain Res.*, 146:269–277.

Hiebert, G.W., Whelan, P.J., Prochazka, A., and Pearson, K.G. (1996). Contribution of hindlimb flexor afferents to the timing of phase transitions in the cat step cycle. *J. Neurophysiol.*, 75:1126–1137.

Jones, C.A. and Yang, J.F. (1994). Reflex behavior during walking in incomplete spinal-cord injured subjects. *Exp. Neurol.*, 128:239–248.

Komiyama, T., Zehr, E.P., and Stein, R.B. (1995). A novel method to evaluate the overall reflex effects of cutaneous nerve stimulation in humans. *EEG Clin. Neurophysiol.*, 97:S179.

Pearson, K.G. (1995). Proprioceptive regulation of locomotion. *Curr. Opin. Neurobiol.*, 5:786–791.

Pearson, K.G. and Collins, D.F. (1993). Reversal of the influence of group Ib afferents from plantaris on activity in medial gastrocnemius muscle during locomotor activity. *J. Neurophysiol.*, 70:1009–1017.

Pratt, C.A. (1995). Evidence of positive force feedback among hindlimb extensors in the intact standing cat. *J. Neurophysiol.*, 73:2578–2583.

Prochazka, A., Sontag, K.-H., and Wand, P. (1978). Motor reactions to perturbations of gait: proprioceptive and somesthetic involvement. *Neurosci. Lett.*, 7:35–39.

Rossignol, S. (1996). Neural control of stereotypic limb movements. In *Handbook of Physiology, Section 12. Exercise: Regulation and integration of multiple systems*. Rowell, L.B. and J.T. Sheperd (eds.), pp. 173–216. American Physiological Society.

Schillings, A.M., Van Wezel, B.M.H., and Duysens, J. (1996). Mechanically induced stumbling during human treadmill walking. *J. Neurosci. Methods*, 67:11–17.

Severin, F.V. (1970). The role of the gamma motor system in the activation of the extensor alpha motor neurones during controlled locomotion. *Biophysics*, 15:1138–1144.

Sherrington, C.S. (1906). *The Integrative Action of the Nervous System*. Oxford University Press, London.

Sherrington, C.S. (1939). *Selected Writings of Sir Charles Sherrington, A Testimonial Presented by the Neurologists Forming the Guarantors of the Journal Brain*. Denny-Brown, D. (ed.), Hamish Hamilton Medical Books, London.

Sinkjaer, T., Jacob, B.A., and Birgit, L. (1996). Soleus stretch reflex modulation during gait in humans. *J. Neurophysiol.*, 76:1112–1120.

Sinkjaer, T., Anderson, B.A., and Nielson J.F. (1996b). Impaired stretch reflex and joint torque modulation during spastic gait. *J. Neurol.*, 243:566–574.

Sinkjaer, T., Toft, E., and Hansen, H.J. (1995). H-reflex modulation during gait in multiple sclerosis patients with spasticity. *Acta Neurol. Scand.*, 91:239–246.

Stein, R.B. (1991). Reflex modulation during locomotion: functional significance. In *Adaptability of Human Gait*. Patla, A.E. (ed.), pp. 21–36. Elsevier Science Publishers.

Stein, R.B. and Kearney, R.E. (1995). Nonlinear behavior of muscle reflexes at the human ankle joint. *J. Neurophysiol.*, 73:65–72.

Stein, R.B., DeSerres, S.J., and Kearney, R.E. (1995). Modulation of stretch reflexes during behavior. In *Neural Control of Movement*. Ferrell, W.R. and Proske, U. (eds.), Plenum Press, New York.

Stein, R.B., Misiaszek, J., and Pearson, K.G. (2000). Function of muscle reflexes in the decerebrate walking cat. *Can. J. Physiol. Pharmacol.*, in press.

Stephens, M.J., and Yang, J.F. (1999). Loading during the stance phase of walking in humans increases the extensor EMG amplitude but not change the duration of the step cycle. *Exp. Br. Res.* 124:363–370.

Wand, P., Prochazka, A., and Sontag, K.-H. (1980). Neuromuscular responses to gait perturbations in freely moving cats. *Exp. Brain Res.*, 38:109–114.

Whelan, P.J. (1996). Control of locomotion in the decerebrate cat. *Prog. Neurobiol.*, 49:481–515.

Whelan, P.J., Hiebert, G.H., and Pearson, K.G. (1995). Stimulation of the group I extensor afferents prolongs the stance phase in walking cats. *Exp. Brain Res.*, 103:20–30.

Yang, J.F. and Stein, R.B. (1990). Phase-dependent reflex reversal in human leg muscles during walking. *J. Neurophysiol.*, 63:1109–1117.

Yang, J.F., Stein, R.B., and James, K.B. (1991a). Contribution of peripheral afferents to the activation of the soleus muscle during walking in humans. *Exp. Brain Res.*, 87:679–687.

Yang, J.F., Fung, J., Edamura, M., Blunt, R., Stein, R.B., and Barbeau, H. (1991b). H-reflex modulation during walking in spastic paretic subjects. *Can. J. Neurol. Sci.*, 18:44–452.

Yang, J.F., Stephens, M.J., Vishram, R. (1998). Transient disturbances to one limb produce coordinated bilateral responses during infant stepping. *J. Neurophysiol.*, 79:2329–2337.

Zehr, E.P., Komiyama, T., and Stein, R.B. (1995). Evaluating the overall electromyographic effect of cutaneous nerve stimulation in humans. *Soc. Neuroscience Abs.*, 21:681.

Zehr, E.P., Komiyama, T., and Stein, R.B. (1997). Cutaneous reflexes during human gait: electromyographic and kinematic responses to electrical stimulation. *J. Neurophysiol.*, 77:3311–3325.

Zehr, E.P., Stein, R.B., Komiyama, T. (1998a). Function of sural reflexes during human walking. *J. Physiol.* 507:305–314.

Zehr, E.P., Fujita, K., Stein, R.B. (1998b). Reflexes from the superficial peroneal nerve during walking in stroke subjects. *J. Neurophysiol.*, 79:848–858.

Commentary: What Is a Reflex?

Gerald E. Loeb

In their chapter, Zehr and Stein use a bit of jargon that I believe has caused some misconceptions about the neural circuitry and functional organization of the spinal cord. During normal, unperturbed locomotion, it is true and, in fact, inescapable that ongoing activity in somatosensory afferents contributes to the total synaptic drive that results in excitation of motoneurons, but it is confusing to call this a contribution of "reflexes." It seems more useful to define a reflex as a short-latency response to a perturbation, leaving open the question of the particular circuitry involved in the reflex and its possible contribution to the unperturbed performance of the same motor task. How might this distinction matter?

The word "reflex" when used to imply neural circuitry implies a level of invariance in that circuitry that is misleading. Some reflex responses to perturbations show remarkable variability from specimen to specimen (Loeb 1993) and may be learned in the sense that the voluntary performance of the locomotor activity is accompanied by descending activity that enables the circuitry required to mediate the reflex response, rather than being enabled by the spinal central pattern generator. This would account for the paradox that many short latency cutaneous reflexes that are active during locomotion and not during stance in normals are also absent during decerebrate locomotion (Duenas et al. 1984).

The word "reflex" should be reserved for processes that are closed-loop, regardless of the nature and location of the circuitry involved. An adaptive response to load (e.g., Stephens and Yang 1996, cited by Zehr and Stein) may well reflect a descending input that is voluntary or open-loop rather than a reflection of what we usually call "reflex circuitry." Consider the increase in respiration that actually precedes rather than follows strenuous use of the musculature.

It is true that the "reflex contribution" to the activation of various motoneuron pools is modulated by the locomotor pattern generator (wherever it resides), but this may suggest to some a misleadingly simple circuit. Monkeys have been trained to produce large changes in the gain of even the simplest circuit responsible for the stretch reflex—the homonymous, monosynaptic projection from muscle spindle primary afferents—but they appear to do so via indirect and distributed effects involving many different spinal circuits (Wolpaw and Lee 1989; Wolpaw and Carp 1993).

The word "reflex input" suggests a degree of dependence on the pathway that may be misleading. Depriving a normal system of a "reflex input" does not necessarily lead to a loss of net synaptic drive (as in Severin 1970, cited by Zehr and Stein). Blockades of fusimotor activity similar to Severin's but in intact rather than decerebrate walking cats produced marked changes in spindle afferent activity with virtually no change in the recruitment of homonymous or synergist muscles (Loeb and Hoffer 1985). Conversely, reflexes may be mediated by circuits that do not contribute motoneuronal excitation during unperturbed behavior but which may be enabled by the descending control system as part of a contingency plan in anticipation of certain perturbations (Cole and Abbs 1987; Cole et al. 1984).

I am sure that Zehr and Stein would not subscribe to any of the misleading implications that I have listed above. By using the venerable word "reflexes" to subsume all of the above noted circuits and phenomena, however, we risk promulgating an overly simplistic "clockwork" model of the spinal cord. This notion has been out of date for 30 years but continues to dominate the presentation of the spinal cord in many textbooks and the thinking of many sensorimotor neuroscientists who study the brain and would prefer the spinal cord to be so obligingly simple.

References

Cole, K.J. and Abbs, J.H. (1987). Kinematic and electromyographic responses to perturbation of a rapid grasp. *J. Neurophysiol.*, 57:1498–1510.

Cole, K.J., Gracco, V.L., and Abbs, J.H. (1984). Autogenic and nonautogenic sensorimotor actions in the control of multiarticulate hand movements. *Exp. Brain. Res.*, 56:582–585.

Duenas, S.H., Loeb, G.E., and Marks, W.B. (1984). A quantitative comparison of hindlimb muscle activity and flexor reflexes in normal and decerebrate cats during walking. *Soc. Neurosci.*, 10:628(Abstract).

Loeb, G.E. (1993). The distal hindlimb musculature of the cat: I. Interanimal variability of locomotor activity and cutaneous reflexes. *Exp. Brain Res.*, 96:125–140.

Loeb, G.E. and Hoffer, J.A. (1985). The activity of spindle afferents from cat anterior thigh muscles. II. Effects of fusimotor blockade. *J. Neurophysiol.*, 54:565–577.

Wolpaw, J.R. and Carp, J.S. (1993). Adaptive plasticity in spinal cord. *Adv. Neurol.*, 59:163–174.

Wolpaw, J.R. and Lee, C.L. (1989). Memory traces in primate spinal cord produced by operant conditioning of H-reflex. *J. Neurophysiol.*, 61:563–572.

18
Fractal Analysis of Human Walking Rhythm

Jeffrey M. Hausdorff, C.K. Peng, Jeanne Y. Wei, and Ary L. Goldberger

1 Introduction

Under healthy conditions, the complex, multilevel locomotor system does a remarkable job of controlling an inherently unstable, multijoint system. During walking, the kinetics, kinematics and muscular activity of gait appear to remain relatively unchanged from one step to the next (Winter 1984; Patla 1985; Kadaba et al. 1989; Pailhous and Bonnard 1992). However, closer examination reveals small fluctuations in the gait pattern, even under stationary conditions (Gabell and Nayak 1984; Yamasaki, Sasaki and Torii 1991; Pailhous and Bonnard 1992). The goal of this chapter is to analyze these subtle step-to-step fluctuations in gait, specifically the duration of the gait cycle, in order to gain insight into the neural control of locomotion. Ultimately, this analysis should improve our understanding of the organization, regulation, and interactions of the entire locomotor system and might also prove clinically useful in the diagnosis and prognosis of gait disorders.

Gait is a process with numerous inputs and outputs. The focus here is on the step-to-step fluctuations in walking rhythm, that is the duration of the gait cycle, the stride interval. We focus on the stride interval (1) because it can be viewed as a final "controlled" output of the neuromuscular control system, and (2) because gait variability may be lowest in the most distal segments of the kinematic chain (i.e., at the foot) (Winter 1984), the time between heelstrikes may function as a gait "clock." An example of the complex fluctuations that occur in the stride interval is shown in Figure 18.1. While considerable "noise" has been observed in the

stride interval, typically measured as the time between heelstrikes of the same foot (Gabell and Nayak 1984; Yamasaki et al. 1991), these fluctuations have not been characterized and their origin is largely unknown.

One possible explanation for these step-to-step variations in walking rhythm is that they simply represent uncorrelated (white) noise superimposed on a basically regular process. This is what one might expect a priori if one assumes that these subtle fluctuations are merely "noise." A second possibility is that there are short-range correlations: the current value is influenced by only the most recent stride intervals, but over the long term, fluctuations are random. A third, less intuitive possibility is that the fluctuations in the stride interval exhibit long-range correlations, as seen in a wide class of scale-free phenomena (Feder 1988; Voss 1988; Bassingthwaighte, Liebovitch and West 1994). In this case, the stride interval at any instant would depend (at least in a statistical sense) on the interval at relatively remote times, and this dependence would decay in a scale-free (fractal-like), power-law fashion.

Defining the nature of these fluctuations is of interest because it may extend the understanding of normal and pathological gait mechanisms. For example, studies in animals have demonstrated that rhythmic movements in general, and locomotor movements in particular, are generated by neural networks often termed central pattern generators (CPGs) (Cohen, Rossignol and Grillner 1988; Collins and Stewart 1993; Strogatz and Stewart 1993; Taga 1994). These networks have been identified in some invertebrates and vertebrates and may be present in humans as well. If the fluctua-

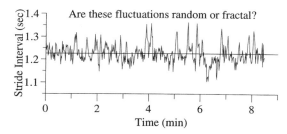

FIGURE 18.1. Stride interval time series of a healthy subject during a walk with constant environmental conditions. While the stride interval appears to be fairly constant, it fluctuates about its mean (solid line) in an apparently unpredictable manner.

tions in the stride interval are uncorrelated or represent only short-range correlations, conventional CPG models could readily be modified to account for these dynamics. However, if the fluctuations indicate long-range correlations and a fractal rhythm, a new theory is needed to account for this scale-free behavior.

2   Fractal Analysis Methods

Before describing the metrics we use to quantitatively characterize the fractal properties of stride interval fluctuations, we briefly explain what we mean by the term fractal. A fractal object is self-similar: it is composed of subunits (and subsubunits) on many levels that (statistically) resemble the structure of the object on many scales (Feder 1988; Bassingthwaighte et al. 1994). The branching of the lung is a biological example of fractal geometry. The concept of fractals can also be applied to temporal processes. Fractal processes that lack a single time scale can generate complex, irregular fluctuations on multiple time scales. However, not all erratic time series are fractal. An important class of fractal time series possess long-range order; fluctuations on one time scale are correlated with fluctuations on multiple time scales. Hence, fractal time series often look irregular and "wrinkly," reminiscent of a coastline or mountain range.

We use two scaling indices that distinguish between fluctuations due to white noise, short-range dependence, and long-range (fractal) correlations.

These indices are based on two previously validated methods: detrended fluctuation analysis (DFA) and power spectral analysis (Bassingthwaighte et al. 1994; Beran 1994; Samorodnitsky and Taqqu 1994; Peng, Havlin, Stanley and Goldberger 1995).

DFA is a modified random walk analysis that makes use of the fact that a long-range (power-law) correlated time series can be mapped to a self-similar, fractal process by simple integration. Methodologic details are provided elsewhere (Hausdorff et al. 1995a; Peng et al. 1995; Hausdorff et al. 1996). Briefly, each integrated time series is self-similar (fractal) if the fluctuations at different observation windows, $F(n)$, scale as a power-law with the window size, n (i.e., the number of strides in a window of observation or the time scale). Typically, $F(n)$ will increase with window size n. A linear relationship on a double log graph indicates that $F(n) \approx n^{\alpha}$, where the scaling index (also called the self-similarity parameter) is determined by calculating the slope of the line relating log $F(n)$ to log n (see Figure 18.2).

A second exponent, β, is determined by finding the negative slope of the line relating the squared amplitude of the power spectrum, log $S(f)$, to log f, where f is the frequency or the inverse stride number. For a process where the value at one step is completely uncorrelated with any previous values (i.e., white noise), $\alpha = 0.5$ and $\beta = 0$. In contrast, persistent long-range correlations are present if $0.5 < \alpha \leq 1.0$ and $0.0 < \beta \leq 1.0$. Pink or $1/f$ noise (i.e., $S(f) \approx 1/f^{\beta}$ where $\beta = 1.0$) lies at the boundary between stationarity ($\alpha < 1.0$) and nonstationarity ($\alpha > 1.0$) and in that case $\alpha = 1.0$. Brown noise, the integration of white noise, has a nonstationary signal (i.e. it has random drifts) and in that case, $\alpha = 1.5$. For stationary data of infinite length, the two scaling indices can be shown to follow the relationship (Peng et al. 1993): $\alpha = (\beta + 1)/2$. An illustration of different types of noise and the results obtained with spectral and DFA analyses methods is shown in Figure 18.2.

Before applying these methods, we wish to stress that α and β depend on the sequential ordering of the fluctuations in the time series, but not on the overall magnitude of the fluctuations (i.e., the variance of the time series). (Indeed, for all the stride interval time series in this study, we confirmed that α was unchanged even if the time series was nor-

FIGURE 18.2. Simulated white, brown and 1/f noise (A) and the DFA (B) and spectral analysis (C). The time series were simulated using standard techniques. (Voss 1988.)

Estimating Scaling Indices α and β

A. Three Simulated Time Series

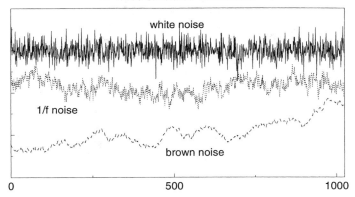

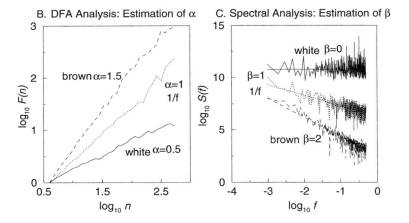

malized to its standard deviation.) Theoretically, a time series can display self-similarity and fractal scaling with relatively small overall variance or large variance, and conversely, a time series can have no correlations while having either small or large overall variance.

3   First Evidence of Fractal Gait Rhythm

We first measured the stride interval in 10 young, healthy men (Hausdorff et al. 1995a). Subjects walked continuously on level ground around an obstacle free, 130-meter long, approximately circular

path at their self-determined, usual rate for about nine minutes. To measure the stride interval, the output of ultrathin, force sensitive switches was recorded on an ambulatory recorder and heelstrike timing was automatically determined (Hasudorff, Ladin and Wei 1995).

A representative stride interval time series is shown in Figure 18.3 (top). First, note the stability of the stride interval; during a nine minute walk the coefficient of variation was only 4%. Thus, as in Figure 18.1, a good first approximation of the dynamics of the stride interval would be a constant. However, fluctuations occur about the mean. The stride interval varies irregularly with some underlying complex temporal "structure." Of note, this structure changes after random shuffling of the data

FIGURE 18.3. Representative stride interval time series before and after random shuffling of the data points (top) and the fluctuation and power spectrum analyses (bottom).

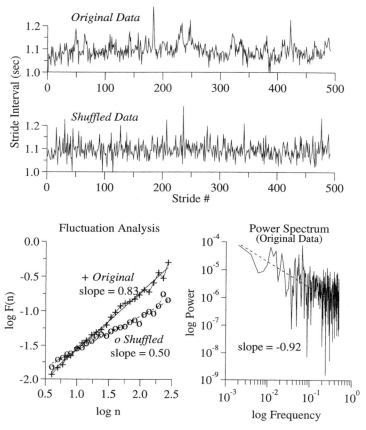

Dynamics During 9 Minute Walk

points, as seen in Figure 18.3, demonstrating that the original structure is a result of the sequential ordering of the stride interval and not a result of the stride interval distribution. Figure 18.3 (bottom left) shows *F(n)* versus *n* plotted on a double log graph, for the original time series and the shuffled time series. The slope of the line relating log *F(n)* to log *n* is .83 for the original time series and .50 after random shuffling. Thus, fluctuations in the stride interval scale as $F(n) \approx n^{.83}$ exhibiting long-range correlations, while the shuffled data set behaves as uncorrelated white noise; $\alpha = .50$. Fig. 3 (bottom right) displays the power spectrum of the original time series. The spectrum is broad band and scales as $1/f^{\beta}$ with $\beta \approx .92$. The two scaling exponents are consistent with each other within statistical error due to finite data length (Peng et al. 1993)

and both indicate the presence of long-range correlations.

4   Modeling the Fractal Gait Rhythm

To investigate this finding of a fractal gait rhythm and the mechanisms that might account for it, we simulated the experimental results of Hausdorff et al. (1995a). Specifically, we were in search of a model that (1) produced a time series whose appearance was similar to that observed experimentally, (2) generated power-law, long-range correlations, and (3) possessed some physiological analogs. Several stochastic and deterministic models were tested. For example, we attempted to reproduce the experimental results by (a) modeling the stride interval time series as the outcome of a second order, deterministic system in combination with some stochastic elements [e.g., a simple "noisy" mass-spring-dashpot model of walking (McMahon and Cheng 1990)], and (b) simulating a "noisy" CPG (an oscillator whose frequency changes randomly). Despite searching over several decades in parameter space, these models were not able to simulate the observed dynamics.

The noisy CPG model fails to generate a fractal time series because it has no memory. To introduce "memory" into the CPG and produce scale-free behavior, we modified the noisy CPG model as follows. Whereas previously, the frequency (mode) of the oscillations could change arbitrarily, we now imposed a rule whereby the CPG mode could change based on some predefined preference where only certain transitions from mode to mode are allowed (possibly with different probabilities). For clarity, we focus here on a simple one dimensional network CPG model, although this model may be generalized in several ways (Hausdorff et al. 1995a). It can be viewed as a network where the nodes represent possible frequencies or stable states. Node transitions will take place from the current node to one of its two neighboring nodes with equal probability. The frequency values at each node of the network are assigned randomly, but once determined, they are fixed in time. The net result is that this model, termed the correlated CPG model, differs from the noisy CPG model in a subtle, but crucial way. The key difference is that oscillator frequencies, once chosen, are now fixed in time. In addition, although the CPG frequency may switch randomly, it switches only to an adjacent node of (randomly) pre-assigned frequency. If this restriction is relaxed (i.e., if the network structure is removed such that each node can switch to any other node, not just neighboring ones), the outcome is the same as that of the original noisy CPG model (i.e., no fractal dynamics). For a time series of an infinite length, the scaling exponent α of the correlated CPG model approaches 0.75, approximating the group mean of the experimental stride interval data ($\alpha = 0.76$). Furthermore, even for finite data, this simple model generates time series indistinguishable from that shown in Figure 18.3, suggesting that some of its features may reflect the underlying physiology.

5 Stability of Healthy Fractal Rhythm

Many natural phenomena are characterized by short-term correlations with a characteristic time scale, τ_0, and an autocorrelation function, $C(\tau)$ that decays exponentially, (i.e. $C(\tau) \approx exp(-\tau/\tau_0)$; relatively quickly) (Van Kampen 1981; Gillespie 1992). In contrast, long-range, fractal correlations have only been observed under unique conditions (Weissman 1988; Beran 1994; Samorodnitsky and Taqqu 1994), for example, when a system is near a critical point of a phase transition (Stanley 1971). In that case, there exists no well-defined correlation length and the autocorrelation function decays as a power-law $(C(\tau) \approx \tau^{-\gamma}$ where $0 < \gamma \leq 1$; slow decay); the current value depends not only on its most recent value but also on its long-term history in a scale-invariant, fractal manner. The establishment of long-range correlations in gait, therefore, raises the possibility of an unidentified mechanism underlying neural control of walking.

To study the stability and extent of these long-range correlations, we asked ten young (ages 18–29 years), healthy men to walk for 1 hour at their usual, slow and fast paces around an outdoor track (Hausdorff et al. 1996). A representative example of the effect of walking rate on the stride interval fluctuations and long-range correlations is shown in Figure 18.4. The locomotor control system maintains the stride interval at an almost constant level throughout the one hour of walking (the standard deviation was 37, 22 and 23 ms, during slow, normal and fast walking, respectively, and the coefficient of variation was less than 3%). Nevertheless, the stride interval fluctuates about its mean value in a highly complex, seemingly random fashion. However, both the DFA and power spectral analysis indicate that these variations in walking rhythm are not random. Instead, the time series exhibits long-range correlations at all three walking rates. The scaling indices α and β remained fairly constant despite substantial changes in walking velocity and mean stride interval.

Consistent results were obtained for all 10 subjects. For all thirty 1 hour trials, α was .95 ± .06 (range: .84 to 1.10). Figure 18.5 illustrates the dependence of α on self-determined walking rate for all 10 subjects. There was a slight dependence of

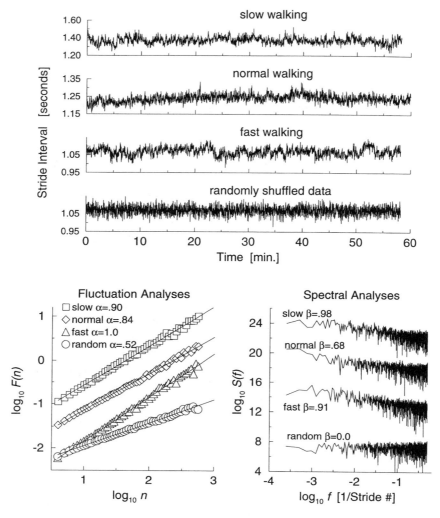

FIGURE 18.4. An example of the effects of walking rate on stride interval dynamics. (Top) One hour stride interval time series for slow (1.0 m/sec), normal (1.3 m/sec) and fast (1.7 m/sec) walking rates. Note the breakdown of structure with random reordering or shuffling of the fast walking trial data points, even though this shuffled time series has the same mean and standard deviation as the original, fast time series. (Bottom) Fluctuation and power spectrum analyses confirm the presence of long-range correlations at all three walking speeds and its absence with random shuffling of the data.

α on chosen walking rate: α was .98 ± .07, .90 ± .04, and .97 ± .05, during the slow, normal, and fast walking trials, respectively. Paired t-tests showed a small, but significant decrease in α at the normal walking rate compared with α at the slow (p < .005) and fast (p < .05) walking rates. Similar results were also observed for the power spectrum scaling exponent. β (e.g., for all thirty trials, β was .93 ± .13). Thus, for all subjects tested at all three rates, *the stride interval time series displayed long-range, fractal correlations over thousands of steps.*

To investigate further the possible mechanisms of this fractal gait rhythm, all 10 subjects were studied under three additional conditions. Subjects were asked to walk in time to a metronome that was set to each subject's mean stride interval (computed from each of the three unconstrained walks). The

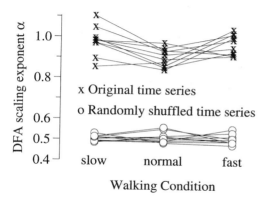

FIGURE 18.5. Dependence of α on self-selected walking rate. x's correspond to α calculated from each original time series and o's to α of a corresponding randomly shuffled time series.

results during metronomic walking were significantly different from those obtained when the walking rhythm was unconstrained. During metronomically paced walking, fluctuations in the stride interval were random and non-fractal in all 30 walking trials.

These findings indicate that the fractal dynamics of walking rhythm are normally quite robust and intrinsic to the locomotor system. The breakdown of long-range correlations during metronomically paced walking demonstrates that supraspinal influences (a metronome) can override the normally present long-range correlations. Since metronomic and free walking utilize the same lower motor neurons, actuators, and feedback, one might speculate further that supraspinal control is critical in generating these long-range correlations.

6 Changes in Fractal Dynamics with Aging and Huntington's Disease

To gain further insight into the basis for this long-term, fractal dependence in walking rhythm, we next investigated the effects of advanced age and Huntington's disease, a neurodegenerative disorder of the central nervous system, on stride interval correlations (Hausdorff et al., 1997). Using detrended fluctuation analysis, we compared the stride interval time series (1) of healthy elderly subjects

($n = 10$) and healthy young adults ($n = 22$), and (2) of subjects with Huntington's disease, ($n = 17$) and healthy controls ($n = 10$).

Figure 18.6 compares the stride interval time series for a young adult and an elderly subject. Visual inspection suggests a possible subtle difference in the dynamics of the two time series. Fluctuation analysis reveals a marked distinction in how the fluctuations change with time scale for these subjects. The slope of the line relating log $F(n)$ to log n is less steep and closer to 0.5 (uncorrelated, white

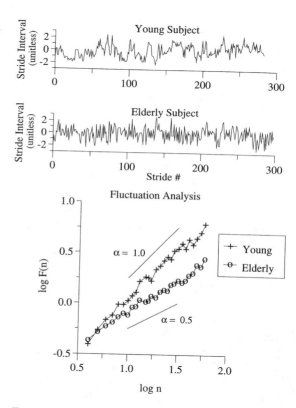

FIGURE 18.6. Example of the effects of aging on the fluctuation analysis of stride interval dynamics. Stride interval time series are shown above and fluctuation analysis below for a 71 year old elderly subject and a 23 year old, young subject. For illustrative purposes, each time series is normalized by subtracting its mean and dividing by its standard deviation. This normalization process highlights any temporal "structure" in the time series, but does not affect the fluctuation analysis. Thus, in this Fig, stride interval is unitless. For the elderly subject, fluctuation analysis indicates a more random and less correlated time series. Indeed, α is 0.56 ($\approx$white noise) for the elderly subject and 1.04 ($\approx$1/f noise) for the young subject.

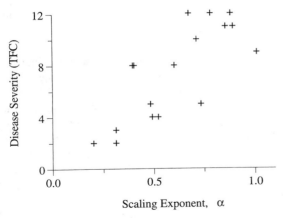

FIGURE 18.7. Relationship between disease severity and degree of stride interval correlations (α) among subjects with Huntington's disease. Disease severity is measured using the Total Functional Capacity (TFC) score of the Unified Huntington's Disease Rating Scale (0 = most impairment; 13 = no impairment). This clinical measure of function has been shown to correlate with positron emission tomography (PET) scan indices of caudate metabolism. (Podsiadlo and Richardson 1991.)

noise) for the elderly subject. This indicates that the stride interval fluctuations are more random and less correlated for the elderly subject than for the young subject. Similar results were obtained for other subjects in these groups as well. α was 0.68 $\pm$ 0.14 for the elderly subjects versus 0.87 $\pm$ 0.15 in the young adults (p < .003).

Interestingly, on average, elderly and young subjects had similar average stride intervals (elderly: 1.05 $\pm$ .10 sec; young: 1.05 $\pm$ .07 sec) and required almost identical amounts of time to perform a standardized functional test of gait and balance (Podsiadlo and Richardson 1991). The magnitude of stride-to-stride variability (i.e., stride interval coefficient of variation) was also very similar in the two groups (elderly: 2.0 $\pm$ 0.7%; young: 1.9 $\pm$ 0.4 %). These results show that while α was different in the two age groups, the gross measures of gait and mobility function of these elderly subjects were not significantly affected by age. Average gait speed of elderly subjects was slightly less than that of the young subjects, however, α was not associated with gait speed ($r = 0.07$; p > 0.7). Furthermore, multiple regression analysis demonstrated that even after adjusting for any potential con-

founders (e.g., speed), age still remained independently associated with α (p < .0005).

The scaling exponent α was also reduced in the subjects with Huntington's disease compared to disease-free controls (Huntington's disease: 0.60 $\pm$ 0.24; controls: 0.88 $\pm$ 0.17; p < 0.005). Moreover, among the subjects with Huntington's disease, α was related to degree of functional impairment ($r = 0.78$, p < .0005; see Figure 18.7).

7   Discussion

Our findings demonstrate that the stride interval time series exhibits long-range, self-similar correlations. This fractal scaling is apparently an intrinsic feature of normal walking rhythm. Scaling exponents obtained using two complementary methods, fluctuation analysis and Fourier analysis, both indicate the presence of long-range, power-law correlations in the gait rhythm during slow, normal and fast walking. With random shuffling of the stride interval, the scaling exponents change to that of an uncorrelated random process. Thus, stride interval fluctuations are not random like white noise, nor are they the outcome of a process with short-term correlations. Instead, the present stride interval is related to the interval thousands of strides earlier and this scaling occurs in a scale-invariant, fractal-like manner.

The presence of a fractal gait rhythm is notable for several reasons:

1. It suggests the presence of a non-trivial long-term dependence ("memory" effect).
2. Such fluctuations are often associated with a nonequilibrium dynamical system with multiple-degrees-of-freedom, rather than being the output of a classical "homeostatic" process.
3. Models of the neural basis of rhythmic motor acts (e.g., CPGs) need to be reexamined to account for this previously unanticipated fractal scaling property.
4. The finding of reduced stride interval correlations with aging and with Huntington's disease parallels the effects of age and disease on the fractal scaling of other processes under neural control (Lipsitz and Goldberger 1993; Belair, Glass, van der Heider and Milton 1995; Iyengar et al. 1996).

Precise elucidation of the factors affecting the fractal dynamics of gait remain to be determined. Nevertheless, we can begin to form an idea of what contributes to this behavior. The breakdown of long-range correlations during metronomically paced walking in the same neuro-mechanical system that produces this fractal behavior during normal walking (i.e., in healthy young adults) suggests: (1) that supraspinal influences can override the normally present fractal rhythm, and (2) that this behavior is not simply a result of the neuro-mechanical interaction of a highly complex system.

The alterations in the fractal dynamics of the stride interval with advanced age and Huntington's disease provide additional evidence. Changes in the fractal rhythm in these populations are not simply attributable to reduced gait speed or increased stride-to-stride variability with aging or disease. When healthy young adults walk slowly, the fractal rhythm is not reduced. Moreover, the magnitude of the stride interval correlations was independent of gait speed and stride-to-stride variability. Apparently, stride interval correlations depend on some aspect of the neuromuscular control system that is not directly related to walking velocity or gait unsteadiness.

Most of the pathologic changes in Huntington's disease are seen in the basal ganglia with a loss of neural projection in the striatum (caudate nucleus and putamen) (Penney and Young 1993). The subjects with Huntington's disease were relatively young (46.3 ± 12.8 yrs) with impairment limited primarily to the central nervous system and were free from concomitant disease and age-related physiological changes. It therefore seems likely that the areas of the cerebrum that are affected by Huntington's disease play an important role in generating stride interval correlations. While some pathologic changes have been found in the cortex (Cudkowicz and Kowall 1990), the primary pathology is in the basal ganglia. Perhaps, the striatal pathology that leads to a decrease in fine motor control in Huntington's disease also impairs the "long-term dependence" and fine control necessary for a fractal gait rhythm.

A reduction in stride interval correlations was also seen in healthy, elderly subjects. This may have been be due in part to comorbidities that we did not detect. Alternatively, the alterations in gait dynamics in the elderly may be due to subtle changes in neural control that were not revealed by our clinical evaluation. These physically fit elderly subjects were free from overt neurological disease. However, even in "normal" elderly adults, there is an age-related decline in dopamine content in the basal ganglia and it has been suggested that age-related changes in gait may result from subtle changes in striatal dopamine mechanisms (Dobbs et al. 1992). The reduction in stride interval correlations may also be a manifestation of these changes in central processing. Either way, it is interesting to speculate why measures of stride-to-stride variability and gross measures of mobility were unchanged in these healthy elderly subjects, while stride interval correlations were diminished. Perhaps the ability to produce a correlated gait depends more sensitively on intact central processing.

8   Future Directions

Future work should likely focus on addressing three general questions:

1. What are the neural control mechanism(s) that generate this fractal gait rhythm?
2. What, if any, are the adaptive benefits of this fractal property?
3. Do the fractal properties of gait rhythm provide information useful for clinical diagnosis and prognosis?

Although we have begun to answer these questions, work in a number of areas will be useful to fully address these questions. For example:

1. Preliminary data indicate the presence of a fractal gait rhythm even during treadmill and blindfolded walking. Further study of these and other walking conditions will be helpful for determining the role of sensory feedback and the environment.
2. Studies that examine subjects with pathologies involving different portions of the extrapyramidal and pyramidal system may provide additional insight into the origins of this fractal behavior and its breakdown in disease, as well as any practical clinical utility.
3. Animal studies could also be useful in identifying the mechanisms and physiologic implications of this behavior. Fractal analysis of rhyth-

mic movements in smaller, less complex animals and in animals with isolated lesions (e.g., decerebrate cats) could provide insight into determining the minimum components (neural and mechanical) necessary for the generation of fractal rhythmic movements.

4. Investigation of other parameters of gait (e.g., neural firing patterns at different locations) and additional dynamical analysis of the stride interval fluctuations could also provide a key towards understanding the origin of the fractal gait rhythm.

5. Whole body modeling and modeling of central pattern generators may be useful in determining to what extent stochastic or deterministic mechanisms underly this property of walking rhythm and the extent to which these models need to be modified to explain the presence of fractal gait.

In summary, our findings indicate that a fractal gait rhythm appears to be a characteristic feature of the normal gait pattern and that the degree of fractal scaling decreases both with advanced age and with Huntington's disease. However, the mechanisms responsible for this scale-free, far from equilibrium behavior and its physiological role remain to be established.

Acknowledgments. We thank P. Purdon, M. Cudkowicz, H. Edelberg, and Z. Ladin, S. Mitchell for their assistance in various phases in this project and H.E. Stanley and J.J. Collins for valuable discussions. This chapter is based largely on previous work (Hausdorff et al. 1995a; Hausdorff et al. 1996; Hausdorff et al. 1997).

This work was supported in part by NIH grants from the National Institute of Aging, the National Institute of Mental Health, and the National Center for Research Resources. We are also grateful for partial support from the G. Harold and Leila Y. Mathers Charitable Foundation, and the National Aeronautics and Space Administration.

References

Bassingthwaighte, J.B., Liebovitch, L.S., and West, B.J. (1994). *Fractal Physiology*. Oxford University Press, New York.

Belair, J., Glass, L., van der Heider, U., and Milton, J., (eds.). (1995). *Dynamical Disease: Mathematical Analysis of Human Illness*. American Institute of Physics Press, New York.

Beran, J. (1994). *Statistics for Long-Memory Processes.* Chapman and Hall.

Cohen, A.H., Rossignol, S., and Grillner, S. (1988). *Neural Control of Rhythmic Movements in Vertebrates.* Wiley & Sons, New York.

Collins, J.J. and Stewart, I. (1993). Hexapodal gaits and coupled nonliner oscillator models. *Biol. Cybern.*, 68:287–298.

Cudkowicz, M. and Kowall, N.W. (1990). Degeneration of pyramidal projection neurons in Huntington's disease cortex. *Ann. Neurol.*, 27:200–204.

Dobbs, R.J., Lubel, D.D., Charlett, A., et al. (1992). Hypothesis: age-associated changes in gait represent, in part, a tendency towards parkinsonism. *Age Ageing*, 21:221–225.

Feder, J. (1988). *Fractals.* Plenum Press, New York.

Gabell, A. and Nayak, U.S.L. (1984). The effect of age on variability in gait. *J. Gerontol.*, 39:662–666.

Gillespie, D.T. (1992). *Markov processes: an introduction for physical scientists.* Academic Press, Boston.

Hausdorff, J.M., Ladin, Z., and Wei, J.Y. (1995). Footswitch system for measurement of the temporal parameters of gait. *J. Biomech.*, 28:347–351.

Hausdorff, J.M., Mitchell, S.L., Firtion, R., et al. (1997). Altered fractal dynamics of gait: reduced stride interval correlations with aging and Huntington's disease. *J. Appl. Physiol.*, 82:262–269.

Hausdorff, J.M., Peng, C.-K., Ladin, Z., Wei, J.Y., and Goldberger, A.L. (1995a). Is walking a random walk? Evidence for long-range correlations in the stride interval of human gait. *J. Appl. Physiol.*, 78:349–358.

Hausdorff, J.M., Purdon, P.L., Peng, C.-K., Ladin, Z., Wei, J.Y., and Goldberger, A.L. (1996). Fractal dynamics of human gait: stability of long-range correlation in stride interval fluctuations. *J. Appl. Physiol.*, 80:1448–1457.

Iyengar, N., Peng, C.-K., Morin, R., Goldberger, A.L., and Lipsitz, L.A. (1996). Age-related alterations in the fractal scaling of cardiac interbeat interval dynamics. *Am. J. Physiol.*, 271:R1078–1084.

Kadaba, M.P., Ramakrishnan, H.K., Wootten, M.E., Gainey, J., Gorton, G., and Cochran, G.V.B. (1989). Repeatibility of kinematic, kinetic and electromyographic data in normal adult gait. *J. Orthop. Res.*, 7:849–860.

Lipsitz, L.A. and Goldberger, A.L. (1993). Loss of 'complexity' and aging. Potential applications of fractals and chaos theory to senescence. *JAMA*, 267:1806–1809.

McMahon, T.A. and Cheng, G.C. (1990). The mechanics of running: how does stiffness couple with speed? *J. Biomech.*, 23:65–78.

Pailhous, J. and Bonnard, M. (1992). Steady-state fluctuations of human walking. *Behav. Brain. Res.*, 47:181–190.

Patla, A.E. (1985). Some characteristics of EMG patterns during locomotion: implications for locomotor control processes. *J. Mot. Behav.*, 17:443–461.

Peng, C.-K., Buldyrev, S.V., Goldberger, A.L., Havlin, S., Simons, M., and Stanley, H.E. (1993). Finite size effects on long-range correlations: implications for analyzing DNA sequences. *Phys. Rev. E*, 47:3730–3733.

Peng, C.-K., Havlin, S., Stanley, H.E., and Goldberger, A.L. (1995). Quantification of scaling exponents and crossover phenomena in nonstationary heartbeat time series. *Chaos*, 6:82–87.

Penney, J.B. and Young, A.B. (1993). Huntington's disease. In *Parkinson's Disease and Movement Disorders*. Jankovic J. and Tolosa E. (eds.), 205–216. Williams and Wilkins, Baltimore.

Podsiadlo, D. and Richardson, S. (1991). The timed "up and go": a test of basic functional mobility for frail elderly persons. *JAGS*, 39:142–148.

Samorodnitsky, G. and Taqqu, M.S. (1994). Stable non-Gaussian random processes: stochastic models with infinite variance. Chapman and Hall, New York.

Stanley, H.E. (1971). *Introduction to Phase Transitions and Critical Phenomena*. Oxford University Press, Oxford and New York.

Strogatz, S.H. and Stewart, I. (1993). Coupled oscillators and biological synchronization. *Sci. Am.*, 102–109.

Taga, G. (1994). Emergence of bipedal locomotion through entrainment among the neuro-musculo-skeletal system and the environment. *Physica. D.*, 190–208.

Van Kampen, N.G. (1981). *Stochastic Processes in Physics and Chemistry*. North-Holland, Amsterdam.

Voss, R.F. (1988). Fractals in nature: from characterization to simulation. In *The Science of Fractal Images*. Peitgen H.O. and Saupe D. (eds.), pp. 21–70. Springer-Verlag, New York.

Weissman, M.B. (1988). 1/f noise and other slow, non-exponential kinetics in condensed matter. *Rev. Mod. Phys.*, 60:537–571.

Winter, D.A. (1984). Kinematic and kinetic patterns in human gait: variability and compensating effects. *Hum. Mov. Sci.*, 3:51–76.

Yamasaki, M., Sasaki, T., and Torii, M. (1991). Sex difference in the pattern of lower limb movement during treadmill walking. *Eur. J. Appl. Phys.*, 62:99–103.

Young, A.B., Penney, J.B., Starosta-Rubinstein, S., et al. (1986). PET scan investigations of Huntington's disease: cerebral metabolic correlates of neurological features and functional decline. *Ann. Neurol.*, 20:296–303.

Commentary:
The Fractal Nature of the Locomotor Rhythm May Be Due to Interactions Between the Brain and the Spinal Pattern Generator

Ranu Jung

Hausdorff et al. present a very intriguing and interesting chapter examining the nature of the human walking rhythm. They show that the usual Gaussian statistical measures, such as mean, standard deviation and variance, which are often used to describe time-series behavior, are not applicable to the human walking rhythm because of the presence of long-range persistence. They find this persistence to be independent of the speed of walking, or the degree of variance.

The long-term correlations observed in normal walking in young adults disappear when the subjects listen to a metronome. The authors suggest that this is because of supraspinal influences on the underlying fractal dynamics. While the effect of the metronome is mediated by supraspinal centers, the stimulus from the metronome is a component of the environment. These studies have demonstrated that *in the presence of a periodic stimulus* the influence of the supraspinal centers is to reduce the fractal component. In the absence of the periodic stimulus, however, the influence of the supraspinal centers may be quite different and, in fact, it may be that the fractal nature of the rhythm is due to the interactions between spinal cord and supraspinal centers.

Investigations in a simple vertebrate, the lamprey, indicate that their exists a spinoreticulospinal loop that profoundly affects locomotor control. Stable "fictive locomotion" can be induced by pharmacologically activating the central pattern generator for swimming in in-vitro spinal cord preparations of the lamprey. However, if the spinal central pattern generator is activated in an isolated-brain-spinal cord preparation the rhythm is profoundly affected and becomes much more variable. This effect is presumably due to the feedforward-feedback connections between the brain and the spinal cord (Vinay and

Grillner 1993; Cohen et al. 1996). The role of these feedforward-feedback brain-spinal cord connections has been examined using computer simulations which show that a change in the balance between the feedforward input from the brain to the spinal cord and the feedback from the spinal cord to the brain can alter the oscillatory frequency (Jung et al. 1996). We have also recently observed the complex effects of brain-spinal cord interactions in conscious lamprey in which the spinal CPG has been activated by excitatory amino acids, before and after spinal cord lesion (unpublished observations). These results are consistent with the data from the in vitro preparation in that the rhythm is much more repeatable in the spinalized animal. Although the nature of the variance in the rhythm obtained from the in vitro preparations and the conscious animals remains to be determined, it is clear that the regularity of the rhythm of the spinal pattern generator is reduced by interactions with the brain.

The authors have explored the possibility that the central pattern generating circuitry within the spinal cord is responsible for the fractal behavior. Only, with a constrained model are they able to obtain a fractal nature in the output time series. It may be that details, such as cellular level properties of neurons which were not present in there models, are essential.

If the human supraspinal-spinal circuitry and interactions are viewed as a dynamical system, then external inputs, such as the metronome, or alterations within intrinsic components of this system could move the dynamical system away from the critical state necessary to express the fractal nature.

The reduction of long-term correlations in Huntington's disease and aging suggest that the intrinsic supraspinal component of the dynamical system may be the basis for the fractal nature of the walking rhythm. In light of such an interpretation, it would not be necessary for the central pattern generating circuitry within the spinal cord to exhibit fractal behavior. While the spinal circuitry may be the primary central pattern generator, the brain provides more than descending tonic drive to regulate frequency. Supraspinal components may be an essential part of a dynamical system that generates a fractal locomotor rhythm. As suggested by the authors, studies conducted in simpler vertebrates may help identify the mechanisms for the fractal nature of the locomotor rhythm. In our laboratory, we are using the lamprey to conduct such investigations. One of the advantages of using the lamprey is that studies can be conducted both in the intact animal and in isolated in vitro preparations.

References

Cohen, A.H., Guan, L., Harris, J., Jung, R., and Kiemel, T. (1996). Interaction between the caudal brainstem and the lamprey central pattern generator for locomotion. *Neuroscience*, 74(4):1161–1173.

Jung, R., Kiemel, T., and Cohen, A.H. (1996). Dynamical behavior of a neural network model of locomotor control in the lamprey. *J. Neurophysiol.*, 75(3):1074–1086.

Vinay, L. and Grillner, S. (1993). The spino-reticulospinal loop can slow down the NMDA-activated spinal locomotor network in lamprey. *NeuroReport*, 4:609–612.

Section V

19
Postural Adaptation for Altered Environments, Tasks, and Intentions

Fay Horak and Art Kuo

1 Definition of Posture and Adaptation

Posture consists of all the musculoskeletal and sensorimotor components involved in controlling the goals of postural equilibrium and spatial orientation. Postural equilibrium involves balancing all the forces acting on the body such that it tends to stay in a desired position or moves in a controlled way. Spatial orientation involves interpretation of sensory information from various sources for a congruent representation of body position with reference to its environment as well as the appropriate positioning of body segments relative to each other and to the environment. To accomplish task goals that require stability and a particular orientation, the constraints of the musculoskeletal system interact with neural control systems involved in multijoint coordination, sensory orientation, and environmental adaptation. Posture is not a static state but a dynamic interaction among many context- and task-specific automatic neural behaviors. Postural adaptation allows changes in postural behavior so that it is optimized for changes in environmental contexts, particular tasks, and subjects' intentions (Horak 1994, 1996).

Coordination of body segments for control of dynamic equilibrium include energy-efficient movement strategies to control the many degrees of freedom involved in whole-body stabilization. A postural movement strategy is the behavioral solution to particular context, task, and intention. Biomechanical constraints inherent in the musculoskeletal system limit the potential movement strategies available for moving body segments for control of equilibrium. Equilibrium in stance involves controlling the position of the body's Center of Mass (CoM) over its limits of stability (Horak and Macpherson 1996). In free stance, the limits of stability consist of the base of foot support but also depends on the range of joint motion and muscle strength and stiffness. When the body is supported, or in contact with stable objects, the base of support includes the external support, changing how the center of body mass can be moved while maintaining stability (McCollum and Leen 1989). The nervous system must adapt to such changes in base of support and initial position as well as to the gradual changes in the mass, strength, and stiffness of segments which occur with development through the lifespan. Coordination of the head with the trunk and legs is particularly important for posture control since the head houses important sensors (vestibular, vision, and neck) for detecting and controlling body motion and balance.

Postural orientation involves interpreting information from the somatosensory, vestibular, and visual systems to determine body alignment, motion, and relative stability. Sensory information cannot be used directly for postural control, without multisensory interpretation, because there is no direct mapping of body CoM and because the body is multisegmented such that sensory reference frames can be altered by changes in body position. For example, turning the head on the body during sway in stance must involve reinterpretation of visual and vestibular information relative to center of body mass motion (Nashner and Wolfson 1974; Lund and Broberg 1983). Multiple channels of information are necessary in order to resolve ambiguities about postural orientation and body motion. The role of individual sensory systems in orientation and equilibrium can change, depending upon the availability of sensory information within a particular environment and depending on the task. For example, when subjects stand on unstable or narrow surfaces, they increase the use of vestibular and visual senses for postural orientation (Horak et al. 1989; Mergner et al. 1997; Fitzpatrick et al. 1994). The quality of sensory information for postural control also depends on how body movements are coordinated—movement strategies—because the sensors are attached to body segments.

In Section V of this book we address postural mechanisms, with a focus on adaptive processes. In this chapter we focus on observed postural strategies within the earth's gravitational field. Chapters 20 and 21 then focus on adaptations due to space flight. In Chapter 22, based on a review of postural data from studies on earth and in space, a new goal for postural control is proposed that is based on an effective weighting of sensory information with respect to an individual's intent, experience, instruction and environment.

2 Multiple Postural Strategies Exist

Studies have demonstrated a continuum of strategies whereby the standing human can maintain equilibrium in the face of a perturbation of the support surface. For example, in response to backward translations of the support surface, forward CoM displacement can be corrected by using an "ankle strategy," "hip strategy," or "stepping strategy," or some combination of these strategies (Nashner and McCollum 1985; Horak and Nashner, 1986; Romos and Starak 1987; Horak et al. 1990; McIlroy and Maki, 1993). The ankle strategy is used to control anterior–posterior sway in quiet stance and in response to small, slow surface translations while standing on a firm, even surface. It controls CoM of the body by exerting torques about the ankle joints to move the center of pressure beyond the CoM. In response to forward sway, a distal-to-proximal sequence of activation ankle, knee and hip extensor muscles rotates the body about the ankle joints with relatively little motion at the hip and knee. The ankle strategy maintains an erect postural alignment while restoring the CoM (Figure 19.1A).

In contrast to the ankle strategy, the hip strategy is used for rapid or large amplitude perturbations, especially under conditions where it is difficult to produce much ankle torque (e.g., standing on a narrow beam or compliant surface or preleaning) but the feet remain in contact with the surface. It consists of flexing the trunk at the hip joints, and, at the same time, counterrotating at the ankle and neck joints (Figure 19.1A). When standing on a surface which does not allow development of surface torque, the hip strategy involves early activation of the hip/trunk flexors such as rectus abdominis and rectus quadratus (and neck flexors), with little activation or coactivation of ankle muscles. When responding to large, fast surface translations when standing on a firm surface, early activation of hip/trunk flexors and neck flexors are added to early activation of the gastrocnemius and soleus muscles.

Like the hip strategy, the stepping strategy is used in response to large and/or fast perturbations and involves moving the base of foot support under the falling CoM (Figure 19.1A) (Horak and Shumway-Cook 1990; McIlroy and Maki 1993b). The first muscles activated for the stepping strategy are the tensor facia lata hip abductor in the initial stance leg and bilateral tibialis anterior muscles. Subjects tend to use a stepping strategy, instead of a hip strategy, in response to perturbations that they have not experienced before or when they receive no instruction about keeping the feet in place (McIlroy and Maki 1993b). A stepping strategy is usually associated with an anticipatory lateral weight shift to unload the stepping leg but may consist of only a step when subjects have inadequate time to prepare to step (McIlroy and Maki 1993a).

Subjects often used combinations of the ankle, hip, and stepping strategies. The ankle and hip strategies actually represent extremes of a response continuum in which subjects may combine hip and ankle strategies in any proportion (Runge et al. 1999). Figure 19.1B shows the extremes of ankle–hip angle changes observed in a subject exposed to repeated backward surface translations while standing on a firm support surface (ankle strategy) or on a narrow beam (hip strategy). When subjects switch from the beam to the normal sur-

FIGURE 19.1. Continuum of ankle–hip strategies. (a) Subject using an ankle, hip, and stepping strategy for postural correction. (b) Continuum of ankle–hip angle relations in response to surface perturbations resulting in forward sway and the associated muscle activation patterns for an ankle, hip, and mixed ankle–hip strategy for dynamic equilibrium in standing humans.

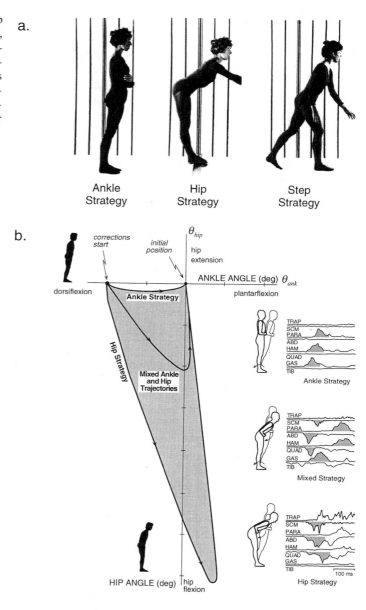

face, their first several trials show considerable hip motion as they gradually change to an ankle strategy and when they switch from the normal surface to the beam, their first several trials show less hip motion than required to accomplish the task (Horak and Nashner 1986); mixed ankle and hip trajectories in Figure 19.1B). This gradual adaptation from one postural strategy to a more optimal one demonstrates that the central nervous system selects a preprogrammed strategy, based on stimulus

characteristics as well as on prior experience and expectations of stimulus and task conditions, and then gradually modifies the strategy using trial and error (Horak and Nashner 1986; Horak 1996; Timmann and Horak 1997).

Strategy selection also depends on the availability of sensory information. Subjects with profound loss of vestibular function are unable to use a hip strategy to stand across narrow beams, in tandem stance, or on one foot, although they use quite normal ankle strategies for postural correction (Horak et al. 1990, 1994). In contrast, reducing somatosensory information in the feet with ischemic pressure cuffs results in increase use of the hip strategy such that subjects appear to be standing on a narrow support surface (Horak et al. 1990). Some elderly subjects with multiple sensory loss use the stepping strategy for postural correction even in response to very small, slow perturbations in which an ankle or mixed ankle-hip strategy is normally used (Horak et al. 1989; Horak 1992). Thus, sensory constraints appear to limit selection or control of postural strategies. Further studies are needed to determine the critical roles of sensory information in postural strategy selection, triggering, and control.

3 Modeling Control of Postural Coordination and Strategy Selection

Since automatic postural responses are not initiated until about 100 ms and a corrective action can take place up to several thousand milliseconds before a subject falls, neural mechanisms have time to modify postural adjustments during the response itself as well as prior to triggering a response. It is unclear, however, specifically what conditions define the contexts for strategy shifts, and what information is necessary to identify these conditions. To address this issue, we examine the dynamics of posture, which govern the possible responses to disturbances and the consequences of those possibilities. We show that the constraints on effective responses vary with factors such as the magnitude of a disturbance and characteristics of the support surface, necessitating a mechanism for identifying when these factors are present, and adjusting postural control accordingly. Because the identification of such factors involves cues which may arise from a variety of sensors, we are led to believe that a confluence of visual, vestibular, and peripheral inputs must contribute to contextual patterns. These patterns, in turn, define appropriate strategies or combinations of strategies for stabilizing the body. We use a control model to show that changes of postural strategies may be related to dynamic reweighting of goals within the nervous system.

A simple rigid body model of body dynamics is used to study the biomechanical constraints on pos-

tural strategies (Kuo and Zajac 1993; Kuo 1995). The body is modeled as a four-segment system confined to the sagittal plane (see Figure 19.2A). The head, arms, and trunk are lumped into a single segment, as are the pairs of feet, shanks, and thighs. Muscles are lumped into 14 groups based on anatomical location of origin and insertion points, and are modeled quasistatically, taking into account force-length characteristics but assuming that shortening velocities are small. Maximum isometric forces, musculoskeletal geometry, and body segment inertial parameters correspond to those of an average adult male. Each muscle produces a passive force dependent on its length, as well as an active force. The active force is a fraction f, specified by activation level, of the maximum possible force, which is dependent on length and, in faster movements, shortening velocity. For a given muscle, each of these active and passive forces corresponds to a set of torques about one or more joints, which are, in turn, mapped into vectors of joint angular accelerations through the dynamical equations of motion.

The net joint angular accelerations ($\ddot{\theta}$) are equal to the sum of the acceleration vectors associated with the active and passive muscle forces, along with terms corresponding to the Coriolis and centripetal accelerations, $V(\theta,\dot{\theta})$, and the effects of gravity, $G(\theta)$. The equation for joint angular accelerations is, in vector form,

$$\ddot{\theta} = M^{-1}(\theta) \cdot R(\theta) \cdot F_l(\theta) \cdot f + M^{-1}(\theta)[R(\theta) \cdot F_{lp}(\theta) + G(\theta) + V(\theta,\dot{\theta})] \quad (19.1)$$

where $M(\theta)$ is the mass matrix and $R(\theta)$ describes muscle moment arms, and F_l and F_{lp} describe muscle length-dependent effects on active and passive forces, respectively. These terms may be condensed into a simpler form

$$\ddot{\theta} = A f + b \quad (19.2)$$

This simplified equation for joint angular accelerations shows that the accelerations are due to one linear term in the controlled inputs and a second term dependent solely on joint positions and velocities.

Although it is impossible to describe motor patterns without knowledge of individual muscle activation levels f, our model of body dynamics makes it possible to restrict the set of motor patterns worthy of consideration. Consider first the action of a single muscle. The net acceleration vector must lie between the accelerations correspond-

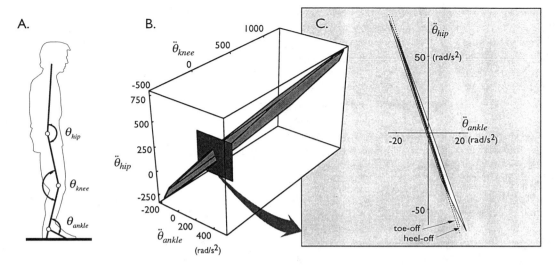

FIGURE 19.2. Feasible accelerations of the body. (A) Definition of body angles. Joint angles are measured positive in extension, referenced with respect to the upright position. (B) Feasible accelerations of the joints are limited to a polyhedron in ankle–knee–hip joint angular acceleration coordinates. (C) When knees are kept straight, accelerations are confined to ankle–hip joint angular acceleration plane. Constraints to keep feet flat on ground, "toe-off" and "heel-off," further limit viable accelerations (dark gray shaded area). Negative joint angles are flexions in this and the following figures.

ing to the upper and lower extremes of muscle activation. For more than one muscle, the net acceleration vector must lie within the set defined by the extremes for all of the muscles. This set, termed the *feasible acceleration set*, describes all possible combinations of joint angular accelerations which may be achieved with all possible combinations of muscle activations (Kuo and Zajac 1993). It embodies the physical constraints because of body dynamics and musculoskeletal geometry.

The feasible accelerations when the body is in upright stance have components of ankle, knee, and hip acceleration, and may be plotted as a polyhedron (Figure 19.2B). We may note immediately that these feasible outputs are confined to a limited region and that some combinations of accelerations can be achieved with much higher magnitude than others. In particular, there is one direction where exceptionally high accelerations may be achieved. This is a consequence of multibody dynamics and musculoskeletal geometry, both of which depend on the configuration of body alignment in stance. However, we have found that for the range of configurations achieved during most postural motions, the feasible acceleration set tends to remain relatively unchanged.

While the maximum accelerations attainable are directly related to physical limitations which are essentially fixed, there are other, more flexible constraints related to a specific motor task. These task constraints may be explicit or implicit. Explicit task constraints include specific instructions to a subject such as to maintain upright stance, avoid stepping, and keep the feet flat on the ground. They act to make particular, physically achievable accelerations incompatible with completion of a task. These type of subject intentions are explicit in the sense that they are relatively obvious or clearly stated requirements of a task. However, the combination of physical and explicit task constraints is often still insufficient to completely specify motor behavior, implying that there remain additional choices to be made. Given the stereotyped nature of postural responses, we must conclude that there are other, implicit constraints imposed by the nervous system, which specify the remaining degrees of freedom.

Implicit constraints are those enforced by default within the nervous system. They reflect task requirements which are not the result of physical or explicit constraints, but which nonetheless occur repeatedly. For example, the tendency for subjects to keep their knees relatively straight when recovering from backward disturbances to the support surface is not due to obvious requirements against knee flexion, and is therefore considered an implicit constraint. Other constraints of this type may be more difficult to uncover, because they affect behavior less directly than explicit constraints. For example, the desire to conserve metabolic energy or otherwise avoid inefficient movements would have consequences which are difficult to define unambiguously. To study these, we must start with reasonable hypotheses which lead to observable behaviors which may be compared with actual behavior. Quantitative models are helpful in this respect, because they be may used to predict the consequences of hypothesized constraints, and therefore to aid in eliminating untenable candidates. In the present study, we will consider conservation of effort to be a reasonable hypothesis, and use models to explore the expected behaviors.

To employ quantitative modeling, it is necessary to express constraints mathematically. Physical limitations have already been expressed in terms of feasible accelerations of the body segments. Many explicit and implicit constraints may also be described as limits on allowable positions or movements, but are often better described with graded levels of desirability. For example, the desire to remain upright does not imply that certain body positions are restricted entirely, but rather that they are relatively undesirable. Similarly, the desire to avoid excess expenditure of energy is best considered as an objective rather than as a limiting constraint. In fact, the language of objective functions (see also discussion on optimization in Chapter 32

and Chapter 35) is better suited to explicit and implicit task requirements because it allows a flexible weighing of constraints. Objective functions can also be used to model rigid constraints by placing essentially infinite costs on violations of these constraints. This approach is also attractive because it makes computation of optimal responses more tractable. In practice, many constraints may be reasonably approximated with objective functions of suitably high cost while improving computational features. We will use the terms "objective" and "constraint" interchangeably.

The first two constraints to be considered are the physical limitations on accelerations and an implicit constraint on knee motion. As discussed above, feasible accelerations of the ankles, knees, and hips are constrained by body dynamics, musculoskeletal geometry, and muscle parameters (see Table 19.1; constraint #1, and Figure 19.2B). An additional knee constraint (#2) is added to prevent motion of the knee during responses to backward disturbances. This implicit constraint arises from the fact that the ideal knee motion for moving the COM backward would involve knee hyperextension, which is prevented by the joint itself. The effect of this constraint is to confine accelerations to the intersection of the ankle-knee-hip accelerations with the plane defined by zero knee acceleration. The result is a set of feasible accelerations of the ankles and hips (see Figure 19.2C). This set allows for much larger magnitudes of acceleration when hip and ankle motion are in the approximate ratio of $-3:1$ (the negative sign indicates that one of the joints must accelerate in flexion while the other moves in extension). This ratio is similar to that as-

TABLE 19.1. Constraints and constraint types restricting possible postural responses.

Constraint	Type	Basis	Quantitative description
1. Feasible accelerations	Physical	Body dynamics and muscle strength	Hard constraints on joint angular accelerations
2. Knees	Implicit	Keep knees straight	Equality constraint on knee angle
3. Heel–toe	Explicit	Instructions to keep feet flat on ground	Objective function in acceleration space, cost on violations to heel–toe constraint
4. COM	Explicit	Instructions to remain standing, avoid stepping	Objective function in joint space, cost on positions near limits of support
5. Upright	Explicit	Instructions to remain upright, avoid stepping	Objective function in joint space, minimum at upright stance
6. Effort	Implicit	Avoid excess expenditure of energy	Objective function in acceleration space, cost on large accelerations scaled by feasible accelerations

sociated with fast, proximal-to-distal activation patterns seen when subjects are given large perturbations or are placed in a shortened base of support. We will, therefore, refer to use of the $-3{:}1$ ratio of hip to ankle motion as a hip strategy (Nashner and McCollum 1985; Horak and Nashner 1986).

Another constraint limiting viable accelerations is the explicit task instruction to keep the feet flat on the ground. This instruction to keep the feet flat on the ground is associated with limitations on the torque that can be exerted on the ground without lifting the heels or toes. These limitations have the effect of further narrowing the region of allowable ankle–hip accelerations due to physical and knee constraints, and enhancing the relative favorability of the $-3{:}1$ combination (see Figure 19.2C). These "heel–toe" constraints also have a significant dependence on body configuration, shifting so as to amplify the limit on ankle plantarflexion when the body is leaned forward, and on ankle dorsiflexion when the body is leaned backward. This dependence causes the viable range of ankle torques to decrease as disturbances become larger, and is the

principal reason for needing alternate movement strategies.

In addition to the heel-toe constraint, there are other explicit objectives to keep the CoM over the base of support and to maintain upright stance, in accordance with demands of typical experimental conditions. Each of these objectives is expressed as a function of the joint angles or accelerations, representing relative graded costs associated with particular body movements or configurations (see Table 19.1). To model the heel-toe constraint, an objective function (#3) is introduced to place high costs on accelerations which will lift the heels or toes (see Figure 19.3A). The COM objective (#4) places high costs on body positions which place the COM far from a neutrally stable position (see Figure 19.3B). There are many of these positions, in which the COM is above the base of support but the body is in a bent position. To guarantee that the body returns to an upright position, an additional objective (#5) must be introduced, which has a minimum at the origin of the ankle–hip joint angle plane, corresponding to the fully upright position (see Figure 19.3C).

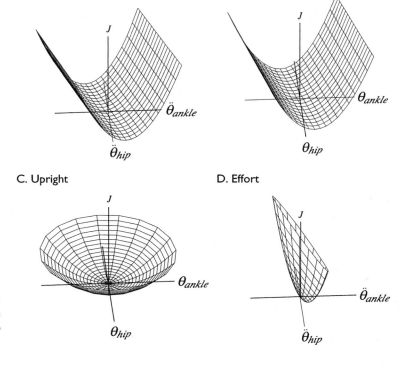

FIGURE 19.3. Objective functions reflecting explicit and implicit constraints. (A) Heel–toe objective penalizes accelerations which tend to lift heels or toes. (B) COM objective places cost on positions in which center-of-mass is near edge of base of support. (C) Upright objective is needed to bring body to upright position. (D) Effort objective places cost on accelerations which require high neural effort, proportional to normalized muscle force.

One additional implicit objective is needed to place a cost on effort (#6). Philosophically, it is appealing to recognize a desire to avoid excess expenditure of energy, but there is also a compelling theoretical justification for this objective. Control theory has shown that with the objectives stated above, the optimum behavior is to exert maximum available muscle forces so as to reach desired body positions as soon as possible. This behavior is known as "bang-bang control" because it is characterized by use of only the high and low extremes of muscle activation, and is not consistent with observed behavior. An objective function (performance subcriterion) on effort is required to eliminate bang-bang control, as well as to distribute forces to multiple muscles (see also Chapter 35). There are many possible ways to place costs on muscle force, metabolic energy expenditure, and other relevant functions, but we introduce a simple quadratic function of neural effort, equal to the percentage of maximum muscle activation (see Figure 19.3D). In practice, many of these possible functions result in similar movements which can only be differentiated through careful measurement of individual muscle forces. For the purpose of studying posture control, the objective on neural effort is adequate for reproducing gross behaviors without addressing the issue of precise distribution of muscle forces.

The multiple objectives defined above are combined into a single unambiguous function by assigning a weight to each individual objective, and summing these terms. If each objective is denoted by J_i with a weight w_i, the overall objective is then

$$J = \sum_{i=1}^{n} w_i J_i \qquad (19.3)$$

where n is the number of objectives (compare to Chapter 35, Winters). The weights place relative priorities on objectives which conflict with each other. For example, there is a trade-off between returning the body to upright and the expenditure of effort—the control behavior is dependent on the relative weighting of the two. This scheme also makes it possible for some objectives to be poorly satisfied or even for some constraints to be violated, in the event that certain competing objectives are weighted sufficiently to override others. For example, disturbances which place the body near the limits of the base of support will cause the COM

and heel–toe objectives to be of paramount importance, and the trade-off may result in the body taking intermediate positions which are not upright, as in the hip strategy.

Using this optimization approach, the problem of selecting appropriate postural responses has been transformed into one of selecting appropriate weightings between competing objectives, which themselves are based upon hypothesized explicit and implicit constraints. Because the weightings are unknown, they must be derived from experimental data. It is also common for the weightings to vary with experimental conditions such as the speed of disturbance and width of the surface. This makes it difficult to test objectives independently of the data used to set the weightings. It is insufficient for a model to reproduce behaviors it has been tuned for; it must also exhibit changes in behavior associated with changes in weights, which are commensurate with changes in experimental conditions.

Rather than test model behavior for weightings on all six of the hypothesized constraints, it is preferable to reduce the number of weightings to a minimal number. One reduction takes place by recognizing that the effort objective (#6), if weighted sufficiently, will effectively enforce the feasible accelerations constraint (#1). The weighting on effort also conflicts with the upright (#3) and COM (#4) objectives, so that increasing its weighting is equivalent to decreasing the magnitudes of the upright and COM weightings, and vice versa. That is, less effort is required to move the COM using the hip strategy than an ankle strategy but the upright objective is compromised. It is, therefore, only necessary to specify the relative, rather than absolute, weightings for COM and upright objectives, reducing another degree of freedom. In addition, the knee constraint (#2) reduces the number of degrees of freedom in the body using an equality constraint rather than a weighting. Finally, simulations show that the effects of the heel–toe (#5) and COM (#4) constraints are both to increase use of the hip strategy, indicating that sufficient weighting on COM eliminates the need for a separate heel–toe constraint.

The result of these reductions is that only two parameters are needed to specify the final objective. The first parameter, termed σ, sets the trade-off between maintaining desired segment positions

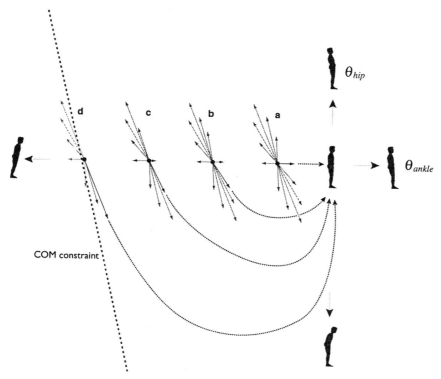

FIGURE 19.4. Interaction between heel–toe constraint and feasible accelerations, plotted in ankle–hip joint angle space. Four different backward perturbations (a,b,c,d) bring body to different forward-leaning positions. From these positions, accelerations of the ankles and hips may be made in different directions. (a) For small perturbations, accelerations may be made in all directions, though −3:1 hip–ankle combination is favored by feasible ac-celeration set. One possible response is to accelerate di-rectly to upright position as shown by dotted line. (b) As perturbations become larger, heel–toe constraint limits direct accelerations toward upright, favors accelerations closer to −3:1 combination. (c) Limitation increases fur-ther, resulting in highly curved path. (d) At extreme body positions, heel–toe and COM constraint limit viable ac-celerations to small set.

and conserving effort. The greater the emphasis on effort, the smaller the magnitude of the response, and the slower the body's return to upright. It is expected that larger disturbances will require larger postural responses to prevent falling. Another in-terpretation of this parameter is as a scale factor for the time axis for the response kinematics. The sec-ond parameter, μ, sets the relative weighting of the upright and COM objectives. It is expected that ex-perimental conditions will affect this parameter, re-sulting in associated changes in postural responses.

Two situations illustrate how these parameters, and hence postural responses, would be expected to change with postural disturbances. A small back-ward translation of the support surface would leave the body in a slightly forward sway disturbed po-sition close to upright with small velocity (see Fig-ure 19.4A). From this position, few constraints ac-tively limit possible accelerations of the ankles and hips. The physically feasible accelerations and the heel-toe constraint make it possible to exert much larger accelerations in the −3:1 combination than in others, but conditions permit the use of small ac-celerations in essentially any direction. The central nervous system is relatively unconstrained by ex-ternal constraints so, by default, other constraints specify the behavior. One possible preference is for the body to take a fairly direct route back to the upright position. In the second situation (Figure 19.4D), a large postural disturbance can place the body in a configuration such that external con-straints are much more important, overriding default constraints and resulting in a different behavior. For example, the perturbation could leave the body near the limit of the base of support and moving at a significant velocity. In this case, the COM and heel–toe constraints greatly limit the viable directions of travel, and a large acceleration may be needed to prevent the COM from moving too far beyond the limits of support. For intermediate situations (Figure 19.4B and 19.4C), the trajectories would be within the two extremes.

This dependence of responses on disturbances is supported by optimal control simulations. Any given set of parameter values corresponds to a single overall quadratic objective function. This objective function in turn defines a set of optimal feedback gains, computed using linear quadratic regulator (LQR) theory, which specify the torque at each joint as a linear function of the angles and angular velocities of all of the joints. Optimal con-

trol theory shows that these gains are constant for a given objective function, and define a family of postural responses dependent on body position and velocity. If the objective function parameters are dependent on postural disturbance, it is expected that different families of responses can occur. This is demonstrated by various behaviors predicted by the model while varying the two weighting parameters. Variation of the effort parameter σ results in changes in the time scale of response, without affecting coordination between the ankles and hips (see Figure 19.5A). Variation of the COM/heel–toe parameter μ results in substantial changes in ankle–hip coordination, with smaller values corresponding to more direct routes from perturbed to upright positions, and larger values corresponding to increased use of the hip strategy. These latter responses begin with accelerations limited by the heel–toe constraints (as described in Figure 19.4).

A. Variations in effort

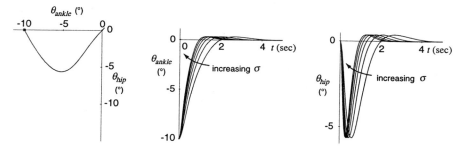

B. Variations in strategy

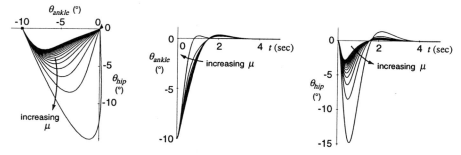

FIGURE 19.5. Effect of weighting variations on simulated responses to backward disturbances. Plots shown are hip versus ankle angle, ankle angle versus time, and hip angle versus time. (A) Increased weight on body positions, σ, results in faster response at the cost of increased ef-

fort, but no change in hip–ankle coordination. (B) Increased weight on COM/heel–toe constraints, μ, results in increased use of hip strategy, in which hip undergoes larger amplitude displacements.

It is interesting to note that even when the COM/heel–toe constraint is given extremely low weight, the body takes a more direct but nevertheless curved route back to the upright position, making partial use of the favored −3:1 combination.

These behaviors are commensurate with those observed in normal human subjects. Horak and Nashner (1986) and Runge et al. (1999) reported use of the hip strategy in responses to large disturbances and when subjects stood on a shortened support surface. Kinematically, this strategy resembles that predicted by the model, making use of the −3:1 hip–ankle combination in the initial response. For small disturbances, the ankle strategy involves decreased hip motion, also as demonstrated by the simulations. As disturbances increase in magnitude, the relative effect of the COM/heel–toe constraint increases. The model predicts that the weighting of the upright and COM objectives, parameter μ, must change as disturbance size changes (see Figure 19.5B and 19.6).

This analysis demonstrates that the various constraints associated with posture control play a role in restricting possible response coordination. The model suggests that a single, fixed set of feedback gains would be insufficient to produce the changes in coordination needed to accommodate the requirements imposed by different perturbations. We, therefore, propose that there exists a mechanism for modulating feedback gains or selecting strategies based on context. Because this modulation depends on identification of various dynamical environmental and neural constraints, its complexity appears to be beyond that normally associated with spinal reflexes. Consideration of dynamical constraints may be equivalent to an internal, neural representation of body dynamics and the environment. Consideration of environmental and implicit and explicit constraints necessarily involves complex processing of sensory and cognitive information involving many levels of neural control.

Another argument for a central role in determining postural strategies is that there is no single sensory input which may be associated with each constraint. Detection of COM position is a function of somatosensory, visual, and vestibular inputs, and detection of the heel–toe constraint most likely depends on somatosensory inputs as well as an evaluation of whether the body is approaching or moving away from that full foot contact. In ad-

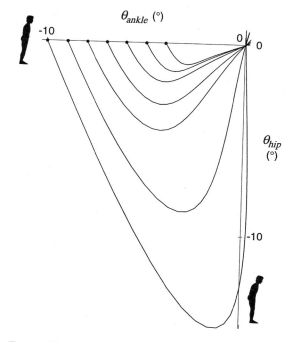

FIGURE 19.6. Predicted responses when model automatically alters strategy based on perturation size and COM/heel–toe constraints. In order to accommodate heel–toe constraint as perturbation increases, model must increase μ, resulting in increased proportion of hip strategy.

dition, instructions to the subject have a cognitive origin, necessitating a flexible means of reweighting goals according to subjects' intentions and external context. Such flexibility cannot be explained by fixed spinally generated activation patterns.

4   Flexible Weighting of Goals Affects Selection and Mixing of Strategies

There are several behaviors related to flexible weighting of goals which remain to be addressed. First, dynamical variables determine contexts for which different strategies are appropriate. These contexts may induce ankle, hip, stepping, and other strategies. Second, the boundaries between these contexts appear to be flexible and may be set within a limited range through instruction, practice, and constraints. Normal subjects may generate different strategies if instructed to, if they have prior experience with the conditions, or if placed on a shortened support surface. This dynamic flexibility is also demonstrated by overriding of constraints (e.g., lifting the heels or stepping) when conditions make enforcement of all constraints impossible. Third, there is evidence that switching between strategies is not entirely discrete. Mixed ankle and hip strategies have been observed experimentally and are predicted by the control model (Nashner and McCollum 1985; Horak and Nashner 1986). When subjects are perturbed nearly to the threshold for stepping, they tend to partially unweight one foot, mixing the beginnings of a step with a hip strategy (McIlroy and Maki 1993a). We propose that these phenomena may be interpreted in terms of combinations of strategies, specified through the flexible weighting of postural objectives.

The effects of perturbation speed and explicit and implicit neural constraints on postural strategy selection in response to sagittal perturbations are illustrated in Figure 19.7. We consider, first, the experimental condition in which subjects are asked to remain upright, avoid stepping, and keep the feet flat on the ground (Figure 19.7A). Starting with small perturbations, a gradual increase in the magnitude of the ankle strategy is observed in response to increasing perturbation. The amount of pure ankle strategy that can be employed is limited physically by the heel–toe constraint, meaning that an alternate strategy must be employed. Given instructions not to step, there is no choice but to mix in the hip strategy. The proportion of ankle and hip strategies changes gradually until the response is nearly a pure hip strategy. However, feasible accelerations and other physical constraints limit the size of perturbation which may be combated with this hip strategy, eventually necessitating a step (or possibly a fall) even though the instructions are to avoid stepping if possible. The addition of the stepping strategy is gradual, initially consisting of unweighting of the heels and stepping foot, but increasing until actual steps are taken, after which step size can increase to accommodate larger or faster perturbations. This progression of behavior is due largely to explicit instructional task constraints, and occurs as long as the central nervous system is able to correctly evaluate the boundaries between the various strategy regimes and choose the appropriate contexts.

In the absence of these explicit task constraints, implicit constraints are responsible for a default behavior. In this case, implicit desires to conserve on effort may cause the central nervous system to choose to use the hip or stepping strategies before physical limits demand, as long as there exists an effort advantage in doing so (see Figure 19.7B). Rather than employing maximal amounts of ankle and hip strategies, the nervous system may prefer to employ the other strategies at lower thresholds, with a potentially small region of hip strategy use. We presume that the nervous system evaluates trade-offs between effort, stability, and many other conditions in setting this default behavior.

It is also possible for explicit constraints to produce other types of behavior. For example, task instructions may call for the subject to step, upon detection of a perturbation of any size. In this case, the ankle strategy may be nearly entirely suppressed, and the response may use the stepping strategy almost exclusively (Burleigh et al. 1995, 1996). Here, the limits on ankle strategy are explicitly set to be as low as possible, and the threshold for stepping is reduced accordingly (see Figure 19.7C).

There are numerous other strategies which may be incorporated as well. These may include lifting the heels or toes, flailing of the arms, and falling, all of which may be necessitated by external conditions, especially when there are no explicit instructions to avoid them.

The flexibility of these behaviors suggests that the central nervous system is able to voluntarily reweight its objectives, evaluate trade-offs, and select the appropriate contexts. It appears that this reweighting can occur dynamically, in order for unexpected and unanticipated disturbances to be counteracted properly. Strategies must be modulated and mixed to meet constraints as demonstrated by mixed ankle and hip strategies and the heel–toe constraint. However, this mixing may occur even when there is no need, as in partial unweighting of one foot during large amplitude hip strategy responses. This may be due to a neural constraint related to context identification. We hypothesize that the central nervous system identifies contexts through pattern matching, and that each pattern is associated with a movement strategy, which could, for example, be encoded as a modulated set of feedback gains. The separation between

FIGURE 19.7. Flexible posture control strategies can be programmed through explicit and implicit constraints. Strategies may be mixed with proportion shown on vertical axis, depending on weighting on constraints, which varies with perturbation magnitude. (A) When subjects are instructed to keep feet flat on ground and avoid stepping, responses consist of ankle strategy for small perturbations. Beyond limits defined by heel–toe constraint, hip strategy must be incorporated, increasing to a maximum at the physical limit on hip strategy. Beyond this point, subjects are forced to step. Stepping strategy may consist of unweighting of one foot for small proportions, in addition to steps of different distances. (B) When subjects receive no instruction, default constraints result in reduced use of hip strategy, with response moving gradually from ankle strategy to stepping. Weightings may be such that strategy shifts occur before external constraints require, if other objectives, such as to reduce effort, are active. (C) Subjects may also be commanded to step immediately, resulting in responses consisting nearly entirely of stepping. (Burleigh and Horak 1995.)

A. Explicitly constrained strategies

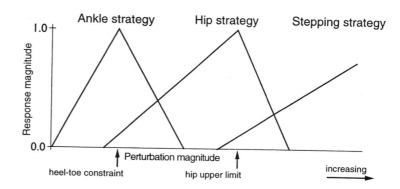

B. Default (implicit) strategies

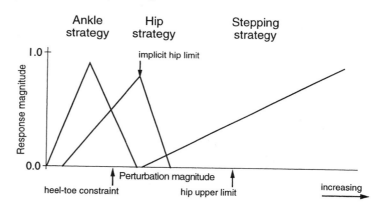

C. Alternate explicit strategies

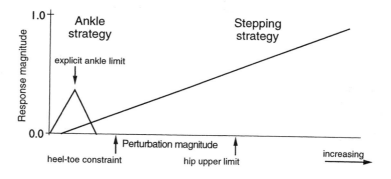

contexts is rarely demarcated unambiguously, meaning that some sensory inputs may partially match more than one pattern, and therefore trigger a combination of strategies, each weighted by the respective closeness of matching.

Of note is that the form of Figure 19.7 bears a resemblance to overlapping fuzzy membership functions, and indeed one approach toward synthesis could involve using fuzzy inference (see also Chapters 22, 40, and 35).

5 Future Work

Our model depends on context identification to determine the appropriate weighting of constraints, and hence, the appropriate mix of postural coordination strategies. We hypothesize that the central nervous system performs context identification through pattern matching. The set of inputs used in this matching process can include a variety of sensory inputs, as well as central set signals which are at least in part generated cognitively. Future work will study the nature of these inputs, and how different combinations of inputs can result in changes in context. We hypothesize that a neural internal representation of the body and its environment is necessary to decode a complex array of inputs for this purpose.

Internal representations of body dynamics are also useful within a feedback control loop for interpreting multiple sensory modalities and estimating body orientation in space. In particular, in conditions of sensory conflict, control systems making use of internal representations tend to be more robust in rejecting spurious input. Models of such control systems may be tested by applying incomplete or incorrect sensory conditions. This approach will make it possible to test the ability of the central nervous system to make certain computations which are attractive from a stabilization standpoint. These computations must then have neurophysiological substrates, making it possible to establish a connection between behavior and the underlying physiology.

References

Lund, S. and Broberg, C. (1983). Effects of different head positions on postural sway in man induced by a reproducible vestibular error signal. *Acta. Physiol. Scand.*, 117:307–309.

Burleigh, A.L. and Horak, F.B. (1996). Influence of instruction, prediction, and afferent sensory information on the postural organization of step initiation. *J. Neurophysiol.*, 75:1619–1628.

Burleigh, A.L., Horak, F.B., and Malouin, F. (1995). Modification of postural responses and step initiation: evidence for goal directed postural interactions. *J. Neurophysiol.*, 72:2892–2902.

Fitzpatrick, R., Burke, D., and Gandevia, S.C. (1994). Task-dependent reflex responses and movement illusions evoked by galvanic vestibular stimulation in standing humans. *J. Physiol.*, 478.2:363–372.

Horak, F.B. (1992). Effects of neurological disorders on postural movement strategies in the elderly. In *Falls, Balance and Gait Disorders in the Elderly.* Vellas, B., Toupet, M., Rubenstein, L., Albarede, J.L., and Christen, Y. (eds.), pp. 137–151. Elsevier Science Publishers, Paris.

Horak, F.B. (1996). Adaptation of automatic postural responses. In *Acquisition of Motor Behavior in Vertebrates.* Bloedel, J., Ebner, T.J., and Wise, S.P. (eds.), 57–85. The MIT Press, Cambridge, Massachusetts.

Horak, F.B. and Diener, H.C. (1994). Cerebellar control of postural scaling and central set in stance. *J. Neurophys.*, 72:479–493.

Horak, F.B. and Macpherson, J.M. (1996). Postural orientation and equilibrium. In *Handbook of Physiology: Section 12: Exercise: Regulation and Integration of Multiple Systems.* Smith, J.L. (ed.), pp. 255–292. Oxford University Press, New York.

Horak, F.B. and Nashner, L.M. (1986). Central programming of postural movements: adaptation to altered support surface configurations. *J. Neurophysiol.*, 55:1369–1381.

Horak, F.B., Nashner, L.M., and Diener, H.C. (1990). Postural strategies associated with somatosensory and vestibular loss. *Exp. Brain Res.*, 82:167–177.

Horak, F.B. and Shumway-Cook A. (1990). Clinical implications of posture control research. In *Balance.* Duncan, P.W. (ed.), pp. 105–111. American Physical Therapy Association, Alexandria.

Horak, F.B., Shupert, C.L., and Mirka, A. (1989). Components of postural dyscontrol in the elderly: a review. *Neurobiol. Aging*, 10:727–738.

Horak, F.B., Shupert, C.L., Dietz, V., and Horstmann, G. (1994). Vestibular and somatosensory contributions to responses to head and body displacements in stance. *Exp. Brain Res.*, 100:93–106.

Kuo, A.D. (1995). An optimal control model for analyzing human postural balance. *IEEE Trans. Biomed. Eng.*, 42:87–101.

Kuo, A.D. and Zajac, F.E. (1993). A biomechanical analysis of muscle strength as a limiting factor in standing posture. *J. Biomech. (Elmsford NY)* 26:137–150.

Kuo, A.D. and Zajac, F.E. (1993). Human standing posture: multijoint movement strategies based on biomechanical constraints. In *Natural and Artificial Control of Hearing and Balance: Progress in Brain Research, Vol. 97.* Allum, J.H.J., Allum-Mecklenburg, D.J., Harris, F.P., Probst, R. (eds.), pp. 349–358. Elsevier, Amsterdam.

McCollum, G. and Leen, T.K. (1989). Form and exploration of mechanical stability limits in erect stance. *J. Motor Behav.*, 21:225–244.

McIlroy, W.E. and Maki, B.E. (1993). Do anticipatory postural adjustments precede compensatory stepping reactions evoked by perturbation? *Neurosci. Lett.,* 164:199–202.

McIlroy, W.E. and Maki, B.E. (1993). Task constraints on foot movement and the incidence of compensatory stepping following perturbation of upright stance. *Brain Res.,* 616:30–38.

Mergner, T., Huber, W., and Becker, W. (1997). Vestibular-neck interaction and transformation of sensory coordinates. *J. Vestib. Res.,* 7:347–367.

Nashner, L.M. and McCollum, G. (1985). The organization of human postural movements: a formal basis and experimental synthesis. *Behav. Brain Sci.,* 8:135–172.

Nashner, L.M. and Wolfson, P. (1974). Influence of head position and proprioceptive cues on short latency postural reflexes evoked by galvanic stimulation of the human labyrinth. *Brain Res.,* 67:255–268.

Nashner, L.M., Shupert, C.L., and Horak, F.B. (1988). Head-trunk movement coordination in the standing posture. *Prog. Brain Res.,* 76:243–251.

Ramos, C.F. and Stark, L.W. (1987). Simulation studies of descending and reflex control of fast movements. *J. Motor Behav.,* 19:38–61.

Runge, C.F., Shupert, C.L., Horak, F.B., and Zajac, F.E. (1994). Possible contribution of an otolith signal to automatic postural strategies. *Soc. Neurosci. Abstr.,* 20: 793.

Runge, C.F., Shupert, C.L., Hovak, F.B., Zajac, F.E., (1999). Ankle and hip postural strategies defined by joint torques. *Gait Posture,* 10:161–170.

Timmann, D. and Horak, F.B. (1997). Prediction and set-dependent scaling of early postural responses in cerebellar patients. *Brain,* 120:327–337.

20
Altered Astronaut Performance Following Spaceflight: Control and Modeling Insights

Dava J. Newman and D. Keoki Jackson

1 Introduction

The mechanisms that the central nervous system (CNS) uses to control posture and movement are the subject of considerable controversy and ongoing study and have historically been treated independently. The premise of this perspective is that functional integration of neuronal structures and biomechanics involved in postural and movement control is essential, and it is the inclusion of muscle and reflex properties that provides the most intriguing research questions. Sherrington (1906) first described this unity as "posture shadowing movement." The synergy of movement, defined as the cooperative nature of many different muscles, sensors, reflexes and circuitry to orchestrate a body movement, deserves further attention. Paillard (1988) points out that "goal directed movements appear to be shaped within a changing context of postural constraints and according to the varied requirements of the environment." The astounding adaptability of human motor control is not well understood, but can be investigated by placing subjects in altered environments (i.e., reduced gravity) and measuring their *adaptive* responses both inflight and upon return to earth gravity.

Human mechanisms for control of posture and motion are normally optimized to perform in earth's gravity (1G) environment (e.g., see Chapter 19). A variety of studies of astronaut performance during and following spaceflight indicate that exposure to microgravity can cause neurosensory, sensory-motor, and perceptual changes with profound effects on the control and dynamics of human motion and posture. The adaptation process involves reinterpretation, modification, or elimination of "normal" motor commands to facilitate motion in microgravity. As mentioned, an additional concern is that of readaptation after return to earth since altered movement control strategies for microgravity are no longer optimized for terrestrial gravity, causing postflight degradation of stability in posture, locomotion, and jumping tasks. The control of jump landings after spaceflight is particularly interesting and is used as an illustrative example in this perspective due to task complexity requiring postural stability, voluntary movement, preprogrammed synergies, and environmental interactions. Experimental data and a model for the overall mechanics of jump landings are highlighted.

Performance decrements in postural stability after spaceflight may result from various changes in the sensorimotor complex. In postflight posture tests, Kenyon and Young (1986) found decrements in standing ability with the eyes closed. Paloski et al. (1993) reported that astronauts demonstrated abnormal postural sway oscillations and drift immediately postflight when the support was sway-referenced to alter ankle proprioception cues (see also Chapter 21). Parker et al. (1985) found direct evidence for the concept of stimulus rearrangement and a reinterpretation of sensory graviceptors during spaceflight. Young et al. (1986) also provided evidence for sensory compensation during spaceflight resulting in interpretation of utricular otolith signals as linear acceleration rather than head tilt, as well as increased dependence on visual cues for perception of orientation. The otolith-spinal reflex, which helps prepare the leg musculature for impact in response to sudden falls, is dramatically reduced during spaceflight and indicates a rapid course of readaptation upon return to earth (Watt et al. 1986). A theory to account for illusions of self-motion experienced by returning astronauts, the otolith tilt-translation reinterpretation, has been proposed (Melvill-Jones 1974; Young et al. 1984; Reschke et al. 1986; Reschke and Parker 1987). Other researchers investigated the role of muscle proprioceptive receptors in posture control showing adaptive sensory-motor responses appropriate for weightlessness (Watt et al. 1985, Roll et al. 1990).

Bryanov and colleagues found distinct decrements in dynamic tasks, such as locomotion and jumping, after spaceflight (1976). Normal gait control partially depends on preprogrammed patterns of muscle activation and continuous monitoring of external sensory input and internal signals. Anecdotal descriptions of astronaut locomotion postflight reveal abnormalities in walking, including adapting a wide stance during gait and difficulty in rounding corners (Homick and Reschke 1977). Chekirda et al. (1971) also reported a "stamping" gait, shift of the body toward the support leg and deviations from straight paths while walking on the first day following spaceflight. Bloomberg et al. (1997) observed alterations in head–trunk coordination during locomotion after spaceflight, contributing to postural and locomotion disturbances. Reschke et al. (1986) used the H-reflex to examine the effect of drops on the sensitivity of the lumbosacral motoneuron pool, which is presumably set by descending postural control signals. In related studies, Watt et al. (1986) reported that all subjects were unsteady postflight, and that one subject fell over backward consistently. H-reflex and tendon reflex (T-reflex) studies postflight indicate that both α and γ motor systems can be altered by spaceflight. Skylab crewmembers showed potentiation in T-reflexes after flight as compared to before (Baker et al. 1977) and Grigoriev and Yegorov (1990) have more recently reported that T-reflexes in crewmembers who spent up to 241 days aboard Mir show decreased thresholds and three-to four-fold increases in amplitude over preflight values. Postflight reflex potentiations support the notion of adaptation within the central nervous system in response to the altered stimulus provided by microgravity.

Impedance control techniques offer important insights into strategies for human motor control.

Impedance control is so named because it relates the dynamic relationship between force and position (Asada and Slotine 1986; Hogan et al. 1987), rather than controlling force and position separately, and takes as a premise that muscles possess spring-like properties. It also permits modulation of the desired apparent "stiffness" and "damping" as seen from the environment. Synergistic coactivation of antagonist muscles serves to modulate the mechanical impedance about a joint without imposing a net torque; increasing antagonist contractions about a joint yields greater stiffness and viscous damping. Grillner (1972) demonstrated that the earliest response of the lower limb muscles to the disturbances encountered in walking is due to the intrinsic properties of muscle, which can be modulated by appropriate coactivation strategies. Hogan et al. (1987) showed increased antagonist coactivation levels for unstable loads, as well as progressive increases in coactivation with increased frequency of perturbations. While studying jumping, McKinley and Pedotti (1992) found preimpact changes in lower-limb configuration for different landing surface compliances implying modulation of endpoint impedance to accommodate surface properties. In a study of space shuttle astronauts, McDonald et al. (1996) cited postflight changes in the phase-plane description of knee joint kinematics during gait as preliminary evidence for changes in joint impedance resulting from microgravity exposure. Musculoskeletal impedance control is most often implemented through feedback loops, but feedforward signals should also be included. Open-loop modulation of impedance is also important since open-loop behavior dominates at high frequencies, characterizing fast events such as jump landing or locomotion heel strike.

Various engineering models can be proposed to reflect the overall mechanics of jump dynamics (see also Chapter 13). The simplest jump landing model, a mass-spring model, accounts for purely vertical motion. The full body mass is concentrated at the center of mass (COM) and supported by a linear spring representing the stiffness of the legs (Figure 20.1A), resulting in a single degree of freedom, second order system in which damping is modeled by a linear velocity dependent element in parallel with the spring. McMahon and Cheng (1990) and Newman (1996) have used similar models to represent the mechanics of human running at

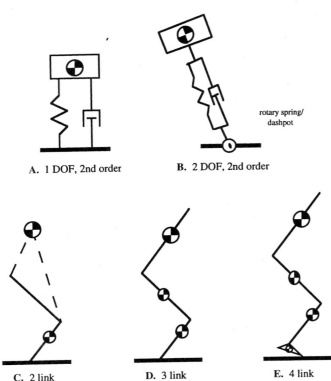

rotary spring/
dashpot

A. 1 DOF, 2nd order   **B.** 2 DOF, 2nd order

C. 2 link   **D.** 3 link   **E.** 4 link

1G and partial gravity, respectively. An enhanced model would incorporate an ankle joint, with its own stiffness and damping characteristics, to gain the horizontal degree of freedom (Figure 20.1B), allowing initial modeling of the fore-aft motion of the mass center, as well as a quantification of the relative dimensions of the stiffness ellipse in Cartesian space. A major shortcoming of both of these models is the constant leg stiffness. Evidence from Greene and McMahon (1979) shows that the leg stiffness decreases as the knee angle increases, due to decreased mechanical advantage. This argument also extends to the other joints (i.e., the impedance and mobility properties of the body are configuration dependent). A more complex approach would incorporate compound inverted pendulum models with rigid segments of fixed lengths and inertial properties connected by hinge joints. The first in this series of models has only two segments: a shank and a thigh, with the COM of the trunk, arms and head fixed relative to the thigh segment (Figure 20.1C). Next, a segment representing the head, arms, and trunk (HAT) is added with a hip joint between the HAT and thigh segments (Figure

20.1D). Because the suggested models are valid only if the foot is flat on the ground, the most complex model proposed here incorporates a fourth segment to describe the foot (Figure 20.1E). This model must take into account the discontinuity in the dynamics in the transition from heel-off to flat-footed stance, as the number of degrees of freedom changes from four to three.

2   Methods and Results

A study was performed to determine the effects of microgravity exposure on astronauts' control of posture and movement in a dynamic jumping task (Newman et al. in review). Astronaut subjects performed multiple jump landings with eyes open and closed from a 30-cm step after 7 to 14 day Space Shuttle missions. Full-body kinematic data was recorded using a video-based motion analysis system. The ankle, knee, and hip joint angles and velocities were computed using the positions of the markers at the toe, ankle, knee, hip and shoulder. A simple second order mass-spring model was used to estimate effective leg stiffness and damping values to complement the kinematic data.

The effect of exposure to an altered gravitational environment on control of lower limb impedance and preprogrammed motor strategies during jump landings was investigated. The joint kinematics of the lower extremity during the jump landings, as well as the kinematics of the whole-body mass center, were of particular interest. Phase-plane plots of hip, knee, and ankle angular velocities versus the corresponding joint angles were used to describe the lower limb kinematics during jump landings. The position of the full-body COM was also calculated in the sagittal plane using an 8 segment body model. The majority of subjects (5 of 9) exhibited expanded phase-plane portraits postflight, with significant increases in peak joint flexion angles and flexion rates following spaceflight (Figure 20.2, top). In contrast, 2 subjects showed significant contractions of their phase-plane portraits postflight (Figure 20.2, bot-

tom). Analysis of the vertical COM motion generally supported the joint angle results. Subjects with expanded joint angle phase-plane portraits postflight exhibited larger downward deviations of the COM and longer times from impact to peak deflection, as well as lower upward recovery velocities. Subjects with postflight joint angle phase-plane contraction demonstrated opposite effects in the COM motion. Computation of the settling time of the COM motion showed a trend in all subjects toward slower equilibration postflight. The joint kinematics results indicated the existence of two contrasting response modes due to microgravity exposure. Most subjects exhibited *compliant* impact absorption postflight, consistent with decreased limb stiffness and damping, and fewer subjects showed *stiff* behavior after spaceflight consistent with increased limb stiffness and damping. The changes appeared to result from adaptive modifications in the control of lower limb impedance.

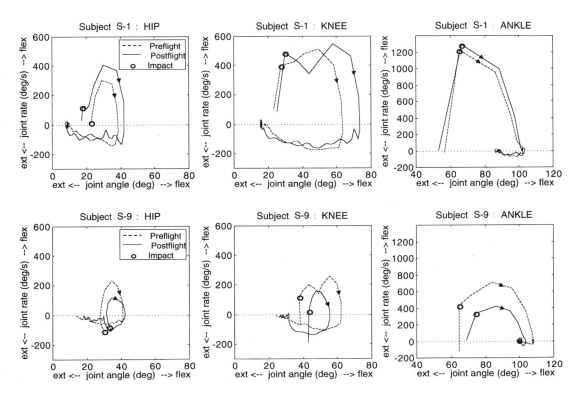

FIGURE 20.2. Comparison of preflight (dashed) and postflight (solid) joint angle phase plane portraits for hip, knee, and ankle. For subject 1 (upper plots), the postflight phase diagram is expanded with respect to the preflight diagram. In contrast, subject 9 (lower plots) demonstrates postflight contraction of the phase portrait in comparison to preflight results.

The lower leg musculature may be thought of as contributing a resistance to joint displacements, or stiffness (modeled as a torsional spring-like element), as well as a resistance to joint angular velocity, or damping (represented by a viscous damper or dashpot). These stiffness and damping elements represent the displacement and velocity dependent components of the joint impedance, respectively. Using this description, the *compliant* group exhibits postflight increases in both peak joint flexion and peak joint rates indicating the reduction in stiffness and damping about the joints following microgravity exposure. In these subjects, increases in joint flexion provide quantitative support for the reports of Watt et al.'s (1986) astronaut subjects that their legs were bending more during drop landings postflight. These changes are also consistent with reductions in the joint torques, and a reduction in the bandwidth of the postural control system as a whole. In contrast, the postflight *stiff* response after returning from spaceflight showed contracted phase-plane plots, indicating increases in limb stiffness and bandwidth of the postural controller.

A number of possible explanations exist for the observed changes in joint impedance during these jump landings including: loss of strength in the antigravity musculature, altered muscle stretch reflexes or vestibular feedback, and changes in modulation of limb stiffness. Because the stiffness and damping that can be exerted about a joint are directly related to the forces in the muscles acting about the joint, significant strength decreases in the antigravity muscles of the legs could well account for the *compliant* synergy. However, the *stiff* synergy exhibits postflight increases in stiffness and damping indicating *increased* joint torques; thus, the results from these subjects undermine the hypothesis that loss of strength alone can account for the observations in this study.

In addition to the muscular commands linked to the flight and impact phases of the jump, the underlying tonic activation in the leg musculature may contribute to the impedance in the lower limbs during jump landing. Clément et al. (1984) found an increase in tonic ankle flexor activity combined with a decrease in tonic extensor activity during spaceflight, that if carried over postflight could lead to a reduction in the stiffness about the ankle joint

against gravitational loads. It is well established that suppression of vestibular function results in depression of the γ_{static} innervation to the leg extensors, causing reduction in extensor tone (Molina-Negro et al. 1980). However, because relative enhancement of the knee flexor was not observed, Clément's group viewed the changes at the ankle as a "subject initiated postural strategy" rather than a functional deafferentation of the otoliths because of microgravity. Regardless of the origin, significant changes in leg muscle tone could well contribute to altered leg stiffness postflight.

Sensory feedback pathways also contribute to the stiffness and damping of the closed-loop postural control system. Feedback quantities that could play a role in the jump landings include postural muscle stretch (modulated through spinal reflexes), vestibular sensing of head orientation and angular velocity, and visual inputs. The stretch reflexes effectively increase the stiffness about the joints by recruiting additional muscle fibers to counteract perturbations to the muscle lengths; the stretch reflexes in concert with Golgi tendon organ force feedback probably serve to modulate the tension-length behavior (impedance) of the muscles. Gurfinkel (1994) reported decreases in the strength of the stretch reflex in tibialis anterior following spaceflight; Kozlovskaya et al. (1981) found amplitude reductions in Achilles tendon stretch reflexes after long-duration flight. Such decreases could have the effect of reducing the stiffness about the leg joints, and hence the stiffness of the leg "spring" supporting the body mass. However, Melvill-Jones, and Watt (1971a,b) demonstrated that the monosynaptic stretch response (occurring approximately 40 ms after forcible dorsiflexion of the foot) did not contribute to gastrocnemius muscle tension. Rather, the development of force was found to correspond to a sustained EMG burst with a latency of 120 ms following the dorsiflexion stimulus, that they termed the "functional stretch reflex." Because the peak joint angle deflections in the jump landing occur only 100 ms after impact, functional stretch reflex activity cannot account for the characteristics of the impact absorption phase.

The delays in the vestibular and visual feedback loops are of comparable or greater duration. Furthermore, the eyes were closed in half of the jumps, without a measurable effect on performance, indicating that vision's effect during the jump landings was minimal. Hence, sensory feedback information from these sources following impact cannot be expected to play a significant role over the jump impact absorption and early recovery phases of jump landings, although they probably contribute to reaching and maintaining the final equilibrium posture.

The limitations on the sensory feedback pathways indicate that the stiffness properties of the lower limbs may be largely predetermined before impact. The stiffness about the joints (and hence the overall leg springiness) is determined by the level of muscle activation, and the overall impedance of the leg to COM motion is also affected by the configuration of the limbs at impact. McKinley and Pedotti (1992) found that the knee extensor muscles (rectus femoris and vastus lateralis) were activated slightly before impact, while the ankle plantarflexors (gastrocnemius and soleus) were continuously active from midflight during jumps. Other investigators (Dyhre-Poulsen and Laursen 1984; Thompson and McKinley 1988) have determined that the timing of the preparatory muscle activation and limb configuration is keyed to the expected time of impact. McNitt-Gray (1993) examined 1 G jump landings and suggested that a common landing strategy was used for different height jumps that was progressively adjusted by selectively changing joint moment gains. For downward stepping and repetitive hopping, Melvill-Jones and Watt (1971b) found that muscular activity commenced from 80 to 140 ms prior to ground contact, and concluded that the deceleration associated with landing was the result of a preprogrammed neuromuscular activity pattern rather than stretch reflex action.

Comparison of the pre- and postflight fits of the mass-spring model indicate that variations in model parameters can adequately predict the alterations in COM motion seen in astronaut jump landings following spaceflight. More specifically, changes in the leg stiffness alone appear to govern the differences in transient response observed upon return to earth. The postflight decreases and increases in the vertical leg stiffness found for these subjects correspond exactly to the classifications of postflight *compliant* and postflight *stiff* made previously on the basis of joint angle kinematics. In the model, decreases in leg stiffness lead to decreases in bandwidth, with slower and less oscillatory time responses. In contrast, increased stiffness results in faster, higher bandwidth performance with greater overshoots. Interestingly, the model fits did not show changes in the leg damping to play a significant role in the postflight differences. This is counterintuitive, because an increase in antagonist muscle activation to raise the limb stiffness might be expected to cause a corresponding increase in the mechanical damping properties of the muscles as well. This leads us to believe that the damping properties of the limbs can be modulated independently of the stiffness, or simply that the damping characteristics are largely constant in the face of large changes in leg stiffness.

3 Summary

In summary, the postflight changes in the kinematics of astronaut jump landings reported here have been attributed to changes in the control of the lower limb impedance due to exposure to the microgravity conditions of spaceflight. The decreased stiffness of the posture control system observed in the *compliant* group of subjects may reflect inflight adaptation to the reduced requirements for posture control in the absence of gravitational forces. On the ground, the nature of the body's compound inverted pendulum structure requires the maintenance of a certain minimum stiffness for stability in an upright position. In space, the body need not be stabilized against gravity, and the control bandwidth and stiffness may therefore be reduced without compromising postural stability. Inflight, an overall reduction in postural stiffness may be observed as reduction in extensor tone and decreases in stretch reflex gain, and may be related to the loss of drop-induced H-reflex potentiation. Compliant postflight behavior may result from a residual decrement in the stiffness of the postural control system following return to earth. In contrast, by using a stiffening strategy postflight, subjects minimize deviations from equilibrium to avoid approaching the boundaries of the cone of stability. Such stiffening in turn requires a commensurate increase in postural control bandwidth.

These observations indicate that the impedance characteristics of the lower limbs are prescribed during the flight phase of the jump, because activation of the leg muscles effectively sets the stiffness about the leg joints prior to impact. Therefore, the changes in joint and mass center kinematics observed in the astronauts postflight were likely due to changes in the preprogrammed muscle activity prior to impact, which sets the limb impedance in an open-loop fashion by controlling the muscle tension-length properties and the limb configuration. The changes observed in the impact absorption phase support the notion that spaceflight contributed to altered neuromuscular activity during the flight phase of the jump, even though EMG records were not available. The presumed alterations in muscle activation patterns following spaceflight could reflect changes in the relative recruitment of antagonist muscles, or differences in the timing of activation (e.g., failure to activate antigravity muscles early enough during the flight phase to stiffen the limbs for impact). The reduced requirements for maintenance of posture under microgravity conditions probably contribute to the changes seen postflight, in concert with decrements in limb proprioception and altered interpretation of otolith acceleration cues.

4 Future Directions

Future research directions include investigating changes in jump landings under different musculoskeletal loading conditions (either unloaded using a suspension system or weighted down) as 1 G analogs to spaceflight adaptation and return to earth-normal gravity. Additional jumping experiments will explore control of limb trajectories and interaction with the landing surface. For example, unrestrained limb trajectories and preprogrammed synergies can be investigated using false platform jumps where the expected landing surface is removed. A comprehensive modeling effort would include physiological models of motor control in addition to the models proposed here to assess the overall dynamics of the task. The following references are recommended (Feldman, 1966; Stark, 1968; Hatze, 1978; Houk and Rymer, 1981; Loeb and Levine, 1990; and Winters, 1995). We hope that new insights from spaceflight data can help

lead to theories that apply across the continuum of gravity to assess not only astronaut performance, but altered posture and movement synergies at 1 G.

Acknowledgments. This work is supported by NASA Grant NGT-51228, NASA Contract NAS1-18690, and NASA NAGW 4336.

References

Allum, J. and Honegger, F. (1992). A postural model of balance-correcting movement strategies, *J. Vest. Res.*, 2:323–347.

Allum, J. and Pfaltz, C. (1985). Visual and vestibular contributions to pitch sway stabilization in the ankle muscles of normals and patients with bilateral peripheral vestibular deficits, *Exp. Brain Res.*, 58:82–94.

Asada, H. and Slotine, J.-J. (1986). *Robot Anal. Cont.*, John Wiley & Sons, New York.

Barker, J.T., et al. (1977). Changes in the achilles tendon reflexes following Skylab missions. In *Biomedical Results from Skylab* (NASA SP-377). Johnston, R.S., Dietlein, L.F., and Washington, D.C. (eds.), pp. 131–135, U.S. Government Printing Office.

Bloomberg, J.J., Peters, B.T., Smith, S.L., Huebner, W.P., and Reschke, M.F. (1997). Locomotor head-trunk coordination strategies following space flight. *J. Vest. Res.*, 7:161–177.

Chekirda, I.F., Bogdashevskiy, A.V., Yeremin, A.V., Kolosov, I.A. (1971). Coordination structure of walking of Soyuz-9 crew members before and after flight. *Kosmicheskaya Biologiya i Meditsina*, 5(6):48–52.

Clément, G., Gurfinkel, V.S., Lestienne, F., Lipshits, M.I., and Popov, K.E. (1984). Adaptation of postural control to weightlessness. *Exp. Brain Res.*, 57:61–72.

Dyhre-Poulsen, P., Laursen, A.M. (1984). Programmed electromyographic activity and negative incremental muscle stiffness in monkeys jumping down. *J. Physiol.*, 350:121–136.

Feldman, A.G. (1966). Functional tuning of the nervous system during control of movement or maintenance of a steady posture. II. Controllable parameters of muscle. *Biophysiology*, 11:565–578.

Greene, P.R. and McMahon, T.A. (1979). Reflex stiffness of man's anti-gravity muscles during kneebends while carrying extra weights. *J. Biomech.*, 12:881–891.

Grigoriev, A.I. and Yegorov, A.D. (eds.) (1990). Preliminary medical results of the 180-day flight of prime crew 6 on space station Mir. Fourth meeting of the US/USSR Joint Working Group on Space Biology and Medicine, San Francisco, California.

Gurfinkel, V. (1994). The mechanisms of postural regulation in man. *Sov. Sci. Rev. Sect. F: Physiol. Gen. Biol. Rev.*, 7(5):59–89.

Hatze, H. (1978). A general myocybernetic control model of skeletal muscle. *Biol. Cybern.*, 28:143–157.

Hogan, N. (1987). Stable execution of contact tasks using impedance control. *IEEE Int. Conf. Robotics Auto.*, pp. 1047–1054.

Homick, J.L. and Reschke, M.F. (1977). Postural equilibrium following exposure to weightless space flight. *Acta Otolaryngologica.*, 83:455–464.

Houk, J.C. and Rymer, W.Z. (1981). Neural control of muscle length and tensions. In *Handbook of Physiology-The Nervous System II*, Vol. 8, Kandel, E.R. (ed.), pp. 257–323.

Kenyon, R.V. and Young, L.R. (1986). M.I.T./Canadian vestibular experiments on the Spacelab-1 mission: 5. Postural responses following exposure to weightlessness. *Exp. Brain Res.*, 64:335–346.

Kozlovskaya, I.B., Kriendich, Y.V., Oganov, V.S., and Koserenko, O.P. (1981). Pathophysiology of motor functions in prolonged manned space flights. *Acta. Astronautica.*, 8(9–10):1059–1072.

Loeb, G.E. and Levine, W. (1990). Linking musculoskeletal mechanics to sensorimotor neurophysiology. In *Multiple Muscle Systems: Biomechanics and Movement Organization*. Winters, J.M. and Woo, S.L.-Y. (eds.), pp. 165–181. Springer-Verlag, New York.

McDonald, P.V., Basdogan, C., Bloomberg, J.J., and Layne, C.S. (1996). Lower limb kinematics during treadmill walking following space flight: implications for gaze stabilization. *Exp. Brain. Res.*, 112:325–334.

McKinley, P. and Pedotti, A. (1992). Motor strategies in landing from a jump: the role of skill in task execution. *Exp. Brain Res.*, 90:427–440.

McMahon, T.A. and Cheng, G.C. (1990). The mechanics of running: how does stiffness couple with speed? *J. Biomech.*, 23(Suppl. 1):65–78.

McNitt-Gray, J. (1993). Kinetics of the lower extremities during drop landings from three heights. *J. Biomech.*, 26(9):1037–1046.

Melvill-Jones, G. (1974). Adaptive neurobiology in space flight. In *Proceedings of the Skylab Life Sciences Symposium* (JSC-09275). Lyndon B. Johnson Space Center, Houston, Texas.

Melvill-Jones, G. and Watt, D.G.D. (1971a). Observations on the control of stepping and hopping movements in man. *J. Physiol.*, 219:709–727.

Melvill-Jones, G. and Watt, D.G.D. (1971b). Muscular control of landing from unexpected falls in man. *J. Physiol.*, 219:729–737.

Molina-Negro, P., Bertrand, R.A., Martin, E., and Gioani, Y. (1980). The role of the vestibular system in relation to muscle tone and postural reflexes in man. *Acta. Otolaryngologica.*, 89:524–533.

Mouncastle, V.B. (1979). *The Mindful Brain*. Edelman, G.M. and Mouncastle, V.B. (eds.), pp. 7–50. The MIT Press, Cambridge, Massachusetts.

Nashner, L. and G. McCollum (1985). The organization of human postural movements: a formal basis and experimental synthesis. *Behav. Brain Sci.*, 8:135–172.

Newman, D.J. (1996). Modeling reduced gravity human locomotion. In Advances in mathematical modeling of biological processes. Kirschner, D. (ed.), *Int'l. J. Appl. Sci. Comput.*, 3(1):91–101.

Paillard, J. (1988). *Posture and Gait: Development, adaptation and modulation*. Amblard, B., Berthoz, A., and Clarac, F. (eds.), Elsevier Science Publishers, Amsterdam, The Netherlands.

Paloski, W. et al. (1993). Vestibular ataxia following shuttle flights: effects of transient microgravity on otolith-mediated sensorimotor control of posture. *Am. J. Otol.*, 14(1).

Parker, D.E. et al. (1985). Otolith tilt-translation reinterpretation following prolonged weightlessness: implications for preflight training. *Aviat. Space Environ. Med.*, 56:601–606.

Reschke, M.F. and Parker, D.E. (1987). Effects of prolonged weightlessness on self-motion perception and eye movements evoked by roll and pitch. *Aviat. Space Environ. Med.*, 58:A153–158.

Reschke, M.F., Anderson, D.J., and Homick, J.L. (1986). Vestibulo-spinal response modification as determined with the H-reflex during the Spacelab-1 flight. *Exp. Brain Res.*, 64:367–379.

Roll, J.P. et al. (1990). Body proprioceptive references in weightlessness as studied by nuscle tendon vibration. In *Proceedings of the Fourth European Symposium on Life Sciences Research in Space* (ESA SP-307). Trieste, Italy, May 28–June 1, pp. 43–48.

Sherrinton, C.S. (1906). The integrative action of the nervous system. Constable, London.

Stark, L. (1968). *Neurological Control Systems*. Plenum Press, New York.

Thompson, H.W. and McKinley, P.A. (1988). Effect of visual perturbations in programming landing from a jump in humans. *Soc. Neurosci. Abst.*, 14:66.

Watt, D.G.D., Money, K.E., and Tomi, L.M. (1986). M.I.T./Canadian vestibular experiments on the Spacelab-1 mission: 3. Effects of prolonged weightlessness on a human otolith-spinal reflex. *Exp. Brain Res.*, 64:308–315.

Watt, D.G.D., Money, K.E., Bondar, R L., Thirsk, R.B., Garneau, M., and Scully-Power, P. (1985). Canadian Medical Experiments on Shuttle Flight 41-G. *Can. Aero. Space J.*, 31(3):215–226.

Winters, J.M. (1995). An improved muscle-reflex actuator for use in large-scale neuromusculoskeletal models. *Ann. Biomed. Eng.*, 23:359–374.

Young, L., Oman, C., Watt, D.G.D., Money, K.E., and Lichtenberg, B.K. (1984). Spatial orientation in weightlessness and readaptation to earth's gravity. *Science*, 225:205–108.

Young, L., Oman, C., Watt, D., Money, K., Lichtenberg, B., Kenyon, R., and Arrott, A. (1986). M.I.T./Canadian vestibular experiments on the Spacelab-1 mission: 1. Sensory adaptation to weightlessness and readaptation to one-g: An overview. *Exp. Brain Res.*, 64:291–298.

Commentary:
Altered Astronaut Performance Following Spaceflight—Control and Modeling Insights

Guido Baroni, Giancarlo Ferrigno, and Antonio Pedotti

Understanding of the mechanism of functional sensorimotor integration made by the CNS remains a fascinating challenge. Despite fifty years having gone by since N.A. Bernstein formulated the theory of the "Physiology of Activity" (e.g., Bernstein 1967), the modernity of many of his assumptions and the presence in the scientific community of detractors and supporters of his thoughts, testify how far we still are from a unifying and irrefutable theory.

Experimental activity in microgravity plays a fundamental role in the frame of motor control investigation. Weightlessness represents an extremely "clean" condition, which cancels all the information provided by those sensory systems, which have their function dependent by the gravity vector. Subjects in the highly altered microgravity environment are subjected to a disabling condition, which requires adaptation processes and the re-calibration of many sensorimotor mechanisms (Massion 1992; Massion et al. 1993).

The quantitative evaluation of a motor task performed by subjects after microgravity exposure normally provides the investigator with evidence of alteration in subjects motor behavior. The interpretation of such evidences is however often a hard challenge. In this case, dynamic jumping task is under discussion. It is not of doubt that jump landing is a complex task, with feed-forward and open loop control mechanisms activated in preparation of ground contact, regulated by anticipatory evaluation of jump height, landing surface features and subject training condition. Ground based studies have pointed out both *compliant* and *stiff* landing strategies, which do not always have consistent correlation with jump height, training, and surface characteristics. Jump landing strategy seems regulated by the trade-off between the biomechanical minimization of loads on lower limbs joints and requirements of equilibrium recovery and maintenance immediately after landing (McKinley et al. 1992). The motor task "simplicity," in the sense intended by Bernstein, is in this case not uniquely identifiable and is probably subjectively established (Bernstein 1967).

A mechanical model is presented by Newman and Jackson for the evaluation of subject adopted strategy for jump landing. The attribution to a specific class represents the starting point for understanding the role that the process of adaptation to microgravity plays for the specific subject's performance after reentry on Earth. Recalibration of the posture regulation mechanism, involving neuromuscular modification of the activity of posture muscle groups, may not be recovered in short time after flight and could explain a reduced subject's skill in counterbalancing gravity loads. EMG evaluation (Clement et al. 1984) are confirmed by the results of human movement analysis recently performed during a 179–days ESA Mission on board the Russian Space Station MIR (Baroni et al. 1998). Having as reference the configuration in weightlessness of lower limbs of a subject freely floating (neutral body posture) (Andreoni et al. 1996), the three-dimensional analysis of one subject's erect posture along a period in microgravity of 113 days shows a progressive recovery of an upright position similar to the one assumed on ground (Baroni et al. 1999).

This is probably obtained by elaborating diversified activation patterns of ankle flexors and extensors, in the frame of a voluntary strategy for the optimization of the whole body dynamic postural control. The repeatability of a comparable erect posture after the nineteenth flight day in both investigated experimental conditions (eyes open and eyes closed) (Baroni et al. 1999) may suggest that the above cited balance strategy assumes the character of a systematic posture regulation mechanism, able to influence the tonic activity of ankle flexor and extensor muscle groups.

Neuromuscular modification is accompanied by reduction of gravity dependent sensory pathways, strongly involved in jump landing performance. If motor learning is considered associated with a process of freeing degrees of freedom (Vereijken 1996), the *stiff* strategy could be interpreted as a consequence of sensory degeneration. Immediately after flight, the subject performs the landing task with reduced or wrongly calibrated information due to from altered sensory pathways (see also Chapter 22). An exaggerated fixation of relationship between joints, with the consequent limitation of joints range of motion can be seen as a way to minimize the disturbances that the voluntary motor task implies, aiming to the essential equilibrium maintenance.

Observed contradictory motor behavior among subjects, and the proposed interpretation of experimental results, underline that jump landing represents one of the many open questions in the field of motor control.

We believe that experimental activity in microgravity represents a basic requirement for our better understanding of human motor control. It is our opinion that many scientists wish to include human motion analysis during many space missions in the International Space Station (ISS) era. The implementation of a reliable permanent motion analysis facility on board of ISS is the necessary premise and is therefore a fascinating technical and scientific challenge. The systematic collection of kinematics data relative to subjects acting within a space module, would allow a more complete characterization of human motor strategies and postural behavior in weightlessness, enriching our knowledge in the field of movement physiology and improving the design of manned orbital working environment.

References

Andreoni, G., Ferrigno, G., Baroni, G., Colford, N.A., Bracciaferri, F. and Pedotti A. (1996). Postural modification in microgravity. In *Proceedings VIth European Symposium on Life Sciences Research in Space*, pp. 50–103. Trondheim, Norway.

Baroni, G., Ferrigno, G., Anolli, A., Andreoni, G., and Pedotti, A. (1998). Quantitative analysis of motion control in long-term microgravity. *Acta Astronautica*, 43(6):131–151.

Baroni, G., Ferrigno, G., Rabuffetti, M., Pedotti, A., and Massion, J. (1999). Long-term adaptation of postural control in microgravity. *Exp. Brain Res.* 128(3):410–416.

Bernstein, N.A. (1967). The Co-ordination and Regulation of Movement. Pergamon Press, Oxford.

Clement, G., Gurinkel, V.S., Lestienne, F., Lipshits, M.L., and Popov, K.E. (1984). Adaptation of postural control to weightlessness. *Exp. Brain Res.*, 57:61–72.

Massion, J. (1992). Movement, posture and equilibrium: interaction and co-ordination. *Prog. Neurobiol.*, 38:35–56.

Massion, J., Gurfinkel, V.S., Lipshits, M., Obadia, A., and Popov, K. (1993). Axial synergies under microgravity conditions. *J. Vestibular Res.*, 3:275–287.

McKinley, P. and Pedotti, A. (1992). Motor strategies in landing from a jump: the role of skill in task execution. *Exp. Brain Res.*, 90:427–440.

Vereijken, B. (1996). Free(z)ing degrees of freedom in motor development. In *Proceeding of International Conference Berntein's Tradition in Motor Control*, p. 134. Penn State University, Pennsylvania.

21
Adaptive Sensory-Motor Processes Disturb Balance Control After Spaceflight

William H. Paloski

1 Introduction

Balance control is a fundamental task performed by the central nervous system (CNS) on a continuous, (normally) subconscious basis, beginning when, as infants, we learn to hold our heads erect. Virtually every voluntary body movement we make, from running and jumping during a sporting event to reaching for a pencil while sitting at a desk, is accompanied by an automatic set of muscle activations that are commanded by the CNS and designed to keep the focal movement from disrupting the postural equilibrium of the body. One of the most important and well-studied balance control tasks is the maintenance of stable upright stance. Patients suffering from degradation of this function have significantly reduced abilities to perform normal activities of daily living.

To control balance during upright stance, the CNS uses state feedback information provided by the visual, vestibular, and proprioceptive sensory organs (Figure 21.1). The receptive fields of these sensors overlap, providing some level of redundancy; however, loss of one of the sense organs, through a disease process or traumatic injury, can not be completely compensated for by the remaining receptors. Nevertheless, the CNS is plastic—it has elaborate mechanisms for adapting to sustained changes in sensory function, motor function, or environmental cues or constraints. Understanding these mechanisms is crucial to designing effective rehabilitative physical therapy regimes for patients and effective countermeasures for returning astronauts.

The mechanisms underlying adaptive changes in balance control are not well understood at present. Most of what is known has been obtained from evaluation of patients with irreversible loss of sensory or motor function. These observations have generally been made after compensation was already complete, so details of the adaptation process were lost. Also, with the possible exception of the recent ground-based reports by Gordon et al. (1995) and Melvill Jones et al. (1996), observations of the recovery of balance control in astronauts following space flight have provided the only available information on the responses of the balance control system in normal human subjects during the CNS adaptation process (e.g., Homick and Reschke 1977; Kozlovskaya et al. 1981; Anderson et al. 1984; Clément et al. 1984; Kenyon and Young 1986; Paloski et al. 1992, 1994; Collins et al. 1995; see also Chapter 20).

Upon insertion into orbit space flight, an astronaut experiences sudden loss of the gravitational stimuli transduced by the vestibular otolith organs and thought to be used by the CNS as the primary vertical spatial reference for controlling upright stance. Over time, the astronaut's CNS adapts to the zero-g environment, compensating for the lost spatial reference. Unfortunately, upon return to Earth, the astronaut's CNS is no longer capable of using the gravity-mediated otolith inputs, and balance control is disrupted by these confusing sensory afferent signals until the CNS again adapts to the new environmental cues.

Studies of the characteristics of balance dysfunction during the immediate postflight (re)adaptation period have led to new insights into the CNS

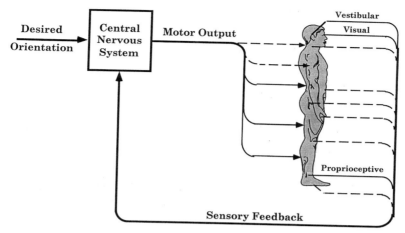

FIGURE 21.1. Sensory-motor control of balance. The central nervous system (CNS) estimates the current state of the body's orientation from sensory feedback provided primarily by the visual, vestibular, and proprioceptive systems. Motor output commands are generated and adjusted by the CNS to minimize differences between desired and estimated body orientations. Desired orientation commands presumably originate in higher level brain centers.

adaptive process. This chapter presents a conceptual model for this adaptive process. Then, based on predictions of the model, results from a simple pre- and postflight experiment that led to surprising insight into balance control limitations during the adaptive process is presented. Finally, the experimental results are discussed and possible future directions suggested.

2 Conceptual Model of CNS Adaptation

A conceptual model of the CNS adaptation to changes in sensory, motor, or environmental characteristics is presented in Figures 21.2 and 21.3. The general control scheme is described in Figure

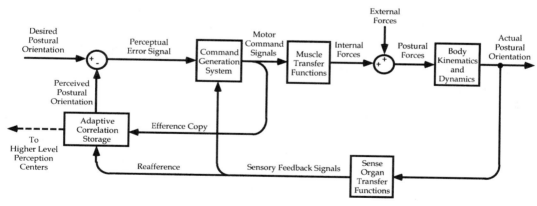

FIGURE 21.2. Adaptive control scheme for maintaining balance. Feedback control of upright postural stability depends on an Adaptive Correlation Storage block in which context-dependent internal models of the sensory, motor, and environmental characteristics are stored. A copy of the multidimensional efferent motor command signal triggers a stored memory (internal model) that is used to predict the multi- dimensional, spatiotemporal sensory afferent information expected to be received (reafference) in response to the commanded motor activity. Comparisons between the expected and actual reafference form postural orientation perceptions that are used to modify the motor command generation and/or inform higher level centers of significant or sustained incongruences between expected and actual reafference.

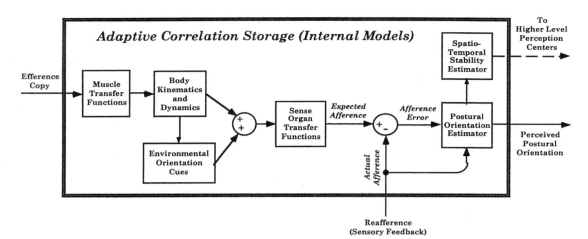

FIGURE 21.3. Detailed scheme for Adaptive Correlation Storage block. Internal models of the motor, biomechanical, sensory, and environmental characteristics are used to predict the reafference expected to be received as a result of the efferent motor command. Differences between the expected and actual afferent signals are used to estimate the true (perceived) postural orientation and the relative postural stability.

21.2, and the contents of the Adaptive Correlation Storage block are elaborated on in Figure 21.3. This conceptual model is based on the efference copy hypothesis proposed by von Holst and Mittlestadt (1950), as modified by Hein and Held (1967). Adaptive responses to persistent changes in sensory, motor, or environmental characteristics are thought to be driven by higher level perception centers and realized through trial and error by altering the internal models to provide predicted reafferent signals that are congruent with the actual reafference received following a voluntary motor command.

This scheme can be used to explain how the loss of gravity-mediated sensory inputs during space flight would disrupt balance control after flight. On-orbit, the sensory input signals resulting from a voluntary body movement would initially be incongruent with those expected to be received based on the terrestrial internal models. Because the incongruence would be sustained, however, adaptive changes would be triggered by higher level centers, leading, through trial and error, to development of new efference copy models optimized for the microgravity environment. These zero-g internal models would not expect sustained changes in vestibular otolith inputs to result from head tilts as would the terrestrial internal models. Immediately after space flight, terrestrial gravity-mediated sensory inputs would return. Initially upon return, the

CNS would continue to use the microgravity efference copy models developed during flight; however, the unexpected afferent signals caused by gravitational stimulation would disrupt sensory-motor control and trigger readaptive modification, leading eventually to a reemergence of the original, terrestrial efference copy models. Thus, while in-flight adaptive changes in the CNS processing of sensory information may optimize central neural control of coordinated body movements in microgravity, they may also maladapt the astronaut for control of coordinated body movements immediately after return to Earth (Reschke et al. 1994; see also Chapter 20).

3   Experimental Methods and Results

The conceptual model suggests that sustained afference errors (Figure 21.3) cause neurobehavioral changes aimed at identifying the source of the error and appropriately adjusting the internal models to minimize it. If true, then new biomechanical strategies may emerge during the adaptation period following a change in sensory, motor, or environmental characteristics. To determine whether postural biomechanics are affected by adaptive neural control, changes in postural sway strategies and head-trunk coordination strategies were examined during the readaptation process following orbital space flight using a simple set of motor control experiments performed by astronauts before flight and on landing day.

The experimental stimuli were administered using a modified computerized dynamic posturography system (Equitest, NeuroCom, International, Clackamas, OR). During each test session, the subject, in stocking feet, stood upright on the force-plate of the posturography system with arms folded across the chest and feet "a comfortable distance apart." The subject was directed to "maintain stable upright posture throughout the test protocol," and was then presented with three sequential trials of backward support surface translation followed by five sequential trials of toes-up support surface rotation, three sequential trials of forward support surface translation, and five sequential trials of toes-down support surface rotation. These types of support surface perturbations have often been used to study postural coordination (e.g., see Chapter 19). Data from the translation trials were analyzed to assess changes in postural sway strategy and head–trunk coordination strategy associated with space flight. Each translation trial lasted for 2.5 seconds. During the first 500 m/sec of the trial, the support surface remained stationary. Then, over the next 500 m/sec, the support surface was translated at a rate of approximately 15 cm/sec. For the final 1500 m/sec, the support surface again remained stationary. Between trials the support surface was slowly translated back to its initial position. The time between trials was normally 3 to 5 sec.

The support surface comprised two foot plates supported by four force transducers which independently sensed the anterior and posterior normal forces by each foot. Throughout each test session, the subject wore a lightweight safety harness and a headset through which continuous binaural acoustic noise was provided to mask auditory spatial orientation cues. Anteroposterior sway excursions of the hip (greater trochanter) and shoulder (C-7/T-1 joint) were monitored by potentiometers attached to the subject using lightweight wooden dowels (sway bars), and the sagittal (pitch) plane angular velocity of the subject's head was monitored using an angular rate sensor (Watson Industries, Inc., Eau Claire, WI) attached to the subject's head set. Outputs from the force transducers, sway bar potentiometers, and angular rate sensor were amplified, digitized at a rate of 103 Hz, and stored

on a Macintosh computer. Segmental body motions were subsequently estimated using a link-segment biomechanical model (Nicholas et al. 1996).

Immediately after flight, the sway responses to the translational perturbations were exaggerated relative to preflight in all subjects. The center of pressure and hip sway trajectories were generally more labile, or underdamped, after flight than before flight, and the learning associated with successive sequential perturbations disappeared in some subjects on landing day. Before flight, specific head–trunk coordination strategies were not systematically observed. After flight, most subjects attempted to minimize head movements with respect to trunk movements during their active response to the support surface translation; however, some subjects attempted to minimize head movements with respect to space.

Results from three sequential forward translation trials for a typical subject are presented in Figure 21.4. Before flight (Figure 21.4: upper left and lower left panels), the changes in hip and ankle flexion angles as well as head and trunk angles (re: space) were limited to less than 5 deg, and little correlation was noted between hip and ankle flexion angles or between head and trunk angles. Thus, no stereotyped postural sway strategy or head–trunk coordination strategy was evident. After flight, the hip flexion response was greatly increased and was negatively correlated with ankle flexion (Figure 21.4: lower right panel), suggesting that the subject was employing the hip sway strategy described by Horak and Nashner (1986) and Nashner et al. (1989) (see also Chapter 19). Surprisingly, however, unlike the classic hip sway strategy, the subject's head was not stabilized in space by counter-rotation at the neck. Instead, the head angular motion also increased and appeared to be directly correlated with trunk angular motion (Figure 21.4: upper right panel), suggesting a head-on-trunk locking strategy similar to that observed by Bloomberg et al. (1996) during postflight locomotion studies in astronauts.

4   Discussion

It has been previously shown that changes in the ability to process vestibular information compromise postural stability control in astronauts immediately after space flight (Paloski et al. 1992, 1994).

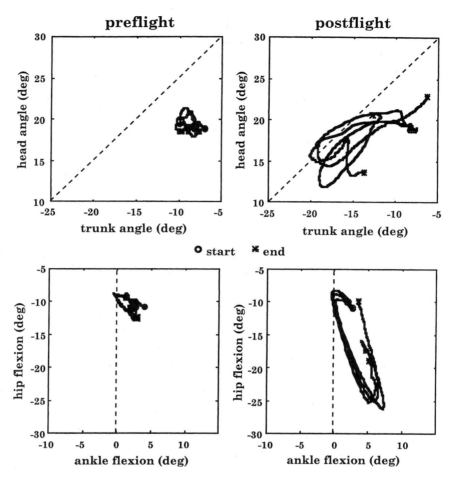

FIGURE 21.4. Response to sudden translations of the support surface before and after space flight. Three sequential perturbations were provided by driving the support surface forward for 500 msec at a rate of approximately 15 cm/sec.

Head and trunk angles were measured with respect to a fixed spatial reference. Hip and ankle flexion angles were defined relative to an idealized body geometry. Negative flexion angles represent extension of the joint.

To better understand the mechanisms underlying this postflight postural ataxia, this study examined the hypothesis that postflight postural biomechanics are affected by adopted head–trunk coordination strategies aimed at minimizing confusing vestibular inputs. This idea first arose from observations of the paradoxical postural behaviors exhibited by crew members immediately after return from space. During quiet stance, those most affected by neurosensory adaptation to microgravity adopted an obvious, seemingly conscious strategy of stabilizing their heads with respect to their trunks, often going so far as to turn their whole bodies (rather than just their heads) to address

speakers at their sides. Paradoxically, these same subjects responded to sudden base-of-support perturbations with exaggerated, underdamped postural responses similar to those observed in cerebellar patients. Subjective observations suggested that these subjects employed both of the head–trunk coordination strategies described by Nashner (1985): during quiet and/or controlled postural activities they appeared to employ the "strapped-down" strategy, while during postural activities requiring more dynamic center-of-mass movements they appeared to employ the "stable-platform" strategy. It was postulated that these paradoxical postural behaviors would result from the reacquisition of graviceptor inputs following space flight, and that specific head–trunk coordination strategies might be adopted to minimize confusing vestibular inputs (see also Chapter 22).

Berthoz and Pozzo (1988) hypothesized that the brain uses a "top down" approach to the control of postural and gait stability. They suggested that a primary constraint on the posture control problem is to stabilize the head relative to the Earth-vertical in order to ensure gaze stability and maintain visual acuity. This novel concept was studied by Reschke et al. (1986) during the drop and landing phases of a Hoffmann reflex experiment flown aboard the Spacelab-1 space shuttle mission. Results from this investigation support the hypothesis that locomotor and postural instabilities experienced by astronauts upon return to Earth are caused by in-flight adaptive acquisition of new top down motor strategies designed to maintain head and gaze stability during body movement in microgravity. The authors concluded that novel and potentially unstable gait strategies may be adopted postflight in an attempt to maintain head stability in the face of conflicting sensory cues during the period of sensory recalibration on reexposure to the terrestrial gravitational environment. More recently, Bloomberg et al. (1996) investigated the effects of space flight on head and gaze stability during locomotion by having astronaut subjects walk and run on a motorized treadmill while visually fixating a stationary target positioned in the center of view. They found significant alterations in head movement control during locomotion following space flight and concluded that these alterations may alter descending control of locomotor function.

Landing day results from the present study showed that most subjects attempted to minimize head movements with respect to trunk movements (strapped-down strategy), often resulting in exaggerated head movements in space. On the other hand, some subjects attempted to minimize head movements with respect to space (stable-platform strategy), often resulting in exaggerated hip motions. Both head-trunk coordination strategies were also observed during a sensory organization test battery performed at the same time as the postflight motor control test battery (Paloski et al. 1992, 1994). Indeed, during the sensory organization testing, some subjects appeared to switch from the strapped-down to the stable-platform strategy within a single 20-second trial.

These findings support the notion that balance control after space flight is driven by a top down program; however, in contrast to Berthoz and Pozzo (1988), the top-down program may be organized around minimizing confounding vestibular inputs rather than stabilizing vision. Because the orientation of the head remains fixed with respect to the torso in the strapped-down strategy observed in many of the subjects in the study, it may have been employed to minimize conflicts between neck proprioceptive and vestibular otolith sensory inputs. Alternatively, it may have been employed to simplify the segmental motor control task by reducing the number of independent body segments to be controlled. A drawback of this strategy is that it would have also complicated the sensory task required to separate gravity from linear head accelerations. The stable-platform strategy observed in some of the subjects in the study may have been employed to simplify the sensory task of separating gravity from linear accelerations by maintaining constant the gravitational component of otolith stimulation. A drawback of this strategy is that it would also have increased the complexity of the segmental motor control task.

As predicted by the conceptual model, it appears that gravity induced otolith inputs are confusing to the posture control system immediately after space flight, and that either to minimize the confusion caused by these inputs or to test the source of the afference error, many of the returning astronauts tried to simultaneously adopt both a stable-platform and a strapped-down head-trunk coordination strategy. Because this was impossible, the subjects were forced to choose one or the other of the strategies. This demonstrates that postflight postural instabilities may result, in part, from new constraints on biomechanical movement caused by the central nervous system adopting strategies designed to aid the adaptation process.

5   Future Directions

Adaptive control models that accurately describe neuro-behavioral responses to altered sensory, motor, or environmental constraints are needed to aid our understanding of the responses to be expected when astronauts, high-performance aircraft pilots, and underwater explorers venture into and return from their unusual working environments. Such models would also be useful clinically to predict responses that patients with sensory-motor pathological conditions may have to various environments, and to guide physical therapists in designing rehabilitation programs to aid CNS adaptation (compensation) in their patients. Such models might also be useful to the robotics industry, in which, as evidenced by the failure of Dante to negotiate the interior of an active volcano crater in Alaska, stable locomotion in ambulatory robots seems to be limited by poor adaptive balance control algorithms.

The conceptual model presented in this chapter may be a good starting point for more robust models in the future, but it is only a conceptual model at this point, and, therefore, it needs to be further validated both experimentally and mathematically. Fortunately, mathematical validation has already begun. Kuo (1995) has developed an optimal control model that mimics certain of the characteristics of the conceptual model. Whether linear control theory is robust enough to capture the salient adaptive characteristics of the human balance control system is not yet known (an alternative conceptual framework is presented in Chapter 22). Further iterations between experimental validation of the model predictions and model modifications to fit the experimental results will be required to answer such questions. Should linear methods be incapable of modelling the system, nonlinear control models and/or neural network approaches may be required.

References

Anderson, D.J., Reschke, M.F., Homick, J.L., and Werness, S.A.S. (1986). Dynamic posture analysis of Spacelab-1 crew members. *Exp. Brain Res.*, 64:380–391.

Bloomberg, J.J., Reschke, M.F., Huebner, W.P., Peters, B.T., and Smith, S.L. (1997). Locomotor head trunk coordination strategies following spaceflight. *J. Vestib. Res.*, 7:161–167.

Clément, G., Gurfinkel, V.S., Lestienne, F., Lipshits, M.I., and Popov, K.E. (1984). Adaptation of posture control to weightlessness. *Exp. Brain Res.*, 57:61–72.

Collins, J.J., DeLuca, C.J., Pavlik, A.E., Roy, S.H., and Emley, M.S. (1995). The effects of spaceflight on open-loop and closed-loop postural control mechanisms: human neurovestibular studies on SLS-2. *Exp. Brain Res.*, 107:145–150.

Gordon, C.R., Fletcher, W.A., Melvill Jones, G., and Block, E.W. (1995). Adaptive plasticity in the control of locomotor trajectory. *Exp. Brain Res.*, 102:540–545.

Hein, A. and Held, R. (1967). Dissociation of the visual placing response into elicited and guided components. *Science*, 158:390–392.

Holst, von E. and Mittlestaedt, H. (1950). *Das Reafferenzprinzip. Naturwissenschaften* 37:464–476.

Homick, J.L. and Reschke, M.F. (1977). Postural equilibrium following exposure to weightless space flight. *Acta Otolaryngologica*, 83:455–464.

Horak, F.B., Nashner, L.M. (1986). Central programming of postural movements: adaptations to altered support-surface configurations. *J. Neurophysiol.*, 55:1369–1381.

Kenyon, R.V. and Young, L.R. (1986). MIT/Canadian vestibular experiments on Spacelab-1 mission: 5. Postural responses following exposure to weightlessness. *Exp. Brain Res.*, 64:335–346.

Kozlovskaya, I.B., Kreidich, Y.V., Oganov, V.S., and Koserenko, O.P. (1981). Pathophysiology of motor functions in prolonged manned space flights. *Acta Astronautica*, 8:1059–1072.

Kuo, A.D. (1996). An optimal control model for analyzing human postural balance. *IEEE Trans. Biomed. Eng.*, 42:87–101.

Melvill Jones, G., Fletcher, W.A., Weber, K., Gordon, C., and Block, E. (1996). Spatial orientation through the feet: an adaptive sensory-motor system? *J. Vestib. Res.*, 6(4S):S15.

Nashner, L.M. (1985). Strategies for organization of human posture. In *Vestibular and Visual Control on Posture and Locomotor Equilibrium*. Igarashi, M. and Black, F.O. (eds.), pp. 1–8. Basel: Karger.

Nashner, L.M., Shupert, C.L., Horak, F.B., and Black, F.O. (1989). Organization of posture controls: an analysis of sensory and mechanical constraints. In *Progress in Brain Research*. Allum, J.H.J. and Hulliger, M. (eds.), 80:411–418. Elsevier.

Nicholas, S.C., Doxey-Gasway, D.D., and Paloski, W.H. (1998). A link-segment model of upright human posture for analysis. *J. Vestib. Res.*, 8(3):187–200.

Paloski, W.H., Bloomberg, J.J., Reschke, M.F., and Harm, D.L. (1994). Spaceflight-induced changes in posture and locomotion. *J. Biomech.*, 27(6):812.

Paloski, W.H., Reschke, M.F., Black, F.O., Doxey, D.D., and Harm, D.L. (1992). Recovery of postural equilibrium control following space flight. In *Sensing and Controlling Motion: Vestibular and Sensorimotor Function*. Cohen, B., Tomko, D.L., and Guedry, F. (eds.), 682:747–754. Ann. NY Acad. Sci.

Reschke, M.F., Bloomberg, J.J., Harm, D.L., Paloski, W.H., and Parker, D.E. (1994). Chapter 13. Neurophysiological aspects: sensory and sensory-motor function. In *Space Physiology and Medicine*, 3rd ed. Nicogossian, A.E., Huntoon, C.L., and Pool, S.L. (eds.). Lea & Febiger, Philadelphia.

Commentary: Adaptive Sensory-Motor Processes Disturb Balance Control After Spaceflight

Robert J. Peterka

Paloski has demonstrated that fairly large changes in postural control properties can occur following space flight. However from the point of view of someone attempting to understand human postural control mechanisms, it is frustrating that very different postflight strategies were adopted by different subjects. What is the source of these differences and how can future experiments be designed to provide more specific insight into the adaptive postural control problem?

It is reasonable to assume that adaptation occurs in order to achieve some goal or set of goals. But what is the goal of postural control during space flight? It may be that job duties in space require different adaptation goals, thus leading to widely varying results among subjects postflight. For example one astronaut may perform most tasks with his/her feet strapped to a support surface requiring orientation with respect to that surface provided by torques exerted against that surface, and requiring altered control because of the loss of gravity. Another astronaut may spend significant time exercising on a treadmill against an artificial gravity provided by elastic cords, therefore requiring control of body position above the base of support and control of joint stiffness similar to that required on earth. A third may perform most tasks while strapped into a chair. These considerations suggest that it may not be possible to gain greater insight into the adaptation process unless the experiences/environment that lead to balance control adaptation can be controlled more precisely.

Assuming we can more precisely control the goal of adaptation, then we are faced with the problem

that a multi-joint system, with feedback available from distributed and partially redundant sensors, can likely achieve a particular adaptation goal by different means. Perhaps we need to place experimental limits on the system both during adaptation and while testing the system. Limits include sensory limits (such as performing experiments with eyes closed, using patients with severe sensory system losses, and using experimental tricks such as sway-referencing to limit sensory motion cues), and motor limits (such as constrained joint motions and altered joint compliance). While these limits in some sense produce an artificial and unrealistic system, a reduced system might provide insights into the system's adaptive behavior, and these insights can be extended later to more realistic systems.

We should explore methods that more precisely control the experimental subject's intent, goals, and capabilities. The simple request that the subject maintain upright stance in response to perturbations may not stress the system enough to identify specific limits placed on balance control by the current adaptive state. Perhaps preadaptation training regimes which emphasize repeatable response patterns to perturbations might enhance our ability to identify specific effects of postural adaptation. For example, train a subject to respond to a support surface translation by bringing his/her body as rapidly as possible to an upright position while maintaining a "strapped-down strategy." Then adapt the system in some specific manner and finally test the subject's ability to perform the well trained response.

Finally, we should consider the possibility that changes in postural control associated with space flight may be a secondary affect. What if some other system (e.g., eye movement control) is driving an adaptive process in order to optimize its performance in the absence of gravity. If there are brain systems shared between the newly adapted system and the postural control system (e.g., some orientation reference system), then postural control properties would be altered even though the adaptation process was not specifically associated with achieving some posture related goal. Experiments that explore this possibility could involve exposure to a stimulus that induces adaptation not specifically related to postural control (e.g., adaptation of visual/vestibular eye movement reflexes in a restrained, seated subject), followed by monitoring of changes in postural control performance.

22
Neuromuscular Control Strategies in Postural Coordination

Ron Jacobs and Anne Burleigh-Jacobs

1 Introduction

It is difficult to grasp how the human nervous system is able to keep up with the requirements of walking, reaching, or even doing something as seemingly simple as standing itself. The nervous system decides what joints need to be moved, exactly when they should be moved and by how much, and then sends the proper series of impulses along nerves to activate the appropriate combination of muscles that leads to a desired movement. Through extensive training, using "oops" and "wow" learning starting at birth and continuing throughout life, the nervous system has learned the sensitivities of how different muscles work together in affecting movement at the various joints. We sometimes forget how hard we have worked to establish more or less automatic and smooth standing behavior. As developed in Chapters 37 and 38, only after a disabling event, for instance a stroke, are we reminded about the complexity of a simple motor task.

Postural coordination is a prerequisite for each motor task. Therefore, in studies of restoration of motor ability following injury or disease, it is important to understand how postural coordination is achieved. Even though postural control has been studied exhaustively, it is still not precisely clear which neuromuscular control strategies are utilized by the nervous system in contribution to postural coordination. To date, the data are obtained from what are called standard postural studies, for instance studying maintenance of stance during: (1) various sensory and cognitive stimuli situations, (2) externally imposed perturbations, (3) self-imposed perturbations, and (4) combinations of 1, 2, and 3 (Ho-

rak and Macpherson 1996; Mulder et al. 1996; see also Chapter 19). Development of postural control models and simulations is strongly dependent on these studies but it seems that we are unable to pool the exhaustive amount of data to create a unified theory on postural control. This chapter discusses:

- A new way of looking at postural control by stating a new definition of the goal of postural control.
- Control strategies that simplify the role of the nervous system in movement and postural control.
- A new way of analyzing and modeling postural control in stance, based on the new definition and using soft computing techniques.

2 Traditional Goal of Postural Control

The study of postural control is greatly simplified by the definition of a *goal* of postural control. The defined goal focuses the direction in which to look for answers to research questions. The defined goal helps interpret changes in output of the nervous system relative to changes in input to the nervous system. The traditional definition, commonly used with respect to postural control, is summarized by Dietz (1996) when he states:

The neuronal pattern evoked during a particular task is always directed to hold the body's center of mass over the base of support. All control mechanisms must, therefore, be considered and discussed in this respect. . . . Only a combination of afferent mechanisms can provide the information needed to control body equilibrium.

This traditional goal of postural control, holding the body's center of mass over the base of support, is limited and only represents a small subset of the many postural control capabilities of our central nervous system (CNS). This limitation can be nicely illustrated when postural control on earth and in space are compared. Traveling from a 1 g to a 0 g environment makes us wonder how we do control our center of mass. In this respect, Gurfinkel and his colleagues have made a very nice observation related to this phenomenon as is stated in Massion et al. (1993):

It is worth noting that the internal representation of the center of mass position with respect to the feet is partly lost during the first days in flight. During this period, erect posture adopted by the "naive" astronaut is inclined forward. In this inclined position, the center of mass will project in front of the subject's feet and would lead to falling under terrestrial conditions.

As the quotation indicates, the traditional goal is limited to 1 g and no longer works in 0 g. Although this is probably one of the best arguments disputing the fact that we are controlling our body center of mass in postural control, several other arguments and questions can be put forward. For instance:

• How does the nervous system calculate body center of mass? We propose that the central nervous system does not, rather body center of mass is a variable used by researchers in an effort to have a self-limited, understandable variable.
• Controlling body center of mass is primarily based on an inverted pendulum model, and implies redundancy of our sensory mechanisms. However, it is well known that control strategies change when one sensory system has been removed or altered (e.g., Nashner et al. 1982; Paulus et al. 1984), suggesting that sensory systems are not simply redundant.

Based on the above reasons, it is time to change gears by defining new experimental paradigms guided by a new defined goal and develop different modeling approaches.

3   New Goal of Postural Control

The question now arises as to whether an alternative goal of postural control can be defined to embrace experimental data and eventually come to a unified theory. We would like to propose an alternative goal for postural control. It is to achieve equilibrium (i.e., effective weighting of all sensory systems) to maintain a stable vertical and horizontal orientation of the body with respect to the in-dividuals intent, experience, instruction, and environment.

The concept of motor control emerging from an interaction between the individual, task, and environment is not new (e.g. Newell et al. 1989; Shumway-Cook and Woollacott 1995). However, such a systems approach is not generally applied to the goal of postural control. But what does the new goal of postural control mean? Three important aspects can be distinguished:

• *The CNS continuously receives afferent information from all peripheral sensory systems.* This suggests that sensory systems are not redundant, but rather act in parallel with the strength of contribution dependent on the motor task. Furthermore, not all sensory information is continuously used to modulate the efferent signals transmitted to the muscles unless a certain threshold value is exceeded. This introduces a certain sloppiness in the postural control system (Collins and De Luca 1993). But as Collins and De Luca (1993) state, this sloppiness may have evolved taking into account time delays of feedback loops and to simplify the task by integrating vast amounts of sensory information when the postural system is not in jeopardy of instability.
• *The CNS continuously adapts, learns and relearns based on all sensory inputs to achieve postural stability.* For instance, adaptation and learning occur when sensory systems have been removed or altered:
    1.   Johansson et al. (1995) demonstrated postural adaptation in erect human stance during maintained galvanic stimulation of the vestibular nerve and labyrinth. The time constant for adaptation was in the range of 40 to 50 s.
    2.   Paloski (1997) showed that after being in space, astronauts learned new postural control strategies to minimize head movements aimed at minimizing confounding vestibular inputs.
    3.   Cole (1995) described a patient with loss of A-fiber type proprioception who was no longer able to stand and walk. Vision, vestibular, and neck proprioception were the only sources of intero- and

exteroception available. After extensive learning periods the patient relearned standing and walking although at slow pace, requiring increased visual reliance and concentration.

- *The central nervous system adaptation and (re)learning of postural control depends on IEIE: intent, experience, instruction, and environment.* For instance:

    1.   Intent (i.e., the voluntary decision to act in a certain way) influences the postural response to an external surface perturbation. For instance, when subjects intend to step forward rather than maintain standing in response to a backward surface perturbation, there is an asymmetrical response of the stance and swing limb gastrocnemius and an early increase in vertical propulsive forces under the initial swing limb (Burleigh et al. 1994; Burleigh and Horak 1996).

    2.   Experience, via prior exposure to testing conditions, has pronounced influences on postural responses. Large displacements of the center of foot pressure and even stepping are common in the first trials of large surface translations. With repeated exposure to the same perturbation, there is a reduced excursion of the center of foot pressure with reduced magnitude of both the stabilizing and destabilizing muscle responses to perturbation (Horak et al., 1989; Maki and Whitelaw 1993).

    3.   Instruction results in different neuromuscular control strategies, for instance when instructed to stand at ease or stand as still as possible (Rothwell 1994). While standing as still as possible, the stiffness at the ankle increases. However, this increase is not caused by co-contraction of muscles since no change in EMG is observed, but is caused by an increase in feedback by a change in the instruction and action of the subject.

    4.   Environmental influences on postural control strategies are nicely illustrated by the experimental results obtained in a 1 g and 0 g environment as described earlier (Massion et al. 1993).

This alternative goal is different from the traditional goal of postural control since it explicitly allows:

- Interpretation of behavior in different environments, e.g. 0 g and 1 g,
- Weighting of all sensory inputs versus the concept of redundancy of sensory systems,
- Modeling and quantification of open-loop and closed-loop control mechanisms, and
- A goal that depends on intent, experience, instruction, and environment.

Our behavior is based on what we have experienced during our lives as being *correct*! Therefore, it is important for our models and experimental paradigms to encompass detection and reweighting of all sensory inputs based on intent, experience, instruction, and environment to modulate efferent signals transmitted to the muscles.

4   Strategies on How the Nervous System Controls Movement

Having the new definition of postural control, it is now important to raise the question: Which control variables are actually important and utilized by the nervous system to develop effective postural control strategies? We would like to argue that control variables and rules identified for *movement control* can be extrapolated to *postural control*. In other words, we do not distinguish posture from movement, but define posture as being the equivalent of movement with reduced velocity. Currently two separate approaches on how the nervous system controls the various movements can be distinguished:

- The first strategy is based on the inverse dynamics approach. It involves a computation of the individual joint torques by some complex representation of muscle skeletal dynamics in the nervous system (e.g., Hollarbach and Atkeson 1987; Kuo 1995; Hatze 1997; van der Helm 1997). This strategy is well adapted to the machine model of biological systems and lends many insights into study of robotics (Roberts and McCollum 1996). However, the actual use of such a strategy by the nervous system is doubtful since it would require too great a calculating load on the nervous system for accurate manipulation of a multijoint body with all the nonlinearities that are involved in motor control.
- The second strategy involves the use of control strategies in which *global control variables* are utilized and movement emerges as a function of these global control variables (e.g., Lacquiniti and Maioli 1992; Gordon et al. 1994; Gottlieb et al. 1996; Jacobs and Macpherson 1996; Jacobs et al.

1996). Rather than computing the individual joint torques, the nervous system actually simplifies the control of multijoint movements by using global control variables. Recently, from neurophysiological experiments, there are several important indications that the nervous system seems to simplify the control of movement in terms of global control variables.

5   Simplified Control Strategies

We argue that the second strategy, use of global control variables, is more representative of nervous system control of movement and posture. For instance, global control variables have been identified in human and animal motor control, utilized in legged robot control, and even proposed for the organization in the spinal circuitry in cats (Table 22.1). Without attempting to give a complete review on simplified control strategies, we will provide only a few examples.

In *human arm control*, Flanagan and Rao (1995) demonstrated that planning of arm movements does not occur at the level of each individual joint of the arm, but at the endpoint of the limb (i.e., the hand). Other studies have suggested that amplitude and direction are independently specified in the planning of a targeted force impulse task (Favilla et al. 1989, 1990). Gordon et al. (1994) further showed that a distinction can be made in nervous system control of extent (i.e., amplitude) and direction of the hand in planar reaching movements. In *human finger control*, Edin et al. (1992) demonstrated that normal and shear forces at the fingertip were independently controlled during precision lifting of an object.

In *animal body control*, lesion studies in the frog have demonstrated that leftward and rightward horizontal components of whole body orienting movements are controlled by distinct circuitry, and that these circuits are independent of those controlling elevation and distance (Masino and Grobstein 1989). Similarly, in *animal head control*, horizontal and vertical components of head movements in the barn owl are controlled by distinct neural circuits in the brainstem (Masino and Knudsen 1990).

In *animal leg control*, Maioli and Poppele (1991) identified an independent control of global variables of limb geometry in standing cats during platform rotations. Limb length and orientation resulted from a parallel processing of multisensory inputs into separate central representations of body tilt. Jacobs and Macpherson (1996) identified a simplified control strategy in standing cats' postural control of balance during platform perturbation experiments. They identified a separate control of force magnitude and direction applied by the cat's paw on the support surface. In addition, they were able to identify two distinct muscle groupings related to force magnitude and direction; flexor and extensor muscles were related to force magnitude control, whereas biarticular muscles were related to force direction control by regulating the difference in knee and hip torque (see also Jacobs and Ingen Schenau 1992). The results of Jacobs and Macpherson (1996) also put forward strong evidence showing that the nervous system does not control at the level of individual joint torques.

In *legged robot control*, Raibert (1986) found that the control of locomotion (hopping) could be greatly simplified by decoupling the process into three parallel control tasks:

TABLE 22.1. Examples of global control variables simplifying movement and postural coordination.

System	Global control variables	Reference
Human arm	Endpoint (i.e., hand)	Flanagan and Rao 1995
	Force impulse amplitude and direction	Favilla et al. 1989, 1990
	Extent and direction	Gordon et al. 1994
Human finger	Normal and shear force	Edin et al. 1992
Frog body	Elevation and distance	Masino and Grobstein 1989
Owl head	Horizontal and vertical movement	Masino and Knudsen 1989
Cat hind limb	Limb length and orientation	Maioli and Poppele 1991
	Force amplitude and direction	Jacobs and Macpherson 1996
Legged robots	Vertical thrust, body attitude, and forward speed	Raibert 1986
Spinal circuitry	Limb length and orientation	Bosco and Poppele 1996
Human body	*Limb length, limb orientation, and trunk attitude*	*Jacobs and Burleigh-Jacobs*

1. control of vertical thrust,
2. control of body attitude, or torque between body and leg,
3. control of forward speed.

This "decoupling strategy" worked well for one-, two-, and four-legged robots. Raibert's approach was a simple alternative to solving the equations of motion based on the mechanical characteristics of the device. For the case of stance (i.e., hopping in place), the forward velocity component drops out and the system is reduced to two controllers, vertical thrust and body attitude. Vertical thrust is similar to force magnitude control in the cat, whereas body attitude is similar to force direction control (Jacobs and Macpherson 1996).

In *spinal circuitry control*, Bosco and Poppele (1996) proposed that even the dorsal spinocerebellar tract activity may represent movement kinematics in a limb-centered reference frame using a two-dimensional coordinate system defined approximately by limb length and orientation. They also suggest that the presynaptic circuitry is responsible for creating and shaping the temporal domain for this coordinate system.

Most of these examples support the idea of a simplification of nervous system control of movements using global control variables. To date, only a few examples are known that support this idea for postural control (e.g., Maioli and Poppele 1991; Lacquiniti and Maioli 1992; Jacobs and Macpherson 1996). However, we would argue again that control variables and rules identified for *movement control* can be extrapolated to *postural control* and

posture is defined as being the equivalent of movement with reduced velocity. In conclusion, it is interesting to note the similarity between the different reports in their description of global control variables. In our opinion, based on these reports, a general separation can be made between limb length (also called amplitude or extent) and limb orientation (also called direction) as global control variables for control of limb movements.

6 Proposed Model Using Simplified Control Strategy for Postural Coordination

Stimulated by multiple reports in the literature (Table 22.1), suggesting global control variables of various motor tasks, we propose a general control strategy that simplifies postural coordination of stance in humans. Our model is not based on the control of body's center of mass, but rather on the control of different global control variables. This is an important difference since the model can now be applied to changes in environment, for instance posture control on earth and in space. Furthermore, our control model is not based on purely inverted pendulum body mechanics where only motion at one joint is controlled, as for instance the ankle. In our model, the degrees of freedom are controlled by using reciprocal and synergistic muscle actions at multiple joints (see later). In our control model, we distinguish three sets of global control variables which act in parallel (Figure 22.1A):

A. Definition of global control variables

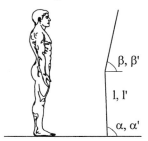

B. Flow diagram of postural control model

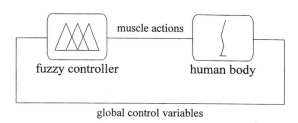

global control variables

FIGURE 22.1. (A) Three sets of global control variables, which act in parallel, are defined: (1) trunk attitude, β, and derivative of β, β'; (2) limb length, l, and derivative of l, l'; and (3) limb orientation, α, and derivative of α, α'. (B) Flow diagram showing that global control vari-

ables go into the fuzzy controller. In the controller are rules that determine the muscle actions, which then influence the body posture dependent on the objective of each of the three sets of control variables.

I. *Trunk attitude*, β, and derivative of β, β';
II. *Limb length*, l (distance between two respective points of the limb), and derivative of l, l';
III. *Limb orientation*, α, and derivative of α, α'.

These global control variables are mechanically coupled but remain sufficiently separate sets of control variables. Each part of the control system behaves as it affects only the one variable it is supposed to control and interactions show up as disturbances in each of the control systems. The global control variables form the basis of the postural control. In the controllers,

I. β and β' determine synergistic actions of trunk flexor muscles which are controlled reciprocally with synergistic actions of trunk extensor muscles,
II. l and l' determine synergistic actions of flexor muscles at each joint which are controlled reciprocally with synergistic actions of extensor muscles at each joint,
III. α and α' determine synergistic actions of biarticular muscles on the anterior side of the limb which are controlled reciprocally with biarticular muscles on the posterior side of the limb.

These control rules should be implemented in a control model for postural coordination of stance. This is not an easy thing to do since in this implementation process we will be confronted with three major problems:

• the relationship between the control variables is nonlinear,
• there is still a lack of precise and detailed data available concerning the control rules,
• our nervous system is by no means precise.

At this time, using conventional techniques, it is still difficult to implement control algorithms and model postural coordination. What is needed is a technique that:

• is capable of approximating nonlinear functions,
• can deal with imprecise data,
• is relatively easy to implement compared to some conventional techniques,
• allows a meaningful and explicit representation of nervous system behavior and is not a black box.

Fuzzy logic can do the job (see Zadeh 1996; Chapter 40). In fuzzy logic, *if-then* control rules determine the relationship between input variables (our global control variables) and output variables

(muscle torque or stimulation) in a meaningful and explicit manner. The control rules in our model are based on experimental findings indicating relations between muscle activation patterns and global control variables (in particular, Jacobs and Ingen Schenau 1992; Jacobs et al. 1993; Ingen Schenau et al. 1995; Jacobs and Macpherson 1996). An important aspect of using fuzzy logic in our application is that all control variables (i.e., sensory inputs) act in parallel, which is in accordance with our new definition of the goal of postural control. In addition, fuzzy logic deals with imprecise data by using natural language and words instead of equations and numbers (Zadeh 1996).

The control strategy is implemented using a four-linked segment model consisting of a trunk, thigh, shank and foot (Figure 22.1B). Uni- and biarticular limb muscles and trunk muscles are represented as torque actuators at each individual joint. The following muscle groups are represented: ankle, knee and hip flexor and extensor muscles in controller I; biarticular leg muscles (i.e., hamstrings, rectus femoris and gastrocnemii) in controller II; trunk flexor and extensor muscles in controller III.

In the control strategy, 36 fuzzy *if-then* rules, based on experimental findings, are implemented. An example of a rule is: "if l is small and l' is negative then make *extensor action* large." In this example of a fuzzy rule, l, l' *and muscle action* are fuzzy control variables. *Small*, *negative*, and *large* are fuzzy sets. A summary of all the *if-then* rules is given in Figure 22.2.

The objective of each of the three parallel controllers is that: controller I (β and β'), maintains maximum limb length and zero velocity; controller II (l and l'), maintains vertical limb position and zero velocity; and controller III (α and α'), maintains vertical trunk position and zero velocity. These three parallel controllers define standing posture with respect to the goal to remain vertical. If the goal objectives change, then simple changes in the rules will meet the new postural requirements.

7 Results of Proposed Model

In our four-linked model of a standing human, the three control mechanisms act in parallel and make corrective and coordinated responses to internal, self-induced perturbations (Figure 22.3). The data shows that the standing human sways about a point of equi-

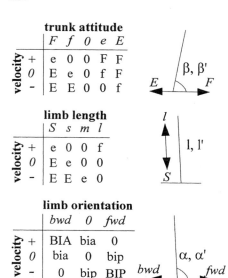

trunk attitude

velocity	F	f	0	e	E
+	e	0	0	F	F
0	E	e	0	f	F
−	E	E	0	0	f

limb length

velocity	S	s	m	l
+	e	0	0	f
0	E	e	0	0
−	E	E	e	0

limb orientation

velocity	bwd	0	fwd
+	BIA	bia	0
0	bia	0	bip
−	0	bip	BIP

FIGURE 22.2. Matrices of each of the three sets of parallel control variables. Dependent on the combinations of the control variables, fuzzy rules determine the muscle actions. For instance, the gray shaded box area corresponds to the fuzzy rule: "if l is small and l' is negative, then make *extensor action* large. (F = large flexion, E = large extension, f = flexion, e = extension; S = smaller, s = small, m = medium, l = large; bwd = backward, fwd = forward; BIA = large action biarticular anterior, BIP = large action biarticular posterior, bia = action biarticular anterior, bip = action biarticular posterior.)

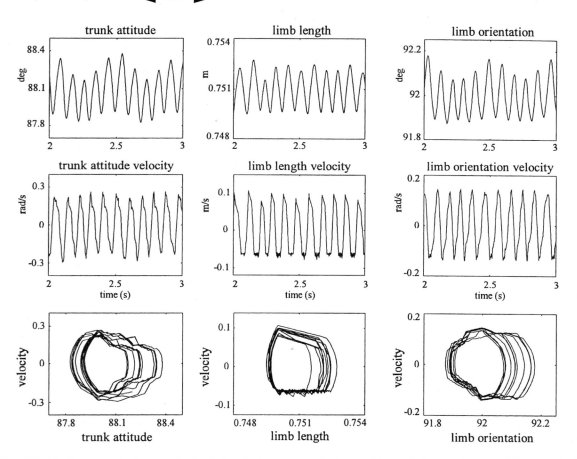

FIGURE 22.3. Results of 1s from a 10s simulation during which a four-linked segment human model maintained independent stance. The three controllers act in parallel to make corrective and coordinated responses to internal perturbations, with small changes in each of the control variables. Trunk attitude, limb length, and limb orientation remain reasonably stable, as shown by the circular pattern of the phase-plane plots.

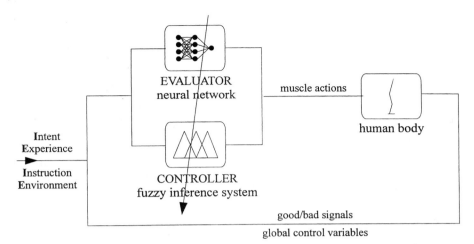

FIGURE 22.4. Flow diagram of an adaptive postural control model. A neural network is included as a good/bad evaluator for optimization of the control task. The evaluator learns and tunes the fuzzy controller based on changes in Intent, Experience, Instruction and Environment (IEIE). Experience ensures that the evaluator learns good behavior. Intent and Instruction provide the objective of the control task. Environment defines the conditions in which the task is performed.

librium with small changes in each of the control variables. Trunk attitude, limb length, and limb orientation controllers are reasonably stable, as shown by the circular pattern of the phase-plane plots.

At this time, the results shown apply only to control of quiet stance. In the future, similar control strategies and rules will be applicable to changes in postural task, environment, sensory inputs, and changes in IEIE. Changes are accomplished by tuning the fuzzy sets and rules in the controller. Tuning utilizes a neural network that will be implemented for optimization of the desired postural task (Figure 22.4). The neural net will act as an evaluator for good/bad actions by the controllers and update the fuzzy control system (Berenji and Khedkar 1992). Having this evaluator of good/bad actions, the model has a built-in internal representation. The use of good/bad evaluation for the internal representation is in contrast to a complex representation of muscle-skeletal dynamics used in other modeling approaches (e.g., Sections III and VII of this book).

8   Future Directions

This chapter discusses the limitations of the traditional definition of the goal of postural control in the way it limits our embracing experimental data and coming to a unified theory on postural control. We introduced an alternative way of looking at postural control by stating a new definition of the goal of postural control. In this new definition, we emphasized the importance of achieving effective weighting of all sensory inputs with respect to IEIE (intent, experience, instruction and environment). The new definition should enable us to embrace experimental data and come to a unified theory of postural control. We also introduced a control model of postural coordination that simplifies nervous control of human stance and is based on experimental findings. This model is based on control of different global control variables and allows changes in postural task, environment, sensory inputs, and IEIE, as is required by our new definition. An important feature of the control model is the use of fuzzy logic, that enables us to model experimental findings in a meaningful and explicit way.

Studying neuromuscular control strategies is eminently important, not only from a scientific point of view to gain better insight in the mysteries of how the nervous system controls movements, but also in studies of restoration of movement after injury or disease. For instance:

• *Artificial stimulation for standing and walking control in paraplegics.* By means of functional electrical stimulation (FES), paralysed muscles can be used functionally. To accomplish this, we must know which control strategies should be used for functional standing and walking.

- *Development of adaptive intelligent limb prosthesis.* The design of an ideal prosthetic limb to optimise functional standing and walking must be based on mechanical models as well as models of neural muscular function.
- *Intervention in rehabilitation, for instance in stroke patients.* Insight into neural control strategies of standing and walking can be applied to effectively improve treatment exercises performed by physical therapists.

Future research work should be directed towards questions that enable us to further develop a unified postural control theory based on our new definition. In our research, we should aim to:

1. Conduct experiments under natural movement conditions and functional postural perturbations to define control variables.

2. Understand our manipulations of sensory inputs or IEIE in our experimental paradigms. With increasing complexity of experimental paradigms, it is important to understand precisely how the output of the nervous system is affected by these manipulations.

3. Further develop models that can encompass and unify experimental findings. Our approach using soft computing techniques (fuzzy logic and neural networks; neuro-fuzzy), is an example of one approach that can be used (see also Chapter 40).

4. Develop theories and models that take into account the CNS development, adaptation, learning over a lifetime of experiences and/or changes in biomechanics and external environment.

References

Berenji, H.R. and Khedkar, P. (1992). Learning and tuning fuzzy logic controllers through reinforcements. *IEEE Trans. Neural Net.*, 3:724–740.

Bosco, G. and Poppelo, R.E. (1996). Temporal features of directional tuning by spinocerebellar neurons: relation to limb geometry. *J. Neurophysiol.*, 75:1647–1658.

Burleigh, A. and Horak, F. (1996). Influence of instruction, prediction, and afferent sensory information on the postural organization of step initiation. *J. Neurophysiol.*, 75:1619–1628.

Burleigh, A., Horak, F., and Malouin, F. (1994). Modification of postural responses and step initiation: evidence for goal-directed postural interactions. *J. Neurophysiol.*, 72:2892–2902.

Cole, J. (1995). Pride and a daily marathon. The MIT Press, Cambridge, Massachusetts.

Collins, J.J. and De Luca, C.J. (1993). Open-loop and closed-loop control of posture: a random walk analysis of center-of-foot-pressure trajectories. *Exp. Brain Res.*, 95:308–318.

Dietz, V. (1996). Interaction between central programs and afferent input in the control of posture and locomotion. *J. Biomech.*, 29:841–844.

Edin, B.B., Westling, G., and Johansson, R.S. (1992). Independent control of human finger-tip forces at individual digits during precision lifting. *J. Physiol.*, 450:547–564.

Favilla, M., Gordon, J., Hening, W., and Ghez, C. (1990). Trajectory control in targeted force impulses. VII. Independent setting of amplitude and direction in response preparation. *Exp. Brain Res.*, 79:530–538.

Favilla, M., Hening, W., and Ghez, C. (1989). Trajectory control in targeted force impulses VI. Independent specification of response amplitude and direction. *Exp. Brain Res.*, 75:280–294.

Flanagan, J.R. and Rao, A.K. (1995). Trajectory adaptation to a nonlinear visuomotor transformation: evidence of motion planning in visually perceived space. *J. Neurophysiol.*, 74:2174–2178.

Gordon, J., Ghilardi, M.F., and Ghez, C. (1994). Accuracy of planar reaching movements. I. Independence of direction and extent variability. *Exp. Brain Res.*, 99:97–111.

Gottlieb, G.L., Song, Q., Hong, D., Almeida, G.L., and Corcos, A.D. (1996). Coordinating movement at two joints: a principle of linear covariance. *J. Neurophysiol.*, 75:1760–1764.

Hatze, H. (1997). Chapter 33: Progression of musculoskeletal models towards large-scale cybernetic myoskeletal models. This book.

Helm, F.C.T. van der (1997). Chapter 32: Large-scaling models: Sensorimotor integration and optimization. This book.

Hollarbach, J.M. and Atkeson, C.G. (1987). Deducing planning variables from experimental arm trajectories: pitfalls and possibilities. *Biol. Cybern.*, 56:279–292.

Horak, F.B. and Macpherson, J.M. (1996). Postural orientation and equilibrium. Handbook of Physiology. Oxford University Press, New York.

Horak, F.B., Diener, H.C., and Nashner, L.M. (1989). Influence of central set on human postural responses. *J. Neurophysiol.*, 62:841–853.

Ingen Schenau, G.J. van, Welter, T., and Jacobs, R. (1995). On the control of monoarticular muscles in multi-joint leg extensions in man. *J. Physiol.*, 484: 247–254.

Jacobs, R. and Ingen Schenau, G.J. van (1992). Control of an external force in leg extensions in humans. *J. Physiol.*, 457:611–626.

Jacobs, R. and Macpherson, J.M. (1996). Two functional muscle groupings during postural equilibrium tasks in standing cats. *J. Neurophysiol.*, 76:2402–2411.

Jacobs, R. and Tucker, C. (1997). Chapter 40: soft computing techniques for evaluation and control of human performance. This book.

Jacobs, R., Bobbert, M.F., and Ingen Schenau, G.J. van (1993). Muscle function of mono- and bi-articular muscles during running. *Med. Sci. Sports Exerc.*, 25: 1163–1173.

Jacobs, R., Koopman, B., Veltink, P., Huijing, P.A., Nene, A., van der Kooij, H., and Grootenboer, H. (1996). A simplified control strategy for postural coordination in human stance. In *Proc. Engineering Foundation Meeting: Biomechanics and Neural Control of Movement IX.* Winters, J. and Crago, P. (eds.), Mt. Sterling Ohio.

Johansson, R., Magnusson, M., and Fransson, P.A. (1995). Galvanic vestibular stimulation for analysis of postural adaptation and stability. *IEEE Trans. Biomed. Eng.*, 42:282–292.

Kuo, A. (1995). An optimal control model for analyzing human postural balance. *IEEE Trans. Biomed. Eng.*, 42:87–101.

Lacquaniti, F. and Maioli, C. (1992). Distributed control of limb position and force. In *Tutorials in Motor Behavior II.* Stelmach, G.E. and Requin, J. (eds.), pp. 31–54. Elsevier, North Holland.

Maioli, C. and Poppele, R.E. (1991). Parallel processing of multisensory information concerning self-motion. *Exp. Brain Res.*, 87:119–125.

Maki, B. and Whitelaw, R.S. (1993). Influence of expectation and arousal on center-of-pressure responses to transient postural perturbations. *J. Vestibular Res.*, 3:25–39.

Masino, T. and Grobstein, P. (1989). The organization of descending tectofugal pathways underlying orienting in the frog, Rana pipiens. I. Lateralization, parcellation and an intermediate representation. *Exp. Brain Res.*, 75:227–244.

Massion, J., Gurfinkel, V., Lipshits, M., Obadia, A., and Popov, K. (1993). Axial synergies under microgravity conditions. *J. Vestibular Res.*, 3:275–287.

Masino, T. and Knudsen, E.I. (1990). Horizontal and vertical components of head movement are controlled by distinct neural circuits in the barn owl. *Nature*, 345: 434–437.

Mulder, T., Nienhuis, B., and Pauwels, J. (1996). The assessment of motor recovery: a new look at an old problem. *J. Electromyogr. Kinesiol.*, 6:137–145.

Nashner, L.M., Black, F.O., and Wall, C. III (1982). Adaptation to altered support and visual conditions during stance: patients with vestibular deficits. *J. Neurosci.*, 2:536–544.

Newell, K.M., Emmerik, R.E.A. van, and McDonald, P.V. (1989). Biomechanical constraints and action theory. *Hum. Mov. Sci.*, 8:403–409.

Paulus, W.M., Straube, A., and Brandt, T.H. (1984). Visual stabilization of posture: physiological stimulus characteristics and clinical aspects. *Brain*, 107:1143–1163.

Peterka, R.J. and Benolken, M.S. (1995). Role of somatosensory and vestibular cues in attenuating visually induced human postural sway. *Exp. Brain Res.*, 105:101–110.

Paloski, W.H. (1997). Chapter 21: Adaptive sensory-motor processes disturb balance control after space flight. This book.

Raibert, M.H. (1986). Legged robots that balance. The MIT Press, Cambridge, Massachusetts.

Roberts, P.D. and McCollum, G. (1996). Dynamics of the sit-to-stand movement. *Biol. Cybern.*, 74:147–157.

Rothwell, J.R. (1994). Control of human voluntary movement—2nd edition. Chapman and Hall, London.

Shumway-Cook, A. and Woollacott, M. (1995). Motor control: theory and practical applications. Williams and Wilkins, Baltimore, Maryland.

Zadeh, L. (1996). Fuzzy Logic = Computing with words. *IEEE Trans. Fuzzy Sys.*, 4:103–111.

Commentary: Neuromuscular Control Strategies in Postural Coordination

David A. Winter

Based on their review of the literature the authors are proposing a new goal of postural control, as opposed to what they describe as the traditional goal of postural control, that is, that of holding the body's center of mass (COM) over the base of support. There is no doubt that such a goal is far too simplistic. The COM displacement and its horizontal velocity must both be considered and during some balance situations (single support phase of gait) the COM is not within the base of support so the simple definition falls down. Their proposed goal is to "achieve equilibrium, i.e., effective weighting of all sensory systems to maintain a stable vertical and horizontal equilibrium of the body with respect to the individuals intent, experience, instruction and environment." As part of their rebuttal of the traditional goal they somehow dismiss the inverted pendulum model and the fact that the CNS cannot calculate the body's COM trajectory.

The inverted pendulum model of balance control can be expressed mathematically and therefore tested (Winter et al. 1996) but the authors have not defined what they mean by a "stable vertical and horizontal equilibrium". From some of their statements they do not appear to understand what an inverted pendulum model can tell them. The simplest version (in the sagittal plane) would have the pendulum pivoting about the ankle joint, and it is easy to predict the horizontal acceleration of the COM as being proportional to (COP-COM) when the COP is the center of pressure controlled by the ankle joint. However, such a model also applies in the frontal plane where rotation takes place about the ankle and hip joints simultaneously and has also been used to predict the horizontal accelerations of the COM during both initiation and termination of gait (Jian et al. 1993). In these latter movements there were several articulations of joints of both limbs. Such a model can also be used to analyze reaction responses after the perturbation has ended. They also state that an inverted pendulum model is one where "only motion at one joint is controlled." This is not so. The simple case described above, in which the inverted pendulum articulated about the ankle joint, does not mean that only the ankle torque is being controlled. Inverse dynamics will show that the knee and hip moments must also be controlled to control the initial loads of their respective pendulums and thereby keep their joints from articulation. In the initiation and termination of gait, Jian et al. (1993) noted many simultaneous motor patterns in both sagittal and frontal planes in order to achieve a controlled unbalancing of the pendulum (during initiation) and a controlled rebalancing of the pendulum (during termination). The inverted pendulum model describes the final integrated output (COM) but does not specify what mechanisms act to control the joint torques to control that COM. These authors are focusing on the rules of those mechanisms.

Section 3 lists four items that they claim can be achieved with their proposed definition and distinguish it from the inverted model. I do not see how the last three in their tests are denied by the use of an inverted model. For example, a simultaneous record of COP and COM from the inverted model would seem to be mandatory to achieve their third aim, that is to model and quantify open-loop and closed-loop mechanisms. Also, the example of gait initiation and termination is a perfect example where the inverted pendulum model can be used to investigate their fourth goal: dependence on intent, experience, instruction and environment. There is nothing in the motor controllers for an inverted pendulum model that cannot be handled by all levels of controllers (passive stiffness, reactive feedback or proactive anticipatory control).

The authors propose a model of control strategies for postural coordination which has three sets of global control variables. This is very interesting, especially as it essentially models the anthropometrics of the body (limb lengths) and their kinematics (displacements and derivatives). They also introduce the idea of fuzzy logic, which would stipulate a set of rules that would state the relationship between their inputs (global control variables) and the output variables (muscle torques). This is commendable but perhaps contradicts their earlier statement that the CNS does not calculate the body COM; their input global control variables are precisely what would be needed to calculate the body COM trajectory and velocity. Thus their proposed model is totally compatible with (or may be the same as) the inverted pendulum model. The fuzzy logic table of rules is entirely compatible with reactive and proactive control but the controller is independent of the biomechanical model on which it acts. For example, their fuzzy logic table might stipulate a hip (torque) strategy when a certain set of kinematics inputs are seen and a different strategy (i.e., ankle) when the kinematics are less dangerous. This is not different from what would be needed in a model that hypothesized that the COM displacement and velocity (in an inverted pendulum model) was the integrated input variable. I would predict that a fuzzy logic controller acting on the COM as an input would produce a very similar set of joint torques as the model they are now proposing. Traditional linear feedback control modeling has been shown to be a subset of fuzzy logic control, thus it is quite possible that traditional modeling of reactive control (to include neural delays and lags due to muscle low pass characteristics) would predict the same motor output responses as is done by the fuzzy logic controller.

In summary, I encourage these researchers to pursue their model but caution them that what they are doing is quite compatible with much of what has gone on before and should not be dismissed. The inverted pendulum model is a biomechanical model that does not specify the control of the motors. They are now modeling the controller which still has to act on a biomechanical model to control some final variable (presumably COM).

References

Jian, Y., Winter, D.A., Ishac, M.G., and Gilchrist, L. (1993). Trajectory of the body COG and COP during initiation and termination of gait. Gait and posture. 1: 9–22.

Winter, D.A., Prince, F., Powell, C., and Zabjek, K.F. (1996). A unified theory regarding A/P and M/l balance during quiet stance. J. Neurophysiol., 75: 2334–2343.

Section VI

Introduction: Neural and Mechanical Contributions to Upper Limb Movement

Jack M. Winters and Patrick E. Crago

Section VI contains the largest number of chapters and represents an extension of an important and lively area of research. The common theme is a consideration of principles underlying movements of the upper limb. In many ways, these nine chapters represent a continuation of the twelve chapters in Section III "Principles underlying movement organization: Upper limb" of the book *Multiple Muscle Systems: Biomechanics and Movement Organization* (Winters and Woo 1990). For that reason, these chapters are repeated within the reference list on this page; they can be obtained as a unit (196 pages total) from the editors, at cost (~$10; contact winters@cua.edu).

There is no introductory chapter for Section VI, in part because of the broad scope of these perspectives and the inherent challenge of synthesis within an area of significant controversy, and in part because of the availability of a considerable number of reviews and target articles of a related nature. These include several chapters from the *Multiple Muscle Systems* book (e.g., an interpretation of the Equilibrium Point hypothesis from Feldman's viewpoint in Chapter 12, considerations for arm path and movement planning in Chapters 14–18, the use of optimal control tools in Chapter 19, and the use of neural networks for study of arm movements, in Chapter 20).

For similar reasons there is also no commentary after these chapters. For the interested reader, several sources of commentary are readily available. In particular, a special issue of the journal *Brain and Behavioral Science*, entitled *Movement Control* (Cordo and Harnad, 1994), has a target article by Bizzi et al. that includes considerable commentary, and an issue on upper-limb movements edited by Flash (1993). For those interested in further discussion of equilibrium point concepts, the book by Latash (1993) should be reviewed. In addition, the recent book entitled *Models in human movement sciences* (Bosch et al. 1998) provides a thought-provoking format that especially addresses issues in dynamical systems and learning.

Chapter 23 starts with a broad perspective that identifies many of the key issues from a neuroscience perspective, with ties to the chapters in Section VII as well as the other chapters in this section. Chapters 24 and 25 deal with the concept of postural primitives, especially as related to a role for lower-level "spinal" force fields. Chapter 26 introduces some key concepts related to movement learning. These three perspectives evolved in part because of past ties with the MIT group led by Emilio Bizzi. Chapter 27 represents some of the evolving work on voluntary movement planning by a long-active researcher, extending ideas presented in Chapters 14 and 15 of the 1990 book, with the concept of a linear synergy rule between joint torques. Chapter 31 develops another hypothesis, the leading-joint hypothesis for multisegment coordination. Chapters 28 and 29 use simulation as a tool to gain insight into movement tracking and equilibrium point concepts. Chapter 30 considers another important area involving upper limb coordination, that of principles underlying grasping.

References

Bosch, M.S.J. Boschker (ed.). (1998). Models in human movement sciences. Amsterdam, Instituut voor Fundamentele en Klinische Bewegingswetenschappen, pp. 45–50.

Cordo, P. and Harnda, S. (1994). *Movement Control.* Cambridge University Press, Cambridge. [see especially Target Article 1, by Bizzi et al]

Flash, T. (ed.). (1993). Special issue of *J. Mot. Behav.,* Vol. 25, #3.

Latashs, M.L. (1993). *Control of Human Movement.* Human Kinetics, Champaign, Illinois.

Winters, J.M. and Woo, S.-L.Y. (eds.). (1990). *Multiple Muscle Systems: Biomechanics and Neural Control of Movement.* Listing of Part III chpaters (Chapters 11–22, respectively). Springer-Verlag, New York.

Hogan, N. and Winters, J.M. *Overview: Principles Underlying Upper Limb Coordination,* pp. 182–194.

Feldman, A.G. and Adamovich, S.V., et al. *The Origin of Electromyogram—Explanations Based on the Equilibrium Point Hypothesis,* pp. 195–213.

Wu, C.-H., Houk, J.C., et al. *Nonlinear Damping of Limb Motion,* pp. 214–235.

Gottlieb, G.L., Corcos, D.M., et al. *Principles Underlying Single-Joint Movement Strategies,* pp. 236–250.

Corcos, D.M., Gottlieb, et al. *Organizing Principles Underlying Motor Skill Acquisition,* pp. 251–268.

Karst, G.M. and Hasan, Z. *Direction-Dependent Strategy for Control of Multi-Joint Arm Movements,* pp. 268–281.

Flash, T. *The Organization of Human Arm Trajectory Control,* pp. 282–301.

Gielen, S., van Ingen Schenau, et al. *The Activation of Mono- and Bi-articular Muscles in Multi-joint Movements,* pp. 302–311.

Seif-Naraghi, A.H. and Winters, J.M. *Optimized Strategies for Scaling Goal-directed Dynamic Limb Movements,* pp. 312–334.

Denier van der Gon, J.J., Coolen, A.C.C., et al. *Self-organizing Neural Mechanisms Possibly Responsible for Movement Coordination,* pp. 334–342.

Crago, P.E., Lemay, M.A., and Lui, L. *External Control of Limb Movements Involving Environmental Interactions,* pp. 343–359.

Meek, S.G., Wood, J.E., and Jacobsen, S.C. *Model-based, Multi-muscle EMG Control of Upper-extremity Prostheses,* pp. 360–376.

23
Maps, Modules, and Internal Models in Human Motor Control

Daniel M. Wolpert and Zoubin Ghahramani

1 Introduction

Neural network models of computation have recently provided a strong foundation from which to formulate computational theories of learning, planning and action (Kawato et al. 1987; Jordan 1995; see also Chapter 34). Here we consider three computational ideas—the generalization properties of function approximators, self-organized modularity, and optimal estimation—and show how these can be used to design and interpret psychophysical experiments which explore the computations involved in motor control.

2 Function Approximation and Generalization

One of the basic problems the motor system is faced with is mapping information between different coordinate systems. For example the visuomotor map must transform visual coordinates (e.g., object location) into coordinates appropriate for action (e.g., the set of joint angles required to reach an object). Formally, such maps are functions from an input space to an output space; the adaptability of these maps throughout life can be considered the process of approximating these functions from a finite set of input-outputs pairs experienced. However, as there are infinitely many possible functions consistent with any finite set of input-output pairs, the problem is ill-posed (Tikhonov and Arsenin 1977). The actual function that arises from approximating the input–output pairs must therefore reflect intrinsic constraints or biases of the function approximator.

The structure and constraints on these maps can be explored by studying their generalization to a limited set of novel input–output pairs (for a review of this technique see Bedford 1989). The paradigm consists of three phases: in the first the output of the mapping is evaluated at many points over the input space, in the second "exposure" phase a novel set of input-output pairs (i.e., local remapping) is introduced, and in the third the mapping is once again evaluated over the input space. Changes in the map between the first and third phases at inputs which were not part of the exposure phase define the way the map generalizes to the exposure.

Two different inferences about the representation of the map can be derived from these studies. The first is the local-to-global nature of the map which can be assessed by the rate of decay of the generalization away from the input-output pairs experienced in the exposure phase. For example, if a map were represented as a look-up table in which corresponding input-output pairs are stored (Atkeson 1989; Rosenbaum et al. 1993), the change in the map would be local—training at one point would simply change the pairing at that point while leaving unaltered previously learned pairings. Alternatively, if the map was parameterized by a small set of parameters, for example, the eye position (Harris 1965) and physical parameters of the arm for the visuomotor map, then changing a parameter for a single input–output pair might produce widespread changes in the map. Intermediate in the range of local-to-global representations are function approximators such as neural network models (for a

review see Hertz et al. 1991). Neural network models are defined by a large number of parameters
(e.g., the weights in the network) that do not necessarily correspond to the physical parameters of the
system. In these models, the structure of the network determines the extent of generalization to a
local novel remapping. Consequently, the extent of
generalization can be used to infer structural properties of the network such as the effective receptive field size of the map. The second inference
which can be derived from generalization studies
is the natural coordinate system of the map. Because a limited remapping does not specify the expected change outside the exposure region the pattern of generalization can be used to assess the
coordinate system which best captures the learning
(e.g., joint vs. Cartesian).

We have examined the generalization of the visuomotor coordinate transformation to novel one-
and two-point remappings (Ghahramani et al.
1997). Subjects' pointing accuracy was assessed in
the horizontal plane before and after exposure to a

novel remapping. During the remapping, concomitant visuomotor input–output pairs were limited to
either one or two points. This was achieved by using a virtual reality-like setup (for details see
Wolpert et al. 1995a) in which direct visual feedback of the hand was replaced by a computer-controlled illuminated cursor projected into the plane
of movement. During the exposure phase, the cursor was only illuminated when the finger was at
one or two points in the workspace—at all other
locations it was extinguished. At those points, a discrepancy was introduced between where the subject had to point to and where the illuminated cursor appeared. Therefore, during this phase only one
or two remapped input–output pairs were experienced. The first column of Figure 23.1 shows the
remappings used. The base of the arrows shows the
inputs—that is, the visual location of the targets—
and the tip of the arrows show where subjects had
to point to see their finger at the target.

The change in pointing behavior over the workspace is shown in the second and third columns of

Remapping   Generalization   Simulation

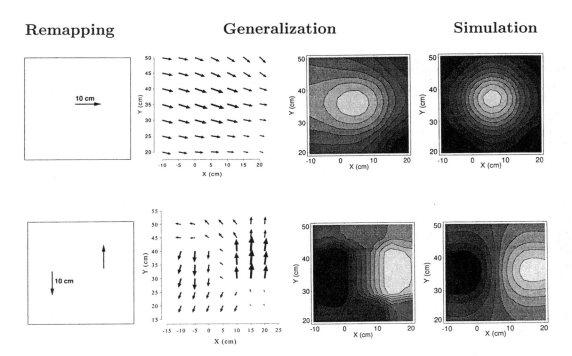

FIGURE 23.1. One point (upper row) and two point (lower
row) generalization of the visuomotor map (average of
eight subjects each) shown over the horizontal workspace
seen from above. From left to right: the remapping during the exposure phase, the induced change in pointing

behavior as an interpolated vector field, adaptation as a
contour plot (light shading represents right for one point
and up for two point), and simulation of an RBF network.

Figure 23.1 for the one and two point remappings. Both these remappings produced most adaptation at the training points with a decaying generalization to points away from the remapped locations. These results rule out a look-up table representation in which only visited areas in the map are adjusted. The generalization seen was also consistent with the mapping being represented in Cartesian space as the generalization was approximately colinear when plotted in Cartesian space (compared, for example, to codings such as eye-rotation).

This generalization behavior was well captured (last column of Figure 23.1) by a Gaussian radial basis function (RBF) network (Broomhead and Lowe 1988; Moody and Darken 1989; see also Chapter 24 and Chapter 25). This model approximates the visuomotor function via a superposition of bases—each unit learns a preferred motor output and the total motor output is the weighted sum of the outputs of each unit. The weighting comes from the Gaussian receptive field assigned to each unit—so that each unit contributes to positions in the input space to which it is best suited. The standard deviation of the receptive field determines the trade off between how closely the model fits the input–output data with how smooth the resulting function is (Poggio and Girosi 1989). In our simulations, a standard deviation of 5 cm (~6° of visual angle) was found to be able match the experimental data—a smaller or larger value led to a more local or global generalization respectively. This study shows that the visuo-motor system is intrinsically biased towards learning smooth mappings, that the effective receptive field size for learning is approximately 6°, and that the coordinate system which best captures the generalization is Cartesian.

3   Modular Decomposition

The principle of "divide-and-conquer," the decomposition of a complex task into subtasks, has been a powerful design tool for computational models of learning (Jacobs et al. 1991; Jordan and Jacobs 1994; Cacciatore and Nowlan 1994). A general strategy for learning is to divide a complex task into simpler subtasks and learn each subtask with a separate module (see also Chapters 19, 34, 35, and 40). This strategy has recently been formalized

into a computational model of learning known as the "mixture of experts," in which the outputs of a set of expert modules are combined by a single gating module (Jacobs et al. 1991). The system simultaneously learns to partition the task into subtasks (the role of the gating module) and to learn each of these subtasks (the role of the experts). The gating module smoothly combines the output of each of the experts based on its estimated probability that each expert will produce the desired output.

A modular decomposition model is shown in Figure 23.2—this represents the simplest instantiation of the hierarchical mixture of experts in which there is only one level and two experts. Consider the task of learning a mapping between some input $\mathbf{x}$ and desired output $\hat{\mathbf{y}}$. Each expert learns a different mapping between this input and the outputs ($\mathbf{y}_1$ and $\mathbf{y}_2$). The contribution of each of these experts to the total system output $\hat{\mathbf{y}}$ is determined by the gating module's output p. The inset shows the relationship between the modulation parameter p as a function of the input (shown here for one dimension). The logistic (or soft-max for multidimensional outputs) form of this relationship can be derived from a probabilistic interpretation in which each expert is responsible for an equal variance Gaussian region, its receptive field, around its preferred input. Therefore p reflects the probability

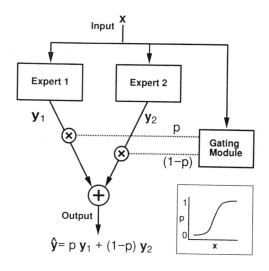

FIGURE 23.2. A modular decomposition model in which each expert learns a different mapping and the outputs are combined by the gating module.

that Expert 1 is the correct module to use for a particular input. The mixture of experts, therefore, partitions the task into subtasks and then combines the output of each expert based on the gating modules expectation that each expert is the appropriate one to use for a particular input.

We have investigated a learning paradigm in which a virtual visual feedback system is used to remap a single visual target location to two different finger positions depending on the starting location of the movement (Ghahramani and Wolpert 1997). Such a perturbation creates a conflict in the visuomotor map of the kinematics of the arm which captures the (normally one-to-one) relation between visually perceived and actual finger locations. One way to resolve this conflict is to develop two separate visuomotor maps (i.e., the expert modules), each appropriate for one of the two starting locations. A separate mechanism (i.e., the gating module) then combines, based on the starting location of the movement, the outputs of the two visuomotor maps. As in previous studies of the visuomotor system (Bedford 1989; Imamizu et al. 1995; Ghahramani et al. 1996), the internal structure of the system can be probed by investigating the generalization properties in response to novel inputs, which in this case are the starting locations on which it has not been trained. The hallmark of a system with modular decomposition would be the ability to learn both conflicting mappings, and to smoothly transition from one visuomotor map to the other in a sigmoidal fashion as the starting location is varied.

Subjects were exposed to two different visuomotor rearrangements at a single visual target location during movements made from two possible starting locations (S2 and S6). Two perturbations, equal in magnitude but opposite in sign, were used, where the sign of the perturbation was determined by the starting location of the movement (Figure 23.3 left). The experimental setup consisted of a virtual visual feedback system in the horizontal plane (described in Wolpert et al. 1995). This setup allowed real-time capture of finger location and the presentation of both targets and visual feedback of the finger in the movement plane. All movements were made in the absence of any visual feedback of the true finger location. The relation between the actual finger location and its visually presented position, represented as a cursor spot, was computer controlled so as to allow arbitrary remappings. Subjects' pointing accuracy, in the absence of visual feedback, for pointing to the target (T) from the seven starting locations (S1-S7) was assessed before and after an exposure to a novel remapping. During the exposure phase, vision of the finger cursor was provided and subjects repeatedly traced out a visual triangle S2-S6-T-S6-S2-T-S2, thereby alternately pointing to the target from S2 and S6. The dotted lines in Figure 23.3 (left) show the path taken by the visual feedback of the finger location and the solid line the actual path taken by the finger. The single visual location T is, therefore, remapped into two distinct finger locations depending on whether the movement starts from S2 or S6.

Figure 23.3 (middle) shows the mean change in pointing behavior (average of 8 subjects) from each starting location (S1-S7 denoted by shading) as a vector with 95% confidence ellipses. Although subjects were unaware of the remappings, they showed significant adaptation when pointing from S2 and

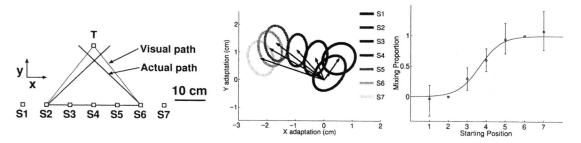

FIGURE 23.3. (Left) The remapping of the visuomotor map dependent on starting location. (Middle) The change in pointing behavior induced from the 7 starting locations. (Right) Estimated mixing proportion p for the 7 starting locations. All error bars and ellipses are 95% confidence intervals.

S6. The adaptation seen for movements from these two points were significantly different ($P < 0.001$), showing that the subjects were able to learn two distinct remappings of the same point in visual space as a function of the starting location. Furthermore, a smooth transition can be seen in the pattern of learning as the starting location is varied between S1 and S7. Figure 23.3 (right) shows the estimated mixing proportion (p in Figure 23.2). The values of p are fixed to be 0 and 1 at starting locations S2 and S6, respectively. The values of p at points other than S2 and S6 capture the form of the generalization as a function of the two learned mappings at S2 and S6. These estimates show a significant modulation over the starting locations ($P < 0.001$), which was captured by a logistic function ($P < 0.001$).

The hypothesis of modular decomposition can be contrasted with models in which a single module combines both the role of the gating module and experts, in our case taking in as inputs both the target and starting locations. It is unclear how such a single module would generalize to new starting locations as this depends crucially on the internal structure of the module. Therefore, the finding that two different maps can be learned for the same point, and that the gating has a logistic relationship, provides evidence that modular decomposition is a feature of visuomotor learning.

4   Optimal Estimation

Although many studies have examined integration among purely sensory stimuli (for a psychophysical review see Welch and Warren 1986) little is known of how sensory and motor information is integrated during movement. When we move our arm in the absence of visual feedback, there are three basic methods the central nervous system can use to obtain an estimate of the current state, the position and velocity, of the hand. The system can make use of sensory inflow (the information available from proprioception), it can make use of integrated motor outflow (the motor commands sent to the arm), or it can combine these two sources of information. Although sensory signals can directly cue the location of the hand, motor outflow generally does not. For example, given a sequence of torques applied to the arm (the motor outflow)

an internal model of the arm's dynamics is needed to estimate the arm's final configuration. To examine whether an internal model of the arm is used we have studied a sensorimotor integration task in which subjects, after initially viewing their arm in the light, made arm movements in the dark (Wolpert et al. 1995b). The subjects' internal estimate of hand location was assessed by asking them to visually localize the position of their hand (which was hidden from view) at the end of the movement. The bias of this location estimate, plotted as a function of movement duration shows a consistent overestimation of the distance moved (Figure 23.4). This bias shows two distinct phases as a function of movement duration, an initial increase reaching a peak after one second followed by a transition to a region of gradual decline. The variance of the estimate also shows an initial increase during the first second of movement after which it plateaus.

We have developed a model of the sensorimotor integration process which can fully account for the experimental results. This model of the sensorimotor integration process which integrates the efferent outflow and the reafferent sensory inflow is based on the optimal state estimation framework from engineering (Goodwin and Sin 1984). We chose to use a Kalman filter, a linear dynamical system that produces an estimate of the location of the hand by using both the motor outflow and sensory feedback in conjunction with a model of the motor system thereby reducing the overall uncertainty in its estimate. The model is a combination of two processes which together contribute to the state estimate. The first, feedforward process (Figure 23.5, upper part) uses the current state estimate and motor command to predict the next state by simulating the movement dynamics with a forward model. The second, feedback process (Figure 23.5, lower part) uses a model of the sensory output process to predict the sensory feedback from the current state estimate. The sensory error, the difference between actual and predicted sensory feedback, is used to correct the state estimate resulting from the forward model. The relative contributions of the internal simulation and sensory correction processes to the final estimate are modulated by the Kalman gain so as to provide optimal state estimates. To accommodate the observation that subjects generally tend to overestimate

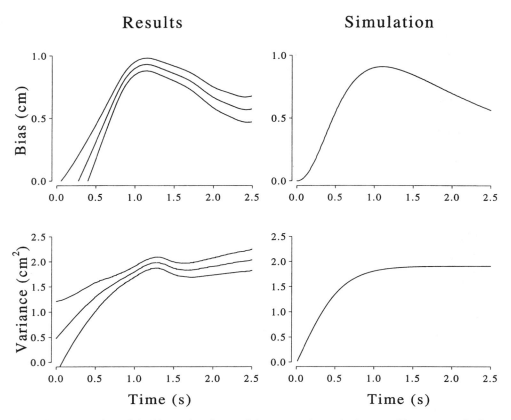

FIGURE 23.4. The propagation of the bias and variance of the state estimate is shown, with outer standard error lines, against movement duration. Also shown are the bias and variance form the simulation of the Kalman filter.

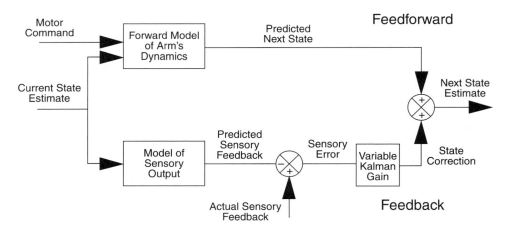

FIGURE 23.5. The Kalman filter model is shown schematically comprising two processes. The first (upper part) uses the motor command and the current state estimate to achieve a state estimate using the forward model to simulate the arm's dynamics. The second process (lower part) uses the difference between expected and actual sensory feedback to correct the forward model state estimate. The relative weighting of these two processes is mediated through the Kalman gain.

the distance that their arm has moved, we set the gain that couples force to state estimates to a value that is larger than its veridical value.

The Kalman filter model demonstrates the two distinct phases of bias propagation observed (Figure 23.4, upper right). By overestimating the force acting on the arm the forward model overestimates the distance traveled, an integrative process eventually balanced by the sensory correction. The pattern of variance propagation is also captured by the model. The variance of the state estimate derives from two sources of variance in the system, the first is the variability in the response of the arm to the motor commands and the second is the noise in the subsequent sensory feedback. Initially, when the hand is in view, the state estimate is assumed to be accurate. The accuracy of the prediction from the forward model component of the Kalman filter depends on the accuracy of the current state estimate (one of its inputs). Therefore during the early part of the movement, when the current state estimate is accurate, the sensorimotor integration process weights heavily the contribution of the forward model to the final estimate. However, in the later stages of the movement, when the current state estimate is less accurate the sensory feedback must be relied upon to correct for inaccuracies in the forward model. In the Kalman filter, the relative weighting shifts from the forward model towards sensory feedback over the first second of movement and then remains approximately constant resulting in the asymptote of the variance propagation.

The Kalman filter model suggests that the peaking and gradual decline in bias is a consequence of a trade off between the inaccuracies accumulating in the internal simulation of the arm's dynamics and the feedback of actual sensory information. Simple models which do not trade off the contributions of a forward model with sensory feedback, such as those based purely on sensory inflow or on motor outflow, are unable to reproduce the observed pattern of bias and variance propagation. The ability of the Kalman filter to parsimoniously model our data suggests that the processes embodied in the filter, namely internal simulation through a forward model together with sensory correction, are likely to be embodied in the sensorimotor integration process.

5 Summary and Future Directions

We have shown that powerful computational principles can be used to guide experimental work in sensorimotor control. Two benefits should accrue from further work within this computational approach. The first is that by comparing the predictions of specifically lesioned models with the data from neurological patients, components of the models can be neurally localized, shedding light both on the computations underlying normal behavior and the deficits in patients. For example, the sensorimotor integration model makes precise predictions regarding the effects of removing the forward model which can be correlated with behavior in neurological patients. The second and more challenging avenue will be the investigation of these models at a neurophysiological level. At the moment, although psychophysics can be used to uncover the computations underlying behavior, neurophysiology will illuminate the way in which these computations are neurally instantiated.

Acknowledgments. We thank the Wellcome Trust. Zoubin Ghahramani was supported by a fellowships from the McDonnell-Pew Foundation and the Ontario Information Technology Research Centre.

References

Atkeson, C. (1989). Learning arm kinematics and dynamics. *Ann. Rev. Neurosci.*, 12:157–183.

Bedford, F. (1989). Constraints on learning new mappings between perceptual dimensions. *J. Exp. Psychol. Hum. Percept. Perform.*, 15:232–248.

Broomhead, D. and Lowe, D. (1988). Multivariable functional interpolation and adaptive networks. *Complex Sys.*, 2:321–355.

Cacciatore, T.W. and Nowlan, S.J. (1994). Mixtures of controllers for jump Linear and non-linear plants. In *Advances in Neural Information Processing Systems 6*, Cowan, J.D., Tesauro, G., and Alspector, J., (eds.), pp. 719–726. Morgan Kaufman Publishers, San Francisco, California.

Ghahramani, Z. and Wolpert, D. (1997). Modular decomposition in visuomotor learning. *Nature*, 386: 392–395

Ghahramani, Z., Wolpert, D., and Jordan, M. (1996). Generalization to remappings of the visuomotor coordinate transformation. *J. Neurosci.*, 16:7085–7096.

Goodwin, G. and Sin, K. (1984). Adaptive filtering prediction and control. Prentice-Hall.

Harris, C. (1965). Perceptual adaptation to inverted, reversed, and displaced vision. *Psychol. Rev.*, 72:419–444.

Hertz, J., Krogh, A., and Palmer, R. (1991). *Introduction to the Theory of Neural Computation*. Addison-Wesley, Redwood City, California.

Imamizu, H., Uno, Y., and Kawato, M. (1995). Internal representations of the motor apparatus: implications from generalization in visuomotor learning. *J. Exp. Psychol. Hum. Percept. Perform.*, 21:1174–1198.

Jacobs, R.A., Jordan, M.I., Nowlan, S.J., and Hinton, G.E. (1991). Adaptive mixture of local experts. *Neural Comput.*, 3:79–87.

Jordan, M.I. (1995). Computational aspects of motor control and motor learning. In *Handbook of Perception and Action: Motor Skills*. Heuer, H. and Keele, S., (eds.), Academic Press, New York.

Jordan, M.I. and Jacobs, R. (1994). Hierarchical mixtures of experts and the EM algorithm. *Neural Comput.*, 6:181–214.

Kawato, M., Furawaka, K., and Suzuki, R. (1987). A hierarchical neural network model for the control and learning of voluntary movements. *Biol. Cybern.*, 56:1–17.

Moody, J. and Darken, C. (1989). Fast learning in networks of locally-tuned processing units. *Neural Comput.*, 1:281–294.

Poggio, T. and Girosi, F. (1989). A theory of networks for approximation and learning. AI Lab Memo 1140, The MIT Press.

Rosenbaum, D., Engelbrecht, S., Bushe, M., and Loukopoulos, L. (1993). Knowledge model for selecting and producing reaching movements. *J. Motor Behav.*, 25:217–227.

Tikhonov, A. and Arsenin, V. (1977). *Solutions of Ill-PosedProblems*. W.H. Winston, Washington, D.C.

Welch, R. and Warren, D. (1986). Intersensory interactions. In *Handbook of Perception and Human Performance. Volume I: Sensory Processes and Perception*. Boff, K., Kaufman, L., and Thomas, J., (eds.), John Wiley & Sons, New York.

Wolpert, D.M., Ghahramani, Z., and Jordan, M.I. (1995a). Are arm trajectories planned in kinematic or dynamic coordinates? An adaptation study. *Exp. Brain Res.*, 103:460–470.

Wolpert, D.M., Ghahramani, Z., and Jordan, M.I. (1995b). An internal model for sensorimotor integration. *Science*, 269:1880–1882.

24

How Much Coordination Can Be Obtained Without Representing Time?

Ferdinando A. Mussa-Ivaldi

1 Introduction

It is a commonplace that to produce even the simplest natural behaviors, the central nervous system must generate complex patterns of motor commands. From the standpoint of control theory, a motor command is a way to achieve a desired goal. In many cases, motor goals may be formulated as movements, that is as temporal sequences of positions to be assumed by a limb or by a limb's endpoint. From the standpoint of information processing, a motor command is a way to encode a desired behavior. Thus, a task such as "grab the cup on the table" may be represented by one of the possible command patterns that our nervous system sets up for its execution.

This chapter discusses to what degree the motor control system must be capable of representing time explicitly when generating a motor command. Intuitively, when we represent an action in abstract terms (e.g., "grab the cup on the table"), we start from a symbolic entity with little or no temporal attributes. Time comes progressively into the picture as the execution details are filled-in ("Move the hand toward the cup, then reach it, then grasp it . . ."). How are these temporal details specified by the process that translates planning into execution? To what extent are they specified implicitly by the dynamical properties of the controlled system? To what extent must they be explicitly supplied by higher planning levels? These questions may be formulated in more specific terms when dealing with the execution of an arm's movement. Should we assume that such a movement is produced by some process defining the coordinated temporal sequence of joint torques (Kawato et al. 1990) or the temporal sequence of the arm's equilibrium positions

(Hogan 1984)? Or, instead, is it possible for the entire movement to be created by a single timeless command, equivalent to state, for example, "move the hand on a straight line from A to B?"

Those who deal with dynamic systems know that time must be explicitly represented when the system under study is connected to something else. Something that one does not know how (or does not want) to describe. This "something else" is seen by the system as the source of a forcing function—a function of time. Conversely, a system whose differential equation does not contain time explicitly is called an "autonomous" system because its behavior is entirely and solely determined by its own state.

So, when we think of a control system as an element that executes some externally planned command, the degree to which time is explicitly represented in the external command reflects the degree of coupling between the control system and the higher stages of motor planning.

The need to minimize the explicit representation of time in the description of a control system is well illustrated by the models of "simple" behaviors, such as the movements of the eye. A typical and well-studied movement of the eye is the saccade: a rapid stepwise transition from an initial to a final fixation point. Studies of oculomotor neuron activities (Robinson 1970) revealed a pattern that has been described as "pulse-step" control: the firing frequency of oculomotor neurons during a saccade is well represented by the superposition of a stepwise function, coding the transition from start to end location, and a rapid and short-lived impulse that drives the eyeball against its own viscous drag. Pulse-step control has a very limited representation of time—the only relevant parameter being the pulse duration—and yet it accounts for the main features of saccadic movements. Given that pulse-like and step-like activity patterns are ubiquitous in brain structures involved with motor control, it is tempting to speculate that a variety of motor behaviors might be commanded in this way. However, there is a major obstacle along this path. The coordination of natural multi-segmental movement has revealed a complex temporal structure.

A number of studies of multijoint arm movements have suggested that even to generate some of the simplest trajectories of the hand, the CNS must orchestrate critically crafted temporal sequences of control signals. Investigations by Morasso (1981) revealed that the CNS seems to be keen in orchestrating complex intersegmental coordinations so as to maintain a simple pattern of hand motions. When subjects are required to make a target reaching movement of the hand in the horizontal plane, they tend to move the hand along straight pathways and with a unimodal "bell-shaped" velocity profile. To generate such simple movements of the hand, one must coordinate precisely timed reversals of one or two joint angles. Morasso's observation suggested that a simple movement of the hand may involve a complex temporal structure of the motor command. This idea is consistent with the hypothesis that limb movements are generated by the CNS as temporal sequences of the limb's equilibrium point (Bizzi et al. 1984; Hoga 1984; Feldman 1986; Flash 1987; Bizzi et al. 1992), also known as virtual trajectories.

In theory, the most direct implementation of a virtual trajectory could be achieved by a control system that, at each time, specifies the static equilibrium of a limb. Such an implementation would require the explicit representation of time—by means of what is functionally equivalent to a "clock." But is such an explicit representation of time a real necessity? Or, is it possible for the temporal order of movements to emerge from the intrinsic dynamics of the motor primitives that are expressed by the output stages of the motor system?

2   Force Fields as Control Primitives

The main point in this perspective is that the mechanics of the motor system determine the spectrum of available control alternatives and, in particular, the degree to which representing time is necessary for achieving coordination. A way to describe the mechanical behavior of a linear system is by specifying its impedance: the relation between externally imposed motions (input) and consequent force (output). Impedance summarizes the behavior of a control system at the interface with its environment (Hogan 1985). It has different components, associated with different time derivatives of motion, such as stiffness (force vs. displacement), viscosity (force vs. velocity), and inertia (force vs. acceleration). These terms can be applied to the analysis of non-linear systems if one limits the scope of the observation to sufficiently small regions of space. Earlier work of Mussa-Ivaldi et al. (1985) applied this local linearization to estimate the stiffness of the hand at a number of equilibrium locations in the horizontal plane.

This local analysis however cannot be extrapolated to the mechanical behavior of the arm across broader regions surrounding the arm's equilibrium location, x_0. One cannot assume that the force vector, F, is related to the hand position, x, by a Hooke's law relation such as

$$F = K(x - x_0).$$

More recent investigations by Shadmehr et al. (1993) revealed that the estimated stiffness of a hand posture decreases with the distance from the equilibrium point. In this study, the spring-like behavior of the arm was explored over a larger range of locations (10 cm) surrounding hand posture and was described as a nonlinear force field (Figure 24.1). Mathematically, the measured fields were reconstructed and represented as linear combinations of Gaussian fields* centered at different workspace locations. This turned out to be an efficient representation as only 25 Gaussian functions were generally sufficient to obtain an approximation with less than 5% error.

Similar force fields were measured by Giszter et al. (1993; see also Chapter 23) in a very different preparation: the spinalized frog. In these experiments the hindlimb of the frog was kept at a number of locations while a brief train of electrical im-

---

*Here, Gaussian field refers to a vector field that is derived as the gradient of a multivariate Gaussian function. Thus, given a real-valued Gaussian, $G(x) = \exp(-x^T K x)$, the corresponding Gaussian field is the vector-valued function $\phi(x) = \nabla G(x) = -Kx \exp(-x^T K x)$.

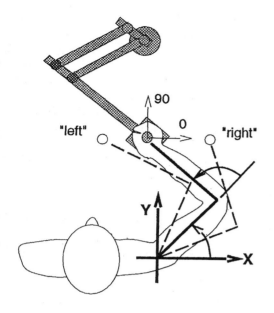

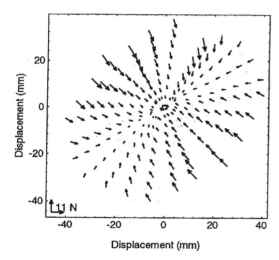

FIGURE 24.1. Force field associated with arm posture. (Top) Experimental set-up. Subjects were asked to keep the handle of a manipulandum either in the "right" or in the "left" position. A set of slow displacements (3–4 cm in 7–10 sec) in different directions were applied by the manipulandum and the resulting elastic forces were measured by a sensor mounted on its handle. (Bottom) A force field measured in the "right" position. (Modified from Shadmehr et al. 1993.)

pulses was delivered by a microelectrode to a site in the lumbar spinal cord. The effect of each stimulation was to elicit a field of forces acting at the endpoint of the ipsilateral hindlimb. When two spinal sites were stimulated at once, the resulting

field corresponded to the vector sum of the fields elicited by each stimulation site (Mussa-Ivaldi et al. 1994). The empirical finding of vector summation suggested that the central nervous system may create a variety of control patterns by the linear combination of independent modules. The problem of representing and implementing different control policies was then reduced to something similar to Fourier analysis of continuous functions. In our case, a variety of control fields, $C(x)$, may be obtained from the superposition of elementary *basis fields*, $\phi_1(x), \phi_2(x), ..., \phi_K(x)$:

$$C(x) = \sum_{i=1}^{K} c_i \phi_i(x)$$

A related computational analysis (Mussa-Ivaldi 1992; Mussa-Ivaldi and Giszter 1992) demonstrated that it is precisely the nonlinearity of these basis fields—the nonlinear dependence of the force upon the limb location—that ensure "expressive power" to this linear summation mechanism.

The force fields measured in biological systems may be formalized as elementary control structures, or in the terminology used elsewhere in this book, primitives (e.g., Chapters 25 and 35). To this end, consider a hypothetical limb operated by K independent modules, each one generating a field of forces over the limb's state space. Let x and x^Y indicate the configuration and the velocity of the limb. A *control field* is a mapping from the limb's state to the output force, $F = \phi(x, \dot{x})$. A set of K modules can be described by providing the corresponding control fields, $\{\phi_1(x, \dot{x}), \phi_2(x, \dot{x}), ..., \phi_K(x, \dot{x})\}$. Based on the empirical finding of vector summation of spinal force fields, we may represent the net control field obtained from a combination of modules as

$$C(x, \dot{x}) = \sum_{i=1}^{K} c_i \phi_i(x, \dot{x}) \qquad (24.1)$$

Here, the parameters c_i are coefficients that tune the intensity of each module's field without altering its functional shape. This net control field drives the passive dynamics of the limb which contain a combination of inertial, viscous and elastic components. The passive dynamics may also be represented as a nonlinear field of forces, $D(x, \dot{x}, \ddot{x})$. Putting together the passive dynamics and the control field one obtains an ordinary differential equation (ODE) that describes the motor behavior of the limb:

$$D(x, \dot{x}, \ddot{x}) = C(x, \dot{x}) \qquad (24.2)$$

This ODE is an equilibrium condition between the fields D and C : a trajectory, $x(t)$, is a solution if and only if, after substituting it for x, $\dot{x}$, and $\ddot{x}$, Eqn. (24.2) becomes identically satisfied. This equation has some variants which reflect different assumptions on the control field. The control field may depend only on position ($C(x)$), or on the full state ($C(x,\dot{x})$), or on position and time $C(x,t)$ or on state and time and so on. Here, I begin by considering the first, simplest case, a case in which time is totally absent from the formulation of the controller.

3 Generation of Movements by Field Approximation

Let us begin by representing a desired sequence of states as a continuous trajectory of the limb's endpoint. A typical problem in classical mechanics is to find the family of solutions (i.e., the trajectories) of a differential equation such as Eqn (24.2). Our goal is the opposite: given a trajectory, we wish to derive a differential equation admitting this trajectory as a solution. We make the hypothesis that the only way for generating different equations is to select the tuning parameters, c_i, of the control field in Eqn (24.1).

It is not generally the case that a desired trajectory may be generated by some allowable setting of the control coefficients. Thus, to produce exactly any desired trajectory is not a feasible goal for the motor control system. A more reasonable goal is to approximate the desired trajectories. Movement approximation may be cast in a rigorous framework by introducing the notion of the *inner product* between two fields, $F(x,\dot{x},\ddot{x})$ and $G(x,\dot{x},\ddot{x})$ along the trajectory $\hat{x}(t)$. To this end, we first define the *restriction*, $F[\hat{x}(t)]$, *of the field $F(x,\dot{x},\ddot{x})$ over the trajectory $\hat{x}(t)$*:

$$F[\hat{x}(t)] \equiv F(\hat{x}(t),\dot{\hat{x}}(t),\ddot{\hat{x}}(t)).$$

This is an operation that maps a trajectory, $\hat{x}(t)$, in the temporal sequence of force vectors generated by the field over this trajectory.

The restriction of a vector field is a useful device for reducing a differential equation to an algebraic equality. By taking the restriction of the

two force fields in Eqn (24.2) over a generic solution, $x(t)$, one obtains

$$D[x(t)] = C[x(t)].$$

Unlike the original differential equation that may or may not be satisfied by any particular setting of the independent variables, $x,\dot{x},\ddot{x}$, the above expression is an algebraic equality that must be identically satisfied for all values of t.

Next, let us define the inner product of two fields, $F(x,\dot{x},\ddot{x})$ and $G(x,\dot{x},\ddot{x})$ along the trajectory $\hat{x}(t)$ as the integral

$$\langle F,G \rangle_{\hat{x}(t)} \equiv \int_{tI}^{tF} F[\hat{x}(t)]^T G[\hat{x}(t)]dt \quad (24.3)$$

Accordingly, the norm of a field along $\hat{x}(t)$ is:

$$\|F\|_{\hat{x}(T)} = \langle F,F \rangle^{1/2}_{\hat{x}(T)}.$$

Following the above definitions, we derive a least-squares approximation of the desired trajectory by minimizing

$$
\begin{aligned}
\varepsilon^2 &\equiv \|D - C\|^2_{\hat{x}(T)} \\
&= \|D\|^2_{\hat{x}(T)} - 2\sum_i c_i \langle \phi_i,D \rangle_{\hat{x}(t)} \\
&\quad + \sum_{i,j} c_i c_j \langle \phi_i,\phi_j \rangle_{\hat{x}(t)}. \quad (24.4)
\end{aligned}
$$

This is the residual of Eqn (24.2) calculated along the desired trajectory, $\hat{x}(t)$. Minimization of ε^2 is achieved by solving for c_i the system of linear equations:

$$\Phi_{j,i} c_i = \Lambda_j \quad (24.5)$$

where

$$
\begin{aligned}
\Phi_{j,i} &= \langle \phi_j,\phi_i \rangle_{\hat{x}(t)} \\
\Lambda_j &= \langle \phi_j,D \rangle_{\hat{x}(t)} \quad (1 \le ij \le K)
\end{aligned}
$$

A relevant aspect of this method is that time is "suppressed" by the projection operation in Eqn. 24.3. The expression for the error, ϵ^2, is quadratic in the control coefficients. Therefore a single global solution is guaranteed to exist.

4 Simulation

The approximation of a desired movement by a summation of static fields has been tested in a simulation of a two-joint planar mechanism (Mussa-Ivaldi and Bizzi 1996). In this case, the controller fields

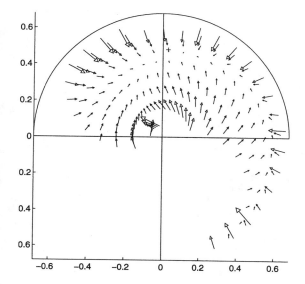

FIGURE 24.2. A Gaussian force field. This is the gradient of a Gaussian potential function over the joint angle space. The cross indicates the equilibrium point (where $F = 0$). (From Mussa-Ivaldi and Bizzi 1996.)

were the gradients of 9 bivariate Gaussian potential functions centered at 9 locations on a 3×3 grid.

One such field is shown in Figure 24.2. When a single controller is activated with a constant input c_i, the limb is attracted toward the equilibrium point of the corresponding force field, $\phi_i(x)$. This attractive force interacts with the nonlinear dynamics of the arm. The result of this interaction is a trajectory $x_i(t)$ which satisfies the differential equation

$$D(x,\dot{x},\ddot{x}) = c_i\phi_i(x).$$

Because of the strong nonlinearity of both the passive dynamics and of the controller, this trajectory will generally follow a curved path with an asymmetric velocity profile (Figure 24.3). The shape of the path and of the velocity profile depend upon the starting location of the hand.

Investigations of equilibrium-point control based on simulation have shown that with a multi-joint limb the setting of a final equilibrium position leads to complex-shaped trajectories, similar to the one shown in Figure 24.3 (Delatizky 1982; Hatsopoulos 1994). These studies suggested that the neural controller must explicitly provide a temporal pattern of intersegmental coordination in order to achieve smooth movements of the hand in the extrapersonal space. Here, I challenge this view and consider the alternative hypothesis that the observed end-point kinematics can be generated by some time-invariant choice of control parameters—that is by a single step in the multidimensional space of these parameters. What makes this idea different from previous studies of final position control is the consideration of an entire nonlinear force-field over the whole limb's workspace instead of a linear controller.

The approximation of three desired hand trajectories is shown in Figure 24.4. The top left panel shows the desired rectilinear paths of the hand. The velocity profile of the desired movement, labeled by the letter T, is on the top right panel. All the desired movements occur on a straight line from start to end position. They last about 800 ms and the tangential velocity profile is expected to follow a symmetric, bell-shaped pattern. The results of the approximation are reported on the lower panels.

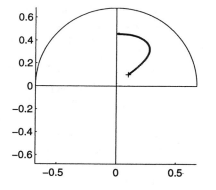
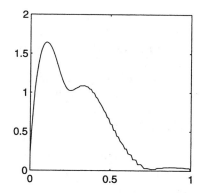

FIGURE 24.3. (Left) Trajectory of the hand following the stepwise activation of the Gaussian controller shown in the previous figure. (Right) Temporal profile of the tangential velocity of the hand. (From Mussa-Ivaldi and Bizzi 1996.)

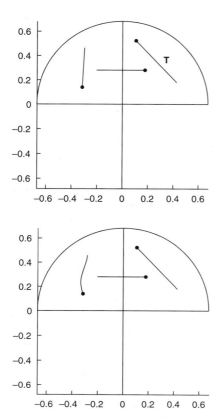

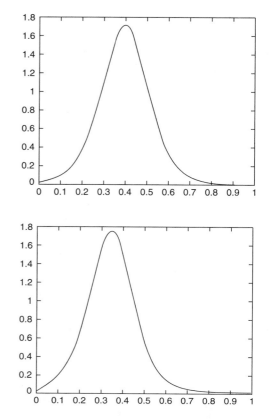

FIGURE 24.4. Approximation of desired trajectories by field summation. (Top left) Desired trajectories of the hand. The small circles indicate the final position. (Top right) Tangential velocity profile for the trajectory indicated by "T". (Bottom panels) Approximation results with 9 Gaussian fields. The variance of these fields was equal to the angular distance between the centers. (Bottom left) Approximating trajectories. (Bottom right) Tangential velocity of the rightmost trajectory (approximation of T). (From Mussa-Ivaldi and Bizzi 1996.)

As remarked by Morasso (1981) the joint kinematics of this movement are rather complex (Figure 24.5). However, the example of Figure 24.5 shows that the limb's kinematics do not necessarily reflect the degree of complexity of the control pattern. In this case the trajectory derives from a control field that does not have any explicit time dependence. The setting of the nine parameters, c_i, may be seen as a combination of synchronous step functions whose mechanical outcome is the force field shown in Figure 24.6 together with three trajectories starting from different initial positions.

This simulation illustrates that, at least in theory, a rich temporal behavior may emerge from timeless control patterns when a system's mechanics is sufficiently nonlinear. While this is an encouraging result, I must also point out three shortcomings of the approximation with time-independent control fields, namely:

1. The approximation failed in some cases. The simulation was not always as successful as shown in Figure 24.4. Occasionally, the minimization of the force error did not lead to minimizing the error between desired and actual trajectories. Preliminary observations suggest that the performance of the approximation may be improved by increasing the number of independent control modules and by varying the centers and the variances of the control fields.

2. The actual trajectories were only partially stable. In our simulation (Figure 24.6) we have found that a combination of static force fields may lead to some degree of stability in the required trajectory: if the initial position of the hand is per-

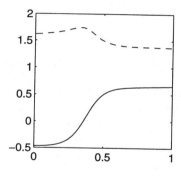

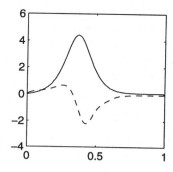

FIGURE 24.5. Joint-angle kinematics obtained from the approximation of the trajectory indicated by "T" in Figure 24.4. (Top panel) Elbow angle (dashed line) and shoulder angle (solid line) vs. time. The abscissa is in seconds and the ordinate in radians. (Bottom panel) The corresponding angular velocity profiles. Note that the elbow goes through a reversal of motion. (From Mussa-Ivaldi and Bizzi 1996.)

turbed, the resulting movement may still converge toward the original unperturbed trajectory. However, (a) the basin of attraction of the desired movement may be narrow and (b) if the initial position is moved forward along the desired movement pathway, the static control field drives the hand toward the final target, without creating a backward initial component. These feature of the simulation are at odds with available experimental evidence. Bizzi et al. (1984) studied the effects of mechanical perturbations applied to the arm of deafferented monkeys during the execution of reaching movements. If, at the onset of a movement, a limb was suddenly displaced towards the intended target by an external perturbation, then at the end of the perturbation, the limb moved back toward the starting point before resuming the original forward movement.

3.   The final field depended upon the starting location. The force field that, in the simulation, drives the arm along the desired trajectory is obviously determined both by the starting and by the ending locations. Intuitively, one may represent the static approximation as a process in which the force fields are tuned so as to mold a single potential landscape that guides the movement along the intended trajectory. Therefore, different trajectories lead necessarily to different landscapes. This feature is not supported by available experimental data (Mussa-Ivaldi et al. 1985; Hocherman et al. 1986) showing that the force field associated with a given arm's posture does not depend upon the movement used to reach that posture.

These shortcomings are sufficient to discard the extreme timeless formulation of this example as a biologically plausible model. Nevertheless, a simple modification of the control system may address these issues by augmenting the set of time-inde-

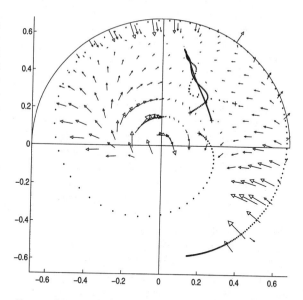

FIGURE 24.6. Outcome of the force field approximation. This force field generates the straight line movement of Figures 24.4 and 24.5. The resulting trajectory is shown here together with two other trajectories obtained by changing the starting location. If, as in these cases, the starting locations belong to the same basin of attraction, the trajectories converge toward the same final location and tend to follow the same path of approach. A trajectory starting outside of the basin of attraction is also shown. (From Mussa-Ivaldi and Bizzi 1996.)

pendent control fields with a set of transient force fields. This modified system of fields is a generalization of pulse-step control.

5   A Kind of Pulse-Step Control

The idea of combining static force fields may be simply modified by partitioning the control field into two components. A static field centered at the final posture, $\Phi(x)$, and a transient time-dependent field, $\Psi(x,t)$, which is present only during the execution of a movement:

$$C(x,t) = \Phi(x) + \Psi(x,t) \qquad (24.6)$$

The static field, $\Phi(x)$, is independent of the movement used to reach equilibrium and the dynamic field, $\Psi(x,t)$, is supposed to vanish at the end of the movement. Therefore, at the end of the movement one is left with a static force field that—as required—depends only upon the final location.

Both fields may be expressed as vectorial combinations of elementary controllers. By combining a set of temporal functions, $u_j(t)$, with the static basis fields one obtains a new set of time-varying fields

$$\Psi_j(x,t) = u_j(t)\phi_{i(j)}(x)$$

that may be used to construct the controller $\Psi(x,t)$ in Eqn (24.6). A particular type of temporal function is a pulse, as it may be approximated by a bell-shaped curve. With such a function, the above time-varying fields would correspond to patterns of force that start from zero amplitude, reach a peak of intensity and vanish. This variation of intensity would not affect the orientation of the force vectors, a feature that has been consistently observed in the active force fields evoked by microstimulation in the spinal frog (Giszter et al. 1993).

Putting together the static and the dynamic components of this distributed control system the net dynamics equation becomes

$$D(x,\dot{x},\ddot{x}) = \sum_i c_i\phi_i(x) + \sum_j d_j\psi_j(x,t) \quad (24.7)$$

This is the vector-field equivalent of a pulse-step controller where the step is given by the static field $\Sigma_i c_i\phi_i(x)$ and the pulse is given by the transient field $\Sigma_j d_j\psi_j(x,t)$.

The coefficients, c_i and d_i may be readily computed following the least-squares approach outlined in Section 3. An important feature of this generalized pulse-step control is that the combination coefficients, c_i and d_i, do not depend upon time. This corresponds to assuming that the supraspinal control does not deal with the fine temporal coordination that is associated with simple movements. Instead, all temporal details may be left to the care of the spinal pattern generators that induce the basis fields $\psi_i(x,t)$. A practical side effect of this computation is the implicit derivation of an equilibrium trajectory (in the form of a time-varying force field) that corresponds to a desired multijoint movement. Once the coefficients c_i and d_i have been computed, the virtual trajectory is defined as the set of locations, $x_0(t)$ that satisfy:

$$D(x_0,0,0) = \sum_i c_i\phi_i(x_0) + \sum_j d_j\psi_j(x_0,t).$$

In this context the virtual trajectory is a necessary outcome of the computational process, rather than an organizing principle.

6   Future Directions

The investigations presented in this perspective indicate that a significant degree of intersegmental coordination may be obtained by control signals with very little temporal structure. Of course, this finding does not prove that actual control signals in the central nervous system should bear any resemblance with the one used in these simulations. However, this study opens a new perspective in two ways. First, it disproves the idea that complex multi-joint kinematics must be the product of complex and critically timed control signals. Studies of neuronal discharge in brain structures such as the motor cortex (Kalaska et al. 1989) and the red nucleus (Miller and Houk 1995) have shown that the pattern of discharge of the higher brain neurons need not to reflect the kinematics of movement.

Second, this work indicates that the mechanics of the control system determines the extent to which time must be explicitly represented by higher centers. The possibility of generating smooth trajectories of the hand by step-wise control signals rests upon the nonlinear features of the control fields. Previous studies have demonstrated the incompetence of linear models to execute simple hand trajectories without utilizing time-varying control functions such as virtual trajectories (Delatizky 1982; Hatsopoulos 1994). This investigation also shows that something as simple as the stepwise activation of static force fields is not compatible with the experimental finding that the force field associated with any given posture does not depend upon the trajectory used to reach that posture. A simple correction of the stepwise model which is consistent with the experimental findings may be obtained by combining a static field with a transient dynamic field. The static field determines the final posture and is the same for all the starting positions. In contrast, the transient dynamic field determines the shape of the trajectory required for reaching the final posture from the starting position. This, again, is just a hypothetical scenario. What is most needed in this scenario is a more plausible mechanical model of the force fields that may be considered as primitive building blocks of multijoint posture and movement. This is a very complex task that can only be approached by some combination of system identification techniques and of prior knowledge about the geometry and mechanics of the musculoskeletal system.

Acknowledgments. This research was supported by ONR grant N00014-95-1-0571 and by NIH grant NS09343.

References

Bizzi, E., Accornero, N., Chapple, W., and Hogan, N. (1984). Posture control and trajectory formation during arm movement. *J. Neurosci.*, 4:2738–2744.

Bizzi, E., Hogan, N., Mussa-Ivaldi, F.A., and Giszter, S.F. (1992). Does the nervous system use equilibrium-point control to guide single and multiple joint movements? *Behav. Brain Sci.*, 15:603–613.

Delatizky, J. (1982). *Final Position Control in Simulated Planar Horizontal Arm Movements.* PhD thesis, M.I.T. Department of Electrical Engineering.

Feldman, A.G. (1986). Once more on the equilibrium-point hypothesis (gamma model) for motor control. *J. Motor Behav.*, 18:17–54.

Flash, T. (1987). The control of hand equilibrium trajectories in multi-joint arm movements. *Biol. Cybern.*, 57:257–274.

Giszter, S.F., Mussa-Ivaldi, F.A., and Bizzi, E. (1993). Convergent force fields organized in the frog's spinal cord. *J. Neurosci.*, 13:467–491.

Hatsopoulos, N.G. (1994). Is a virtual trajectory necessary in reaching movements? *Biol. Cybern.*, 70:541–551.

Hocherman, S., Bizzi, E., Hogan, N., and Mussa-Ivaldi, F.A (1986). Target acquisition and maintenance in two joint arm movements. In *Sensorimotor Plasticity, Theoretical and Clinical Aspects*. Schmidt, R.S. and Jeannerod, M. (eds.). Editions INSERM, Paris.

Hogan, N. (1984). An organizing principle for a class of voluntary movements. *J. Neurosci.*, 4:2745–2754.

Hogan, N. (1985). Impedance control: an approach to manipulation: Parts i, ii, iii. *ASME J. Dynamic Sys., Measurement and Control*, 107:1–24.

Kalaska, J.F., Cohen, D.A.D., Hyde, M.L., and Prud'homme, M.A.. (1989). A comparison of movement direction-related versus load direction-related activity in primate motor cortex, using a two-dimensional reaching task. *J. Neurosci.*, 9:2080–2102.

Kawato, M., Maeda, Y., Uno, Y. and Suzuki, R. (1990). Trajectory formation of arm movement by cascade neural network model based on minimum torque-change criterion. *Biol. Cybern.*, 62:275–288.

Miller, L.E. and Houk, J.C. (1995). Motor coordinates in primate red nucleus: preferential relation to muscle activation versus kinematic variables. *J. Physiol.*, 488:533–548.

Morasso, P. (1981). Spatial control of arm movements. *Exp. Brain Res.*, 42:223–227.

Mussa-Ivaldi, F.A. (1992). From basis functions to basis fields: using vector primitives to capture vector patterns. *Biol. Cybern.*, 67:479–489.

Mussa-Ivaldi, F A. and Bizzi, E. (1997). Learning Newtonian mechanics. In *Self-Organization, Computational Maps and Motor Control*. Morasso, P. and Sanguineti, V. (eds.), Elsevier, Amsterdam pp. 191–238.

Mussa-Ivaldi, F.A. and Giszter, S.F., (1992). Vector field approximation: a computational paradigm for motor control and learning. *Biol. Cybern.*, 67:491–500.

Mussa-Ivaldi, F.A., Giszter, S.F. and Bizzi, E. (1994). Linear combinations of primitives in vertebrate motor control. *Proc. Natl. Acad. Sci. USA*, 91:7534–7538.

Mussa-Ivaldi, F.A., Hogan, N., and Bizzi, E. (1985). Neural, mechanical and geometrical factors subserving arm posture in humans. *J. Neurosci.*, 5:2732–2743.

D.A. Robinson. (1970). Oculomotor unit behavior in the monkey. *J. Neurophysiol.*, 33:393–403.

Shadmehr, R, Mussa-Ivaldi, F.A., and Bizzi, E. (1993). Postural force fields of the human arm and their role in generating multi-joint movements. *J. Neurosci.*, 13:45–62.

25
Augmenting Postural Primitives in Spinal Cord: Dynamic Force-Field Structures Used in Trajectory Generation

Simon Giszter, Michelle Davies, and William Kargo

1 Introduction

The perspective presented here builds on the idea that the central nervous system (CNS), and particularly the spinal cord, is initially organized into, or equipped with, a set of neural/ biomechanical primitives. These primitives can then be used to construct novel movements, or perhaps molded by learning and development into new sets that support new behaviors. The primitives provide a bootstrap system for motor control and avoid the problems of learning movements from a tabula rasa condition. They could be likened to the idea of deep structures proposed in neurolinguistics. We will discuss the use of primitives in movement and new data related to their organization.

1.1 Ill-Posed Problems

Understanding multijoint control of a limb is often complicated by the degrees of freedom problem. The CNS must solve this to control and plan trajectories. The task for the CNS is to locate a single solution from a vast domain of solutions. This problem is one reason why tabula rasa ideas of motor learning are largely untenable. Ill-posed problems pertain in kinematic planning, choice of joint torques and stiffnesses, and choice of muscle activations. Some of these problems are solved (or at least circumvented) in the spinal cords of lower vertebrates. Kinematic solutions to ill-posed problems have been shown to occur in spinal cord in reflex scratching or wiping behaviors in 'spinal' turtles and frogs (frogs and turtles with the spinal cord isolated from descending controls, Fukson et al. 1980;

Berkinblitt et al 1984,1989; Mortin et al. 1985; Giszter et al. 1989; Sergio and Ostry 1993). These animals also show motor equivalence: a spinal frog will use different motor patterns to achieve the same task goal depending on limb state and context. Finally, the data also supported the idea of reusable reflex units in these behaviors (Berkinblitt et al. 1984, 1989).

1.2 Force-Field Solutions: Spatially Organized Forces

Bizzi et al. (1991) examined the organization of muscle responses and end-point forces associated with microstimulation of small collections of spinal interneurons (10–1000), or associated with reflex behaviors (Giszter et al. 1993). This work applied the ideas developed by Feldman, Ostry, and coworkers, and Bizzi, Hogan and Mussa-Ivaldi (see Bizzi et al 1992 for discussion), in human and monkey extending the ideas to the spinal cord circuitry. As a result of these studies, a motor primitive has been suggested, based on combined physiological and biomechanical data (Giszter et al. 1992). This unit has been called a "force-field primitive." A force-field, in this context, is a function that maps forces generated in the limb to the limb's configuration.

Experimentally, force fields are measured by recording the forces generated as a result of stimulation of either the skin or of the central nervous system (CNS). Forces are recorded while holding an animal's limb immobile in a location. The measurement is then repeated at each of an array of positions, for the same stimulation. In this way a sampled map of force-position pairs is constructed.

This map is a summary of effects of (a) the passive and active isometric mechanical properties of the individual muscles, (b) the configuration and mechanics of the skeletal system transforming these forces, and (c) the effects of neural feedback pathways and circuit controls altering activity. These factors interact to generate the measured end-point translational force in the limb, following the stimulation. The force-field description might be used to summarize or predict the motion and interactions of the end-point of the leg with the environment, once motion based effects on the field are accounted for.

The main findings of this isometric analysis of spinal cord were: (1) There were only a few convergent force-fields in frog lumbar spinal cord, which could be revealed by microstimulation. (2) An individual force field was scaled in magnitude through time following stimulation but its structure did not alter. (3) Combination of two different convergent force fields elicited in parallel could usually be represented as a vector summation of the component fields. (4) There was an ordered topography of spinal cord regions eliciting different force fields. (5) The force-field structures elicited by microstimulation resembled those measured during reflex behaviors. It was therefore hypothesized, based on microstimulation, that the collection of spinal force-field primitives could form a basis for the construction of arbitrary force fields or movements by recruiting them in combinations. It was further hypothesized that such recruitment might be performed either by spinal pattern generators or by descending pathways. The findings supported the idea of a biomechanical or force-field primitive organized in spinal circuitry and closely allied to components of reflex behaviors. This perspective is a modern interpretation of Sherrington (e.g., Sherrington 1961).

1.3   Primitives as a Basis for Synthesis

The discovery of vector summation of force fields elicited by microstimulation, and the subsequent elaboration of this finding to show that isometric end-point force fields could often be summed in a multijoint limb (Bizzi et al. 1991; Giszter et al. 1992; Mussa-Ivaldi et al. 1995), suggested a mechanism with which to construct patterns of motion, interactions with the environment, and postural

control. This was given formal structure by Mussa-Ivaldi (1992). Briefly, Mussa-Ivaldi showed that arbitrary fields were readily approximated using a linear combination of conservative and circulating radial vector field primitives (see also Chapter 24). The coefficients of these field primitives could be found by least squares or minimum norm methods.

Mathematically the force field F(x) relating force to position x can be approximated by using k basis fields $q_i(x)$ and control parameters c_i:

$$F(x) = \Sigma_{i=1}^{k} c_i q_i(x) \qquad (25.1)$$

To approximate arbitrary smooth fields, the k basis fields $q_i(x)$ are subdivided into two groupings representing the decomposition of an arbitrary smooth field into a conservative field portion and a circulating field portion. These two groupings consist of $k/2$ conservative or irrotational fields and $k/2$ solenoidal or circulating fields. Thus

$$F(x) = C(x) + R(x) = \Sigma_{i=1}^{k/2} q_i(x) + \Sigma_{i=k/2+1}^{k} q_i(x) \qquad (25.2)$$

Where $C(x)$ is a pure conservative field and R(x) a pure solenoidal field.

Given a set of j sampled (or planned/desired) force vectors P_i at locations x_i the task of a planner is to find the control vector c which minimizes the error e given by:

$$e = \Sigma_{i=1}^{j} [F(x_i) - P_i]^2 \qquad (25.3)$$

Depending on the number of samples (j) the minimum norm, an exact, or a least square approximate solution can be found with a particular set of k basis fields.

Mussa-Ivaldi and Giszter (1992) applied the vector field approximation to the motor system in the nonredundant limb. It was possible that serially redundant manipulators might pose difficulty for the use of this class of models in many biological systems. However, Gandolfo and Mussa-Ivaldi (1993) tested how far the mechanism of end-point defined basis field summation might apply to serially redundant planar linkages. Their results suggested that in serially redundant planar linkages, near vector summation may occur among a large fraction of randomly chosen control primitives. The situation for redundant manipulators operating in three or higher dimensions of the end-point tool may be more complex, and the limitations of vector summation of end-point fields are not as well understood.

In principle, arbitrary field structures and field time courses can be generated by the techniques of basis field approximation in non-redundant systems (Mussa-Ivaldi 1992). The clear advantage of this framework is that several of the processes necessary to actual movement execution can be supported by basis field approximation (e.g., Mussa-Ivaldi and Gandolfo 1993). However this flexibility also implies that planning the details of these processes must be deferred to other mechanisms: it is not a direct outcome of the force-field framework.

A salient point relevant to the basis field approach is the absence of circulating fields or non conservative fields in the biological data. This observation suggests either that the classes of fields approximated by the biological system are more limited than the fields in Mussa-Ivaldi (1992), or that additional undiscovered mechanisms must exist. It has been argued that by avoiding circulating fields and simulating passive systems the biological limb may guarantee stability in passive environments (Colgate and Hogan 1989).

1.4 Issues in Understanding Spinal Cord Primitives

Several issues arise in considering the force field approach, some of which will be addressed with new experimental data below. Foremost among these is the lack of understanding of the relationships between the force-field primitives and other spinal mechanisms. Primitives could represent general purpose premotor elements. For example, they could be the site of convergence of signals from descending, pattern generator and coordinating systems (see also Chapter 35). Alternatively, they could be elements associated exclusively with spinal pattern generators (see Chapter 12) or reflexes (see Chapter 29). Finally, it is also possible that different force-field types could have very different structural underpinnings and interneuronal control in the CNS. There is very little experimental data on how the static force fields measured in frogs and rats under isometric conditions relate to the dynamic force fields underlying the reflex movements of spinal animals.

Data strongly suggested that the set of force-field primitives form an important component of the spinal control of reflex behaviors (Giszter et al. 1993). The observation was made that the types of force fields revealed by microstimulation corre-

sponded closely to force fields that could be measured during wiping, turning and flexion behaviors elicited in spinal frogs. Further, robust translational force fields of similar structure and properties could be measured in limbs allowed to express excess degrees of freedom during the same spinal behaviors (Figure 25.1). The stability of field structures through time was also a feature in reflex behaviors. It is clearly important to understand how force-field primitives are related to, and used in these reflex behaviors. This is, perhaps, the most central issue in understanding how force-field primitives are used and how they should be related to other spinal and descending mechanisms. We have begun to examine this important aspect of the force-field controls in the spinal cord by focusing on spinal responses to different imposed mechanical environments.

The remainder of this perspective will address the role and the control of force-fields during specific identified reflex movements such as wiping. It is likely that force and velocity feedback from the limb are used in movement, but the role of force and velocity feedback in reflex force fields has to date remained unexplored. We are beginning to examine 3-dimensional (3D) translational end-point force-field structure and temporal evolution generated by the proximal 4 degrees of freedom (DOF) of the limb during movement.

2 Problem of Primitives in Movement

The force-field primitive was initially described using isometric techniques. This had several advantages in the short term. The measurements were uncomplicated by velocity dependent muscle properties or afferent feedback. However, these same issues bedevil simple application of the experimental force-field data to movement. Several issues complicate analysis of movement and must be incorporated in the consideration of primitives during motion. Foremost among these are the complexity of the multijoint musculoskeletal system and its behavior during movement.

2.1 Individual Muscle Properties

Muscle structure and properties differ widely (e.g., moment arms, passive elastic, tendon, fiber composition, pinnation). The contributions of most of

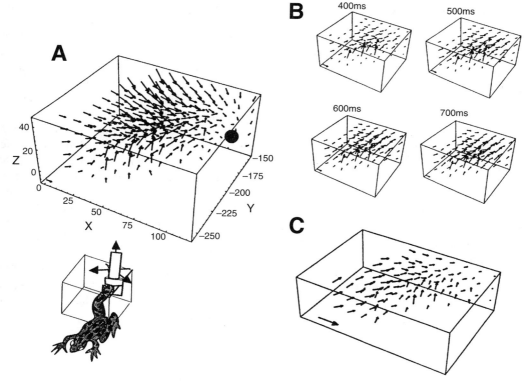

FIGURE 25.1. Three-dimensional isometric force fields supporting wiping behaviors. Data was collected with a gimbal allowing the frog to choose the 4DOF limb orientation but preventing end-point translation. (A) The maximum (or peak) total force field is shown interpolated with the equilibrium position (black ball) indicated. Forces converge to a location caudal and along the midline of the frog's body axis. The wiping phase supported by this force field terminates in a similar position. Distance scales are in millimeters, the calibration force vector (lower left corner) is 1 N. (B) Invariance of active fields. Four active fields after removal of baseline forces and normalization are shown. The normalization opera- tion adjusted the fields so the average vector magnitude was constant (thus the scaling). After this operation it can be seen that the underlying active force pattern is constant over time. This could be confirmed using similarity measures. Time shown is relative to stimulus onset. (C) Active forces in a second frog stimulated weakly. In this frog the free limb moved briefly and weakly and did not overcome gravity. Nonetheless, the orientation of the force patterns shows little variation when compared to force patterns in B. Where applicable, correlation of active wiping fields was greater than 0.9. Similarity between frogs was strong for all three behaviors and their associated force fields.

these are well estimated in the isometric measures, but there are locally varying properties relevant to movement which are complex (Joyce et al. 1969). For example, local to a position the stiffness of the muscle varies with generated force, and this variation is not directly examined in a way relevant to limb dynamics in the isometric force-field framework (e.g., Cecchi et al. 1986)). Most challenging in attempting to apply isometric measures to dynamic limb control are the force-velocity properties of muscle, and hysteresis in these properties (Atevaldt and Crowe 1980; see Chapter 2 and Chapter 8). Several properties of muscle are asymmetric and differ in lengthening versus shortening contractions. In the multijoint limb a mix of these may come into play in the moving limb, distributed across several joints (see Winters et al. 1988; Chapters in Sections II and III of this book).

2.2 Afferent Feedback Effects

Feedback has been invoked as a means of controlling the musculoskeletal plant properties so as to provide a simpler, and more linear, interface to the

CNS (e.g., Nichols and Houk 1975; Houk 1979; Hoffer and Andreassen, 1981). In the context of single muscles, afferent feedback may help linearize the muscle responses to provide a more uniform stiffness. This effect is particularly significant in sudden stretch.

In the context of the multijoint limb the examination of single muscle responses based on homonymous feedback is insufficient. The interactions among muscles and joints and the dynamic interactions during movement suggest higher order behavior of the limb must be controlled by the reflex connections in spinal cord, and two-joint muscles figure prominently in this control. Examination of these properties and the effects of heteronymous connections has been addressed both experimentally and theoretically. In general, these results might be summarized as vector fields and related to ideas of primitives although this has not been a focus of such work thus far. In Loeb et al. (1994) it was shown that reflex feedback contributed to isometric force-field stiffness in microstimulation experiments, as might be expected from the known effects of homonymous feedback.

2.3   A Force-Field Approach to Reflex Movements

There is great difficulty in making completely satisfying top down and bottom up links in multijoint motor control. The range of properties to synthesize run from individual muscle fiber properties and cellular physiology of the spinal cord on the one hand to whole limb behavior and properties on the other. Rather than attempting to synthesize measurements of individual muscle properties and limb mechanics (e.g., by integrating data such as Lieber et al. 1991,1992 into a whole limb model; see also review in Chapter 8), the force-field approach measures whole limb properties. To begin to understand how primitives relate to whole limb dynamic behavior, we have continued this approach while trying to be sensitive to the issues and complexity of the preceding discussion, which may confound such a direct method. During a movement the muscles act to accelerate and decelerate the limb inertia, and to overcome gravity, friction, viscosity or other environmental resistance. The types of contractions that have been examined in muscle physiology which most closely approximate the dynamic con-

ditions the muscle experiences during normal movement are termed auxotonic: contractions against an elastic load. We devised experiments which allowed us to examine the isometric force that is generated in a limb at the end of a movement through an elastic environment. At this point the whole limb and individual contributing muscles were generating force under isometric conditions at equilibrium, and the muscle to end-point force transformation was not changing. However, the impact of the environmental resistance and motion related afference on force-field evolution through the preceding motion was present. Auxotonic whole muscle and single fiber work in frogs suggests the muscle state and behavior at the end of an auxotonic contraction is close to that in an isometric contraction (Iwamoto et al. 1990). Thus the results of this approach were directly comparable to previous force-field work and able to extend our ideas of spinal force-field primitives in a direct way. Their implications and scope will be discussed below.

3   Analysis of Primitives in Movement: A First Attempt

3.1   Simulated Environments

We used the PHANToM robot, a small haptic interface robot, as the core element in our experiments. The PHANToM has a spatial resolution of 400 dpi and a maximum force of 10 N, 12 bit force resolution and a dynamic range of 100:1. The robot was attached to the spinal frog using a clamp comprised a light aluminum modification to the three DOF gimbal which forms the standard mechanical interaction port on the PHANToM. The clamp positioned the frog's limb so that the calf ran through the intersection of the three gimbal axes. The PHANToM was controlled using a laboratory developed software control loop and host program, which was derived from Sensable Devices code. The program ran on a 90MHz Pentium PC. The PHANToM robot's software control loop ran at 2.5KHz. Cartesian ankle position and applied Cartesian force were obtained from encoders and software calculation each control cycle, using specific algorithms based on the experimental protocol chosen. The force and position data from the control loop were subsampled and stored at 150Hz.

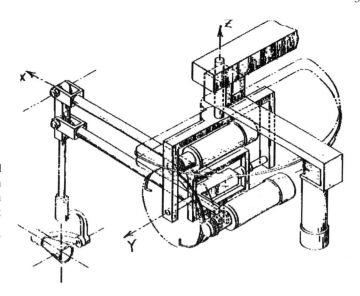

FIGURE 25.2. The PHANToM robot used in our experiments. The thimble shown was replaced with a light aluminum clamp which placed the ankle of the frog at the intersection of the gimbal axes. The robot gimbal allowed the 4DOF proximal to attachment to alter orientation freely at this point but generated translational forces at the end-point based on Cartesian position.

The stationary (isometric) force responses of the phantom and the spatial resolution were previously fully tested and calibrated against an ATI 3/10 force sensor. With the robotic device we were able to present various artificial environments to the frog's leg. Here we will discuss the effects of cartesian isotropic elastic fields. We applied constant pre-specified linear stiffness fields to the frog using the PHANToM. In our experiments we varied the stiffness and the equilibrium coordinates of these applied fields. The apparatus and experimental setup are illustrated in Figure 25.2 and 25.3.

3.2 Experiment Design

Ten spinal frogs were stimulated to elicit the wiping reflex or flexion withdrawal while the limb was in a constant isotropic stiffness environment provided by the robot. To avoid the various complications due to the dynamics of the robotic linkage, the dynamics of the frog's limb and the velocity dependence of muscle force generation, we examined the forces and electromyographic (EMG) activity of the spinal frog when the frog's limb had come to active equilibrium with the robot's force-field. This is illustrated in Figure 25.3. When the limb and robot are stationary and in equilibrium for hundreds of milliseconds with near constant EMGs, the limb state and that of individual muscles is closely equivalent to the state in previous isometric measurements (Iwamoto et al. 1990). The contractions underlying the movements described here represent near auxotonic conditions for the shortening muscles. Inertial and interaction torques in the limb and robot need not be considered, and we assume that active viscosity controls and the velocity dependent properties of muscle are negligible at rest. The equilibrium between robot and frog was always achieved at a similar latency following stimulation in a given frog.

Figure 25.4 shows the active vectors through time during wiping for a variety of starting configurations and different compliant environments. The equilibrium was usually maintained for at least 100 ms. In some frogs such as that shown, when using prolonged stimulation trains we also observed that the equilibrium could be sustained for a prolonged period. With shorter trains the latency of the peak excursion and equilibrium was also remarkably similar from frog to frog, and without regard to the manipulation of the limb's mechanical environment. This observation was critical in enabling subsequent experiments and analysis. The equilibrium isometric responses, because they occurred at the same time following stimulation, could be directly compared with one another and used as samples from force fields that were generated in particular environments. This allowed us to

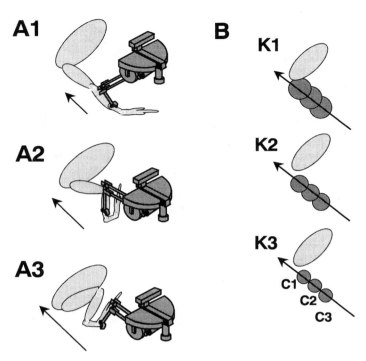

FIGURE 25.3. The method of use of the robot. The frog was stimulated and moved along a trajectory through the elastic field generated by the robot until the frog and robot were in equilibrium. This indicated in A1-3, the magnitudes of interaction forces generated during movement are drawn beneath each cartoon. At A3 the frog and robot are in equilibrium and isometric conditions are re-established after a period of motion. B1-3 shows the experimental design. The robot's field equilibrium was moved along a line towards the target of the reflex movement (*c1-c3*) and the stiffness of the elastic field was varied at each location in a geometric progression (*k1-k3*).

do the following: (1) construct isometric force fields generated at fixed latency in different environments, (2) examine the interaction of initial limb configuration and environment on peak force production.

3.3   Possible Experimental Outcomes

During the unperturbed hind-limb wiping reflex or flexion withdrawal the frog's limb moves to specific locations. The robot's elastic field presents the frog's limb with progressively increasing resistance controlled by the experimenter. The perfect isometric experiment represents one extreme of this continuum: an infinitely stiff environmental field. We chose to distinguish several possible outcomes of the manipulation of the limb's external mechanical environment:

a. *Task achievement.* In this outcome, the frog overcomes the robot's finite resistance and achieves the task goal (this result is suggested by mass perturbation experiments of Schotland et al. (1993) in which frogs coped with large mass changes successfully), as for example, the behavior of a PID controller might.

b. *Field structure is constant.* In this outcome, the frog maintains, increases, or decreases, the magnitude and stiffness of the reflex force fields in relation to environmental interactions by maintaining or simply scaling the isometric force fields.

c. *Novel field structures are produced.* In this outcome, the frog behavior involves new force fields as a result of the motion and interactions with environment or cannot be simply explained in the existing framework. Outcome (a) is actually a special subset of this outcome.

Each of the first two outcomes has specific testable predictions. Outcome (a) predicts motion to the wiping behavior target point in all lower stiffness environments. Outcome (b) is more complex but can predict the relationship of force and position in different stiffness environments and how equilibria will vary based on the robots equilibrium location: To repeat the limb and robot are stationary and in equilibrium for hundreds of milliseconds with near constant electromyograms and the limb state is equivalent to the state in previous isometric measurements. Inertial and interaction torques in the limb and robot need not be considered, and

FIGURE 25.4. Preliminary two-dimensional data in a compliant mechanical environment. Sample data from a long train experiment are shown. The isometric force field measured using our standard technique is shown below. The frog moves to a stable plateau when moving against the environment. By relating the position of equilibrium to the isometric field we determined that the forces generated after moving in the compliant environment were only 20% of those generated in purely isometric conditions despite their sustained isometric plateaus in responses. Note that the experiment presented here was performed with a mechanical spring and ATI 3/10 sensor, not with the robot, although results are in agreement in both arrangements.

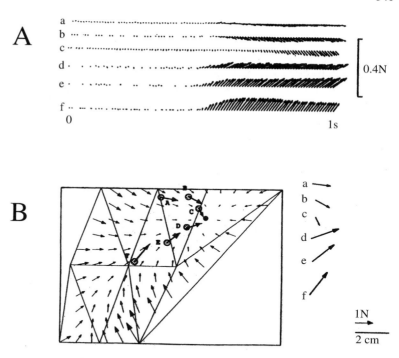

we assume that active viscosity controls and the velocity dependent properties of muscle can be ignored. The mechanical equilibrium of frog and robot occurred at approximately the same time period following stimulus onset regardless of the environmental force-field conditions we imposed. In previous isometric measurements in spinal frogs the reflex force-field stiffness was relatively constant across the workspace (in contrast to human arm measurements). Thus, if the outcome represents a scaling of the isometric field we can predict the equilibrium interaction forces at different locations. For example, if the frog spinal cord made no adjustment of activation based on the *history* of dynamic interaction of the limb and environment we would expect that the magnitude of equilibrated interaction forces should be predicted by the stiffness of the environment, the stiffness of the frog's force field (estimated from the previous isometric measurements) and the environments' equilibrium location. We expected a relationship for interaction force $F(x,k)$ as follows:

$$F(x,k) = \frac{k_1 \cdot k_2 \cdot x}{k_1 + k_2} \qquad (25.4)$$

where k_1 is the frog's reflex force-field stiffness, k_2 the robot stiffness, and x the separation of robot and frog equilibria. For a force-field time evolution that is independent of the environment we thus expected a sigmoidal dependence of interaction force on k_2 or $ln(k_2)$ at a single location, and a multiplicative effect of variations of location on interaction force. We observed such a sigmoidal relationship in a deafferented frog. If the frog's field stiffness, k_1, is regulated by the frog simply and uniformly in relation to an imposed environment stiffness k_2, then Eq. (25.4) can be modified accordingly to describe the data.

3.4 Roles of Position and Resistance

Our data provided a snapshot of the end result of the dynamic evolution of force-field structure in a compliant environment. We found that in the ranges of stiffnesses and positions that we utilized, and over a range of frog stiffness estimates, the magnitude of reflex interaction forces measured after a period of dynamics in the robot imposed environment differed from those predicted in outcomes (a) or (b) above. We never saw outcome (a) in elastic environmental fields. In afferented frogs the elastic fields prevented achievement of the target position despite this lying within the frogs' force generation capabilities. Nor did we observe data confirming outcome (b). In afferented frogs the interaction force modulated linearly with $ln(k_2)$ and with x. In the ranges of test parameters that we used we had predicted a multiplicative doubling or tripling of the steep sigmoidal portion of the force magnitude to $ln(k_2)$ slope over the range of equilibrium test positions for a preserved force-field structure. In afferented frogs the slope of the dependence on $ln(k_2)$ we actually observed was independent of the external equilibrium test location, although the intercept of this relationship did vary systematically with the environment's equilibrium location. The relationships of the magnitude of the interaction force to the environmental stiffness and to the environment's equilibrium location thus appeared to be independent and additive (not multiplicative). This result is not consistent with a simple scaling of the original force-field structure. Our data indicated that, while the frog may be at the same position and limb configuration and at the same latency following stimulation, very different isometric force levels were exerted depending on the

position at which the limb started from and the environment through which it moved.

Figure 25.5 summarizes the key features of these results diagramatically; data are described in fuller detail elsewhere.

We also examined whole three-dimensional force fields collected at the time of peak force in different stiffness environments. Despite a substantial uniformity of magnitude such fields exhibited a convergent force pattern oriented similarly

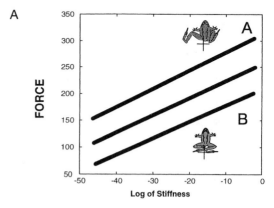

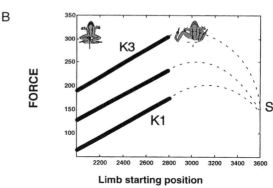

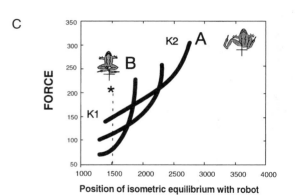

FIGURE 25.5. A summary of recent experiments with the PHANToM. The effects of varying starting position and environment stiffness on peak (isometric) force are shown in A and B diagrammatically. A cartoon frog shows how the limb position is varying in each panel. A shows the effect of ln stiffness is linear and of similar slope at different positions (A extended, B flexed). B shows force varies with configuration linearly in a given stiffness environment (e.g. *K1* or *K3*) until the singularity in extension is approached (S). C shows the effect of these influences together on peak forces, and the point in space at which they are achieved. Each curve represents peak force and its position as stiffness varies. Thus curve A represents position force pairs at equilibrium starting from extended position A. *K2* end of the curve represents the high stiffness environment end and *K1* the low stiffness environment end. The frog may thus be at the same time after identical stimuli and at the same point in space (dotted line and asterisk) with different force levels depending on the history of environment passed through and initial starting position. These relationships can be expressed as the sum of two fields with different properties and controls (Figure 25.6).

to the isometric fields observed previously. We have found that force orientation appears to be unaffected by the environment stiffness.

3.5 Descriptions of the Reponses as Force-Field Primitives: The Dynamic Primitive

In our data we found it was possible to decompose the dynamic force-field [$F(x,t,k)$] and its modulation in an environment of stiffness k into two force-field components, $f_1(x)$, a static primitive, and $f_2(x)$, a dynamic primitive with time evolutions $A(t)$ and $B(t)G(k)$ respectively:

$$F(x,t,k) = A(t)f_1(x) + B(t)G(k)f_2(x) \quad (25.5)$$

The temporal evolution of the whole force field always plateaued or peaked at approximately the same time following stimulation regardless of environment and/or extent of movement. This strongly suggests there is an intrinsic timing regulation in the production of the component force-fields during reflex behavior (i.e., in parameters $A(t)$ and $B(t)$ above). It is not clear how tightly the timing and force generating circuitry described by Eq. (25.5) is bound together. It is conceivable these represent separable elements.

Eq. (25.5) represents an extension of the original scheme of Eq. (25.1). However it is also still possible that the field during motion and the dynamic field primitive represented by $f_2(x)$ cannot generally be described in the form used in Eq. (25.5), but must be represented as a single function of both position, interaction and time:

$$F(x,t,k) = A(t)f_1(x) + B(t)f_2(x,k,t) \quad (25.6)$$

In either formulation the first field [$f_1(x)$] is of the original convergent elastic field type. The second [$f_2(x,k,t)$] is a force field which is modulated in amplitude by interaction with the environment during motion (represented here as the environment's stiffness k). In the experiments described here it is convergent in pattern but has uniform force magnitude with a force discontinuity or very abrupt change in force at the equilibrium position. We have chosen to term this second field a "dynamic primitive" because (1) it was detected only following limb dynamics and (2) it is modulated as a result of dynamic interactions with the environment. Further, the data suggest the control during

dynamics may be analyzed in the force-field primitive framework if Eq. (25.5) or a closely related formulation still pertains in future experiments.

4   An Emerging Scheme

The experiments described here suggest a scheme of spinal organization very similar to that which has been proposed by many previously, dating back to Sherrington and Brown. The scheme we favor is that of a modular system comprised of timers and phaselocking devices, and sets of dynamic and postural primitives (i.e., stable assemblies of synergies and reflexes with a biomechanical coherence). The timers are responsible for generating specific timing, specific phasing and perhaps quantized information processing, synchronized decisions and quantized actions in reflex, repetitive, and voluntary behaviors. The system of primitives are parameterized to organize postural and dynamic force responses in relation to limb state, history, and current action under the control of the local timers and/or descending control or supervision. The primitives also provide a means for synthesizing novel actions and dynamic responses. They could be used to this end by descending controls and learning mechanisms. The control of force based on configuration, and based on interaction currently appear to be separable. This system allows a very flexible construction of new movements but provides the bootstrap for learning systems, the intrinsic stability and the set of base behaviors that allow animals such as frogs to function rapidly and effectively with minimal experience. Regarding equilibrium point ideas it is worth noting that a fixed point or equilibrium is defined in the "dynamic" primitive field, but the "dynamic" primitive field is not elastic. Regarding pattern generation and reflex behavior, models based on spreading activation, relaxation oscillators and primitives reproduce many aspects of reflex behaviors in a biologically satisfying way, and can be tuned to improve performance (Giszter 1993,1994).

In the context of motor learning the addition of dynamic primitives to the postural primitives is significant. The dynamic primitives represent low or zero stiffness force fields.

However, the dynamic fields discovered thus far are mostly conservative, and thus do not in them-

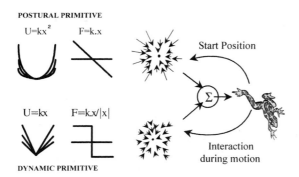

POSTURAL PRIMITIVE

$U=kx^2$ $F=k.x$

$U=kx$ $F=k.x/|x|$

DYNAMIC PRIMITIVE

Start Position

Σ

Interaction
during motion

FIGURE 25.6. A suggested scheme of spinal cord organization. The scheme consists of two types of primitives, static and dynamic, which are driven by reflexes, pattern generators and descending controls. Each primitive has a different responsiveness to configuration and environment and a different potential well.

selves extend motor learning repertoires greatly away from the purely conservative class of fields. This scheme is shown in Figure 25.6. The nonconservative behavior of, for example, locomotion is generated in this scheme by the driving interactions of oscillators with the primitive system. The likelihood is also that some primitives and their relationship to the CPG in mammals show plasticity, and can be modified by descending control.

5 Conclusion and Future Directions

In summary, the data discussed here suggest force-field primitives are organized in the spinal cord. The identification of elements of motor organization is clearly a critical ingredient of motor control understanding (Windhorst 1990). These primitives have now been shown to be relevant to multijoint 3D reflex movements. They have also been extended to more dynamic contexts. This effort has revealed a new force-field organization which, we suggest, is a dynamic primitive. Further, these elements may eventually be of clinical significance. It has been shown elsewhere that the primitives are laid out in ways that may make them available as targets for functional electrical simulation (Bizzi et al 1990; Giszter, Loeb Mussa-Ivaldi and Bizzi, 1994, forthcoming), or perhaps be suitable synaptic targets for nerve regeneration. Data described here shows that in the spinal cord removed from descending controls reflex proprioceptive feedback strongly regulates the muscle activation underlying behavioral force fields and these compensations may be described simply as precisely timed and

systematic force-field effects (cf. Chapter 29). However, it is important to note that the static data published previously and the measurements following movement that are described here represent isometric "snapshots" of two aspects of the spinal cord mechanisms that contribute to the force-field regulation used in complex multijoint limb motions. The task ahead is to understand the control of force-field primitives throughout movement, in different environments, and to understand the circuitry that underpins these control processes.

Acknowledgments. Supported by NIH R29-NS34640. Drs. Neville Hogan, Terence Sanger, Sandro Mussa-Ivaldi, Emilio Bizzi, Stefan Schaal, Maja Mataric, Tamar Flash, and Daniel Wolpert, and attendees of the Biological and Artificial Motor Systems Workshops have provided invaluable advice and discussion at various stages.

References

van Attevaldt, H. and Crowe, A. (1980). Active tension changes in frog skeletal muscles during and after mechanical extension *J. Biomech.*, 13:323–331.

Berkinblitt, M.B., Feldman, A.G., Fukson, O.I. (1989). Wiping reflex in the frog: movement patterns, receptive fields, and blends. In *Visuomotor Coordination: Amphibians, Comparisons, Models, and Robots.* Ewert, J.-P., Arbib, M.A. (eds.), pp. 615–630. Plenum, New York.

Berkinblitt, M.B., Zharkova, I.S., Feldman, A.G., and Fukson, O.I. (1984). Biomechanical aspects of the wiping reflex cycle. Biophysics, 29:530–535.

Bizzi, E., Hogan, N., Mussa-Ivaldi, F.A., Giszter, S.F. (1992). Does the nervous system use equilibrium-point control to guide single and multiple joint movements? *Behav. Brain Sci.*, 15:603–613.

Bizzi, E., Mussa-Ivaldi, F.A., and Giszter, S.F. (1991). Computations underlying the execution of movement: a novel biological perspective. Science, 253:287–291.

Cecchi, G., Griffiths, P.J., and Taylor, S. (1986). Stiffness and force in activated frog skeletal muscle fibres. *Biophys. J.*, 49:437–451.

Colgate, E. and Hogan, N. (1989). An analysis of contact instability in terms of passive physical equivalents. *IEEE Proc. Int. Conf. Robot. Automat.*, pp. 404–409.

Fukson, O.I., Berkinblit, M.B., and Feldman, A.G. (1980). The spinal frog takes into account the scheme of its body during the wiping reflex. Science, 209: 1261–1263.

Gandolfo, F. and Mussa-Ivaldi, F.A. (1993b). Vector summation of end-point impedance in kinematically redundant manipulators Proc 1993. *IEEE/RSJ Int. Conf. Intell. Robots and Sys.*

Giszter, S.F. (1993). Behavior networks and force fields for simulating spinal reflex behaviors of the frog. Second international conference on the simulation of adaptive behavior. The MIT Press, Cambridge MA.

Giszter, S.F. (1993). Force-fields in spinal behaviors: models versus experiments. *Soc. Neurosci. Abstr.*, 19:(1)146.

Giszter, S.F. (1993). Behavior networks and force fields for simulating spinal reflex behaviors of the frog. *2nd. International Conference on the Simulation of Adaptive Behavior*, (from Animals to Animats 2), 172–181. The MIT Press.

Giszter, S.F. (1994). Reinforcement tuning of action synthesis and selection in a virtual frog. *Third International conference on the Simulation of Adaptive Behavior*, (from Animals to Animats 3), 172–181. The MIT Press.

Giszter, S.F., McIntyre, J., and Bizzi, E. (1989). Kinematic strategies and sensorimotor transformations in the wiping movements of frogs. *J. Neurophysiol.*, 62:750–767.

Giszter, S.F., Mussa-Ivaldi, F.A., and Bizzi, E. (1993). Convergent force fields organized in the frog spinal cord. *J. Neurosci.*, 13:467–491.

Giszter, S.F., Mussa-Ivaldi, F.A., Bizzi, E. (1993). Movement primitives in the frog spinal cord. *Neural Systems: Analysis and Modeling*. Eeckman, F.H. (ed.). Kluwer, Boston, MA. pp. 431–446.

Giszter, S.F., Mussa-Ivaldi, F.A., Loeb, E., and Bizzi, E. (1994). Systematic mapping of force response and muscle activity in the lumbar spinal cord of the frog. *Soc. Neurosci. Abstr.*

Grillner, S. and Wallen, P. (1985). Central pattern generators for locomotion, with special reference to vertebrates. *Ann. Rev. Neurosci.*, 8:233–261.

Hoffer, J.A. and Andreassen, S. (1981). Regulation of soleus muscle stiffness in premamillary cats: intrinsic and reflex components. *J. Neurophysiol.*, 45:267–285.

Houk, J.C. (1979). Regulation of stiffness by skeletomotor reflexes. *Ann. Rev. Physiol.*, 41:99–114.

Iwamoto, H., Sugaya, R., and Sugi, H. (1990). Force-velocity relation of frog skeletal muscle fibers shortening under continuously changing load. *J. Physiol.*, 442:185–202.

Joyce, G.C., Rack, P.M.H., and Westbury, D.R. (1969). The mechanical properties of cat soleus muscle during controlled lengthening and shortening movements. *J. Physiol.*, 204:461–474.

Lieber, R.L. and Shoemaker, S.D. (1992). Muscle, joint and tendon contributions to the torque profile of frog hip joint. *Am. J. Physiol.*, 263:R586–R590.

Lieber, R.L., Leonard, M.E., Brown, C.G., and Trestik, C.L. (1991). Frog semitendinosus tendon load-strain and stress-strain properties during passive loading. *Am. J. Physiol.*, 261:C86–C92.

Loeb, E., Giszter, S.F., Borghesani, P., and Bizzi, E. (1993). The role of afference in convergent force fields elicited in the frog spinal cord. *Somatosens. Mot. Behav.*, 10:81–95.

Lombard, W.P. and Abbot, F.M. (1907). The mechanical effects produced by the contraction of individual muscles of the thigh of the frog. *Am. J. Physiol.*, 20:1–60.

Mortin, L.I., Keifer, J., Stein, P.S.G. (1985). Three forms of the scratch reflex in the spinal turtle: movement analyses. *J. Neurophysiol.*, 53:1501–1516.

Mussa Ivaldi, F.A. (1992). From basis functions to basis fields: using vector primitives to capture vector patterns. *Biol. Cybern.*, 67:479–489.

Mussa-Ivaldi, F.A. and Gandolfo, F. (1993). Networks that approximate vector valued mappings. *IEEE Conf. Neural Networks*, pp. 1973–1978.

Mussa-Ivaldi, F.A. and Giszter S.F. (1992). Vector field approximation: a computational paradigm for motor control and learning. *Biol. Cybern.*, 67:491–500.

Mussa-Ivaldi, F.A., Giszter S.F., and Bizzi E. (1994). Linear superposition of primitives in motor control. *Proc. Natl. Acad. Sci.*, 91:7534–7538.

Mussa-Ivaldi, F.A., Hogan, N., and Bizzi, E. (1985). Neural, mechanical, and geometric factors subserving arm posture in humans. *J. Neurosci.*, 5:2732–2743.

Nichols, T.R. and Houk, J.C. (1975). Improvement in linearity and regulation of stiffness that results from actions of stretch reflex. *J. Neurophysiol.*, 39:119–142.

Ostry, D.J., Feldman, A.G., and Flanagan R.F. (1991). Kinematics and control of frog hindlimb movements. *J. Neurophysiol.*, 65:547–562.

Schotland, J.L. and Giszter, S.F. (1993). Interaction of

stimulus strength and configuration in frog wiping behaviors. *Soc. Neurosci. Abstr.*, 19:(1)146.

Schotland, J.L. and Rymer, W.Z. (1993). Wiping and flexion reflexes in the frog: I Kinematics and EMG patterns. *J. Neurophysiol.*, 69:1725–1735.

Schotland, J.L. and Rymer, W.Z. (1993). Wiping and flexion reflexes in the frog: II Response to perturbations. *J. Neurophysiol.*, 69:1736–1748.

Sergio, L. and Ostry, D.J. (1993). Three-dimensional kinematic analysis of frog hindlimb movement in reflex wiping. *Exp. Brain. Res.*, 94:53–70.

Shadmehr, R. and Mussa-Ivaldi, F.A. (1993). Geometric structure of the adaptive controller of the human arm. AI Memo 1437, CBCL Memo 82. The MIT Press, Cambridge, MA.

Shadmehr, R., Mussa-Ivaldi, F.A., and Bizzi, E. (1993). Postural force fields of the human arm and their role in generating multijoint movements. *J. Neurosci.*, 13:45–62.

Sherrington, C. (1961). The integrative action of the nervous system. Yale University Press, New Haven, CT.

Smeraski, C., Kargo, W., Davies, M.R., Miya, D., Mori, F., Shibayama, M., Tessler, A., Murray, M., and Giszter, S.F. (1996). Synergies and motor mechanisms in normal rats and rats with transplants placed into neonatally transected spinal cords. (submitted).

Windhorst, U.R. (1991). Group report: what are the units of motor behavior and how are they controlled? In *Motor Control: Concepts and Issues*. Humphrey, D.R. and Freund, H.-J. (eds.), pp. 101–120. John Wiley & Sons, New York.

Winters, J., Stark, L., and Seif, Naraghi, A.-H. (1988). An analysis of the sources of musculoskeletal system impedance. *J. Biomechan.*, 21:1011–1025.

26
Learning and Memory Formation of Arm Movements

Reza Shadmehr and Kurt Thoroughman

1   Introduction

Learning a motor task is characterized by a gradual transition from a high demand on attention to the task becoming automatic and nonattentive. Studies that have recorded limb movements during learning of a motor task have shown that this increase in automaticity of movements is accompanied by key kinematic features:

1. Stiffness of the limbs decreases (Milner and Cloutier 1993), as evidenced by a decreased coactivation of the muscles and an increased compliance in response to a perturbation.
2. Movements become smoother (Hreljac 1993), as evidenced by a reduction in a cost function that scales with the jerkiness of the movement (second derivative of velocity).
3. Motion of the joints become decoupled (Vereijken et al. 1992), as evidenced by a reduction in the cross-correlation between patterns of joint rotations.

The central hypothesis is that these kinematic features result from the formation of motor memory: the content of motor memory is called an "internal model" of the task. Formation of the internal model allows the nervous system to reduce the dependence of the motor program on the visual and proprioceptive feedback. This leads to a reduction in the attention requirements of the task and increases the reliance of the motor output on an internal model that predicts motor patterns that should be produced in order to execute a desired movement.

Although this description of an internal model brings to mind learning of complex motor skills, it is equally valid for simple tasks. This can be illustrated by an example. If one is asked to rapidly pick up an empty bottle of milk that has been painted white, the arm exhibits a flailing like motion. This is an indication that in programming the motor output to the muscles, the nervous system predicts and attempts to compensate for the mechanical dynamics of the preceived full bottle. In a control theory framework, the internal model (IM) is an association from a desired trajectory for the hand to a set of muscle torques (Shadmehr and Mussa-Ivaldi 1994a). Because in principle this map is unique for the objects and tools which we have learned to interact with, "motor memory" may be thought to contain, at least in part, a collection of IMs in which visual information serves as an identifying cue that allows for binding of an appropriate association (i.e., recall). We learn these IMs with experience (Gordon et al. 1992), and they are an integral part of our ability to interact with the objects and systems in our environment. Yet we know little about the neural substrate of motor memory or the processes that culminate in its formation. The objective of this chapter is to review as well as present some new results on psychophysics of learning to make arm movements, and then put these results in perspective of what we know about memory systems of the brain.

2   Learning Novel Dynamics

As one learns to control the arm to perform a novel motion, the motor output begins to rely on an IM. The evidence for this comes from two directions.

In the first approach, changes in the motor output have been recorded as some aspect of a well-learned task has been altered (Milner and Cloutier 1993; Gottlieb 1994). For example, Gottlieb (1994) trained subjects to make elbow movements against various loads. He found that after a number of practice movements, the EMG had changed in such a way as to suggest that the motor output was being preprogrammed before the onset of movement based on the expected behavior of the load. Milner and Cloutier (1993) quantified changes in EMG during practice and observed that initially, when presented with a novel load, subjects tended to co-contract antagonist muscles, increasing stiffness of the arm. An increase in stiffness is a reasonable way of dealing with an unknown load: the arm will

show less deviations from the desired kinematic behavior for a given disturbing force. With practice, there was a decrease in the level of co-contraction. This decrease paralleled an increase in the reliance of the motor output on a pattern of muscle torques which specifically compensated for the imposed load.

In the second approach, the formation of the internal model has been inferred by what is termed "after-effects": Novel forces were imposed on the arm by having subjects move a manipulandum (Figure 26.1) (Shadmehr and Mussa-Ivaldi 1994a; Sanes 1986), or by positioning a subject at the center of a rotating room (Lackner and Dizio 1994) so that strong coriolis forces acted on the arm during multijoint reaching movement. The imposed forces

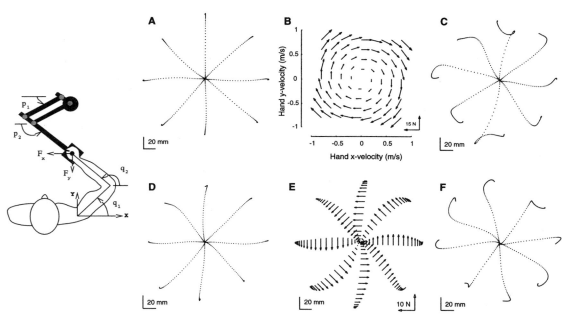

FIGURE 26.1. The robot manipulandum and the experimental setup. The manipulandum is a very low friction, planar mechanism powered by two high-performance torque motors. The subject grips the handle of the robot. The handle houses a force transducer. The video monitor facing the subject displays a cursor corresponding to the position of the handle. A target position is displayed and the subject makes a reaching movement. With practice, the subject learns to compensate for the forces produced by the robot. (A) Hand path of a typical subject in the null field (i.e. no forces being produced by the robot). (B) An example of a force field produced by the robot, $F = B\dot{x}$, where is the hand velocity vector. (C) Resulting hand path of an untrained subject in the field. (D) Hand path after 300 movements in the field. The trajectory in the field converges to the trajectory observed in the null field. (E) Forces produced by a typical trained subject to counter the effect of the force field as a function of hand position for each movement. These forces are the projection of the forces measured at the interaction point between the subject and robot onto a direction perpendicular to the direction of target. (F) While training in the field, random targets are presented with null field conditions. The result are aftereffects. The points in all hand paths are 10 msec apart.

perturbed the trajectory of the arm and required the subject to use visual and proprioceptive feedback to correct errors in hand path (Figure 26.1C). With practice, subjects were able to make accurate and smooth reaching movements without visual guidance (Figure 26.1D). It is suggested that the IM functions as a mapping from a desired arm movement (i.e., plan) to a prediction of the forces that will be encountered during the movement (Shadmehr and Mussa-Ivaldi 1994b). As a consequence, because of the reliance of the motor output on the IM, removal of the imposed forces should lead to after-effects: in the absence of forces, the hand trajectory should be a mirror image of that observed before adaptation (one can describe a mathematical model and make exact predictions of this trajectory, see Shadmehr and Mussa-Ivaldi 1994a). This prediction has been experimentally confirmed (Figure 26.1F).

We have recently developed a tool that gives a fairly direct measure of the IM constructed by the subject's motor system (Shadmehr et al. 1995). The idea is to measure the change in the mechanical impedance of the arm (i.e., how the arm's neuromuscular system reacts to a perturbation).

Consider the equations of motion for the robot-human system of Figure 26.1:

$$I_r(p)\ddot{p} + G_r(p,\dot{p})\dot{p} = E(p,\dot{p}) + J_r^T F \quad (26.1)$$

$$I_s(q)\ddot{q} + G_s(q,\dot{q})\dot{q} = C(q,\dot{q}) + J_s^T F \quad (26.2)$$

where I and G are inertial and coriolis/centripetal matrix functions, E is the torque field produced by the robot's motors, i.e. the environment, F is the force measured at the handle of the robot, C is the controller implemented by the motor system of the subject, $q*(t)$ is the reference trajectory planned by the motor control system of the subject, J is the Jacobian matrix describing the differential transformation of coordinates from endpoint to joints, q and p are column vectors representing joint positions (e.g. q_1 and q_2) of the subject and the robot (Figure 26.1), and the subscripts s and r denote subject or robot matrices of parameters. In the null field, that is, $E = 0$ in Eq. (26.1), assume that a solution to this coupled system is $q = q*(t)$, that is, the arm follows the reference trajectory (typically a straight hand path with a Gaussian tangential velocity profile). Let us name the controller which ac-

complishes this task $C = C_0$ in Eq. (26.2). When the robot motors are producing a force field (i.e., $E \neq 0$), the arm's motion converges back to the reference trajectory if the new controller in Eq. (26.2) is

$$C = C_1 = C_0 + J_s^T J_r^{-T} \hat{E}$$

The internal model composed by the subject is $C_1 - C_0$, i.e. the change in the controller after some training period. We can estimate this quantity by measuring the change in the interaction force along a given trajectory before and after training. If we call these functions F_0 and F_1, then we have:

$$F_0(q,\dot{q},\ddot{q},q*(t)) = J_s^{-T}(C_0 - I_s\ddot{q} - G_s\dot{q}) \quad (26.3)$$

$$F_1(q,\dot{q},\ddot{q},q*(t)) = \\ J_s^{-T}(C_0 + J_s^T J_r^{-T} \hat{E} - I_s\ddot{q} - G_s\dot{q}) \quad (26.4)$$

The functions F_0 and F_1 are impedances of the subject's arm before and after training in a field. By approximating the function $F_1 - F_0$, we have an estimate of the change in the output of the human arm's adaptive controller, which we have defined to be the internal model. In order to measure F_0, we had the subjects make movements in a series of environments. The environments were unpredictable (no opportunity to learn) and their purpose was to perturb the controller about the reference trajectory so we could measure F_0 at states neighboring the reference trajectory. Next, the environment in Figure 26.2A was presented and the subject given a practice period to adapt. After training, F_1 was estimated in a similar fashion as F_0. The difference between these two functions was calculated along all measured arm trajectories and the results were projected onto the hand velocity space. The resulting pattern of forces were interpolated via a sum of Gaussian radial basis functions, and are shown in Figure 26.2B. This is the change in the impedance of the arm and estimates the input–output property of the internal model that was learned by this subject. We found that subjects learned to change the effective impedance of their arm in a way that approximated the imposed force field.

We also recorded EMG activity during learning of the reaching task. Figure 26.3 shows the root-mean squared EMG from four arm muscles during movements in the null field (no forces) and those in field of Figure 26.1B. The particular movement shown is

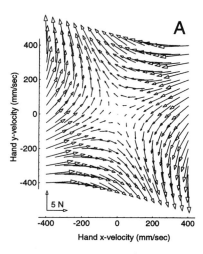

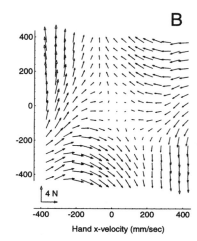

FIGURE 26.2. Quantification of the change in impedance of a subject's arm after learning a force field. (A) The force field produced by the robot during the training period. (B) The change in the subject's arm impedance after the training

to a target at 90 degrees (at 12 o'clock). The force field would tend to push the hand in a clockwise direction, causing shoulder and elbow extension. After practice, the learned response is an increase in activation of all muscles, but this increase is particularly strong in the elbow and shoulder flexors. In effect,

the subjects learn to make a movement which is primarily an elbow extension with strong activation of the elbow and shoulder flexors. The EMG activity of the biceps suggests that for this movement, practice results in generation of compensatory flexor torques that correlate with hand velocity.

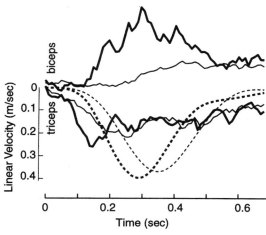

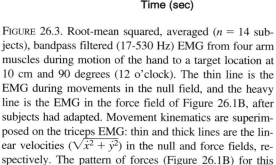

FIGURE 26.3. Root-mean squared, averaged ($n = 14$ subjects), bandpass filtered (17-530 Hz) EMG from four arm muscles during motion of the hand to a target location at 10 cm and 90 degrees (12 o'clock). The thin line is the EMG during movements in the null field, and the heavy line is the EMG in the force field of Figure 26.1B, after subjects had adapted. Movement kinematics are superimposed on the triceps EMG: thin and thick lines are the linear velocities ($\sqrt{\dot{x}^2 + \dot{y}^2}$) in the null and force fields, respectively. The pattern of forces (Figure 26.1B) for this

movement imposes an extension torque on the elbow and shoulder joints. The torque will be proportional to hand velocity. Effect of training is an increase in the activity of all muscles, but the magnitude of the increase is particularly large for the biceps and the ant. deltoid (elbow and shoulder flexors). With training, subjects learn to significantly increase activity of the flexors for a movement which is essentially an extension of the elbow joint. Note that after adaptation, biceps EMG is essentially proportional to hand velocity, compensating for the imposed force field.

3 Representation of Memory Changes with Time

Once the pattern of muscular activity that was necessary to make smooth reaching movements in the force field was learned, it became available for recall: performance in the same environment was significantly improved (compared to initial practice) when tested 24 to 48 hours later (Shadmehr et al. 1995). In comparison, performance in an untrained environment remained at naive levels. Recent experiments show that the improvement in performance persists without further practice for at least 5 months (Shadmehr and Brashers-Krug 1997). This suggests that practicing arm movements with a novel mechanical environment sets in motion processes that result in long-term motor memories.

Unfortunately, we know little about the processes that culminate in motor memory formation (Salmon and Butters 1995; Halsband and Freund 1993). However, a feature of memory across the animal kingdom is that it continues to develop long after practice has stopped: in general, memory appears to functionally progress from a short-lived fragile form to a long-lasting stable form (DeZazzo and Tully 1995). For example, a person who has been knocked unconscious will have memory loss for the events that occurred just before the blow. Severely depressed patients who undergo electric shock therapy show a significant (and sometimes transient) reduction in their ability to recall items they recently learned (Squire et al. 1984). The progression to long-term memory, which is referred to as consolidation, is a time dependent process that is initiated during the practice session but continues long after completion of practice. The time during which information becomes consolidated has been used to functionally define short-term memory (Fuster 1995).

Does formation of motor memory progress from a short-term, fragile form to a long-term stable form? Is there a distinct short-term motor memory phase? Until recently, there was little evidence to support the notion of motor memory consolidation. For example, severely depressed patients receiving electric shock therapy 24 hours after acquiring a visuo-motor skill showed no loss of the skill (Squire et al. 1984). However, the same patients had little recollection of having practiced the task

before. In other words, the intervention appeared to have stopped the consolidation process of one type of information (the memory of the episode), but not of the visuo-motor memory. This and a lack of evidence regarding a functional transformation of motor memory from a fragile to a stable form had suggested that the distinction between short-term and long-term memory did not apply to learning of visuo-motor skills (Squire 1987).

However, we know that representation of memory of certain skills also changes with time: in a perceptual discrimination task, it was observed that subjects rapidly improved with training, and continued to improve at a slower rate after completion of practice and without further training (Karni and Sagi 1993). The fast learning took place with the presentation of the stimulus, but further, slower learning took place hours after the end of the training session and was critically dependent on a component of sleep (Karni et al. 1994). In another set of experiments it was shown that motor memories are not permanent and may be vulnerable to experimental intervention: Lewis and coworkers (Lewis et al. 1951; Lewis and Miles 1956) demonstrated that association of visual stimuli to specific motor actions could be learned and subsequently "unlearned" when a second task required subjects to associate similar visual stimuli to different motor actions. However, these studies did not investigate whether the vulnerability of the original learning changed with the passage of time.

If a newly acquired motor memory gradually (and without further practice) becomes stable with time, then one would expect that an appropriate intervention will eventually have little effect on the ability to recall a previously learned IM. Using the paradigm of learning to control arm movements in a force field (Figure 26.1), this idea was recently tested. We asked subjects to practice movements in field B_1, and then gave them a period of rest, varying from 5 minutes to 24 hours (during which subjects were free to do what they wished). After this period of rest, subjects trained in field B_2, where $B_2 = -B_1$, (i.e., the forces in B_2 were in the opposite direction of B_1). We then tested for retention of the skill acquired in practicing in B_1 some days later (Figure 26.4). Our results showed that retention of the IM for B_1 could be disrupted when a second, anti-correlated to the first, IM for B_2 was learned (Brashers-Krug et al. 1995). We theorized

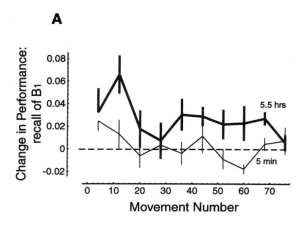

A

B

FIGURE 26.4. Performance during the test of recall for B_1 as a function of temporal distance between learning of $B1$ and $B2$. $B1$ was tested for recall one week after B_1 and B_2 were learned. The performance measure is a correlation between a typical hand trajectory of the subject in the null field before introduction of the forces (as in Figure 26.1A) and the trajectory in the field. (A) Mean improvement in performance $\pm$ SE for two groups of subjects. Thin line is for the group ($n = 9$) that practiced in B_2 five minutes after completion of practice in B_1. Thick line is for the group ($n = 10$) that practiced in B_2 5.5 hours after B_1. (B) Ability to recall B_1 is significantly dependent on temporal distance between B_1 and B_2. Each bar is the mean $\pm$ 95% confidence interval of change in performance as measured for a target set during the recall test vs. during initial practice.

that this disruption occurred because the adaptive system was attempting to associate the same visual target to two very different muscular force patterns: learning of IM_2 was causing an "unlearning" of IM_1. However, if IM_2 was learned beyond a critical time window (approximately 4–5 hours) after acquisition of IM_1, it had little effect on recall of IM_1 (Brashers-Krug et al. 1996; Shadmehr and Brashers-Krug 1997). In other words, within hours, the representation of IM_1 became gradually less vulnerable to the "intervention" caused by learning of IM_2. This suggested that the memory representation of IM_1 rapidly underwent a process of consolidation.

The main mechanism currently believed to underlie memory formation in the central nervous system is long-lasting changes in synaptic efficacy (Bliss and Collingridge 1993). A prominent example of synaptic plasticity is long-term potentiation (LTP) or depression (LTD), both of which have been observed in the motor areas of the cortex (Asanuma and Keller 1991) and cerebellum (Castro-Alamancos et al. 1995). With regard to neural basis of consolidation, it has been shown that after inducing LTP

(in the hippocampus), certain low frequency stimuli can de-potentiate the synapse (Fujii et al. 1991), effectively reducing the synapse's efficacy to near baseline levels. These stimuli, however, are only effective if they are given with in a small time window after potentiation of the synapse: 20 minutes after induction of LTP, the low-frequency stimuli depotentiate the synapse by 70%, while at 100 minutes, the depotentiation is only at 30%. This suggests that one mechanism for consolidation might be changes in the resistance of LTP or LTD to events that might reverse the potentiation. Recently, it was reported that a molecule that participates in long-term synaptic remodeling of Purkinje cells in the cerebellum was highly expressed only during the 1 to 4 hours after completion of a motor learning task. This suggests that within a short time window after completion of practice of a motor task, there may be critical events taking place in representation of the learned skill in the central nervous system. The time course of these neurophysiological changes are similar to the functional changes that we have observed in our subjects.

References

Asanuma, H. and Keller, A. (1991). Neuronal mechanisms of motor learning in mammals. *Neuroreport*, 2:217–224.

Bliss, T.V. and Collingridge, G.L. (1993). A synaptic model of memory: long-term potentiation in the hippocampus. *Nature*, 361(6407):31–39.

Brashers-Krug, T., Shadmehr, R., and Bizzi, E. (1996). Consolidation in human motor memory. *Nature*, 382: 252–255.

Brashers-Krug, T., Shadmehr, R., and Todorov, E. (1995). Catastrophic interference in human motor learning. In *Advances in Neural Information Processing Systems*. Tesauro, G., Touretzky, D.S., and Leen, T.K. (eds.), vol. 7, pp. 19–26. The MIT Press, Boston.

Castro-Alamancos, M.A., Donoghue, J.P., and Connors, B.W. (1995). Different forms of synaptic plasticity in somatosensory and motor areas of the neocortex. *J. Neurosci.*, 15:5324–5333.

DeZazzo, J. and Tully, T. (1995). Dissection of memory formation: from behavioral pharmacology to molecular genetics. *Trends Neurosci.*, 18:212–218.

Fujii, S., Saito, K., Miyakawa, H., Ito, K., and Kato, H. (1991). Reversal of long-term potentiation (depotentiation) induced by tetanus stimulation of the input to the CA1 neurons of guniea pig hippocampal slices. *Brain Res.*, 555:112–122.

Fuster, J.M. (1995). *Memory in the Cerebral Cortex: An Empirical Approach to Neural Networks in the Human and Nonhuman Primates*. The MIT Press, Cambridge, MA.

Gordon, A.M., Forssberg, H., Johansson, R.S., Eliasson, A.C., and Westling, G. (1992). Development of human precision grip. III. Integration of visual size cues during the programming of isometric forces. *Exp. Brain Res.*, 90:399–403.

Gottlieb, G.L. (1994). The generation of the efferent command and the importance of joint compliance in fast elbow movements. *Exp. Brain Res.*, 97:545–550.

Halsband, U. and Freund, H. (1993). Motor learning. *Cur. Opin. Neurobiol.*, 3:940–949.

Hreljac, A. (1993). The relationship between smoothness and performance during the practice of a lower limb obstacle avoidance task. *Biol. Cybern.*, 68:375–379.

Karni, A. and Sagi, D. (1993). The time course of learning a visual skill. *Nature*, 365(6443):250–252.

Karni, A., Tanne, D., Rubenstein, B.S., Askenasy, J.J.M., and Sagi, D. (1994). Dependence on rem sleep of overnight improvement of a perceptual skill. *Science*, 265:679–682.

Lackner, J.R. and Dizio, P. (1994). Rapid adaptation to coriolis force perturbations of arm trajectory. *J. Neurophys.*, 72:299–313.

Lewis, D. and Miles, G. H. (1956). Retroactive interference in performance on the star discrimeter as a function of amount of interpolated learning. *Percept. Mot. Skills*, 6:295–298.

Lewis, D., McAllister, D., and Adams, J. (1951). Facilitation and interference in performance of the modified Mashburn apparatus: I. The effects of varying the amount of original learning. *J. Exp. Psychol.*, 41: 247–260.

Milner, T.E. and Cloutier, C. (1993). Compensation for mechanically unstable loading in voluntary wrist movement. *Exp. Brain Res.*, 94:522–532.

Salmon, D.P. and Butters, N. (1995). Neurobiology of skill and habit learning. *Curr. Opin. Neurobiol.*, 5: 184–190.

Sanes, J.N. (1986). Kinematic and endpoint control of arm movements is modified by unexpected changes in viscous loading. *J. Neurosci.*, 6:3120–3127.

Shadmehr, R. and Brashers-Krug, T. (1997). Functional stages in the formation of human long-term motor memory. *J. Neurosci.*, 17:409–19.

Shadmehr, R. and Mussa-Ivaldi, F.A. (1994a). Adaptive representation of dynamics during learning of a motor task. *J. Neurosci.*, 14(5):3208–3224.

Shadmehr, R. and Mussa-Ivaldi, F.A. (1994b). Computational elements of the adaptive controller of the human arm. In *Advances in Neural Information Processing Systems*. Cowan, J.D., Tesauro, G., and Alspector, J. (eds.), vol. 6, pp. 1077–1084. Morgan Kaufmann, San Francisco.

Shadmehr, R., Brashers-Krug, T., and Mussa-Ivaldi, F.A. (1995). Interference in learning internal models of inverse dynamics in humans. In *Advances in Neural Information Processing Systems*. Tesauro, G., Touretzky, D.S., and Leen, T.K. (eds.), vol. 7, pp. 1117–1124. The MIT Press, Boston.

Squire, L.R. (1987). *Memory and Brain*. Oxford University Press.

Squire, L.R., Cohen, N.J., and Zouzounis, J.A. (1984). Preserved memory in retrograde amnesia: sparing of a recently acquired skill. *Neuropsychologia*, 22:145–152.

Squire, L.R., Slater, P.C., and Chace, P.M. (1975). Retrograde amnesia: temporal gradient in very long-term memory following electroconvulsive therapy. *Science*, 187:77–79.

Vereijken, B., van Emmerik, R.E.A., Whiting, H.T.A., and Newell, K.M. (1992). Free(z)ing degrees of freedom in skill acquisition. *J. Mot. Behav.*, 24:133–142.

27

What Do We Plan or Control When We Perform a Voluntary Movement?

Gerald L. Gottlieb

1   Introduction

The quantitative analysis of multiple degree of freedom movements is a relatively recent practice in motor control. In the early 1980s, Morasso, Lacquaniti, and Soechting published studies of arm reaching that identified certain distinctive kinematic characteristics (Morasso 1981; Soechting and Lacquaniti 1981; Lacquaniti et al. 1982). Morasso noted (p. 224) that "the common features among the different reaching movements are the single-peaked shape of the hand tangential velocity and the [straight] shape of the hand trajectory." Soechting and Lacquaniti further noted that these properties were unaffected by changes in the load held in the hand or by the intended speed of movement. These properties of straightness and "bell-shaped" velocity profiles have become defining features of unconstrained human reaching movements, even though Hollerbach (1982) noted that movements in the sagittal plane tended to be more curved than those in the horizontal plane. A model which captures many of these kinematic features in a parsimonious way, the minimum jerk trajectory, was proposed by Hogan (1984). This is widely used, although it is important to appreciate that it is a *description* of the movement trajectory and cannot be the exclusive basis for *planning* the trajectory.*

We argue here for an alternative to kinematic planning. Voluntary movement is accomplished by the execution of motor programs for planned forces and corresponding EMG patterns in the muscles. However, rather than directly solving inverse dynamic equations, the CNS uses relatively simple coordination rules among muscles and joints that greatly simplify the problem of finding muscle activation patterns to satisfactorily approximate our kinematic goals. The well-known kinematic features of movements result from a trial and error tuning of force profiles based upon visual and kinesthetic feedback. This chapter presents illustrative data for this hypothesis. Some data have been presented at greater length in Gottlieb et al. (1996a), Gottlieb et al. (1996b), and Gottlieb et al. (1997).

2   Methods and Results

Our studies have been of undisturbed movements of the arm in a sagittal plane. In one experiment, twelve targets 20 cm from center were positioned at the hours of a clock. Subjects were instructed be both fast and accurate (Gottlieb, et al. 1996c). The positions of infrared light emitting diodes taped over joint centers at the shoulder, elbow, wrist and on the finger tip were recorded using an OPTO-TRAK 3010 (Northern Digital) system. The sampling rate was 200 per second.

---

*Jerk magnitude is an inverse function of movement time. Hence, to minimize the jerk of any movement of finite duration, additional constraints must be added. For example, minimizing the weighted sum of jerk and movement time or simply specifying movement time allows the computation of a minimum jerk trajectory under those constraints. Any rule for movement planning must specify the additional constraints, not just the minimum jerk criterion, for the plan to be implemented by neural structures.

Ten movements were made to each target. Trajectories were aligned on 5% of their peak tangential velocity and averaged. The joint torques were computed by inverse dynamical equations, given in Eq. (27.1) in simplified form.* The angles of the upper arm θ_s and forearm θ_e are defined relative to vertical. The interior angle of the elbow joint is given by $\phi = \theta_e - \theta_s$. The net muscle torques at the elbow and shoulder are given by t_e and t_m. Subscripts l and u refer to lower and upper arm segments. The mass and length of the segments are defined by m and l, r is the location of the segment

center of mass, and g is the acceleration of gravity.

$$t_e = \begin{array}{l} I_l \ddot{\theta}_e \\ + r_l l_u m_l \cos \phi \ddot{\theta}_s \\ + r_l l_u m_l \sin \phi \dot{\theta}_s^2 \\ + r_l m_l \sin \theta_e g \end{array} \qquad (27.1)$$

$$t_m = \begin{array}{l} (I_u + l_u^2 m_l)\ddot{\theta}_s \\ + (r_l l_u m_l \cos \phi)\ddot{\theta}_e \\ - r_l l_u m_l \sin \phi \dot{\theta}_e^2 \\ + (r_u m_u + l_u m_l)\sin \theta_s g \\ + Elbow\ Torque \end{array} \qquad (27.2)$$

Figure 27.1 illustrates movements in four of the twelve directions. All are fairly straight with bell shaped velocity profiles. The dynamic joint torques† are all biphasic pulses. They are usually relatively symmetrical, deviating most from symmetry when the pulses are small as they are for the elbow torque at 2 and 9 o'clock and shoulder torque at 5 o'clock. These characteristics of the dynamic torque patterns, independent of the load or speed of the hand (Gottlieb et al. 1996b), were noted by Soechting and Hollerbach (Soechting and Lacquaniti 1981; Hollerbach 1982) at about the same time that the distinctive kinematic properties of the movements were first described. They have received far less comment but are no less important than kinematic straightness and smoothness.

3 Discussion

The first point to stress is that linear synergy, like straight paths, bell-shaped velocity profiles, and biphasic torque pulses, is simply an observation of natural behavior. No one of these four mechanical features is a "constraint" that the motor control system is obligated to exhibit. We can change them. However we usually do not unless there are external constraints on performance that make it necessary. Furthermore, we probably cannot change the four features individually and independently be-

---

*The net muscle torque (the difference between the torques of the flexors and extensors) is equal and opposite to weighted sum of motion and gravitational dependent components. The simplified form shown in Eqn 27.1 omits wrist rotation and shoulder translation which were included in our computations. The coordinates we have used are in terms of upper and forearm segment angles (q_s and q_e) in an inertially fixed reference frame. In a joint angle frame of reference, the shoulder angle is equivalent to q_s (because we assume the trunk does not rotate) and the elbow angle is the difference between the segment angles ($\phi = \theta_e - \theta_s$). Motion dependent torques have sometimes been further subdivided into net and interaction components. In this nomenclature, the "net" torque at a joint is inertial and proportional to the acceleration at that joint. The interaction torque is the sum of "inertial," centripetal and Coriolis components. The inertial interaction component of a joint is proportional to the weighted sum of the accelerations at all the other joints, the centripetal to the weighted sum of the squares of the velocities of all other joints, and the Coriolis to the weighted sum of the velocity cross products of all pairs of joints. Our form of the equations has no cross products so Coriolis and centripetal components are combined in the terms that are proportional to the squared velocities. Our equations have "net" and "inertial" components but they are not individually identical to the components when written in terms of joint angles.

The reason for this digression is to suggest that the labeling of components may be of relevance in terms of the physics, but is useful only for mathematical convenience; it is "book keeping." For motor control, the CNS does not distinguish between interaction and net components. It generates muscle forces that produce motion. If those forces are wrong (due to error or pathology), then movements go awry but this cannot be attributed to the failure of the CNS to properly compensate for any one motion dependent component or the other (see for example, Bastian et al. 1996; Sainburg et al. 1995). It is a failure to produce "correct" torques on the left side of the equations (a torque because of muscle activation by the CNS) that leads to undesired consequences on the right side of the equation, not the other way around.

---

†Dynamic torque is torque in which the gravitational terms have been omitted. These can be computed from Eqn. 27.1 with the gravitational constant g set to zero. We have suggested that the creation of a stable equilibrium state is dealt with independently of the dynamical problem of propulsion of the limb between such states (Gottlieb 1996).

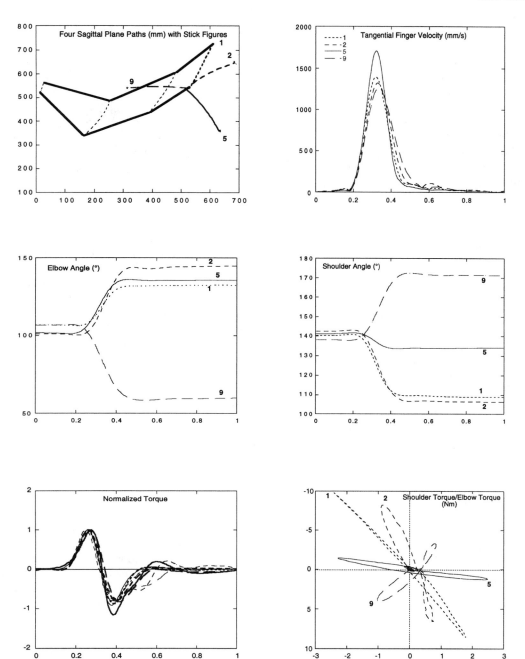

FIGURE 27.1. This figure illustrates four movements out of a series of 12, performed in different directions. The top two panels show that the kinematics are typically straight (a stick figure of the arm at the initial and final positions of the 1 o'clock movement are drawn on the left) with bell-shaped tangential velocity profiles. The middle panels show that the joint angles change uniformly with the distance moved and that very different directions of movement may occur with similar motions of one or the other joint. The bottom left panel shows that the amplitude normalized dynamic muscle torques are all very similar in their temporal features. Heavy lines are shoulder, thin lines are elbow. On the right, the plot of one joint torque vs the other results in narrow profiles with long axis that rotate with the direction of movement.

cause they are related by true constraints, Newton's laws of motion. Those laws, for movements involving more than a single joint, are too complex to allow us to easily intuit the consequences of an arbitrary pattern of joint torques or to perform inverse dynamics "in our heads." It is their obvious complexity that makes us look askance at the idea of the brain "solving" equation 1 to generate torque patterns from a planned trajectory. Thus we look at linear synergy as a "discovered" rule that the CNS uses to coordinate the torques across multiple joints. This rule, along with biphasic patterns constitutes a torque based movement "plan" that leads to fairly straight, smooth motion.

It is clear from Eq. (27.1) that the net muscle torque at one joint cannot be determined solely from the motion at that joint. Conversely, the motion at a joint cannot be determined exclusively by the torque at that joint. For example, elbow torques at 2 and 9 o'clock are small and almost the same while the elbow excursions are near maximal but in opposite directions. The relationship between muscle torque and direction of motion is less complex however, than it might seem thus far.

It is not sufficient to describe the behavior of individual joints. We must also describe the relationship between them. We have proposed that the relationship between joint torques can be approximated by a simple linear equation.

$$t_s = K_d\, t_e \qquad (27.3)$$

We call this rule *linear synergy* (compare to the arm angular velocity scaling law of Chapter 8.3). Figure 27.1 also shows the dynamic elbow and shoulder torques for the four directions of movement, all scaled so that their first peaks have magnitudes of $+1$. Torques at both joints are biphasic, highly synchronized pulses that have nearly simultaneous peaks and zero crossings. Also plotted are the dynamic shoulder torques, plotted versus dynamic elbow torques. These form tight ellipses or figure eights rather than straight lines because of imperfect synchrony. The correlation coefficients between the two joints are greater than 0.92 for all four directions. These figures imply that the direction of movement can be controlled by the relative sizes of the torques at the two joints while preserving their individual and nearly identical temporal patterns. Preserving the relative sizes but changing the magnitude and timing properties of

the biphasic pulses alters movement speed or distance or accommodates changes in the inertial load at the hand. The trajectories of these four movements are extremely simple both in external (Cartesian) and internal (joint angle) frames of reference. It might be asked whether the simple torque relationship of Eq. (27.3) is only possible with equally simple kinematics. To address this we have looked at a variety of other movements (and see Figures 1 and 2 of Almeida et al. (1995) and Figure 2B of Gottlieb et al. (1996b).

Figures 27.2 and 27.3 show pairs of individual movements. In Figure 27.2, one is a reach out to a target similar to the 1 o'clock movement in Figure 27.1. The other is a reversal movement in which the subject reached out to the same point and immediately returned to the initial position, similar to the movements studied in (Sainburg et al. 1995). In the latter we a have triphasic torque pattern at the elbow. Nevertheless, there is linear synergy between elbow and shoulder torques (and wrist torques which are not illustrated). In Figure 27.3 we have shown two point-to-point movements with no reversal of the finger tip but have chosen the initial and final points such that there is a reversal of one joint. Again there is linear synergy between shoulder, elbow and wrist torques.

The fact that these simple torque patterns lead to the observed kinematics is not obvious and in fact, is surprising. It prompted three questions. The first was had we made a programming error? Linear synergy has been independently reproduced in three other laboratories making that unlikely. The second is how general a rule is linear synergy? Work presented and cited here only begins to answer this.

The third is what do the four observed features of natural movements imply about the planning and control of those movements? Morasso also noted (p. 224) that "As a consequence [of the kinematic features], one may hypothesize that the central commands which underlie the observed movements are more likely to specify the trajectory of the hand than the motion of the joints. . . ." In light of the features of the torque described above and the fact that the appearance of straightness appears to outweigh physical straightness (Wolpert et al. 1994; Flanagan and Rao 1995), we do not consider this argument compelling. If movement planning is done in that way, then it remains to be demonstrated how such plans are executed.

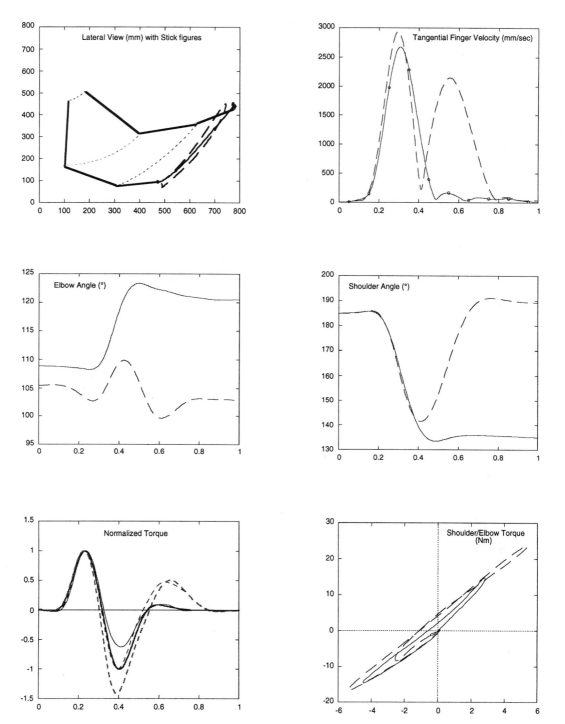

FIGURE 27.2. Two individual movements to the same target are illustrated. In the second movement (dashed lines), the hand did not remain at the target but promptly returned to its original position. Both movements are relatively straight but the latter has a double-peaked velocity profile. In this latter movement, there are reversals in the angular trajectories of both joints and the muscle torques have a triphasic rather than biphasic pattern. Heavy lines are shoulder, thin lines are elbow. Linear synergy between joint torques is preserved however.

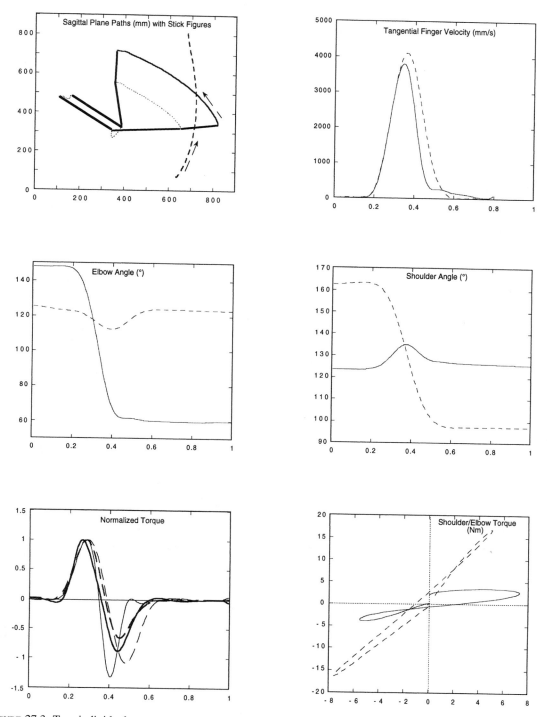

FIGURE 27.3. Two individual movements are illustrated. The relative straightness of the of the hand paths requires a reversal of one or the other joint midway through the movement. Nevertheless, torques at both joints remain biphasic and linear synergy persists. Heavy lines are shoulder, thin lines are elbow. The largest deviation from linear synergy is seen for the shoulder reversal, a movement in which the linear component of the shoulder torque would be quite small.

Straight, smooth movements require the CNS to produce appropriate muscle torques at the joints. The direct solution of the inverse dynamical equations is a possible but unlikely way to accomplish this. The use of force optimization criteria has been suggested (Uno et al. 1989) but the generality of this approach over the work space has not been established. The often discussed λ-equilibrium models assume that central commands from higher neural centers can be expressed in terms of a variable λ, which has units of length and is the threshold of a muscle's tonic stretch reflex. The difference between λ and concurrent feedback from muscle proprioceptors defines a state in space, an "equilibrium point" (EP) at which all muscle and external torques are balanced. Voluntary movement is a matter of shifting this EP from the initial to the final position and allowing lower level neuromuscular mechanisms, especially those of the spinal cord, to drive the muscles to their new EPs. Feldman's λ-model (λ_F) posits a central command that shifts monotonically between its initial and final values, is "independent of current external conditions" and expressed in terms of "positional frames of reference" to control of both single and multiple muscles and degrees of freedom (Feldman and Levin 1995). Latash has described a variant version of this hypothesis (λ_L) (Latash and Gottlieb 1991a, 1991b) in which the virtual trajectory of the EP is more complex (but see also (Gomi and Kawato 1996)).

The central commands of both EP models include a moving EP or virtual trajectory command ($R(l)$) and a coactivation command ($C(l)$) plus others that affect reflex gain and reciprocal inhibition. In the λ_F model, R moves monotonically and quickly (relative to the movement itself) from the initial to the final EP. The changes in dynamical torques needed to accommodate changes in distance, speed, load and direction are reflex driven and require appropriate changes in C (and the others) but we do not yet know many details of how this command is planned. Some of our other reservations concerning the λ_F-hypothesis can be found in [(Gottlieb 1995) and http://nmrc.bu.edu/MCL/Lambda_Thread/lambda.html]. The λ_L version proposes that for fast movements R has an "N-shape" which begins and ends at the same EPs as the R of the λ_F model but has intervening extrema. In both versions, muscle activation and ultimately muscle

force (F) is graded by the difference between R and the actual limb position (x) and scaled by C according to Eq. (27.3) [see for example Latash (1992) or Section 3.4 of Latash and Zatsiorsky (1993), Feldman and Levin (1995, St-Onge et al. (1993) for a more elaborate development that includes the velocity of motion].

$$F = f(C)(x - R) \qquad (27.4)$$

This equation is not incompatible with biphasic torque pulses or linear synergy. The extrema of R (in the λ_L version) create the dynamic biphasic torque components while the end points of R establish the static torques required to balance external forces such as gravity which appear in Eqn (27.1). Thus, controlling in terms of R is similar to controlling in terms of F, combining dynamic and static components, and requiring the parallel and compatible planning of C at both joints. One difference is that the execution of an R/C plan for Eq. (27.4) would appear to require knowledge of a variable (x) that is only available by sensory feedback in order to create linear synergy* while explicit planning in terms of F does not.

We conclude that movement planning is done explicitly in terms of muscle torques (or more precisely, the muscle activation patterns). Actual muscle torques will differ somewhat from those plans because of the compliant properties of the peripheral neuromuscular system. Actual EMG patterns will differ somewhat because of reflexes but feedback from muscle proprioceptors makes only a modest contribution (Gottlieb 1996) to the execution of well-planned movements. Movement emerges from the compliant interaction between muscles and their load according to Newtonian mechanics. This planning is based on a learned, internal model of limb and load dynamics that requires only a few parameters to create a biphasic torque pattern that matches the task (Gottlieb 1993).

In addition to the torque plan, we also have a trajectory plan, that is an "expectation" of what the trajectory should be, and by which we evaluate the kinematic outcome. This may be described as a tra-

---

*Linear synergy of torque (or F in terms of Eq. (27.3)) implies a linear relation between (x-R) across joints that seems difficult to achieve. Only R is a controllable variable while x, a feedback variable, differs with the joint and the task.

jectory of an EP (which exists by virtue of neuro-muscular compliance) that is similar to the actual movement. If the movement deviates from what we expected because we planned poorly or were inter-fered with, segmental reflexes offer some compen-sation but we will not go straight to and may even miss our targets if the dynamic trajectory was im-portant as in throwing for example. If the learned model was wrong (Shadmehr et al. 1993), it will be revised, given sufficient practice, until the trajectory is restored by a new set of joint torques (e.g., see Figure 26.3, in which the EMG patterns after prac-tice have changed dramatically; cf. Chapter 35, where such learning is viewed as an optimization process). If the movement appears to deviate from what we expected because of experimental perver-sity (Wolpert et al. 1994; Flanagan and Rao 1995; see also Chapter 23) then we modify both our force and our kinematic plan but not our model until the movement appears satisfactory. Generating curved movements requires deviation from linear synergy but often little more than small changes in the tim-ing of one or the other joint torque are sufficient. Purposeful deviation from linear synergy with a precise kinematic objective may be more difficult to accomplish and require extensive practice.

4   Future Directions

There are a number of questions that arise from this work.

1. How generally is linear synergy used by the nervous system? Knowledge of which movements violate it might tell us something about what are grace and skill and how we acquire them, and how this hypothesis compares to others. [Other hypoth-esized rules for upper limb movements include the leading joint hypothesis of Chapter 31, the ratio of joint angular velocities in Chapter 39, and the neuro-optimization approach of Chapter 7.4.

2. How might we experimentally distinguish be-tween force and kinematic planning? As we indi-cated above, given the compliant nature of the pe-ripheral neuromuscular system, it is difficult to infer from the outside what the unmeasurable cen-tral command might be. The same movement can, in principle, be produced by either one.

3. Does sensory feedback play an important role in the performance of well planned movements? One problem with this question is deciding what "important" means.

4. What is the effective limb compliance during movement? These compliant properties mean that the movement plan (in whatever terms it is formu-lated) will under many circumstances differ from the actual trajectory.

Question 1 is of general interest, regardless of the mechanisms responsible for it. Questions 2 to 4 are addressing the issue of how or if we can decide be-tween EP hypothesis based models and force based models (see also Chapter 29). The two EP versions make very different predictions about the location of the EP during a fast movement but both make strong demands on sensory feedback during the move-ments. Both versions require an appropriate degree of limb stiffness, for example, $f(C)$ in Eq. (27.3) that must be specified along with R for every task and which remains to be independently measured during fast movements. Our force model would qualita-tively agree with the λ_L version about where the tra-jectory of the EP but does not assume a major role for feedback or require any particular value of limb compliance to produce the joint torques. At present, the reader may look at the data base that has been published and apply Occam's razor.

References

Almeida, G.L., Hong, D.H., Corcos, D.M., and Gottlieb, G.L. (1995). Organizing principles for voluntary movement: extending single joint rules. *J. Neuro-physiol.*, 74(4):1374–1381.

Bastian, A.J., Martin, T.A., Keating, J.G., and Thach, W.T. (1996). Cerebellar ataxia: abnormal control of interaction torques across multiple joints. *J. Neuro-physiol.*

Feldman, A.G. and Levin, M.F. (1995). The origin and use of positional frames of reference in motor control. *Behav. Brain Sci.*, 18:723–806.

Flanagan, J.R. and Rao, A.K. (1995). Trajectory adapta-tion to a nonlinear visuomotor transformation: evi-dence of motion planning in visually perceived space. *J. Neurophysiol.*, 74(5):2174–2178.

Gomi, H. and Kawato, M. (1996). Equilibrium-point con-trol hypothesis examined by measured arm stiffness during multijoint movement. *Science*, 272:117–120.

Gottlieb, G.L. (1993). A computational model of the sim-plest motor program. *J. Mot. Behav.*, 25(3):153–161.

Gottlieb, G.L. (1995). Shifting frames of reference but the same old point of view. *Behav. Brain Sci.*, 18:758–759.

Gottlieb, G.L. (1996). On the voluntary movement of compliant (inertial-viscoelastic) loads by parcellated control mechanisms. *J. Neurophysiol.*, 76(5); (in press).

Gottlieb, G.L., Song, Q., Hong, D., Almeida, G.L., and Corcos, D.M. (1996a). Coordinating movement at two joints: a principal of linear covariance. *J. Neurophysiol.*, 75(4):1760–1764.

Gottlieb, G.L., Song, Q., Hong, D., and Corcos, D.M. (1996b). Coordinating two degrees of freedom during human arm movement: load and speed invariance of relative joint torques. *J. Neurophysiol.*, 76(5):3196–3206.

Gottlieb, G.L., Song, Q., Almeida, G.L., Hong, D., and Corcos, D.M. (1997). Directional control of planar human arm movement. *J. Neurophysiol.*, 78:2985–2998.

Hogan, N. (1984). An organizing principal for a class of voluntary movements. *J. Neurosci.*, 11:2745–2754.

Hollerbach, J.M. (1982). Computers, brains and the control of movement. *TINS*, 5:189–192.

Lacquaniti, F., Soechting, J.F., and Terzuolo, C.A. (1982). Some factors pertinent to the organization and control of arm movements. *Brain Res.*, 252:394–397.

Latash, M.L. (1992). Independent control of joint stiffness in the framework of the equilibrium-point hypothesis. *Biol. Cybern.*, 67:377–384.

Latash, M.L. and Gottlieb, G.L. (1991a). An equilibrium-point model for fast single-joint movement. I. Emergence of strategy-dependent EMG patterns. *J. Mot. Behav.*, 23:163–178.

Latash, M.L. and Gottlieb, G.L. (1991b). Reconstruction of shifting elbow joint compliant characteristics during fast and slow voluntary movement. *Neuroscience*, 43(2/3):697–712.

Latash, M.L. and Zatsiorsky, V.M. (1993). Joint stiffness: myth or reality? *Hum. Move. Sci.*, 12:653–692.

Morasso, P. (1981). Spatial control of arm movements. *Exp. Brain Res.*, 42:223–227.

Sainburg, R.L., Ghilardi, M.F., Poizner, H., and Ghez, C. (1995). Control of limb dynamics in normal subjects and patients without proprioception. *J. Neurophysiol.*, 73(2):820–835.

Shadmehr, R., Mussa-Ivaldi, F.A., and Bizzi, E. (1993). Postural force fields of the human arm and their role in generating multijoint movments. *J. Neurosci.*, 13(1):45–62.

Soechting, J.F. and Lacquaniti, F. (1981). Invariant characteristics of a pointing movement in man. *J. Neurosci.*, 1(7):710–720.

St-Onge, N., Qi, H., and Feldman, A.G. (1993). The patterns of control signal underlying elbow joint movements in humans. *Neurosci. Lett.*, 164:171–174.

Uno, Y., Kawato, M., and Suzuki, R. (1989). Formation and control of optimum trajectory in human multijoint arm movement—minimum torque-change model. *Biol. Cybern.*, 61:89–101.

Wolpert, D.M., Ghahramani, Z., and Jordan, M.I. Perceptual distortion contributes to the curvature of human reaching movements. (1994). *Exp. Brain Res.*, 98:153–156.

28
Simulation of Multijoint Arm Movements

Evert-Jan Nijhof and Erik Kouwenhoven

1 Introduction

Arm movements are so common that we easily forget how complicated they are. Even a simple movement like reaching for a cup of coffee requires fine-tuned muscle activation patterns. Not only the magnitudes of the individual muscle contractions, but also their timing are of key importance in getting one's hand to the cup.

Over the last two decades, numerous subjects were observed while performing goal-directed arm movements and their hand trajectories* were recorded. From these observations it became clear that hand paths are more or less straight (or slightly curved if significance is attributed to the small deviations of the straightness) (e.g., chapters in Winters and Woo 1990). Moreover, the velocity profiles are smooth and show only one maximum. This implies that subjects smoothly accelerate their hand achieving a maximum velocity somewhere around the middle of the path and subsequently decelerate smoothly again towards the target.

There is no doubt that experiments like the ones described above give deeper insight into human motor behavior and control. In particular, natural behavior can be studied qualitatively as well as quantitatively simply by watching and analyzing how people move under normal conditions. Abnormal movements are much harder to assess; either normal subjects have to be forced to move ab-

normally or one has to recruit subjects with some form of handicap or motor disorder. As an alternative, motion simulations might yield comparable or even superior information about the motor system. It is possible to simulate conditions that are difficult to realise in real life, like dissecting a muscle or immobilizing an entire bone segment. In this way researchers can gain deeper insight into the field of motor control.

Even if wrist and finger motions are not taken into account, the description of arm movements is still extremely complicated. The influences of more than 25 muscles across the shoulder joint and more than 20 muscles across the elbow joint have to be considered, as well as five bone structures. Moreover, the shoulder is not a fixed joint with respect to the torso, since the scapula slides over the rib cage as soon as the humerus is moved around. Nevertheless, various authors have tried to model the dynamical behavior of the human arm or part of it. Models of the shoulder mechanism, for example, have been developed by Karlsson and Peterson (1992) and by Van der Helm (1994), and the elbow joint has been modeled by Winters and Stark (1985, 1988). Van Dijk (1978) built a lumped-muscle model of both shoulder and elbow joints similar to the model presented here.

In this chapter we present a lumped-muscle model of the arm which is simple and comprehensible. The primary objective of the model is to serve as a tool in the process of understanding human motor control; it is not intended to predict the most accurate arm trajectories. The chapter is arranged as follows: first, we shall give details about various parts of the model like the bone structures and

---

*The term *path* refers to the spatial positions of the hand only, whereas the term *trajectory* also incorporates the temporal aspects.

the muscles. Then, we shall focus on how synergistic muscles can be lumped. In the final part we compare the results of some typical simulations with actually performed arm movements.

2 Methods

2.1 Numerical Model

We wanted a model that is simple and relatively easy to comprehend, yet capable of generating realistic arm movements. In addition, it should be capable of working in a *forward* mode where hand trajectories are calculated on the basis of muscle activation patterns, as well as in an *inverse* mode where muscle activation patterns are calculated from (measured) hand trajectories.

2.1.1 Workspace

We met the first of these design criteria by restricting the model to two dimensions and by excluding gravity and any wrist motions. This left us with a two-segment model: an upper arm that can rotate in a frictionless shoulder joint and a lower arm that is connected to the upper arm by an elbow joint. The lower arm is defined as a combination of the forearm and the immobilized hand. Both segments can move freely in the horizontal workspace at shoulder level. The part of the workspace that can be reached by the tip of the index finger is indicated by the gray area in Figure 28.1. Note that shoulder horizontal adduction (hereafter termed flexion) angles range between −30° and 140° and elbow flexion angles between 0° and 150°.*

2.1.2 Muscles

To generate realistic arm movements, we included muscles in the model. The muscle model is based on the extensive work of Winters and Stark (1985, 1988). A muscle is represented by three components, as is schematically depicted in Figure 28.2 (see also Chapter 1 and Chapter 8): an active, contractile element (*CE*) that generates force by shortening and two passive elements in the form of a parallel elastic element (*PE*) and a series elastic element (*SE*).

---

*Full extension of shoulder and elbow corresponds to 0°.

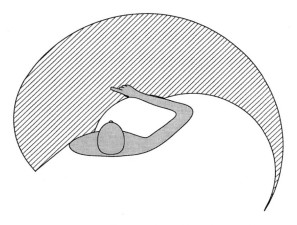

FIGURE 28.1. Part of the horizontal workspace that can be reached by the tip of index finger. Shoulder angles range between −30° and 140° and elbow angles between 0° and 150°.

The amount of muscle contraction is controlled by an activation signal *A*, a continuous signal between 0 and 1. Two first-order filters have been incorporated representing the excitation-activation coupling dynamics. The activation dynamics T_a can be seen as a simplified, mathematical description of the complex active state dynamics of the myofibrils inside the muscle filaments. The activation dynamics is split into two time sequences: activation that occurs in the order of 5 to 10 ms and deactivation that takes about five times longer. The input of the activation dynamics is the excitation *E*, a signal that can be regarded as a lightly filtered and rectified *EMG*. All the neural dynamics like re-

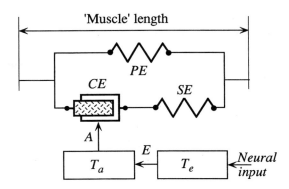

FIGURE 28.2. Classical Hill model of a muscle. The input is a neural signal between 0 and 1 and the output is the length of the muscle.

cruitment, decruitment, firing rate, etc. have been incorporated in a time delay T_e of the order of 40 ms between the neural input and the excitation signal E.

2.1.3 Active Element

The contractile element CE generates force once it has been activated. This force is determined by the activation A, the force-velocity relation (F_v) and the force-length relation (F_l) by (see also Chapter 8)

$$F = A\, F_v F_l \qquad (28.1)$$

Various models that fit the experimental force-velocity relation have been proposed (see Winters 1990 for a review). In our model, we have used the well-known hyperbolic Hill relation between contractile force F_v and muscle shortening velocity v_{ce}. This relation can be written in the form

$$F_v^c = F_m \frac{a_f(1 - v_{ce}/v_m)}{a_f + v_{ce}/v_m} \qquad (28.2)$$

where the dimensionless parameter a_f of order 1 defines the steepness of the hyperbolic shape. The other two parameters determine the intercepts of the velocity axis and of the force axis. They represent the maximum, unloaded contraction velocity v_m and the maximum, isometric force F_m, respectively. Both parameters scale with activation A, having their maximum values at maximum activation $A = 1$. For a lengthening muscle, we used a similar relation (Winters and Stark 1985), which, after imposing the appropriate boundary conditions at $v_{ce} = 0$, reads

$$F_v^e = F_m \frac{f_\infty(a_f + 1)\dfrac{v_{ce}}{v_m} + (f_\infty - 1)a_f}{(a_f + 1)\dfrac{v_{ce}}{v_m} + (f_\infty - 1)a_f} \qquad (28.3)$$

where the relative maximum eccentric force $f_\infty = F_\infty/F_m$ is taken to be 1.3. In Figure 28.3 graphic representations of Eqs. (28.2) and (28.3) are shown for various values of the neural activation A and $a_f = 0.35$.

Muscle force depends on the length of the muscle, a fact described by its force-length relation. Since the muscle spans a joint, it is more natural and convenient to express this relation as a torque-angle relation which includes changes in the moment arm of the muscle with joint angle. We have

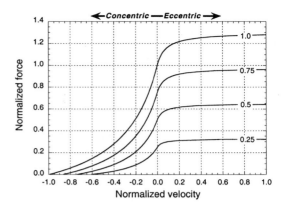

FIGURE 28.3. Normalized muscle force versus normalized contraction speed (Eqns. 28.2 and 28.3, with $a_f = 0.35$ and $f_\infty = 1.3$) for various values of the neural activation. Negative velocities correspond to concentric contractions, positive velocities to eccentric contractions.

fitted experimental data of shoulder and elbow muscles (Kulig et al. 1984) with a Gaussian curve

$$T(\theta) = T_m \exp[-((\theta - \theta_{Tmax})/\theta_f)^2] \qquad (28.4)$$

where θ_{Tmax} denotes the joint angle at maximum torque T_m and θ_f is the Gaussian shape-parameter.

2.1.4 Passive Elements

Muscles are connected in series and in parallel with soft tissue like tendon, skin, blood vessels and so on. The series elastic element (SE) models the tendon between muscle and bone as a nonlinear spring. The parallel elastic element (PE) works as a limiter. By generating a progressively increasing restoring torque it prevents the bone segment from passing beyond a certain angle. The soft tissue tends to have quasistatic mechanical properties in which the stiffness (i.e., the slope of the force-extension curve) increases fairly linearly with force over the primary range. This results in a first-order differential equation for the elastic force. Solving the equation leads to

$$F(\Delta x) = F_m(e^{a_e \Delta x/\Delta x_m} - 1)/(e^{a_e} - 1) \qquad (28.5)$$

where Δx is the muscle extension relative to the rest (i.e., zero force) length, F_m is the maximum force of the elastic element, a_e is a shape parameter where a higher value results in a steeper curve, and Δx_m is the extension at F_m.

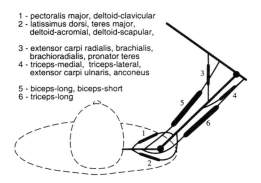

1 - pectoralis major, deltoid-clavicular
2 - latissimus dorsi, teres major,
     deltoid-acromial, deltoid-scapular,

3 - extensor carpi radialis, brachialis,
     brachioradialis, pronator teres
4 - triceps-medial, triceps-lateral,
     extensor carpi ulnaris, anconeus

5 - biceps-long, biceps-short
6 - triceps-long

FIGURE 28.4. Arm model consisting of six lumped muscles and two segments. Indicated are the synergistic muscles that are lumped into one effective muscle.

In addition to being elastic, muscles also exhibit viscous behavior, that is, they show an increase in repulsing force with velocity. Viscosity can be introduced either into the parallel elastic element or into the force-velocity relation. We opted for the latter and included a small viscosity of 0.2 Nms/rad in Eqs. 28.2 and 28.3.

2.2   Muscle Lumping

Because muscles can only pull and not push, at least four are required to operate a two-joint arm model (Figure 28.4): one pair of antagonistic, mono-articular muscles across the shoulder joint (1 and 2) and another pair across the elbow joint (3 and 4). A more extended version of the model might also incorporate a biarticular pair running from the shoulder to the lower arm (5 and 6).

From anatomy it is known that several muscles act as synergists during a particular movement, in which each of them has its own anatomical and physiological characteristics. This brings us face to face with one of the major problems we have to solve in order to develop a *manageable* muscu-loskeletal model: How can we lump the different synergistic muscles in the lowest number of effective muscles?

One of the first assumptions we make relates to the recruitment order of the synergists. Are they innervated simultaneously or is there a recruitment order whereby, for example, the larger muscles with large moment arms are recruited first? On the

basis of our experiences with synergistic elbow muscles like the m. brachialis and the m. brachioradialis (Tax et al. 1990) and medial and lateral heads of the m. triceps (Van Groeningen and Erkelens 1994), we assumed that innervation was simultaneous. In these studies that focused on recruitment in particular, no significant differences were found in the recruitment order of the two synergists. The synergistic muscles are therefore lumped only on the basis of their relative torque contribution about the joint.

The anatomical data of the shoulder muscles are derived from the extensive work of Wood et al. (1989a,b) and those of the elbow muscles partly from An et al. (1981) and partly from a human skeleton model. Because we are interested in arm movements in the horizontal plane at shoulder level, we consider only those muscles across shoulder and elbow joints which generate significant horizontal adduction/abduction of the humerus and flexion/extension of the forearm bones. Vertical humeral abduction/adduction and pronation/supination of the forearm are not taken into account. This leaves us with 19 muscles to consider (Table 28.1): 4 shoulder flexors, SFs (*deltoid—clavicular part* (DELC), *pectoralis major—abdominal part* (PMJA), *—clavicular part* (PMJC), *—sterno-costal part* (PMJS)), 4 shoulder extensors (*deltoid—acromial part* (DELA), *—scapular part* (DELS), *latissimus dorsi* (LATD), *teres major* (TMAJ)), 4 elbow flexors, EFs (*brachioradialis* (BRAD), *brach-ialis* (BRAC), *pronator teres* (PROT), *extensor carpi radialis* (ECRD)), 4 elbow extensors, EEs (*triceps—lateralis* (TRIA), *—medialis* (TRIM), *anconeus* (ANCO), *extensor carpi ulnaris* (ECUL)), 2 biarticular flexors, BFs (*biceps—long head* (BILH), *—short head* (BISH)) and 1 biarticular extensor (*triceps—longum* (TRIO)).

For each of the 19 muscles, at least 8 parameters have to be specified: the muscle's origin and insertion coordinates, the muscle's moment arm(s) with respect to the joint(s) and the maximum muscle strength. The origin and insertion coordinates cannot be taken directly from the work of Wood et al. (1989b) because these authors presented the data with the arm in an arbitrary orientation. After a coordinate translation and several rotations about the humeral head we ended up with the upper arm and the forearm in the horizontal plane. Figure 28.5 shows the linear trajectories of the 19 muscles to-

TABLE 28.1. Anatomical data of the 19 muscles involved. The coordinate system refers to Figure 28.5, where the origin is at the center of the humeral head, the X-axis runs to the right, the Y-axis upwards and the Z-axis out of the plane.

Muscle	Ty	Or	In	O_x (cm)	O_y (cm)	O_z (cm)	I_x (cm)	I_y (cm)	I_z (cm)	R_s (cm)	R_e (cm)	PCSA (cm^2)	T_s (Nm)	T_e (Nm)	T_s (%)	T_e (%)
DELC	SF	C	H	−3.4	2.2	2.8	12.4	1.0	1.4	3.9		4.5	9		13	
PMJA	SF	T	H				6.3	1.7	1.1	4.8		3.9	9		14	
PMJC	SF	C	H	−11.8	7.6	0.0	9.4	1.7	0.9	8.2		5.2	21		32	
PMJS	SF	T	H				6.9	1.7	1.1	6.6		4.5	15		22	
DELA	SE	S	H	1.0	−1.5	2.3	15.3	0.6	1.3	−1.0		13.5	−7		11	
DELS	SE	S	H	−4.7	−6.0	0.8	13.2	−0.1	1.2	−6.1		3.9	−12		10	
LATD	SE	T	H				5.5	0.9	−1.2	−1.9		12.9	−12		20	
TMAJ	SE	S	H	−4.0	−8.3	−8.9	6.2	0.4	−1.0	−4.2		5.8	−12		20	
BRAD	EF	H	R	23.0	−1.1	0.8	29.1	22.6	−2.8		4.2	1.5		6		11
BRAC	EF	H	U	16.1	0.0	0.1	31.8	3.5	−1.3		2.1	7.0		13		25
PROT	EF	H	R	28.7	−0.5	0.4	30.6	7.5	0.1		1.6	3.4		5		10
ECRD	EF	H	M	27.8	−1.1	1.4					2.9	5.3		14		27
ANCO	EE	H	U	30.6	−1.1	1.0	33.0	0.8	−0.8		−1.1	2.5		−3		6
ECUL	EE	H	M	30.3	−1.4	1.7					−1.3	3.4		−4		10
TRIA	EE	H	U	13.8	−1.9	0.8	32.6	−1.9	−1.1		−2.0	6.0		−11		27
TRIM	EE	H	U	20.5	−1.7	−1.9	32.6	−1.9	−1.1		−2.0	6.1		−11		27
BILH	BF	S	R	−1.1	1.9	1.3	31.0	4.4	−0.1	1.9	3.4	2.5	4	8	7	15
BISH	BF	S	R	−3.2	3.3	0.8	31.0	4.4	−0.1	4.6	3.4	2.1	9	6	13	13
TRIO	BE	S	U	−1.4	−2.8	−2.8	32.6	−1.9	−1.1	−3.2	−2.0	6.7	−19	−12	31	30

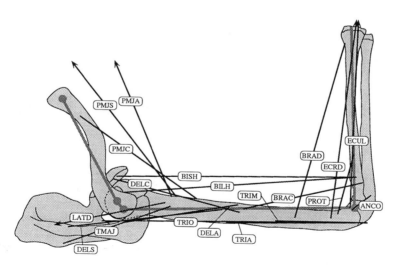

FIGURE 28.5. The linear trajectories of the 19 flexor/extensor muscles across the shoulder and elbow joints together with the 5 bone segments involved. The arrows represent the 5 muscles that only partially attach to the sketched bones.

gether with the 5 bone segments involved. Note that the origins of the PMJA, PMJS and LATD and the insertions of the ECUL and ECRD are somewhat obscured because they do not originate or insert on one of the sketched arm segments. In Table 28.1, the anatomical data are summarized, where *Or* and *In* denote origin and insertion, and *C, T, H, R, U* and *M* stand for clavicle, thorax, humerus, radius, ulna and metacarpal bone. The origins and insertions are given in the coordinate system of Figure 28.5, where the upper arm is 30° flexed and the forearm 90° flexed. In Figure 28.5 the *X*-axis runs to the right, the *Y*-axis upwards, the *Z*-axis out of the plane and the origin is the center of the humeral head. (Note that the model itself uses another, body-fixed, coordinate system.)

The effective moment arms R_s and R_e can be calculated from Wood et al. (1989b) and An et al. (1981) and are reproduced in Table 28.1 where a positive value represents a flexion moment arm and a negative value an extension moment arm.

It is virtually impossible to determine the maximum strength of an individual muscle in vivo. As a first-order approximation it is common to take the maximum strength proportional to the physiological cross-sectional area (PCSA) of the muscle (Crowninshield and Brand 1981). For the ratio *k* between muscle strength and PCSA, values in the range 40–100 N/cm^2 have been proposed (Crowninshield and Brand 1981). In order to obtain a valid constant based on our data we calculated the total torque T_j of a muscle group about either the shoulder or the elbow joint *j* according to

$$T_j = k_j \sum_i (PCSA_i \cdot R_{ji}) \qquad (28.6)$$

where R_{ji} denotes the moment arm of muscle *i* and the summation runs over all the contributing muscles. Note that we used two constants *k*, since shoulder and elbow data are obtained by different techniques. The maximum torques can be compared to results reported in existing literature[*] on maximum voluntary torques about the shoulder joint (Williams and Stutzmann, 1959) and elbow joint (Singh and Karpovich 1968). In our model we use $k_s = 50$

---

[*]For an extensive review on muscle forces, see Kulig et al. (1984).

TABLE 28.2. Skeleton parameters of the model

Parameter	Unit	Upper	Lower
Mass	kg	1.82	1.43
Length	cm	30.9	33.3
Mass center*	cm	13.5	16.5
Moment inertia†	kgcm^2	510	575

*Proximal from joint.
†With respect to joint.

N/cm^2 and $k_e = 90$ N/cm^2. In the last two columns of Table 28.1, the relative torque of a particular muscle about a joint is given. For example, the clavicular head of the deltoid (DELC) contributes 13% to the total shoulder flexion torque.

Finally, the individual insertions, origins and moment arms of the synergistic muscles across a particular joint contribute to their lumped version an amount that is proportional to their relative torque contribution according to

$$X = \sum_i (X_i \cdot T_i) / \sum_i T_i \qquad (28.7)$$

where X_i denotes one of the quantities of muscle *i* and the summation runs over all the synergistic muscles.

2.3 Skeleton Parameters

The anthropometric data listed in Table 28.2 are taken from the standard work of Winter (1979). The length values are scaled to the upper arm length of Wood et al. (1989b) and a body mass of 65 kg is assumed.

3 Simulations

At first glance it might seem obvious that the hand path of a point-to-point movement is straight, simply because it is the shortest path between two points in space. From a control point of view, however, it is as difficult to generate a straight path as an arbitrarily curved path (Hollerbach and Atkeson 1987). Although horizontal reaching movements are undoubtedly straight in nature, small but significant and reproducible deviations from straightness can be observed (e.g., Miall and Haggard 1995). The following questions arise: What is the

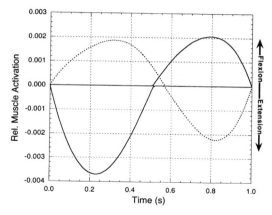

FIGURE 28.6. Activation patterns of elbow (solid) and shoulder (dotted) muscles for the midsagittal movement.

3.1 Accuracy of Activation Patterns

In the first series of simulations, effects of "errors" in the muscle activation patterns are studied. To get a feeling of what kinds of activations are required to generate a movement, we used a stripped version of the model, where unnecessary cocontraction is avoided and the torques of the bi-articular muscles are incorporated into the monoarticular ones. Figure 28.6 shows the activation patterns of the remaining four muscles as the result of an inverse simulation. The input hand path was a minimum jerk trajectory in the midsagittal plane from starting position $(-0.179, 0.293)$ to target position $(-0.179, 0.431)$. (Positions are specified in a right-handed, shoulder-centered coordinate system, where the X-axis runs through both shoulders, and the Y-axis points away from the body. As the speed of the movement, we used a value that is quite often observed in our experiments, although the conclusions still hold for faster movements.) The solid lines denote the elbow muscle activations and the dashed lines the shoulder muscle activations. These four patterns will be used as the input for the subsequent (forward) simulations.

Figure 28.7 shows the simulated hand paths if the activation signals from Figure 28.6 are scaled in amplitude (a type of sensitivity analysis). The right panel represents the movement from $(-0.179,$

origin of this curvedness? And is it in the motor planning or in the execution of that motor program. For example, "miscalculations" of the muscle activations generate nonstraight hand paths. Misjudgment of the payload at the hand (or simply the inertia of the hand itself) also yields incorrect muscle activation patterns which results in curvedness.

Simulations with the described model were performed to study the effects of "errors" in muscle activations patterns and the effects of misjudging the payload at the hand on the shape of the resulting hand paths.

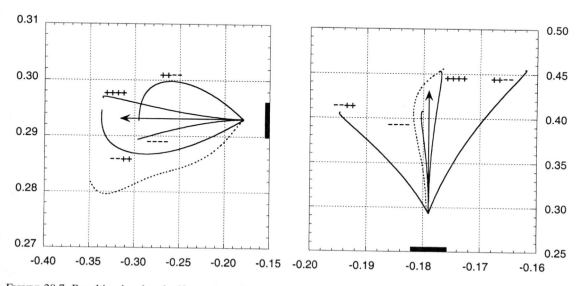

FIGURE 28.7. Resulting hand paths if muscle activations are 20% enlarged (+) or 20% reduced (—). Dotted curves show the paths of the actually measured hand movements.

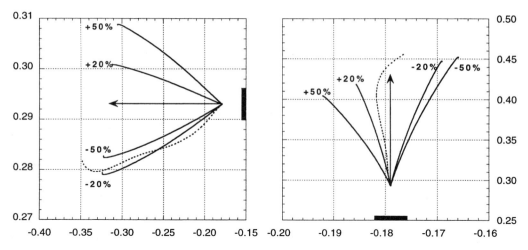

FIGURE 28.8. Resulting hand paths if payload at the hand is misjudged by ±20% or ±50%.

0.293) to (−0.179, 0.431) and the left panel from (−0.179, 0.293) to (−0.320, 0.293). The legends beside the curves denote whether an activation is scaled by +20% or by −20%, in the order elbow flexor, elbow extensor, shoulder flexor, shoulder extensor. The arrows show the hand paths resulting from the original activations (i.e., the input paths of the inverse simulations) and the dotted curves give the averaged paths produced by a subject in complete darkness. The black bars indicate the magnified axes on their original scale. It is obvious that errors in the amplitude of the activation patterns give rise to curved hand paths, often with small hooks at the end. Note that these end hooks have been reported to occur experimentally too, although during much faster movements (e.g. van Sonderen et al. 1988).

3.2 Misjudgment of Payload

Effects of misjudging the payload at the hand is shown in Figure 28.8. First, activation patterns in the inverse mode have been calculated with a payload of 0.5 kg at the hand and subsequent forward calculations with modified loads as indicated in the figures yield the paths shown. Once again, paths were curved if the payload is misjudged, although the shapes are quite different from those depicted in Figure 28.7.

4 Discussion

From an anatomical point of view, the (human) shoulder and elbow mechanisms are already ex-

tremely complicated with more than 45 muscles and 5 bones. Therefore it is really challenging to try to model multijoint arm movements. Two roads can be followed; either one develops a model that is as accurate as possible with all muscles involved (e.g., Van der Helm 1994) or one simplifies the musculoskeletal system as far as possible without losing the essential characteristics. Since our design objective was to have a model that helps us to understand planar *hand* movements and was not to predict the most accurate *arm* positions, we took the latter road and developed a lumped-muscle model with 2 degrees of freedom, 6 effective muscles, and 2 bone structures. It is obvious that this is a really gross approximation. In particular taking the shoulder joint as a grounded hinge joint neglects all the excursions of the scapula and clavicle and undoubtedly results in incorrect humerus positions. However, realistic hand positions are still possible if small corrections in the elbow angle are allowed.

As an example of the kind of simulations that can be performed with the model, we presented point-to-point hand movements that resulted from imperfect muscle activation patterns. Of course, errors in muscle activation amplitude and misjudgment of the payload at the hand give rise to curved hand paths which do not end at the target. However, the absolute target errors are only centimeters on a travel distance scale of nearly 15 cm. In order to make a hand movement toward a fixed target it is therefore unnecessary for the motor apparatus to specify the activations patterns very accurately, say to within 1%. A small final correction based on visual feedback is sufficient to arrive at the target. If such a correction cannot be performed due to a lack of visual feedback, as was the case for the experimental data shown, end-point errors and path curvednesses are of the same order of magnitude as the simulations presented. Moreover, the scatter in individual hand paths performed in complete darkness is comparable to the scatter due to "errors" in activation patterns as is shown in this chapter. It is therefore not unreasonable to state that hand trajectories are only fuzzily specified. It cannot be concluded from these simulations whether the motor commands themselves are not very accurate or whether it is just their execution.

5   Future Directions

Motor control is still a rather unexplored area simply because it is so hard to obtain objective measurements. Unintentional interventions by the subject might obscure the experimental outcomes. Simulations do yield objective outcomes and will become more and more important as a tool for understanding normal and abnormal motor behavior. It remains to be seen whether a complete model, with all the muscles and bones included, is the holy grail that should be sought. Even the simplest inverse problem with two muscles and one degree of freedom already becomes an underdetermined problem if co-contraction is allowed. Adding more muscles requires additional constraints and/or optimization criteria in order to distribute joint torques over the synergistic muscles (see also Chapter 35). For a model with more than 40 muscles this is quite a challenging task. The same argument holds for a forward model: How can realistic activation patterns be specified for a large number of muscles if one cannot measure them properly in vivo?

In our opinion building a musculo-skeletal model should be a bottom-up process (see also Chapter 32). The optimal model should, at least, have the proper kinematics, i.e., all bone structures and joints should be included with all the possible degrees of freedom. Then a minimum number of (effective) muscles should be included in order to operate the joints. More muscles and/or ligaments can be added to this basic model in a later stage, depending on the kind of question one wants to answer.

Acknowledgments. The authors would like to thank Dr. J.J. Boessenkool for providing the recorded hand trajectories and Prof. Dr. C.J. Erkelens for his stimulating discussions.

References

An, K.N., Hui, F.C., Morrey, B.F., Linscheid, R.L. and Chao, E.Y. (1981). Muscles across the elbow joint: a biomechanical analysis. *J. Biomech.*, 10:659–669.

Crowninshield, R.D. and Brand, R.A. (1981). A physiologically based criterion of muscle force prediction in locomotion. *J. Biomech.*, 14:793–801.

Hollerbach, J.M. and Atkeson, C.G. (1987). Deducing planning variables from experimental arm trajectories: pitfalls and possibilities. *Biol. Cybern.*, 56:279–292.

Karlsson, D. and Peterson, B. (1992). Towards a model for force predictions in the human shoulder. *J. Biomech.*, 25:189–199.

Kulig, K., Andrews, J.G., and Hay, J.G. (1984). Human strength curves. *Exerc. Sport Sci. Rev.*, 12:417–466.

Miall, R.C. and Haggard, P.N. (1995). The curvature of human arm movements in the absence of visual experience. *Exp. Brain Res.*, 103:421–428.

Singh, M. and Karpovich, P.V. (1968). Strength of forearm flexors and extensors in men and women. *J. Appl. Physiol.*, 25:77–189.

Tax, A.A.M, Denier van der Gon, J.J., and Erkelens, C.J. (1990). Differences in the coordination of elbow flexors muscles in force tasks and movement tasks. *Exp. Brain Res.*, 81:567–577.

Van der Helm, F.C.T. (1994). A finite element musculoskeletal model of the shoulder mechanism. *J. Biomech.*, 27:551–569.

Van Dijk, J.H.M. (1978). Simulation of human arm movements controlled by peripheral feedback. *Biol. Cybern.*, 29:175–186.

Van Groeningen, C.J.J.E. and Erkelens, C.J. (1994). Task-dependent differences between mono- and bi-articular heads of the triceps brachii muscle. *Exp. Brain. Res.*, 100:345–352.

Van Sonderen, J.F., Denier van der Gon, J.J., and Gielen, C.C.A.M. (1988). Conditions determining early modification of motor programmes in response to changes in target location. *Exp. Brain Res.*, 71:320–328.

Williams, M. and Stutzmann, L. (1959). Strength variation through the range of joint motion. *Phys. Ther.*, 39:145–152.

Winter, D.A. (1979). *Biomechanics of Human Movement*. John Wiley & Sons, New York.

Winters, J.M. (1990). Hill-based muscle models: a systems engineering perspective. In *Multiple Muscle Systems*. Winters, J.M. and Woo, S.L.Y. (eds.). Springer Verlag, New York.

Winters, J.M. and Stark, L. (1985). Analysis of fundamental movement patterns through the use of in-depth antagonistic muscle models. *IEEE Trans. Biomed. Eng.*, BME-32:826–839.

Winters, J.M. and Stark, L. (1988). Estimated mechanical properties of synergistic muscles involved in movements of a variety of human joints. *J. Biomech.*, 21:1027–1041.

Winters, J.M. and Woo, S.L.Y. (1990). *Multiple Muscle systems*, Part III: Principles underlying movement organization: upper limb. Springer Verlag, New York.

Wood, J.E., Meek, S.G., and Jacobsen, S.C. (1989a). Quantitation of human shoulder anatomy for prosthetic arm control—I. Surface modelling. *J. Biomech.*, 22:273–292.

Wood, J.E., Meek, S.G., and Jacobsen, S.C. (1989b). Quantitation of human shoulder anatomy for prosthetic arm control—II. Anatomy matrices. *J. Biomech.*, 22:309–325.

29
Planning of Human Motions: How Simple Must It Be?

Frans C.T. van der Helm and A.J. (Knoek) van Soest

1 Introduction

Human beings are able to perform a tremendously large set of motions. These motions vary from cyclic (e.g., walking, biking), to explosive (e.g., jumping, hitting, throwing) to positioning tasks (e.g., reaching). The way these motions are accomplished is still the subject of debate among psychologists, neurophysiologists, biomechanists, and others.

Often humans are compared with robots. Whereas the accomplished tasks can be similar, the system underneath is strikingly different. Human limbs contain highly nonlinear elements as multiple degree-of-freedom joints, muscles and sensory organs, and moreover the system has kinematic and actuator redundancies. Motions are controlled by the central nervous system (CNS). It is very likely that the CNS takes advantage of (or at least takes into account) the nonlinearities of the peripheral system. The specific characteristics of the peripheral system are learned in years of trial and error, and practicing of skilled motions. Generally, robots are much simpler in construction, avoiding actuator or kinematic redundancies. Robot motions are generated using a trajectory planner and a position (and sometimes velocity) feedback control. This is in contrast to humans, where the time-delays in the feedback pathways would make a position-controlled feedback control system very slow, and therefore a feedforward component is required to control motions.

The goal of this chapter is to set up and discuss a general scheme for control of positioning tasks. The problem to be addressed is to what extent the CNS is capable of complex planning. For instance, does the CNS take all nonlinearities into account, or is planning by the CNS likely to use simplifying strategies ('primitives') like the equilibrium point and/or force fields discussed in Chapter 24 and Chapter 25? Consequences of these simplifications are analyzed from a control engineering viewpoint. What is gained and what is lost by a simplified planning strategy?

From this perspective, we make a distinction between the planning of a motion and the execution of the motion. Whereas the planning of motions takes place in the CNS, both the CNS and the peripheral system are involved in execution. For optimal planning, knowledge about the execution part is necessary.

2 Approach

Analyzing human motions by looking at separate components will only reveal a limited view. It is better to use an integrative approach in which all the components (CNS and the peripheral system) are represented with their specific dynamic features. In Figure 29.1, a block scheme is shown for the planning and execution of human motions (Rozendaal 1997). The scheme is developed with control engineering conventions in mind (i.e., it can be used for quantative analysis and computer simulation right away). The blocks contain linearized representations of nonlinear phenomena, in order to use available linear analysis tools. However, the essence is maintained: It provides insight in the dynamics of feedback and feedforward planning and

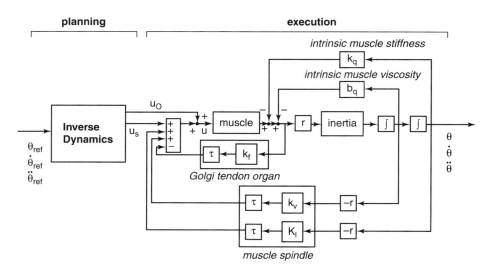

FIGURE 29.1. A block scheme of the motion control of a human arm. Input is a reference trajectory; output is the actual trajectory. In the inverse dynamic block a feedforward signal u_0 is calculated from the reference trajectory, as well as a setpoint signal u_s. The muscle block contains the muscle dynamics, transferring the neural input signal (α-activation) to muscle force. In the inertia block muscle force is transferred into accelerations, subsequently twice integrated to obtain the actual position. In the in-trinsic muscle stiffness block the force-length relation is represented, and in the intrinsic muscle viscosity the force-velocity relation is represented. Proprioceptive feedback can be divided into force feedback (represented by a gain k_f and a time-delay τ_d) in the Golgi tendon organ, and length and velocity feedback (represented by a gain k_l and k_v, respectively, and a time-delay τ_d) in the muscle spindle. The moment arm is represented by r, transferring force to moment, and angle to length.

motion execution. The scheme is similar to that presented in Chapter 11. The execution part contains well-known components of the musculoskeletal system (muscles, inertia, viscosity and stiffness of the skeletal system, muscle spindles, Golgi tendon organs).

Planning is done in a feedforward part, which contains some sort of internal model of the arm and the environment. This takes place presumably in the cerebellum. Input to the planning part is the desired motion or goal of the task. The output of the planning part to the execution part can be conceptually divided into a neural input to the muscle u_0 and a setpoint signal u_s. Actually, u_s can be regarded as a prediction of the sensory signals based on the reference trajectory. If the actual sensory signals match (processed in interneurons?), no error signal is present and no neural input signal to the α-motor neuron is generated. At the level of the α-motor neuron the neural signals of u_s and u_0 are combined, and a single neural signal to the muscle results. The conceptual distinction between feedforward signal u_0 and setpoint u_s is important. In a servocontrol loop, u_s would function as the reference signal, representing the desired position. The speed of the servocontrol loop depends on the dynamics of its components. However, the speed of the whole system can be increased by adding another signal (u_0) to the reference signal, in which the (slow) dynamics of the system are compensated for. The forward path of the execution part contains the muscle dynamics and the skeletal dynamics including the geometrical aspects. From this scheme it is clear that if the feedforward part u_0 results from a perfect inverse model of the muscle and skeletal dynamics, the actual motions will be exactly equal to the desired motions, and no feedback is needed. In other words, the contribution of the feedback path will be cancelled out by its prediction u_s, which is generated by the same perfect inverse model.

However, in reality some sort of perturbations will always be present, and therefore feedback loops are necessary for stability and/or robustness. Two sorts of feedback are present (i.e., originating from the intrinsic muscle properties (force-length, force-velocity relationships) and originating from the proprioceptive feedback (muscle spindles and Golgi Tendon Organs (GTO)). The muscle force-length and force-velocity relationships result in stiffness and damping of the skeletal system. The dynamics of the proprioceptive feedback pathways are dominated by the time-delay caused by transportation of signals along the nerves and the processing of information in the CNS. Typically, these time-delays are estimated to be about 30 to 40 ms for arm muscles, and 30 to 50 ms for leg muscles. It is well known from control theory that time-delays in feedback loops result in a decreased bandwidth (that is, only low-frequency reference signals (desired motions) can be tracked if a pure feedback control would have been applied. Simulations revealed that the upper bandwidth is limited to 2 to 3 Hz, whereas humans are capable of motions of about 6 to 8 Hz. It must be concluded that fast motions are generated with help of the feedforward planning part, and are not under feedback control!

The details of the scheme in Figure 29.1 are discussed in Chapter 11. Here, only the most important features for the motion planning will be presented. The setpoint u_s, necessary as a representation of the reference trajectory, is compared with signals derived from the proprioceptive sensors as muscle spindles and GTO's, and an error signal results. This signal is combined together with the feedforward drive u_0 at the level of the α-motor neuron. If in absence of the feedforward drive u_0, the setpoint u_s contains only position information (which would be the case if only a desired position would be input to the planning part), the executive system is effectively a *position feedback* control system, without a contribution of the velocity and force feedback. However, from a control engineering point of view, this is very unlikely. The skeletal system will act as a double integrator (muscle force is transferred into acceleration, twice integrated to obtain position), where the stiffness and viscosity might be low when the muscles are not activated. Combined with the activation dynamics in the muscle (calcium uptake/release) and the time-delays in the loop, the phase lag of the feedback loop is below 180 degrees starting at very low frequencies on. In other words, only position feedback will result in a barely stable or unstable system.

However, to prevent drifting whenever the de-

sired endpoint is reached, position feedback information is very important. To increase the performance of the feedback system, velocity and force feedback are necessary: The phase lag will then be decreased and the frequency bandwidth will be increased (see Figures 29.2A and 29.2B). This implies that the planning part must produce velocity and force setpoints as well, which can only result from an extended inverse kinematic planning module, incorporating the desired velocity and acceleration.

3   Simplified Planning Strategies

As argued in the previous section, the optimal feedforward planning exploits a perfect model of the neuromusculoskeletal system (i.e., assumes perfect knowledge of muscle and skeletal dynamics). Intrinsic and proprioceptive feedback will compensate not only for external perturbations, but also for imperfections in the internal model (model noise). Any planning strategy that employs a feedforward component that is derived using a simplified internal model *must rely on feedback loops* to achieve the desired performance, because the feedforward component does not result in the desired accuracy even in the absence of external perturbations. Why would one devise motor control theories that entail simplifications of the scheme presented above? Two main reasons seem to drive the attempts to devise such theories. First, it is argued by many that the CNS is unlikely to be able to form a full model of the effector system. Second, it is argued that in many cases we are not interested in having a perfect control system, as long as the goal of the task to be performed is attained. The first argument seems to be losing ground, given recent advances in the field of neural networks. Thus, the second argument forms the main motivation for investigating the extent to which simpler control schemes are applicable to specific domains of motor control.

In particular, when considering positioning tasks, it may be argued that as long as the goal (here the desired position at time infinite) is attained, full control over the end-point trajectory is not necessarily required. However, it is important to have a clear view of the simplifications introduced in such control schemes, to analyze the behavioral limitations

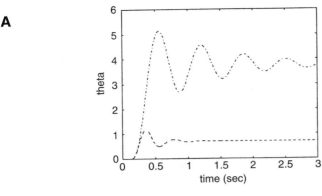

A

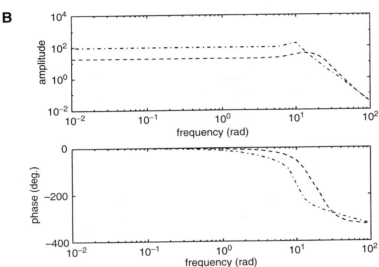

B

FIGURE 29.2. The transfer function from input signal u_s to output signal θ, for position and velocity feedback (-.-.), and for position, velocity, and force feedback (- - -). It can be seen that the addition of force feedback results in faster and better damped behaviour. (A) Simulations in the time-domain, response to a unit step. (B) Simulations in the frequency domain.

that follow from these simplifications and to check if the motor control theory is in accordance with general control principles. Only in this way will it be possible to judge for which type of tasks the scheme will indeed result in adequate behavior. In this chapter, we attempt to contribute to this endeavor by outlining how two motor control schemes in which the concept of equilibrium points is central fit into the general scheme outlined above. In particular, we will consider the λ theory (e.g., Feldman and Levin 1995), and force-field theories (e.g., Bizzi et al., 1991; Mussa-Ivaldi and Giszter 1992; Giszter et al. 1993; Chapter 24, Chapter 25).

3.1 The λ Model

Where does the λ model (e.g., Feldman 1966; Latash 1993; Feldman and Levin 1995) fit into the

general scheme presented? More precisely, which simplifications are applied to the general scheme in that model? In the classical form of the λ hypothesis, the feedforward component is absent, and the specified λ represents a setpoint for a servotype controller (i.e., supraspinal neural input signal u_s). Furthermore, the only feedback channel considered was position feedback arising from spindles. At the level of individual muscles, the controlled variable, termed λ, was conceptualized as the threshold length for the stretch reflex—the length above which muscle activation would occur because of the input from muscle spindles coding for muscle length. It was assumed that λ for agonist and antagonist are independently controlled.

At the level of joints, it has been argued that a change of variables yields a more insightful representation; at this level, the controlled variables are reformulated into the so-called R and C commands, the R (reciprocal) command specifying the equilibrium angle for the joint in the absence of external forces, and the C (cocontraction) command specifying joint stiffness (e.g., Feldman et al. 1990). In our scheme, R would represent the position setpoint in the absence of external forces, whereas C would represent the amount of co-contraction obtained by the difference in the position setpoints for the agonist and antagonist. According to Feldman, positioning movements are controlled by letting R change at maximum rate from the initial endeffector position to the desired endpoint.

In the absence of feedback delays, a control strategy based on just specification of λ would suffice to obtain perfect behavior, as exemplified by robot control, where excellent positioning behaviour is obtained using only position feedback control. Given the presence of latencies however, position feedback results in adequate tracking only for low-frequency behavior. Even then the gain of the feedback loop must be modest if the system is to behave stably. That is, in the presence of latencies λ control may eventually take the system to the desired end point, but control of the movement taking the end-effector there will be very slow. It is generally accepted that a pure time delay exists on the order of 35 ms, representing the time it takes for the signal to travel up and down the extremity. In addition, a phase shift results from the intrinsic dynamics of the muscle and the transition from muscle force (resulting in acceleration) to position by a double integration. In models in which these dynamics are explicitly accounted for (e.g., the general scheme presented), this phase lag is automatically included in the model. In models in which muscle activation is assumed to be directly (i.e., algebraically) coupled to muscle force, however, the phase lag is not fully taken into account. How large is this phase lag? Interestingly, the literature regarding this phenomenon largely describes this phase lag as a time delay between EMG and joint torque, the delay being referred to as Electro-Mechanical Delay. Based on crosscorrelation of EMG and joint torque during 2 Hz sinusoidal isometric contractions of the quadriceps group, values in the order of 90 ms have been recorded.

The combined dynamics of muscles and skeletal system already results in a considerable phase lag. Adding the phase lag of the time-delay of the position feedback system, the frequency bandwidth is about 0.5 Hz and allows only for a position feedback gain of about 0.001. Effective servocontrol behavior with only position feedback is then impossible. Since humans are able to move up to 6 to 8 Hz, the limited frequency bandwidth is a severe objection against the original formulation of the λ theory.

In the current version of the λ theory (Feldman and Levin 1995), velocity feedback is also considered. In particular, muscle activation is regulated not only through the specification of $\lambda(t)$ (see above), but depends also on the specification of μ. It is suggested by Feldman that the ultimate λ that regulates muscle activation, referred to as $\lambda*$, equals λ plus μ times the sensed movement velocity. In terms of our control scheme, μ represents the gain of velocity feedback; the setpoint velocity is assumed to be fixed and equal to zero. In contrast to the specification of λ, the specification of μ does not follow from the desired kinematics. It is assumed to be an independent variable that is directly specified by higher centers in the nervous system. As in any control system, if the controller gain μ is too high, an unstable feedback loop can result.

Velocity feedback clearly will improve system behavior, i.e. increase the frequency bandwidth; force feedback will increase the bandwidth even more (see Figure 29.2). If μ is assumed not to be constrained, adequate specification allows reproduction of somewhat richer behavior (e.g., triphasic pattern in positioning), though the bandwidth of the feedback-controlled system will remain limited. If we assume that μ is indeed freely specified (adjustable feedback gain of the velocity feedback loop), optimal specification increases the bandwidth of the system to about 2.3 Hz. Adding force feedback increases the bandwidth to about 3 Hz. Even with the combined force-velocity-position feedback, the system does not show fast-tracking behavior. It is concluded that a feedforward input signal is necessary in addition to the setpoint signal in order to account for the fast motions observed in human beings.

In conclusion, time delays and muscle dynamics are the main determinants of the quality of the behavior obtainable from the combination of position setpoint and gain control and velocity gain control that in our view represents the λ model. In both the

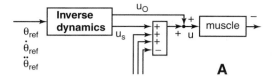

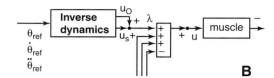

FIGURE 29.3. In A the setpoint signal u_s is compared to the proprioceptive signals coming from the muscle spindles and Golgi tendon organ. The feedforward signal u_0 is added to the resulting error signal. In effect, this is similar to the

scheme in B. Here, the feedforward signal u_0 is added to the setpoint signal u_s before the summing point (α-motor neuron). This is in agreement with the concept that λ would contain feedforward information: $\lambda = u_0 + u_s$.

theoretical and the simulation work on the λ theory, remarkably little attention is paid to the value to be assigned to these delays.

What about external and inertial forces? In the description of the λ model, it is frequently acknowledged that resulting movements emerge from the interplay of central commands, feedback, and external conditions. If external and inertial conditions are taken into account, some part of λ contains the feedforward planning signal: $\lambda = u_0 + u_s$, where u_0 is the feedforward signal, calculated using inverse kinematics or inverse dynamics (see Figure 29.3). However, it appears to be assumed that the primary controlled variable is still planned without taking these external and inertial forces into account. If the system would be very stiff, this results only in slight deflections of the intended movement. However, it has been shown experimentally that joint stiffness is not that high. Katayama and Kawato (1993) showed that typical N-shaped motion patterns of the EP are required if the hand is to follow a straight line from start to endpoint. Furthermore, the stiffness resulting from proprioceptive feedback is limited if instability is to be avoided. Thus, when λ is planned irrespective of external conditions, large deflections may arise; if external forces *are* taken into account, this would require a representation of the effect of those forces on the system, reintroducing to a large extent the computational problems (i.e., inverse dynamic calculations) that the theory was meant to dispense with. An extreme example of the effect of external forces discussed by Van Ingen Schenau et al. (1995), concerned a contact control task where a subject seated at a horizontal table is required to move a heavy object along a straight line in front of the body. Inverse kinematical analysis of this task shows that it requires extension of the elbow. Inverse dynamical analysis reveals that the net joint

torque at the elbow is flexing. Thus, due to the external force, the joint torque required actually opposes the required joint movement. In terms of the equilibrium point for the joint, this would mean that during the entire movement, the elbow equilibrium angle is to lag behind the actual elbow angle!

In conclusion, it is our view that the λ theory is appealing in that it overcomes many of the computational problems. From the above analysis, however, we conclude that it is applicable only to low-frequency behavior in the absence of external forces. If relatively fast motions with inertial load are performed, feedforward planning (i.e. input signal u_0) is necessary.

3.2 Force Fields

A more recent group of equilibrium point theories may be referred to as force field theories (e.g., Bizzi et al. 1991; Mussa-Ivaldi & Giszter 1992; Giszter et al. 1993). Let's first summarize the starting point of these theories. They are all inspired by experiments in frogs, in which it is found that stimulation of specific points in the spinal cord results in movement of the paw to a specific point in space, irrespective of the initial state of the paw. Only a small number of points of stimulation in the spinal cord have been identified that are associated with such equilibrium points. By definition, associated with each of these stimulation points, a potential energy field may be constructed that has a single minimum at the identified equilibrium position. Here, these potential energy fields are usually represented in the form of force fields (first derivative of the potential energy field), indicating what force the endeffector exerts on a static manipulandum as a function of position. Based on the identification of this small number of equilibrium points and associated force fields, it is suggested that movement is generated by activating a number of these force fields in such intensity, that their sum defines an equilibrium point at the intended endpoint of the positioning movement.

Is it remarkable in itself that stimulation of specific points in the spinal cord generates equilibrium points of the endeffector? In our admittedly naive view, it is not at all! Given the fact that we have a damped system with a number of elastic components (e.g., muscle's force- length relation; length feedback), it is natural that a fixed input will take the limb to some equilibrium point. We would expect that many of these equilibrium points are partly determined by passive elastic structures such as joint capsules, ligaments, and parallel elastic elements in muscles. In fact, the only thing that would surprise us is if it is indeed the case that the number of equilibrium points is as small as presently suggested. In our view, it is too early to provide a definite answer to the question of how many stimulation points that result in a equilibrium position of the paw actually exist.

In the experimental studies, it has not yet been investigated to what extent these force fields have dynamic properties. That is, to what extent the force associated with a certain position depends on things other than that position, in particular on time. From the fact that the stiffness associated with these force fields is too high to be generated by intrinsic muscle properties, at least as muscle's force-length relation, we conclude that proprioceptive feedback loops with their corresponding latencies must contribute to the generation of the force field. As a result, we do not expect the force fields to be an *instantaneous* function of position. Rather, at a given stimulation, the force generated at a particular position will depend directly on supraspinal stimulation and feedback contributions.

At present very little has been said in the literature about the way in which the inputs to these alleged stimulation points are to be generated to produce a movement to a given target. Do these inputs depend just on the desired end position? Are the movement speed and external forces accounted for? If they do indeed depend on all of these factors, then the simplification obtained may be limited.

Where does the idea that movement may be controlled through activation of force fields fit into our scheme? As at present the mechanisms underlying the generation of the force fields are not well de-

fined, a final answer is impossible. It is clear, however, that the implementation includes many of the lower level feedback processes included in our scheme. Thus, we would suggest that the force field approach implies that it is not inputs onto individual muscles that are controlled, but rather that somewhere higher in the hierarchy, fixed combinations of such inputs are activated. If indeed the number of combinations is limited, this inevitably will result in a less rich movement repertoire, at the advantage of a lower-dimensional input vector. It may well be that execution of positioning tasks, the goal of which fits very well in the context of force fields, is indeed simplified by this approach at an acceptable cost; we suggest however that in tasks where it is not the aim to end at a specified point in space with zero velocity, control on the basis of combinations of force-fields may even complicate control.

4   Future Directions: Toward an Inverse Dynamic Model for the Planning of Motions

Why would evolution opt for a simpler control strategy? Is it beyond the capacity of the central nervous system to include the inertial properties and knowledge of external conditions in the planning strategy? Many hypotheses about motion control employ a kind of internal model. Years and years of practice make it very likely that the properties of the limb are well known, but also a fair estimate of the environment can be made. The muscle activation patterns and resulting motion are completely different if a heavy suitcase is lifted instead of a light suitcase. Hence, expected mass *does* affect the planning strategy.

What role do kinematics play in movement planning? In experiments it has been shown that the hand moves along a straight line from start to endpoint (see also Chapter 27 and Chapter 29). Dynamic optimization techniques would certainly predict that the optimal trajectory (from an energy point of view) is often a curved line, due to the inertia of the arm. Does this mean that the motion is planned in kinematic variables? An explanation could be that the motion is learned under visual control (i.e., the "error function" is indeed in end-

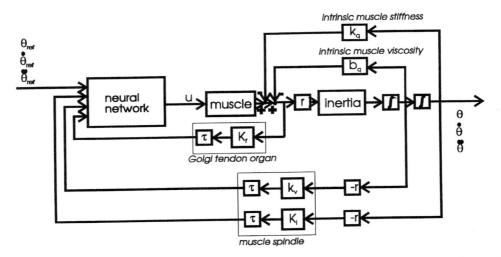

FIGURE 29.4. A block scheme for the learning and control of human arm motions. The feedforward controller together with the feedback controller are combined in one neural network. (Stroeve 1999A,B)

point coordinates). Thus, studying simple, artificial motions (like hands moving in a plane) only show part of the capacity of the CNS in motion planning. Whenever a motion is truly learned and optimized, e.g. in sports, hardly any straight line motions remain (e.g., see Chapter 31).

In a kinematically redundant linkage, the joint positions are not specified by the endpoint alone. In a forward dynamic optimization, indeed an optimal motion pattern can be found. Nonetheless, it is possible that the CNS exploits simple kinematic relations to reduce the redundancy of the system (e.g., see Chapter 27, Chapter 31; Chapter 39). For example, if the arm (shoulder and elbow) is considered as a 5 DOF system, and the endpoint (without orientation specified) has 3 DOF, the simplification would not be too far off if the 5 DOF are mapped to a 3 DOF kinematic subspace. However, such a "kinematic" simplification *could hardly be* distinguished from a dynamic optimization, because it is learned under dynamic conditions.

Artificial neural networks are definitely a simplification of the biological neural network, in the capacities and sheer numbers of the neurons. Stroeve (1999A,B) built a neural network for the combined feedforward and feedback control of a 2 DOF-6 muscle musculoskeletal model of the arm (see Figure 35.C2). Muscle length, velocity, and force feedback, including time-delays, are input to the network, as well as the desired trajectory, needed

for the feedforward calculations (compare to Figure 29.1, where the in- and output signals are the same). The neural network had 32 nodes in the hidden layer, and was trained by Backpropagation Through Time. He showed that this neural network can indeed "learn" the inertial properties of the arm and environment, and use this knowledge for a accurate feedforward planning of the motion (Figure 29.4).

An internal model of the arm and environment containing the inverse dynamics of the system would be the optimal feedforward model. Imperfections of the optimal feedforward model (model noise) can be adjusted for by intrinsic muscle properties and proprioceptive feedback. Simplifying strategies as the EP hypothesis or force fields can be regarded as a subset of the optimal feedforward model. As long as it has not been shown that human motions are structurally deviating from the optimal performance (and it has not), there is no need to assume that simplifying strategies are exploited. Certainly, the central nervous system has more than enough "computational" power to incorporate the dynamic properties in addition to the kinematic properties!

References

Bizzi, E., Mussa-Ivaldi, F.A., Giszter, S.F. (1991). Computations underlying the execution of movement: a novel biological perspective. *Science*, 253:287–291.

Feldman, A.G. (1966). Functionable tuning of the nervous system with control of movement or maintenance of a steady posture: 2. Controllable parameters of the muscle. *Biophysics*, 11:565–578.

Feldman, A.G. and Levin, M.F. (1995). The origin and use of positional frames of reference in motor control. *Behav. Brain Sci.*, 18:723–806.

Giszter, S.F., Mussa-Ivaldi, F.A., and Bizzi, E. (1993). Converging force fields organized in the frog's spnal cord. *J. Neurosci.*, 13:467–491.

Ingen Schenau, G.J., van Soest, A.J., Gabreels, F.J.M., and Horstink, M.W.I.M. (1995). The control of multijoint movements relies on detailed internal representations. *Human Movement Sci.*, 14:511–538.

Katayama, M. and Kawato, M. (1993). Virtual trajectory and stiffness ellipse during multi-joint movement predicted by neural inverse models. *Biol. Cyber.*, 69:353–362.

Latash, M. (1993). *Control of Human Movement.* Human Kinetics, Champaign, Il.

Mussa-Ivaldi, F.A. and Giszter, S.F. (1992). Vector field approximation: a computational paradigm for motor control and learning. *Biol. Cybern.*, 67:491–500.

Rozendaal, R.H. (1997). *Stability of the Shoulder: Intrinsic Muscle Properties and Reflexive Control.* PhD thesis Delft University of Technology, Delft, The Netherlands.

Stroeve, S. (1999a). Impedance characteristics of a neuromusculoskeletal model of the human arm. I: Postural control. *Biol. Cybern.*, 81:475–494.

Stroeve, S. (1999b). Impedance characteristics of a neuromusculoskeletal model of the human arm. II: Movement control. *Biol. Cybern.*, 81:495–504.

30
Biomechanics of Manipulation: Grasping the Task at Hand

Aram Z. Hajian and Robert D. Howe

1 Introduction

From most of the biomechanics literature on the upper limb, you might not suspect that arms end in hands that interact with real objects. Much research has focused on the role of the central nervous system in controlling limb motion, so tasks are reduced to free trajectories (e.g., Chapter 27) or interactions with abstract "force fields" (see Chapter 24 and Chapter 25). The arm is often characterized by a single end-point position or force, or the velocity or torque at each joint. This approach has produced a number of interesting insights, and it promises to account for important aspects of the underlying structure of the motor control system. It does not, however, address many of the mechanisms responsible for our ability to perform real tasks.

Hands mediate a large majority of our mechanical interactions with the world, and an understanding of their characteristics is essential to explain manipulation skills. In this direction, physical therapists, hand surgeons, and rehabilitation engineers have succeeded in relating the functional physiology of the hand to specific task properties. Unfortunately, this empirical knowledge does not elucidate the neural and active muscular basis for the observed behavior. The absence of such an explanation is not surprising, since hands are singularly complex appendages, with many joints, a complicated musculature, high enervation density, and unique reflexes. Many tasks involve the use of multiple fingers in parallel and in opposition, and fingers often slide and roll as an object is manipulated. Just instrumenting such tasks to record position and force information is challenging, and analyzing the resulting data is even more so.

Some of the characteristics of our interactions with the environment are atypical of traditional physical systems analysis. Before and after we contact objects, the equations of motion which govern the dynamics change across a short time scale. The environment is often unknown and varying, and thus, difficult to precisely model. In addition, the mechanics of the operations such as grasping, twisting, lifting, and assembling are poorly understood or quantified. These difficulties notwithstanding, recent experiments have begun to illuminate some of the sensing and control strategies that make human hands dexterous. Results from robotic analysis and experimentation also provide insight into the key issues.

From this perspective, we begin with a discussion of the issues raised in the study of dexterous manipulation. Next, we review unique aspects of the physiology of the hand, including muscular, skeletal, and neural components. The following section examines work on the various segments of manipulation tasks, beginning with the initial contact against the finger, and proceeding through grasping and various types of manipulation. As an illustrative example of one method of analyzing task execution, we discuss issues and strategies we have observed during human drumming. We conclude with a discussion of important open questions and promising areas for future research.

2 Functional Characteristics of Hands

Hands are not like five miniature arms: they have several unique mechanical and neural features that distinguish them as the primary effectors of con-

382

tact. The enormous complexity is evident from the kinematics; we have over twenty degrees of freedom in each hand. The glabrous (hairless) skin of the hands has a very high density of specialized cutaneous mechanoreceptors. In addition to our prehensile characteristics, such as the opposable thumb and fingers which can curl to grasp an object tightly, we have specialized reflexes for manipulation. One example is the slip reflex, a control mechanism that acts to increase grip force on the detection of incipient slip of an object from grasp. As a result of their small relative size, as well as their mechanical design, hands have a higher bandwidth for movement and control than other parts of the body. One aspect of that mechanical design is the fact that many of the hands' muscles are not adjacent to the joint that they actuate, but are located in the forearm and act through long tendons passing through the wrist. This solves the problem of powering many joints in close proximity, but complicates the internal kinematics and dynamics for control purposes.

Our hands have many other unique features. We take advantage of the parallel chains in the kinematic structure of our hands to apply opposition forces to form a grasp, a primary method of interaction with the environment. The dynamic range in force control is perhaps eight orders of magnitude from the most precise to the most powerful—microsurgeons can sense and control forces as low as a few mg while power lifters can hoist weights as high as hundreds of kg.

Manipulation skill is dependent on extensive tactile afferent information; people become clumsy when deprived of touch information through numbness due to cold or anesthetics (Johansson and Westling 1984). Each hand contains about 17,000 mechanoreceptive units (Vallbo and Johansson 1984). Our tactile sensory experience is built from a variety of sensors responding to a number of physical parameters, including skin curvature, deformation, and acceleration.

3   From Contact to Task Analysis

Analyzing neuromuscular control of the hand in manipulation tasks is difficult because of the number of factors involved—from both the hand and

the task. This complexity becomes clear in considering the progression typical of many precision manipulation tasks. Significant interaction begins with the first contact between the fingers and the object to be manipulated, when cutaneous receptors begin to report the object's properties and location. These neural signals often modify the motor commands as the object is securely grasped in a configuration appropriate to the task. The object is then manipulated in some fashion by the fingers, and interactions can occur both between the fingers and object, and between the object and other surfaces in the environment. This section reviews research results that pertain to each phase of this process, and outlines some of the outstanding issues.

3.1   Contact

There are two key aspects of the first contact against the finger tips: sensation and mechanics. High enervation density allows exact localization of the object relative to the fingers, which can be essential for precise grasping, especially with poor visual information. These receptors also detect surface features and texture, which are particularly important if the fingers are to slide or roll over the object surface during the task. This sensing process can be remarkably fast: the local coefficient of friction is sensed in the first 200 milliseconds after contact in some tasks (Johansson and Westling 1984). The fact that contact occurs at a variety of places and in a variety of patterns on the fingers is markedly different from the usual biomechanics treatment of "limbs," where loading is almost always applied at the end point.

For small-scale precision tasks that use fingertip grasps, the passive mechanics of the fingertips can play an important role. Recent experiments have measured the force-displacement-velocity relationship of the fingertip pulp, as well as the pressure distribution across the fingertip, during indentation by a flat surface (Pawluk and Howe 1999a). The force increases exponentially with displacement through the first few Newtons of contact (Figure 30.1). This pattern is observed in many soft biological materials, but it is perhaps surprising here because the contact area is also changing. This force-displacement relationship results in a stiffness increasing with the applied load, just as observed for muscles, although the mechanism here

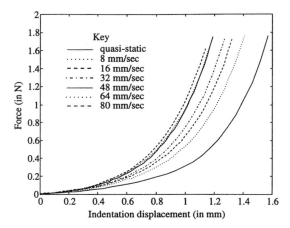

FIGURE 30.1. Force vs. displacement at various indentation velocities. (From Pawluk and Howe 1999a)

is completely passive. One advantage of this relationship is that at low force levels the stiffness is near-zero, so that as the finger first makes contact with the object, small position errors produce minimal disturbance forces. As the object is securely grasped and the applied force increases, the contact stiffens, resulting in little motion in response to disturbances.

For contact with a flat surface, the pressure distribution across the finger tip pad is approximately parabolic (Figure 30.2), and reasonably well modeled by the Hertz theory that describes contact of elastic spherical solids. These passive mechanics affect both the sensory system (the mechanoreceptor population must operate across a huge range of local pressures) and the mechanical interactions (the object will press, roll, and slide across the finger)

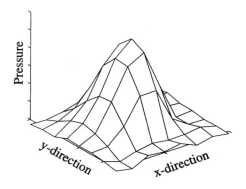

FIGURE 30.2. Static pressure distribution across the human index finger pad. Grid spacing is 2 mm.

(Pawluk 1996b). The details of these relationships among finger pad properties, manipulation mechanics, and tactile sensing have yet to be addressed.

3.2 Grasping

Following contact, many tasks involve the grasping of an object. These require selecting specific configurations of the fingers with respect to objects, consistent with constraints dictated by the object, the hand, and the task. This problem has received considerable attention from both empirical and analytical perspectives. Studies of human grasping have often focused on categorizing observed grasps; perhaps best-known are the six grasps defined by Schlesinger (1919): cylindrical, fingertip, hook, palmar, spherical, and lateral. Such a categorization leads to associating grasps with object shapes. Thus a ball suggests a spherical grip while a cylinder suggests a wrap grip. However, when people use objects in everyday tasks, the choice of grasp is dictated less by the size and shape of objects than by the tasks to be accomplished. Even during the course of a single task with a single object, the hand adopts different grips to adjust to changing force conditions.

Napier (1980) suggested that grasps should first be categorized by function rather than appearance. In Napier's scheme, grasps are divided into power grasps and precision grasps. Power grasps are useful where considerations of stability, security, or high forces predominate, and are characterized by large areas of contact on the fingers and the palm. Precision grasps are used when sensitivity and dexterity are paramount, and the object is held with the tips of the fingers and thumb. Cutkosky and Howe (1990) have developed a more detailed taxonomy that includes both shape and functional divisions (Figure 30.3). It is important to note that although this type of taxonomy organizes the range of human grasps, the correspondence between tasks and grasps is not well explained.

In robotics research, considerable attention has been devoted to a complementary approach: mechanical analysis of the grasping process. These studies attempt to model the interaction between object and hand from first principles. For example, contact forces, kinematics, or compliance of the hand–object system are often included in a model. However, in order to make analyses tractable, many

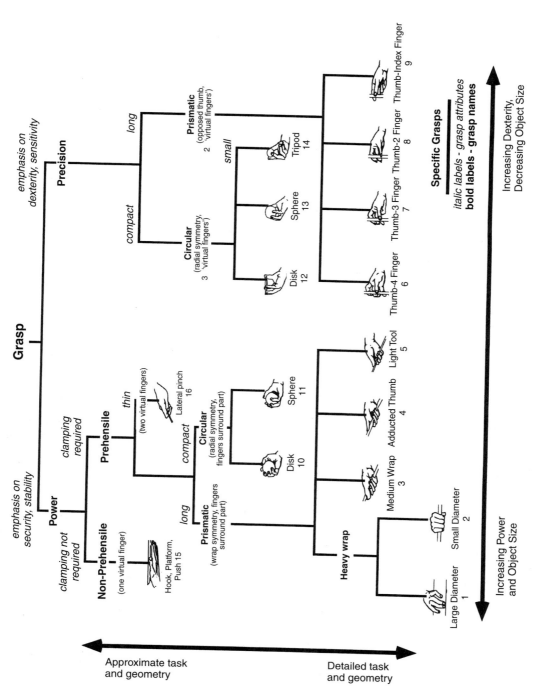

FIGURE 30.3. Grasp taxonomy. (from Cutkosky and Howe 1990)

important factors are not included. Shimoga (1996) reviewed the diverse techniques used in these models. Grasp choice is often formulated as an optimization problem, with objective functions that include dexterity, precision, sensitivity, power, and stability; various methods are used to evaluate these measures from task requirements and hand and object geometry. Unfortunately, there has been little experimental validation of these analyses, largely because of the lack of multifingered robotic hands with sufficient functional dexterity to execute the grasps. These approaches to grasp planning usually ignore the role of real-time sensory feedback during grasping, which is clearly an important component of human grasping.

Recently, grasp analysis has been applied to human manipulation, with interesting results. Tendick et al. (1993) constructed a model of the kinematics of the pencil grasp often used for writing and other precision tasks, specifically for analysis of surgical suturing. The results indicated that at the tool tip, effective stiffness is low in the direction of needle motion, where mobility is important, and higher in the perpendicular directions, where resistance to disturbances is important. This analysis is significant for its use of analytical tools to relate the properties of the hand and the task, and it helps to explain the ubiquitous choice of this grasp for suturing and many other tasks. Buttolo (1996) performed similar analyses, and confirmed the results through experimental measurements of grasp stiffness during manipulation of a pen.

These analyses also underscore a number of essential differences between the hand and the arm. Although the arm has more than the six degrees of freedom required to arbitrarily position and orient an object in space, the hand is superior for many tasks. Factors in this precedence include the lower inertia of the fingers, the parallel rather than serial configuration (which can, among other things, provide high stiffness or sensitivity), and the ability to configure the grasp to produce a wide range of object impedance values and compliance center locations. Many tasks take advantage of the complementary capabilities of the arm and the hand, by using the hand to generate the desired stiffness and precise mobility in certain directions, and using the arm to move the hand so these directions are correctly aligned for the task. These considerations show that the fingers work together to create capabilities that are not explained by traditional single limb analyses.

3.3 Manipulation

Manipulation can involve many different mechanical interactions between the hand and object. For tasks that use power grasps, manipulation is often accomplished by the arm, while precision grasps often involve rolling and sliding of fingers, and compliant changes in contact force direction and magnitude. Little biomechanics research has appeared on the more complex of these operations, due to the evident difficulties in measurement and analysis. A number of robotic analyses of manipulation with rolling and sliding have appeared in recent years (Cole et al. 1992; Bicchi and Sorrentino 1995; Chang and Cutkosky 1995). Although these consider only simplified situations, one important lesson is that the equations describing these systems are greatly affected by the assumed contact conditions. This means that the mechanical interactions at the fingertips, including compliance, friction, and damping, can have a large role in determining the behavior of the hand–object system.

Johansson and Westling have performed a series of experiments that illustrate some of the sensing and control mechanisms for dealing with contact conditions. These experiments also elucidate some of the functions of touch sensing in the performance of manipulation tasks (Johansson and Westling 1988). Nerve signals from mechanoreceptors were monitored as subjects grasped and lifted specially instrumented objects. The measured grasp force was always near the minimum required to avoid slipping, despite large variations in object weight and coefficient of friction of the grasping surface. Signals from finger tip nerve endings that indicated the earliest stages of slip were followed by a reflexive and unconscious increase in the grasp force that prevented further slipping. Other nerve signals indicated the making and breaking of contact between the fingers and the object, and between the object and the table. These signals are apparently used by the central nervous system to trigger motor commands for successive phases of the manipulation task (Johansson and Westling 1988).

Lehman and his colleagues have investigated bimanual tasks, which reveal some of the same types of control strategies (Lum et al. 1992; Reinkens-meyer et al. 1992). This work describes methods for controlling grasp force based on anticipated load changes and accelerations. As with individual hands, these studies show that mechanisms required for parallel and opposition configurations are significantly different from the serial chains of single arms and legs.

3.4 Impedance

Mechanical impedance is widely used in biomechanics research as a convenient way to characterize the interaction between limbs and externally applied task forces and motions (Hogan 1990). Dexterous manipulation often capitalizes on the kinematic complexities of the hand to vary the impedance of manipulated objects. This is accomplished by selecting an appropriate grasp configuration and by regulating grasp force, and a wide range of impedances can be obtained in this way. This strategy is particularly useful in tasks where the grasped object interacts with other surfaces in the environments. As an analytical tool, impedance also allows us to reduce the complexity of the system under investigation by considering only the effective end-point impedance at the interface between the finger and the manipulated object, or between the object and the environment.

This section analyzes drumming, as an example of a task where the impedance of a grasped object is modulated by changing grasp force. Playing a drum roll relies on impedance variation to compensate for slow response times. By adjusting the effective stiffness of the hand, the passive dynamics of the hand–drumstick system can be modulated, and thus the drum roll frequency. Using a simplified model of the complicated hand kinematics, we demonstrate that the drumming frequency varies with the pinch grasp force.

Humans use many different techniques to produce fast, continuous drumming. Here the focus is on the standard double stroke roll practiced by trained drummers, which underscores the passive dynamic interaction of interest and is straightforward to measure and analyze. The double stroke roll begins with one hand striking the drum with the stick, which is then allowed to bounce once. As this stick is retracted, the other hand brings its stick onto the drum for the next double stroke. This al-ternating sequence of double bounces is repeated for the duration of the roll in establishing a steady drumming frequency well above the rate possible if each stroke is actuated directly by the hand.

To analyze this process, we must reduce its kinematic complexity: as in virtually every manipulation task involving the hands, the number of links and degrees of freedom is unwieldy. During the double bounce, the arm and hand are basically motionless, and the dynamics of the fingers-stick-drum system govern the passive bounce.

Based on our observations (Hajian and Howe 1996), a simple lumped-element second order model can represent the fingers-stick-drum system during the interaction (Figure 30.4). The drum head is represented by a massless spring and damper, while the stick is modeled as an equivalent translational mass at the tip, including contributions from the finger and thumb mass. This stick-and-finger mass is coupled to the rest of the hand through the variable joint impedance of fingers, represented by a spring and damper. The hand position is essentially fixed during the interactions with the drum and is then retracted across longer time scales by wrist and/or arm motion. While the stick is in contact with the drum, the model on the

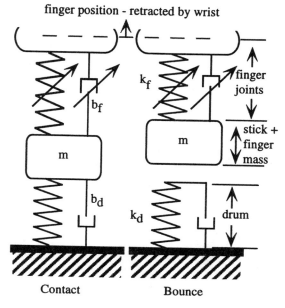

FIGURE 30.4. Lumped element model of hand-stick-drum system during drum roll bounce.

left side of Figure 30.4 is valid; while the stick is in free flight during the passive bounce, the right side pertains.

Drumming frequency can be described by the bounce interval, the time between the two contacts with the drum head. Neglecting damping, the model of Figure 30.4 suggests that the bounce interval should be proportional to the natural frequency $\omega_{bounce} = \sqrt{k/m}$, where k is the stiffness of the fingers, and m is the effective mass of the finger-stick system. Since muscle stiffness increases in proportion to the force generated, modulation of the grasp force between the finger and thumb allows the drummer to control k (Hajian and Howe 1996). This relationship shows how voluntary modification of grasp force can directly control drumming frequency.

To experimentally validate this model, we recorded grasp forces and drum head impacts as skilled drummers played double stroke rolls at a range of frequencies. Measured bounce durations ranged from 30 to 140 msecs, and grasp forces from 3 N to 36 N. Figure 30.5 shows typical data for one drummer for strokes with peak impact forces between 34 N and 42 N. The square of the bounce frequency, $\omega_{bounce} = \left(\frac{2\pi}{t_{bounce}}\right)^2$, is plotted against the grip force, measured at the thumb. From the relationship above, the stiffness should be proportional to ω_{bounce}^2, and the measurements are in fact grouped along the best fit line. Overall, the data shows the expected correlation between grasp force

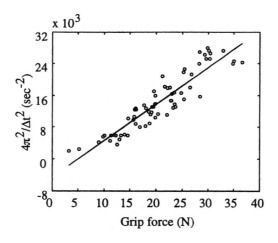

FIGURE 30.5. Bounce frequency squared $(\omega)^2$ vs. grasp force.

level and drumming frequency predicted by the model. Thus, we demonstrate a simple example where grasp force, and consequently impedance variations, act to vary the parameters (drumming frequency) of a task.

Unfortunately, as successful as impedance characterization is, both for hands as well as other dynamic systems, it does not capture all of the intricacies of real tasks. Execution of dexterous tasks involves learning and memory, adaptive control, complex sensing, and neural signal processing—all beyond the realm of an explanation rooted in impedance. Nevertheless, impedance analysis does encapsulate real dynamic interaction and provides a sense of quantitative comparison to known, analyzed, passive systems.

4   Future Directions

Focusing on hands and tasks will raise a different set of questions than studies of the arm. This begins at the smallest scale, where the mechanics of distributed contact are important. Rolling and sliding are frequently used in precision manipulation, but little is known about the mechanics of the interaction, or the role of the compliant finger tip pad. Sensory information about parameters such as contact location and fingertip torque are clearly important, but how these parameters are derived from afferent signals, and how they are incorporated into control for task execution is unclear. Even for the links that have been demonstrated between cutaneous afferent signals and motor responses, little is known about the gating and magnitude of the coupling.

Coordination among the fingers is essential in many tasks, but the mechanisms are largely unknown. Many finger joints are actuated by several muscles, and some hand muscles serve multiple fingers. Hand surgeons have considerable empirical knowledge of the functional aspects of this kinematic dependence. Organizational patterns in motor control and sensing may lend insight into multifinger coordination. The motor and sensory requirements for tasks in general are not well known (McPhee 1987), and thus it is difficult to analyze task execution and infer biomechanical function. Conceivably, it may be sufficient to define only a few constituent subtasks to understand the essence of manipulation and task execution. Establishing a task taxonomy or a grouping of constituent actions on a biomechanical (as opposed to therapeutic) basis might help simplify the analysis of real tasks.

Much work on the upper limb has focused on general principles, such as impedance regulation and the equilibrium point hypothesis, because of their potential to explain a broad range of behavior. In contrast, work on task execution by the hand suggests that skillful manipulation relies on a number of very diverse mechanisms. For example, grasp force control is adapted to local friction conditions in a pinch grip, and bimanual tasks use feedfoward compensation for changes in loading. There seems to be little in common among these various strategies, despite their importance to skillful task execution. The fact that some of them are evident at an early stage in development (Forssberg et al. 1992) suggests that they are an intrinsic part of the motor control system rather than learned behaviors. Further attention to the role of hands in task execution may reveal more of these mechanisms, and the common substrates upon which they are built.

References

Bicchi, A. and Sorrentino, R. (1995). Dexterous manipulation through rolling. *Proc. 1995 IEEE Int. Conf. Robotics and Automat.*, 1:452–457.

Buttolo, P. (1996). Characterization of human pen grasp with haptic displays. Ph.D. dissertation, University of Washington, Department of EE, June.

Chang, D.C. and Cutkosky, M.R. (1995). Rolling with deformable fingertips. *Proc. 1995 IEEE/RSJ Int. Conf. Intell. Robots Sys. Hum. Robot Inter. Cooperat. Robots*, 2:194–199.

Cole, A.A., Hsu, P., and Sastry, S.S. (1992). Dynamic control of sliding by robot hands for regrasping. *IEEE Transact. Robotics Automat.*, 8(1):42–52.

Cutkosky, M.R. and Howe, R.D. (1990). Human grasp choice and robotic grasp analysis. In *Dextrous Robot Hands*. Iberall, T. and Venkataraman, S.T. (eds.), ch. 1, pp. 5–31. Springer-Verlag, New York.

Forssberg, H., Eliasson, A.C., Kinoshita, H., Westling, G., and Johansson, R.S. (1995). Development of human precision grip. IV. Tactile adaptation of isometric finger forces to the frictional condition. *Exp. Brain Res.*, 104(2):323–330.

Hajian, A.Z., Sanchez, D.S., and Howe, R.D. "Drum roll: Increasing bandwidth through passive impedance modulation." *Proc. 1997 IEEE Intl. Conf. Robotics and Autom.* 2294–2299.

Hogan, N. (1990). Mechanical impedance of single and multi-articulate systems. In *Multiple Muscle Systems: Biomechanics and Movement Organization*. Winters, J.M. and Woo, S.L.-Y. (eds.), Chapter 9, pp. 149–164. Springer-Verlag, New York.

Johansson, R.S. and Westling, G. (1984). Roles of glabrous skin receptors and sensorimotor memory in automatic control of precision grip when lifting rougher or more slippery objects. *Exp. Brain Res.*, 56:550–564.

Johansson, R.S. and Westling, G. (1988). Programmed and triggered actions to rapid load changes during precision grip. *Exp. Brain Res.*, 7:72–86.

Lum, P.S., Reinkensmeyer, D.J., Lehman, S.L., Li, P.Y., and Stark, L.W. (1992). Feedforward stabilization in a bimanual unloading task. *Exp. Brain Res.*, 89(1): 172–180.

McPhee, S.D. (1987). Functional hand evaluations: a review. *Am. J. Occup. Ther.*, 41(3):158–163.

Napier, J.R. (1980). *Hands*. Princeton University Press, Princeton, New Jersey.

Pawluk, D.T.V. and Howe, R.D. (1999a). "Dynamic contact of the human fingerpad against a flat surface." *ASME J. Biomech. Eng.*, 121(2):605–611.

Pawluk, D.T.V. and Howe, R.D. (1999b). "Dynamic lumped element response of the human fingerpad." *ASME J. Biomech. Eng.*, 121(2):178–184.

Reinkensmeyer, D.J., Lum, P.S., and Lehman, S.L. (1992). Human control of a simple two-hand grasp. *Biol. Cybern.*, 67(6):553–564.

Schlesinger, G. (1919). Der Mechanische Aufbau der Kunstlichen Glieder. In *Ersatzglieder und Arbeitshilfen fur Kriegsbeschadigte und Unfallverletzte*. Borchardt, M. et al. (eds.), pp. 321–699. Springer, Berlin.

Shimoga, K.B. (1996). Robot grasp synthesis algorithms: a survey. *Int. J. Robotics Res.*, 15(3):230–266.

Tendick, F., Jennings, R.W., Tharp, G., and Stark, L. (1993). Sensing and manipulation problems in endoscopic surgery: experiment, analysis and observation. *Presence*, 2(1):66–81.

Vallbo, A.B. and Johansson, R.S. (1984). Properties of cutaneous mechanoreceptors in the human hand related to touch sensation. *Human Neurobiol.*, 3:3–13.

31
A Principle of Control of Rapid Multijoint Movements

Natalia V. Dounskaia, Stephan P. Swinnen, and Charles B. Walter

1 Introduction

It is widely recognized in robotics that the main problem associated with control of multijoint mechanical systems is the interaction between movements of adjacent segments of the system. Movement at one joint produces torques that act at all other joints, and therefore, the joints cannot be controlled independently from each other. The importance of this problem for human movement study was previously noted by the neurophysiologist N. Bernstein. He asserted that the control strategy used by the CNS in multijoint limb movements fully exploits the mechanical forces due to segmental interactions, enhancing the efficiency of muscular forces. According to Bernstein, interactive forces are used for movement control "in such a way as to employ active muscle forces only in the capacity of complementary forces" (Bernstein 1967, p.109). Following his writings, questions pertaining to the way in which interactive and other passive forces influence joint movements and how they subserve movement control have frequently been discussed in the literature (e.g., Hollerbach and Flash 1982; Jöris et al. 1985; Kaminski and Gentile 1989; Schneider et al. 1990; Gordon et al. 1994; Ulrich et al. 1994; Virji-Babul and Cooke 1995). Some recent studies in development also emphasize the importance of mechanical factors during learning (Thelen et al. 1993; Konczak et al. 1995). In this work, acquisition of movement patterns is considered as active exploration of the match between mechanical features of the infant's body and the task.

In spite of an increased appreciation for the fact that the CNS takes interactive torques into account in its control strategies, concrete mechanisms for the employment of these torques have not been proposed. In the present work, a hypothesis is offered that describes a control strategy applicable to multijoint rapid movements, that is, to movements during which reactive forces are particularly powerful and, therefore, can not be neglected.

2 Formulation of the Leading Joint Hypothesis

It is assumed that during rapid multijoint limb movements there is a difference between control principles applied to different joints in the limb. One joint of the limb plays a predominant role in movement production. Its movement grossly determines motion of the whole limb, and therefore, this joint can be referred as the "leading" one. Muscles governing the leading joint give rise to its consequent acceleration and deceleration and, if the movement is discrete, fixation near the desired final position (the well-known tri-phasic pattern). This motion generates passive torques that determine to a large extent the motions of all the other joints. Therefore, the latter are "subordinate" joints in some sense. Activity of muscles surrounding these joints differs from the activity of the leading joint muscles. The subordinate joint muscles serve to correct or compensate for the effect of passive torques generated by the leading joint motion and to adjust movement to the task requirements. These muscles can intensify the effect of passive torques during one portion of movement and suppress it during another portion, depending on the required coordination pattern. Nevertheless, the effect of the motion at the leading joint forms the basis upon which the whole limb movement is built up.

Naturally, the presented brief formulation of the leading joint hypothesis (LJH) poses more questions than it answers. To elucidate the principles that are implied by the LJH, an illustrative example of two-joint movement will be presented in the following section. Additionally, examples of some movements whose description is available in the literature will also be considered briefly.

3 Joint Coordination During Elbow-Wrist Movements and Other Illustrative Examples

The presented data are part of a study that is reported in detail in Dounskaia et al. (1998). Only a brief description of the experimental design and some results of the data analysis will be given here. Cyclical, planar elbow–wrist movements were performed at 5 frequency plateaus (0.5, 1.1, 1.75, 2.4, and 3.05 Hz) in two planes: sagittal, where the forearm was in a supine position, and horizontal, where the forearm was in a neutral position. Three patterns of movement were studied: (1) a continuous

sequence of simultaneous flexion and extension at both joints (*unidirectional* pattern); (2) a sequence of elbow flexion accompanied by wrist extension and vice versa (*bidirectional* pattern); (3) oscillation at the elbow with the wrist relaxed (*wrist-relaxed* pattern). The extreme joint positions for the first two patterns are shown schematically in Figure 31.1.

Similar movement types were analyzed by Kelso et al. (1991). They showed that an increase in cycling frequency affected stability of the bidirectional (antiphase) pattern and finally lead to its deterioration. The authors suggested that loss of the bidirectional pattern was accompanied by transitions to the unidirectional (in-phase) pattern. They interpreted the result in terms of a coordinative dynamics approach that did not rely on mechanics. The manipulation of cycling frequencies across a broader range in the present study enabled us to reveal that not only the bidirectional, but also the unidirectional pattern was affected by an increase in frequency. In particular, we demonstrated that features of both patterns tended to become similar to those characterizing the wrist-relaxed pattern at the highest cycling frequency plateaus. The data indicated that the deformation of both the unidirectional and bidirectional patterns was largely the result of the inability of the neuromuscular system to

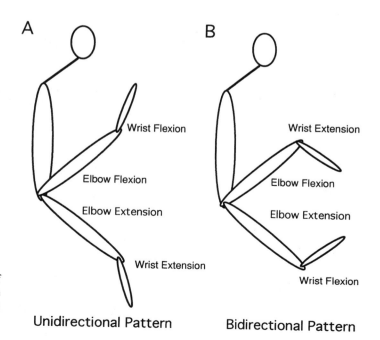

FIGURE 31.1. Schematic representation of the peak elbow and wrist positions for the unidirectional (A) and bidirectional (B) patterns.

cope with passive torques which arose during motion at the joints and which increased when cycling frequency increased.

Several converging lines of evidence support this conclusion. First, the movement amplitude, EMG level and EMG timing analyses provided convincing evidence that the elbow movement was highly independent of movement pattern. Accelerations and decelerations at the elbow rhythmically alternated according to activity of the elbow antagonist pairs during all types of movement at all frequency plateaus. Because this stable elbow performance was accompanied by a wide spectrum of different wrist movements, it is reasonable to conclude that elbow movement was not much affected by wrist motion. Moreover, the uniformity of elbow control across all three movement patterns allows the conclusion that differences between the patterns arose from differences in wrist control. This difference can be observed during movements at low and moderate frequency plateaus, where all three movement patterns were successfully performed. Relative phase between the elbow and wrist movements (confined between 0° and 180°) was used as a measure of successful performance (Schmidt et al. 1992). During successful performance, relative phase is about 180° in the bidirectional pattern and 0° in the unidirectional pattern. To detect deterioration of the patterns, the threshold value 90° of relative phase, averaged across cycles, was used. The analysis of relative phase indicated that the second frequency plateau (1.1 Hz) was the highest

plateau at which both patterns preserved their specific features in most trials (Figure 31.2).

To reveal the control strategies applied to the wrist during successful performance, torques exerted at the wrist were analyzed and compared with EMG records for a representative subject. Torques were calculated using a simplified dynamic model of a two-joint planar system (Putnam 1993). This model allows an estimation of the net torque (NT) acting at the wrist and its two components: gravitational torque (GRT) and inertial torque (IT). The latter arises from elbow motion. Based on these data, generalized muscle torque (GT) can be computed as a difference GT = NT − IT − GRT. The GT includes active and passive components. The latter arises from muscle elasticity, and the restraining effects of periarticular tissues. This component will be further referred to as restraining torque.

Typical examples of torques, angles, and EMG activity in the wrist observed for the three patterns in the horizontal plane at the second frequency plateau are shown in Figure 31.3A-C. The top panel of each figure represents changes in the elbow and wrist angle. The latter was measured as a relative angle between forearm and hand. The middle panel shows the NT and its two components, IT and GT. The GRT was not calculated, because it was assumed to be negligible during horizontal movements. The bottom panel shows smoothed, rectified and band-pass filtered EMG records. As suggested by the bottom panel of Figure 31.3A, re-

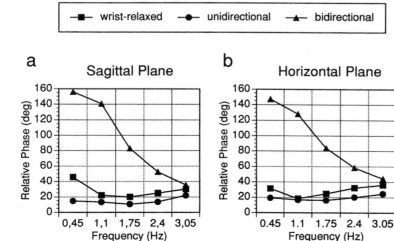

FIGURE 31.2. Plots of mean relative phase against frequency plateaus obtained for the three coordination patterns: unidirectional, bidirectional and wrist-relaxed. The data are the result of averaging continuous relative phase across all trials of the same conditions. The two panels correspond to movements in (a) the sagittal plane and (b) the horizontal plane.

laxation of the wrist muscles was quite successful during the represented wrist-relaxed movement. At the same time, the GT was substantial in this movement. This suggests that the dominating part of the GT was its passive component (i.e., restraining torque). Thus, during wrist-relaxed movements, the wrist was moved predominantly by two types of passive torques that acted in antiphase: inertial torque dependent on elbow motion and passive restraining torque.

Contrary to the wrist-relaxed movement, wrist muscle activity was substantial during unidirectional and bidirectional movements, as shown by the EMG plots in Figure 31.3B and 31.3C. However, a comparison of EMG activity with the GT reveals that passive restraining torque was also dramatically involved in the GT during these two patterns. Indeed, during the unidirectional pattern, the periods of maximal EMG activity (roughly marked by the shadowed areas) took place during transi-

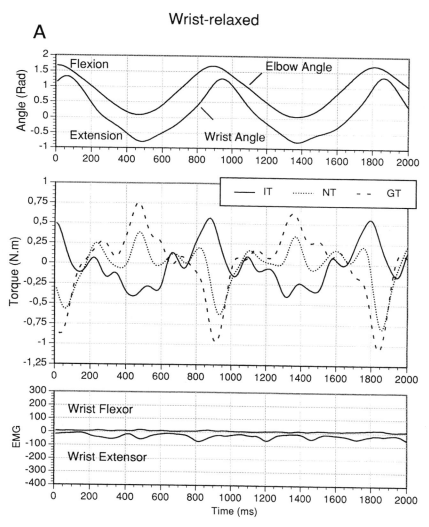

FIGURE 31.3. Representative fragments of joint angles, torques and rectified, filtered EMG records associated with wrist movements in the three patterns: (A) wrist-relaxed, (B) unidirectional, and (C) bidirectional. All movements were performed in the horizontal plane at the second fre-quency plateau. For the wrist extensor, an increase in EMG is plotted as a downward deflection. The positive direction corresponds to flexion, and the negative direction corresponds to extension in all panels. The shadowed areas indicate periods of most extensive wrist muscle activity.

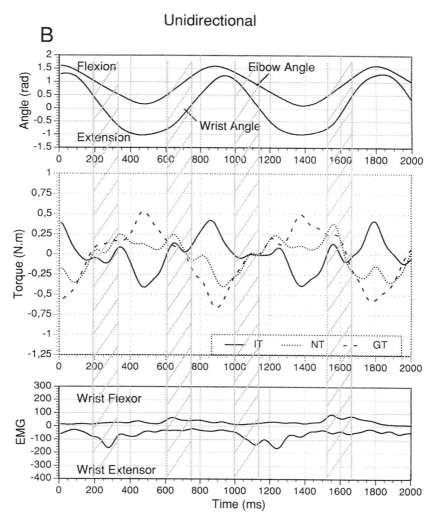

FIGURE 31.3. Continued.

tions of the wrist from one extreme position to another. However, GT reached its peak values at wrist reversals. This implies that the active component of the GT did not substantially participate in wrist control near the wrist extreme positions. Instead, the restraining component of the GT was responsible for the wrist reversals, coping with antiphase inertial torque (notice that both these torques are passive). The role of wrist muscle activity was limited to assisting passive motion during wrist peak-to-peak transitions, which resulted in slight decreases of relative phase in comparison with the wrist-relaxed pattern.

The GT and IT acted in the same direction (i.e., they were in-phase) during bidirectional movement (Figure 31.3C). Muscle activity was high at the wrist reversals and just prior to them, and its action was opposite to the action of the GT and IT. This suggests that the function of wrist muscle activity was to cope with both passive torques (inertial and restraining) in the vicinity of wrist peak positions. Since the active and passive components composing the GT had opposite signs, the GT in Figure 31.3C represents the difference between both. Inasmuch as the resulting sign of the GT was opposite to the sign of EMG, it was the sign of the

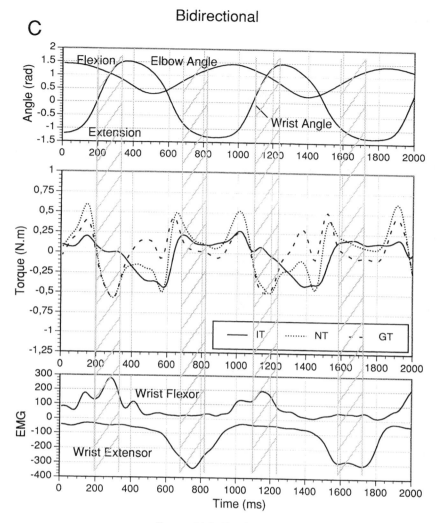

FIGURE 31.3. Continued.

passive component, and hence, passive torque, that substantially surpassed the active torque during the course of motion. An exception was the portion of movement in the vicinity of wrist peak positions where active torque neutralized passive torque. This resulted in the valleys observed between the pairs of GT peaks of the same sign in Figure 31.3C. Thus, although passive torques played a significant role in wrist control during bidirectional movements, they were less compatible with the desired movement pattern than during unidirectional movements. Passive torques were largely responsible for transitions of the wrist from one extreme position

to another, but they opposed the desired movement in the vicinity of the extreme positions and had to be coped with by active muscle torque during these movement epochs. Passive torques reached maximal values during these movement periods, and therefore, muscle activity had to be relatively high to suppress them (notice the higher EMG levels in the bidirectional than in the unidirectional pattern in Figure 31.3B and 31.3C).

The data shown above was typical of successful performance observed at the second frequency plateau (1.1 Hz). Starting from the third frequency plateau (1.75 Hz), performance deteriorated, and this was particularly evident with respect to the bidirectional pattern: disruption of the movement was accompanied by dramatic changes in relative phase values. However, the relative phase analysis did not provide convincing information about changes in the unidirectional pattern, because only slight (although significant) increases in relative phase were observed under scaling-up frequency conditions (see Figure 31.2). Therefore, two other characteristics were computed to study the evolution of the patterns. First, relative phase profiles within a cycle provided an additional tool to track

the evolution of all three movement patterns. To display relative phase profiles, continuous relative phase obtained for each trial was averaged and plotted against 32 equal sections of elbow phase, each spanning 11.25°. Figure 31.4 shows the plots of averaged relative phase profiles computed for movements in the sagittal plane at two frequency plateaus: the second plateau (the left column) and the fifth plateau (the right column). The left three panels demonstrate that relative phase showed different profiles across the different movement patterns at the second frequency plateau. As cycling

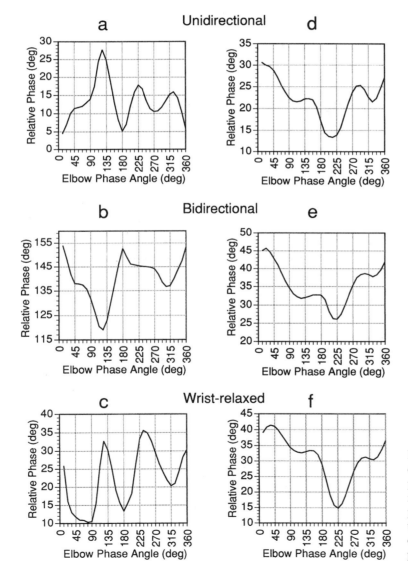

FIGURE 31.4. Relative phase within-cycle profiles. (a–c): Relative phase profiles averaged across sagittal movements at the second frequency plateau for the three movement patterns. (d–f): The same characteristics obtained at the fifth frequency plateau.

frequency increased, the profiles of the three patterns became very similar to each other, as can be observed from the right panels.

Second, timing of peak EMG activity was determined for three muscles (biceps, the wrist flexor and wrist extensor) as portions of cycle duration. Elbow peak flexions were used to divide trials into cycles. Biceps timing did not display any difference across the three patterns. A different result was obtained for the wrist muscles. Their averaged EMG timing data are shown in Figure 31.5 by means of a two-dimensional representation. Each line in the figure connects five points representing the frequency plateaus. The coordinates of each point are: wrist extensor timing in the horizontal axis, and wrist flexor timing in the vertical axis. The direction of increase in frequency is indicated by black arrows next to each graph. In each part of Figure 31.4, the graphs of the three movement patterns originate in different areas of the two-dimensional space at low frequency plateaus, but they gradually converge, virtually meeting each other at the fifth frequency plateau. This suggests that peak activity of the wrist muscles occurred at different cycle epochs in different movement patterns, but this movement characteristic became similar for all three patterns at the highest frequency plateau.

In summary, the three movement characteristics (i.e., relative phase, relative phase profiles and EMG timing) clearly demonstrate that both actively controlled movement patterns displayed their specific features at lower frequency plateaus, but were no longer distinguishable from the wrist-relaxed pattern at higher frequency plateaus. The most reasonable explanation for this fact is a drastic role assigned to passive torques during elbow–wrist movements. These torques emerged as a result of elbow movement. At moderate cycling frequencies, they served as a basis upon which active torque generated by wrist muscles could build a desired pattern. However, as cycling frequency increased, the neuro-muscular structures responsible for the active component of wrist control were apparently unable to properly intervene with passive torques which, therefore, became a dominating factor in wrist movement.

From the perspective of the LJH formulated at the beginning of this chapter, the elbow functioned

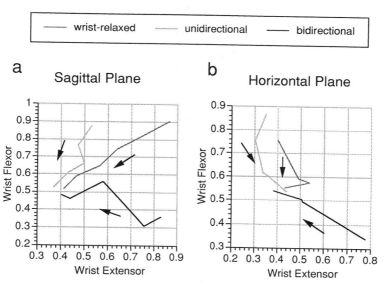

FIGURE 31.5. Two-dimensional representation of mean wrist EMG timing as a proportion of cycle duration. The wrist extensor timing is plotted on the horizontal axis, and the wrist flexor timing is plotted on the vertical axis. The three graphs in each panel correspond to wrist timing during the unidirectional, bidirectional and wrist-relaxed patterns across the five frequency plateaus. The black arrows indicate the direction of frequency increase. In both the sagittal (a) and horizontal (b) planes, the three graphs converge to the same region of the two-dimensional space with increases in cycling frequency. The timing values are confined between 0 and 1 and are represented in arbitrary units.

as a leading joint and the wrist as a subordinate joint during the three movement patterns. The leading role of the elbow is seen from the fact that elbow control was relatively independent from wrist movements, providing stable stereotypical elbow oscillations across all three coordination patterns. Additionally, elbow movement gave rise to passive torques that contributed substantially to the net torque exerted at the wrist. Conversely, the subordinate features of the wrist are clearly displayed not only by high variability of its movements, but also by the strategy applied for its control. This strategy consisted of generation of active torques which, interplaying with passive torques in different movement epochs, provided different movement patterns; it also has implications to the coordination of grasping (the topic of Chapter 30). This control principle yielded successful performance at moderate frequencies. It was economical and efficient, because it allowed generation of rapid, full-range wrist movements with the least expenditure of muscle power because of a maximal substitution by passive effects. However, this control strategy had some restrictions that became explicit at the highest frequency plateaus, where coping with passive torques was possibly beyond the possibilities of the controlling and actuating structures.

The type of subordination between joints suggested by the LJH can be observed not only during the elbow–wrist movements described above, but also in many rapid upper and lower-limb movements described in the literature. For example, it has been noted in studies of free motion of the lower extremities during walking, running and kicking that hip muscle torque directly controls accelerations and decelerations of the thigh during most movement epochs (Miller 1980; Putnam 1991, 1993; Ulrich et al. 1994). Interactive torques resulting from thigh motion play a substantial role in lower leg motion. In terms of the LJH, hip movement is leading and knee movement is subordinate in these types of motion. The validity of the leading-joint control principle in its application to the hip–knee mechanical linkage is supported by the data presented by Phillips et al. (1983). The authors conducted simulations of movement at the hip and knee during the swing phase of running and showed that mechanical interactions generated by the proximal segment motion alone produced large increases in distal segment speed without an aid from active

control at the distal joint. The control principle formulated in the LJH is also evident in upper extremity movements during throwing. In this case, the leading role is obviously assigned to the shoulder, because the upper arm moves mostly in accordance with the activity of the shoulder muscles, whereas the forearm movement evolves largely from the action of passive torques arising from shoulder movement (Feltner 1989; Putnam 1993).

This section has attempted to demonstrate how the control mechanism briefly formulated in Section 2 as the LJH is implemented in concrete multijoint movements. The next section considers the LJH in more detail.

4   Some Questions Pertinent to Different Aspects of the LJH

4.1   Is the Leading Joint Always a Proximal One?

The proximal joint was suggested to be a leading joint in all examples considered in the previous section. Many studies on sport movements report the action of interactive forces exerted by proximal segments on distal segments (see for instance, Alexander and Haddow 1982; Jöris 1985; Putnam 1993). The proximal-to-distal effect of interactive forces can be accounted for by the peculiarities of mechanical structure of the human body. Bernstein (1967, 1984) hinted at this phenomenon a long time ago, when drawing attention to the difference between the controllability of the proximal and the distal limb segments. In particular, he noted that there is a difference between the amount of muscle tissues governing movement of the proximal and distal limb segments, and between the moments of inertia of these segments (Bernstein 1984). He concluded that because of these differences, muscle control affected the proximal segment far more easily than the distal segment. "In order to be perceptible in this latter system, the effector impulse either must be very strong or must coincide exactly with the moment when conditions in the distal system are particularly favourable for its appearance" (Bernstein 1984, p.200). The LJH perfectly converges with these considerations; it is also at least consistent with the linear synergy between joint torques that is developed in Chapter 27.

However, the effect of interactive forces is not necessarily proximal-to-distal in all types of multijoint movements. Studies of paw-shake response in the cat provide an example of distal-to-proximal effects. It has been demonstrated that the ankle joint flexions and extensions are primarily caused by moments created by the ankle muscles. Motion in the knee and hip results from mechanical interaction between the ankle and the other joints, and muscle activity at these two joints is directed at suppressing this motion (Hoy et al. 1985; Smith and Zernicke 1987). Thus, the ankle is a leading joint during paw-shake, and the knee and hip are subordinate. The hip and knee muscles act to counterbalance movement of the distal segment.

The LJH does not imply any restrictions on which joint can be a leading one. When movement at a joint generates mechanical interactions dominating in the whole limb and when the movement at this joint is in accordance with activity of the muscles spanning it, the latter joint is a serious candidate for a leading joint. What the LJH suggests is that during each epoch of a rapid movement there is one joint whose movement generates dominating interactive torques and which, therefore, can be considered as a source of mechanical interactions participating in control of movement at the other joints. The nature of multijoint structures allows creating different mechanical effects, providing a rich diversity of human movements. Revealing the leading joint in each concrete movement may help to understand in what way the multijoint mechanical structure of the limb is used for organization and control of this particular movement.

4.2 How Can the Leading Joint Be Determined?

It is presently difficult to formulate general rules for detection of the leading joint. A diversity of multijoint movement types should be analyzed from the perspective of the LJH to generate such rules. Nevertheless, the currently available findings allow us to put forward several preliminary considerations that may be of help in searching for the leading joint during multijoint movement.

It is hypothesized that the main feature of the leading joint is that it undergoes minimal interactional effects from the other joints. Accordingly, the association between muscular activity around the leading joint and the kinematic changes at this joint during multijoint movement should be similar to that observed during single-joint movement. This implies that the agonist muscle group at the leading joint provides concentric torque at the beginning of the movement, resulting in joint acceleration. The subsequent eccentric torque of the antagonistic muscle group brakes the joint motion. However, for some multijoint movements, the "single-joint movement pattern" at the leading joint may not be very distinct, because of the reciprocal reaction of subordinate joint muscles to the undesirable effect produced by the leading joint motion. Resisting the passive joint movement, muscle activity at the subordinate joints also indirectly resists the leading joint movement and thus influences its kinematics to some extent. Nevertheless, it is reasonable to expect that an increase in angular velocity at a joint in response to its agonist activity during the initial stage of movement is a reliable characteristic of the leading movement. This relation between muscle activity and joint kinematics in proximal joints has been reported in studies of many upper and lower limb movements such as overarm throwing, walking, running and kicking at different speed levels (Miller 1980; Feltner 1989; Putnam 1991, 1993). Conversely, because interactive torques acting at the subordinate joints can be high from the very beginning of movement (due to high angular acceleration at the leading joint), it is typical of subordinate joints that the initial direction of their movement opposes action of their muscles (e.g., Putnam 1993).

Although the predominant influence of one segment motion on motions of other segments is clearly apparent at the start of many multijoint movements, this may not always be the case. In particular, dominating features of the leading joint can vanish if movement is not sufficiently fast. Accordingly, a second approach that may help in detecting the leading joint is by manipulating movement speed. Higher angular velocities and accelerations induce higher interactive torques, making motion-dependent effects more explicit. This was the experimental strategy used in the analysis of elbow–wrist movements presented in the previous section. As cycling frequency increased, the pattern of elbow movement remained essentially the same, but wrist movement underwent drastic changes so that both controlled patterns of movement acquired features of the wrist-relaxed pattern at the highest frequencies. Accordingly, the leading role of the elbow became more explicit under these experimental manipulations. Unfortunately, the technique of movement speed scaling has its limitations, because increases in speed may cause changes of the leading joint during some movements. This is supported, for example, by observations made by Meulenbroek et al. (1993), who demonstrated that the contribution of different arm segments to back-and-forth drawing movements depends on movement frequency.

Finally, the calculation of various torque components composing net torques which act on limb segments can aid in revealing the role of interactive torques during each movement epoch. This method is based on the development of dynamic models of the limb and is widely used in biomechanical studies (see, for example, Hoy et al. 1985; Feltner 1989; Schneider et al. 1990; Putnam 1991, 1993). However, in applying this method, it is important to realize that it only yields approximate estimations of torque levels due to the inevitable incompleteness of any model. Combining kinetic analysis with EMG analysis is strongly recommended. This can help not only to verify modeling results by comparing computed muscle torque with real muscle activity, but also to evaluate which portion of generalized muscle torque arises from active control, and which portion of this torque arises from its passive component. The important role of the latter was revealed in the analysis of elbow-wrist movements (see Section 3), even at moderate cycling frequencies (1.1 Hz). This suggests that the impact of passive torques in the generalized muscle torque has often been underestimated.

4.3   What Is the Nature of the Control of the Leading Joint?

The control mode applied to the leading joint may become clear when the goal of the joint motion is better understood. One feasible goal of the leading joint movement is to produce movement at the leading joint itself as it is, for example, during the paw-shake response. In this case, interactions affecting other joints are an artefact of the leading joint movement. More often, however, the primary goal of the leading joint movement is to exert interactive torques that produce movement in subordinate

joints. In this case, the leading joint movement does not imply high accuracy or modifiability. Leading joint control is likely to be gross, being responsible mostly for the rate of the whole movement, whereas its current displacement is determined with significant tolerance. This suggests that the leading joint can essentially be controlled in a feedforward mode without use of concurrent afferent feedback (Hoy et al. 1985; Smith and Zernicke 1987; Hasan and Stuart 1988). Different contemporary approaches to the issue of uniarticular movement control can account for the acceleration-deceleration movement at the leading joint (Wallace 1981; Bizzi et al. 1984; Feldman 1986). Most of these models suggest relatively simple control strategies: one or two control parameters have to be set before a voluntary movement is initiated. Thus, a series of stereotypic movements of the leading joint can be performed "automatically" with similar values of the control parameters.

4.4   What Is the Nature of the Control of the Subordinate Joints?

Contrary to the leading joint control, which is supposed to be dominantly open-loop, we hypothesize that the control of the subordinate joints is likely to be closed-loop with afferent information playing a critical role. During the considered elbow-wrist movements (see Section 3), the elbow performed virtually the same movement across all movement patterns. The patterns differed according to the way in which wrist muscle activity intervened with elbow motion-dependent dynamics. This suggests that the primary goal of subordinate joint muscle activity was to interplay with motion-dependent torques by superimposing active torques that changed passive motion of the joint, in accordance with the task. But motion-dependent torques are highly mutable. They are modulated within each cycle, in accordance with angular velocity and acceleration of the leading joint. In addition, their pattern is not invariant from cycle to cycle, because the leading joint motion is not an ideal periodic oscillation. The requirement for the wrist muscles to operate in mutable, and, to some extent unpredictable dynamic conditions, suggests that proprioceptive feedback, signaling motion-dependent information, is essential for the regulation of wrist muscle activity. The same conclusion was made in connection with the paw-shake response during which knee and hip muscle activity was highly dependent on interactive torques generated by ankle motion (Hoy et al. 1985; Hasan and Stuart 1988). Supportive observations are also reported in recent studies with patients suffering severe proprioceptive deficits as a result of large-fiber sensory neuropathy (Sainburg et al. 1993; Ghez et al. 1995; Gordon et al. 1995).

The use of kinesthetic afferent information does not rule out the possibility that subordinate joint muscle activity is preplanned to a large extent. Some support for this idea has been provided through comparison of movements of deafferented patients with and without visual feedback (Sainburg et al. 1993; Ghez et al. 1995). It was demonstrated that there was a significant improvement in performance when patients were able to view their limbs prior to movement. Converging evidence was reported by Koshland and Hasan (1994), who studied pointing movements in the horizontal plane. It was shown that during unconstrained movements, initial muscle activity at the wrist always functioned to resist the reactional effect of proximal segment motion on the wrist. When constraints were imposed on movement, such as instructions to subjects to achieve a different final orientation of the hand for the same target location, the choice of muscle to be activated first was the same as in the unconstrained conditions, independently of whether flexion or extension at the wrist was required at the final position. These data support the idea of preprogrammed motion for the subordinate joints, which, however, can be modified "on-line".

To summarize, it is hypothesized that the pattern of subordinate joint muscle activity is roughly preplanned. However, interactive torques cannot be predicted in detail. Therefore, ongoing movements of the subordinate joints are possibly corrected with the use of proprioceptive feedback, due to the strong dependence of these movements on highly mutable interactive forces and to accuracy requirements which are usually imposed on the subordinate joint movement. Thus, the control of multijoint movements is a combination of open-loop and closed-loop control modes which can be implemented without use of the dynamic model of the limb and solution of the inverse dynamic problem.

4.5 Learning from the LJH Perspective

It has just been argued that the pattern of both leading and subordinate joint muscle activity is roughly preplanned in each movement, even though feedback information is essentially used for control of multijoint movements. The "outline" of the basic pattern of activity may be learned and may then be used as an approximation of the activity required for the actual movement.

The LJH simplifies the learning problem, because the control based on the use of the leading joint may be constructed, roughly speaking, in two steps. First, the gross movement of the leading joint may be learned with other joints relatively fixed. This stage probably includes exploration of joint interactions and search for the leading joint. Then, movement of the subordinate joints can be gradually learned on the background of interactive torques produced by motion at the leading joint. This simplified structure of the learning process is consistent with Bernstein's formulation (1967). According to his theory, unskilled performers decrease the number of degrees of freedom at the beginning of learning. With this purpose, they "freeze out" a portion of the degrees of freedom. Following Bernstein, as a skill is elaborated, the frozen degrees of freedom are gradually released. Towards the end of the learning course, the organism learns to use reactive forces, which generates great economy of muscle activity. Although empirical verification of the Bernstein's ideas about skill acquisition has not been the focus of many studies in motor behavior, limited work has provided support for this form of learning process (McDonald et al. 1989; Schneider et al. 1989; Vereijken et al. 1992; see also Chapter 26).

The interpretation of the learning process within the framework of the LJH suggests that a critical stage is a gradual release of the relatively rigid control at the subordinate joints against the background of passive effects generated by the leading joint movement. Gradual increases in passive torques are accompanied by the generation of patterns of active control through musculature at the subordinate joints. Because the total suppression of passive torques would be very energy consuming, only those movement types are initially developed that are highly compatible with passive effects and that require only their partial suppression. At the same time, the control of the leading joint can require updating with respect to the condition of subordinate joint movement. This learning process may result in a rough mapping between intended types of movements, defined in the extrinsic coordinates, and simple sequences of neural commands that elicit the associated muscle activation patterns (Thelen et al. 1993; Loeb 1995). Visual and proprioceptive information can help to update these patterns for specific goal-directed movements.

To summarize, the LJH implies that the learning process for the leading joint is fundamentally different from the learning process for the subordinate joints, because the latter proceeds under much more complicated dynamic conditions. However, the complexity of acquisition of the leading joint motion should not be underestimated. The leading joint muscle activity is responsible not only for the leading joint motion, but also indirectly for passive motion imposed on the subordinate joint(s). From the perspective of LJH, the problems that are solved during each multijoint movement are comparable to those solved by acrobats performing somersaults on the back of a galloping horse. Both the horse and the acrobat have to be well trained. However, this is not enough, because the performance will only be successful, if the acrobat concentrates attention on his movements and those of the horse, and effectively uses available feedback information.

5   Conclusions

Different types of constraints underlie the CNS control strategies. We contend in this chapter that mechanical properties of the human body are among the most important factors defining principles of movement organization. The human body is a linked mechanical system moving in a physical environment and obeying the laws of general mechanics. Clearly, the CNS has to adjust its control signals to mechanical factors to allow a successful operation of the human body in the environment. N. Bernstein was the first neurophysiologist who perceived the importance of this problem, indicating that the CNS employs joint interactions to provide efficient and economical control of human limbs (Bernstein 1967). However, no concrete control mechanism has been proposed

whereby the neural structures deal with human body mechanics and even benefit from them. In this work, we have attempted to formulate a control principle for rapid multijoint movements. However, the notion "rapid movement" is a conventional one. The effect of inertial forces, even during moderate movement speeds, should not be underestimated. The analysis of elbow–wrist movements (see Section 3) demonstrated that inertial forces dramatically affect wrist movement, even for a cycling frequency of 1.1 Hz, which intuitively cannot be considered as "rapid" movement (see also Chapter 29). Another point worth mentioning to is that passive effects in the subordinate joints are not an artefact of leading joint movement, but they are often *deliberately* generated by the muscles at the leading joint. From the LJH perspective, the purpose of the leading joint movement is often not the movement at the leading joint itself, but the generation of interactive torques that put the subordinate joints in motion. This implies that the leading joint is sometimes used as an intermediate mechanical link for transmission of control inputs from the powerful leading joint muscles to the unspanned subordinate joints.

Finally, we would like to emphasize once again that the LJH does not require the inverse dynamics problem to be solved to produce multijoint movements. From the LJH perspectives, each type of skilled movement is a result of a learning process which is, at the same time, a process of movement organization. The learning problem is facilitated, because learning of control for the leading and subordinate joints can be organized in a successive mode. The LJH approach also reduces complexity of the problem concerning storage of learned patterns in memory, because the most essential information pertains to the control signals providing leading joint movement, which often is a simple acceleration–deceleration pattern. As for the subordinate joints, their control is to some extent online with the use of afferent feedback, and therefore must not be stored in memory in great detail.

6   Future Directions

Even though it is in its infancy, the leading joint approach may have promising applications to a diversity of motor control and learning issues. One of the most urgent directions of study is an analysis of different types of human movements for the purpose of revealing the leading joint and the influence of its movement on movement at the other joints. On the one hand, such work would further elaborate the LJH, generating more profound insights about the features of leading and subordinate joint movements and allowing a better understanding of the mechanical effects employed for movement control. On the other hand, these analyses would help us to understand how the studied movements are organized and what factors influence joint coordination in each movement pattern.

Another potentially fruitful direction of investigation is associated with the problem of excessive degrees of freedom. Following the LJH, the set of possible patterns of joint coordination is limited for a limb. The effect of joint mechanical interaction imposes constraints on the possible movement combinations afforded by the joints. There is good reason to believe that only those coordination patterns are acquired that are in agreement with the interactive torques produced by movement at the leading joint. Although this factor essentially decreases the number of possible movement patterns, a significant freedom in choice of movement can still exist for many movement tasks. Thus, the LJH does not fully eliminate the question of how the choice is made from a set of possible movements. It is expected that some other factors impose constraints on movement performance in addition to mechanical factors. For example, requirement to move the end-point along a straight line in the visual space can be such an additional factor during some reaching movements (Wolpert et al. 1995). Revealing the factors that influence movements, together with those originating from mechanical properties of the human limbs, would help us to understand the specifics of each movement type.

Finally, research should focus more on issues of motor learning and development. Frequently performed movements, such as reaching, walking, sports movements, and so on, become automatic as a result of intensive learning that sometimes takes years. To understand the principles of control applied to skilful movements, it is necessary to track the development of these principles in individuals who acquire new skills. The LJH opens a way to trace all stages of learning from the very beginning of practice until high performance levels are obtained and to reveal the principles underlying movement organization.

References

Alexander, M.J. and Haddow, J.B. (1982). A kinematic analysis of an upper extremity ballistic skill: the windmill pitch. *Can. J. Sport Sci.*, 7:209–217.

Bernstein, N. (1984). Biodynamics of locomotion. In *Human Motor Actions: Bernstein Reassessed*. Whiting, H.T.A. (ed.), pp. 171–222. North-Holland, Amsterdam.

Bernstein, N. (1967). *The Co-ordination and Regulation of Movements*. Pergamon Press, Oxford, United Kingdom.

Bizzi, E., Accornero, N., Chapple, W., and Hogan, N. (1984). Posture control and trajectory formation during arm movement. *J. Neurosci.*, 4:2738–2744.

Dounskaia, N.V., Swinnen, S.P., Walter, C.B., Spaepen, A.J., and Verschueren, S.M.P. (1998). Hierarchical control of different elbow–wrist coordination patterns. *Exp. Brain Res.*, 121:339–354.

Feldman, A. G. (1986). Once more for the equilibrium point hypothesis (l model). *J. Mot. Behav.*, 18:17–54.

Feltner, M. E. (1989). Three-dimensional interactions in a two-segment kinetic chain. Part II: Application to the throwing arm in baseball pitching. *Int. J. Sport Biomech.*, 5:420–450.

Ghez, C., Gordon, J., and Ghilardi, M.F. (1995). Impairments of reaching movements in patients without proprioception. II. Effect of visual information on accuracy. *J. Neurophysiol.*, 73:361–372.

Gordon, J., Ghilardi, M.F., and Ghez, C. (1995). Impairments of reaching movements in patients without proprioception. I. Spatial errors. *J. Neurophysiol.*, 73: 347–360.

Gordon, J., Ghilardi, M.F., Cooper, S.E., and Ghez, C. (1994). Accuracy of planar reaching movements. II. Systematic extent errors resulting from interactional anisotropy. *Exp. Brain Res.*, 99:112–130.

Hasan, Z. and Stuart, D.G. (1988). Animal solutions to problems of movement control: the role of proprioceptors. *Ann. Rev. Neurosci.*, 11:199–223.

Hollerbach, J.M. and Flash, T. (1982). Dynamic interactions between limb segments during planar arm movement. *Biol. Cybern.*, 44:67–77.

Hoy, M.G., Zernicke, R.F., and Smith, J.L. (1985). Contrasting roles of inertial and muscle moments at knee and ankle during paw-shake response. *J. Neurophysiol.*, 54:1282–1294.

Jöris, H.J.J., Edwards van Muyen, A.J., van Ingen Schenau, G.J., and Kemper, H.C.G. (1985). Force, velocity and energy flow during the overarm throw in female handball players. *J. Biomech.*, 18:409–414.

Kaminski, T. and Gentile, A.M. (1989). A kinematic comparison of single and multijoint movements. *Exp. Brain Res.*, 78:547–556.

Kelso, J.A.S., Buchanan, J.J., Wallace, S.A. (1991). Order parameters for the neural organization of single, multijoint limb movement patterns. *Exp. Brain. Res.*, 85:432–444.

Konczak J., Borutta M., Topka H., and Dichgans J. (1995). The development of goal-directed reaching in infants: hand trajectory formation and joint torque control. *Exp. Brain Res.*, 106:156–168.

Koshland, G.F. and Hasan, Z. (1994). Selection of muscles for initiation of planar, three-joint arm movements with different final orientations of the hand. *Exp. Brain Res.*, 98:157–162.

Loeb, G.E. (1995). Control implications of musculoskeletal mechanics. In *Proceedings of Annual International Conference IEEE-EMBS*, 17:1393–1394 (#6.2.2.1).

McDonald, P.V., Emmerik, R.E.A. van, and Newell, K.M. (1989). The effects of practice on limb kinematics in a throwing task. *J. Mot. Behav.*, 21:245–264.

Meulenbroek, R.G.J., Rosenbaum, D.A., Thomassen, A.J.W.M., and Schomaker, L.R.B. (1993). Limb-segment selection in drawing behaviour. *Q. J. Exp. Psychol.*, 46A:273–299.

Miller, D.I. (1980). Body segment contributions to sport skill performance: two contrasting approaches. *Res. Q. Exerc. Sport*, 51:219–233.

Phillips, S.J., Roberts, E.M., and Huang, T.C. (1983). Quantification of intersegmental reactions during rapid swing motion. *J. Biomech.*, 16:411–417.

Putnam, C.A.A. (1993). Sequential motions of body segments in striking and throwing skills: descriptions and explanations. *J. Biomech.*, 26:125–135.

Putnam, C.A.A. (1991). Segment interaction analysis of proximal-to-distal sequential segment motion patterns. *Med. Sci. Sports Exerc.*, 23:130–144.

Sainburg, R.L., Poizner, H., and Ghez, C. (1993). Loss of proprioception produces deficits in interjoint coordination. *J. Neurophysiol.*, 70:2136–2147.

Schmidt, R.C., Treffner, P.J., Shaw, B.K., Turvey, M.T. (1992). Dynamical aspects of learning an interlimb rhythmic movement pattern. *J. Mot. Behav.*, 24:67–84.

Schneider, K., Zernicke, R.F., Schmidt, R.A., and Hart, T.J. (1989). Changes in limb dynamics during the practice of rapid arm movements. *J. Biomech.*, 22:805–817.

Schneider, K., Zernicke, R.F., Ulrich, B.D., Jensen, J.L., Thelen, E. (1990). Understanding movement control in infants through the analysis of limb intersegmental dynamics. *J. Mot. Behav.*, 22:521–535.

Smith, J.L. and Zernicke, R.F. (1987). Predictions for neural control based on limb dynamics. *Trends Neurosci.*, 10:123–128.

Thelen, E., Corbetta, D., Kamm, K., and Spencer, J.P. (1993). The transition to reaching: mapping intention and intrinsic dynamics. *Child Dev.*, 64:1058–1098.

Ulrich, B.D., Jensen, J.L., Thelen, E., Schneider, K., and Zernicke, R.F. (1994). Adaptive dynamics of the leg movement patterns of human infants: II. Treadmill stepping in infants and adults. *J. Mot. Behav.*, 26:313–324.

Vereijken, B., Van Emmerik, R.E.A., Whiting, H.T.A., and Newell, K.M. (1992). Free(z)ing degrees of freedom in skill acquisition. *J. Mot. Behav.*, 24:133–142.

Virji-Babul, N. and Cooke, J.D. (1995). Influence of joint interactional effects on the coordination of planar two-joint arm movements. *Exp. Brain Res.*, 103:451–459.

Wallace, S.A. (1981). An impulse-timing theory for reciprocal control of muscular activity in rapid, discrete movements. *J. Mot. Behav.*, 13:144–160.

Wolpert, D.M., Ghahramani, Z., and Jordan, M.I. (1995). Are arm trajectories planned in kinematic or dynamic coordinates? An adaptation study. *Exp. Brain Res.*, 103:460–470.

Section VII

32
Large-Scale Musculoskeletal Systems: Sensorimotor Integration and Optimization

Frans C.T. van der Helm

1 Introduction

A wide variety of musculoskeletal systems have been described in the last few decades. A musculoskeletal model in its simplest form can be described as one segment with a one degree-of-freedom (DOF) hinge joint, moved by two muscles (agonist–antagonist). Simple models obviously are not representative of any of the joints in the human body, but their goal is to gain insight in the *principles* underlying the role of the components and the control of human movements. Even the simplest 1-DOF models show *actuator redundancy*, that is, there are two actuators (muscles) controlling one DOF. In the context of musculoskeletal systems this is a necessity, since muscles can only exert pulling forces. However, the muscles can also co-contract: while the net moment does not change, the forces of the muscles increase. For an excellent overview refer to Winters and Stark (1987).

Models of increasing complexity have been developed to investigate other features of the musculoskeletal system. 2-DOF models, like a planar description of the shoulder and elbow joint in the horizontal plane, have been used to analyze the role of biarticular muscles, and the control of the endpoint stiffness (i.e. the resistance of the hand to unexpected perturbations), (Hogan 1985; Flash 1987; van Zuylen et al. 1988; Van Ingen Schenau 1989). 3-DOF models are frequently encountered for the (planar) representation of the hip, knee and ankle joint for the analysis of walking (Pedotti et al. 1978), and jumping (e.g., Pandy et al. 1990; Pandy and Zajac 1991). 3-DOF models are interesting for the control of planar arm movements, since they

can provide the simplest form of *kinematic redundancy*. Kinematic redundancy means that there are more DOF present in the joints than at the endpoint. In such a case, if the endpoint is fixed, like the hand holding a handle, the arm configuration still can change.

At the other end of the continuum of increasing complexity, large-scale musculoskeletal models are found. A perfect definition of a truly large-scale model is difficult, but a common feature is that they have a detailed representation of the joints and all muscles present. The usual *goal* of large-scale musculoskeletal models is *to gain insight into the function of specific muscles* for certain tasks and into the resulting behavior. Because of the sheer size of the models they tend not to lend themselves to deriving principles of motor control, but these principles can be recognized in the simulation results. Moreover, very often it is a good test for motor control principles if they are still applicable if the system complexity increases. For example, many motor control theories use the notion of agonist–antagonist and reciprocal inhibition. However, in a complex system like the shoulder, no pure agonist–antagonist pairs can be found. The function of muscles is very position-dependent. Large-scale musculoskeletal systems pose a very complex system to control. There are many interactions between the control inputs, resulting from inertial dynamics, polyarticular muscles, kinematic and actuator redundancies.

A model is a simplification, an abstraction of the real system, meant as a tool to analyze the system. If a model would be as detailed (and as complex!) as the real system, it becomes very difficult to analyze. The simplest model that serves the research goal is the best model. However, a model is only as good as its weakest link. The weakest link can be the omission of important features, but also the representation of detailed properties from which the parameters can not be obtained accurately. A very precise representation of particular details can be a waste of effort if there is only a crude estimate available of other parameters. In a large-scale model, it is desired to represent all important parameters: muscle properties, particular nonlinear properties, geometry, and so on. A difficult step is the validation of large-scale models. It requires the measurement of input, output, and internal signals, which is for most signals impossible in the human body. Strictly, a model is as good as it is validated, which would mean for most large-scale models that their predictions should be used with precautions. However, in the effort of building a large-scale model and in the (sensitivity) analysis of the results, much is learned about the important features of the system.

Large-scale musculoskeletal models contain many, many parameters. These parameters can be distinguished as inertia parameters of the segments (mass, rotational inertia), geometry parameters (joint rotation centers, muscle attachments, ligament attachments, etc.), and muscle parameters (e.g., physiological cross-sectional area, optimum length, maximal contraction velocity, etc; see also Appendix A1). Geometry and inertia parameters are typically derived from cadaver measurements, muscle parameters from a wide range of controlled experiments using a variety of species ranging from humans to frogs.

Currently, large-scale musculoskeletal models are generic, representing a "typical" adult male rather than any specific individual. Almost by definition, generic models are used for a general analysis of the musculoskeletal system, like predictions for a large group. Model predictions may depend to a large extent on the specific morphology present in the model. Morphological parameters are usually not an average of a large set of data, but indeed the data of just one or a few cadavers. It could be very well that there is a mutual dependency between some parameters (e.g., if one muscle is smaller another muscle is larger). For most joint systems these interrelationships and their effects are simply unknown. Individual models would require parameters derived for the individual subject itself. Currently there is no complete set of parameters available for a living human being (see Appendix 1 for a summary of what is available). If a generic model is used for individual predictions, it could be that due to unknown interrelationships and/or imprecise data, the predictions are off.

If the aim of large-scale musculoskeletal models is to on one hand gain insight into the function of specific muscles, and on the other hand use models that are as simple as possible without missing important characteristics of the system, then there is an area of tension in making choices necessary in developing a model. These choices include

- Model structure,
- Parameter estimation: kinematic and dynamic parameters,
- Optimization,
- Validation, and
- Sensitivity analysis.

The goal of this chapter is to discuss each of these choices, and assess to what extent they affect the quality of the large-scale model.

2 Model Structure

Modeling starts with the choice of the model structure. Physically, the model structure is the set of properties that are included in the model. Mathematically speaking, it is the set of equations connecting the input variables with the output variables. Properties of the neuromusculoskeletal system are

- The linkage system (bones, joints, ligaments) with the associated mass and viscoelastic properties,
- The actuator system containing the muscles,
- The sensory system with the muscle spindles and Golgi tendon organs, and
- The control system at various levels of the central nervous system (CNS).

In Figure 32.1 the interrelations between these systems are shown in a block scheme. Undisputedly, all these components are important for human movements. However, in the *study* of human movement one can see a "bottom-up" approach of models including more and more features of higher levels (see Table 32.1). This is accompanied by a

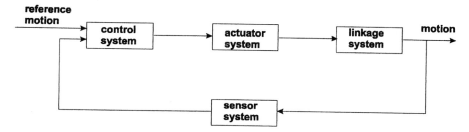

FIGURE 32.1. Block scheme of a neuromusculoskeletal system with the linkage system (bones, joints, ligaments), actuator system (muscles), sensor system (muscle spindles, Golgi tendon organs, etc.), and the control system (central nervous system).

change of input/output variables and detail of analysis, but not necessarily by an increase of the validity of the predictions. A model is as good as its weakest link. If more features are incorporated but their model structure or parameters are poor, the predictions can even be worse!

2.1 Equations of Motions, Forward and Inverse Dynamic Simulations

The equations of motions of a mechanical system (musculoskeletal systems no exception) can be written as a non-linear set of differential equations:

$$\underline{M}_q(\dot{q}, q; t) = I(q) \quad \ddot{q}(t) + B(\dot{q}, q)\dot{q}(t) + K(q) + \underline{M}_{ex} \quad (32.1)$$

where q is the vector of generalized coordinates (rotational and translational DOF), M_q is the vector with net generalized forces (joint moments and joint forces in case of rotational and translational DOF, respectively), I, B, and K are the inertia, viscosity, and stiffness matrices. $\mathbf{M}_{ex}$ is the vector of external moments and forces, generally consisting of contact forces and gravity. M_q consists of all sources contributing to the net generalized forces, like muscles and ligaments.

Passive forces (ligaments, capsule) are incorporated into K and B; gravitational forces are incorporated in M_{ex}. The generalized forces and moments are in fact the result of the muscle forces F_m multiplied by their moment arms J_{mq} (n_q x n_m; n_q is the number of DOF and n_m is the number of muscles) for each DOF q:

$$\underline{M}_q(\dot{q}, q; t) = J_{mq}(q) \cdot \underline{F}_m(u(t); \dot{q}, q) \quad (32.2)$$

In *any* musculoskeletal model n_m is larger than n_q for the simple fact that muscles can only exert traction forces: there is *actuator redundancy*. In physiological terms, there are more muscles present that DOF, meaning that there can be an infinite number of muscle force combinations to result in the same (submaximal) motion. Generally, a unique solution is found by optimization of a certain criterion in which a measure of muscle "effort" is within the criterion.

In a forward dynamic simulation, M_q is given or can be calculated from the neural input using muscle models, and $q(t)$ is calculated by integration of the differential equations. In an inverse dynamic simulation, $q(t)$ and its derivatives are given, or measured, and the net generalized forces $\underline{M}_q$ can be calculated. In the latter case, the system of differential equations becomes a system of algebraic equations, which is much easier to solve!

TABLE 32.1. Input–output variables for models with increasing levels of complexity. For variables between parentheses, it has not (yet) been shown that it is feasible to calculate these.

Model structure	Forward dynamic simulations		Inverse dynamic simulations	
	Input variables	Output variables	Input variables	Output variables
Linkage system	Net moments	Motion/external forces	Motion/external forces	Net moments
Plus actuator system	Neural input	Motion/external forces	Motion/external forces	Neural input/muscle forces
Plus sensor system	Supraspinal signals	Motion/external forces	Motion/external forces	(Supraspinal signals)
Plus control system	Reference motion	Motion/external forces	Motion/external forces	(Desired motion)

Muscle force is the result of the neural input $u(t)$ to the muscle, and a number of non-linear differential equations describing the muscle dynamics [e.g., see Chapter 8, Chapter 10, and Chapter 11).

2.2   Linkage System

The linkage system contains the bones, intermediate joints, and ligaments. A thorough choice of the components of the linkage system is important, because the *kinematic structure* of the model is involved. Two bones can move with respect to each other by virtue of the joint in between. In principle, a bone has six DOF of motion with respect another bone: *three rotations and three translations*. The motions of the joint are limited by passive structures like the articular surfaces and the ligaments. These passive structures pose *restraints* to the joint motions: though motions in many directions are possible, often these motions will be very small. For example, cartilage can be compressed only a few millimeters. Most of the time these small motions are neglected, and the motion is said to be *constrained*. For each constrained direction the number of DOF diminishes by one.

Traditionally, joints have been studied (e.g., by Fick 1911) by comparing them with standard idealized revolute joints: hinges, spherical joints, ellipsoidal joints, and saddle joints. In the human body, joints can be found with motions that resemble these standardized revolute joints. The glenohumeral joint of the shoulder and the hip joint behave approximately as spherical joints, only permitting three rotational DOF. The elbow joint and finger joints can be regarded as hinge joints with one rotational DOF. The first thumb joint and the ankle joint resemble a saddle joint, having two nonintersecting rotational axes. But many times one can not derive the potential motions from the shape of the articular surfaces. For example, the knee joint displays a complex combination of translations and rotations, resulting from an interaction between the articular surfaces and the knee ligaments. The sternoclavicular joint between the sternum and the clavicle almost subluxates at the end of the arm elevation.

Ligaments call for special attention when modeling. A typical stress–strain relation for a ligament has a slack length (below which no force is transmitted), a nonlinear toe region when the ligament is stretched just beyond slack length, and a reasonably linear region for higher strains. The maximal strain above slack length is small, only about 3 to 6% for most ligaments, but reaching about 50% for some such as the ligamentum flavum in the spine. The high stiffness of ligaments result in a high sensitivity of the ligament forces for the joint motions. For inverse dynamic simulations it is unusual to include ligaments (e.g., assumed to be in "slack" region). In forward dynamic simulations the stiffness of the ligaments can be a major factor in selecting the magnitude of the integration step size.

Ligaments can be distinguished into intracapsular and extracapsular. An intracapsular ligament is a local thickening of the connective tissue, of which it is sometimes difficult to distinguish the transition with the capsule. They are often best included in the model as restraints to the joint excursion (e.g., restraining elbow motion between full extension and 140° flexed), which has the advantage of having less impact on the integration step size. Extracapsular ligaments, like the cruciate ligaments in the knee and the coracoclavicular ligament in the shoulder, are normally represented as separate structures with one line of action, since their function is of interest. However, Wismans (1980) showed that the knee motion is hypersensitive to the location of the attachment sites of the cruciate ligaments. A simple knee model with a fixed or moving rotation axis is often preferable above a ligamentous model which shows weird motion patterns!

2.3   Actuator System

Muscles are the actuators of the musculoskeletal system. They generate forces in the system, but also generate power, i.e. increase of decrease of energy, thereby enabling the human body to do "work." Muscles consists of origin and insertion tendons, and a muscle belly in between. Their shape can be very complex, with curved muscle fibers, pennation angles, bi- and multipennate, and so on. Modeling muscles, which is the focus of many chapters in this book (see especially Chapter 3), involves two steps:

1. Determination of the muscle line of action and/or muscle moment arm,
2. Representation of the muscle dynamics.

2.3.1 Determination of the Muscle Line of Action and/or Muscle Moment Arm

A muscle line of action is defined as a straight or curved line along which the muscle force is acting (see also Chapter 1). The muscle force is the result of the summation of the contribution of all individual muscle fibers. Hence, assuming one muscle line of action is already a simplification. There have been some attempts to represent the whole muscle body using finite element models, which allows for deformations of the muscle shape, change of pennation angle, and so on (Van Leeuwen & Spoor 1993). However, these models are difficult to build for each individual muscle and would require a surplus of computation time when connected in a musculoskeletal model.

There are several ways to estimate the muscle line of action or muscle moment arm. The simplest way is to take a straight line from origin to insertion, and calculate the moment arm as the distance to the rotation center. However, for some muscles this is an impermissible simplification (see Figure 32.2A). Another approach is to determine the centroid line of a muscle. The centroid line is defined as the central line through the muscle belly, often estimated from MRI or CT images (Jensen and Davy 1975), and the muscle moment arm is estimated relative to the joint rotation center. An apparent problem in this approach is that the moment arm can be estimated, but the direction of the muscle force is still unknown. A force acting along a curved line will generate reaction forces at the structures underneath, forces that must be known for a force equilibrium (see Figure 32.2B). A third

approach is to use bony contours, around which the muscles are wrapped (Högfors et al. 1987; Winters and Kleweno 1993; Van der Helm 1994). The muscle line of action is calculated as the shortest line between origin and insertion, curved around a bony contour in between. The *effective* muscle line of action is between the tangent point at the bony contour and the origin (or insertion if applicable). The moment arm can be calculated as the distance of this line to the joint rotation center. An advantage is that a proper representation of the forces acting in the system is provided, that is, a good estimate of the joint reaction forces can be calculated (see Figure 32.2C). A fourth approach is to calculate the effective moment arm r as the derivative of the muscle length l to joint angle (q^{th} DOF):

$$r = \frac{\partial l}{\partial q} \qquad (32.3)$$

By thorough measurements of the muscle length and joint angle, the instantaneous moment arm as a function of joint angle can be calculated (Spoor et al. 1990). A similar approach is used by Delp et al. (1990) to calculate the effective moment arm for a complex kinematic structure like the knee.

For muscles with large attachment sites, one muscle line of action is not sufficient to represent the mechanical effect of the muscle. Usually, representative attachment points are estimated by eye during the cadaver measurements (Brand et al. 1982; Högfors et al. 1987; Johnson et al. 1996). Van der Helm and Veenbaas (1991) showed an approach in which the required number and location of the muscle lines of action could be calculated afterward.

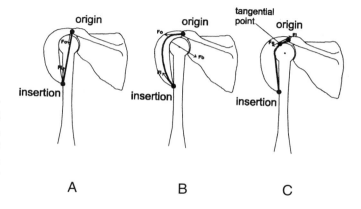

FIGURE 32.2. Modeling the muscle line of action. (A) Straight line approach; (B) Centroid line approach; (C) Bony contour approach. F_o is the force at the origin, F_i is the force at the insertion or pseudoinsertion (tangential point), F_b is the missing force for the force equilibrium of the muscle itself.

A   B   C

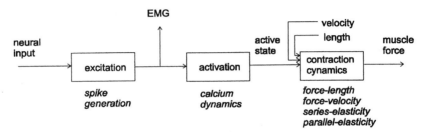

FIGURE 32.3. Muscle model block diagram according to Winters and Stark (1985). The "excitation" block is a linear first-order model describing a filtering of (infinite bandwidth) neural input. The "activation" block is a non-linear first-order model describing the fast increase in calcium concentration and the slow decrease by the calcium pump. The "contraction dynamics" block contains the contractile element, series-elastic element, and parallel-elastic element.

2.3.2 Representation of the Muscle Dynamics

As shown in Figure 32.3, muscle dynamics can be divided into two stages: activation dynamics (from neural input to "active" state ($\sim$ intracellular calcium concentration)) and contraction dynamics (from active state to muscle force). Although complex biochemical reactions take place in these transitions (see Chapter 2), they are necessarily (for computational efficiency) lumped into rather crude descriptive models. These descriptive models are reasonably good in representating major dynamic events in force generation, but can not be used to study the internal biochemical sequence of events.

Basically, there are two types of muscle models (see also Chapters 1, 2, and 8): Hill-type models and Huxley-type models. Hill-type models use descriptive functions for the force-length and force-velocity relationship of the muscle fiber and muscle tendon (see Chapters 10, 11, 28, 29, 33, 43). Huxley-type employ models of dynamic cross-bridges to determine the force-velocity relation. For connection to tendons and the skeleton, the latter still use descriptive force-length curves for the fiber and tendon. Huxley (1957) represented the cross-bridge dynamics with a partial differential equation which is difficult and time-consuming to simulate. Zahalak (1981) simplified this partial differential equation to a system of ordinary differential equations using the zero-, first-, and second-order approximation.

Virtually all muscle models used in large-scale musculoskeletal modelling are Hill-type models (Hatze 1981; Winters and Stark 1985; Otten 1988;

Zajac 1989). Hatze (1981; Chapter 33) exploits the most complex muscle model, containing the ordered recruitment of motor units and neural spike frequency as inputs, and a very detailed descriptive representation of the inner muscle (e.g., 9 motor units of different size and type). Other models (e.g., Winters and Stark 1985; Otten 1988; Zajac 1989) are reasonably similar in the way that the non-linear muscle features (force-length relation, force-velocity relation and series- and parallel-elastic elements) are represented. However, the weakest point in the muscle models is not the detail of muscle dynamics represented, but the very limited way that the actual *parameters* of the model can be obtained. It is suggested that the most important parameters for the functioning of a muscle are the muscle moment arm, the physiological cross-sectional area (PCSA) and the optimum length (defining the force-length excursion of the muscle). Other parameters defining the series-elastic and parallel-elastic properties, and subtleties of the force-velocity relation, can be reasonably guessed. In my opinion, the scarce knowledge about the parameters does not warrant the use of complex muscle models, because these models do not really improve the quality of the predictions.

2.4 The Sensory System with the Muscle Spindles and Golgi Tendon Organs

Virtually all existing musculoskeletal models, simple or large-scale, are open-loop models without any proprioceptive feedback. Yet, although one would rarely build a robot without any feedback mechanism, in musculoskeletal modeling incorporating in feedback is often considered a nuisance! Without feedback, any perturbation can result in a deviation from the intended motion. Within the musculoskeletal system, two different "feedback" mechanisms can be distinguished: that due to intrinsic muscle properties [e.g., length and velocity "feedback" in Figure 32.3; see also Chapter 10 (in which this is called effect) "preflexes"; Chapter 11, Chapter 2, and Chapter 7] and proprioceptive feedback from muscle spindles and Golgi tendon organs (GTO), Chapter 8. The intrinsic muscle properties are normally reasonably represented in musculoskeletal models, but it is useful to conceptualize this as a source of feedback and that the stability of the whole mechanism depends in part on these properties.

Muscle spindles are small (about 8–9 mm) sensory units attached in parallel to extrafusal muscle fibers (see Figure 32.4). In a muscle spindle, a sensory region (around which an afferent nerve ending is wrapped) is found in between two intrafusal muscle regions. If the sensory part is stretched, the afferent nerve will fire. Muscle spindles contain two type of sensory fibers (see also Chapters 8 and 11): nuclear chain fibers and nuclear bag fibers. Nuclear chain fibers are often thought of a mainly length sensors, and nuclear bag fibers are mainly velocity sensors (with some acceleration sensitivity as well). Type II afferent nerve fibers contain information from the nuclear chain fibers, whereas type Ia afferent nerve fibers contain information from both the nuclear bag and nuclear chain fibers. In Chapter 11, signals from nuclear bag and nuclear chain fibers are hypothesized to add in a competitive way.

In Figure 32.4 there are four inputs to the muscle spindles: Two mechanical inputs (length and velocity) and two neural inputs (γ_s (static) to the nuclear chain intrafusal fibers and γ_d (dynamic) to the nuclear bag intrafusal fibers. If the whole muscle is lengthened, the sensory part will be stretched accordingly, and hence function as a length sensor. If the intrafusal fiber is activated, the sensor will be stretched and an afferent signal will be generated as well. In fact, as long as a significant level of γ activation lasts, there will be afferent output. Hence, the hypothesis in the earlier days that the γ-activation would be a kind of position reference, and the muscle with spindle would function as a servosystem, can not be true, because there will always be a (error) signal present during γ-activation.

A few attempts have been made to incorporate the spindle dynamics into the muscle model, the most well-known by Hasan (1983) and Gielen and Houk (1987). Schaafsma et al. (1991) and Winters (1995, see also Chapter 7) built a reflexive dynamic muscle model (muscle with spindle feedback) based on representation of the spindle dynamics by

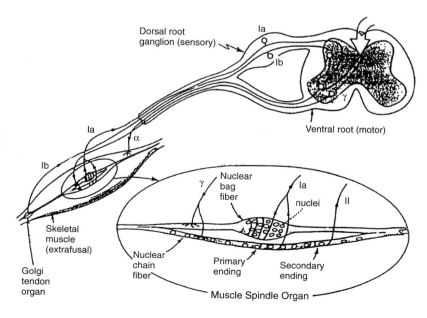

FIGURE 32.4. Overview of the spinal cord, efferent motor-neurons to the extra- and intrafusal muscle fibers, the muscle spindle with the static nuclear chain and dynamic nuclear bag fibers, the Golgi tendon organ, and the afferent Ia (length and velocity information), II (length information) and Ib (force information) fibers. (Adapted from Matthews 1972.)

a separate muscle model of the intrafusal fibers. Such models are able to capture some of the classic force transient and quasistatic responses seen in single cat muscles (e.g., Nichols and Houk 1976; Hoffer and Andreasson 1981; Hulliger 1984). Because of activation (and hence force) changes with position perturbation, it also predicts quasistatic behavior similar to the elbow flexion–extension "do not intervene" paradigm of Feldman (1966). Recently, Rozendaal (1997) and Van der Helm and Rozendaal (Chapter 11) presented an integral approach to the musculoskeletal feedback system, incorporating force feedback from the Golgi tendon organ and length and velocity feedback from the muscle spindle (see Figure 11.1). They used this model to gain insight into optimum force, length and velocity feedback gains for one DOF systems. With time, such models should play an important role for larger-scale systems.

2.5   Control System Design

To summarize, at present most musculoskeletal models are open-loop, and the neural input to the muscles can be compared with the α-motor neuron signal. No control system is specified, only that an optimal "α motor neuron signal" is sought. As soon as proprioceptive feedback is added to the model, the structure of the control system becomes much more well defined, and the neuromusculoskeletal control system can be assumed to have two functions:

1. Disturbance rejection,
2. Motion generation.

The neural input signal to a musculoskeletal system with proprioceptive feedback can be compared with a supraspinal signal, for instance from the cerebellum. The α-motor neuron is just a node (adding/subtracting point?) in the control loop, and the neural input to the muscles a combination of feedback input and supraspinal input. For disturbance rejection, a setpoint signal with the desired length (position), velocity and acceleration should be available (although with large-scale systems, this may become challenging), resulting in a compensatory neural input (activation) of the muscles. For motion generation, an optimal neural input signal must be generated, assuming full knowledge of the desired motion trajectories and muscle and

skeletal dynamics. In fact, Chapter 29 suggests that a kind of inverse dynamic optimization is performed; Kawato's group has contributed significantly to this type of approach, using neural networks, and has tied it to learning (e.g., Kawato et al. 1987; Chapter 34).

3   Parameter Estimation: Kinematic and Dynamic Parameters

Because the aim of large-scale musculoskeletal models is to gain insight into the function of specific muscles for certain tasks, the parameters in the model should be as accurate as possible. Unfortunately, the parameters are almost always obtained via human cadavers, whereas the motions, forces, and EMG signals to compare the predictions are obtained from other subjects. Errors due to differences in morphology are inevitable. Necessary parameters are inertial parameters, geometrical parameters and muscle parameters, which are preferably all measured in the same cadaver.

Inertial parameters, that is, the mass and rotational inertia, and center of gravity, can be calculated from antropometric measures using regression equations like studies of Dempster (1955), Clauser et al. (1969), Hinrichs (1985), and Yeadon and Morlock (1989). Errors can be up to about 10%, especially for such larger segments as the trunk.

Geometrical parameters comprise the position coordinates of the joint rotation centers, the muscle attachments, ligament attachments, some bony contours around which muscles are wrapped, and so on. These data are normally obtained from cadavers, though at the moment much research is focused on deriving these data from living subjects using MRI. The quality of the model depends on the quality of the morphological data (see Appendix 1 for an updated summary that extends Yamaguchi et al. 1990). Therefore, it is surprising that for the lower extremity only one study is known in which the position of all muscle attachments have been measured (Brand et al. 1982). Many modeling studies use these data. In the upper extremity a quite different situation is encountered. In the shoulder region, each research group starts with measuring their own geometric data (Högfors et al. 1987; Wood et al. 1989; Veeger et al. 1991; Van der Helm et al. 1992; Johnson et al. 1996), whereas only a few large-scale models exist (Karlsson and Peterson 1992; Van der Helm 1994a,b). In the elbow region more data are found (An et al. 1981; Murray et al. 1995), and for the wrist and hand there is Brand et al. (1981). It must be concluded that no complete data set exists for the whole upper extremity. Figure 32.5 shows an example of how the geometry of a muscle is modeled.

The data needed for a muscle model depends on the muscle model that is used, although all essentially use the same data. Zajac (1989) proposed a muscle model which five basic parameters: maximal muscle force, optimal muscle length, tendon length, pennation angle, and fiber composition. Winters and Stark (1985) developed a muscle model in which the nonlinearities of the key constitutive relations (contractile element force-length and force-velocity, series and parallel elements) are described through dimensionless "shape" parameters. Winters and Stark (1988) describe how the key

parameters can be derived from morphological measurements, such as PCSA, tendon length, optimum length, pennation angle, mass (or volume) of the muscles, using loosely derived "regression" equations from a wide range of measurements of many species. Winters and Stark (1987), based on sensitivity analysis, found that the most sensitive parameters tend to be task-specific. Thunissen (1993) suggests that PCSA and optimum length are the most sensitive parameters for a muscle. Hatze's muscle model uses many more parameters, some of which cannot be recorded in a morphological study (Hatze, 1981; Chapter 33).

Klein Breteler et al. (1997) is the first study to measure muscle optimum length for *all* muscles of the shoulder, and work is in progress to obtain these data for the elbow and wrist using the same cadaver.

It is suggested that the weakest link in musculoskeletal modelling is the quality of *morphological data*, and *not* the model structure included in the muscle *model*.

4   Optimization

In the optimization procedure used here, an optimal *input signal* is estimated given a certain output or performance criterion. The distinct process of using input–output error minimization (e.g., least squares error criterion) as part of system identification, is addressed in Chapter 9 and Ljung (1987). Here we start with a model (complete with internal structure and parameters), and the output signal is calculated, given this model and its input signals.

As noted previously, in any musculoskeletal model there are more muscles present than strictly necessary to make a certain motion. In other words, there exist an infinite number of solutions (sets of muscle forces) to make the same motion. Optimization seeks the *optimal* solution, where optimal is defined as the extremum of a certain criterion. The next thing to do is to compare the *optimal* solution with experimental data. Then, potential differences can be explained, for example, by a critical look at the model ("Are all important parameters included?", "Is the model structure too simplified?") and by a critical look at the data ("Is motor control aimed at the 'optimal' solution, or is the solution 'good enough'?", "Is the motion really well-

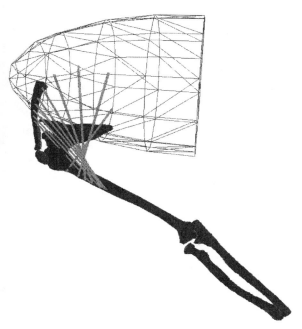

FIGURE 32.5. An example of the representation of a muscle with a large attachment site by multiple lines of action: M. pectoralis major. (Van der Helm and Veenbaas 1991.)

trained?"). Another important step is to analyze the optimal solution itself. Why is this particular solution picked out by the optimizing algorithm? A very helpful tool in this step is the *sensitivity analysis*. By slightly changing parameters, the effect on a certain behaviour of the model simulation can be calculated.

In accordance with the way that the motion equations (Eqs. (32.1) and (32.2)) are evaluated, in optimization one should distinguish the *forward dynamic optimization* and the *inverse dynamic optimization*. Zajac and Winters (1990) and Winters (Chapter 35) provided an excellent overview of potential optimization criteria. In general, the terms in the optimization criterion can be divided into task-related terms J_{task} and load-sharing related terms J_{ls}. Sometimes, 'damage preventing' terms (J_{damage}) as high joint loading or pain are introduced:

$$J_{pc} = J_{task} + J_{ls} + J_{damage} \qquad (32.4)$$

Task-related terms are, for example, deviations from the endpoint position reached, maximal hand speed, maximal kinetic energy of the body center of gravity, minimal time (time-optimal), and so on. Load-sharing criteria are only necessary if the task-related criteria do not specify a maximal effort (maximal hand speed, maximal kinetic energy of the body center of gravity, time-optimal) but a sub-maximal task (achieving a certain endpoint). Then, multiple solutions are possible and a load-sharing criterion will favor one of them. In a load-sharing criterion normally it is desired to minimize the total energy that is spend on the task. A physiological objection to such a criterion might be that there are no direct energy sensors in the body, and that one muscle does not care how another muscle is loaded. A more mathematical objection is that the states of all muscle models used in large-scale musculoskeletal models are *not informative* enough to calculate the energy *consumed* by the muscle. Hence, minimization of the energy consumed by the muscles is approximated by, for example, minimization of the squared muscle forces ($\min\sum_i (F_i)^2$), squared muscle stresses ($\min\sum_i (F_i/\text{PCSA}_i)^2$) or maximal muscle stress ($\min(\max_i(F_i/\text{PCSA}_i))$). Because Hatze (1981) exploits the most detailed muscle model, he has the best approximation of muscle energy consumed, using approximations of activation

heat, maintenance heat, shortening heat, work done and viscosity effects.

Happee (1992) had an interesting improvement, in which he used as a criterion the minimization of the squared muscle active state **a**, weighed by the muscle volume ($\min\sum_i V_i.(a_i)^2$). It can be argued that the muscle active state is an approximation of the calcium concentration (the calcium pump into the sarcoplasmatic reticulum accounts for roughly 50% of the energy used) and that it is related to the rate of cross-bridge binding and release (the other major energy consuming factor). Muscle active state **a** is dynamically related to muscle force ($F/F_{max}(l_{mus},v_{mus})$, the 'relative' muscle force with muscle length l_{mus} and velocity v_{mus} accounted for). The volume term V (PCSA x l_{mus}) results from the consideration that the consumed energy is linearly related with the 'amount of muscle' being activated (i.e., a twice as much volume of muscle consumes twice as much energy with the same activation).

4.1 Inverse Dynamic Optimization

In the inverse dynamic situation, the "task" is given, that is, to mimic the recorded motion as close as possible. With the position (q), velocity ($\dot{q}$) and acceleration ($\ddot{q}$) and external forces F_{ex} given, the motion equations act as linear equation constraints in the optimization procedure. Only load-sharing terms are incorporated in the optimization criterion.

It is easy to see that if the position and velocity of the musculoskeletal system are known (e.g., by motion recordings), the motion equations are only a function of the time instant they are evaluated. In other words, the optimal solution in terms of muscle forces can be calculated for each time-step separately. The only constraint is the muscle dynamics. Muscle force can not rise from zero to maximal in one instant, or neither can it decrease to zero in one instant. Therefore, the optimal solution for muscle forces must be constrained using a representation of the (inverse) muscle dynamics. Several solutions can be found for this problem.

1. The most trivial (and historically most common) solution is that only (quasi)static situations are analyzed! Muscle dynamics can be neglected and muscle force can range from zero to maximal isometric force.

2. Happee (1994) introduced an inverted muscle model. If the muscle force is known at a certain time, and also the dynamic muscle states, the minimal and maximal allowable muscle force for the next time-step can be calculated from the minimal and maximal neural input. These boundaries on the muscle force are used in the optimization procedure. The estimated muscle force calculated in the optimization procedure will be between these boundaries. Then, the neural input, as well as an update on the muscle states, needed to produce this muscle force can be calculated using an inverted muscle model. Next, the same procedure can be started for the next time-step. A disadvantage is that no "anticipation" of neural activation can occur, that is, early activation of a muscle if it is foreseen that later on this muscle must exert very high muscle forces. However, this is only a limitation in very fast motion with maximal effort.

3. With an application in human walking, Thunisse (1993) developed an algorithm in which the muscle dynamics are taken into account by forward integration of the differential equations and backward integration of Lagrange multipliers. His algorithm truly computes the most efficient muscle activation through time by "backwards propagating the need for activation," and thereby anticipating large muscle forces needed in short time.

4. Recently, Yamaguchi et al. (1995) described an algorithm for directly obtaining a known motion, but by using forward dynamic integration of the differential equations. At a certain time (with all states known), one by one the neural input variables are activated, and the resulting accelerations (for all $\ddot{q}$) are calculated. Then, these neural input variables are weighed (with a load-sharing criterion) to result in the desired accelerations (from the motion recordings). In this algorithm there is no anticipation (in that respect it is similar to option 2), but in using only forward integration a stable solution is ascertained (see Section 4.4).

4.2   Forward Dynamic Optimization

In a forward dynamic optimization, the algorithm has to solve the same problem that the CNS is faced with: finding a neural input to the muscles which will result in the desired motion, taking care of the *kinematic* and *actuator* redundancies, and being reasonably energy-efficient. At first, given the task

one may not have a clue what could be the best solution. By analyzing the optimal solution, one learns about the most important properties of the musculoskeletal *model* (and maybe the *system*).

In a forward simulation the nonlinear differential equations ($\underline{\dot{x}} = f(\underline{x}, \underline{u}(t))$) are integrated with input variables $\underline{u}(t)$ and initial values $x(t = 0)$. The objective of the optimization is to find an input $\underline{u}(t)$ which results in a certain motion with a minimal (maximal) value for the optimization criterion J_{pc}. The optimization criterion J_{pc} is calculated over the whole time window in which the motion occurs. In the optimization algorithm an update of $\underline{u}(t)$ is calculated, which results in a lower value of J_{pc}. Two types of algorithms have been used (of which there are hundreds of examples and variations within the engineering optimization literature):

1. Gauss–Newton (gradient-based) methods, in which the first $\dfrac{\partial\,Jpc}{\partial\mathrm{u}(t)}$ and sometimes the second derivatives $\dfrac{(\partial\,Jpc)^2}{\partial\mathrm{u}(t)^T\underline{\partial}\mathrm{u}(t)}$ of the optimization criterion with respect to $\underline{u}(t)$ are calculated. Then, the search direction can be updated in the direction of approximated minimum of a second-order function. Normally, the derivatives are calculated numerically, which involves many forward dynamic simulations. For example, Pandy et al. (1994) state that about 95% of CPU time is consumed by calculating the derivatives. They used about 77 *hours* CPU on a Cray supercomputer in order to optimize one walking cycle, with a lower extremity model. One can view this algorithm as a blind person standing on a hill searching for the valley, turning around on his spot, sensing the steepest descent, and making small steps in this direction. The traps are obvious. Progress is slow, and there is a risk of a local minimum (a small valley but not the deepest one we are interested in).

2. Random-search methods (e.g., Bremerman 1973; Winters and Stark 1988; various genetic algorithms), in which certain search directions are randomly (or quasirandomly) evaluated, and each promising one is followed until the deepest point. A disadvantage can be a large number of evaluations without progress. An advantage is that the tedious calculation of derivatives is avoided. One can view this algorithm as a blind

person making relatively big steps in random directions, and as soon as he feels he is stepping downhill, this search direction is continued until the slope goes up again. The risk of local minima is smaller, since big steps are taken anyway. For large number of variables a random-search method may converge faster.

Of note are that a wide variety of heuristic approaches have been put forward (especially in engineering), some of which try to combine the best features of random and gradient-based approaches.

Forward dynamic optimization is very computer-time consuming, and therefore has not often been applied to a large extent on large-scale musculoskeletal models. Pandy et al. (1994) has probably performed the most extensive optimization effort ever, by optimizing the lower extremity musculoskeletal model for a walking cycle. Hatze (1981; Chapter 33) has developed a very large-scale model with 15 segments and 240 muscles, and a very detailed muscle model. However, in the optimization stage presented in Chapter 33 the model is reduced to a three-link model with eight muscles.

4.3   EMG Driven Models

The problem of deriving the proper input signals $u(t)$ is a severe problem in the use of forward dynamic simulations using large-scale musculoskeletal systems. Therefore, numerous studies have attempted to use EMG signals as an input to the forward dynamic simulation. Typically, these studies have an identification stage and a prediction stage. In the identification stage, unknown parameters relating EMG to muscle force, or EMG to joint torque, are estimated in a regression equation using motion and EMG recordings of a subject. Then, in the prediction stage, these regression equations are used to predict the motion from a novel recording of the subject.

These models have been reasonably successful, for static situations (Laursen 1997) and for dynamic situations (Chapter 34). The study in Chapter 34 is probably one of the most general in this area, using a neural network model to relate the input EMG with the output joint torques. As little as possible a priori knowledge of the musculoskeletal model is incorporated. However, a disadvantage of such a pure black-box approach is that little is *learned*

about the musculoskeletal system modelled in terms of the function of specific muscles, forces inside the system, etc. Maybe there is a better future for the grey-box approaches, such as used by Laursen (1997). He used data about the muscle moment arms from a musculoskeletal model (Karlsson and Peterson 1992) and fitted EMG input to the joint torques, using these moment arms. Hence, muscle *forces* are a accessable variables inside the system, and many other biomechanical properties can be analyzed.

4.4   Quasistatic Stability

Probably the most neglected topic in optimization of musculoskeletal models is stability. Stability can be defined as a robustness against perturbations (i.e., the property that the system returns to its original position after a perturbation). Clearly, this definition has its origin in quasistatic postural tasks, where the objective is to stay in place. During motions, stability is less well defined. Getting back to the original position at the time of perturbation is not really an option, but getting back to the reference trajectory and/or arriving at the goal-position anyway might be more useful definitions. Quasistatically, a system is stable if the potential energy increases for *all* possible perturbations of the DOF. Potential energy in a musculoskeletal system is stored in the masses in the field of gravity, and in the spring-like behavior of muscles and passive structures. Sources of potential energy can be categorized by type (i.e., due to conservative forces (E_v) or due to strain in spring-like elements (E_s)), and by location, that is, internal to the musculoskeletal system (muscles, passive structures, segmental masses) or external (external constant forces and external springlike forces). Five "sources of potential energy" can be discerned (see Figure 32.6):

E_{sm}: due to spring-like behavior of muscles

E_{sp}: due to spring-like behavior of passive skeletal structures

E_{se}: due to spring-like behavior of environmental structures

E_{vg}: due to gravitational forces of body segmental mass

E_{ve}: due to constant forces applied at end-point

Spring-like behavior of muscles originates from a number of sources. For very small amplitudes of

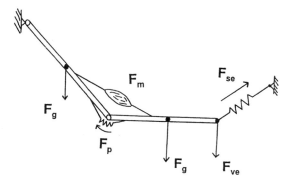

FIGURE 32.6. Five sources of potential energy: Spring-like forces like muscle, passive tissue (ligaments), and external contact forces; conservative forces, like gravity and external weight. For any combination of weight and spring constants, the mechanism will find an equilibrium point.

perturbation (a few millimeters), the cross-bridge stiffness (extension of cross-bridges inside the sarcomeres) come into play. For larger amplitudes, the derivative of the muscle force-length relation is equivalent with the muscle stiffness. It is generally acknowledged that if the muscle is on the negative slope of the force-length curve (above optimum length), that the stiffness is negative and instability could occur (in the absence of passive stiffness). Beyond the scope of a quasistatic analysis, the reflexive feedback from the muscle spindles will increase the 'stiffness' of the muscles (see Section 2.4). This will be incorporated into a dynamic analysis.

One can distinguish two strategies to increase the stiffness of a joint: co-contraction and increasing

the gains of the feedback loops. Co-contraction is not very energy efficient, but is functional for all frequencies of perturbation. Proprioceptive feedback is energy efficient because it requires only compensation if a perturbation is present. However, because of the time delays in the feedback loops it will only be effective below certain frequencies (e.g., 3 Hz according to Chapter 11).

One can imagine in Figure 32.6 that after a small perturbation the arm will return to its equilibrium position. However, this is only true if the springs are sufficiently stiff, that is, sufficient potential energy is stored during the perturbation, *more* than compensating for the decrease of potential energy with respect to the gravity field (Van der Helm and Winters 1994).

Table 32.2 presents the first and second derivative of the potential energy for spring-like sources (muscles, ligaments, external contact forces) and conservative sources (gravity, external weight). The first derivatives result in the *static* equilibrium equation:

$$F_s.r_s + F_h.r_h = 0 \qquad (32.5)$$

for all spring-like forces F_s and conservative forces F_h, where r_s and r_h are the moment arms, respectively. In other words, Eqn. (32.5) describes the moment and force equilibrium, but there is no guarantee that this equilibrium is *stable*. For example, if the arm is elevated to the vertical position, theoretically no muscle forces are needed for the static equilibrium. However, everyone will agree that muscle forces are needed to prevent the arm from falling down. Antagonistic muscle forces will com-

TABLE 32.2. The Potential Energy for spring-like source and for conservative sources. It should be noted that $r_s = \dfrac{dl}{dq}$ where r_s is the moment arm of the spring-like force with respect to the qth DOF. x_s is the displacement of the spring with respect to the equilibrium position, and x_h is the displacement of the point of application f_h with respect to an arbitrary point

	Spring force: $k_s.x_s = -F_s$	Constant force: f_h
Potential energy E_p	$E_p = 0.5.k_s.x_s^2$	$E_p = m.g.x_h = -F_h.x_h$
1st derivative (static equil.)	$\dfrac{\partial E_p}{\partial q} = k_s.x_s.\dfrac{dx_s}{dq} = -F_s.r_s$	$\dfrac{\partial E_p}{\partial q} = m.g.\dfrac{dx_h}{dq} = -F_h.r_s$
2nd derivative (stable equil.)	$\dfrac{\partial^2 E_p}{\partial q^2} = k_s.r_s^2 - F_s.\dfrac{dr_s}{dq}$	$\dfrac{\partial^2 E_p}{\partial q^2} = -F_s.\dfrac{dr_s}{dq}$

pensate each others moment, but together they will
increase the stiffness of the joint.

In the last row in Table 32.2 the second derivative
contributions to the potential energy is shown. If the
second derivative is positive, this source of potential
energy contributes to a stable equilibrium. It is interesting to note that for springlike forces the spring
constant times the squared moment arm will *stabilize* the system (as long as the spring constant is positive!), and that the second term (the change of moment arm due to the change of DOF x_q) is generally
destabilizing the system (see Figures 32.7 and 32.8).

In mechanical terms, an equilibrium is stable if

$$dxq \cdot \frac{\partial^2 Ep}{\partial xq^2} \cdot dx_q \geq 0 \qquad (32.5)$$

for any combination of small perturbations dx_q. Introducing the apparent "joint" stiffness matrix K_q,
consisting of the contributions of the different
sources of potential energy:

$$K_q = \frac{\partial^2 Ep}{\partial xq^2} \qquad (32.6)$$

Stable equilibrium implies that all eigenvalues of
K_q must be positive since the eigenvectors of K_q
are orthonormal and cover the whole space of perturbations:

$$V_q^T \cdot K_q \cdot Vq = Dq; \quad \forall dq \in Dq : dq \geq 0 \qquad (32.7)$$

where V_q is the eigenvector matrix and D_q is the
diagonal matrix with eigenvalues d_q.

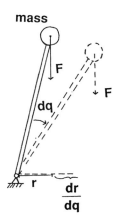

FIGURE 32.8. The apparent *joint* stiffness resulting from
a conservative force is equal to the force F_h multiplied
by the change of the moment arm dr/dq.

Generally, when a solution is found in an inverse
dynamic optimization problem, only requirements
for static equilibrium are included. There is then *no
guarantee* that this solution is a stable solution without adding the constraints of Eq. (32.7). Hence, the
inequality constraints of Eq. (32.7) should be added
in the optimization procedure (Van der Helm and
Winters 1994). Additional muscle activity, typically
of agonistic muscles, is necessary for stability.

A forward dynamic optimal solution is always stable, though since normally additional muscle activity is penalized by the load-sharing criterion, the solution will tend to be borderline stable: little
additional muscle effort is spent for stability. This
has two disadvantages. First, it is not likely that human beings operate at the verge of instability, so the
optimal solution is underestimating the effort spent
for stability. Second, the search for an almost unstable solution will hit every now and then unstable
regions of the input space, which will give no information and could destruct the optimization algorithm. Therefore, it is advocated to include a certain
level of stability in the optimization, for example, by
requiring that the eigenvalues of K_q are above a certain (positive) value instead of just being positive.

4.5 Dynamic Stability

The analysis of Section 4.4 applies to the quasistatic case without proprioceptive feedback. Unfortunately, the stability analysis for a nonlinear dynamic system requires very complex mathematics,

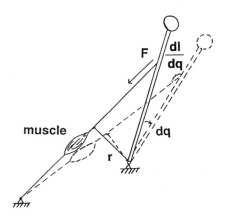

FIGURE 32.7. The apparent *joint* stiffness resulting from
a spring-like force is the sum of the spring constant k_s
multiplied by the squared moment arm r_s^2, and the force
F_s multiplied by the change of the moment arm dr/dq.

such as the Lyapunov approach, and has not yet been demonstrated for *any* musculoskeletal model. A step in the right direction might be the analysis of the *linearized* model, which can be done in any point in the state space. It is feasible to implement proprioceptive feedback loops and the accompanying gains like presented in Figure 32.1 into a large-scale musculoskeletal model (Rozendaal 1997; Chapter 11). The procedure starts with a linearization of the model to make a linear state-space description of the muscles, skeletal system and the interaction between both. Time-delays can be incorporated using a second-order Padé approximation, adding two states per time-delay. This results in Eq. (11.11):

$$
\begin{bmatrix} \dot{exc} \\ \dot{act} \\ v_{ce} \\ \dot{p}_1 \\ \dot{p}_2 \\ \dot{q} \\ \ddot{q} \end{bmatrix} = \begin{bmatrix} A_{11} & A_{12} & A_{13} \\ A_{21} & A_{22} & A_{23} \\ A_{31} & 0 & A_{33} \end{bmatrix} \cdot \begin{bmatrix} exc \\ act \\ l_{ce} \\ p_1 \\ p_2 \\ q \\ \dot{q} \end{bmatrix} \quad (32.8)
$$

where (exc, act, l_{ce}) are the muscle states, (p_1, p_2) are the states of the Padé approximation, and $(q, \dot{q})$ are the states associated with the skeletal system. The subsystems can be interpreted as follows: A_{11} is the linearized muscle model, A_{12} is the input of delayed information of the states q and $\dot{q}$ (proprioceptive feedback) and the nondelayed information of the states q and $\dot{q}$ (intrinsic muscle stiffness and viscosity), A_{21} (not a full matrix) is the force feedback (a function of the length of the series-elastic element: $l_{se} = l_{mus} - l_{ce}$), A_{22} is the approximation of the Padé filter, A_{23} is the length and velocity feedback, A_{31} is the representation of the muscle forces acting on the skeletal system, no time-delayed effects (proprioceptive feedback is unidirectional), and A_{33} denotes the second-order nature of inertial systems.

The state space matrix A is a function of the length, velocity and force feedback, as an approximation represented by simple gains k_l, k_v, and k_f, respectively. For a stable system it is required that the eigenvalues of A are smaller/equal to zero; the feedback gains must be optimized to achieve this. Whenever the optimal gains are assessed for a sequence of positions (the linearization allows for

transient states q and $\dot{q}$), a stable forward dynamic simulation is approximated. Generally, during motion the gains will be overestimated since the effect of the transition to a new working point in state space is not incorporated.

Rozendaal (1997) showed that this approach is feasible for large-scale systems such as a three-dimensional model of the shoulder (Van der Helm 1994a,b). *A priori* assumptions were that the loop gain $k_f.F_{max}$ is fixed, and the relation $k_v/k_l = 0.1$. These assumptions were justified in the analysis of simpler one and two DOF models. The optimization criterion weighted the speed of the response with the oscillating behavior (Rozendaal 1997).

In the future this approach might be useful to start forward dynamic simulations. In an inverse dynamic simulation the neural input to the muscles can be calculated, without incorporating any stability constraints and/or co-contraction. Then, the appropriate proprioceptive feedback gains can be established and the complete musculoskeletal model including feedback loops can be simulated forward dynamically.

5 Validation

The validation of musculoskeletal models is severely hampered since the access to internal signals in the human body is limited. For a true validation, the input and output signals, and the states of the *system* should be recorded, where the output signals should be compared with the predicted output from the *model* based on simulations with the recorded input signal. However, the neural input signals to the muscles, let alone for the supraspinal signals, can not be measured. The resulting motion can be measured, but since many ensembles of muscle forces can result in the same motion, this could not be regarded as a satisfying validation. Muscle forces can not be directly recorded, except for a few exceptions with human volunteers with buckle transducers on the Achilles tendon (Komi 1990), and animal experiments where strain gages were inserted in tendons, and calibrated after the animal was sacrificed.

EMG signals are the easiest to measure internal signals, and are therefore used for comparison with model results. The EMG-force relation can be established under very well defined circumstances, but the estimated parameters can not be generalized, and thus need to be assessed for each new subject (Hof and Van der Berg 1981). Heckathorne and Childress (1981) showed that the EMG depends on the contraction velocity and muscle length, which can only be assessed using a musculoskeletal model. Van der Helm (1994[a]) compared four different optimization criteria with EMG recordings and concluded that only the *magnitude* of the muscle force depends on the criterium, but not the fact whether or not a muscle was active. Because the relation between force amplitude and EMG amplitude is unknown, EMG *amplitude* can not be used to validate musculoskeletal models.

However, other aspects of EMG may be useful for validation. Comparison of the onset of muscle activation and the onset of EMG is a good indication that the dynamic features of the musculoskeletal model are correct (e.g., Van Soest et al. 1993; Happee and Van der Helm 1995). From the *isometric* EMG (where length and velocity effects are not important) the principal action of a muscle can be estimated. For example, in a static position forces at the elbow are exerted in all directions perpendicular to the humerus. The range of EMG activity (in polar coordinates) can then be compared with the range of muscle forces (Laursen 1997).

Because all forces in the mechanical model are known, each force which can be measured can be compared with the prediction. For example, Pandy et al. (1993) used the predicted foot reaction forces during standing up from a chair for comparison with measurements. However, it is fair to say that this is not a *complete* validation of the model, because the reaction forces depend to a large extent on the mass distribution of the person and to a lesser extent on the particular muscular coordination to generate the motion. Pandy et al. (1990) and Soest et al. (1993) compared the predicted optimal jumping height, muscular coordination strategy and motion patterns with the performance of skilled jumpers, and concluded that their models were predicting these features reasonably well.

Other internal signals have been used for comparison, like the muscle pressure (Jarvholm et al. 1991). In the near future, it will likely be possible to measure the muscle heat exchange and the muscle metabolics (phosphate turnover) using MRI. That would really show if the energy criterion is a proper performance criterion for submaximal tasks.

6   Future Directions

A new era of musculoskeletal modeling has arrived, because it is acknowledged that the muscle spindle and Golgi tendon feedback properties are important for the execution and stabilization of human motions, and as such are implemented in musculoskeletal models. Rozendaal (1997) used a rough description of these sensors (static gains!), but could show a remarkable resemblance with experimental values for the joint impedance. He concluded that feedback was the most important factor in the resulting impedance, much more important than the intrinsic muscle properties. It is likely that the focus will shift from very precise muscle models to good models of the muscle spindle and Golgi tendon organs (e.g., Winters, 1995; Chapter 7). The *control* properties of the CNS will be investigated also, next to the usual feedforward neural drive as currently is the state of the art.

It is likely that 1-DOF and 2-DOF models will continue to be used for investigating certain *principles* underlying these morphological structures and control. However, the role of large-scale musculoskeletal models will become more and more important. Any proposed control structure must be able to deal with the complexity of a three-dimensional system, in which muscles have multiple functions in exerting force, generating energy, stabilizing joints and compensating for each others unwanted moments about redundant DOF. It is well-known that control becomes many dimensions more complex if it has a multivariable structure, in which each signal interacts which the other signals.

Until now, one neural drive to the motor-unit/muscle has been defined. Incorporating the feedback properties will shift the nature of this drive from the α-motor neuron input towards a supraspinal signal activating the motor neuron *including* its feedback loops. In addition, *other input signals* to the musculoskeletal system must be defined and incorporated in the optimization, e.g. the γ-motor neuron activation. Chapter 11 showed that one of the functions of the γ-activation is to adjust the gain of the length and velocity feedback loops, and shift the *relative* gain of these loops; thereby the reflexive impedance (stiffness, viscosity and inertia effects) is directly dependent on these gains. The force feedback is very effective in compensating for the slow muscle activation dynamics. The role of the CNS in generating *feedforward as well as feedback control* signals will be investigated using *neuromusculoskeletal* models and a whole new set-up of experiments.

It is concluded that in the next decade the *paradigm of large-scale neuromusculoskeletal models* will shift from assessing muscular function to control issues dealing with the complex coordination between these muscles.

References

An, K.N., Hui, F.C., Morrey, B.F., Linsscheid, R.L., and Chao, E.Y. (1981). Muscles across the elbow joint: a biomechanical analysis. *J. Biomech.*, 14:659–669.

Anderson, F.C., Ziegler, J.M., Pandy, M.G., and Whalen, R.T. (1993). Dynamic optimization of large-scale musculoskeletal systems. Abstracts XIV ISB congress Paris.

Brand, R.A., Crowningshield, R.D., Wittstock, C.E., Pederson, D.R., Clark, C.R., and van Krieken, F.M. (1982). A model of lower extremity muscular anatomy. *J. Biomech. Eng.*, 104:304–310.

Bremermann, H. (1970). A method of unconstrained global optimization. *Math. Biosc.*, 9:1–15.

Clauser, C.E., McConville, J.T., and Young, J.M. (1969). Weight, volume and center of mass of segments of the human body. AMRL-TR-69–70, Wright Patterson Air Force Base, Ohio.

Delp, S.L., Loan, J.P., Hoy, M.G., Zajac, F.E., Topp, E.L., and Rosen, J.M. (1990). An interactive, graphics-based model of the lower extremity to study orthopaedic surgical procedures. *IEEE Trans. Biomed. Eng.*, 37:757–767.

Dempster, W.T. (1955). Space requirements of the seated operator. WADC-TR-55–159. Wright Patterson Air Force Base, Ohio.

Flash, T. (1987). The control of hand equilibrium trajectories in multi-joint arm movements. *Biol. Cybern.*, 57:257–274.

Gielen, C.C.A.M. and Houk, J.C. (1987). A model of the motor servo: incorporating nonlinear spindle receptor and muscle mechanical properties. *Biol. Cybern.*, 57:217–231.

Happee, R. (1994). Inverse dynamic optimization including muscular dynamics, a new simulation method applied to goal directed movements. *J. Biomech.*, 27:953–960.

Hasan, Z. (1983). A model of spindle afferent response to muscle stretch. *J. Neurophysiol.*, 49:989–1006.

Hatze, H. (1981). Myocybernetic control models of skeletal muscle. University of South Africa Press, Pretoria.

Hatze, H. (1997). Progression of musculoskeletal models towards large-scale cybernetic myoskeletal models. Section VII.

Hill, A.V. (1938). The heat of shortening and the dynamic constants of muscle. *Proc. R. Soc.*, (London Ser. B.) 126:136–195.

Hinrichs, R.N. (1985). Regression equations to predict segmental moments of inertia from antropometric measurements: an extension of the data of Chandler et al. (1975). *J. Biomech.*, 18:621–624.

Hoffer, J.A. and Andreasson, S. (1981). Regulation of soleus muscle stiffness in premammillary cats: areflexive and reflex components. *J. Neurophysiol.*, 45:267–285.

Högfors, C., Sigholm, G., and Herberts, P. (1987). Biomechanical model of the human shoulder—I: muscle model. *J. Biomech.*, 20:157–166.

Huxley, A.F. (1957). Muscle structure and the theories of contraction. *Prog. Biophys. Mol. Biol.*, 7:257–318.

Ingen Schenau, G.J. van (1989). From rotations to translation: constraints on multi-joint movements and the unique action of bi-articular muscles. *Hum. Mov. Sci.*, 8:301–337.

Jensen, R.H. and Davy, D.T. (1975). An investigation of muscle lines of actions about the hip: a centroid approach vs. the straight line approach. *J. Biomech.*, 8:103–110.

Johnson, G.R., Spalding, D., Nowizke, A., and Bogduk, N. (1996). Modelling the muscles of the scapula: Morphometric and coordinate data and functional implications. *J. Biomech.*, 29:1039–1051.

Karlsson, D. and Peterson, B. (1992). Towards a model for force predictions in the human shoulder. *J. Biomech.*, 25(2):189–199.

Katayama, M. and Kawato, M. (1993). Virtual trajectory and stifness ellipse during multi-joint movement predicted by neural inverse models. *Biol. Cybern.*, 69:353–362.

Klein Breteler, M., Spoor, C.M., and Van der Helm, F.C.T. (1997). Measuring muscle and joint geometry parameters for a shoulder model. *Proc. First Conference of the International Shoulder Group*. Delft, The Netherlands.

Koike, Y. and Kawato, M. (1997). Estimation of posture and mmovement from surface EMG signals using a neural network model. Section VII.

Komi, P.V. (1990). Relevance of invivo force measurements to human biomechanics. *J. Biomech.*, 23:23–34.

Laursen, B. (1997). Shoulder muscle force predictions, comparison of two models. *Proc. First Conference of the International Shoulder Group*. Delft, The Netherlands.

Ljung, L. (1987). System identification: theory for the user. Prentice-Hall, New Yersey.

Matthews, P.B.C. (1972). Mammalian muscle receptors and their central actions. Arnold, London.

Murray, W.M., Delp, S.L., and Buchanan, T.S. (1995). Variation of muscle moment arms with elbow and forearm position. *J. Biomech.*, 28(5):513–525.

Nichols, T.R. and Houk, J.C. (1976). Improvements in linearity and regulation of stiffness that results from actions of stretch reflex. *J. Neurophysiol.*, 39:119–142.

Otten, B. (1988). Concepts and models of functional architecture in skeletal muscle. *Exerc. Sport Sci. Rev.*, 89–137.

Pandy, M.G. and Zajac, F.E. (1991). Optimal muscular coordination strategies for jumping. *J. Biomech.*, 24:1–10.

Pandy, M.G., Zajac, F.E., Sime, E., and Levine, E. (1990). An optimal control model for maximum height human jumping. *J. Biomech.*, 23:1185–1198.

Pedotti, A., Krishnan, V., and Stark, L. (1978). Optimization of muscle-force sequencing in human locomotion. *Math. Biosc.*, 38:57–76.

Rozendaal, L.A. (1997). Stability of the shoulder: intrinsic muscle properties and reflexive control. PhD thesis Delft University of Technology, The Netherlands.

Schaafsma, A., Otten, B., and Van Willigen, J.D. (1991). A muscle spindle model for primary afferent firing based on a simulation of intrafusal mechanical events. *J. Neurophysiol.*, 65:1297–1312.

Soest, A.J. van, Schwab, A.L., Bobbert, M.F., and Ingen Schenau, G.J. van (1993). The influence of the bi-articularity of the gastrocnemius muscle on vertical-jumping achievement. *J. Biomech.*, 26:1–8.

Spoor, C.W., Leeuwen, J.L. van, Meskers, C.G.M., Titulaer, A.F., and Huson, A. (1990). Estimation of instantaneous moment arm of lower leg muscles. *J. Biomech.*, 23:1247–1259.

Thunissen, J. (1993). Muscle force prediction during human gait. PhD Thesis University of Twente, The Netherlands.

Van der Helm, F.C.T. (1994a). A finite element musculoskeletal model of the shoulder mechanism. *J. Biomech.*, 27:551–569.

Van der Helm, F.C.T. (1994b). Analysis of the kinematic and dynamic behavior of the shoulder mechanism. *J. Biomech.*, 27:527–550.

Van der Helm, F.C.T. and Winters, J.M. (1994). Optimizing workspace postures for a large-scale shoulder-arm system: neuro-mechanical mapping and 'field' possibilities. *Proc. 13th Southern Biomed. Eng. Conf.*, pp. 243–247. Bethesda.

Van der Helm, F.C.T. and Veenbaas, R. (1991). Modelling the mechanical effect of muscles with large attachment sites: application to the shoulder mechanism. *J. Biomech.*, 24:1151–1163.

Van der Helm, F.C.T., Veeger, H.E.J., Pronk, G.M., Van der Woude, L.H.V., and Rozendal, R.H. (1992). Geometry parameters for musculoskeletal modelling of the shoulder mechanism. *J. Biomech.*, 25:129–144.

Van Leeuwen, J.L. and Spoor, C.W. (1993). Modelling the pressure and force equilibrium in unipennate muscles with in-line tendons. *Phil. Trans. R. Soc.*, (Lond. B.) 342:321–333.

Van Zuylen, E.J., Gielen, C.C.A.M., and Denier van der Gon, J.J. (1988). Coordination and inhomogeneous activation of human arm muscles during isometric torques. *J. Neurophysiol.*, 60:1523–1548.

Veeger, H.E.J., Van der Helm, F.C.T., Van der Woude, L.H.V., Pronk, G.M., and Rozendal, R.H. (1991). Inertia and muscle contraction parameters for musculoskeltal modelling of the shoulder mechanism. *J. Biomech.*, 24(7):615–629.

Wismans, J.S.H.M. (1980). A three-dimensional mathematical model of the human knee. PhD thesis, Eindhoven University of Technology, The Netherlands.

Winters, J.M. (1995). How detailed should muscle models be to understand multi-joint movement coordination? *Hum. Mov. Sci.*, 14:401–442.

Winters, J.M. and Stark, L. (1985). Analysis of fundamental human movement patterns through the use of in-depth antagonistic muscle models. *Trans. Biomed. Eng.*, BME-32(10):826–839.

Winters, J.M. and Stark, L. (1987). Muscle models: what is gained and what is lost by varying model complexity. *Biol. Cybern.*, 55:403–420.

Winters, J.M. and Stark, L. (1988). Estimated properties of synergistic muscles involved in movements of a variety of human joints. *J. Biomech.*, 21(12):1027–1041.

Wood, J.E., Meek, S.G., and Jacobsen, S.C. (1989). Quantification of human shoulder anatomy for prosthetic arm control—I. Surface modelling. *J. Biomech.*, 22:273–292.

Yamaguchi, G.T., Moran, D.W., and Si, J. (1995). A computationally efficient method for solving the redundant problem in biomechanics. *J. Biomech.*, 28(8):999–1005.

Yeadon, M.R. and Morlock, M. (1989). The appropriate use of regression equations for the estimation of segmental inertia paramters. *J. Biomech.*, 22:683–689.

Zajac, F.E. (1989). Muscle and tendon: properties, models, scaling and application to biomechanics and motor control. *CRC Crit. Rev. Biomed. Engng.*, 17:359–419.

Zajac, F.E. and Winters, J.M. (1990). Modelling musculoskeltal movement systems: joint and body segmental dynamics, musculoskeletal actuation, and neuromuscular control. In *Multiple Muscle Systems*. Winters, J.M. and Woo, S.L.Y. (eds). Springer-Verlag, New York.

33
Progression of Musculoskeletal Models Toward Large-Scale Cybernetic Myoskeletal Models

Herbert Hatze

1 Introduction

In recent years, it has been recognized by many researchers (e.g., Lehman 1990) that the study of oversimplified models of both the skeletal and the muscular subsystem may produce totally unrealistic and even fundamentally erroneous results (see also Chapter 7). This has prompted some investigators to extend and refine musculoskeletal models toward more complex and biologically realistic ones. A discussion of the problems associated with the development of controllable large-scale musculoskeletal models, the main features of the latter, and some computer simulation results will be the main topic of the present perspective. It will be organized as follows: Subsection 1.1 gives a general definition and brief description of types of models, followed by a discussion of model complexity in Subsection 1.2. Section 2 is devoted to a description of the procedures typically applied to large-scale myoskeletal modeling and demonstrated by the development of the specific three-dimensional large-scale cybernetic myoskeletal model whose figural appearance is shown in Figure 33.1. Section 3 presents some simulation responses of this model relating to neural control sensitivity phenomena, while Section 4 contains anticipated future directions with regard to the evolution of large-scale neuromyoskeletal models.

1.1 General Definition and Types of Models

If we accept the *general definition of a model* as being the "*abstract representation of selected at-*

tributes of a real object, event, or process" (Hatze 1981a), then humankind has excelled in building models since the stone ages. Each picture of a hunting scene painted on the wall of a cave by prehistoric people is a (graphical) model representation of a real event. At a later stage, humans discovered that models of continuous processes could be quite useful in forecasting future events. A typical such example is Newton's third law of motion. It allows one, for instance, to compute the position $z(t)$ of a freely falling stone (air resistance neglected) at each instant, t, of time by simply double integrating the second order ordinary differential equation $\ddot{z} = -g$, subject to the initial conditions $\dot{z}(t_o) = \dot{z}_o$, $z(t_o) = z_o$. The first type of purely empirical prediction models is called *inductive* and is essentially a black box with empirical input–output relations, while the second type (e.g., Newton's law) is of general validity since it has been generated by *deductive* reasoning. (Other categorizations of models are, of course, also possible—see, for instance, Jacoby and Kowalik 1980). From this it follows and is, in fact, a *maxim in modeling that deduction should always be carried as far as possible*. It ensures minimization of uncertainty in the modeling procedure and the incorporation into the model structure of *as many known properties* of the modeled system as possible. These fundamentals of modeling techniques apply in particular to the modeling of the human musculoskeletal system, which is defined to be a "*skeletal assemblage whose elements are acted upon by myo-(muscular) actuators*" (Hatze 1980a).

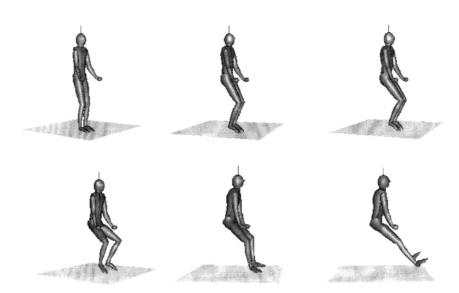

FIGURE 33.1. Six configurations (left to right, top to bottom) of a kinematic animation sequence showing the figural appearance of a typical large-scale 3-dimensional, 17-segment, 42 degrees-of-freedom, 240 muscle myoskeletal model, suspended from its head and executing a double-legged kicking motion.

1.2 Complexity of Musculoskeletal Models: The Importance of Biological Realism

Complexity is currently a hotly debated issue and the reader is referred to the literature for an overview of musculoskeletal models of varying structure and an exhaustive discussion on purpose-dependent model complexity (e.g., Hatze 1980a; Lehman 1990; Loeb and Levine 1990; Yamaguchi 1990; Zajac and Winters 1990). It should be mentioned at this point that model complexity is not a well-defined term. Rosen (1988), for instance, expresses an extremely radical view by defining a complex system as one that "possesses a potentially infinite number of descriptions" and cannot be simulated at all. He quotes the central nervous system as an example. We shall adopt a less restrictive view and *define a complex musculoskeletal model as one that incorporates in its structure all macrobiologically relevant features of the biosystem, and whose simulation responses deviate from the corresponding behavior of the real (natural) system only by a prescribed margin, for all conceivable modes of operation.* This automatically implies that such a system model is also a *general-purpose* model.

As is evident from the above definition, the topic of model complexity is closely related to that of biological realism, especially with respect to muscle models and neuronal networks. An illustrative example in this regard is the investigation of Lehman (1990) into the relationship between musculoskeletal model input and muscle model complexity. Lehman studied flexion–extension movements of the wrist using a single-joint skeletal model driven by one equivalent flexor and one equivalent extensor muscle. Three different types of muscle models were employed and the active state for each equivalent muscle was the model input. The wrist angle was the output function. The latter trajectory was also observed experimentally together with the surface EMGs of both flexor and extensor muscles. Lehman then solved the inverse problem for each of the three muscle models to determine the respective active state input functions for the observed wrist angle trajectory. He found *profound qualitative and quantitative* differences between the respective inferred active state inputs (Lehman 1990, Figure 6.1), concluding that "as (muscle) model complexity increases, so does the similarity between the inferred active state and electromyograms." Although active state and EMG are only indirectly related, this example nevertheless convincingly demonstrates that even for a comparatively simple one-degree-of-freedom model containing two myoactuators, *the choice of muscle model complexity is of crucial importance*, with the linearized Hill muscle model producing the worst and totally unrealistic results.

The above remarks and the example cited lead to the formulation of the following *procedures governing the creation of general purpose (complex) musculoskeletal models* (Hatze 1980; 1981b):

1. *Identification* of the human musculoskeletal assemblage as an interlinked multibody system acted upon by various internal active and passive force generators (skeletal muscles, ligaments, and so on) and by external forces, and consisting of nonrigid segments that are connected by joints, each of which has basically six degrees of freedom. *Utilization* of deductive reasoning (laws of mechanics) to establish the functional relationships that govern the dynamical behavior of the system.

2. Collection and utilization of as much *reliable a priori information* as possible on the *biological structure* and the *functional behavior* of the individual system components (skeletal assemblage, skeletal muscle).

3. *Creation of submodels* of the various system components that are *as complex as feasible*, using deductive reasoning wherever applicable, and accounting fully for all nonlinearities inherent in the substructures. It is always possible to reduce model complexity if so required but the reverse process is much harder to accomplish. Reasonable simplifications should be introduced at this stage, avoiding all modifications of biological basic properties.

4. *Development of the intended musculoskeletal model structure* by appropriately combining the skeletal (executor) and muscular (myoactuator) submodels to a single, general-purpose system model.

5. *Validation* of the complete model by establishing its ability to predict system responses that agree, within specified limits of accuracy, with the corresponding responses as observed on the real biosystem (Hatze 1981b). Modern methods for large-scale simulation model validation are readily available (Murray-Smith 1995) and establish the credibility of the model.

The rationale behind the creation of this type of model is the fact that the interactions between the skeletal and muscular components of the whole system are so complicated that the influence on the functional behavior of the whole model of unrealistic simplifications and undue linearizations of the model substructures is largely unpredictable. This has been strikingly demonstrated by the investigations of Lehman cited above.

2 Large-Scale Myoskeletal Modeling

The large-scale property of a myoskeletal model is determined by the morphology of the skeletomechanical submodel and by the dimensions of the skeletomechanical and myoactuator state space. Based on the dimensions of existing models, we shall *define a myoskeletal model to be large-scale if its skeletal morphology is three-dimensional, it comprises at least 15 segments, its skeletomechanical state space dimension exceeds a value of 76 (38 configurational coordinates plus 38 first derivatives) and that of the myoactuator state space a value of 92.*

We shall not indulge here in detailed descriptions of the numerous simple planar musculo-skeletal models that are currently in use, as such desriptions can be found in various chapters of Winters and Woo (1990). Rather, we shall concentrate on some important aspects relating to the development of complex spatial, large-scale cybernetic myoskeletal models.

2.1 Selection of Hominoid Morphology

A *hominoid* is an anthropomorphic mathematico-geometrical model of the segmented human body (Hatze 1980a). Its morphology (i.e., the number and shapes of its segments, and the types of the joints), defines the number, f, of skeletomechanical degrees of freedom of the (unconstrained) model. The first obstacle encountered in considering body segmentation is the definition of *sharp intersegmental boundaries* which in the real biosystem do not exist but are fuzzy and change during segment motion. It is, however, possible to circumvent these difficulties by appropriately defining the intersegmental boundaries as surfaces generated by tracing out the average layers of the fuzzy sets of the respective boundary particles (Hatze 1980b).

The next problem facing the modeler is the *non-rigidity of the body segments*. It is well known that parts of segments (muscles, connective tissue, organs, and so on) execute movements relative to the skeleton. In addition, other factors such as breathing, nonstationary joint axes, changes in the distribution of body fluids, etc. also contribute to the nonrigidity problem. There have been attempts to account for some of these effects by attaching so-called wobbling masses to rigid-rod segment models. Gruber (1987), for instance, used such a three-link wobbling-mass body model to predict reaction forces and moments in the knee and the hip during vertical-jump landings. She found tremendous differences between the predictions of the wobbling-mass and the rigid-segment model, amounting in some cases to a few thousand percent! The hip moment, for example, would show a positive peak of 1300 Nm during the first 0.01 s of the motion for the rigid-segment model, while the corresponding value for the wobbling-mass model was virtually zero (Gruber 1987, Figure 26d). Even more incredible were the predictions of the horizontal knee reaction forces: A positive peak of about 1000 N during the first 0.01 s for the rigid-segment model and a sign reversal to a negative peak of about −230 N for the wobbling-mass model (Gruber 1987, Figure 26a). No attempt was made to explain these excessive discrepancies or to validate the models in any way. Instead, the wobbling-mass model predictions were accepted as realistic and correct. Although this study again demonstrates the limitations and dangers associated with the use of oversimplified models (the 3-link model of Gruber contained no feet, among other shortcomings), an effort has at least been made to somehow incorporate into the model the soft-tissue behavior of segmental substructures. However, such endeavors are doomed to failure from the outset, for the following reason. For most segments, the soft-tissue substructures consist predominantly of muscles which change their consistency during action from very soft in the relaxed state to almost rigid in the completely contracted state. In the course of a motion, the various muscles attached to a specific segment may change their contractive state, and hence their consistency to anything between these two extremes. In addition, they also act to a certain extent on other soft tissue structures such as blood vessels and connective tissue. All of these effects combine in such a way as to continuously vary the consistency of the segmental soft-tissue substructures during motion, in a virtually unpredictable manner. Moreover, the maximal excursions relative to the skeleton of the soft-tissue structures are comparatively small, even in the relaxed muscle state. Taking all of these arguments into account, the *rigid-body approximation* for segments *appears justifiable* (see also Zajac and Winters 1990, p.125), especially since the shape fluctuations, exomorphic differences, varying density distributions, and asymmetries of individual segments are all accounted for (Hatze 1980b).

Although there is little dispute about the fact that head, upper arms, forearms, upper legs, lower legs, and pelvis constitute, to some extent, segmental units (once the rigidity assumption has been accepted), the situation is not so clear cut for segments like the neck, shoulder, thorax, abdomen, hands, and feet. Indeed, these body parts can not really be called segments, since they consist of highly intricate bony structures connected to various types of soft tissue. Consider, for example, the thorax. The bony structure consists of the spine and the ribs (the shoulders are regarded separate segments), while the inner organs, muscles, connective tissue, skin, and so on, constitute the soft tissue components. A reasonable division of the thorax into subsegments is virtually impossible. Even if each of the thoracic vertebrae would be considered a disk-shaped subsegment, there would be no corresponding soft-tissue structures that could be regarded as belonging to that disk. Yet the thorax is obviously a flexible skeletal structure that can bend and rotate in all directions. Here, the modeller is in a real predicament: He can either attempt to create a complex and fairly realistic thorax model comprising some 340 interconnected hard- and soft-tissue subsegments and face the gigantic task of combining these with the remaining segments to form the complete skeletal subsystem, or he can consider the abdominothoracic complex as a single (rigid) segment, assuming that all rotational degrees of freedom of this pseudo-segment reside in a (pseudo)ball-and-socket joint located in the abdomino-pelvic region. For simulating gross body motions not involving internal thorax motions the single-segment thorax model will suffice for most purposes, while for detailed investigations into the responses of the spine to stresses during lifting tasks, the complex thorax model would be appropriate. Similar arguments apply to the segmental modeling of the neck and the abdomen.

A somewhat different approach is required for the hands and the feet. These are terminal segments especially designed for manipulating, respectively contacting, the environment. They are also special in the sense that their bony content is comparatively large and of a rather complicated structure. Because hands and feet are the segments predominantly responsible for the mechanical communication of the hominoid with its surroundings, they are the ones who most frequently experience the environmental contact forces. We shall have to say more about this later. For the moment we shall postulate that these segments can be modeled as single units as far as the segmental dynamics is concerned, but can subsequently be extended to more complex structures without having to modify the dynamical equations of motion.

Based on the above considerations we are lead to conclude that *a skeletomechanical model comprising 17 segments* (head–neck, abdominothoracic, abdominopelvic, two shoulders, upper arms, forearms, hands, thighs, lower legs, and feet) might be appropriate for simulating a large class of gross body motions. This model is certainly large-scale (see also below) but only semicomplex, for reasons pointed out above.

2.2 Selection of Dimensionality: From Planar to Spatial Models

It would be hard to dismiss the argument that spatial (3D) hominoids are more realistic than planar (2D) ones. In fact, there exists not a single human motion that is really planar but only motions which may be considered to possess *predominantly* planar components. Why then are the overwhelming majority of musculoskeletal models in use today of the planar type and therefore unrealistic? A partial answer is to be found in the fact that, until fairly recently, the skeletodynamical (differential) equations of motion were much easier to derive for 2D-hominoids than for 3D models. This has changed with the advent of programs (such as NEWEUL, AUTOLEV, or SYMBA, see Yamaguchi 1990) for

the automatic generation of the equations of motion by symbolic manipulation. There are, however, other obstacles. To characterize the inertial properties of a segment, only three quantities are required for planar models (mass, moment of inertia, distance of center of mass location), while the corresponding number for spatial models is at least seven (mass, three principal moments of inertia, three coordinates for center of mass location) and may increase if there are nonzero products of inertia (segment axes not principal), or if the segment is not symmetrical (deviation angles between original and principal axes, see Hatze 1980b). In addition, the interpretation of the various passive and active torques generated at a specific joint in relation to the configurational joint coordinates, as well as the modeling of the muscle architecture becomes considerably more difficult for 3D hominoids. These are, however, problems that can be overcome as will be demonstrated later. Thus, realistic *spatial* skeletomechanical models exhibiting a semicomplex morphological structure are the choice of today, although the transition from planar to spatial structures is not trivial.

2.3 Definition of Spatial and Segmental Axis Systems and of Skeletomechanical Configuration Coordinates

Once a decision on the morphology of the hominoid and its dimensionality has been made, generalized coordinates have to be defined that unambiguously determine the configuration of the skeletomechanical model at any point in time. In order to do this, a Cartesian axis system has to be fixed in space, and *local Cartesian axis systems* must be attached to the model segments. This is not a trivial task because of the irregular segment shapes. In the present context we shall not enter into extensive discussions of this issue but merely note that for technical reasons, it is most advantageous to fix the origins of the segmental axes at the segmental mass centroids, and to orientate the axes such that they constitute principal axes of inertia (Hatze 1983).

The next step is the definition of the *skeletomechanical configuration coordinates*. Their number and type depends on the geometry of model joints that connect the model segments. It has already been pointed out in Subsection 1.2 that each human joint has, in principle, six degrees of freedom (three rotational and three translational). In many cases, however, the excursions in the direction of some of these six degrees are minute so that they may be ignored for most practical purposes, thereby effectively reducing the number of degrees of freedom to be considered. The problem is now to decide whether or not a certain articular degree of freedom is to be regarded as irrelevant. Consider, for instance, the knee joint which, at first sight, would appear to be a simple hinge joint. The truth is that it possesses a nonstationary axis of rotation that changes its position and orientation as a function of both, the angular excursion and the joint load. However, these changes only slightly affect the dynamics of the postulated hinge joint motion of the lower leg but do have a significant influence on the length- and moment-arm functions of the active and passive torque-generating structures spanning the knee joint. Hence the extremely complicated kinematic structure of the natural joints does not necessarily have to be incorporated in all detail in the inertial part of the equations of motion but must be taken into account in the part containing the kinematics of the driving torques of the system. On these premises, it turns out that the configuration of the 17-segment hominoid introduced in Subsection 2.1 and displayed in Figure 33.2 can be described by 42 generalized coordinates q_i, of which 3 are linear (q_1, q_2, q_3) and 39 angular. The skeletomechanical state space dimension of this model is therefore 84, i.e., it is large-scale.

2.4   Hominoid Dynamics: Formulation of Skeletomechanical Equations of Motion

In Subsection 1.2 it has already been mentioned that, with the restrictions discussed above, the human skeletal system may be regarded as an interlinked multibody system and that deductive reasoning should be applied whenever appropriate. This implies that formalisms for the generation of equations of motion for 3D multibody systems may be used for establishing the hominoid dynamics description. Such formalisms may be based on Newton's method, Langrange's equations, Kane's method, or D'Alembert's principle. We shall use

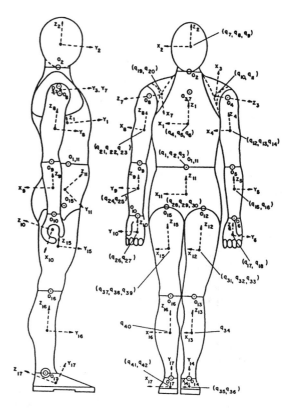

FIGURE 33.2. Lateral and anterior view of 17-segment hominoid. The local segment coordinate systems and configurational coordinates $q_1,...,q_{42}$ are also shown. (Adapted from Hatze 1983)

the latter for the present 17-segment hominoid, in the form suggested by Wittenburg (1977).

Let the vector of configurational coordinates be defined by $\mathbf{q}:=(q_1,q_2,...,q_{42})^T$ and the corresponding generalized velocities by $\dot{\mathbf{q}}:=\boldsymbol{\chi}=(\chi_1,\chi_2,...,\chi_{42})^T$. It can then be shown (Hatze 1981c) that the 84-dimensional first-order differential system governing the motion of the skeletomechanical subsystem is given in vector form by

$$\dot{\mathbf{q}} = \boldsymbol{\chi}, \qquad\qquad\qquad \mathbf{q}(t_o) = \mathbf{q_o},$$
$$\dot{\boldsymbol{\chi}} = \mathbf{A}^{-1}(\mathbf{B} + \mathbf{Q}^M(\boldsymbol{\mu}) + \mathbf{Q}^L), \ \boldsymbol{\chi}(t_o) = \boldsymbol{\chi_o}, \quad (33.1)$$

where $\mathbf{A}$ denotes the (state-dependent) inertia matrix and $\mathbf{B}$ is given by

$$\mathbf{B} = \mathbf{B}(\mathbf{p},\mathbf{F},\bullet), \qquad\qquad (33.2)$$

with $\mathbf{p}$ denoting the 42×17 matrix of angular velocity base vectors, $\mathbf{F}$ the matrix of external force vectors acting at the respective segments, and the dot indicates other variables not relevant to the present discussion. Of special interest are the matrices $\mathbf{F}$ of external force vectors, $\mathbf{Q}^L$ of internal passive articular moments, and $\mathbf{Q}^M(\boldsymbol{\mu})$ of muscle moments. We shall discuss each of these quantities separately.

2.5   External Forces Acting on Hominoid Segments: The Special Case of Environmental Impact Forces

The present model permits the definition of three externally active force vectors (air and water resistance, environmental contact forces) together with their points of application, for each of the 17 model segments. Difficulties arise, however, if these vectors are not known a priori, as is the case with environmental impact forces, whose magnitudes and directions are the result of the dynamical interaction of the hominoid with its surroundings. In most instances, the terminal segments hands and feet are involved. In this case, two different approaches are possible: Either instantaneous changes in the generalized velocities occuring upon impact are computed followed by the calculation of the constraint forces as described in Hatze and Venter (1981), or all potential environmental contact points on the terminal segments are modeled as nonlinear viscoelastic structures as suggested by Hatze (1980a, p. 378). The first technique mentioned has been implemented in practice (Hatze, 1981d), is the computationally more efficient one but is physiologically unrealistic. The second method is biologically realistic but requires extremely complex models of the terminal segments. It does, however, have the added advantage of preserving the state space dimension without sacrificing the availability of information on the impact and constraint force components. Without going into details, we shall only mention here that work in this direction is in progress, utilizing the fact that motions of skeletal substructures (such as bones) of terminal segments relative to the segments' inertial coordinate systems may be regarded as dynamically irrelevant and therefore be treated as kinematic phenomena. There is no doubt that this sort of microstructural modeling of terminal segments will be the approach of the future and will substantially contribute to the large-scale status of complex myoskeletal models. Existing models of this kind are still in their infancy (see, e.g., Scott and Winter 1993) so that this modeling discipline represents a tough challenge to biomechanists.

2.6   Passive Joint Torques and Articular Boundaries

The next quantities to be discussed are the passive joint torques $\mathbf{Q}^L$ appearing in Eq. 33.1. These torques provide the natural restrictions for the range of joint motion and originate from passive structures like ligaments, cartilage, joint capsules, connective tissue, and the parallel elasticities of uni- and biarticular muscles that either span the joint in question or are located within that joint. With the exception of the passive torques generated by biarticular muscles, which are treated separately, it is, in general, extremely difficult if not impossible to separate the contributions of the individual structural elements to the total externally observable torque. Some modeling aspects relating to passive moments occuring across *one-dimensional* (hinge) joints have been discussed by Zajac and Winters (1990 p. 127). However, the major interest today is in multidimensional torque models. Such a comprehensive, three-dimensional, multivariate model of passive human joint torques together with experimental and computational methods for the identification of its parameters has recently been developed and published by the author (Hatze 1997). It is the first model of its kind that is applicable to joints with three (or fewer) angular degrees of freedom and fully accounts for all (nonlinear) interrelationships that exist between the individual moment functions. In addition, it permits the direct determination of the articular boundaries of the joint and is therefore especially suited for the inclusion in large-scale myoskeletal models such as the present 17-segment hominoid. As an example, Figure 33.3 depicts one of the two passive torque functions and the articular boundaries of a specific human elbow joint.

2.7   Modeling the Myodynamics and Myocybernetics of Skeletal Muscle

The last quantity appearing in Eq. 33.1 and not yet discussed is the matrix $\mathbf{Q}^M(\boldsymbol{\mu})$ of muscle moments which are functions of the myostate vector $\boldsymbol{\mu}$. Very detailed expositions of all facets of contemporary skeletal muscle modeling can be found in a num-

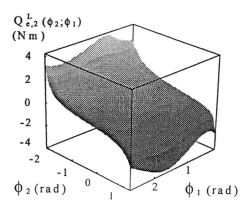

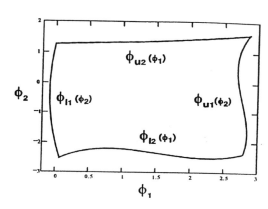

FIGURE 33.3. Passive elastic torque function $Q_{e,2}^L$ $(\phi_2;\phi_1)$ of a human elbow joint for rotations ϕ_2 about the longitudinal forearm axis, for a set of fixed flexion angles ϕ_1 (left), and articular boundary functions of the same joint (right). Viscous effects were not considered here.

ber of chapters of Winters and Woo (1990) and need therefore not be repeated here. Of special importance in the present context is the contribution of Lehman (1990), already mentioned in Section 1, in which the critical dependence of musculoskeletal model predictions on muscle model complexity was clearly demonstrated. In the present subsection we shall therefore restrict our attention to the description of the salient features of the complex myocybernetic control model of skeletal muscle used in the present large-scale 17-segment myoskeletal model. The basic features of the model are described in detail in Hatze (1977, 1978, 1980c, 1981b, and 1990). In constructing the model, the following principles were observed: (1) The model should be as biologically realistic as feasible and account for all nonlinearities and behavioral peculiarities inherent in the myostructures; (2) It should be capable of predicting myocontractive phenomena that were not used in the construction of the model; (3) It should adequately mimic the discrete motor unit structure of skeletal muscle since numerous myodynamic properties are recruitment-rankorder-dependent; and (4) It should, first and foremost, contain *control parameters that reflect the actual neural controls motor unit recruitment and individual motor unit firing rates, especially with a view to possible extensions of large-scale myoskeletal to complex neuromyoskeletal models.* Because of the latter feature which emphasizes the suitability of the model for neural control applications, it has been termed

a "myocybernetic control model" (see also Hatze 1981b).

A certain muscle to be modeled is regarded as consisting of *9 model motor units*, each having its own specific excitation dynamics, size, and contractile characteristics. The common contraction dynamics is determined by the length change of the whole muscle contractile element. The *single muscle myostate* μ' comprises the normalized motor unit population n, the 9 normalized motor unit calcium ion concentrations γ_α, $\alpha=1,...,9$, the normalized contractile element length ξ, and the normalized stretch potentiator ζ. (If fatiguing effects are to be included, the additional state variable κ and its differential equation is required).

Hence $\mu' := (n,\gamma_1,\gamma_2,...,\gamma_9,\xi,\zeta)^T$ and the 12 *single-muscle myostate equations* are given by

$$\dot{n} = \hat{n}z(t), \qquad\qquad n(0) = n_o, \quad (33.3)$$
$$\dot{\gamma}_\alpha = m_\alpha[w_\alpha(n)v_\alpha(t) - \gamma_\alpha], \ \alpha = 1,...,9,$$
$$\qquad\qquad\qquad\qquad \gamma_\alpha(0) = \gamma_{o\alpha},$$
$$\dot{\xi} = f_\xi(n,\gamma_\alpha,\xi,\zeta,F^{SE}(\xi,q)) \qquad \xi(0) = \xi_o,$$
$$\dot{\zeta} = f_\zeta(n,\gamma_\alpha,\xi,\zeta,F^{SE}(\xi,q),\dot{\xi},t) \quad \zeta(0) = 1,$$

where $z(t)$ and $v_\alpha(t)$ are the normalized recruitment and motor unit stimulation rates respectively, $w_\alpha(n)$ are switching functions, and $f_\xi(\bullet)$ and $f_\zeta(\bullet)$ are highly nonlinear functions of their respective arguments. The remaining symbols denote constants. The first of these 12 differential equations describes the *recruitment* dynamics, the 9 subsequent ones the *excitation dynamics*, and the two last ones the *contraction dynamics* of the muscle.

The *normalized muscle output force* across the series elastic element is then given by

$$F^{SE}/\overline{F} = \frac{\left[\exp\left(\frac{\sigma}{\alpha\lambda_{To}}\left\{\ell - (\overline{\lambda}^2\xi^2 - d^2)^{\frac{1}{2}} - \lambda_{To}\right\}\right) - 1\right].\sin\overline{\Theta}}{[e^\sigma - 1][1 - (d/\overline{\lambda}\xi)^2]^{\frac{1}{2}}}$$

(33.4)

where $\overline{F}$ is the maximum isometric force, ℓ the (generally curved-path) length of the muscle and hence a function of skeletomechanical coordinates q_i, and the remaining symbols denote constants. (For fusiform muscles, $d \equiv 0$ and $\overline{\Theta} \equiv \pi/2$). For biarticular muscles, the passive parallel elastic force F^{PE} must be added to F^{SE} since it has not yet been accounted for, as was the case with uniarticular passive muscle forces via the passive joint torques $\mathbf{Q}^L$. The muscle moments $\mathbf{Q}^M(\boldsymbol{\mu})$ appearing in (33.1) are obtained by cross-multiplying the respective moment arm vectors with the corresponding active and passive muscle force vectors, and scalar-multiplying the result by the matrix $\mathbf{p}$ of angular velocity base vectors. Curved paths of muscle action centroids are fully accounted for. Methods for the *identification of most model parameters* appearing in the above equations are presented in Hatze (1981b and 1981e), while validation procedures for this model can be found in Hatze (1980c and 1990). It should be emphasized that the present (fairly complex) myocybernetic muscle model captures all known macrobiological morphological and functional major characteristics of skeletal muscle and tendon, such as the exponentially growing motor unit sizes, their rank-ordered and time-delayed recruitment, their varying contraction times, the changing properties of their force-velocity relations including the lengthening mode; the history-dependence of stretch-potentiation, the length-dependence of the contractile force, the nonlinear stress-strain relation of tendinous tissue, and so on. If a muscle model does not incorporate these characteristics, and is therefore not capable of correctly reproducing the contractile phenomena, its inclusion into a comprehensive large-scale myoskeletal model will necessarily lead to incorrect simulation responses.

To complete the description of the whole myoac-

tuator subsystem model, we need to combine the differential equations of all M (maximally 240) muscles that are included in the model. To this end, we define the *myostate* $\boldsymbol{\mu}$ *of the complete myoactuator subsystem* consisting of M model muscles in terms of the single-muscle myostates as $\boldsymbol{\mu} := (\boldsymbol{\mu}'_1, \boldsymbol{\mu}'_2, ..., \boldsymbol{\mu}'_M)^T$. The *complete myodynamics* of the system may then be concisely represented as (see also Hatze, 1980a)

$$\dot{\boldsymbol{\mu}} = g(\mathbf{q}, \boldsymbol{\mu}, \mathbf{v}(t), \mathbf{z}(t)), \qquad \boldsymbol{\mu}(0) = \boldsymbol{\mu}_o \quad (33.5)$$

where $\mathbf{v}(t) := (v_{a1}(t), ..., v_{aM}(t))^T$, $\alpha = 1, ..., 9$, and $\mathbf{z}(t) := (z_1(t), ..., z_M(t))^T$. Equations (33.1) and (33.5) are coupled via the state vectors $\boldsymbol{\mu}$ and $\mathbf{q}$, and together constitute the complete dynamical description of the present cybernetic large-scale myoskeletal model *with open-loop control input vector* $(\mathbf{v}(t), \mathbf{z}(t))^T$. The dimension of its skeletomechanical state vector is 84, that of its myostate vector has a value of (maximally, but without the fatigue variable) 2880 corresponding to 240 muscles, and that of its control vector a value of (maximally) 2400. Truly a large-scale myoskeletal model.

3 Computer Simulation of Large-Scale Myoskeletal Models

The computer simulation of large-scale systems is not a trivial task. Because of space limitations it is, however, not possible to discuss the numerous problems and pitfalls associated with the practical implementation and execution of the computerized version of the present and similar large-scale myoskeletal models. Instead, we shall proceed to the presentation of some remarkable simulation results that are pertinent to the subsequent discussion.

3.1 Myoskeletal Model Responses: Insensitivity of Specific Skeletal Motions to Random Perturbations of Neural Control Inputs

As a practical application, the present large-scale model was used to shed some light on how, and to what extent, random (and deterministic) perturbations of control inputs affect the resulting motion of the system. The computerized version

of the combined differential system (33.1) and
(33.5) as well as all auxiliary algorithms required
for its integration was used in the form of the
BIOMLIB® computer program HOMYOS to perform the simulation. Eight leg muscle groups in
the model were activated, four of which are biarticular: m. iliopsoas (left and right), m. rectus
femoris (l. and r.), vasti group (l. and r.), and the
hamstring group (l. and r.). The segmental, articular, myodynamic, and morphometric input parameter values were determined from a healthy, 23-
year old male subject. Where necessary, data for
the left and right segments and muscles were
equalized to guarantee left-right symmetry of the
motion, and a mass of 10 kg was attached to each
foot in order to slow down the otherwise too rapid
kicking movement. The hominoid was suspended
from its head and the pelvic segment was constrained in all directions by appropriately dimen-
sioned visco-elastic external forces, as shown in

Figure 33.1. The unperturbed (nominal) neural
control input functions were of bang-bang form
and chosen such as to produce the symmetric double-legged kicking motion of the vertically suspended hominoid depicted in Figure 33.1. Simulation time was 0.5 s.

Gaussian random and deterministic constant
perturbations of neural interspike intervals $\tau_{\alpha j}$,
$\alpha = 1,...,9$ and recruitment rates z_j of the j-th muscle were used to produce deviations from the nominal (unperturbed) controls. The measure of perturbation used for all variables y is the *normalized
variance* $\Delta^2 = \sigma^2/\sigma_\omega^2 = T^{-1} \int_0^T [y(t) - y^o(t)]^2 dt/\sigma_\omega^2$,
over the simulation time interval T; $y^o(t)$ is the
nominal value of the variable in question and σ_ω^2
the corresponding reference variance of chaotic excursions (Hatze 1995).

The results of this simulation were remarkable
indeed and are partly summarized in Figure 33.4.

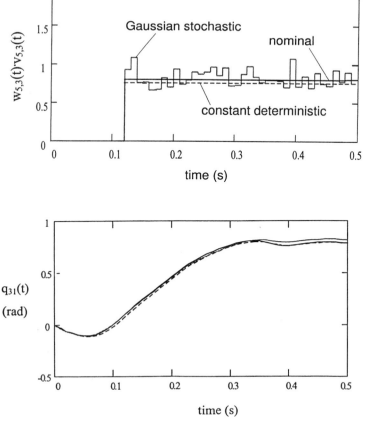

FIGURE 33.4. (Top) Nominal, Gaussian
stochastic, and constant deterministic
control function $w_{5,3}(t) \cdot v_{5,3}(t)$ of the
5th motor unit in muscle 3 as a representative example of the remaining 71
controls. (Bottom) Left hip angle $q_{31}(t)$
(flexion–extension) corresponding to
the above controls. The second and
third hip angles and the knee angle are
not shown.

(A condensed account of this study is given in Hatze (1996) while a detailed account will appear elsewhere.)

As can be seen from Figure 33.4, *large perturbations of the neural control inputs result in comparatively small variations of the corresponding coordinate trajectories*, with standard deviation ratios for random perturbations ranging from 7.03 to as much as 23.87. It should, however, be emphasized that these results pertain to this particular kicking motion only, although they possess some generality.

The observed hyposensitivity of skeletal motion to neural control perturbations has its origin partly in the muscular and partly in the skeletal subsystem. Among the muscular causes of this phenomenon are the comparatively sluggish reaction of the intramuscular calcium ion concentration to variations of the stimulation rate, the nonlinearity of the active-state function, and the influence of recruitment rank-order dependent motor unit properties. A further reason for the occurance of the hyposensitivity phenomenon is, of cource, the smoothing action of the inertia of the skeletomechanical system components which act like a low-pass filter.

The *most important implications* of these findings are the facts that comparatively chaotic neural controls may produce well coordinated motions of the skeletal subsystem and that the neuromusculoskeletal inverse dynamics problem is highly ill conditioned and does, in fact, belong to the class of so-called physically (and mathematically) incorrectly posed problems which do not, by definition, possess unique solutions. In practice this means that, in general, the reconstruction of neural control inputs that correspond to an observed motion is not possible with any reasonable confidence. A similar argument holds true also for the inference from motion data of muscular joint torques. These statements are strongly supported by the frequently made experimental observation of considerably varying EMG records that correspond to repetitiions of one and the same stereotyped motion. Also, DeLuca and Erim (1994, Figure 1B) have demonstrated that mean firing rates of motor units in the isometrically contracting tibialis anterior muscle may vary simultaneously by as much as 14% without significantly affecting the muscle force level.

4   Future Directions: The Evolution of Large-Scale Neuromyoskeletal Models

Virtually all complex musculoskeletal models in use today are open-loop control models, as far as the neural inputs are concerned. The present cybernetic myoskeletal model is no exception. This implies that the input signals to the model muscles are either provided by the user, or are determined by some algorithm, such as an optimization procedure. This is acceptable for myoskeletal models but would be unrealistic for models claiming to be *neuro*myoskeletal ones. It is no secret that the motor units of the muscular system are not directly controlled by the central nervous system (CNS) but via a complex, hierarchical network of intricate neuronal interconnections, including numerous sensory feedback loops. Indeed, as is well known, the muscle itself contains several sensory organs (two types of muscle spindles, Golgi tendon organs) that provide afferent information which is utilized in the control circuitry (e.g., see Chapter 2 and Chapter 32). Additional proprioceptive information is provided by receptors embedded in ligaments, joints, and articulating membranes. As Zajac and Winters (1990, p. 131) have pointed out, modeling efforts directed at peripheral neuromotor circuitry are scarce, mainly because of the uncertainties associated with the immense complexity of the neuronal interconnections existing already at this (comparatively low) level of neuronal organisation. There are, nevertheless, some attempts to model the dynamics of sensory organs such as the muscle spindles (for a brief review, see Chapter 8, Zajac and Winters 1990, pp. 131–134 and Loeb and Levine 1990), or even direct muscular control (e.g. Abbas and Chizeck, 1995).

There is, however, no question that despite the almost prohibitive difficulties, the modeling (and understanding) of the neural control of skeletal motion will become the main challenge in this field in the future. There is also little doubt that such efforts will be greatly facilitated by the availability of biologically realistic *cybernetic* models of skeletal muscle that contain (a reasonable number of) model motor units which can be recruited and which, in principle, can all be controlled individually by separate stimulation rates emanating from *model motoneurons* situated in the respective *model motoneuron pool*. In fact, the "common-drive phenomenon" of motoneuron pool function, as described by De Luca and Erim (1994) with its potential of dramatically reducing neural control complexity, can be modeled only by employing a multi-motor-unit type muscle model. It is my firm belief that, in conjunction with the skeletal muscle's *proprioceptive feedback mechanisms*, the modeling of the *common- drive phenomenon of motoneuron pool function* is the next step to be taken in the direction of creating comprehensive *neuromyoskeletal* models.

To conclude, it should be mentioned in this context that the *overriding principle governing the execution of any movement task is, in all likelihood, the teleology of the process, that is, the tendency of the neuromyoskeletal system to act goal-directed, thereby optimizing certain task-specific performance criteria* (see also Chapter 35).

References

Abbas, J.J. and Chizeck, H.J. (1995). Neural network control of functional neuromuscular stimulation systems: Computer simulation studies. *IEEE Trans. Biomed. Eng.*, 42:1117–1127.

De Luca, C.J. and Erim, Z. (1994). Common drive of motor units in regulation of muscle force. *Trends Neurosci.*, 17:299–305.

Gruber, K. (1987). Entwicklung eines Modells zur Berechnung der Kräfte im Knie- und Hüftgelenk bei sportlichen Bewegungsabläufen mit hohen Beschleunigungen. Ph.D.-dissertation, Laboratory of Biomechanics, ETH Zürich.

Hatze, H. (1977). A myocybernetic control model of skeletal muscle. *Biol. Cybern.*, 25:103–119.

Hatze, H. (1978). A general myocybernetic control model of skeletal muscle. *Biol. Cybern.*, 28:143–157.

Hatze, H. (1980a). Neuromusculoskeletal control systems modeling—a critical survey of recent developments. *IEEE Trans. Auto. Control*, AC-25:375–385.

Hatze, H. (1980b). A mathematical model for the computational determination of parameter values of anthropomorphic segments. *J. Biomech.*, 13:833–843.

Hatze, H. (1980c). Discrete approximation of a continuous myocybernetic model of skeletal muscle. Special Report SWISK 18, CSIR, Pretoria.

Hatze, H. (1981a). *Mechanische Prinzipien von Bewegungsabläufen.* University of Vienna, Vienna.

Hatze, H. (1981b). *Myocybernetic Control Models of Skeletal Muscle.* University of South Africa Press, Pretoria.

Hatze, H. (1981c). HOMSIM: a simulator of three-dimensional hominoid dynamics. Special Report SWISK 23, CSIR, Pretoria.

Hatze, H. (1981d). A comprehensive model for human motion simulation and its application to the take-off phase of the long jump. *J. Biomech.*, 14:135–142.

Hatze, H. (1981e). Estimation of myodynamic parameter values from observations on isometrically contracting muscle groups. *Europ. J. Appl. Physiol.*, 46:325–338.

Hatze, H. (1983). Computerized optimization of sports motions: an overview of possibilities, methods and recent developments. *J. Sports Sci.*, 1:3–12.

Hatze, H. (1990). The charge-transfer model of myofilamentary interaction: prediction of force enhancement and related myodynamic phenomena. In *Multiple Muscle Systems.* Winters, J.M. and Woo, S.L-Y. (ed.), pp. 24–45. Springer, New York.

Hatze, H. (1995). The extended transentropy function as a useful quantifier of human motion variability. *Med. Sci. Sport Exerc.*, 27:751–759.

Hatze, H. (1996). A three-dimensional multivariate model of passive human joint torques and articular boundaries. *Clin. Biomech.*, (in press).

Hatze, H. and Venter, A. (1981). Practical activation and retention of locomotion constraints in neuromusculoskeletal control system models. *J. Biomech.*, 14:873–877.

Jacoby, S.L.S. and Kowalik, J.S. (1980). *Mathematical Modeling with Computers*, pp. 3–21. Prentice-Hall, New Jersey.

Lehman, S.L. (1990). Input identification depends on model complexity. In *Multiple Muscle Systems.* Winters, J.M. and Woo, S.L-Y. (ed.), pp. 94–99. Springer, New York.

Loeb, G.E. and Levine, W.S. (1990). Linking musculoskeletal mechanics to sensorimotor neurophysiology. In *Multiple Muscle Systems.* Winters, J.M. and Woo, S.L-Y. (ed.), pp. 165–181. Springer, New York.

Murray-Smith, D.T. (1995). Advances in simulation model validation: theory, software and applications. In: *Eurosim '95.* Breitenecker, F. and Husinsky, I. (ed.), pp. 75–84. Elsevier, Oxford.

Rosen, R. (1988). The epistemology of complexity. In *Dynamic Patterns in Complex Systems.* Kelso, J.A.S., Mandell, A.J., and Schlesinger, M.F. (ed.), pp. 7–30. World Scientific, New Jersey.

Scott, H.S. and Winter, D.A. (1993). Biomechanical model of the human foot: kinematics and kinetics during the stance phase of walking. *J. Biomech.*, 26:1091–1104.

Winters, J.M. and Woo, S.L-Y., (ed.) (1990). *Multiple Muscle Systems.* Springer, New York.

Wittenburg, J. (1977). *Dynamics of Systems of Rigid Bodies*, Teubner, Stuttgart.

Yamaguchi, G.T. (1990). Performing whole-body simulations of gait with 3-D, dynamic musculoskeletal models. In *Multiple Muscle Systems*. Winters, J.M. and Woo, S.L-Y. (ed.), pp. 663–679. Springer, New York.

Zajac, F.E. and Winters, J.M. (1990). Modeling musculoskeletal movement systems: joint and body segmental dynamics, musculoskeletal actuation, and neuromuscular control. In *Multiple Muscle Systems*. Winters, J.M. and Woo, S.L-Y. (eds.), pp. 121–148. Springer, New York.

Commentary: Does Progression of Musculoskeletal Models Toward Large-Scale Cybernetic Models Yield Progress Toward Understanding of Muscle and Human or Animal Movement?

Peter A. Huijing

In this chapter, Hatze argues that deductive reasoning should be the basis of scientific effort. This means that theory forming should not be tried on an ad hoc basis. Few scientist will argue with that point. However, in order not to fall back in the ancient argument concerning scientific methodology that pits deductive reasoning against empirical observation, it should be kept just as clear that experimental confrontation of model results with reality is as essential for scientific progress. In my personal experience it has always been the combination of these two factors that leads to most progress in understanding and chances for practical application. It also prevents a situation arising that in a modeler's mind reality will be replaced by the truth of a limited model. Initially people are usually very well aware of limitations built by them into the model but after some time this becomes less explicit and one may tend to forget.

What do these views on deductive reasoning

mean for musculoskeletal modeling? Hatze draws the conclusion that deduction should be carried as far as possible in modeling. He contributes in a very positive way to scientific argument by defining what a complex musculoskeletal model should be. It should incorporate *all macrobiologically relevant* features with a well defined margin of error for *all conceivable modes of operation*. Thus the strive for preferably three dimensional general-purpose models is opened.

Application of this view has led Hatze's work to be unique at least in one respect: In his models more aspects of muscular physiology are represented than in most models. Examples are effects of muscle architecture, series elasticity, and non-linear effects of submaximal activation which he tried to take into account. Such contributions are recognized most frequently only implicitly by modellers who follow the general approach of his methods. In that light it is surprising that Hatze's attempts to deal with non-linear muscular properties depending on activation are not generally followed. This is probably not related to possible limitations of the approach used (see Chapters 5 and 6) but to a general lack of awareness of the importance of the principle of the approach.

In contrast, sometimes Hatze's contributions are not recognized as such, or are even condemned explicitly by such statements that these methods "render little practical value" (Baratta and Solomonow 1992). Even though controversy is essential for scientific progress, controversy should be handled in such a way that it invites rather than precludes discussion. Otherwise we run the chance of contributing to disappearance from general knowledge of very valuable insights, that have been shown to have occurred in myology in the past.

Despite Hatze's recognized contributions to the methods of muscle and movement modeling, it should be realized there are many known aspects of muscle physiology that are not incorporated in his models, even though they can be considered to be relevant at the macrobiological level. An example is the effects of potentiation by myosin light chain phosphorylation, which allows a muscle to exert a force which is much closer to its maximal value that one would expect from its actual calcium ion dependent level of activation. Nor are variation of length of sarcomeres in series within the fibers of a muscle described which lead to sizable alteration of length force characteristics after previous shortening (Edman et al. 1993, Meijer et al. 1997). In the older models (e.g., Hatze 1981) the properties of the muscle are described as lumped fiber characteristics, variation of properties within parts of a muscle are generally not considered. The present model seems to have been brought somewhat closer to biological reality, even though it does not seem to incorporate a effects of a primary distribution of fiber mean sarcomere length.

I am not arguing that all these factors should be incorporated in any model (actually the contrary is the case; see below) but models that are claimed to be comprehensive should contain such effects.

Actually it is my opinion that it is presently impossible to incorporate such effects for human muscles in an adequate way simply because even though we know about general mechanisms, we actually have very little knowledge about realistic parameter values.

Audu and Davy (1985) and also Barrata and Solomonow (1992) argued that computational cost of comprehensive models would be limiting. Clearly such effects of cost has decreased significantly in the last decade. However in my view the limiting factor is the proposed wide scope itself, coupled to the fact that biomechanists tend to expect too much from their modelling efforts. For example, predictive models in a clinical setting are expected to be able to predict even individual results, despite the fact that we are not able to model classical physiological characteristics of isolated in situ muscles from experimental animals very well without actually using variables obtained for individuals (e.g., Bobbert et al. 1990; van Ingen Schenau et al. 1988; van Zantwijk et al. 1996).

In my view (see also Huijing 1995, 1996, 1997, Chapters 5 and 6), the goals for which the model was constructed should also be involved in the selection of features of the model which should be constructed as simple as possible for a number of reasons. This will allow an intuitive contact with interactions of elements that take place within the model and we will stand a small chance of being able to find adequate values for model variables also for human muscles.

For example, it does not seem useful to model all relevant molecular events in many relevant muscles if a major purpose is to understand aspects of intermuscular coordination of human or animal move-

ment. Instead, it seems more profitable to construct a model of the musculoskeletal system with relatively simple myoactuators for well defined tasks, so that simple ideas can be tested using the model and confronted experimentally with reality. As an example I can point to development of an idea in my immediate surroundings: that power can transported effectively by polyarticular muscles from joint to joint. This idea was (re-?)discovered by van Ingen Schenau and coworkers on the basis of unpublished data and very simple model calculations for a very limited number of muscles around knee and ankle. It was clear that the muscles could not generate the power that was found experimentally at the ankle during a jump, and thus the search for other sources of power was started. This type of information must have been hidden within the complex hominoid models, but was never discovered—probably because the complexity of the model prevented the seeding of the idea, or alternatively optimization of models removed the whole feature from the model. Following the idea of transport of energy a still very simple model was constructed by Bobbert (Bobbert et al. 1986) to develop a quantitative idea of this concept.

In such cases involving application of very simple models, the models should not be expected anymore to predict reality but instead should be considered as helpful instruments to determine which questions to ask in further experiments. In other words a vision on direction and goals is essential as a prerequisite to both modelling and experimentation. Confrontation of model output as well as reality with very simple idea's is a valuable way to progress. If these idea's are too simple it will become clear and usually directions can be found regarding the source of the deviation.

In contrast, during the *Deer Creek Conference* (*Biomechanics and Neural Control of Movement*) Jack Winters argued that at least somewhat more complicated models may be necessary in some cases to develop the necessary vision referred to above: A sensitivity analysis may provide some guidance towards relative importance of variables incorporated in the model. Also Hatze (1997) states that it is much easier to simplify models than it is to make them more complex. Even though performing sensitivity analysis of a given model is a sensible thing to do, it may also lead away from insights since an almost natural temptation would be to manipulate values of variables in order to increase agreement with experimental results. Apart from that danger I feel it is uneconomical to build larger models than necessary in order to reduce them later.

Rather than constructing comprehensive models, in muscular modeling our efforts should better be directed at important and fairly new issues such as trying to replace the paradigm of separation of static (isometric) and dynamic (shortening velocity) characteristics, as that here are indications that even sustained isometric contractions remain dynamic in time particularly at higher muscle lengths.

If my views on modeling of muscle and movement are incorrect, it should be fairly easy for proponents of comprehensive modelling to indicate what the contributions to either understanding of muscles, movement and locomotion or changes of practice of application in one of many fields of human movement should be attributed to this type of model. I would invite any modeler to make clear to his or her audience what the value of a particular approach is. Doing that will also allow creation of room for explicit considerations of limitations of a particular approach. In such a discussion a defense can be presented for the exclusion of certain aspects of muscular properties.

It would have looked a lot better for any member of the modeling community if it could shown that for example the results of Rack and Westbury (1969) regarding submaximal activation were considered, but not included in the model for reasons specified.

Particularly with respect to increased understanding, there are many examples where using simple models has been effective. Changes brought about directly in fields related to human movement are more rare but do exist.

References

Audu, M.L. and Davy, D.T. (1985). The influence of muscle model complexity in musculoskeletal motion modelling. *J. Biomed. Eng.*, 107:147–157.

Baratta, R.V. and Solomonow, M. (1992). The dynamic performance model of skeletal muscle. *Crit. Rev. Biom. Eng.*, 19:419–454.

Bobbert, M.F., Ettema, G.J., and Huijing, P.A. (1990). Force length relationship of a muscle tendon complex: experimental results and model calculations. *Eur. J. Appl. Physiol.*, 61:323–329.

Bobbert, M.F., Huijing, P.A., and van Ingen Schenau, G.J. (1986). An estimation of power output and work done by human triceps surae muscle-tendon complex in jumping. *J. Biomech.*, 11:899–906.

Edman, K.A.P., Caputo, C. and Lou, F. (1993). Depression of tetanic force induced by loaded shortening of frog muscle fibers. *J. Physiol.*, 466:535–552.

Hatze, H. (1981). Myocybernetic control models of skeletal muscle. University of South Africa Press, Pretoria.

Hatze, H. (2000). Progression of musculoskeletal models toward large cybernetic myoskeletal models. In *Biomechanics and neural control of movement.* Winters, J.M. and Crago, P.E. (eds.), pp. 425–437. Springer, New York.

Huijing, P.A. (1995). Parameter interdependence and succes of skeletal muscle modelling. *Hum. Mov. Sci.*, 14:443–486.

Huijing, P.A. (1996). Important experimental factors for skeletal muscle modelling. *Eur. J. Morph.*, 34:47–54.

Huijing, P.A. (1998). Muscle, the motor of movement: properties in experiments, modelling and function. *J. Kinesiol*, 8:61–77.

Meijer, K., Grootenboer, H.J., Koopman, H.F.J.M., and Huijing, P.A. (1997). Fully Isometric Length-Force Curves of Rat Muscle Differ From those During and After Concentric Contractions. *J. Appl. Biomech.*, 13:164–181.

Rack, P.M.H. and Westbury, D.R. (1969). The effects of length and stimulus rate on tension in the isometric cat soleus muscle. *J. Physiol.*, 204:443–460.

van Ingen Schenau, G.J., Bobbert, M.F., Ettema, G.J., de Graaf, J.B., and Huijing, P.A. (1988). Simulation of rat EDL force output based on intrinsic muscle properties. *J. Biomech.*, 21:815–824.

van Zandwijk, J.P., Bobbert, M.F., Baan, G.C., and Huijing, P.A. (1996). From twitch to tetanus: performance of excitation dynamics optimized for a twitch in predicting tetanic muscle forces. *Biol. Cybern.*, 75:409–417.

34
Estimation of Movement from Surface EMG Signals Using a Neural Network Model

Yasuharu Koike and Mitsuo Kawato

1 Introduction

Over the years, the neurophysiology and biomechanics of muscle systems have been investigated quite extensively in order to characterize the relations between muscle activity (EMG) and various dynamical and/or kinematic aspects of the ensuing movement behavior. There have been numerous efforts to correlate the duration, magnitude and timing of phasic EMG bursts with the amplitude, duration, and maximum speed of limb motion (Gottlieb, Corcos, and Agarwal 1989; Brown and Cooke 1990; Karst and Hasan 1991). Although the complexity of musculoskeletal systems has made it difficult to reconstruct movement accurately from EMG signals, this goal is central to efforts to model motor control mechanisms of the central nervous system (CNS) computationally.

For example, quantitative dynamic models of the arm and muscle force generation have been used to predict muscle tension and/or motion from EMG signals (Akazawa, Takizawa, Hayashi, and Fujii 1988; Wood, Meeks, and Jacobsen 1989,Winters 1990; Clancy and Hogan 1991). Typically, these models have been based on the spring-like properties of muscles: muscle tension can be derived by controlling muscle length and activation level (Rack and Westbury 1969). It was hoped that piecemeal examination of the basic dynamical parameters would result in progressively better quantitative models of the musculoskeletal system (i.e., muscle model, neural model, skeletal system model with variable muscle moment arms, Lagrangian dynamics model of the arm). To this end, our group earlier proposed a 6-muscle human arm model

(Katayama and Kawato 1993) and a 17-muscle monkey arm model (Dornay, Uno, Kawato, and Suzuki 1992). A problem with this approach, however, is that assumptions have to be made at each step about the largely unknown nonlinear properties of the musculoskeletal and nervous systems.

The aim of the current study was to construct a complete forward dynamics model (FDM) of the human arm that affords accurate estimation of movement trajectories from input of physiological signals such as muscle EMG. To achieve this, we used an artificial neural network that learned nonlinear functions relating physiological recordings of EMG signals to simultaneous measurement of two-joint planar movement trajectories. Previously, we have used surface EMG signals to estimate: (1) joint torques under isometric conditions in the horizontal plane (Koike et al. 1993), (2) three-dimensional posture (three degrees-of-freedom at the shoulder and one degree-of-freedom at the elbow) (Koike and Kawato 1994a) and (3) joint angular acceleration with subsequent reconstruction of movement trajectories in the horizontal plane (Koike and Kawato 1994b).

The model developed here incorporated the following as domain-specific information: (1) the relationship between the EMG input signal and quasitension, (2) the dynamics of the arm, and (3) nonlinear muscle properties. To implement (1), a network was prepared to act as a temporal FIR filter. For (2), the physical parameters of the subject arm were calculated from the measured 3D shape of the arm, and the arm dynamics were described by Lagrange equations. To efficiently implement (3), various nonlinear properties of the musculoskeletal system were obtained through training of the modular network.

Section 2 of this chapter describes how these three domains of knowledge were incorporated into our model. Section 3 outlines the procedures used to collect and process the experimental data. Section 4 shows simulation results for joint torque estimation and trajectory formation using the obtained neural network. Section 5 discusses the reliability of the model and the directions of further development.

2   The Model That Estimates Trajectories from Surface EMG Signals

Figure 34.1 compares a view of information flow in a human (A) to the computational procedure adopted in this paper (B).

Figure 34.1(A) shows a process of transformation from motor commands to a trajectory. The CNS first sends a command to the muscles, causing them to exert tension. There exists nonlinear relationships between muscle-exerted tension and motor commands.

Muscle tension is related to motor command (firing rate) through a sigmoid function (Rack and Westbury 1969; Mannard and Stein 1973). This nonlinearity is caused not only by the firing-rate–tension relationship but also by recruitment of α-motor neurons. Moreover, there are two nonlinear relationships: between muscle tension and muscle length, and between muscle tension and muscle contractile velocity (Fig.1(A)) [Basmajian and De-Luca 1985; see also Chapters 1, 2, 7, 8, 10, and 33]. One is called the length-tension curve, which describes how muscle tension increases with length even if the motor command does not change. The other is called the velocity-tension curve, and describes muscle tension decreases with contractile velocity for a constant motor command.

The joint torque is then calculated from the muscle tension and muscle moment arms. The distance between the joint axis and the force action line of the muscle is the muscle moment arm. The moment arm changes nonlinearly depending on the joint angle and because muscles wrap around other muscles, bones and connective tissues (e.g., see Chapter 32). Joint torque is produced as the difference between agonist and antagonist muscle torques, which depends on the muscle tension and the mus-

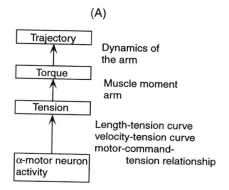

(A)

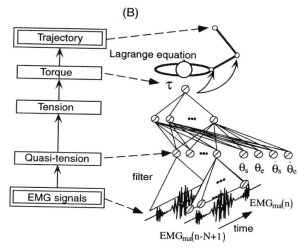

(B)

FIGURE 34.1. The comparison of the information flow in the organism (A) to the computational procedure adopted in this paper (B). In (A), control signals from the CNS are sent to each muscle via α motor neurons. The signals activate the muscles (muscle tension), the contraction causes joint torques, and then the arm moves along movement trajectories according to the arm dynamics. In (B), we can measure EMG signals and trajectory: they both have a double-line box around them. EMG signals, though temporally and spatially distorted, reflect the motor commands fed to the muscles. Since we can not measure motor neuron activity directly, though not ideal, we will treat the low pass filtered EMG activity as a substitute of the firing rate of motor neurons.

cle moment arm. Arm dynamics exist between the joint torque and the joint angle, velocity, and acceleration. Understanding the dynamics, however, is difficult because of the presence of complex, nonlinear interaction forces among the moving joint segments.

Our aim has been to construct a *forward model* of the human arm using data obtained by physiological measurements. This model took EMG signal as input and produced end point trajectories as output. Figure 34.1(B) shows the current procedure employed to compute end point trajectories from EMG signals. We have been able to measure the behavior of EMG signals and trajectories of the hand, elbow, and shoulder (measured quantities are shown with a double-line box around them). We treat EMG signals as a record of the motor commands to the muscles, since we can not directly measure the motor neuron activity. Though not ideal, EMG activity is a reasonable reflection of the firing rate of a motor neuron. Actual EMG activity was transformed by a linear, second-order low pass filter. The transformed signal is called "quasitension," because it seems to be highly correlated with the true muscle tension. We also used a neural network with a modular architecture to convert quasitension to estimated dynamic torque. If a single, multilayer network performs different tasks under different occasions, there will generally be strong interference effects which lead to slow learning and poor generalization. If we know in advance that a set of training cases may be naturally divided into subsets that correspond to distinct subtasks, interference can be reduced by using a modular architecture. In the physiological view point, this division is natural. Muscle has nonlinear properties, such as the length-tension and velocity-tension relationships. In the case of movement, both nonlinear properties have to be considered. In the case of posture control, however, the velocity-tension relationship may not need to be considered. It is also widely known that dynamic characteristics of spinal and supraspinal reflex loops are very different between movement and posture maintenance. In this part, the neural network learned nonlinear muscle properties, such as the length-tension curve and the velocity-tension curve, and nonlinear properties of musculoskeletal systems, such as the muscle moment arms. We did this by using actual torques as teaching signals and actual joint kinematics as additional inputs. The actual joint kinematics were obtained from the measured Cartesian trajectories of the joints. The teaching torque signal was computed using the actual joint angle kinematics and the measured physical parameters of the arm using inverse dynamics equations. Finally, the estimated torques which were outputs of the neural network and the actual joint kinematics were used to calculate joint angle acceleration using the forward dynamics equations. These angular accelerations were integrated to predict the next-time-step joint state, and end point trajectories were estimated using forward kinematics equations. In this manner, we were able to construct a forward model that transformed EMG signals at the current time step to end point trajectories at the next time step. The following details each step.

2.1 The Relationship Between EMG Signals and Quasitensions

Surface EMG signals are spatiotemporally convoluted action potentials of the muscle membranes, and involve not only descending central motor commands but also reflex motor commands generated from sensory feedback signals. There have been considerable efforts to estimate muscle force from surface EMG signals (Basmajian and DeLuca 1985; Akazawa et al. 1988; Wood et al. 1989; Clancy and Hogan 1991). From these previous studies in medical electronics and biological engineering (Basmajian and DeLuca 1985), it can be expected that low-pass-filtered EMG signals (quasitension) reflect the firing rate of α motor neurons, because high-frequency components of EMG reflect the shape of individual action potential while low-frequency components reflect the firing-frequency of motor nerve fibers. In neurophysiological studies, it was found that a second-order low-pass filter was sufficient for estimating muscle forces from the nerve impulse train (Mannard and Stein 1973). The relationship between the EMG input signal and $\hat{T}$ (quasitension) the output signal can be represented as an FIR (Finite Impulse Response) filter,

$$\hat{T}(t) = \sum_{j=1}^{n} h_j \cdot EMG(t - j + 1) \quad (34.1)$$

where, h_j is the filter, EMG represents EMG signals, $\hat{T}$ represents "quasitensions" and j is the index to discrete time. EMG is actually the digitally rectified, integrated and filtered signal. The second-order frequency response of the filter $H(s)$ is represented as follows.

$$H(s) = \frac{\omega_n^2}{(s^2 + 2\zeta\omega_n + \omega_n^2)} \quad (34.2)$$

where ω_n and ζ denote natural frequency and damping coefficient, respectively. The impulse response of the function in equation 2 is

$$h(t) = a(e^{-bt} - e^{-ct}) \tag{34.3}$$

The coefficients h_j in Eq. 34.1 can be acquired by digitizing $h(t)$ with the given coefficients a, b and c.

2.2 The Relationship Between Quasitensions and Joint Torques

Each joint torque was estimated from quasitension, joint angle, and velocity using an artificial neural network model with a modular architecture as shown in Figure 34.2. for the shoulder network.

The modular architecture consists of two types of networks: expert and gating networks (Jacobs and Jordan 1991; Nowlan and Hinton 1991). Two modular shoulder and elbow networks were used to estimate the two joint torques respectively in order to improve the accuracy of the torque estimation.

2.3 Modular Learning

We briefly illustrate the modular learning algorithm which was proposed by Jacobs and Jordan (1991) and Nowlan and Hinton (1991), and used in this study. The j-th output of the gating network, g_j, is calculated by the following soft-max function

$$g_j = \frac{e^{s_j}}{\sum\limits_{i=1}^{N} e^{s_i}} \tag{34.4}$$

where s_i is calculated from the input signals to the gating network. N denotes the number of outputs. The total output of the modular network is as follows.

$$\tau = \sum\limits_{i=1}^{N} g_i \hat{\tau}_i \tag{34.5}$$

where $\hat{\tau}_i$ is the output of the i-th expert network.

The gating and expert networks are trained to maximize the following log-likelihood function.

$$\ln L = \ln \sum\limits_{i=1}^{N} g_i e^{-(\tau - \hat{\tau}_i)^2/2\sigma_i^2} \tag{34.6}$$

where σ_i is the variance scaling parameter of the i-th expert network.

The adaptation rules of the weights in the gating network are derived from the partial derivative of Eq. (34.6) by applying the chain rule.

$$\frac{\partial \ln L}{\partial s_i} = h_i - g_i \tag{34.7}$$

where h_i is defined by the following equation corresponding to the posterior probability.

$$h_i = \frac{g_i e^{-[\tau - \hat{\tau}_i]2/2\sigma_i^2}}{\sum\limits_{j=1}^{N} g_j e^{-[\tau - \hat{\tau}_j]2/2\sigma_j^2}} \tag{34.8}$$

Similarly, the adaptation rules of the weights in the expert networks are derived from the partial derivative of Eq. (34.6) by applying the chain rule.

$$\frac{\partial \ln L}{\partial \hat{\tau}_i} = h_i(\tau - \hat{\tau}_i)/\sigma_i^2 \tag{34.9}$$

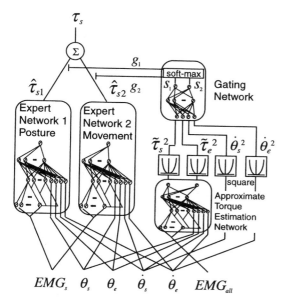

FIGURE 34.2. Structure of the artificial neural network which estimates the shoulder joint torque using a modular architecture. EMG_s are EMG signals of muscles related to the shoulder movement. EMG_{all} are EMG signals of all muscles. $\hat{\tau}_s^2$ and $\hat{\tau}_e^2$ are the square of the approximate torque of the shoulder and elbow respectively calculated by the approximate torque estimation network. $\hat{\tau}_{s1}$ and $\hat{\tau}_{s2}$ are shoulder joint torques estimated by the expert network 1 and 2 respectively. See (Eq. 34.4) for g_1 and g_2.

2.4 A Neural Network That Estimates Each Joint Torque

In Figure 34.2, each expert network estimated shoulder joint torque. For the case of elbow joint torque, the same modular architecture was used except that the expert input signals were EMG_e: the EMG signals of muscles related to the elbow joint movements. The expert network 1 estimated joint torques $\hat{\tau}_{s1}$ mainly during posture control, and the expert network 2 estimated joint torques $\hat{\tau}_{s2}$ mainly during movements. This division of their roles was first attained by pretraining and further refined by the automatic modular learning algorithm.

The gating network switched the expert networks by judging whether the arm moved or not. To judge whether the arm moved or not, the square of the angular velocity and torques which change faster than velocity signals, were used. To calculate this torque input, an approximate torque estimation network was prepared at the input side of the gating network (Figure 34.2).

Each expert network consisted of a four-layer network as shown in Figure 34.3.

The first-layer inputs of this four-layer network were the EMG signals recorded from some of the ten muscles over a 0.5 second interval. The EMG signals from double joint muscles, related single joint muscles, the joint angle and the joint angular velocity of the elbow and shoulder were the expert network inputs. The number of units in the second layer was 11 for the shoulder expert network, and 9 for the elbow expert network. The calculation of the "quasitension" from the EMG signals was implemented in the expert network between the first and the second layers. In a strict sense, "quasitension," a linearly filtered EMG signal can not represent muscle tension. Because the FIR filter is linear, the nonlinear muscle properties found in the motor-command-tension, length-tension, and velocity-tension curves are not represented between the first and second layers. Thus, the network learns these nonlinear properties between the second and the fourth layers. The second-layer inputs were the joint angles and joint angular velocities of the elbow and shoulder, as well as the quasitensions. The third layer consisted of 30 hidden units. The fourth, the output layer, estimated the joint torque. Activation functions, relating the weighted sum of synaptic inputs to the output of an artificial neuron model, of only the third layer are the nonlinear sigmoid function.

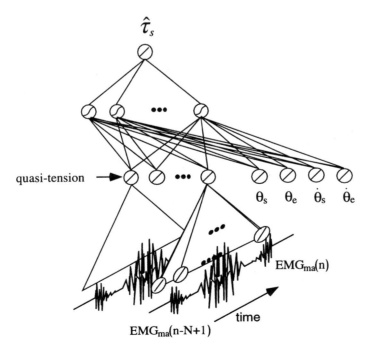

FIGURE 34.3. One of the expert neural network which estimates the shoulder joint torque. The approximate torque estimation network also has a similar structure except that it has two output units for the shoulder and elbow torques.

The gating network consisted of a three-layer network. The first-layer inputs were the square of each joint torque and joint velocity. Thus, the number of units in the first-layer was four (2*2). The second-layer consisted of 10 hidden units. The third, the output layer, consisted of two units which calculate s_j in equation (34.4) corresponding to two expert networks ($j = 1, 2$). Again, only the second layer units are nonlinear. The outputs of the gating network are g_1 and g_2 defined in Eq. (34.4).

The approximate torque estimation network also consisted of a four-layer network like the expert networks shown in Figure 34.3. The first-layer inputs were the EMG signals from all the 10 muscles over a 0.5 second interval. The second-layer inputs were the joint angles and joint angular velocities of the elbow and shoulder, as well as the 10 quasitensions. Thus, the number of units in the second layer is 14. The number of units in the third layer is 30. The fourth, the output layer, consisted of two units which estimated shoulder and elbow joint torques $\tilde{\tau}_s$ and $\tilde{\tau}_e$. Again, only the third layer is nonlinear. This network had less accuracy than the expert networks, but it could provide sufficiently good information to judge whether the arm moved or not. The training method of the approximate torque estimation network was standard. The actual torques τ_s and τ_e are given as the teaching signals and the objective function is defined as the squared sum of the difference between real and estimated joint torques.

2.5 The Relationship Between Joint Torques and Trajectories

In this chapter, we deal with horizontal planar movements of the shoulder joint and the elbow joint (flexion-extension) at the shoulder level. Therefore the controlled object is the two-link system comprised of the upper arm (link 1) and forearm (link 2) shown in Figure 34.4.

We use the following dynamics equations for two-joint horizontal movements of the upper arm and the forearm.

$$(I_1 + I_2 + 2M_2 L_1 l_{g2} \cos(\theta_e) + M_2 L_1^2)\ddot{\theta}_s$$
$$+ (I_2 + M_2 L_1 l_{g2} \cos(\theta_e))\ddot{\theta}_e$$
$$- M_2 L_1 l_{g2}(2\dot{\theta}_s + \dot{\theta}_e)\dot{\theta}_e \sin(\theta_e) = \tau_s$$
$$(I_2 + M_2 L_1 l_{g2} \cos(\theta_e))\ddot{\theta}_s + I_2 \ddot{\theta}_e$$
$$+ M_2 L_1 l_{g2} \dot{\theta}_s^2 \sin(\theta_e) = \tau_e \qquad (34.10)$$

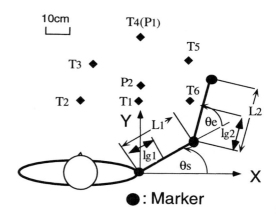

10cm

●: Marker

FIGURE 34.4. Experimental settings and definitions of joint angles and link physical parameters of the two-link arm model. L_1: length of upper arm, L_2: length of forearm, l_{g1}: distance from the center of mass of upper arm to the shoulder joint, l_{g2}: distance from the center of mass of forearm to the elbow joint, θ_s: shoulder joint angle, θ_e: elbow joint angle. Black circles show 5 points where the subject exerted isometric hand forces in Experiment 1. Black diamonds show the start, via- and target points of movements in Experiment 2. White squares show 23 points where postures are maintained in Experiment 3.

where τ, θ, $\dot{\theta}$, $\ddot{\theta}$ represent the joint torque, joint angle, velocity and acceleration, respectively. M_i, L_i, l_{gi}, I_i represent the mass, length, distance from the center of mass to the joint axis, and rotary inertia around the joint for each link, respectively.

When the problem is to find the joint motion corresponding to a known sequence of input torques, the transformation (34.10) is referred to as *forward dynamics*. If the initial conditions (joint angles and velocities), and the control signals (joint torques from the initial time to the final time) are given, then the time course of θ and are obtained by numerical integration of the dynamics equations (34.10).

When the problem is to find the joint torques corresponding to the desired time sequence of joint angles, the transformation (34.10) is referred to as inverse dynamics. In the experimental procedure of this paper, to calculate the joint torques from measured trajectories, the dynamics equations (34.10) are also used. In the case of forward dynamics, the information flows from the right side to the left side of equations (34.10), and in the case of inverse dynamics, the information flows from the left to the right.

3   Experimental Procedures

3.1   Experiment 1: Isometric Force Generation

One healthy subject, 29 years old, participated in this study. The seated subject's shoulder was restrained by a harness. In the first experiment, to analyze the relationship between EMG signals and quasitension, the forces generated at the hand under isometric conditions and surface EMG signals were measured.

His wrist was secured by a cuff and supported horizontally using the beam which was attached to a force-torque sensor. The subject was trained first to exert a hand force of about 50% maximum. The subject exerted isometric hand forces in two different directions: forward and backward, left and right, at five different locations (θ_e, θ_s) of (30°, 110°), (40°, 80°), (50°, 90°), (60°, 100°) or (70°,

70°) indicated by the "black circles" in Figure 34.4. These trials lasted for seven seconds and were of various rates of force production. At each 5 positions, the subject tried 2 times in each direction. Thus, the rate of the hand-force change was intentionally varied and the peak magnitude was roughly controlled.

The hand force was measured by a force-torque sensor and filtered at an upper cut-off frequency of 130 Hz in hardware. These signals were first sampled at 2000 Hz with 12-bit resolution, and were then re-sampled at 200 Hz. The positions of the hand, elbow and shoulder were recorded at 400 Hz using the OPTOTRAK position sensing system. The shoulder and elbow joint angles were calculated from those position data. These joint angles data were digitally filtered at an upper cut-off frequency of 10 Hz by a Butterworth filter. Then, these signals were re-sampled at 200 Hz. The shoulder and elbow joint torques were calculated from the measured hand force within the horizontal plane (two-degree-of-freedom) multiplied by the transpose of the Jacobian of the coordinate transformation.

EMG signals were recorded from the following 10 muscles shown in Figure 34.5. For flexion/extension of the shoulder joint, the deltoid-clavicular

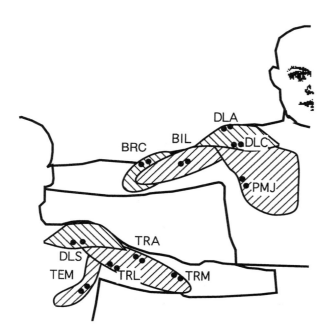

FIGURE 34.5. Electrode positions in EMG measurement. See text for muscle name abbreviations.

part (DLC), deltoid-acromial part (DLA), deltoid-scapular part (DLS), pectorals major (PMJ), and teres major (TEM) were measured. For double-joint muscles, the biceps-long head (BIL) and triceps-long head (TRL) were measured. For flexion/extension of the elbow joint, the brachialis (BRC), triceps-medial head (TRM), and triceps-lateral head (TRA) were measured.

The EMG signals were recorded using a pair of silver-silver chloride surface electrodes, in a bipolar configuration. The electrodes each had a 10-mm diameter and were separated by approximately 15 mm. Test maneuvers were used to verify electrode placement. Each signal was sampled at 2000 Hz with 12-bit resolution. This signal was digitally rectified, integrated for 0.5 ms (EMG_{ave}), sampled at 200 Hz, and finally, filtered (25-ms moving average window). This signal was denoted EMG_{ma}.

$$EMG_{ma}(t) = \frac{1}{5}\sum_{i=-2}^{2} EMG_{ave}(t - i) \quad (34.11)$$

The EMG_{ma} signals were used as the input signals in Eq. (34.1) (i.e. EMG).

3.2 Experiment 2: Movement Generation

These measurements of arm positions and EMG signals were simultaneously continued during movements and maintenance of posture using the same method as Experiment 1. Again, the subject's wrist was secured by a cuff and supported horizontally. In Figure 34.4, the target positions are indicated by the "black diamonds." T_1 to T_6 are starting and ending positions, and P_1 and P_2 are via points (see Uno, Kawato, and Suzuki 1989). The subject was asked to produce five different unrestrained point-to-point movements between the five targets, i.e. $T_3 \rightarrow T_6$, $T_2 \rightarrow T_6$, $T_1 \rightarrow T_3$, $T_4 \rightarrow T_1$, $T_4 \rightarrow T_6$; movements were repeated in the opposite direction. Then, the subject made via point movements between two targets in the horizontal plane. Two cases, $T_3 \rightarrow P_1 \rightarrow T_5$, $T_3 \rightarrow P_2 \rightarrow T_5$; were tested in both directions. The movement durations ranged from about 600 ms to about 800 ms. Each of the 14 movements consisted of 10 trials. During movement, joint angular velocity and acceleration were computed using numerical differentiation. The joint torques were calculated from the trajectories using the dynamics Eq. (34.10), be-

TABLE 34.1. Parameters of the human arm.

	Link 1 upper arm	Link 2 forearm
L_i [m]	0.241	0.367
L_i [m]	0.256	0.315
l_{gi} [m]	0.104	0.165
M_i [kg]	1.02	1.16
I_i [kg m^2]	0.0167	0.0474

cause dynamical torques can not be measured directly during movement.

3.2.1 The Physical Parameters of the Subject Arm

The physical parameters of the arm of a human subject were calculated from the 3D shape of the human arm. First the shape of the subject arm was scanned in 3D space by the Cyberware Laser Range Scanner®. Then, assuming a uniform material with a specific gravity of 1.0, the mass, the center of mass, and the rotary inertia were calculated from the cubic volume. The density of water is a good approximation both for soft and hard tissues. Table 34.1 shows the estimated physical parameters of the subject arm for the Eq. (34.10).

3.3 Experiment 3: Posture Maintenance

In Experiment 3, the subject produced co-contraction of muscles while maintaining the same posture without exertion of force at 23 points over the workspace indicated by the "white squares." Thus, the net torques generated were zero. Three trials at each point lasted for six seconds, and were of various co-contraction levels.

4 Simulation Results

4.1 Joint Torque Estimation Using an Artificial Neural Network Model

To train the network, the popular back-propagation algorithm (Rumelhart, Hinton, and Williams 1986), in conjunction with the steepest ascent method was examined first. Because its rate of convergence is slow, we used the Kick-Out method (Ochiai and Usui 1994) in which learning rates are adjusted, according to the rate of increase in the objective function during the last few steps.

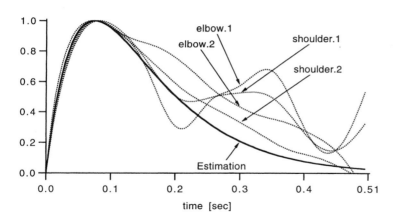

FIGURE 34.6. Impulse response of the second-order temporal filter which determines the quasitension from EMG. The ordinate scale is arbitrary with the peak response of 1.0.

4.2 Estimation of the Weights Between the First and Second Layers (Filter)

To specify the relationship between EMG signals and quasitension, joint torques under isometric conditions measured in Experiment 1 were first estimated from surface EMG signals using a simple, nonmodular four-layer neural network such as shown in Figure 34.3. The network was trained with the odd-numbered trials from Experiment 1, and the even-numbered trials were used for a cross-validation test. The training employed 10,000 sample points from 5.0 seconds * 10 trials * 200 Hz sampling rate (10,000 = 5 * 10* 200). The weights between the first layer and second layer after learning are shown in Figure 34.6.

The dotted lines for shoulder.1 and elbow.1 indicate weights obtained from a previous study using the same subject (Koike et al. 1992). The dotted lines for shoulder.2 and elbow.2 indicate the weights obtained this time. The coefficients of equation (3) were estimated from shoulder.1 and elbow.1 using the least squares error method for 0.25 sec. A comparable calculation for 0.5 sec yielded weights which were less stable and variable for different joints. The solid line in Figure 34.6 shows the resulting impulse response with a = 6.44, b = 10.80, and c = 16.52 in Eq. (34.3). Using these coefficients, the isometric torques were estimated accurately. Because the coefficient of determination (square of the correlation coefficient between actual torques and estimated torques) for the test data was 0.897, and, moreover, shoulder.2 and elbow.2 which were obtained from the present experiment, fit the estimated impulse response well, we can conclude that the obtained filter was

reliable. The coefficients a, b, and c of the filter were fixed when the torques were estimated during movement in the next step.

Figure 34.7 shows EMG signals EMG_{ma} calculated from Eq. (34.11), and quasitension $\hat{T}$ given by Eq. (34.1). We can see that the quasitension signal (smooth curve) lags about 100 ms behind the EMG_{ma} signals.

4.3 Estimation of the Weights Between the Second and Fourth Layer (Nonlinear Transformation)

Next, data from Experiments 2 and 3 explained in Subsections 3.2 and 3.3 were used to finally determine the weights between the second and fourth layers. Thus, nonlinear properties of the musculoskeletal system were determined from the data of movements and posture maintenance without exerted forces. The network was trained with the odd-numbered trials from Experiment 2, and the even-numbered trials were used for a cross-validation test. The training employed 35,000 sample points (2.5 seconds * 70 trials * 200 Hz sampling rate).

The data from the first and third trials for posture control from Experiment 3 were also used to train the network. In this case, the target dynamic torque is zero because no movement or exerted force was generated in Experiment 3. The second trial was spared for a performance test. The training employed 50,600 sample points (5.5 seconds * 46 trials * 200 Hz sampling rate). The test employed 25,300 sample points (5.5 seconds * 23 trials * 200 Hz sampling rate). The learning was broken off before the error in the test data began to increase (cross validation method; (Wada and

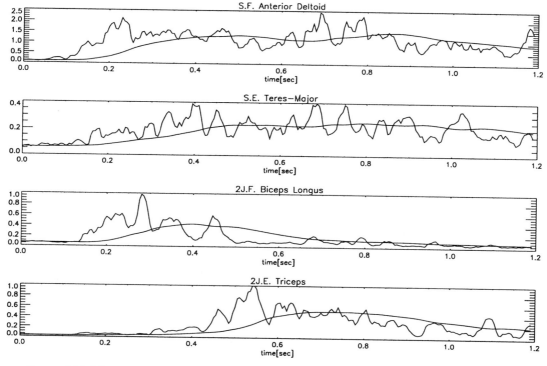

FIGURE 34.7. Measured EMG signals and quasitension for the four muscles (DLC, TEM, BIL, TRL).

Kawato 1992). This is the standard way of cross validation to avoid "overlearning" in which synaptic weights are tuned too much to the training data and generalization capability is lost. This was routinely used for all training sessions.

Before modular learning, the approximate torque estimation network and the gating network were trained using both movement and posture data. Expert network 1 was pretrained using the posture maintenance data from Experiment 3. In contrast, the expert network 2 was pre-trained using the movement data from Experiment 2. After pre-learning, the modular learning was done with the values of 0.05 for both σ_1 and σ_2 in Eq. (34.6) using both moving and stationary position data. The purpose of pre-training of the gating, expert and approximate torque estimation networks is to lay "seeds" in those network weights before the usual modular learning takes place. Good initial synaptic weights obtained by this pretraining greatly enhanced automatic division of two expert networks as well as dramatically reduced the overall learning time.

Figure 34.8 shows one example of the estimation result of the joint torques for the shoulder and elbow (upper traces), and the output of the gating network for test data (lower traces).

Prediction was made at each time step from the position, velocity and EMG data from the test set. As far as the torque is concerned, the dotted line is for the actual torque, and the solid line is for the network output. For the output of the gating network, the solid line corresponds to Expert 1, and the dotted line corresponds to Expert 2. Overall for test data from Experiments 2 and 3, the determination coefficient of dynamic torque was 0.887. Thus, the dynamic torques were reliably predicted by our proposed network. Expert 1's output corresponds to "posture," and expert 2's output corresponds to "movement." From the lower trace of Figure 34.8, we can assert that the gating network switched the expert networks correctly for both the stopping and moving conditions.

4.4   Trajectory Formation

The trajectories were calculated from the initial position and velocity, and the continuous EMG signals for point-to-point movements and via point movements. This was done in the following recur-

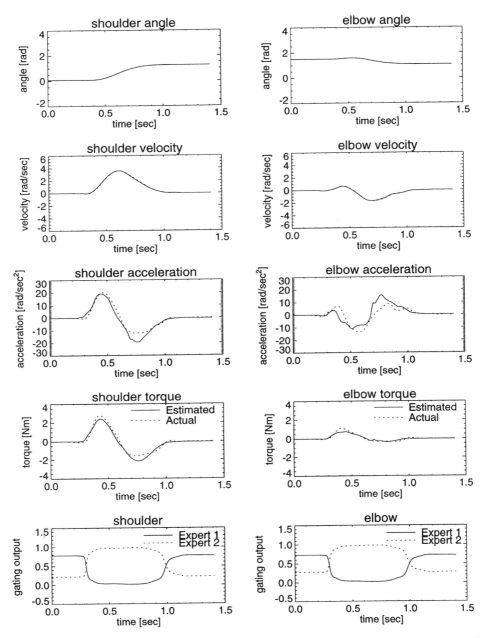

FIGURE 34.8. Estimation results of joint torques for shoulder (left) and elbow (right). Dotted curves show the actual torque and solid curves show the estimated torque in the upper row; solid curves show results for Expert 1 and dotted curves show results for Expert 2 in the lower row.

sive way. These two steps were repeated until the end of the recording duration.

Step 1. "At each time step, the dynamic torque was predicted by the neural network model from the position and velocity values at the current time step and the past 500 ms of EMG data. Then, this predicted torque was used as the control input to the dynamics Eq. (34.10)."

Step 2. "Numerical integration of Eq. (34.10) by Euler's method from the current values of the position, velocity and torque provides the next step value of position and velocity."

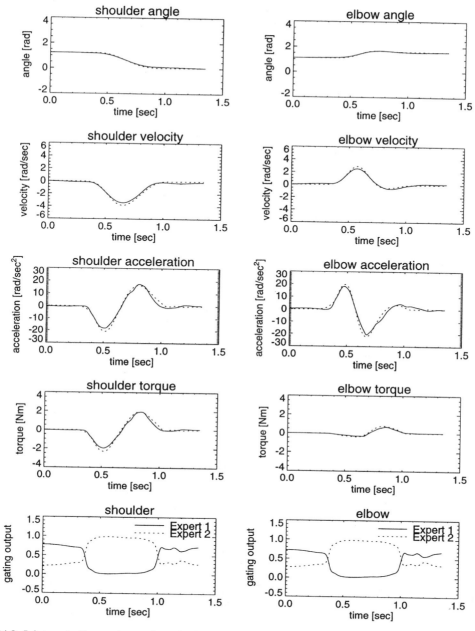

FIGURE 34.9. Joint angle (1st row), angular velocity (2nd row), angular acceleration (3rd row), and joint torque (4th row) predicted during point-to-point movement. Dotted curves show actual values and solid curves show estimated values in the upper 4 rows. The bottom row shows the outputs of the gating network. Here, solid curves show the output of the gating network for Expert 1 and dotted curves show the output of the gating network for Expert 2.

Figure 34.9 shows one example of the simulation results of trajectory generation for T_3 to T_6.

In descending order, the joint angle, angular velocity, angular acceleration, torque, and output of the gating network are shown. The left column corresponds to the shoulder and the right one corresponds to the elbow. In the upper 4 rows, the solid curve is the network output, and the dotted curve is the experimental data. In the bottom row, the solid curve is the output for Expert 1, and the dotted curve is the output for Expert 2. Similar to the one-step prediction described before, the gating network switched the expert networks correctly for both the stopping and moving conditions. It should also be noted that at the start and end of a movement, the output of the gating network began to change in advance of the velocity change, allowing the expert network output to follow.

Figure 34.10 shows trajectories on the *XY* plane. Overall for test data shown in Figure 34.10 from Experiment 2, the coefficient of determination for position data predicted from initial conditions of position and velocity and EMG time course is 0.948. Therefore, even though there was a gradual accumulation of error because the angle and angular velocity at the next time step were recursively calculated by summing the predicted accelerations with the current angular velocity, the trajectories were reconstructed accurately. Some trajectories on the *XY* plane were slightly different from actual trajectories, because of the error accumulation. There is, however, almost no significant error for the joint angle. This is the first demonstration that multijoint movements and posture maintenance can be fairly accurately predicted from multiple surface EMG signals while allowing complicated via-point movements as well as co-contraction.

5 Discussion

Joint torques and then human arm movements have been estimated from surface EMG signals using a four-layer artificial neural network with a modular architecture. In the implementation, we took account of the following domain-specific knowledge: (1) the relationship between the EMG input signal and quasi-tension, (2) the dynamics of the arm, and (3) nonlinear muscle properties. To implement (1), a network was prepared to work as a temporal FIR filter between the first and second layer. For this filter, we found about a 100 ms lag between EMG signals and quasitension. Soechting and Roberts (1975) reported the natural frequency of the impulse response relating EMG to force of human muscle was 2.5 Hz. Moreover, Bawa and Stein (1976) reported that the natural frequency of the impulse response for human soleus muscle was around 2 Hz. These natural frequencies correspond to about 60 to 100 ms delay between EMG signals and muscle tension. Bennett (1993) also pointed out this low-pass property of muscles and reported delays of approximately 60 ms to 90 ms between surface EMG signals and human arm muscle tension. To implement (2), the physical parameters of the subject arm were calculated from the measured 3D shape of the arm, and the arm dynamics were described by the Lagrangian equations. Furthermore, some nonlinear properties of the musculoskeletal system were obtained by training the neural network; expert networks were trained separately from training data focusing on movement or posture control to efficiently implement (3). There are several reasons for using two expert networks. For example, from the physiological view point, the use of muscles are different depending on whether the arm moves or not. When the arm is moving, the relationship between velocity and tension has to be considered. %In the case of the posture control, however, the velocity-tension %relationship does not need to be considered. In the case of the posture control, however, the length-tension relationship is mainly considered. It is also widely known that dynamic characteristics of spinal and supra-spinal reflex loops is very differ-

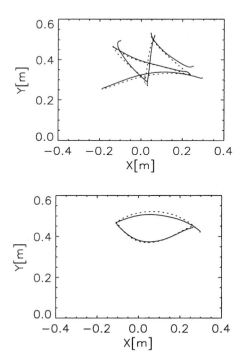

FIGURE 34.10. Calculated trajectories on *XY* plane. Dotted curves show actual and solid curves show estimated paths.

ent between movement and posture maintenance (Brooks 1981). The approximate torque estimation network was added to calculate joint torques to provide the useful information for the gating network.

Until now, mainly qualitative descriptions have been made regarding the relationship between movements and EMG, such as recognizing registered movement patterns from surface EMG signals (Suzuki and Suematsu 1969; Mori, Tsuji, and Ito 1992). In this paper, however, trajectories were estimated quantitatively from surface EMG signals. The complexity of musculoskeletal systems makes it difficult to reconstruct movement trajectories accurately from EMG signals. We use point-to-point movements or via pint movements in the horizontal plane, which almost covered the workspace. Also, the link dynamics is not simple because of the presence of complex interaction forces among moving link segments.

The comparison of the estimated and measured trajectories is a severe test of the goodness of the model because it is essential to estimate the shoulder and elbow joint torques not only qualitatively but also quantitatively to reconstruct trajectories accurately. We have also confirmed that each joint torque was accurately reconstructed by comparing the estimated torque waveforms and data torques as shown in Figure 34.8. Moreover the model was examined by using the test data which were not used for training: this is further confirmation of the generalization capability of the model.

The constructed forward dynamics model can serve as a fundamental tool for the computational study of multi-joint arm movements. Other than this scientific use, several engineering applications might also be possible. For example, by using the network, EMG signals could be used as human interface inputs to control a "virtual arm" in a virtual reality environment. A further possibility is for the motor command produced by a minimum-muscle-tension-change model (Uno, Suzuki, and Kawato 1989) based on the neural network forward dynamics model to be applied to a paralyzed limb.

6 Future Directions

Regarding computational studies of motor control based on the acquired forward dynamics model, our future work includes (1) calculating of virtual trajectories to critically examine the virtual trajectory

hypothesis (see Koike and Kawato (1993) for preliminary results), (2) learning the inverse dynamics model, and (3) examining a minimum-motor-command-change model (Kawato 1992).

References

Akazawa, K., Takizawa, H., Hayashi, Y., and Fujii, K. (1988). Development of control system and myoelectric signal processor for bio-mimetic prosthetic hand. *Biomechanism*, 9:43–53.

Basmajian, J.V. and De-Luca, C.J. (1985). *Muscles Alive*. Williams & Wilkins, Baltimore.

Bawa, P. and Stein, R. (1976). Frequency response of human soleus muscle. *J. Neuropysiol.*, 39:788–793.

Bennett, D. (1993). Electromyographic responses to constant position errors imposed during voluntary elbow joint movement in human. *Exp. Brain Res.*, 95:499–508.

Brooks, V.B. (1981). Section 1: The nervous system, volume II. Motor control. In *Handbook of Physiology*. American Physiological Society, Maryland.

Brown, S. and Cooke, J. (1990). Movement related phasic muscle activation. I. Changes with temporal profile of movement. *J. Neurophysiol.*, 63:455–464.

Clancy, E.A. and Hogan, N. (1991). Estimation of joint torque from the surface EMG. *Annual International Conference of the IEEE Engineering in Medicine and Biology Society* 13:0877–0878.

Dornay, M., Uno, Y., Kawato, M., and Suzuki, R. (1992). Simulation of optimal movements using the minimum-muscle tension-change model. In *Advances in Neural Information Processing Systems 4*. Moody, J.M., Hanson, S.J., and Lippmann, R.P. (eds), pp. 627–634. Morgan Kaufmann, San Meteo.

Gottlieb, G.L., Corcos, D.M., and Agarwal, G.C. (1989). Organizing principles for single-joint movements I. A speed-insensitive strategy. *J. Neurophysiol.*, 62:342–357.

Jacobs, R. and Jordan, M. (1991). A competitive modular connectionist architecture. In *Advances in Neural Information Processing Systems 3*. Moody, J.M., Hanson, S.J., and Lippmann, R.P. (eds), pp. 767–773. Morgan Kaufmann, San Meteo.

Karst, G.M. and Hasan, Z. (1991). Timing and magnitude of electromyographic activity for two-joint arm movements in different directions. *J. Neurophysiol.*, 66:1594–1604.

Katayama, M. and Kawato, M. (1993). Virtual trajectory and stiffness ellipse during multijoint arm movement predicted by neural inverse models. *Biol. Cybern.*, 69:353–362.

Kawato, M. (1992). Optimization and learning in neural networks for formation and control of coordinated movement. In *Attention and Performance, XIV*. Meyer, D. and Kornblum, S. (eds.). Cambridge, MIT Press.

Koike, Y. and Kawato, M. (1993). Virtual trajectories predicted from surface EMG signals. *Soc. Neurosci. Abstr.*, 19:543.

Koike, Y. and Kawato, M. (1994a). Estimation of arm posture in 3D-space from surface EMG signals using a neural network model. *IEICE Transactions Fundamentals D-II*, 4:368–375.

Koike, Y. and Kawato, M. (1994b). Trajectory formation from surface EMG signals using a neural network model. *IEICE Transactions* (D-II), J77:193–203, (in Japanese).

Koike, Y., Honda, K., Hirayama, M., Gomi, H., Bateson, E.-V., and Kawato, M. (1992). Dynamical model of arm using physiological data. *Technical Report of IEICE NC91-146* pp.107–114, (in Japanese).

Koike, Y., Honda, K., Hirayama, M., Gomi, H., Bateson, E.V., and Kawato, M. (1993). Estimation of isometric torques from surface electromyography using a neural network model. *IEICE Transactions D-II* 6:1270–1279, (in Japanese)

Mannard, A. and Stein, R. (1973). Determination of the frequency response of isometric soleus muscle in the cat using random nerve stimulation. *J. Physiol.*, 229:275–296.

Mori, D., Tsuji, T., and Ito, K. (1992). Motion discrimination method from EMG signals using statistically structured neural networks. *Technical Report of IEICE NC91-*143:83–90, (in Japanese).

Nowlan, S. and Hinton, G. (1991). Evaluation of adaptive mixtures of competing experts. In *Advances in Neural Information Processing Systems 3*. Moody, J.M., Hanson, S.J., and Lippmann, R.P. (eds.), pp. 774–780. Morgan Kaufmann, San Meteo.

Ochiai, K. and Usui, S. (1994). Kick-out learning algorithm to reduce the oscillation of weights. *Neural Networks*, 7:797–807.

Rack, P. and Westbury, D. (1969). The effects of length and stimulus rate on tension in the isometric cat soleus muscle. *J. Physiol.*, 217:419–444.

Rumelhart, D., Hinton, G., and Williams, R. (1986). Learning representations by back-propagating errors. *Nature*, 323:533–536.

Soechting, J. and Roberts, W. (1975). Transfer characteristics between EMG activity and muscle tension under isometric conditions in man. *J. Physiol.*, 70:779–793.

Suzuki, R. and Suematsu, T. (1969). Pattern recognition of multi-channel myoelectric signals by learning method. *Japanese J. Med. Elec. Biol. Eng.*, 7:47–48, (in Japanese).

Uno, Y., Kawato, M., and Suzuki, R. (1989). Formation and control of optimal trajectory in human multijoint arm movement: minimum torque-change model. *Biol. Cybern.*, 61:89–101.

Uno, Y., Suzuki, R., and Kawato, M. (1989). Minimum-muscle-tension-change model which reproduces human arm movement. *Proceedings of the 4th Symposium on Biological and Physiological Engineering*, 299–302, (in Japanese)

Wada, Y. and Kawato, M. (1992). A new information criterion combined with cross-validation method to estimate generalization capability. *Sys. Comput. Japan*, 23:92–104.

Winters, J.M. (1990). Hill-based muscle models: a systems engineering perspective. In *Multiple Muscle Systems*. Winters, J.M. and Woo, S.L.-Y. (eds.), pp. 69–93. Springer-Verlag, New York.

Wood, J., Meek, S., and Jacobsen, S. (1989). Quantitation of human shoulder anatomy for prosthetic arm control-II. anatomy matrices. *J. Biomech.*, 22:309–325.

Commentary:
What Can We Learn from Artificial Neural Networks About Human Motor Control?

A.J. (Knoek) van Soest

In this interesting chapter by Koike and Kawato, the authors start out by defining the problem of control of positioning tasks as a series of transformations, ultimately transforming the desired end effector position into accelerations of the extremity. They focus on a part of this cascade of transformations; in particular on the mapping from neural input to muscles (alpha motoneuron activation) to their kinetic output (joint torques). In considering this mapping, they take surface electromyograms as the operational measure of neural input, and attempt to predict the movement that results from this EMG. The most interesting part is the way in which they approach the problem of mapping EMG into net joint torques. Taking an approach inspired by artificial neural networks, they construct an ANN that takes EMG at current and past time, current joint angles and current joint angular velocities as its input and produces the resulting net joint torques at the current time as its output. Subsequently, using standard techniques, they calculate accelerations and numerically integrate these in order to obtain position and velocity at the next point in time. This general approach is worked out for a restricted class of movements: two degrees of freedom arm positioning movements in two-dimensional space. For this class of problems, it is shown that it is indeed possible to come up with an ANN that does the job satisfactorily. As correctly claimed by the author, this is the first actual demonstration that multijoint movements can be predicted from surface EMG signals and initial state of the extremity.

The second part of this chapter starts to consider arm movements in 3D space. For reasons that are not quite clear, the authors take a different approach here. They construct, again using an ANN, the map from EMG to static arm position. When this map is subsequently applied to nonstatic data, the obvious result is obtained that the position predicted by the model does not match the current position—obvious because during movement nonzero acceleration of the arm is required and thus the torque produced by gravity and net joint torque produced by the muscles cannot be in equilibrium. It might have been more interesting to attempt to extend the approach used in 2D to 3D while retaining the ordering of inputs and outputs of the model.

Returning to the first part of the chapter, it must be noted that there have been some earlier attempts to predict the output of the motor system from measured EMG. In particular, Bobbert et al. (1986) were able to predict net joint torque at the ankle during vertical jumping on the basis of EMG data obtained from soleus and gastrocnemius muscles. That and similar studies were less ambitious than the chapter, however, for several reasons. First, because in tasks such as vertical jumping there is little fine-regulation of EMG amplitude; the EMG resembles bang-bang control. Secondly and probably more importantly, in those studies, rather than using the predicted torques to drive a forward dynamic simulation and comparing the resulting kinematics with the actual kinematics, the measured kinematics were considered as an input to the model and the comparison made was between the predicted net joint torque and the net joint torque calculated on the basis of inverse dynamics. Such an approach is easier to be successfully applied because it prevents the accumulation of error in the kinematics over time.

Taking a skeptical position, it might be argued

that because ANNs can approximate any mapping, all that is shown by Koike and Kawato is that for the type of task studied, a deterministic relation exists between the neural input and skeletal state (position and velocity) on the one hand, and the joint torque on the other. However, their contribution is in our view more valuable than that. In particular, it shows that although ANNs can theoretically be shown to be able to approximate any mapping, it takes a lot of physical insight into the problem to actually set up an ANN that does the job. In the case of the authors, it apparently required a modular approach with separate networks for static and dynamic situations, with a gating network deciding which of the two were to be activated; furthermore, pretraining of the modules apparently was essential to obtain convergence in a reasonable time.

If we accept that the structure of the ANN used in this chapter was inspired by physical knowledge of the problem to be solved, the question then arises as to whether this approach is specific to the type of task studied or if it is more generally applicable. Two ways to address this question seem to be open. Most importantly, it may be empirically investigated by extending the approach to more complex tasks in a more complex (i.e., 3D) space. In our view, the extension to 3D made in the current contribution is not very helpful, because an essentially different model is set up. We would like to urge the authors to extend their approach to more complex situations. As an alternative, a first impression of the generalizability of their approach may be obtained analytically, by confronting knowledge from muscle physiology about the variables that influence muscle force to the inputs in their model. From muscle physiology, the most important determinants of muscle force are known to be neural input, fiber length and fiber shortening velocity. At a more refined level, contraction history has an influence on force production (e.g., force potentiation). At first sight, the primary determinants are all present in the input to their model. The only thing that can be noted is that in their model the kinematic input is provided in terms of joint angles, which does not relate unequivocally to muscle fiber length due to tendon elasticity. However, neglecting tendon stretch is unlikely to be a significant problem in the majority of tasks. Thus, it would be our a priori expectation that the approach is indeed generalizable to more complex situations.

An interesting way to test if the physiologically well-known force-length-velocity relation of muscle is actually represented in the neural network would be to clamp the EMG and vary kinematical inputs. On that basis, the torque-joint angle-joint angular velocity relation embodied in the ANN could be obtained and compared against data that may be obtained from dynamometer experiments.

Finally, we wish to comment on the relevance of the approach to motor control issues. Throughout the chapter it is suggested that EMG may be considered as a variable controlled by higher centers in the CNS. Although descending signals are a major determinant of alpha motoneuron activity, there is always a feedback contribution to that activity. In other words, EMG is not the reflection of supraspinal commands; in the terminology of Feldman (Feldman and Levin 1995), it cannot be a controlled variable. According to Feldman, it is furthermore possible that the same EMG level can result in static equilibrium in different positions, a suggestion that casts doubts on the approach taken in the second part of the chapter. From a mathematical point of view, given the fact that both the gravitational torque and the muscular torque are nonlinear functions of joint angle, it may indeed be that these torque-angle functions intersect at more than one point. Thus the possibility cannot be excluded that more than one stable equilibrium position corresponds to a particular level of EMG.

The suggestion following from the final part of the chapter is that EMG may be considered to define an equilibrium position towards which the extremity is attracted due to the intrinsic stiffness of muscle. This suggestion is very close to the so-called alpha-model of equilibrium point control proposed by Bizzi (e.g., Bizzi et al. 1992). It is well known, however, that often the intrinsic stiffness of muscle is neither sufficient to guarantee stability at the equilibrium point nor adequate to generate fast movements. In fact, it was argued by van der Helm and van Soest (Chapter 29) that even when the enhancement of stiffness that results from (time-delayed) position feedback is included, production of fast movements is still problematic. Thus we would expect that if Koike and Kawato would go on to calculate EMG-based equilibrium point trajectories and corresponding stiffnesses (and we hope they do!), they would find relatively low stiffnesses and thus in case of fast movements equilib-

rium point trajectories that are in no way related to the actual trajectory to be produced.

References

Bizzi, E., Hogan, N., Mussa-Ivaldi, F.A., and Giszter, S. (1992). Does the nervous system use equilibrium point control to guide single and multiple joint movements? *Behav. Brain Sci.*, 15:603–613.

Bobbert, M.F., Huijing, P.A., and Ingen Schenau, G.J. van (1986). A model of the human triceps surae muscle-tendon complex applied to jumping. *J. Biomech.*, 19: 887–898.

Feldman, A.G. and Levin, M.F. (1995). The origin and use of positional frames of reference in motor control. *Behav. Brain Sci.*, 18:723–806.

Commentary:
What's the Use of Black Box Musculoskeletal Models?

Sybert Stroeve

In this chapter, Koike and Kawato show that a neural network system can form a fairly accurate mapping from joint angles, angular velocities and relevant EMG-signals to joint torques for arm movements in the horizontal plane. Forward dynamic calculation of end-point trajectories based on these estimated joint torques demonstrates that the errors in this estimation process are small enough to ensure relatively small deviations between the calculated and the actual end-point trajectories. What do we learn from these results? Beyond the fact that it is possible to construct such a mapping, it is hard to reveal any new clues about biomechanics or movement control from the presented calculations. A drawback of the application of a neural network as a model is that it contains no physically interpretable parameters, as are present in for example Hill-type musculoskeletal models. If an adapted model structure which includes the Kelvin–Voight muscle model is applied, as suggested by the authors, it is questionable whether model parameters can be identified well, given the large number of model parameters. Still, it might lead to a more interpretable model. Another method to increase the transparency of the neural network model is linearization of the network during a trajectory (Stroeve 1999). This means that the sensitivities of the joint torques for changes in the EMG-signals are calculated along a trajectory.

Although new insight in the mechanisms of movement control is not offered by the suggested combination of a black box model of muscular torques and a white box model of skeletal dynamics, it presents a useful tool for movement control research, provided that a valid model has been attained. The achieved neural network model was validated for movements which were not in the training set to the neural network optimization, but were similar to those movements. Fairly accurate predictions could now be made, indicating that the variance of the neuromuscular control characteristics between movement trials was small and its basic characteristics were well captured by the neural network. Since the system was not tested for movements with another set of kinematic constraints (as starting points, velocities), it is not clear whether the system captured a general blueprint of the forward dynamics. Given the nonlinear characteristics of the musculoskeletal system, it may be expected that the model only comes up with accurate predictions if the applied movements have about the same characteristics as the movements presented during training. The presented results concern movements with durations ranging from about 600 to 800 ms. It is to be expected that the estimation accuracy of the neural network model is reduced if it would be trained with a larger variety of movement durations. In that case a larger number of expert networks might improve the estimations.

A valid black-box musculoskeletal model may be of use for motor control research in several ways. It might, for instance, be used to calculate the load-sharing problem for a series of criterion functions. Comparison of the results of the load-sharing optimizations with measured EMG signals can support a sound choice of the criterion function. As suggested by the authors, a black-box model can be applied to evaluate hypothesized neuromuscular control schemes as equilibrium-point control (Feldman 1986), feedback error learning (Kawato et al. 1987) or other neural control schemes. In general, a black-box musculoskeletal model may be useful for evaluation of all kinds of neural control principles as long as understanding of the effects of characteristics of the musculoskeletal system on the achieved neural control is not aimed at.

References

Feldman A.G. (1986). Once more on the equilibrium-point hypothesis for motor control. *J. Mot. Behav.*, 18: 17–54.

Kawato M., Furukawa K., and Suzuki R. (1987). A hierarchical neural network model for control and learning of voluntary movement. *Biol. Cybern.*, 57:169–185.

Stroeve S. (1999). Impedance characteristics of a neuromusculoskeletal model of the human arm, I Posture control. *Biol Cybern.*, 81:475–494.

35
Study Movement Selection and Synergies via a Synthesized Neuro-Optimization Framework

Jack M. Winters

1 Introduction

The question of whether or not human movement is "optimized" has generated considerable discussion over the years. In part, this evolving controversy is based on misconceptions regarding the optimization framework.

The purpose of this chapter is to develop a more formalized foundation for neuro-optimization, and in the process make the case for why this will become increasingly important as we move toward studying more realistic larger-scale systems, and addressing the movement selection process.

2 Motivation

As motivation for this topic, consider the following observations:

- We are *goal-directed* creatures. Most tasks we perform are directed toward some aim. Often the primary goal can be formulated via natural language. This is especially true for tasks of research interest (e.g., via instructions to subjects, athletic performance, tasks by persons with disabilities).
- The engineering tool for mathematically handling goal-directed problems is optimization.
- The genuine divergence in design strategy between biological organisms and human-designed machines, such as robotic manipulators, leads naturally to the utilization of optimization as a tool for selecting between alternative biosolutions.
- Optimization can be thought of as a practical tool that maps the multidimensional mechanics/con-

trols problem to a more intuitive one that focuses on causality between goals, performance, and strategy sensitivity to key model parameters or environmental conditions (Hogan and Winters 1990).

- There are virtually always secondary (implicit) criteria (e.g., related to effort, stability, energy, safety) that are an intrinsic part of the neurocontrol infrastructure.
- The optimization process solves *redundancy* problems (kinematic, actuator) as a natural byproduct of meeting primary (explicit) and secondary (implicit) goals.
- *Learning* can be viewed as a search process (Jang et al. 1995). The optimization framework ties fully to evolving connectionist neural net (CNN) concepts, and indeed adaptive CNNs are often used as algorithms for solving optimization problems. Both conventional and connectionist (neural, genetic) algorithms normally try to drive scalars (e.g., called evaluation, value, fitness, energy, cost functions) toward extrema, often by working on approximations of gradients. Also, reinforcement learning approaches such as Adaptive Heuristic Critic networks (Barto et al. 1983) are an approximation of the classic optimization algorithm called Dynamic Programming (Werbos 1990).
- Unlike in the normal optimization problem, where all that ultimately matters is the final solution (i.e., algorithm as tool), in neuro-optimization the *process* of converging toward a solution may itself be a part of the research because it relates to the learning process.
- In contrast to layered artificial CNNs, the brain is a *highly organized structure* (e.g., distinct white and gray matter areas). The computational neuro-

control architecture appears inherently distributed, modular, and hierarchical.

- Precise ("crisp") neuro-information of high certainty is the exception. Posture and movement strategies must be selected, implemented, and evaluated in real time. This suggests a role for nonanalytic, "soft computing" approaches such as neuro-fuzzy classification and control (e.g., see Chapter 40).

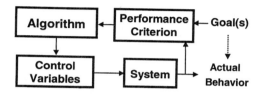

FIGURE 35.1. Structure of optimization problem.

3 Foundations and Key Terminology

3.1 "Redundancy" and "Synergy"

The chapters in Section VII are concerned with more realistic, larger-scale models, and tasks *where alternative solutions are possible*. It is useful to define several working definitions:

- *kinematic redundancy*: more joint degrees of freedom (i.e., generalized joint coordinates) than endpoint coordinates,
- *actuator redundancy*: at least twice as many muscle actuators as joint degrees of freedom,
- *muscle synergy*: available neurocontrol sequence primitives (e.g., for posture, rhythmic patterns).

Notice that redundancy expands the realm of viable solutions, while synergy narrows it. While Brown and Loeb (Chapter 10) prefer "overcomplete" to "redundancy" because there are likely good reasons for virtually every muscle and joint configuration within an organism, the above definitions are useful. "Synergy" is here given a narrower, more operational definition than is common in the neurophysiology literature, one that appears consistent with the "primitive" concept (e.g., see Chapter 25 and Chapter 24) and subtask modules (e.g., Chapter 19, Chapter 23, and Chapter 34).

3.2 Formulation of the Optimization Problem

The classic constrained nonlinear optimization problem can be considered to consist of four distinct parts (Figure 35.1) (Hogan and Winters 1990):

- a *system* to be controlled, fully characterized by a set of bounded states x_s (this can be thought of as providing a set of dynamic nonlinear constraints),
- a scalar *performance criterion* (J_{pc}), that defines

the task goal, and what constitutes successful completion of this goal (often includes multiple subcriteria),

- a set of *control (design) variables* (x_c) available to be modulated (e.g., pulse width), that map to control system inputs (e.g., neuromotor inputs to muscle),
- an *algorithm* that extremizes the performance criterion (e.g., gradient-based or random-based optimizer; some type of connectionist network).

Operationally, the problem definition is:

Minimize $J(x_c)$, by modulating x_c, subject to a set of inequality/equality/dynamic constraints $g_i(x_s, x_c, x_e, t)$, with bounds (hard or soft) on x_s and x_c

where the vector x_e represents any external coupling (e.g., a contact force). When the problem involves designing/refining a system or structure by modifying *model (design) parameters*, it is called *optimal design*—of interest to engineers and to bioscientists studying longer-term evolutionary/adaptive mechanisms (e.g., see Chapter 13). When it involves determining the "best" strategy to complete a task by modifying control inputs, given a defined system, it is called *optimal control*—of interest here. However, there is no reason why the two cannot be combined (e.g., simultaneous modulation of bicycle design parameters and neurocontrol signals).

The constraints can normally be thought of as due to a dynamic system model, of the form:

$$\frac{dx_s(t)}{dt} = \Phi(x_s(t), x_c(t), x_e(t)); \quad x_{c\text{-}low} \le x_c \le x_{c\text{-}high}$$

$$y(t) = \Psi(x_s(t), x_c(t), x_e(t)); \quad x_{s\text{-}low} \le x_s \le x_{s\text{-}high}$$

$$(35.1)$$

This model simply captures the laws of physics. In words, the algorithm searches for the solution (i.e., values for the control variables) from a set of alternative solutions that best satisfies the performance criterion, given the physical system to be controlled.

3.3 Limitations of Inverse Thinking

We must distinguish between *inverse optimization* (IO) approaches and *forward optimization* (FO) approaches (for a more detailed treatment, see Winters (1995a) or Chapter 32). IO is commonly used to solve the actuator redundancy problem, often called the load-sharing problem, after an inverse dynamic analysis was performed using *available* experimental data to estimate joint moments. Typical inputs are joint moments and kinematics, and outputs are muscle forces (or perhaps muscle activation). This normally involves solving algebraic equations at slices of time, and minimizing an effort-type criterion, such as muscle stress. Such approaches have been especially prevalent for studies of gait and of low back loads during lifting tasks (see Winters 1995a for review). Notice that since kinematics are already prespecified, mechanical stability is not normally an issue, nor are the actual performance goals of the task. While often useful within the context of estimating muscle load distribution, IO techniques described within the biomechanics literature as "optimization" have tended to give optimization a bad name, and are not related to the type of problem that must be addressed by the actual neuromotor system. We will not consider such techniques further (although see Chapter 32 for newer, more innovative methods for IO).

In contrast, in FO the "goal" is embedded as part of the performance criteria, neither postural kinematic configurations or movement paths are prespecified, and the system is defined by differential equations that describe the laws of physics. This is what control engineers mean by optimal control. For larger-scale systems, finding analytical solutions is normally impossible, and even finding numerical solutions can prove challenging.

4 Developing a Neuro-Optimization Foundation

4.1 Performance Criterion for Posture/Movement

The practical role of the performance criterion is not only to specify the goal(s) but also to help facilitate the algorithmic process of selecting from a set of viable solutions that would meet the specified task goal(s) reasonably well. We now define the more general optimization performance criterion for human movement, in *three structural forms*: summed subcriteria, "satisficing" multicriterion, and "soft neurofuzzy multicriterion."

The "system" can be either a musculoskeletal (MS) model or a neuromusculoskeletal (NMS) model. If one's "system" is a MS model, it is critical that J_{PC} include multiple subcriteria. The reason is simple: the CNS is a sophisticated and distributed structure, and considerable infrastructure is dedicated to critical ancillary functional needs (e.g., maintaining overall stability, moving safely with low pain, and using relatively low effort)—thankfully, most sensorimotor CNS structures do not "turn off" and let a handful of cells control the motoneuronal pools.

4.1.1 Conventional: Summed Subcriteria and MS Model

Our first approach assumes a single global scalar criterion (e.g., Zajac and Winters 1990; Winters 1995a, 1996), with weights for several types of subcriteria:

$$J_{pc} = w_{task} \sum_{i=1}^{nj} (J_{pc_{task}})^{pi} + [J_{intrinsic}] \; ; \quad (35.2)$$

where $[J_{intrinsic}] =$

$$\left[w_{eff} \sum_{l=1}^{nl} (J_{pc_{eff}})^{pi} + w_{pain} \sum_{k=1}^{nk} (J_{pct_{pain}})^{pk} + \right.$$
$$\left. w_{stab} \sum_{m=1}^{nm} (J_{pc_{stab}})^{pm} + w_{dyn\text{-}syn} \sum_{n=1}^{nn} (J_{pc_{dyn\text{-}syn}})^{pn} \right]$$

where each $J_{pc\text{-}i}$ includes a mathematical expression (often premultiplied by a local relative weight w and raised to a power p), and n_l, n_k, n_m, and n_n represent the number of muscles, tissues, available stability strategies for a given postural configuration, and dynamic macrosynergy coordination restraints, respectively.

Here $J_{pc\text{-}task}$, describes the primary *task goal*, and thus for our purposes is related to the instructions given to the subject. For example, a common goal in research studies is to track a reference trajectory (smooth or with target jumps) with minimal error:

$$J_{pc_{task}} = \sum_{i=1}^{n} \|x_{i_{ref}} - x_i\|^p \qquad (35.3)$$

where x_i is normally an end point (e.g., hand), but could be a joint or a mapped location (e.g., on screen), and typically $p = 2$. Another classic case is for $J_{pc\text{-}task}$ to be minimum time (e.g., for saccadic eye movements) (e.g., Winters et al. 1984; Clark and Stark 1975). In sports, common goals relate to performance outcome, such as jump height or the release velocity (and direction) for an object being thrown or kicked (e.g., Hatze 1981). For more involved tasks, this task goal can also be considered as several task subgoal modules, which could work in parallel or be gated on (i.e., have nonzero weights) sequentially.

The intrinsic subcriteria under $J_{intrinsic}$ represent the various implicit CNS phenomena, here including:

- $J_{pc\text{-}eff}$, represents penalty for *neuromuscular effort or energy*; the subcriteria in most common use is the sum of the n_m muscle stresses. This helps capture the reality that for whatever the reasons, there are intrinsic CNS mechanisms (mostly at lower CNS levels) that tend to pull motoneuronal signals back toward lower sustained rates (e.g., 5–10% of maximum). It can include neural effort (i.e., penalizing neurocontrol inputs), muscle stress (force per cross-sectional area), muscle force, muscle energy dissipation, muscle activation, muscle volume, etc. (Winters, 1995a; see also Chapter 32). Although these "costs" tend to be synergistic, when used in tandem with competing "reference" task criteria, mildly different strategies will emerge (Seif-Naraghi and Winters 1990).

- $J_{pc\text{-}pain}$ represents the profoundly strong drive toward preventing self-injury of tissues (with a certain margin of safety), and especially penalty (high weight) for using tissues that are causing pain or discomfort (Winters 1995a). Here "tissues" include joint loads (e.g., shear stress in the low back or knee).

- $J_{pc\text{-}stab}$, represents a profoundly strong intrinsic CNS drive towards maintenance of a level of relative mechanical stability. An interesting way of viewing J_{stab} is to associate it with a scalar related to *potential energy* (E_{pe}), which when near equilibrium is in turn related to mechanical stability (Winters and Van der Helm 1994; Winters 1995a). If we assume that muscles possess spring-like properties, a formalized mathematical framework emerges that states that for a steady posture under a certain set of control variables to be sta-

ble, $E_{pe}(\boldsymbol{x}_{x\text{-}q})$ must be a minimum, where $\boldsymbol{x}_{s\text{-}q}$ are the states associated with the generalized joint coordinates. While the pure mechanics can be formulated via equality ($\partial E_{pe}/\partial \boldsymbol{x}_{x\text{-}q} = 0$) and inequality ($\partial^2 E_{pe}/\partial \boldsymbol{x}_{x\text{-}q}^2 > 0$) constraint equations (Winters and Van der Helm, 1994; see also Chapter 32, here we address the more realistic case a desired "comfort" level of relative stability (e.g., eigenvalues of the apparent stiffness $\boldsymbol{K}_a = \partial^2 E_{pe}/\partial \boldsymbol{x}_{x\text{-}q}^2 >$ positive constant), a more reasonable alternative is to consider it a *soft neuroconstraint*. The event-driven criteria of Chapter 19 (related to ankle and hip-centered strategies) also fall into this category.

- $J_{pc\text{-}dyn\text{-}syn}$ represents penalty for violating well-documented macro neurodynamic synergies in movement or muscle coordination patterns, such as the intrinsic tendency toward frequency phase-locking between body segments during tasks such as finger tapping or locomotion. In line with the writings of Bernstein (e.g., 1967), known synergies (i.e., available neurosequence primitives) should be used.

If in Eq. (35.2) we let the task be tracking a target, and the secondary subcriteria (terms within brackets in Eq. 35.2) be simply related to input cost ("effort"), we obtain the form used for the classic optimal control problem, for which a rich theory exists for simpler lower-order plants (e.g., Athans and Falb 1966), including applications to human movement (Clark and Stark 1975; Nelson 1983; Oguztorelli and Stein 1983; Seif-Naraghi and Winters 1990). Additionally, for the special case of letting the power n equal 2 and assuming a linear plant, Eq. (35.2) then relates to the classic optimal linear quadratic regulator (LQR) problem, which again includes human movement applications (Loeb and Levine 1990).

It is important to realize that some subcriteria are *competing* while some are *complementary*. For instance, in the classic optimal tracking case $J_{PC\text{-}task}$ (output performance) and $J_{PC\text{-}eff}$ (e.g., input "energy") are competing, and often $J_{pc\text{-}effort}$ and $J_{pc\text{-}stab}$ are competing subcriteria in the sense that effort-oriented subcriteria tend to lower neuromotor drives, while concerns for stability often require that there be cocontraction of muscles so as to raise certain stiffnesses above critical levels.

Here, however, we generalize the ancillary subcriteria concept, more related to numerical (vs. analytical) optimization: $J_{implicit}$ sculpts the J_{PC} -x_c space so as to help *select* between viable solutions, while also affecting (usually positively) the rate of convergence (Seif-Naraghi and Winters 1990). Because these ancillary terms are intended to represent intrinsic neural processes, it is often useful to consider them as pre-set (i.e., task-independent, but at least for $J_{pc\text{-}stab}$ configuration dependent). Thus in contrast to some criticism, they can be viewed as making the process as more realistic, rather than more complex!

Often the simple existence of an effort-type subcriteria—even with low relative weight—helps solve the redundancy problem by identifying ("selecting") from the set of feasible solutions that which best minimizes effort. Thus in some cases final solutions are relatively insensitive to the value of the weights—the mere presence is enough. Yet in many other situations the magnitude of weights does influence performance—and indeed is a mechanism for *scaling movements* when subcriteria are competing. For instance, the basic "speed sensitive" and "speed insensitive" scaling strategies documented in Gottlieb et al. (1990) can be captured by changing the relative weight between competing subcriteria related to fast performance ($J_{PC\text{-}task}$) versus the effort ($J_{PC\text{-}eff}$)—as the relative penalty for effort w_{effort}, grows, movements automatically scale with speed ("speed sensitive") via primarily pulse height modulation (Seif-Naraghi and Winters 1990; Winters and Seif-Naraghi 1991). Similarly, "speed insensitive" pulse-width modulation is seen with changes in the reference trajectory magnitude or in system inertia. Ironically, this subcriteria-based approach provides more realistic control behavior than forcing certain output speeds via a single reference trajectory (a "task" subcriteria)—the optimal solution then involves excessive cocontraction and "pull-pull" behavior (Seif-Naraghi and Winters 1990). And let us be honest: our commands to subjects (and more importantly their interpretations) all involve fuzzy natural language qualifiers—"move slower," "move with less effort," and so on. Even the overused "fast and accurate" instruction involves subjective interpretation.

So why include a "pain/discomfort/injury" subcriteria? This author has recently had anterior cruciate ligament and rib injuries. Both resulted in modified neurocontrol strategies, for tasks ranging from gait to reaching. To simulate this via FO, one can: (1) modify certain system parameters (e.g. to capture changes in tissue mechanics); and (2) add penalty for painful loading of the involved tissues. The result: alternative coordination strategies.

The above approach is certainly computationally complex, yet remarkably intuitive to a user because of its direct cause–effect nature. As such, this type of J_{PC} , in conjunction with FO, *provides a interactive tool for non-biomechanists (e.g., medical experts, students) to participate in what would otherwise be a remarkably complex neuromechanical problem.* Thus rather than dealing directly with the dynamic equations representing the system, one can instead focus on the transformed problem of the form:

$$y(t) = \Psi(J_{pc}) = \Psi(\sum_i w_i J_{pci}) \qquad (35.4)$$

where the output performance $\mathbf{y}(t)$ is now *causally related to the objectives*, as expressed by the subcriteria. A complex mechanics problem is mapped into a more intuitive problem for users (e.g., experts, students) that focuses on causality between the prescribed w_i's of $\mathbf{J}_{pc}$'s and subsequent observed performance.

However, there are three possible concerns: (1) w's depend on subjective expert judgment, (2) the process is not modular; and (3) the optimization structure is "open-loop" rather than adaptive.

4.1.2 Multicriterion Optimization: "Satisficing" Performance

If we choose to instead view subcriteria as a set of *independent objectives*, to somehow be assimilated, we have what is sometimes called the Pareto optimality problem for multicriterion optimization, which involves determining a *"satisficing"* (noninferior) solution from a set of viable alternatives (March and Simon 1958; Sakawa 1993). The general method can be formulated as (Cichocki and Unbehauen 1993):

$$\min\left[\sum_{k=1}^{n} \{w_k J_{pck}(x_c)\}^p\right]^{1/p};$$

$$(1 \le p \le \infty); \quad \sum_{k=1}^{n} w_k = 1; \quad w_k \ge 0 \qquad (35.5)$$

where if $p = 1$, we have the standard additive weighting method (with J_{PCi} becoming subcriteria), while if p is infinity, we have the minimax method (with J_{PCi} thought of as a set of objectives). Notice that we have made a critical step toward combining "apples and oranges" objectives. Hence why this type of approach is becoming popular in engineering design (e.g., Fan et al. 1992). Conceptually, one works on one criteria J_{PCi} at a time, but drives toward an optimal solution that often switches (or "gates") between criteria—not unlike a swimmer whose coach has her focus on the weakest aspect of her stroke, then move to another aspect. This allows intangible factors (e.g., pain in a certain joint, a history of heart problems) to influence one's strategy, yet from within a *mathematical* framework. This is one way to solve the problem of selecting "compromises" among multiple objectives, such as the types of postural perturbation responses presented in Chapter 19. Sakawa (1993) employs a fuzzy nonlinear programming formulation in which J_{PC} and any constraints are "softened" via a mapping to a strictly monotone membership function on the interval $<0,1>$.

A special type of the multicriterion optimization, called the goal programming method (Sakawa 1993), assumes that the human decision maker (HDM) has *known goals* for each of the scalar objectives. We then minimize an scalar "error" associated with each objective (i.e. ultimately the sum of all these errors would approach zero). It is as if there are a set of problems, each solved by a reasonably separate module (i.e., in *parallel*), and potentially using a different structural architecture.

This is a framework for bringing together concepts within the literature. For instance, in the Feedback-Error Learning approach of Kawato's group (e.g., Kawato, 1990; Gomi and Kawato, 1993; Chapter 34, see Figure 34.C1) connectionist neural network (CNN) modules "learn" by gradually transitioning from a feedback-centered control to more efficient and effective control by CNNs, and do so via parallel "expert modules" for tracking and posture. In essence, actual mechanics (via feedback error) serves an implicit teacher for capturing aspects of inverse/forward dynamics and posture. A limitation of their approach (and most others, e.g. Chapter 29) is that the "goal" (reference) is typically prespecified kinematics (e.g. joint angles in Figure 35.C1 and 35.C2, in commentary). This kinematocentric (vs. task-centered) approach limits its extension; yet through their modular struc-

ture we see more clearly the link between the optimization and learning processes.

4.1.3 Performance Criteria Synthesis via a "Soft Computing" Framework

A disadvantage of the form of Eq. (35.2) is that it lacks a convenient methodology for switching (gating) between tasks (e.g., based on "events" defined by expert rules) or refinement of relative weights (e.g., through adaptive processes). More importantly, its not the optimal interface for engaging human decision makers. In Chapter 42, Crago et al. refers to "expert rules" for neuroprosthesis control by finite state methods, while in Chapter 39, Popovic and Popovic refer to "nonanalytic" control approaches that emphasize the need for rule-based hierarchical multilevel control that match decision dynamics to system dynamics. In Chapter 19, Horak and Kuo define certain perturbation response synergies, but then note that for some perturbations the neuromotor response appears to be a combination of several strategies. Then in Chapter 40, Jacobs and Tucker develop the concept of soft computing, and especially bottom-up neurofuzzy systems for evaluation and control of human performance. Earlier we suggested that a degree of "relative" stability (a "soft," though mathematically precise, concept) made more behavioral sense than the "hard" constraint of absolute stability. We see a trend. In dealing with actual human behavior, *performed by humans interpreting instructions and interpreted by humans performing science*, boundaries are rarely crisp, but in some sense fuzzy. Here fuzziness refers to the inherent deterministic ambiguity (vagueness) associated with human inference.

Thus as a third approach, we develop a "soft computing" framework that encompasses both the "additive subcriteria" (Subsection 4.1.1) and "multiobjective" (Subsection 4.1.2) approaches as special cases, while also embedding expert inference directly into the optimization problem. Three key points provide the motivation for a fuzzification of the J_{PC} aspect of the optimization problem:

1. Eq. (32.2) can be viewed as determining, *in parallel*, a set of local criteria—fuzzy set theory provides an intriguing (and computationally formalized) foundation for *synthesizing information* to obtain a more robust index of performance.

2. Because the process of combining subcriteria involves a Human Decision Maker (HDM), normally an "expert" heuristically selecting relative weights (merit) or coming up with rules, there is an imprecise ("fuzzy") nature to the problem that ties to human thinking and natural language.

3. Fuzzy logic and CNNs are complementary technologies for designing intelligent systems (Lin and Lee 1996). The field of integrated neuro-fuzzy systems is growing rapidly, for a pragmatic reason. While CNNs tend to be "black boxes" (with meaning not associated with weights), adaptive fuzzy nets can use rule production algorithms that better interact with HDMs (see also Chapter 40).

The concept of utilizing fuzzy inference and weights for multiobjective optimization problems is not new (Bezdek 1981; Sakawa 1993; Jang et al. 1995) and, indeed, is well developed. Its roots lie in a classic paper on decision-making in a fuzzy environment written by theoretical pioneers Bellman and Zadeh (1970), who gave us Dynamic Programming and fuzzy set theory, respectively.

Consider the following mathematical framework for utilizing fuzzy expert rules to perform modular optimization with fuzzy gating, here for two rules, two variables (y_1, y_2) and two objectives (see also Figure 35.2):

if (y_1 is A_1) and (y_2 is B_1) then $J_{PC} = f(x_c, y, w)$;
$$0 \leq w \leq 1$$
if (y_1 is A_2) and (y_2 is B_2) then $J_{PC} = f(x_c, y, w)$;
$$0 \leq w \leq 1$$
$$(35.6)$$

where here the input y_i could be, for instance, a system output related to performance, a muscle state variable, a weight, the output of another fuzzy rule (i.e., its "firing" level), or the result of an interactive menu involving the user. The arrows indicate the flow of the fuzzy inference process (see Chapter 40 for a description of fuzzy sets and rules). Although there are many forms for fuzzy inference systems, in Figure 35.2 we utilize a Sugeno fuzzy model (Takagi and Sugeno 1985; Jang et al. 1995). Notice that this is essentially just a nonlinear input–output mapping in which internal inference occurred via fuzzy mathematics, and a "defuzzification" process is employed to obtain J_{PC}. Such approaches provide a more friendly linguistic inter-

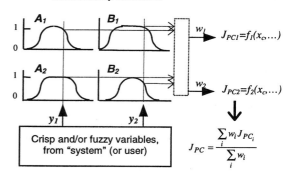

FIGURE 35.2. Structure for determining a "soft" criteria, assuming a simple Sugeno fuzzy model, which yields a crisp scalar output that here is the performance index J_{PC} (see also Jang et al. 1995). A_i and B_i are typically linguistic phrases. Combinations of "and" and "or" inference is possible, with the weights w_i's giving the relative importance of the local J_{PCi} module associated with the respective rule(s).

face for users. Equally importantly, they include the approaches formulated in the prior two sections as special cases: (1) if a criterion to the right (e.g., J_{PC1}) is viewed (if desired) as a full performance criterion as in Eq. 35.2 with polynomial powers of one (two), then we have what is called the first-(second-)order Sugeno fuzzy model—we are simply summing weighted subcriteria, with no rules; and (2) if the fuzzy if–then rules are allowed to be crisp (a special case), the satisficing multicriterion form emerges. A key advantage is that this framework can naturally include *conditional fuzzy expert rules as modules*, which ties it to conditions that trigger behavior based on event recognition, such as the ankle or the hip responses discussed in Chapter 19, and thus to event-driven applications of optimization.

In formalizing this process, it is useful to note that any of the following can be fuzzy or crisp: *data* ("facts"; for simulations these are normally crisp), *mapping* (e.g., fuzzification via membership functions), the set of conditional *rules* which relate one or more premises (antecedents) to consequences (the special case of a crisp rule can serve as a "gate" toggle, or fuzzy singleton), *merit indices*, and the *synthesis process* (here defuzzification so as to yield J_{PC}). There can also be series–parallel structure—"firing" may depend on the state of other "expert" modules (a series arrangement), and indeed on crisp "gating" as well as fuzzy inference.

As formulated so far, this is a top-down, expert-driven *fuzzy subsystem that determines J_{PC}*, with the fuzzy infrastructure providing a degree of expert knowledge representation, uncertainty tolerance, and robust operation. The optimization structure remains intact.

A logical extension is to use CNNs to implement this fuzzy system. As developed in Subsection 4.3 in Chapter 40, the idea behind neurofuzzy systems is to take advantage of the best of each: the adaptive learning capabilities of CNNs and the fuzzy interface to HDMs. Implementation is straightforward, and indeed the package ANFIS (Adaptive Neuro-Fuzzy Inference Systems) is part of the Matlab Fuzzy Toolbox (Jang et al. 1995). Candidates for tuning are at all levels of Figure 35.2: membership functions, fuzzy rules, or the weights that are part of the defuzzification process. A word of caution: most of the vast efforts on neuro-fuzzy learning use "bottom-up" tuning in which a CNN passes on its adaptive "knowledge"—the system "creates" fuzzy if-then rules and modifies weights that subsequently can in theory be interpreted (and modified) by a HDM (e.g., see Chapter 40; Lin and Lee 1996). However, here our approach differs in that such bottom-up tuning is used very selectively (e.g., to refine secondary weights or rules), and is subservient to top-down expert reasoning.

4.2 Moving Toward Modular Neuro-Algorithms

We now shift our focus from the determination of J_{PC} to the form of the algorithm that must find a solution.

4.2.1 Severe Limitations of Analytical Approaches

When the system model is very simple (e.g, low-order, linear), a rich theory of analytical optimal control (e.g., see Athans and Falb 1969; Reklaitis et al. 1983) is available that can be used to determine solutions as a function of the simple performance criteria and the specified initial and final conditions (see Nelson 1983; Seif-Naraghi and Winters 1990). However, here we are only interested in reasonably sophisticated models, and thus

numerical methods are necessary to find optimal solutions. Also, end conditions often need not be precise.

To date, most advanced optimization studies have chosen the "system" to be a musculoskeletal (MS) model, with our control parameters defining an EMG-like input that is essentially the CNS. It is an "open-loop" type of approach, which we justify by noting that we are in fact obtaining the control signal for the "final common pathway." This has served us well for a certain class of tasks, using reasonably sophisticated nonlinear MS models: fast, aggressive, propulsive activities. Thus it works well for study of fast eye and head movements (e.g., Clark and Stark 1975; Lehman and Stark 1979), simple fast arm tracking tasks, with and without loads and with various "instructions" (e.g., Seif-Naraghi and Winters 1990), kicking and jumping (Hatze 1981; Chapter 33), jumping (e.g., Pandy 1990), and walking (e.g., Yamaguchi 1990).

Van der Helm (Chapter 32) and Hatze (Chapter 33) suggest that there is a great need for including closed-loop mechanisms into the process, and several approaches for implementing feedback are proposed that could be applied to larger-scale NMS models (e.g., Chapter 7 and Chapter 11). However, classic engineering LQR optimal control theory is best suited to linear systems without loop time delays, which does not normally provide a good approximation to NMS systems. Perhaps the most heroic effort in this area is that of Loeb and Levine (1990), who applied this approach to cat locomotion. Here we assume more generic numerical algorithms, and acknowledge that there is no reason why control parameters cannot include both feedforward and feedback variables, and the "system" cannot include lower neurocircuitry (e.g., data on postural responses [see Chapter 19]) suggest learned, coordinated synergy responses, which to this author is best dealt with via the fuzzy-optimization framework of Subsection 4.1.3).

4.2.2 Foundations of the General Nonlinearly Constrained Numerical Optimization Problem

Optimization algorithms range from stochastic search with polynomial line search features (e.g., Bremermann 1970; Lehman and Stark 1982) or with genetic algorithms (e.g., Goldberg 1989; Lin and Lee 1996) to a remarkable variety of gradient-based optimization packages (e.g., see Reklaitis 1983). A detailed exposition is beyond the scope of this chapter (see also Chapter 32). However, all involve a search of the J_{PC}-x_c (control) space for a minimum by using an equation like:

$$x_c^{(k+1)} = x_c^{(k)} + \delta d_k \qquad (35.7)$$

where the two key parameters are the search direction vector d_k for step k, and the scalar step size δ (which is variable for many algorithms). For gradient-based methods, d_k is a function of the local gradient in the J_{PC}-x_c space (first-order effects). A more general second-order form is (Cichocki and Unbehauen 1993):

$$m(t)\frac{d^2 x_c}{dt^2} + b(t)T(x_c,t)\frac{dx_c}{dt} = \\ -\nabla J_{pc}(x_c) + z(t)x_{rand} \qquad (35.8)$$

where m and b are real-valued positive functions for $t \geq 0$, T is a symmetric positive definite matrix, and x_{rand} is a random vector. Here we have purposely utilized parameters that are commonly associated with classical mechanics to help give a feel for dynamic search behavior—normally m is relatively small and multidimensional search behavior is "overdamped," yet history dependent. Many special cases emerge:

- steepest-decent: $T = I$ (identity matrix), $m = z = 0$
- Newton's: $T = \nabla^2 J_{pc}(x_c)$ (Hessian), $m = z = 0$, $b = 1$
- Levenberg–Marquardt: $T = \nabla^2 J_{pc}(x_c) + v(t)I$, $m = z = 0$, $b = 1$
- purely random: $m = b = 0$, $\nabla J_{pc}(x_c)$ not calculated

Using pure steepest-decent searching can be remarkably slow, yet calculating the actual Hessian matrix is very computationally expensive, and not done in practice. The better gradient-based algorithms generally involve some type of clever approximation of the inverse Hessian matrix—together, these allow "superlinear" convergence when the J_{PC}-x_c space has a friendly shape. In essence, the gradient gets one fairly close, then the quadratic nature of the second-order effect helps converge to the final solution. Often hybrid approaches are utilized that take advantage of

strengths of specific algorithms or of domain-specific (here J_{PC}-x_c) knowledge; the most common being to use stochastic "global" search (e.g., genetic algorithms) to get close, with local gradient-based approaches used to converge more quickly to the local extrema.

The default algorithm that is currently used by this author in the JAMM software package, called Feasible Sequential Quadratic Programming or FSQP (Panier and Tits 1993; Zhou and Tits 1990–1996), has been found to be especially robust, and computationally efficient (with superlinear convergence). The algorithm solves minimax problems of the form:

$$\min\{\max(J_{pc_k}(x_c))\}:g_i(x_c) \leq 0; h_j(x_c) = 0\} \quad (35.9)$$

subject to i inequality constraint equations, j equality constraints, and k performance criteria (each of which can include subcriteria).

4.2.3 Neural Networks for Optimization

Although a simplification, there are three classic ways that connectionist networks can be trained (i.e., learn via searching): (1) supervised learning, where an "error" or "energy" signal (criterion) is minimized, often using some type of backpropogation algorithm (Werbos 1974, 1990) to perform credit assignment (essentially an approximation of gradient search within the weight space); (2) conventional unsupervised ("self-organizing") learning (e.g., pattern classification or clustering); and (3) reinforcement learning (both supervised and unsupervised learning are possible). Algorithms for the first of these are usually variations of first-order (gradient-based) searches, only in the error-weight space, and consequently convergence is often slow (but robust). The latter appears the most biologically plausible (Barto 1994; Klopf et al. 1993), and ties to "approximate" optimal control (Barto et al. 1995).

CNNs can be biologically inspired, and indeed should be more than just the typical "blind" multi-layers of cells that we typically associate with brute-force input–output CNN optimization. Consider a set of CNN algorithmic modules, as in Figure 35.3. Now assume that it is possible for CNNs to take on a holistic structural design that more intimately takes into account the biomechanical system and/or the task. The most computationally effi-

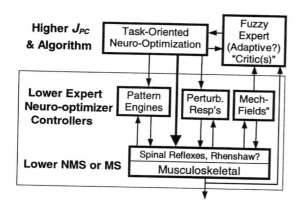

FIGURE 35.3. A modular CNN-MS neuro-optimization structure, including examples of lower neuro-optimization controller modules that have specific structures and are reasonably autonomous (e.g., own reinforcement learning capabilities), but with their relative contribution being task specific. That in dashed could ultimately become the "system."

cient CNNs designed for engineering control applications tend to take on a form in which certain *synaptic weights are associated in some way with model parameters or synergies*, and the internal cells are similar in function to Lagrange multipliers (Cichocki and Unbehauen 1993). For instance, one can formulate a scalar "energy" function (i.e., criteria) to be minimized via a learning algorithm that adjusts the weights. Many algorithms are available (e.g., Barto 1983; Grossberg 1988; Brown et al. 1990; Klopf et al. 1993), the most notable being a few that are inspired by the cerebellum (Berthier et al., 1993; Barto, 1983), and Hebbian learning (Hebb, 1949) algorithms with the form (Amari 1991):

$$\frac{dw_{ij}}{dt} = -\mu(\alpha w_{ij} - \beta_i\, u_j) = -\mu\left(\frac{\partial E(w_{ij})}{\partial w_i}\right) \quad (35.10)$$

where u_j is the *j-th* input, w_{ij} is the *ij-th* weight, μ is the learning rate coefficient, α is the forgetting factor, and β_i is an overall learning signal (notice the product relation between u_i and β_I, i.e. when they fire simultaneously, w_i increases). This form can be considered as a gradient optimization process for a suitably chosen energy (Lyapunov) function $E(w_{ij})$. The challenge is determining β_i, which via its product relationship with the input provides the co-occurance (local reinforcement) feature. More global reinforcement signals may require separate CNNs, and relates to concepts such as using "Adap-

tive Critics" (ACs) as ongoing measures of performance/utility (Werbos 1990; Barto 1994) that can be used for adaptive long-term synaptic gain depression/potentiation.

4.3 Aim to Enlarge the "System Model" to Include Lower Neuro-Optimization Modules

So far we have followed the structure of Figure 35.1, and separately considered the process of determining scalars such as J_{PC} from the "algorithm," and implicitly assumed the existence of a "system" and a "control variable to system" mapping. The implicit assumption was of a MS system model. Yet Figure 35.3 considers the possibility of *enlarging the model of the "system"* to include lower neurocircuitry (i.e., NMS models), and primitives (see also Chapter 7). Then higher-up J_{PC} may be simplified because certain ancillary subcriteria are automatically dealt with via adequate neuro-optimization subsystems that could interface, for instance, via neurofuzzy inference. For instance, there is evidence that recurrent inhibition via Rhenshaw cells helps drive down sustained levels of activity— so does penalty for "effort" (J_{eff}). By letting the "system" include *modular neuro-optimizers* that help to automatically minimize lower-level ancillary subcriteria, the "higher" J_{PC} is less concerned with details (i.e., closer to cognitive thought and global planning). Furthermore, our "control parameters" may become more general (e.g., command signals such as suggested by Feldman et al (1990)). The key principle: proactively plan neural structure *within modules with targeted performance (sub-) objectives*, and with weights that carry meaning.

4.3.1 Examples of Lower Neuro-Optimization Modules

Here we consider frameworks for developing neurooptimization "primitives" in the three areas in Figure 35.3.

Mechanical "Fields"

Many practically useful local learning algorithms are a variation of Eq. (35.10). For instance: supervised "least mean squares" and unsupervised "potential learning" algorithms (Amari 1991) can be framed to locally minimize estimates of potential energy, and neuro-algorithms can automatically perform Principal Component Analysis or PCA (an input pattern correlation technique popular in engineering) using robust subspace Hebbian-based learning algorithms that include the leaky decay term of Eq. (35.10) (e.g., Oja and Karhunen 1985; Sanger 1989) or both Hebbian and anti-Hebbian (i.e., lateral inhibition) local connections (Rubner and Schulten 1990). The result is an ordered set of basis functions that could be used to estimate eigenvectors for "input" matrices, obtained efficiently. Additionally, within engineering, CNNs are now used to implement various augmented Lagrange multiplier methods that solve the regularization of "ill-conditioned" systems (Cicikocki and Unbehauen 1993). Furthermore, CNN-based exterior penalty function methods are now available that efficiently turn constraints (here due to laws of mechanics) into implicit scalar subcriteria.

Several practical applications emerge. Winters and Van der Helm (1994) show that to address mechanical stability for redundant systems, one needs to minimize the potential energy E_{PE} in the generalized joint coordinate (rather than end-point) space. Let us assume "field" modules where we let local criterion J_{PC} be E_{PE}, and assume a local CNN *learning rule* that either uses unsupervised local reinforcement (Hebbian-type) learning or "supervision" via the laws of mechanics (applied to the musculoskeletal-environment system). The key to forming useful "posture modules" then becomes choosing our CNN structure carefully. If one takes conceivably available sensor information as inputs and associates Lagrange multipliers within the classic constrained neuro-optimization formulation described in Cichocki and Unbehauen (1993), weights associated with "stiffness" and "compliance" can be made to emerge. Thus meaningful concepts such as "apparent stiffness fields" and "relative stability" can be attained within the CNN modules. Alternatively, CNNs performing PCA, as described previously, can be used to automatically generate sets of basis eigenvectors from evolving data that relate to relative elastostatic stability.

In Chapter 25 and Chapter 24 (Mussa-Ivaldi), the concept of convergent spinal force fields in the end-point space (in their case preselected) is put forward, in their case using radial basis functions (Poggio and Girosi 1992). Such functions are can be viewed as a special case of a Sugeno fuzzy inference system,

for instance with Gaussian membership functions (Jang et al, 1995). Thus an alternative framework is to study whether the experimental data could be predicted by neurofuzzy optimization modules, perhaps self-organized.

Pattern (Synergy) Engines

Coordinated synergies, of research interest since at least the 1930s (e.g., Bernstein 1967), can evolve because of voluntary movement or as coordinated responses to perturbation. A pragmatic first step is EMG synergy analysis. Here PCA has been used to provide a set of basis functions that succinctly represent EMG data clusters or control functions for specific movement tasks (e.g., Shiavi and Griffin1981; Patla 1985; Wotten et al 1990); as noted previously this type of analysis can now be efficiently done using modular Hebbian CNNs. In our own current work on upper-limb EMG synergy analysis for multiple tasks, we are trying out neurofuzzy models (e.g., Sugeno) for this purpose (Winters 1996).

The "divide and conquer" concept of decomposing complex tasks into subtasks, for which some evidence is available, seems consistent with this approach, and has been used for computational models of learning (see Chapter 23 and Chapter 34). In a recent study using dynamic recurrent CNNs for EMG-to-kinematic systems identification for both tracking ("figure 8's" by hand) and propulsive (rising from chair) tasks, we subdivided a certain previously successful CNN structure into "postural" and "inertial" subnetworks trained by different signals (experimental position and acceleration, respectively), added Gaussian "spatial distances" between neurons within subnetworks, and then compared the performance of trained (optimized) networks for different subnetwork sizes (Draye et al. 1997). We were able to get better performance and faster convergence with rather small (e.g., 10-neuron) modular networks, and furthermore found internal self-selection into "tonic" and "phasic" neurons (identifiable both by their task behavior and their properties). When added to a previous finding from FO studies of better task performance and algorithmic convergence when using nonlinear versus linearized models of a same system for the same task goal (Seif-Naraghi and Winters 1990), optimization algorithms seem to be telling us that a useful FO approach for synergy identification is to use smaller modular CNN structures that control nonlinear systems models.

Neuro-optimization modules can be small and specific. The phase-resetting central pattern generator of Chapter 45, and indeed CPGs in general are here deemed as local CNNs that provide a wonderful example of an efficient structural design in which only a few neurons are utilized.

Perturbation Responses

Perturbation-driven modules make special sense both for training and as event-driven synergy responses. For instance, in the concept of internal expert models (Gomi and Kawato, 1993; Chapter 34), perturbation (feedback) "error" (here kinematically centered) between the actual and internal models drives internal model training (see also Figure 34.C1, in commentary). Furthermore, it is proposed that the postural response "strategies" described by Horak and Kuo (Chapter 19) within an optimal control context seem ideally suited for being considered as neurofuzzy optimization modules that work in parallel, "learn" via trial and error reinforcement, and "fire" when conditions (initial posture, type of perturbation, state of body) are satisfied by an event-driven "neuro-fuzzy critic" (see Subsection 4.1.3).

4.3.2 Neuro-Optimization Module Interplay

A concern regarding the proposed neuro-optimization "system modules" may be that they would be unwieldy and untestable. Not true. Certain tasks bring out certain hypothesized modules. The lower within the structure they are, the more likely controlled experiments can be formulated that bring one or two modules to the surface (e.g., a postural purturbation). Can we mix "supervised" and "unsupervised" low-end modules? Absolutely, as long as the "supervisor" is either the physical world (i.e., laws of mechanics played out through MS and environment interplay) or global reinforcement "critic" signals. Error between expected performance and actual performance (e.g., one view of muscle spindle signals) help refine internal models of NMS-environment subsystems, and physical laws in effect can function as external teachers (Werbos 1994). The algorithmic modules will, in a loose sense, "learn" causal quasi-mechanics (or at least ap-

proximate Jacobians) through reinforcement. Yet it is argued here that it learns within a *task-centered* universe, that is, the brain really does not know mechanics; it just merrily goes about the modular neuro-optimization *process*, working on extremizing local and global scalars. Let us focus more on this process.

Can module interplay be successfully managed? In this author's opinion, the appropriate heuristic hierarchical structure is a *neurofuzzy* formulation, for instance as described in Subsection 4.1.3 for J_{PC} generation. Indeed, the approach in Chapter 34 of using expert neural network modules, gated appropriately, is a special (crisp) case of a neurofuzzy approach. Through the use of fuzzy rules, both parallel and series-parallel architectures evolve naturally. Although not normally thought of this way, the key issues really are two: (1) the determination and management of generated scalars; and (2) information synthesis within a fuzzy infrastructure.

A key need, however, is identifying ongoing *global* reinforcement signals, which brings us back to the heuristic expert fuzzy critic (Winters 1996) or heuristic adaptive critic (Barto 1994; Werbos 1994, Jang et al. 1997) concepts, the challenge of dealing with temporal patterns, and a focus on neural architectures, such as the role of the climbing fibers on the cerebellar Purkinje cells.

5 Discussion

5.1 Why the "System" Should Include the "User"

This chapter proposes an infrastructure for studying movement selection for skilled, realistic tasks. Assume that the key purpose of a large-scale neuromechanical model is to gain insight into human performance. A corollary is that while model creation and validation are useful byproducts, these should not be the end products. "Insight," by definition, is a human attribute. Thus the overall "system" is really *the human using the model as well as the model itself.* Thus, the human–computer interface matters! What are the tools for gaining insight? While plotting of key variables, visualization and animation are all important, their value is limited if the human expert is involved in a passive (vs. interactive) data analysis process. The tool of numerical optimization (along with systematic sensitivity analysis) is intrinsically interactive in that "what if" questions evolve naturally, and the focus is on *understanding causality*.

5.2 Design Larger-Scale System Models

This perspective has focused on the macroscopic topic of structural design. Within engineering design, often the earliest decisions on structural design impact profoundly on the subsequent development. The challenge of determining an appropriate system formulation is tied to the research/clinical problem being addressed. Three concerns usually impact on this formulation:

- The desired form for the model (e.g., desired control inputs to the model, and outputs from the model).
- The task(s) under study.
- Practical issues, such as the type of algorithm being used to get a solution, the available computational capabilities, and the desired generality of the result.

It is suggested here that to understand principles underlying the saga of human coordination, we need to look at large-scale models that include kinematic and actuator redundancy, then study realistic (nonoverconstrained) tasks that promote strategy selection. When appropriate, we also should use integrated neuromuscular elements within the larger model (see Chapter 7), and remember that a model is only as good as its weakest link (for the tasks under study). Most everyday movements are transitions in posture (controllable temporary instability). Equilibrium Point theories in particular beg for larger scale extension. Especially intriguing are the subconscious coordination *synergies* away from the "prime movers" (e.g., anticipatory) or cognitive focus of the subject.

5.3 Movement Scaling via Problem Formulation

A common criticism of numerical optimization (vs. both analytical optimal control solutions and trained CNNs) is that each "run" only creates one solution, and that the new "knowledge" does not really transfer to other tasks. A trained CNN generalizes, performing reasonably well for related tasks. This is an insightful, astute concern. Yet at the same time, it is not really justified.

There are several approaches for addressing this problem. One is to determine the overall J_{pc} as the sum of a collection of the J_{pc}'s from a *sequence of tasks*, and then let the control parameters be fitting equations of the "real" control parameters plotted as a function of the scaling entity (e.g., target amplitude change). For instance, as conceptualized in Figure 35.4 and in Winters et al. (1984), a single optimization "run" consisted of a target sequence with "jumps" (e.g., between 2 and 40°), and control pulse parameters for "time-optimal" saccadic eye movements within this sequence were obtained. The x_c control parameters were actually fitting parameters of the various pulses. The end result is that a user could specify any initial and final position within about ±50°, and be assured of getting a nearly time optimal "saccade" that was within about a degree or so of the desired location (i.e., within the region of high resolution on the fovia). While such scaling for the eye is clearly much simpler than for limb movements, where not just magnitude but also speed, 3D location, load, and so on, could scale, the principle is evident. A second approach is to perform a set of individual runs and then "fit" the solution space (e.g., via regression equations or splines). Then if a fine-tuned optimization is desired, one can use the predicted "fit" as the initial conditions ($x_c(0)$) for the search, and most likely the algorithm will converge very quickly. A third approach is to use assume a modular neurofuzzy architecture, and then use the ap-

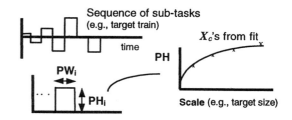

FIGURE 35.4. Conceptualization of approach for scaling, here for fast movements: the neurocontrol input is assumed to have a pulse-like control structure, and a set of optimal pulse heights and/or widths are determined for a series of subtasks in which some performance measure (e.g., target size) changes.

proach of Figure 35.2 (a fuzzy if–then rule architecture) to determine J_{PC}'s and associated controllers—the advantages of trained CNNs are then evident.

5.4 Neuro-Optimization in Context

Humans/animals are optimized to do many things well, none perfectly. This suggests that the primary use of larger-scale optimal control is *to study criterion-control model causality*, that adding performance subcriteria related to "effort/safety" is always necessary, and that tight final bounds on algorithmic convergence may have become overemphasized.

An intriguing model of human coordination is a set of optimization neuromodules working in parallel and series-parallel. The solution is then "satisficing," yet with *all modules involved in an ongoing neuro-optimization process.* Some modules are related to the primary task (but please do not call them motor programs!), some to embedded intrinsic needs ("effort," "safety," "stability," "no pain," etc.). All weigh in with their 2 cents worth, occasionally screaming for a greater weight within a neurofuzzy environment. Using parts of the cerebellum as an analogy, some require advanced near real-time sensor fusion to get reinforcement signals (e.g., AC concept), others anticipate needs during the planning stage, or orchestrate synergies via "primitive" neurobasis functions.

The key point: the neural learning *process* (whether using gradient-based, Hebbian, genetic algorithms, etc.) is literally (both conceptually and mathematically) an ongoing optimization *process*, with the "solutions" being mildly suboptimal (satisficing) for a certain collection of posture and movement tasks of interest. This is fully consistent with the way that genetic (optimum search) algorithms work (see commentary to Chapter 13), and furthermore with the concept of not "over-learning" (see subsection 4.3 in Chapter 34) so as to avoid overtuned CNN solutions that lose some of their generalization capability (and are thus suboptimal in the larger sense of movement robustness).

Has this "neuro-optimization" framework been so generalized as to lose its optimization base? No, because in all cases, regardless of implementation, the conceptual and structural foundation retains the four key structural parts of the optimization prob-lem: performance measure(s), search algorithm(s), controls to be adjusted, and a system model.

6 Future Directions

In comparison to the engineering community, where optimization tools are an integral part of the design of modern technology, optimization is a rarely used tool for NMS analysis. Whether or not this will change is an open question. This author believes it will, but that the drive will be research associated with *problem solving* (where motivation for deliverables is high) rather than hypothesis testing. There is a track record here. Optimal control evolved during the 1960s when the need for new theory to control missiles and aircraft was high, design optimization during the 1980s as industry embraced computer-aided engineering, and CNNs for signal processing and optimization during the late 1980s and early 1990s because they could solve difficult problems. During the 1990s, neurofuzzy CNNs and modular CNN designs have become areas of focus. Here are some predictions:

- With neuro-optimization, we will get away from focusing on static results of a single "final optimized end result," and more on: (1) the dynamic neuro-optimization *process* (i.e., learning over time); and (2) issues in *scalability* and *generalizability*. Thus the static use of optimization, so necessary in the past and yet so instrumental in causing a distaste for optimization among many scientists, should start to fade.

- As noted eloquently by Robinson (1994), there are true challenges in attributing *meaning* to individual neuron firing. Traditional layered CNNs, as brute-force black-box problem solvers (e.g., three layers with backpropagation), have little role to play—one might as well use advanced numerical optimization algorithms because convergence is usually quicker. Neurofuzzy approaches and modular CNNs with specific objectives will change this.

- *Neuro-optimization's growth through neuroengineering (based in problem solving).* Paul Werbos, discoverer of the CNN backpropagation algorithm (Werbos 1974), views the human system as continually striving to solve optimization problems and feels that the most challenging problem is to uncover AC networks that provide ongoing global estimates of utility (merit, performance) useful to *local* neural processing elements. Approximate Dynamic Programming (a neuro-optimization algorithm for CNNs) can be used to evolve toward sophisticated solutions, for example, by using reinforcement learning (Werbos 1990, 1994). This author agrees, but prefers a more rules-oriented neurofuzzy foundation that can more fully pull in HDM problem-solving expertise (e.g., fuzzy expert critics) (Winters 1996). The ideal "intelligent neurosystem" will use both "top-down" (HDM rules-oriented) and "bottom-up" (clustering data to refine rules) approaches, with gradual transfer toward distributed neuro-optimization modules. Such a process should be able to predict trends in *learning* (e.g., "co-contraction" during early learning stages as described in Chapter 26, a trend this author has often observed during the convergence of numerical algorithms, often as they "overlearn").

- *Clinical applications.* Currently the level of utilization of advanced MS models by nonresearchers is practically nonexistent, despite many excellent efforts (e.g., see Chapter 36). Once human–computer interfaces are more interactive, smarter, and fundamentally assistive to experts, optimization will emerge as the mechanism that finally pulls clinicians, surgeons, coaches, and therapists into the realm of computational neuromechanics. Why? Because by remapping of the original problem from the world of 3D inertial dynamics, nonlinear actuators and state equations into a more intuitive world that focuses on causality between performance criteria, key model parameters, and predicted performance, "what–if" questions can be addressed interactively. As computational speed continues to improve and interactive Windows-based interfaces for professionals evolve, it is predicted that interactive optimization will become an important assistive tool. The combination of fuzzy systems and neuro-optimization is especially intriguing.

References

Abbas, J.J. and Chizeck, H.J. (1995). Neural network control of functional neuromuscular stimulation systems: computer simulation studies. *IEEE Trans. Biomed. Eng.*, 42:1117–1127.

Amari, S. (1991). Mathematical theory of neural learning. *New Generat. Comput.*, 8:281–294.

Athans, M. and Falb, P.L. (1966). Optimal control: an introduction to the theory and its applications. McGraw-Hill, New York.

Barto, A.G. (1994). Reinforcement learning control. *Curr. Opin. Neurobiol.*, 4:888–893.

Barto, A.G., Bradtke, S.J., and Singh, S.P. (1995). Learning to act using real-time dynamic programming. *Artif. Intell.*, 72:81–138.

Barto, A.G., Sutton, R.S., and Watkins, C.J.C.H. (1983). Neuronlike elements that can solve difficult learning control problems. *IEEE Trans. Sys., Man, Cybern.*, 13:835–846.

Bellman, R.E. and Zadeh, L.A. (1970). Decision making in a fuzzy environment." *Man. Sci.*, 17:141–164.

Bernstein, N. (1967). *The Co-ordination and Regulation of Movements.* New York: Pergamon.

Berthier, N.E., Singh, S.P., Barto, A.G., and Houk, J.C. (1993). Distributed representation of limb motor programs in arrays of adjustable pattern generators. *J. Cogn. Neurosci.*, 5:56–78.

Bezdek, J.C. (1981). *Pattern Recognition with Fuzzy Objective Function Algorithms.* Plenum Press, New York.

Brown, Kairiss, E.W., and Keenan, C.L. (1990). Hebbian synapses: biophysical mechanisms and algorithms. *Ann. Rev. Neurosci.*, 13:475–511.

Cichocki, A. and Unbehauen, R. (1993). *Neural Networks for Optimization and Signals Processing.* Wiley & Sons, New York.

Clark M.R. and Stark, L. (1975). Time-optimal behavior of human saccadic eye movement. *IEEE Autom. Control*, AC-20:345–348.

Draye, J.-P., Winters, J.M., and Cheron, G. (1997). Self-selected modular recurrent neural networks with postural and inertial subnetworks applied to complex movements. *Biol. Cybern.*, (submitted)

Fan, M.K.H., Tits, A.L., Zhou, J.L., Wang, L.-S., and Koninckz, J. (1992). *Consol-OptCAD User's Manual.* University of Maryland.

Feldman, A.G., Adamovich, S.V., Ostry, D.J. and Flanagan, J.R. (1990). The origin of electromyograms—Explanations based on the equilibrium point hypothesis. In *Multiple Muscle Systems.* Winters, J.M. and Woo, S.L-Y. (eds.), Chapter 12, pp. 195–213. Springer-Verlag, New York.

Gottlieb, G.L., Corcos, D.M., Agarwal, G.C., and Latash, M.L. (1990). Principles underlying single-joint movement strategies. In *Multiple Muscle Systems.* Winters, J.M. and Woo, S.L-Y. (eds.), Chapter 14, pp. 236–250. Springer-Verlag, New York.

Grossberg, S. (1988). Nonlinear neural networks: principles, mechanisms, and architectures. *Neural Networks*, 1:17–61.

Hatze, H. (1981). A comprehensive model for human motion simulation and its application to the take-off phase of the long jump. *J. Biomech.*, 14:135–142.

Hebb, D.O. (1949). *The Organization of Behavior.* John Wiley & Sons, New York.

Hogan, N. and Winters, J.M. (1990). Principles Underlying Movement Organization: Upper Limb and Single Joint. In *Multiple Muscle Systems: Biomechanics and Movement Organization.* Winters, J.M. and Woo, S.L-Y. (eds.), Chapter 9, pp. 182–194. Springer-Verlag, New York.

Jang, J.-S.R. and C.-T. Sun (1995). Neuro-fuzzy modeling and control. *Proc. IEEE*, 29 p., March.

Jang, J.-S.R., Sun, C. and Mizutani, E. (1997). *Neuro-Fuzzy and Soft Computing. A Computational Approach to Learning and Machine Intelligence.* Prentice Hall, Upper Saddle River, New Jersey.

Kawato, M. (1990). Computational schemes and neural network models for formation and control of multi-joint arm trajectory. In *Neural Networks for Control.* Miller, W.T., Sutton, R.S., and Werbos, P.J. (eds.), pp. 197–228. The MIT Press, Cambridge.

Klopf, A.H., Morgan, J.S., and Weaver, S.E. (1993). A hierarchical network of control systems that learn: modeling nervous system function during classical and instrumental conditioning. *Adapt. Behav.*, 1:263–319.

Kosko, B. (1997). *Fuzzy Engineering.* Prentice Hall, Upper Saddle River, New Jersey.

Lehman, S.L. and Stark, L. (1982). Three algorithms for interpreting models consisting of ordinary differential equations: sensitivity coefficients, sensitivity functions, global optimization. *Math. Biosci.*, 62:107–122.

Lehman, S.L. and Stark, L. (1989). Simulation of linear and nonlinear eye movement models: sensitivity analysis and enumeration studies of time optimal control. *J. Cybern. Inform. Sci.*, 2:21–43.

Lin, C.-T. and Lee, C.S.G. (1996). *Neural Fuzzy Systems. A Neuro-Fuzzy Synergism to Intelligent Systems.* Prentice Hall, Upper Saddle River, New Jersey.

Loeb, G.E. (1984). The control and response of mammalian muscle spindles during normally executed motor tasks. *Exerc. Sport Sci. Rev.*, 12:157–204.

Loeb, G.E. and Levine, W.S. (1990). Linking musculoskeletal mechanics to sensorimotor neurophysiology. In *Multiple Muscle Systems.* Winters, J.M. and Woo, S.L-Y. (eds.), Chapter 10, pp. 165–181. Springer-Verlag, New York.

March, J.G. and H.A. Simon (1958). *Organizations.* John Wiley, New York.

Nelson, W.L. (1983). Physical principles for economies of skilled movements. *Biol. Cybern.*, 46:135–147.

Oguzturelli, M.N. and Stein, R.B. (1983). Optimal control of antagonistic muscles. *Biol. Cybern.*, 48:91–99.

Oja, E. (1989). Neural networks, principal components, and subspaces. *Int. J. Neural Systems*, 1:61–68.

Pandy, M.G. (1990). An analytical framework for quantifying muscular action during human movement. In *Multiple Muscle Systems.* Winters, J.M. and Woo, S.L-Y. (eds.), Chapter 42, pp. 653–662. Springer-Verlag, New York.

Panier, E.R. and Tits, A.L. (1993). On combining feasibility, descent and superlinear convergence in inequality constrained optimization. *Math. Program.*, 59:261–276.

Patla, A.E. (1985). Some characteristics of EMG patterns during locomotion: implications for the locomotor control process. *J. Motor Behav.*, 17:443–461.

Reklaitis, G.V., Ravindran, A., and Ragsdell, K.M. (1983). *Engng. Optimizat.*, John Wiley & Sons, New York.

Robinson, D. (1994). Implications of neural networks for how we think about brain function. *Brain Behav. Sci.*, pp. 42–53.

Sakawa, M. (1993). *Fuzzy Sets and Interactive Multiobjective Optimization.*, Plenum Press, New York.

Sanger, T.D. (1989). Optimal unsupervised learning in a single-layer linear feedforward neural network. *Neural Networks*, 2:459–473.

Seif-Naraghi, A.H. and Winters, J.M. (1990). Optimal Strategies for Scaling Goal-Directed Arm Movements." In *Multiple Muscle Systems.* Winters, J.M. and Woo, S.L-Y. (eds.), Chapter 19, pp. 312–334. Springer-Verlag, New York.

Shiavi, R. and Griffin, P. (1981). Representing and clustering electromyographic gait patterns with multivariate techniques. *Med. Biol. Engng. Comput.*, 21:573–578.

Werbos, P.J. (1974). *Beyond Regression: New Tools for Prediction and Analysis in the Behavioral Sciences.* PhD. Thesis, Harvard University.

Werbos, P.J. (1990). A menu of designs for reinforcement learning over time. In *Neural Networks for Control.* Miller et al., (eds.), The MIT Press.

Werbos, P.J. (1994a). The brain as a neurocontroller: new hypotheses and new experimental possibilities. In *Origins: Brain and Self-Organization.* Pribram, K. (ed.), Erlbaum.

Werbos, P.J. (1994b). Circuits for optimization in biological motor systems. *Proc. 13th Southern Biomed. Eng. Conf.*, pp. 247–250, Washington.

Winters, J.M. (1995a). Concepts in neuro-muscular modelling. In *3-D Analysis of Human Movement.* Allard et al. (eds.), Chapter 12, pp. 257–292. Human Kinetics.

Winters, J.M. (1995b). How Detailed Should Muscle Models be to Understand Multi-Joint Movement Coordination? *Hum. Mov. Sci.*, 14:401–442.

Winters, J.M. (1996). Intelligent synthesis of neuromusculo-skeletal signals using fuzzy expert critics. *SPIE*

Smart Sensors & Actuators for Neural Prosthesis, Vol. 2718, pp. 456–468. San Diego, California.

Winters, J.M. and Seif-Naraghi, A.H. (1991). Strategies for goal-directed fast movements are byproducts of satisfying performance criteria. *Behav. Brain Sci.*, 14:357–358.

Winters, J.M. and Van der Helm, F.C.T. (1994). A field-based musculoskeletal framework for studying human posture and manipulation. In 3D., *Proc. Symp. on Modeling and Control of Biomed. Sys.*, IFAC, pp. 410–415, March, Galveston.

Winters, J.M., Nam, M.H., and Stark, L. (1984). Modeling Dynamical Interaction Between Fast and Slow Movement: Saccadic Eye Movement Behavior in the Presence of the Slower VOR. *Math. Biosci.*, 68:150–187.

Wooten, M.E., Kadaba, M.P., and Cochran, G.V.B. (1990). Dynamic electromyography I: Numerical representation using principal component analysis. *J. Orthop. Res.*, 8:247–258.

Yamaguchi, G.T. (1990). Performing whole-body simulations of gait with 3_D, dynamic musculoskeletal models. In *Multiple Muscle Systems*. Winters, J.M. and Woo, S.L-Y. (eds.), Chapter 43, pp. 663–680. Springer-Verlag, New York.

Zadeh, L.A. (1965). Fuzzy sets. *Infor. Cont.*, 8:338–353.

Zajac, F. and Winters, J.M. (1990). Modeling musculoskeletal movement systems: joint and body-segment dynamics, musculotendon actuation, and neuromuscular control. In *Multiple Muscle Systems: Biomechanics and Movement Organization*. Winters, J.M. and Woo, S.Y. (eds.), Chapter 8, pp. 121–148. Springer-Verlag, New York.

Zhou, J.L. and Tits, A.L. (1990–1996). User's Guide for FSQP Version 3.4: A Fortran code for solving constrained nonlinear (minimax) optimization problems, generating iterate satisfying all inequality and linear constraints. *Technical Report SRC TR-92-107r4*, Systems Research Center, University of Maryland.

Commentary: Can Neural Networks Teach Us the Way We Learn?

Frans C.T. van der Helm

This chapter gives an understanding of the optimization of musculoskeletal models, especially as applied to forward dynamics simulations. In the optimization criterion, two factors must be considered: defining the task (J_{pc_task}), for example, a goal such as time optimality, and in case of a *submaximal* task, the inclusion of a load-sharing term (J_{pc_effort}). It is interesting to categorize the other terms in the criterion presented in this perspective. The pain criterion (J_{pc_pain}) has not been used very often yet. It is difficult to assess what mechanical events (stress, strain) and what structures (tendon, muscle fibers, etc.) are involved. However, in itself it can be considered a subcategory of J_{pc_task}. The task is to avoid as much pain as possible, and motion and muscle activation will be adapted accordingly. The stability term (J_{pc_stab}) is also a subcategory of J_{pc_task}. In a forward dynamic simulation, the final solution is always stable. However, stability requires "effort," and the final solution is likely to be borderline stable, because effort is penalized in the load-sharing term. It is a good idea to incorporate a minimal level of stability in the final solution, which is quite realistic. Human motions are reasonably robust against perturbations. Top sport motions might be borderline stable again. Through training the least amount of energy possible is spent on stability. Think about baseball pitchers dislocating their shoulder when they release the ball during their pitching motion! The potential energy approach (E_{pe}) is a quasistatic approach to stability, and therefore only applicable to postural tasks. During motion, stability is more difficult to define. One way to check stability is to linearize the (nonlinear) state equations of the musculoskeletal model, in which the skeletal and muscle dynamics, and eventually the proprioceptive feedback dynamics, are incorporated:

nonlinear state equations: $\dot{\underline{x}} = f(\underline{x}, \underline{u})$

where $\underline{x}$ is the state vector and $\underline{u}$ is the system input vector.

Linearized state equations: $\dot{\underline{x}} = A\underline{x} + B\underline{u}$

where A is the system state matrix and B is the input matrix. For stability of a *linear* system it is required that all (real parts of) the eigenvalues of A must be negative. The system should be linearized in every state encountered during the motion and checked for stability. This is a too strict test for stability, because the system might pass unstable states as long as it ends in a stable state-space region. And it is not even a guarantee for stability, because one assumption in linearization is that only *small* perturbations to the equilibrium state will be present. For a system in fast motion, this is certainly not the case. There are stability criteria for nonlinear dynamic systems, like the Lyapunov criterion. However, these criteria become very complex for musculoskeletal systems and have, to my knowledge, never been applied.

Things become more difficult when proprioceptive feedback pathways are incorporated into the system, and the input signal shifts from neural input (α-motor neuron drive) to the muscles toward a supraspinal neural input to the lower level control unit incorporating the α-motor neuron and its reflexive pathways and interneurons. Then the task of the central nervous system becomes twofold. On the one hand, an appropriate supraspinal signal must be generated to accomplish the task. This signal must consist of a feedforward drive, anticipating on the musculoskeletal system properties and environmental conditions, and of (neural) setpoint signals ($\sim$force, velocity, length) for comparison with the proprioceptive feedback signals from the muscle spindles and Golgi tendon organ. On the other hand, the CNS must generate the γ-motor neuron signal in order to set the gain of the spindle feedback, and can potentially adapt the feedback gains on a central level (spinal cord) or even apply a "dynamic" control algorithm, like a PID (Proportional-Integrating-Differentiating) controller. The supraspinal input signal can be optimized quite comparable with the open-loop optimization algorithms. Interpretation is more difficult because the input signal consists of a feedforward and feedback drive, and is modulated by the feedback loops. To what extent the CNS can change its dynamic control properties is still unknown. In manual control

experiments (McRuer et al. 1965), it is accepted that humans can adapt their controlling behavior from a integrator to a gain factor to a differentiator. There is no reason to assume why on lower control levels these properties could not exist.

What are the consequences for a neural network approach in the optimization of input signals and control signals? Firstly, it should be established what role the NN plays in the control. Is it just generating supraspinal input signals, or is it also acting as a controller and/or adapting the feedback properties? If it is generating supraspinal input signals, why should a CNN network be more efficient and more able to take into account the dynamic properties of the musculoskeletal systems than conventional algorithms? A more interesting approach is to use artificial NN as a nonlinear controller, simultaneously generating the feedforward drive as well as the feedback drive. This approach has been applied by Kawato et al. (1987) in the Feedback-Error-Learning approach (see Figure 35.C1), in which a inverse model, generating the feedforward drive, is learned by minimizing the feedback signal. Recently, these models have been advanced such that also the feedback controller is updated (Gomi and Kawato 1993).

A similar approach has been shown by Stroeve (1996), who incorporated *one* CNN controller for the feedforward and feedback drive (see Figure 34.C2). He used a backpropagation-through-time (BTT) algorithm for learning, and also adequately incorporated the feedback time delays because of to neural signal transport. Severe drawbacks of both Kawato and Stroeve are the slow convergence and large computational burden during the learning

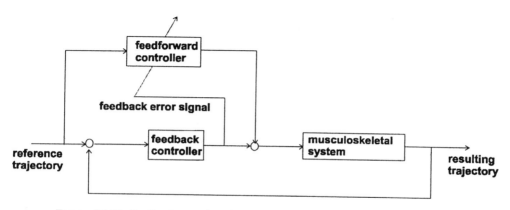

FIGURE 35.C1. Feedback error learning CNN structure used by Kawato's group.

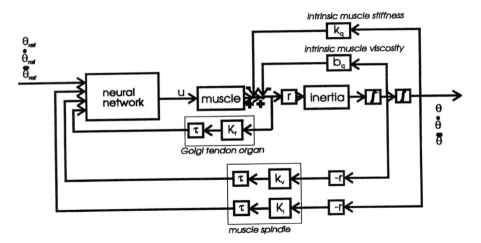

FIGURE 35.C2. CNN feedforward and feedback controller used by Stroeve.

phase. Because of the computational burden these CNN have not been applied yet to large-scale musculoskeletal models. Is there any reason why a connectionist NN could avoid these drawbacks?

I am skeptical about how fast the NN world will improve, so that really large-scale dynamic problems can be solved. In addition, what did we learn when we found a large-scale CNN that can do the same trick as the CNS, but is probably just as complex as the CNS?

Still, I certainly would like to join the optimistic view of Jack Winters about the potential applications, so that novice users can enter their opinion on the weighing of the competing performance criteria, and get interpretable results. That is certainly a goal to strive for, one way or another!

References

Gomi, H. and Kawato, M. (1993). Neural network control for a closed-loop system using feedback-error-learning. *Neural Networks*, 6, 933–946.

Kawato, M., Furukawa, K., and Suzuki, R. (1987). A hierarchical neural network model for control and learning of voluntary movement. *Biol. Cybern.*, 57:169–185.

McRuer, D.T., Grahan, D., Krendel, E.S., and Resener, W. (1965). Human pilot dynamics in compensatory systems—Theory, model and experiments with controlled element and forcing function variations. Report AFFDL TR-65-15, Wright Patterson AFB, Ohio.

Stroeve, S.H. (1996). Learning control of a human arm model. Abstract book Engineering Foundation Conference: Biomechanics and neural control of movement. Deer Creek, Ohio.

36
Clinical Applications of Musculoskeletal Models in Orthopedics and Rehabilitation

Scott L. Delp, Allison S. Arnold, and Stephen J. Piazza

1 Introduction

Persons with gait abnormalities frequently undergo surgical procedures to improve the alignment of their limbs, increase the efficiency of movement, and prevent the progression of deformity (Bleck 1987; Gage 1991). For example, persons with cerebral palsy who walk with excessive flexion of the knees often have their hamstrings lengthened to increase knee extension and improve the crouched posture. Excessive plantarflexion of the ankle during walking is frequently treated by surgical lengthening of the Achilles tendon. Stiff-knee gait, which may arise from abnormal activity of the rectus femoris muscle, is treated by transfer of this muscle to the posterior side of the knee. Rotational deformities are corrected by derotational osteotomies, procedures in which bones are divided and realigned to restore more normal limb rotation.

Although musculoskeletal surgeries can improve posture and walking, the outcomes are unpredictable and sometimes unsuccessful. Theoretically, the mechanics of walking may be improved by first identifying the muscles that are responsible for disrupting normal movement and then altering muscle forces (via tendon lengthenings) and moment arms (via osteotomies and tendon transfers) to achieve a proper balance of joint moments. However, current diagnostic and treatment methods do not allow the offending muscles to be identified easily, nor do they enable the functional consequences of treatments to be predicted prior to surgical intervention. We believe that the design of improved treatments will proceed more effectively if musculoskeletal models are developed that help explain the underlying mechanics of movement abnormalities and the functional consequences of surgical interventions.

A wide variety of musculoskeletal models have been developed to study clinical problems. Some simple models focus on musculoskeletal geometry and enable the lengths and moment arms of muscles to be estimated before and after surgery (e.g., Buford and Thompson 1987). More complex models have been created that characterize muscle force generation, limb dynamics, and CNS control (e.g., Yamaguchi and Zajac 1990; Schutte et al. 1993; see also Chapter 32). In the development of a musculoskeletal model for a specific clinical application, it is important to consider the level of complexity needed for the analysis. Based on our investigations, we have identified three types of models (kinematic, static, and dynamic) that can be used to answer a variety of clinically relevant questions.

A *kinematic* model can be used in conjunction with experimental measurements of joint angles to examine muscle lengths during normal and pathologic movements. Knowledge of muscle lengths is important because a "short" muscle that restricts movement can often be surgically lengthened. However, classifying a muscle as shorter or longer than normal is not a straightforward task. The lengths of multijoint muscles, such as the hamstrings, during movement depend on several factors, including the angular excursions of multiple joints and the geometric relationships of the muscles and bones. A model that accurately represents musculoskeletal geometry can be used to identify which muscles are shorter than normal during movement.

A *static* model that includes force generation can be used to examine how the isometric force-generating capacities of muscles are altered by surgeries, such as tendon lengthenings, tendon transfers, and osteotomies. These procedures are intended to produce a more normal balance of the moments about the joints during movement, but are not always successful. Hence, quantification of surgical effects is needed to design more effective procedures.

A *dynamic* model is an essential tool for understanding the motions produced by muscles. Clinical assessments of muscle function during movement are often based upon the electromyographic (EMG) activity of a muscle and the joint moments to which the muscle contributes. However, the activity and moment arms of a muscle are not the only factors that determine the motion produced by the muscle. A muscle that crosses one joint has the potential to accelerate other joints through dynamic coupling, and biarticular muscles may accelerate joints in a direction that opposes their joint moments (Hollerbach and Flash 1982; Zajac and Gordon 1989). Therefore, an analysis of muscle function during movement must consider the pattern of muscle activity, the moments generated by the muscles, *and* the coupled dynamic equations of the system.

This chapter presents three examples in which musculoskeletal simulations of varying complexity have been used to address clinical problems. Our goal is to briefly summarize our experience using musculoskeletal models; it is not intended to be a comprehensive review. Each example begins with an introduction to the clinical question, followed by a description of the model, and discussion of selected results. Although it was necessary to exclude many details from this perspective, readers are referred to journal publications for additional information. The perspective concludes with a brief discussion of some of the limitations of our current musculoskeletal simulations. Consideration of these limitations suggests areas for future research.

2   Kinematic Analysis of Muscle Lengths During Crouch Gait

Crouch gait, one of the most common movement abnormalities among persons with cerebral palsy, is characterized by excessive flexion of the knee during stance. Short hamstrings are thought to be

the cause of crouch gait, and persistent crouch is often treated by surgically lengthening the hamstrings. Although hamstring lengthenings usually decrease stance phase knee flexion, weakening of the hamstrings may cause several problems, including decreased knee flexion during swing and increased hip flexion during stance (Ray and Ehrlich 1979; Thometz et al. 1989).

We have used a model of the lower extremities (Delp et al. 1990) to estimate the lengths of the hamstrings and other multijoint muscles during normal gait and crouch gait to gain insight into the possible causes of this movement abnormality. The goal of this analysis was to determine if the hamstrings are indeed short during crouch gait, as is often inferred from static clinical measures of hamstrings length, or whether the hamstrings might be of normal length, or longer than normal, due to the excessive hip flexion that frequently accompanies excessive knee flexion in these patients.

All subjects who participated in this study underwent gait analysis at Children's Memorial Medical Center in Chicago. A VICON motion analysis system (Oxford Metrics, Oxford, England) was used to determine three-dimensional kinematics of ten unimpaired subjects and fourteen subjects with crouch gait (knee flexion > 20° in one or both limbs during the entire stance phase). All of the subjects with crouch gait had a diagnosis of spastic cerebral palsy, were over age seven, had no previous surgery, and walked without orthoses or other assistance.

The joint angles measured in the gait analysis were used in connection with the musculoskeletal model of the lower extremities to estimate the changes in hamstrings and psoas lengths that occur over the gait cycle (Figure 36.1). The musculoskeletal model consists of three-dimensional representations of the bones and muscle-tendon paths and kinematic descriptions of the joints (Delp et al. 1990). A single muscle representation of the hamstrings was constructed using a weighted average of the muscle attachment coordinates for the semitendinosus, semimembranosus, and the long head of the biceps femoris, with physiologic cross sectional area as the weighting factor. A model of the psoas was constructed using an average origin on the anterior lumbar spine, via points near the pelvic brim, and an insertion point on the lesser trochanter. The hip model was assumed to be a ball-and-socket

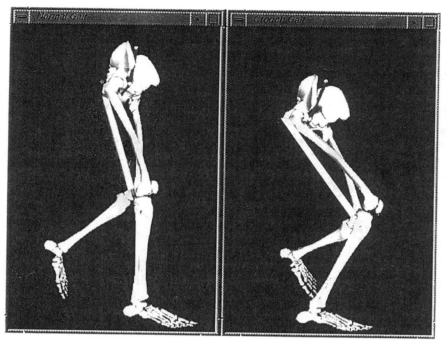

FIGURE 36.1. Graphics-based model of the lower extremity depicting (A) normal and (B) crouch gait at 16% of the gait cycle (midstance). The hamstrings and psoas paths used to estimate the muscle-tendon lengths are highlighted. (From Delp et al. 1996.)

joint. The knee model includes kinematic constraints to represent the motion of the tibiofemoral and patellofemoral joints. The ankle was represented as a revolute joint.

Calculated muscle lengths were normalized by the muscle length at anatomical position and then plotted over the gait cycle for each subject. The normalized muscle lengths for each crouch gait subject were compared to the muscle lengths estimated from the gait kinematics measured in the unimpaired subjects.

Analysis of the muscle lengths showed that a minority of the subjects had hamstrings that were shorter than normal during walking (e.g., subject 1 in Figure 36.2A). Most subjects had hamstrings that appeared to be of normal length or longer despite subjects' excessive knee flexion during stance (e.g., subject 2 in Figure 36.2A). This occurred because the hamstrings span the hip as well as the knee, and the excessive knee flexion was accompanied by excessive hip flexion in most subjects. This result is consistent with the findings of Hoffinger et al. (1993) and suggests that lengthening of the ham-

strings may be inappropriate in some patients with crouch gait.

Although a small number of subjects had psoas muscles that were of nearly normal length (e.g., subject 1 in Figure 36.2B), almost all subjects with crouch gait had psoas muscles that were shorter than normal throughout the gait cycle (e.g., subject 2, in Figure 36.2B). This suggests that the coupling between the knee and hip, via the hamstrings, may be an important factor contributing to excessive knee flexion. If subjects' hip flexion were reduced, then perhaps their knee flexion would improve, since decreasing excessive hip flexion may decrease hamstrings tension and enable knee extension.

This example demonstrates that a musculoskeletal model that includes kinematic constraints describing the joints can be used in conjunction with three-dimensional measurements of joint angles to help identify which muscles are short during movement. Our previous work has shown that static measures of muscle length, which are typically performed by physical therapists through manipulation

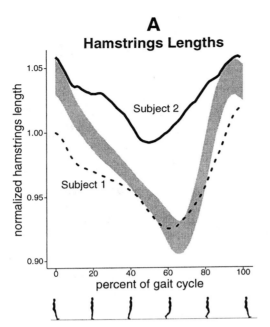

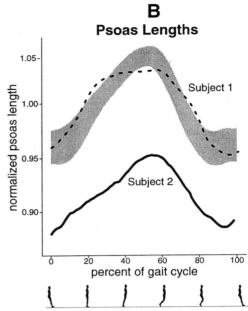

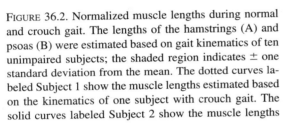

FIGURE 36.2. Normalized muscle lengths during normal and crouch gait. The lengths of the hamstrings (A) and psoas (B) were estimated based on gait kinematics of ten unimpaired subjects; the shaded region indicates ± one standard deviation from the mean. The dotted curves labeled Subject 1 show the muscle lengths estimated based on the kinematics of one subject with crouch gait. The solid curves labeled Subject 2 show the muscle lengths estimated from the kinematics of another subject with crouch gait. Note that the hamstrings are longer than normal (above the shaded region) and the psoas is shorter than normal (below the shaded region) for Subject 2. Muscle lengths are normalized by their lengths at the anatomical position (0° hip flexion, abduction and rotation, and 0° knee flexion).

of the limb, are not correlated well with muscle lengths during movement (Delp et al. 1996). This suggests that static measures of muscle tightness are not related to hamstring lengths during walking and may not be highly relevant when deciding whether a patient's hamstrings need to be lengthened. By contrast, analysis of muscle lengths during movement may help distinguish subjects who do have short hamstrings from those who do not have short hamstrings, and may provide a more rational basis for deciding who should have hamstring lengthening surgery.

3   Static Analysis of Muscle Forces After Tendon Surgery

Persistent plantarflexion of the ankle, termed equinus gait, is a common deformity in stroke and cerebral palsy patients. Equinus gait is frequently caused by shortening of the fibers (i.e., contracture) of the triceps surae. Either isolated contracture of the gastrocnemius or combined contracture of the gastrocnemius and soleus may be present. Surgical lengthening of the gastrocnemius aponeurosis or the Achilles tendon is commonly performed to treat equinus deformities, and several clinical studies have evaluated the effectiveness of these procedures (Sharrard and Bernstein 1972; Lee and Bleck 1980; Graham and Fixsen 1988; Olney et al. 1988; Etnyre et al. 1993; Rose et al. 1993). When only the gastrocnemius is contracted, gastrocnemius aponeurosis lengthening is usually successful in restoring the normal range of ankle motion and maintaining the moment-generating capacity of the plantarflexors (i.e., plantarflexion strength). However, tendo-achilles lengthening, the procedure that is commonly used to treat combined contracture of the gastrocnemius and soleus, is less effective. If the Achilles tendon is not lengthened enough, passive plantarflexion moment continues to cause ankle equinus after surgery. By contrast, if the

Achilles tendon is lengthened too much, the active force-generating capacity of the muscles can be greatly reduced, resulting in disabling muscle weakness.

The force-generating capacity of a muscle after tendon lengthening is influenced by the architecture of the muscle tendon complex (the lengths and arrangement of muscle fibers (see Chapter 3; Delp and Zajac 1992). Because the gastrocnemius and soleus have different architectures, one would expect these muscles to respond differently to tendon lengthening. These effects are difficult to quantify in clinical studies because individual muscle forces cannot be measured without invasive techniques. However, a musculoskeletal model that accounts for the differences in muscle architectures can be used to examine how tendon lengthenings affect the force- and moment-generating characteristics of these muscles.

We developed a musculoskeletal model to examine the tradeoff between restoring range of ankle motion and maintaining plantarflexion strength in cases of combined contracture of the gastrocnemius and soleus (Delp et al. 1995). We first developed a model that represents the normal moment-generating characteristics of the major muscles crossing the ankle. The model of normal muscles was then altered to represent contracture of the gastrocnemius and soleus. The effects of gastrocnemius aponeurosis lengthening and tendo-achilles lengthening were simulated by altering the tendon lengths of the contracted muscles. The theoretical effectiveness of the simulated surgeries was evaluated based on their ability to produce normal passive and active moment-generating characteristics about the ankle.

The musculoskeletal model used in this example includes a model of the isometric force-generating property of each muscle-tendon complex (Delp et al. 1990). The length-tension characteristic of each muscle-tendon complex was derived by scaling a generic model of muscle and tendon (Zajac 1989) with four parameters (peak isometric force, optimal muscle fiber length, tendon slack length, and pennation angle; see Figure 36.3). Values for these parameters were derived from data collected in anatomical studies (Wickiewicz et al. 1983; Friederich and Brand 1990).

The passive plantarflexion moment computed with the model of normal muscles (i.e., the moment

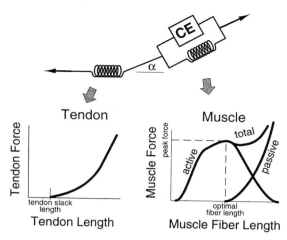

FIGURE 36.3. Muscle-tendon model. The length-tension property of muscle is represented by an active contractile element (CE) in parallel with a passive elastic element. Total isometric muscle force is assumed to be the sum of muscle force when it is inactive (passive) and when it is maximally excited (active). The muscle is in series with tendon, which is represented by a nonlinear spring. Pennation angle (a) is the angle between tendon and muscle fibers. Tendon slack length is the length of tendon at which force initially develops during tendon stretch. This model was scaled to represent each muscle by specifying the muscle's peak force, optimal fiber length, tendon slack length, and pennation angle based on data collected in anatomical experiments.

produced by the muscles when they are inactive) corresponds closely to measured passive moments (Siegler et al. 1984) (Figure 36.4A). Total plantarflexion moments computed with the model (the isometric moment generated by the muscles when they are fully activated, including passive moment) also correspond closely to maximal plantarflexion moments measured (Sale et al. 1982), with the knee extended (Figure 36.4B) and flexed (Figure 36.4C). Because the gastrocnemius and soleus provide such a large percentage of the plantarflexion moment and are the only muscles affected by the tendon lengthening surgeries, this analysis focuses on the moments generated by these muscles.

Light microscopy of muscle biopsies taken from children with cerebral palsy have suggested that the number of sarcomeres in a fiber was decreased in cases of contracture (Tardieu et al. 1982; Tardieu and Tardieu 1987). Ziv et al. (1984) reported that muscle fiber lengths were reduced by 45% in spas-

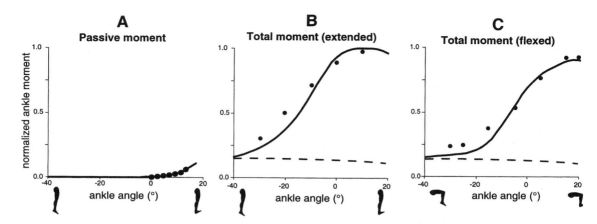

FIGURE 36.4. Ankle moment vs. ankle angle for the normal model. Passive moment (A) computed with the model (solid curve) is compared to the passive plantarflexion moment reported by Siegler et al. (1984) (dots) with the knee extended. Total moments computed with the model (solid curve) are compared to the maximum isometric moments reported by Sale et al. (1982) (dots) with the knee extended (B) and flexed 90°(C). The

dashed curves in B and C represent the sum of the moments produced by tibialis posterior, peroneus longus, peroneus brevis, flexor hallucis longus, and flexor digitorum longus. The difference between the solid and dashed curves in B and C therefore represents the moment produced the triceps surae. Experimental and computed moments are normalized by the maximum active moment with the knee extended.

tic mice. Based on these data, we characterized contracture of the triceps surae by decreasing the optimal fiber lengths of the gastrocnemius and the soleus by 45%. In the computer model, the decrease in fiber lengths resulted in a substantial increase in the passive moment about the ankle. With the knee extended, the computer model of muscle contracture had a passive moment onset at 30° of plantarflexion; this is consistent with the observations of Tardieu et al. (Tardieu and Tardieu 1987), who reported that the average onset of passive moment occurred at approximately 30° of plantarflexion in children with equinus, when the knee was in extension.

Gastrocnemius aponeurosis lengthening was simulated by elongating the tendon of the gastrocnemius while the soleus tendon remained unaltered. Tendo-achilles lengthening was simulated by elongating the tendons of both the contracted gastrocnemius and the contracted soleus.

In the simulation, neither gastrocnemius aponeurosis lengthening nor tendo-achilles lengthening alone was an effective treatment for combined contracture of the gastrocnemius and soleus. Gastrocnemius aponeurosis lengthening did not decrease the excessive passive moment developed by the contracted soleus. Lengthening of the Achilles tendon

2 cm restored the passive moment developed by the contracted triceps surae to slightly greater than normal (Figure 36.5A). However, this increase in tendon length decreased the moment-generating capacity of triceps surae substantially (Figure 36.5B).

The simulations suggest that the active and passive moment-generating characteristics of the contracted triceps surae can be restored more effectively by independent lengthening of the contracted gastrocnemius and soleus. After the Achilles tendon was lengthened 1 cm and the gastrocnemius aponeurosis was lengthened 1 cm, the total moment-generating capacity was less than normal (Figure 36.5B), but greater than after tendo-acilles lengthening alone (cf., dashed and dot-dashed curves in Figure 36.5B). The passive moment generated by the triceps surae after combined tendo-achilles lengthening and gastrocnemius aponeurosis lengthening remained greater than normal (Figure 36.5A).

These results indicate that independent lengthening of the contracted gastrocnemius and soleus accounts for differences in the architecture of these muscles and may be a more effective means to restore range of motion and maintain plantarflexion strength when combined contracture of the gastrocnemius and soleus is present.

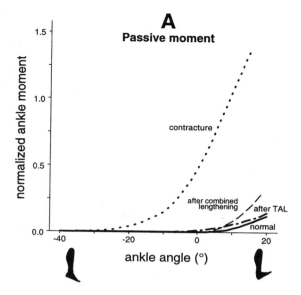

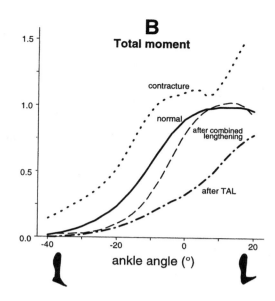

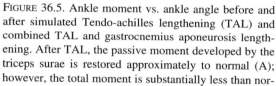

FIGURE 36.5. Ankle moment vs. ankle angle before and after simulated Tendo-achilles lengthening (TAL) and combined TAL and gastrocnemius aponeurosis lengthening. After TAL, the passive moment developed by the triceps surae is restored approximately to normal (A); however, the total moment is substantially less than nor-mal after TAL (B). After combined lengthening the passive moment developed by the triceps surae is greater than normal (A). However, the total moment is nearly normal after combined lengthening (B). Moments were normalized by the maximum active moment of the normal triceps surae.

4 Dynamic Analysis of Muscle Action in Stiff-Knee Gait

Many persons with cerebral palsy walk with de-creased knee flexion in the swing phase, or stiff-knee gait. Stiff-knee gait is often attributed to the knee-extending action of the rectus femoris muscle, which frequently shows prolonged and increased activation in persons with stiff-knee gait (Perry 1987; Suther-land et al. 1990; Damron et al. 1993). Analysis of the rectus femoris is complex, however, because the hip flexion moment it produces has the potential to increase knee flexion while the knee extension mo-ment it produces acts to decrease knee flexion. We developed a dynamic simulation of the swing phase of normal gait to study the influence of muscle forces on knee flexion and to examine how abnormal acti-vation of the rectus femoris affects knee flexion (Pi-azza and Delp 1996).

The simulation calculates the motions of the joints from muscle excitation patterns (Figure 36.6). The excitation inputs to the simulation were derived from the intramuscular EMG recordings

(expressed as a percentage of maximum EMG) re-ported by Perry (1992) for normal gait. These sig-nals determined muscle activations via first-order activation dynamics (Zajac 1989). Each of twelve muscle-tendon actuators was assumed to obey the same normalized force-velocity and force-length curves, and all muscles were assumed to have a maximum shortening velocity of 10 optimal fiber lengths per second as suggested by Zajac (1989) for muscles of mixed fiber type. The normalized force-length-velocity characteristics were scaled to represent each individual muscle by the four pa-rameters described above: optimal fiber length, maximum isometric force, pennation angle, and tendon slack length.

Five segments were represented in the lower ex-tremity model: the pelvis, thigh, patella, shank, and foot of the right leg. The joints connecting the seg-ments permitted motions in the sagittal plane only. The hip and ankle joints were modeled as friction-less revolutes, but the knee joint model included both the rolling and sliding of the femoral condyles on the tibial plateau and the patellofemoral kine-matics as described by Delp et al. (1990). Transla-

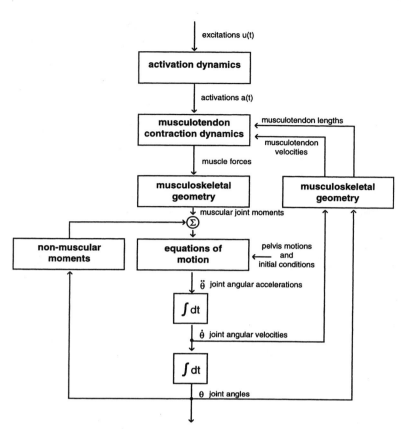

FIGURE 36.6. Block diagram for the simulation of swing phase. Muscle activations were determined from muscle excitation inputs, approximated from EMG recordings. The force generated by each muscle-tendon actuator was a function of its activation, length, and velocity. Total applied joint moments were calculated by multiplying muscular forces by moment arms and adding the angle-dependent joint restraint moments (representing ligaments). Hip joint center translation and pelvic tilt were prescribed from measured gait analysis data. The equations of motion were integrated in time to determine the kinematic output of the simulation. (Modified from Piazza and Delp 1996.)

tion and tilt of the pelvis were prescribed based on experimental measurements of pelvis motion, leaving only three degrees of freedom: flexion–extension of the hip, knee, and ankle. The equations of motion for the system were integrated forward in time to determine the swing phase motions. The results of the simulation were compared to experimental measurements to test the accuracy of the simulation (see Piazza and Delp 1996 for these comparisons).

We performed a "one-at-a-time" factorial analysis (Hogg and Ledolter 1987) to determine the capability of each of the joint moments and initial joint angular velocities to diminish peak knee flexion in swing. Muscular joint moments equal to twice the standard deviation of the joint moment were added to or subtracted from the normal simulation joint moment at each time step during the simulation. The standard deviations of the muscular joint moments were approximated by averaging the standard deviations calculated by Winter (1991) over the duration of the swing phase.

To clarify the role of the rectus femoris during

swing, simulations were performed (1) with the rectus femoris actuator removed from the model and (2) with an exaggerated excitation input to the rectus femoris actuator. For the latter simulation, the excitation input to the rectus femoris actuator continued throughout swing phase at 30% of its maximum level (as opposed to a 0.03 sec burst at 5% of maximum for the normal simulation).

As expected, the simulation demonstrated that either increasing knee extension moment or decreasing knee flexion velocity at toe-off decreased peak knee flexion in swing. Decreasing hip flexion moment also caused a substantial decrease in peak knee flexion. The rectus femoris was found to play an important role in regulating knee flexion during swing phase. A simulation of swing performed with the rectus femoris actuator removed resulted in excessive knee flexion, suggesting that the knee-extending action of the rectus femoris in early swing is important for normal knee flexion during walking (Figure 36.7). Conversely, overactivity in rectus femoris inhibited knee flexion in the simulation.

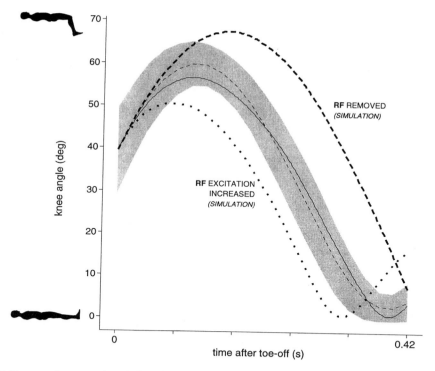

FIGURE 36.7. Knee angle versus time for simulations performed with the rectus femoris (RF) actuator removed and with increased and extended excitation of the rectus femoris. The dashed curve with shading represents measured mean joint angle ± one standard deviation for normal swing. The solid curve within the shaded area represents the knee angle for a simulation performed with all muscle-tendon actuators intact and supplied with normal excitations. Removal of the rectus femoris both increased and delayed peak knee flexion; increased rectus femoris activity limited swing phase knee flexion. (From Piazza and Delp 1996.)

These findings confirm that reduced knee flexion during the swing phase (stiff-knee gait) may be caused by overactivity of the rectus femoris. The simulations also suggest that weakened hip flexors and stance phase factors that determine the angular velocity of the knee at toe-off may be responsible for decreased knee flexion during swing phase. These factors should be considered along with rectus femoris activity before surgery is performed on the rectus femoris in an attempt to correct stiff-knee gait.

An unexpected finding in our investigation of muscle-induced knee accelerations was that the gastrocnemius muscle produced a knee *extension* acceleration between 25% and 60% of swing while simultaneously producing a knee *flexion* moment. Though inactive, the gastrocnemius generated force during the simulation when it was passively stretched. Other biarticular actuators produced accelerations opposite in direction to that of their joint moments: the hamstrings produced a hip flexion acceleration in mid-swing and rectus femoris produced a hip extension acceleration in early swing. Muscle function is often inferred from the moment arms and EMG activity of a muscle. However, the joint accelerations caused by biarticular muscles presented in this example demonstrate that an accurate assessment of muscle function during movement must account for the force-generating properties of the muscle, the musculoskeletal geometry, and the coupled nature of the system dynamics.

5   Discussion and Future Directions

The examples presented above have used generic musculoskeletal models, which are based on measurements of bone geometry, muscle-tendon paths, joint kinematics, muscle architecture, and muscle activations made on a relatively small number of subjects. We have modified the generic models to study bone deformities (Arnold et al. 1997), tendon transfers (Delp et al. 1994b), joint reconstructions (Delp et al. 1994a), muscle pathology (Delp et al. 1995), and abnormal control (Piazza and Delp 1996). Although these studies have begun to provide some clinically useful guidelines, the models have not been widely used to make patient-specific treatment decisions. We believe that the limitations of current models must be reduced and the accuracy with which models represent individuals with neuromuscular pathology must be tested before simulations can be used to make individual treatment decisions.

The accuracy with which a generic model can be transformed to represent the geometry of an individual must be determined to improve the utility of musculoskeletal models. For example, we have used a generic musculoskeletal model to analyze the lengths of muscles in subjects of different sizes, some of whom have bone deformities. Muscle lengths were normalized (e.g., Figure 36.2) in an effort to account for the differences in size, but is not clear how variations in size and the presence of deformities affect the accuracy of muscle length calculations. Future work is needed to determine how musculoskeletal geometry varies among subjects and to understand how bony deformities affect the lengths and moment arms of the muscles. This is especially important because the results of kinematic, static, and dynamic simulations are sensitive to the accuracy with which the musculoskeletal geometry is represented. The development of biomechanical models from medical images, such as MRI, can provide highly accurate representations of the musculoskeletal geometry of individual subjects (Smith et al. 1989). If comparisons between image-based models and transformed, generic models show that current transformation methods are not sufficient, then new methods of transforming the generic models should be developed, or more efficient methods to create models from medical image data should be derived.

The muscle model used in the static and dynamic simulations presented here must also be further tested. While this model represents many features of muscle force generation, it does not account for changes that may occur in disease states. For instance, the model does not account for complexities associated with activation of spastic muscle, such as potential alterations in recruitment or rate modulation (Tang and Rymer 1981). Although we have attempted to account for the decrease in muscle fiber lengths that may occur with contracture, the muscle model does not include the effects of muscle-tendon remodeling, such as alterations in the peak force of a muscle (Williams and Goldspink 1978) or changes in the elasticity of tendon (Woo et al. 1982). Muscle-tendon models that account for the alterations in force-generating characteristics that occur with pathology, surgery, and other treatment modalities are needed to verify the accuracy of existing simulations and to enhance the value of new ones.

It is also important to consider whether a musculoskeletal simulation is of sufficient complexity to answer the clinical question being posed. For example, "short hamstrings" are not the only cause of crouch gait. Crouch gait could arise from a multitude of other causes including weak plantarflexors, weak hip extensors, or difficulties with balance (Rab 1991). More complex models that allow exploration of these other possible causes may be needed to determine the most appropriate treatment.

Perhaps the most profound limitation of the models presented in this chapter is their exclusion of CNS control. The static analysis of equinus gait does not allow one to analyze how tendon lengthening may affect muscle force production through its influence on neural control. The dynamic simulation of stiff-knee gait was performed open-loop; the synthesized motions have no ability to affect the muscle excitations through reflexes, as occurs in vivo. Certainly, the incorporation of accurate representations of sensory-motor control into dynamic simulations of abnormal movements is one of the most important challenges faced by motor control researchers.

We believe that extensively tested computer models of the neuromusculoskeletal system are necessary for explaining the causes of movement abnormalities and the functional consequences of surgical procedures; this information is essential for developing improved treatment plans. Ultimately, prospective clinical studies that compare simulation results to clinical results are needed to determine if the insights gained from musculoskeletal models can indeed improve treatment outcomes. This is the most important test of the clinical utility of musculoskeletal models.

Acknowledgments. We would like to thank Abraham Komattu, Claudia Kelp-Lenane, Peter Loan, Carolyn Moore, Lisa Schutte, Rosemary Speers, Kim Statler, Steve Vankoski, and Tony Weyers for help in the anatomical experiments, computer modeling, and data analysis. We are also grateful Eugene Bleck, Norris Carroll, Luciano Dias, Sylvia Ounpuu, Jacquelin Perry, George Rab, and Felix Zajac for their helpful comments related to movement deformities and musculoskeletal modeling. We thank Idd Delp for help with the illustrations. This work was supported by the Whitaker Foundation, The Baxter Foundation, NSF Grant BCS-9257229, the United Cerebral Palsy Foundation, and NIH R01 HD33929.

References

Arnold, A.A., Komattu, A.V., and Delp, S.L. (1997). Internal rotation gait: a compensatory mechanism to restore abduction capacity decreased by bone deformity? *Dev. Med. Child Neurol.*, 39:40–44.

Bleck, E.E. (1987). *Orthopaedic Management in Cerebral Palsy.* Philadelphia: Mac Keith Press.

Buford, W.L. and Thompson, D.E. (1987). A system for three-dimensional interactive simulation of hand biomechanics. *IEEE Trans. Biomed. Eng.*, 34:444–453.

Damron, T.A., Breed, A.L., and Cook, T. (1993). Diminished knee flexion after hamstring surgery in cerebral palsy: prevalence and severity. *J. Pediatr. Orthop.*, 13:188–191.

Delp, S.L. and Zajac, F.E. (1992). Force- and moment-generating potential of the lower extremity muscles before and after tendon lengthening. *Clin. Orthop. Rel. Res.*, 284:247–259.

Delp, S.L., Komattu, A.V., and Wixson, R.L. (1994a). Superior displacement of the hip in total joint replacement: effects of prosthetic neck length, neck-stem angle, and anteversion angle on the moment-generating capacity of the muscles. *J. Orthop. Res.*, 12:860–870.

Delp, S.L., Ringwelski, D.A., and Carroll, N.C. (1994b). Transfer of the rectus femoris: effects of transfer site on moment arms about the knee and hip. *J. Biomech.*, 27:1201–1211.

Delp, S.L., Statler, K., and Carroll, N.C. (1995). Preserving plantarflexion strength after surgical treatment for contracture of the triceps surae: a computer simulation study. *J. Orthop. Res.*, 13:96–104.

Delp, S.L., Arnold, A.S., Speers, R.A., and Moore, C. (1996). Hamstrings and psoas lengths during normal and crouch gait: implications for muscle-tendon surgery. *J. Orthop. Res.*, 14:144–151.

Delp, S.L., Loan, J.P., Hoy, M.G., Zajac, F.E., Topp, E.L., and Rosen, J.M. (1990). An interactive, graphics-based model of the lower extremity to study orthopaedic surgical procedures. *IEEE Trans. Biomed. Eng.*, 37:757–767.

Etnyre, B., Chambers, C.S., Scarborough, N.H., and Cain, T.E. (1993). Preoperative and postoperative assessment of surgical intervention for equinus gait in children with cerebral palsy. *J. Pediatr. Orthop.*, 13:24–31.

Friederich, J.A. and Brand, R.A. (1990). Muscle fiber architecture in the human lower limb. *J. Biomech.*, 23:91–95.

Gage, J.R. (1991). *Gait Analysis in Cerebral Palsy.* Mac Kieth Press, New York.

Graham, H.K. and Fixsen, J.A. (1988). Lengthening of the calcaneal tendon in spastic hemiplegia by the White slide technique. *J. Bone Jt. Surg. [Br].*, 70-B:472–475.

Hoffinger, S.A., Rab, G.T., and Abou-Ghaida, H. (1993). Hamstrings in cerebral palsy crouch gait. *J. Pediatr. Orthop.*, 13:722–726.

Hollerbach, J.M. and Flash, T. (1982). Dynamic interactions between limb segments during planar arm movement. *Biol. Cybern.*, 44:67–77.

Lee, C.L. and Bleck, E.E. (1980). Surgical correction of equinus deformity in cerebral palsy. *Dev. Med. Child Neurol.*, 22:287–292.

Olney, B.W., Williams, P.F., and Menelaus, M.B. (1988). Treatment of spastic equinus by aponeurosis lengthening. *J. Pediatr. Orthop.*, 8:422–425.

Perry, J. (1987). Distal rectus femoris transfer. *Dev. Med. Child Neurol.*, 29:153–158.

Perry, J. (1992). *Gait Analysis.* Thorofare, New Jersey. SLACK.

Piazza, S.J. and Delp, S.L. (1996). The influence of muscles on knee flexion during the swing phase of gait. *J. Biomech.*, 29:723–733.

Rab, G. (1991). Consensus on crouched gait. *The Diplegic Child*, pp. 337–339. American Academy of Orthopaedic Surgeons. Rosemont, Illinois.

Ray, R.L. and Ehrlich, M.G. (1979). Lateral hamstring transfer and gait improvement in the cerebral palsy patient. *J. Bone Joint Surg.*, 61-A:719–723.

Rose, S.A., DeLuca, P.A., Davis, R.B., Ounpuu, S., and Gage, J.R. (1993). Kinematic and kinetic evaluation of the ankle after lengthening of the gastrocnemius fascia in children with cerebral palsy. *J. Pediatr. Orthop.*, 13:727–732.

Sale, D., Quinlan, J., Marsh, E., McComas, A.J., and Belanger, A.Y. (1982). Influence of joint position on ankle plantarflexion in humans. *J. Appl. Physiol.*, 52:1636–1642.

Schutte, L.M., Rodgers, M.M., and Zajac, F.E. (1993). Improving the efficiency of electrical stimulation-induced leg cycle ergometry: an analysis based on a dynamic musculoskeletal model. *IEEE Trans. Rehab. Eng.*, 1:109–125.

Sharrard, W.J. and Bernstein, S. (1972). Equinus deformity in cerebral palsy—a comparison between elongation of the tendo calcaneus and gastrocnemius recession. *J. Bone Jt. Surg. [Am].*, 54-B:272–276.

Siegler, S., Moskowitz, G.D., and Freedman, W. (1984). Passive and active components of the internal moment developed about the ankle during human ambulation. *J. Biomech.*, 17:647–652.

Smith, D.K., Berquist, T.H., An, K.N., Robb, R.A., and Chao, E.Y.S. (1989). Validation of three-dimensional reconstructions of knee anatomy: CT vs MR imaging. *J. Comput. Assist. Tomogr.*, 13:294–301.

Sutherland, D.H., Santi, M., and Abel, M.F. (1990). Treatment of stiff-knee gait in cerebral palsy: a comparison by gait analysis of distal rectus femoris transfer versus proximal rectus release. *J. Pediatr. Orthop.*, 10:433–441.

Tabary, J.C., Tabary, C., Tardieu, C., Tardieu, G., and Goldspink, G. (1972). Physiological and structural changes in the cat's soleus muscle due to immobilization at different lengths by plaster casts. *J. Physiol.*, (Lond). 224:231–244.

Tang, A. and Rymer, W.Z. (1981). Abnormal force-EMG relations in paretic limbs of hemiplegic human subjects. *J. Neurol. Neurosurg, Psychiatry*, 8:690–698.

Tardieu, C. and Tardieu, G. (1987). Cerebral palsy—mechanical evaluation and conservative correction of limb joint contractures. *Clin. Orthop. Rel. Res.*, 219: 63–69.

Tardieu, C., Huet de la Tour, E., Bret, M.D., and Tardieu, G. (1982). Muscle hypoextensibility in children with cerebral palsy: I Clinical and experimental observations. *Arch. Phys. Med. Rehabil.*, 63:97–102.

Thometz, J., Simon, S., and Rosenthal, R. (1989). The effect on gait of lengthening of the medial hamstrings in cerebral palsy. *J. Bone Joint Surg.*, 71-A:345–353.

Wickiewicz, T.L., Roy, R.R., Powell, P.L., and Edgerton, V.R. (1983). Muscle architecture of the human lower limb. *Clin. Orthop.*, 179:275–283.

Williams, P.E. and Goldspink, G. (1978). Changes in sarcomere length and physiological properties in immobilized muscle. *J. Anat.*, 127:459–468.

Winter, D.A. (1991). *Biomechanics and control of human gait.* Waterloo, Ontario, Canada: University of Waterloo Press.

Woo, S.S.-Y., Gomez, M.A., Woo, Y.-K., and Akeson, W.H. (1982). Mechanical properties of tendons and ligaments. II. The relationships of immobilization and exercise on tissue remodeling. *Biorheology*, 19:397–408.

Yamaguchi, G.T. and Zajac, F.E. (1990). Restoring unassisted natural gait to paraplegics via functional neuromuscular stimulation: a computer simulation study. *IEEE Trans. Biomed. Eng.*, 37:886–902.

Zajac, F.E. (1989). Muscle and tendon: properties, models, scaling, and application to biomechanics and motor control. *CRC Crit. Revs. Biomed. Eng.*, 17:359–411.

Zajac, F.E. and Gordon, M.E. (1989). Determining muscle's force and action in multi-articular movement. *Exerc. Sport Sci. Rev.*, 17:187–230.

Ziv, I., Blackburn, N., Rang, M., and Koreska, J. (1984). Muscle growth in normal and spastic mice. *Develop. Med. Child Neurol.*, 26:94–99.

Commentary:
Comments on Clinical Applications of Musculoskeletal Models in Orthopedics and Rehabilitation

Andy Ruina

This chapter describes an attempt to gain understanding of the workings and pathologies of the human lower body by the creation and refinement of computer simulation models. To date, the models described are mostly concerned with the geometry, kinematics, and mechanics of muscle, tendon and bone. The work shares the hopes and possible weaknesses of all attempts in science to use simulation to advance knowledge.

The main hope presented is that sufficiently accurate simulation can allow the testing of prosthetic or surgical corrections. But the complexities the authors talk about adding to their simulations seems to put them in the large club which, in the back of its mind, seeks the impossibly distant goal of the perfect simulator. What if they had it? They do. They have reality, and reality is its own perfect simulator. It is available to us here and now. But having a real person before you and understanding that person in any useful sense of the word are two different things. As simulations become more realistic, they tend to inherit the complexity and inaccessibility of reality.

That computer experiments do not have problems with the human-subjects committee is only an advantage if there are useful experiments that one could perform on human subjects. Would there be much to learn by varying the musculotendon lengths and thus musculotendon strengths (via force-length changes) in a given person's leg? Would one do it if it were easy and caused no pain or damage? Maybe and maybe not. Even if so, and one wanted to substitute a simulation one would have to trust the actually imperfect computer model.

One observation, repeated in different forms several times in the chapter, is that a biarticulate muscle's torque on one joint does not determine even the sign of that muscle's contribution to the rotation (statics) or rotational acceleration (dynamics) of that joint. Anyone doing surgery on a tendon should probably understand this observation and its kinematic correlates. It should hardly need be mentioned. Do complex computer models of the type presented here add to the generation of this type of basic insight or only serve to more convincingly demonstrate it?

Finally, the large efforts presented emphasizing accurate internal geometry seem to be at the expense of modeling full motion cycles. It might be that more insight could be had by studying more complete motions with, if not neurological models, at least with reasonable optimization criteria. More specifically, one might trust the authors attempt to simulate a walking cure if they could simulate healthy and pathological walking.

Despite these possible shortcomings, it is certainly possible that work of the type presented will help make people walk better by direct patient-specific simulations or, more likely, by the intuition that the modelers gain and share as they develop their general simulations.

Section VIII

37
Human Performance and Rehabilitation Technologies

Jack M. Winters, Corinna Lathan, Sujat Sukthankar,
Tanja M. Pieters, and Tariq Rahman

1 Introduction

Sections VIII and IX of this book differ from the previous sections in that they are tied more closely to applied research, especially as related to rehabilitation. This seems appropriate. When addressing the significance of our work, most of us include a statement that our research will ultimately help enhance the quality of life of certain types of persons with disabilities. Often the motivation behind our claim is that the increased knowledge obtained from our collective basic research will ultimately lead to technological or therapeutic innovation that will benefit society. This concept has deep roots that go back to the influential writings of Vannevar Bush, a U.S. presidential science adviser during the 1940s, who helped spawn dramatic increases in government-sponsored research infrastructure (e.g., the creation of NSF and NIH), and subsequently in the number of research-oriented scientists and engineers within most developed societies. Yet indications are that an ever greater part of the research pie will be targeted toward applied research that is more directly associated with tangible societal aims (if government funded) or products (if privately funded). Clearly there is a broad societal aim for higher quality health care at reduced costs. This must be done within an aging society that is going through dramatic technological change.

Section VIII focuses on human performance, particularly as related to technologies and approaches with application to the field of rehabilitation. This chapter starts with a brief synopsis of the past, then presents some principles related to human-technology interfaces and finally provides a selective review of some technological areas that appear to hold special promise for the future. Subsequent chapters are written by scientists with strong theoretical backgrounds who have purposely chosen to focus on innovative tools and approaches to science that might more readily yield technologies that are applicable within the rehabilitation environment.

2 Background

2.1 Basic Terminology and Classification

An engineer or research scientist working within a modern rehabilitation medicine environment quickly learns the importance of terminology; Table 37.1 provides definitions for some key terms in the rehabilitation research field.

Rehabilitation technologies of interest here can be roughly classified into three types (see also Table 37.2):

- *Assistive devices* used by persons with disabilities.
- Technologies that assist with *evaluation/diagnosis/assessment* by an appropriate expert (or team).
- Technologies that enhance *therapy*.

These need not be mutually exclusive. For instance, the robot technologies described in Chapter 38 are proposed as *therapeutic* devices that have an embedded capacity for *objective evaluation*.

TABLE 37.1. Key terminology.

Categories
Impairment: A loss or abnormality of structure or function at the organ level.
Disability: Restriction or lack of ability to perform an activity in a manner considered normal, manifested in the performance of daily tasks; the functional consequences of impairments.
Handicap: A disadvantage resulting from an impairment or disability, that limits or prevents the fulfillment of a role that is normal for the affected individual.

Clinical research assessment
Validity: The extent to which a measure actually measures what it is intended to (and not something else), that is, the degree to which evidence supports the inferences that are made from a score—types include *content* (are all relevant concepts or items within the domain of interest represented?), *criterion* (is it consistent with "gold standard?"), *predictive* (does it predict important events in the future, e.g. clinical outcome?), *construct* (does it correlate with theoretical construct it was designed to measure?); *concurrent* (does it correlate with events occurring at the time, or other scales?); *ecological* (is it meaningful or useful in the person's real life, outside of the clinical setting?).
Reliability: The extent to which a score or measure is free of random error or noise, that is, the degree to which an instrument is consistent, reproducible, and repeatable when administered by appropriately trained individuals; types include: *internal consistency* (measure same concept or factor?), *test–retest, interrater consistency*.
Bias: Systematic error in a measure, usually associated with subjective self-reporting or stakeholder-reporting, or attitudinal measures—classic examples are *socially desirable* response and *acquiescent* response.
Score sensitivity: Ability of measure(s) to *detect* true cases, to *discriminate* between meaningful differences, and to indicate *changes* in the true attribute being measured.

Rehabilitation research can be roughly categorized into four classes:

- *Basic* research (aimed at basic knowledge, e.g., most of the material in Sections II through VII).
- *Clinical* research (often controlled clinical trials aimed at effectiveness, efficacy, etc.).
- *Applied* research, aimed at developing new technologies (e.g., neural prostheses).

TABLE 37.2. Classification scheme for rehabilitation and human performance technologies.

Assistive devices
- Mobility devices (e.g., wheelchairs [manual, powered], scooters, walkers, crutches).
- Manipulation aids (e.g., alternative utensils and tools, reachers).
- Control interface devices (e.g., joysticks, mouthsticks, pointers).
- Orthotic/prosthetic devices (e.g., upper and lower limb prosthetics, orthotic braces).

Evaluation technologies
- Conventional medical exams (e.g., musculoskeletal, neurological).
- Imaging (e.g., X-rays, MRI, CT scans).
- Hand-held instruments (e.g., goniometers, grip tensometers).
- Measurement via controlled exercise equipment (e.g., joint torque, velocity, power).
- Advanced motion analysis technologies (3D motion, force platforms, EMG).
- Tele-evaluation (e.g., between home and medical facility).

- *Concurrent* "just in time" research, aimed at performing the research necessary to transfer evolving (high-investment) societal technologies into the rehabilitation sector.

The last of these has been an area with little federal rehabilitation research investment, yet a strong argument can be made that it can yield the highest return on investment. One simple example can make this point. It is widely acknowledged that the technology that has most profoundly impacted on the quality of lives of persons with disabilities in developed nations is the personal computer. Indeed, during the 1980s many of the key companies involved in this remarkable societal transformation (including Apple, IBM, and Microsoft) were quite helpful in facilitating the use of their products by persons with disabilities, more as a company policy priority than as a research investment. This facilitated third-party products, and as a consequence the number of persons with disabilities whose lives have been enhanced is remarkably large despite relatively modest federal and private investment in this area. Because research investment within rehabilitation, while significant and mildly growing (e.g., funding via the new *National Center for Medical Rehabilitation Research* within NIH), is small relative to societal technological forces, this chapter focuses on research approaches and emerging rehabilitation technologies that are consistent with expected technological innovation within society.

2.2 Historical Perspective

Historically, the fields of rehabilitation and of ergonomics/human factors have developed somewhat independently, so we will start by considering them separately. They are now more tied together, with the common bond being human performance assessment. This section focuses on trends within the United States, with apologies.

2.2.1 Pre-1970s Historical Perspective

Rehabilitation Technologies

Several historical perspectives are available that address the evolution of *rehabilitation engineering* as a field (Childress 1985; Galvin 1991; Cook and Hussey 1995). Its initial roots lay in the areas of *prosthetics and orthotics*, and in basic assistive technologies such as canes and wheelchairs. The initial two driving motivational factors were societal desires to help: (1) "handicapped" war veterans with injuries return to work; and (2) "crippled" children. In 1917, after World War I, the U.S. Congress enacted the Vocational Rehabilitation Act to provide services to return disabled veterans to civil employment, and in 1943 amendments to this act strengthened it.

A period of relatively rapid growth started after World War II (University of California 1947), and by the 1950s, significant improvements had been made in basic orthotic bracing for children, and in both lower (e.g., SACH foot, UC-BL ankle and knee) and upper limb (body-powered cable) prosthetics. The challenges of lower-limb prostheses literally helped expand basic theories on 3D mechanisms (Radcliffe 1960; Suh and Radcliffe 1967). In upper limb prosthetics, the development of bifunctional power transmission systems for routing residual proximal body power (e.g., shoulder shrug) to a distal hand and/or elbow location(s) via effective body harnesses and Bowden cables (Pursley 1955) proved to provide a remarkably functional and robust interface that remains the system of choice for most amputees (LeBlanc 1987). Ironically, millions of bicycles now use Bowden cable technology that was developed for prosthetics. During this time the Veterans Administration hospital

system was put in place, and formalized training programs started in orthotics and prosthetics. Shriner's Hospitals for Crippled Children, which started in 1922, initiated a significant research program in the early 1960s.

The late 1950s and 1960s also saw a flurry of activity and excitement related to use of external power, using both pneumatic (conventional and braided McKibbon artificial muscles—Schute 1961; Gavrilovic and Maric 1969) and electrical actuators (Groth and Lyman 1961). These efforts, however, remained primarily a research laboratory activity. One of the key motivating factors was the thalidomide drug problem, in which a significant number of children in Europe and Canada were born with limb abnormalities. During the 1960s and 1970s, Simpson's group (e.g., Simpson and Smith 1977) developed some remarkably effective whole-arm pneumatic limbs that, while discontinued for practical reasons, produced self-generated coordinated movements that remain essentially unparalleled in contemporary high-level prosthetic systems (Heckathorne 1990). Insights from these efforts led to the concept of extended physiological proprioception, or EPP (Simpson 1974, Doubler and Childress 1984).

During this era, other technologies improved, such as wheelchairs and canes, but often only incrementally once the initial product was on the market, with the market dominated by a few companies which sold only a few models. For instance, the conventional folding X-frame wheelchair was patented by H. Everest and H. Jennings in 1937, who went on to form Everest and Jennings, Inc., which remains the largest international producer of wheelchairs and during the late 1950s helped electrically powered wheelchairs enter the mainstream.

Human Factors and Ergonomics

This research activity can be documented as early as the 1920s, tied primarily to the fields of industrial psychology and industrial engineering, especially as related to attempts to improve worker efficiency. During World War II, the focus shifted to personnel safety and the "man-machine system." Because of technological advances, high performance was expected under conditions of higher demand. The risk of poor design in aircraft cockpit display became a matter of life and death. During the 1950s and 1960s, growth was driven primarily by the needs of the military, the automotive industry, and the space program. In the 1950s, Human Factors was officially named as a field of research in the United States, whereas in Europe, it was already called Ergonomics (Christensen et al. 1988). Today the terms are relatively interchangeable, although ergonomics tends to be more widely used within the biomechanics community.

By-products of this activity included new anthropometric and ergonomic data, especially for young adult males (e.g., Dempster 1955; Barter 1957), and a wealth of neurocognitive and neuromotor studies of target tracking, especially focusing on speed-accuracy-target size tradeoff issues that came under the umbrella of Fitts Law (Fitts 1954).

Human Performance

Human performance assessment has also been around for a long time, especially as related to gait (e.g., Weber and Weber 1836; Carlet 1872; Braune and Fisher 1895; Bernstein 1935; Glanville and Kreezer 1937; reviewed in Lamoreux 1971). However, quantitative gait analysis really emerged after World War II, with the early motivation being lower limb prosthetic development. By the early 1950s, the Biomechanics Laboratory at the University of California, Berkeley and San Francisco, had developed motion analysis capabilities (e.g., Levins et al. 1948) and 3D force platforms (Cunningham 1950; Cunningham and Brown 1952), and even had used inverse dynamics (with a distal-to-proximal progression) to estimate joint forces and moments (Bressler and Frankel 1950; Bressler et al. 1957). Indeed, it is perhaps fair to say that the knowledge of normal gait that was presented in Eberhart et al. (1954) is not that far removed from the current state of the art. They also had a strong understanding of isolated muscle properties (Ralston et al. 1949; Ralston 1953), and of EMG and EMG-force relations (Inman et al. 1948, 1952); the early analysis of shoulder kinematics by Inman et al. (1944) is still considered the standard. However, for practical purposes, the quantification of human performance was not yet widely available, except in a few research laboratories, and the analysis process was very tedious.

This time period also saw the emergence of 'systems engineering' approaches to neurocontrol of movement by pioneers such as Lawrence Stark (e.g., Young and Stark 1963; Stark 1968); control systems concepts such as optimal control, predictive control, sampled-data control systems, variable feedback, and adaptive control were all applied to biological systems during the 1960s, often with a focus on pathological populations.

2.2.2 Explosive Growth During the 1970s and 1980s

Rehabilitation Field

A remarkable level of growth in this field occurred during the 1970s and 1980s, triggered by a combination of policy issues, health care prioritization and resources, and technological innovation. These years saw the following:

- The emergence of *disability activism* and civil rights, which starting in 1973 helped produce evolving legislation [ranging in the United States from the Rehabilitation Act of 1973 (reasonable accommodation in employment and education; the right to an education for all children) to the Americans with Disabilities Act of 1990].
- Remarkable growth in the number of rehabilitation *professional specialties* and in the number of persons entering such fields as physical therapy, occupational therapy, and special education (there are now about 80,000 physical therapists and 50,000 occupational therapists in the United States).
- Remarkable increases in health care expenditures, well above the rate of inflation [within the United States from $247/person (6.3% of gross national product) in 1967 to $3,510/person (16.4% of GNP) in 1996].
- The development in the late 1970s of RESNA (the Rehabilitation Engineering (and Assistive Technology) Society of North America; the phrase in parentheses was added more recently and represents the fact that now under 40% of the membership are engineers).
- Profound changes within society in its *attitudes* [e.g., a "handicapped person" became a "person with a disability," who in turn was a "consumer" or "customer"; the concept of *independent living centers* evolved (Budde 1990), as did the role of technology to enable independence].

- The number of available *products* expanded dramatically, with technological choices emerging for disabled consumers (e.g., more companies making wheelchairs, with a larger variety of options), and product databases (e.g., ABLEDATA (http://www.abledata.com/, which includes information on over 22,000 products, and information resources (e.g., NARIC, National Rehabilitation Information Center in Silver Spring, MD) were created to provide persons with disabilities with better access to information.
- Funding for rehabilitation research expanded dramatically [e.g., within the United States, 16 Rehabilitation Engineering Research Centers were funded by what is now called the National Institute on Rehabilitation and Disability Research (NIDRR), with several others funded through the Veteran Administration (VA) system].

By the late 1980s, clinically based rehabilitation engineers were less involved in orthotics/prosthetics provision and had moved into such areas as seating and positioning, home/worksite accommodation, and customized alteration of assistive technology. However, these are all areas where as technology improves and customization becomes more embedded into products, other providers such as occupational therapists (OTs) can be adequately trained to provide and service technology. This is reflected in the new RESNA-sponsored Assistive Technology Practitioner (ATP) and Assistive Technology Suppliers credential exams, which are open to a broad range of rehabilitation health providers; this in turn implies that the niche for clinical rehabilitation engineers in these traditional areas may shrink (Winters 1995a).

Human Factors and Ergonomics

This field also expanded, especially during the 1980s, driven in part by the considerable (and steadily growing) cost of occupational (work-related) injury ("worker's comp"), especially as related to low back pain and repetitive motion disorders such as carpal tunnel syndrome. Many companies came to realize that front-end prevention through re-design of work environments and/or job tasks was cost-effective. Thus ergonomists had the dual mission of *enhancing performance* while also minimizing the *risk of injury*. By 1990 computer packages (e.g., *Mannequin* from BCA; *3D Static

Strength Program from the University of Michigan; *Jack* from the University of Pennsylvania (Monheit and Badler 1991); became available. They dramatically enhanced the ability to perform customized ergonomic analysis in a more timely fashion. Additionally, anthropometric databases started to broaden, including data on populations beyond just young adult males.

The term Human Factors Engineering is now commonly used to describe the application of knowledge about human characteristics to the design of the physical aspects of systems and equipment. This field received increased attention in the 1970s and 1980s, in part due to high visibility accidents and media coverage of "human error." Another trend was the increasing prevalence of computer technology and the attention to user friendly computer hardware and software. And finally the concept of *usability* became inherent to designing human-technology interfaces that optimize human performance.

Human Performance

The 1970s and 1980 saw the emergence of human performance assessment as an objective science, driven by the wide availability of 3D motional analysis systems and exercise equipment with sensors, and improved filtering and computational capabilities (e.g., see Chapters 1–7 in Allard et al. 1995). Such technological improvements continue, although driven more by the resources of these relatively small companies than by government investment (*Workshop Report on Gait Analysis on Medical Rehabilitation*, Bethesda, 1997).

2.3 Present Challenges

Most existing rehabilitation devices were designed as engineering technologies, with the human-technology interface based on ad hoc, intuitive, trial-and-error approaches that often took advantage of the remarkable level of adaptability of the human system to a variety of interfaces. Often this works reasonably well, with rather appropriate technologies evolving over time, due to collective input from heath care providers, users (consumers), and market-driven forces. Thus, wheelchairs have reasonably standardized sizes and interface options, and ergonomic guidelines have emerged for customizing a wheelchair to the needs of a specific person.

Yet there are problems. For instance, it is well accepted that the level of *abandonment* of assistive devices is high, around 30%, and that in many cases this is the result of poor interface technology and the lack of adequate research (Philips and Chao 1994). Indeed, few of the 22,000 products listed in ABLEDATA have ever gone through a formalized clinical trials process. Furthermore, product development based on intuition and interaction with health providers may have its limitations: in contrast to the 1970s and 1980s, fewer innovative technologies—whether assistive devices or for functional evaluation or therapy—have entered the rehabilitation field during the 1990s. Perhaps even more importantly, there are identifiable needs within large populations that appear not to be well met, especially populations with broad-based *neurological or neuromuscular impairment* in which a range of functional limitations are apparent because of the complexity of the pathology. Consider stroke patients, the primary focus of Chapter 38. In the United States alone, more than one half million people have a stroke each year. In 1992, at the National Rehabilitation Hospital (NRH), the typical stroke patient spent over 30 days in the hospital; in 1996 it was down to 20 days, and dropping. Yet there have been few changes in the evaluation and therapeutic aspects of the rehabilitation process for stroke patients, and typically *no type of objective (sensor-based) evaluation* is part of this process. Why?

One answer is that perhaps we have gone about as far as we can developing technologies by intuition. This suggests a need for: (1) transferring insights from basic biomedical research into this area; and (2) developing more effective interfaces between technology and its end user (whether this be a person with disability or a clinician). This is exactly what Chapters 38 to 40 address.

It is also crucial that we, as scientists, be aware of issues in product development—Chapter 41 provides a timely treatment on this process. Of note is that in the past, research scientists and engineers have often created technologies that have not been appropriate, for a variety of reasons (e.g., directed toward too small a target consumer population to be cost effective, too large and not cosmetically pleasing, not easily repaired or serviced, not prac-

tical). Part of the challenge is that, as seen throughout this book, neuromusculoskeletal systems are complex and possess a remarkable ability to adapt to environmental conditions. There is nonetheless a sense among many of us that as we come to better understand human performance and human-technology interfaces, we will succeed in transferring our scientific knowledge into innovative, appropriate new twenty-first century technologies. Section VIII is about this challenge.

2.4 Future Opportunities

Predicting the future is difficult, and a detailed discussion is beyond the scope of this chapter. However, there are several trends that are relevant:

- Society in developed countries is *aging*, and the impact on the rehabilitation field will likely be large (Galvin 1991). For instance, within the United States, the number of persons over 65 will approximately double between 1990 and 2030, and the number of people above 85 years is expected to increase five-fold. In recognition of this challenge, a number of special federal research initiatives now address key areas affecting older adults; this is likely to expand in the future. Geriatric rehabilitation will be a growing field.
- *Home care* is by far the fastest growing segment of the health care market (e.g., from 1994 to 1995, home health benefit payments via Medicare grew 26%). All indications are that this will continue, as health care becomes more decentralized.
- The current private investment in *telecommunications and interactive computer technologies* is staggering, and all indications are that profound changes are on the horizon that will affect all aspects of modern society. One targeted area is "intelligent" interactive personal assistants for specific categories of professionals. Another area with considerable investment capital is the entertainment industry, which now drives not only "virtual reality" technologies but also future 3D motion analysis company innovation. The considerably smaller biomechanics/rehabilitation sector should be able to take advantage of such societal investment.
- *Objective evaluation and outcomes assessment* approaches and technologies have emerged as a public funding priority. Although less certain be-

cause of the growth of managed care in the United States, a window is open for adoption and promotion of objective assessment when it is cost effective (e.g., reduces hospital stays; prevents illness).

Although personal perspectives on what the future may hold will clearly vary (and is always a risk), Table 3 in Winters (1995a) suggests that by the year 2010: tele-evaluation and video televisits, supported by home visits by professionals with interactive expert computer assistants, will be the norm; much of physical/occupational therapy will be automated, customized, and interactive; and postformable orthotic braces will be fabricated on the fly, with built-in 'smart' safety warning and tremor management capabilities. Whether or not these predictions happen in this form, the key point is the following: Within the near future, available technological capabilities will be so good that that our ability to effectively utilize these technologies will be *limited by our knowledge of human movement and the human-technology interface*. This suggests a special need for human performance research, and especially a focus on *diffuse neurological pathologies* that lead to multifaceted neuromotor impairment affecting many coordination strategies; hence Chapters 38 to 40.

3 Principles for Designing Appropriate Human-Technology Interfaces

Section VIII focuses on how scientific insights could impact on the rehabilitation field. This implies a consideration of the human-technology interface, and on technologies that could be used for improved diagnosis and therapeutic intervention. However, first we need to develop a scientific perspective on this interface and identify certain conceptual principles for designing and evaluating such interfaces. Interfacing of humans with technology in order to achieve performance goals is not new; technology has always played an important role in the harmonization of humans and their environments. The challenge in the rehab community is to develop or apply *enabling* technology to assist the performance of individuals in their environment.

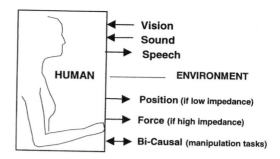

FIGURE 37.1. Information transmission modalities for human-technology interfaces.

3.1 Overview of the Interface

Figure 37.1 provides a more conventional conceptual overview of the key sensorimotor modes for human-technology interface. The bottom three arrows represent information exchange at an interface such as the hand. The first (top) of these represents a mode where the output is position, and the environment provides a low impedance (i.e., as the hand position moves, the applied force changes little). Examples of this would be hand tracking movements, performed without mechanical contact or while using a lightly loaded joystick. This type of paradigm has been used often in neuromotor research, and most of the studies in Section VI fall into this category. Notice that it allows a unidirectional view of the human-technology interface, with conceptually clean sensory inputs and motor outputs (hand position). For the second of these arrows, force is the output, typically applied under isometric conditions, for example with the hand against a stiff surface or immobilized by apparatus or an orthoses. The third case, bicausal interfaces, represents manipulation tasks (i.e., dynamic interaction with an environment). Performance thus depends on the properties of the environment; we now consider this more general case further.

3.2 Bi-Causal Mechanical Interfaces and EPP

The purpose of neuromotor activity is typically *to manipulate the environment* in some way, usually via a direct mechanical interface. In this section we develop some key concepts for mechanically coupled systems, then link this to the important concept of Extended Physiological Proprioception (EPP). Issues such as contact instability and contact impedance modulation then becomes important, and a theoretical framework for these concepts has been established (Hogan 1984, 1990). It is worth briefly developing the theoretical foundation using a simple example.

Following Hogan (1990), we utilize a parallel mass-dashpot-spring second-order linear model of the human to develop these principles:

$$(ms^2 + bs + k)x(s) = cu(s) - f \quad (37.1)$$

where m is generalized mass (due to a limb segment), b is viscosity, k is stiffness (due to muscles), x is a generalized position, u is a controllable input, f is an environmental (applied) force, and s is the Laplace operator.

Now let us assume a desire to regulate position by generating a control signal proportional to the time integral of any deviation between x and a reference value r, while also allowing f to be a variable external mass m_e (e.g., that of an object to be picked up), that is,

$$u(s) = \frac{g}{s} r(s) - x(s); \qquad f = m_e s^2 x(s) \quad (37.2)$$

where g is a feedback gain. We then have, after substitution and simple manipulation:

$$\frac{x(s)}{r(s)} = \frac{cg}{(m + m_e)s^3 + bs^2 + ks + cg} \quad (37.3)$$

and to be stable, the following constraint on the gain:

$$g < \frac{bk}{c(m + m_e)} \quad (37.4)$$

For conventional control, we normally would want g to be reasonably high (i.e., approach the inequality). But then what happens when g is initially set with m_e zero or low, but then m_e rises when an object is picked up? Often some type of sustained pathological chattering; this phenomena of destabilization upon contact is often called *contact (or coupled) instability*. Although analysis becomes considerably more elaborate for more realistic higher order, nonlinear systems, the principle is quite general (Hogan 1990).

To introduce the concepts of *passivity* and *coupled stability*, we define the *driving point impedance z* (relating the velocity v to f) for two possible

feedback schemes, integral (z_i, u_i) and proportional (z_p, u_p) action:

$$z_i = \frac{f(s)}{v(s)} = -(ms^2 + bs + k); \quad u_i(s) = \frac{g}{s}(r - x)$$

$$z_p = \frac{f(s)}{v(s)} = -\left(ms + b + \frac{k}{s}\right); \quad u_p(s) = g(r - x)$$

$$(37.5)$$

Most objects that the human interacts with are passive, that is, while they may store energy (e.g., potential energy because of masses and springs), the recoverable energy is less than that stored because energy is lost across the dashpot b. For systems under sinusoidal driving input, net power must then be absorbed over each cycle and thus the phase angle between v and f must lie between $\pm 90°$. In the top case of Eq. 37.5, this is not satisfied (it is between 0 and $-180°$), while in the bottom it is. The point is the following: *dynamic interaction profoundly affects stability*. Of course in reality this dynamic interface is highly nonlinear. For instance, the human can change the transient stiffness at the hand by an order of magnitude in a fraction of a second, via co-contraction of muscles (Winters 1990). Notice also that this principle extends *to interfaces between body parts* (e.g., mechanics between the torso and shoulder regions is influenced by arm configuration and hand contact load).

This helps illuminate an important concept: the inherent nature of biosystems toward holistic "inclusion" as a dynamic control strategy. This concept is synergistic with that of EPP, first coined by Simpson (1974), who was a pioneer in the design of pneumatic upper limb prosthetic limbs in which residual torso motion was used to control multiple prosthetic joints by a combination of one-to-one cable coupling and external power assist. The observation is this: Humans possess a remarkable capacity to, with sufficient practice, utilize a device such as a golf club or pencil as if it were *an extension of their own bodies*. Technologies that allow such inclusion via predictable (usually one-to-one) mapping and yield control to the human are considerably more likely to be accepted; this could be coined *assistive control*. An experienced golfer does not miss the ball despite the fact that deviation of a few degrees at the wrist would create a large deviation at the end of the club, and an experienced hockey player uses the stick within an environment with dramatic changes in impedance in various directions. With experience, the technology becomes a subconscious extension of the body. What it takes for this to occur is now fairly well understood. A key requirement for such "inclusion" is intimate bicausal (energetic) mechanical interaction in which there is instant *position (and velocity) and force (and its derivative) mapping across the interface*. The product of force and velocity is power, and in essence there is instantaneous power transfer.

In contrast, conventional control systems approaches utilizing inherently unidirectional information flow have met with marginal success. This helps explain why, despite considerable effects by distinguished research groups, most upper limb prosthetic users still prefer conventional body-powered prostheses to externally powered, especially for the hand (terminal device) interface (Michael 1986; LeBlanc 1987). In a very real sense, they can subconsciously "feel" a cup or egg in the grip via the bicausal mapping to the shoulder, while requiring little (if any) cognitive attention. Attempts at providing a more artificial mapping of sensor information by other means (e.g., by vibration or electrical "minishocks" of a skin location) have met with limited success, requiring significant cognitive effort by the user.

Why is this so important? Because not only does it capture a key aspect of human performance, but also is remarkably consistent with the modern approach to rehabilitation engineering: design of assistive devices with a purpose of *assisting human performance* by using technology to *enhance* the person's abilities. This is the foundation of what is often called *appropriate technology*; it is a way of thinking. It applies not only to assistive devices, but also to orthotics and prosthetic devices, and to therapeutic intervention. To summarize: While engineers have had limited success in designing control interfaces that *"take over"* certain neuromechanical functions, technologies that *assist or enhance* human function through careful mechanical interface design have high potential.

3.3 Human-Computer Interfaces

With the rapid evolution in computer technology and the increasing use of computers in the clinical environment, human-technology interaction has become almost synonymous with human-computer

interaction, with applications in areas such as virtual reality (VR) also involving mechanical interfaces that then directly couple to the computer. Yet despite the exponential growth in the number of computer-interfaced clinical projects and systems, relatively little attention has been paid to the human factors issues associated with such systems. Human-computer interaction (HCI) and interface design is a growing field with the recognition that the human-technology interface is a key to optimizing human performance.

3.3.1 Models of the Human Information Processor

By understanding the human as an information processor, we can obtain some basic principles for human-technology interface design. A simple model of the human processing system is shown in Figure 37.2. This is an idealized model of human information-processing capabilities in which the user is described in terms of three processors: perceptual, cognitive, and motor (Card et al. 1983, 1986). Sensory information is gathered from the interface, primarily from the visual system, but increasingly more from auditory and kinesthetic as interfaces advance. This sensory information flows into working memory through the perceptual processor where it can then be accessed by the cognitive processor to decide if and how to act on it. Finally the motor processor executes any actions decided upon.

The information flow through the perceptual processor to working memory is considered to be

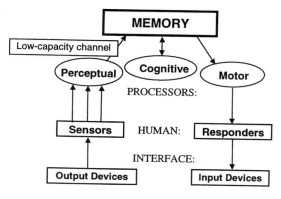

FIGURE 37.2. Simple model of the human processing system.

a low-capacity channel. In other words, there is a limited ability of a person to pay attention to all of the sensory input available. From the point of view of human-technology interfacing, the key is not to overload channel capacity. A successful interface design will take into account user capabilities. Specifically that there are finite resources available and appropriate allocation of attentional resources can improve performance of tasks. As activities change and new tasks are initiated, there will be dynamic distributions of attentional resources. Understanding the distribution of attentional resources will aid designers by providing predictions of human performance.

Two specific examples of interface design constraints coming from this simple model, that will enhance human performance, are as follows:

1. Decrease the demand on attentional resources, for example, visible controls and options eliminate the need to recall options from memory;
2. Improve the quality of the data, for example, a sharp image will require fewer perceptual resources than a blurry image.

With computers and visuomotor control, another critical aspect is the display update (sampling) rate. This problem of the effects of discrete information and time delays has long been of special interest to the researchers addressing teleoperator performance in virtual environments, especially as related to head-mounted displays, which use stereoscopic display, head movement tracking, and powerful graphics computers. It has generally been found that subjects can perform reasonably well with low frame rates, such as 10 f/s for inexperienced subjects and about 2 f/s for trained subjects (Liu et al. 1993), yet for "immersion" experiences the update rates must be considerably higher, approaching that of television (50 Hz in Europe, 60 Hz in the United States).

Another area of considerable research activity has been the relation between visual processing and eye–head movements. Humans only see 1 to 2° of arc with any clarity, and routinely make fast, near time-optimal saccadic eye movements to sample their environment (Clark and Stark 1975). Where humans choose to look is itself a fascinating problem in information processing, and the term scanpaths is often used to document this (Noton and Stark 1971; Stark and Ellis 1981).

3.3.2 Quantifying Performance in Complex Environments

The use of computers as interfaces to achieve performance objectives is becoming more and more prevalent. For example, simulated environments or virtual reality are now being used in clinical environments. Whether the simulated environment is for training, or the actual interface for executing a task, *performance measures* are needed to evaluate the effectiveness of the simulated interface.

Complex simulations such as those afforded by virtual reality applications (discussed below) pose particular challenges for quantifying performance because of the inherent difficulties in designing and evaluating the interface. Here humans participate within a multisensory surrounding environment that for advanced applications provides congruent visual and motor frames of reference. Human computer interface design principles need to be systematically applied to advanced interfaces while keeping in mind the task goals to enable evaluation.

Complex tasks have inherent evaluation difficulties because of their tendency to be multimodal nature. This, combined with executing the task in a simulated environment, ensures that quantifying task performance will not be trivial. For example, simple measures of task execution time generally will be insufficient to characterize performance; some combination of sensorimotor, biomechanical, and cognitive measures will be needed. However, this may not always be limiting if clever solutions can be obtained for collecting assessment data from a more structured subset of the therapy session.

4 Objective Evaluation: A Tool for Diagnosis, Treatment Planning, and Outcomes Assessment

By objective evaluation, here we refer to using sensor-based measures as part of the evaluation process. Here we consider three possible uses for evaluation: (1) diagnosis; (2) treatment planning; and (3) outcomes assessment.

4.1 Expert Diagnosis

One of the most challenging and important aspects of medicine is diagnosis. Historically, considerable biomedical engineering research and development effort has gone into diagnostic technologies, especially designed towards the needs of physicians. Here we focus on the utilization of human movement analysis to diagnose conditions of the neuro-musculoskeletal (NMS) system.

4.1.1 Diagnosis and 3D Motion Analysis: The Challenge

Within the area of 3D human movement analysis, the most successful clinical application to date has been gait assessment of children with cerebral palsy; this has been the foundation for most financially successful clinical gait laboratories. In a recent NIH-sponsored workshop on Gait Analysis (Stanhope et al., 1997), the highest priorities were all related to research directed toward establishing clinical efficacy. The reason in part is that at present, quantitative functional sensorimotor assessment is rarely part of the clinical diagnostic process.

Can this change? Yes, but it is suggested here that to be successful, the motion analysis results must interface more effectively to the Clinical Decision Maker (CDM). A CDM is a busy professional who must already synthesize a wide variety of "apples and oranges" information, while also resisting information overload. In making decisions, CDMs are used to the following types of information: (1) patient charts (which often include subjective scales represented by a single number, and objective single-number data such as range of motion); (2) visual images (e.g., X-rays), and (3) phrases from colleagues (e.g., radiographic diagnosis, nurse's comments).

Now contrast this to the classic type of information that results from a 3D human movement study—sets of curves as a function of time (e.g., motion, force, EMG). Consider a "typical" task of 1 sec duration, with smoothed and filtered 8 EMG collected at 100 Hz (many would consider this a low sampling rate), 3 limb segments (18 channels) of motion data at 50 Hz, and 6 channels of contact force at 100 Hz. That is 2,300 raw numbers collected in 1 sec, and this typically expands dramatically as more measures (e.g., velocity, moments, power) are computed. If only gait is of interest, this may be manageable, with suitable training of appropriate CDMs, but in general it is quite a challenge—there is information overload.

So what is the solution? For certain localized pathophysiological conditions, for example, such sports-related orthopedic impairments as a torn ligament, simple objective measures (e.g. torques) during simple tasks (e.g., single-joint isometric, concentric and eccentric contractions) may be sufficient. Yet what about persons with neuromotor impairment or orthopedic dysfunction that affect multisegment coordination, where no one type of movement task typically is sufficient to capture the functional limitations? The CDM, virtually out of necessity, usually bases decisions on very simple, reduced information from many tasks (e.g., personal interaction and chart data). Thus we see divergence between clinical practice and our conventional mode for 3D quantitative analysis (Winters 1995a).

Two points need to be made. First, there is a need for greater focus on the process and form for providing information to CDMs. This includes approaches for more aggressively reducing and synthesizing data, and better interfaces for presentation. More fundamentally, any added objective information should enhance the performance of the CDM, and it should be possible to objectively document this enhancement. This suggests that the 'system' under study should include the CDM as well as the actual data.

Second, the CDM cannot be an expert on everything, especially as related to 3D human movement. An "intelligent computer assistant" (ICA) or an "interactive personal assistant" (IPA) of some sort is needed that could, in near-real time, use embedded clinical and engineering interpretive expertise to aggressively reduce data (Winters 1996). An example of this approach is the [PI]/ASSET (Principal Investigator in a Box]/Automation and Support System for Expert Tele-science) technology, developed in a collaborative effort by NASA-AMES and MIT (Franier et al. 1994; Groleau 1994). The purpose of this well-tested technology was to impart some of the knowledge of research scientists into an ICA that was designed to assist astronauts in performing research in space under time pressure. The modular software, which includes components of both event-driven and user-driven programming, can troubleshoot for equipment failure, help coach astronauts through the experimental protocol, and evaluate data in near-real time. It only provides assistance and suggestions, and in particular looks for data that does not meet *a priori* expectations (and is thus "interesting"). This provides an existance proof on how the combination of an interactive ICA and an expert system in the background (in this case NASA's CLIPS) can serve as an effective personal assistant. For CDMs, an appropriate extension would be fuzzy expert systems, since approximate (vs. crisp) reasoning better describes the typical CDM process (see Chapter 40), and in particular linguistic inference from human experts.

Imparting knowledge is itself nontrivial, for two key reasons. First, experts often do not agree on how to interpret the same medical data (and even in the relative importance of various types of information), and thus reaching consensus can prove challenging. This is a classic problem in medical informatics. Second, extracting and assembling knowledge from experts can prove challenging; here advances in interactive authoring and multimedia technologies prove helpful.

4.1.2 A Role for Soft Computing

Soft computing (SC), the focus of Chapter 40, differs from conventional (hard) computing in that it is tolerant of imprecision, uncertainty, vagueness, and partial truth. Its guiding principle is to exploit such tolerance to achieve tractability, robustness and low solution cost. The principal constituents of SC are (Zadeh 1994): fuzzy logic, neural network theory, and probabilistic reasoning (which subsumes genetic algorithms, belief networks, and chaos theory). Current approaches focus on combining these techniques to take advantage of the best of each. Here we briefly summarize these approaches (see also Chapter 40).

A classical fuzzy expert system consists of data ("facts") and a reasoning process (rule database) that may involve crisp or fuzzy inference (a crisp fact or rule is a subset of a fuzzy fact or rule, i.e., it has a "hard" boundary). A key advantage of a fuzzy implementation is that facts can be a mixture of precise and imprecise data, and heuristic rules can synthesize "apples and oranges" data; most real-world CDM's must address multiple, often conflicting goals, with decisions ideally be based in part on available *objective* data and in part on *subjective* value judgments (Winters 1996). This would seem to be a critical feature for any IPA de-

signed for medical diagnosis. Of note is that fuzzy expert system algorithms, such as FuzzyCLIPS (a fuzzy extension of NASA's CLIPS by the Research Council of Canada), can have a sophisticated inference engine that differs considerably from serial programming. It can be classified as a "top-down" approach to system design in the sense that the system is "programmed" by experts (i.e., it has captured some of their knowledge and heuristic reasoning capability), but does not possess a mechanism for adaptive learning based on collected data.

A remarkable level of R&D activity is being directed to neurofuzzy systems (e.g., Lin and Lee 1996; Jang et al. 1997; Kodko 1997; see also Chapter 40). Neurofuzzy designs have the advantage of being able to synthesize both top down programming (as with FESs) and *bottom-up learning from data sets* (as with neural networks). The key advantage to neurofuzzy architecture is that an attempt is made to not just learn but also associate meaning to evolving data patterns. This is the concept behind EMG 'synergy analysis" in Chapter 40, and what we coined a "synergy engine" (Winters 1996; see also Chapter 35). Another advantage is that once the network has "learned" through training (e.g., normal movement patterns or certain classic pathological patterns), it has the potential to be computationally efficient at recognizing unexpected (and thus "interesting") behavior. Finally, any mature implementation with a significant number of rules (whether via "top-down" programming, "bottom-up" learning, or both) has an inherently modular structural design.

The box entitled Soft Computing Critic (SCC) in Figure 37.3 and 37.4 represents the process of synthesizing converging modules, and requires some development. In the area of neuro-engineering, the concept of an Adaptive Critic (AC) has emerged, which essentially has the key role of synthesizing incoming information to provide an ongoing assessment of performance by neurocontrol networks involved in an optimization/learning process (Werbos 1990); it ties conceptually to reinforcement learning (Barto et al. 1984). In the context of medical diagnosis, Winters (1996) utilized the term Fuzzy Expert Critic (FEC) to help emphasize the importance of close top-down intimacy with a CMD and of robust system performance, and the value of fuzzy inference as a means for dealing with the information compression and prioritization

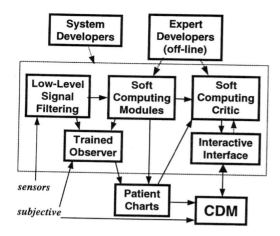

FIGURE 37.3. Schematic for an interactive personal assistant for a Clinical Decision Maker (CDM) that synthesizes objective (sensor-based) data with more conventional measures via soft computing. Notice that the 'system' includes both the ICA and the trained observer. For certain applications the CDM may also be the trained observer.

process. But "SCC" would appear more general, and can include nonanalytic reasoning methods such as the rule production approach of Chapter 39.

A key point is that the final output to the CDM needs to avoid information overload by *interactively assisting* the CDM. Additionally, the final

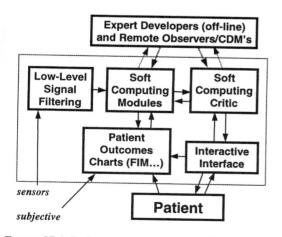

FIGURE 37.4. Patient-centered interactive flow diagram. In contrast to Figure 37.3, there is more of an emphasis on interactive system learning via patient responses and on performance changes over time, less on top-down expert developers.

end-product must *augment* existing clinical practice. In our own system development, we are focusing on SCC analysis of human coordination patterns that could complement or augment subjective expert observation and self-report, as follows:

- The addition of a prioritized set of linguistic phrases (e.g., via Mamdani neurofuzzy filters), often supported via certain selected curves (e.g., unusual synergies), and
- Some reduced numbers that could be added to existing chart data (e.g., via Sugeno neurofuzzy filters).

Ultimately, neuromusculoskeletal simulations, such as envisioned in Chapter 36 or Chapter 35, could be embedded as an available option, but these may have limited diagnostic value to most CDMs.

Finally, it should be noted that the evaluation of performance should non be separated from insights from neuro-control, because this, after all, is what the system is attempting to accomplish. It is generally agreed that the neurocontrol strategies used by humans are remarkably sophisticated, and poorly understood, especially for large-scale systems (e.g., see Chapter 32). Another way of addressing the need for quantitative assessment has been the development of technologies to measure human movement. As Jacobs and Tucker state in Chapter 40, human movement can be quantified in terms of kinetic, kinematic, and electromyographic measures, which are obtained with such technologies as force plates, EMG electrodes, and cameras (e.g. "Vicon") or other electromagnetic tracking (e.g., Flock of Birds). Quantitative strength testing and training devices, such as Cybex and Kincom, have been widely used in rehabilitation and sportsmedicine. In Chapter 38, Reinkensmeyer et al. discuss the potential benefit of robotic devices that could apply therapeutic patterns of forces to stroke patients' limbs, and which could more objectively quantify specific pathophysiological mechanisms, spontaneous recovery, functional ability, and therapy dosage. Several examples of robotic devices are discussed, including bimanual rehabilitators, a reaching guide, a Mirror Image Movement Enhancer (MIME), and MIT-MANUS.

We have been developing, through joint CUA-NRH efforts, a portable evaluation system (PES) as part of our telerehabilitation thrust. The PES utilizes three different sensor technologies: EMG surface electrodes, Flock-of-Birds motion analysis, and six-axis force sensor. It also uses a larger battery of upper limb tasks, some quite similar to standard musculoskeletal exams, some related to reaching or ADL tasks. It is designed to complement standard clinical scales, assisting in rehabilitation evaluation and diagnosis (e.g., stroke patients). The purpose of such evolving systems is to help synthesize the vast amount of data collected from the multiple sensors, and reduce it quickly into concise results that are meaningful and useful to a clinician. As an example of the types of linguistic phrases that could be of value (Winters 1996):

- Check the taping/strapping for the [. . .] EMG—data suggests it is coming loose!"
- "Muscles [. . .] and [. . .], normally synergists, were not! (click to see curves)."
- "[All/,most/[name]] muscle excessively active [esp. during [initiation/clamping] phase,] [and especially for [strength/reaching/manipulation/ADL] tasks]."
- "For [. . .] tasks, the elbow stayed [adducted/abducted], abnormally [close to/far from] the body."

The first of these is related to interactive technical trouble-shooting, and would be addressed by the trained observer. The last three are more related to the CDM process. Notice the inherent divergence between the scientist who feels her/his information is valuable and cannot be so dramatically reduced into a few linguistic phrases, and the busy, overworked CDM.

4.2   Treatment Planning

Virtually all that was said above for diagnosis also applies here. The main distinction is that for treatment planning the CDM often has an even better idea of the information that she/he needs. This could allow even more refined and customized ICA, one that could find specific niches within the marketplace. The obvious application is for gait, which is a reasonably stereotyped task and one that has been the focus of hundreds of research studies; however, to date quantitative gait analysis does not have strong foothold within the CDM community, except for children with Cerebral Palsy at selected gait laboratories with high levels of expertise.

4.3   Outcomes Assessment

Three specific functional objectives are generally associated with medical rehabilitation (Granger and Brownscheidle 1995):

1. Outcome (benefits patients received from program).
2. Progress (type and degree of improvement).
3. Efficiency (amount of staff and institutional resources expended per functional gain by patient).

There is a clear trend in medical practice today toward making patient (and product) evaluation more objective and quantitative. This trend is commonly termed "outcomes assessment," and refers to the measurement, monitoring, and management of outcomes (Davies et al. 1994). *Outcomes measurement* refers to systematic and quantitative observation of outcome indicators; *outcomes monitoring* refers to the repeated measurement over time of outcome indicators in a manner that permits causal inferences about what produced the observed patient outcomes; and *outcomes management* pertains to the use of information and knowledge gained from outcome monitoring to achieve optimal patient outcomes through improved clinical decision making and service delivery.

There are several reasons that help to explain the increased interest in outcomes in the United States (Davies et al. 1994):

1. Public policy has targeted health care as the #1 priority for *reform* in the United States. The measurement of outcomes on a national scale can help establish standards of effectiveness for a basic package of universal health benefits.
2. *Purchaser demand*: Purchasers are looking for demonstrable value of the health care dollars they spend.
3. *Quality assurance*: The introduction of continuous quality improvement (CQI) into health care has the potential to help integrate outcomes assessment more closely into the care delivery process.
4. *Clinical research*: Recognition of importance of outcome studies.
5. *Computerized medical record*: Affords database capability for patient information.

In the specific field of rehabilitation medicine, outcomes assessment may be particularly important. As stated in the *Executive Summary* of the *Report of the Task Force on Medical Rehabilitation Research*: "the Task Force concluded that three overriding needs are critically important to the field's progress:

- Develop meaningful quantitative measures of impairments, disabilities, and handicaps, and of the outcome of rehabilitation interventions. . . .
- Develop standards and guidelines for the design and application of evaluative tools.
- Evaluate the effectiveness of existing and emerging rehabilitation procedures."*

One way in which the recognized need for outcomes assessment has been addressed is in the work of a task force formed in 1983 by the American Congress of Rehabilitation Medicine and the American Academy of Physical Medicine and Rehabilitation. One result was the establishment of the Uniform Data System for Medical Rehabilitation (UDSmr) at the State University of New York at Buffalo. An important component of the UDSmr is a standardized test to measure level of disability called the Functional Independence Measure (FIM), which has been used for tens of thousands of patients from hundreds of facilities, and has been shown to be adequately reliable and valid; it is used in Chapter 38. Perhaps of most importance here is that the FIM meets not only the demand for a quantitative assessment tool, but also for uniform, standardized tools. Uniformity in functional assessment allows comparability of results and over time between providers, thus increasing the likelihood for improving the processes of care (Granger and Brown 1995) and helping establish cost–benefit guidelines.

Another issue that the human movement community needs to be aware of is the concerted move towards outcomes assessment that more fully involves *patient-centered* (vs. CDM-centered) assessment. At first glance this suggests less use of objective (sensor-based) measures. However, perhaps the opposite is true. If we assume that a patient's self-assessment is related to her/his ability to perform key tasks, then through bottom-up reinforcement learning (via SC) approaches, correla-

tions can be established between task-based global performance measures and measured parameters. The output of such a system could be linguistic statements or questions to the patient that complement conventional measures such as FIM. The patient could associate a degree of truth with each statement or answer. The result could be a synthesized assessment that is influenced by the patient response; Figure 37.4 provides a possible patient-centered flow diagram. Of note is that this same type of background infrastructure may also be of value when evaluation is combined with automated therapy, which is our next topic.

5 Technologies to Assist Therapy and Evaluation

Automation of therapy may have a bright future if it can be shown to reduce costs, be convenient and reasonably "fun" (and thus enhance user compliance), allow more frequent objective evaluation, and more generally *enhance outcomes*.

For instance, current treatment for persons with neurological impairment, such as stroke, involves a good deal of focused therapist time. Because of cost constraints, the amount of time devoted to treatment is dropping. Furthermore, evaluation and outcomes measures are remarkably subjective, and not normally based on objective measures (e.g., EMGs, 3D motion), which could be used to document indices that relate pathology to function and performance. This need for objective measures helps provide the motivation for Chapter 38, as well as other activities in robotics and virtual reality that are the focus of this section. Selling this concept of automated therapy with embedded objective evaluation measures to the clinical community, however, may take a sustained effort.

5.1 Virtual Reality (VR) Technologies

The emergence of the computer as a clinical tool and the progress made in sensor and display technologies creates an unique opportunity for conducting clinical human performance measurement research. Of particular interest are developments made in computer simulation technologies such as

---

*1990 Report of the Task Force on Medical Rehabilitation Research, Hunt Valley, MD.

Virtual Reality (VR). VR involves the integration of various sensors, displays, and computers to allow users to interact with artificial computer environments in a reasonably natural and synergistic manner. The VR infrastructure includes the following:

- *Graphics workstation and software.* This is the heart of the VR system and is a primary determinant of a VR system's level of performance. The workstation can vary from IBM PC compatibles with special graphics boards to high end graphics workstations such as the Silicon Graphics multiprocessor Onyx platforms. VR computing software residing on the graphics computer is responsible for coordinating the actions of the various sensors, processing user inputs, and defining the nature of the user virtual world interaction.
- *Sensors* to measure, in *real time*:
  - Position/orientation of any prescribed body segment (e.g., electromagnetic sensors such as the Flock of Birds [Ascension Technologies], and the Polhemus Fastrack, and ultrasonic trackers such as the Logitech 3D mouse; eye tracking systems).
  - Hand posture/ finger flexion measurement tools (e.g., instrumented gloves such as the 5DT Glove [General Reality Inc.] and Cyberglove [Virtual Technologies]).

Interactivity and *immersion* are the cornerstones of VR (Burdea and Coiffet 1994). A key is integrating seemingly disparate technologies into a coherent unit, thereby making more intuitive human computer interaction possible. Applications of VR technology include simulation, training, and visualization.

The use of VR technology in the military and space arenas is well documented. The military uses the technology to train its troops on equipment operation as well as tactics (Johnston 1987; McDonough 1993). NASA uses VR to train and concurrently evaluate the performance of their astronauts for extra vehicular activities.

The applications of VR in medicine are being investigated by several commercial and academic research organizations. These applications include surgical training, surgical planning of complex procedures (Pieper et al. 1992; Rosen 1994), the use of image guided techniques for navigation during surgery, and rehabilitation (Reddy et al. 1994; Suk-

thankar 1996). Surgical simulators can be used to facilitate the learning of fine and gross motor skills crucial to the performing of surgical procedures, which can then be (hopefully) transferred into "real" world situations. Surgical simulators with embedded performance measurement tools can be used for medical accreditation or for training surgeons on innovative surgical procedures. In such cases simulation technologies such as VR can be used for *skills acquisition and enhancement*, an important aspect of rehabilitation in individuals with impaired motor function.

Four types of learning effected in virtual environments include procedural, perceptual motor skill, navigational, and conceptual learning (Wickens and Barker 1995). It is important for such learning experiences to be closed loop with the trainee interacting in a natural manner with the information presented to her/him in the simulator. This helps the trainee to effectively transfer these learned motor skills into the "real" world.

Individuals with compromised motor function, occurring as a primary or secondary result of their stroke or other form of neurological insult, have to reacquire motor skills to function in the real world. Often such persons become discouraged when functioning in a "real world." By creating an immersive environment *based on their abilities and therapeutic needs* where "success" is more achievable and "therapy" is reasonably "fun," the therapeutic value of rehabilitation regimen can be maximized. Bimanual rehabilitators, such as the ones discussed in Chapter 38, can be coupled with immersive display setups used in VR for skills acquisition.

VR-based therapeutic programs can be used to increase patient motivation and thereby help accelerate the recovery for patients. Several VR-based therapeutic programs have been developed and tested with patients suffering from stroke and other disabilities at the NRH (Sukthankar 1996). The goal of these programs has been to motivate individuals to induce motion in their affected extremities. Position sensors mounted on the affected limb measured the position and posture achieved by the patient during the therapeutic task. This position/orientation data was then mapped into actions in the virtual world. Thus, for example, wrist flexion could be mapped to control the throttle on a virtual car. Though the early results from these program evaluations are positive, formal studies are needed to validate the efficacy of such VR based therapy programs. A potential advantage of the VR based therapy approach is the *ability to embed real-time diagnostics into the therapeutic routines.*

Apart from neuromotor therapy and diagnosis, VR-based therapy programs have also been used in the field of neuro-behavioral psychology. Hodges et al. (1993) and Lamson (1994) have demonstrated the use of VR in the treatment and management of acrophobia. Kijima et al. (1994) have similarly explored the use of VR as a virtual sandbox for use in psychotherapy. Tesio (Burdea and Coiffet 1994) used a VR based system to train ataxic individuals to use proprioceptive rather than visual feedback to maintain balance. The variation of the patient's center-of-pressure measured using external sensors was used quantify the patient's recovery process. In this system, distorted visual cues were presented to the user in the head mounted display in the form of exaggerated rotations or inverted dynamics thereby encouraging the subjects to rely more on the proprioceptive feedback to maintain their balance. This application exemplifies one of the advantages of using VR based simulations for studying visuo-motor systems in the human body: The ability to objectively control the visual stimulus in the VR display field gives experimenters an additional avenue for testing hypotheses about the interaction between the vestibular and ocular control systems.

VR can be used as a visualization tool to help in the design of ADA compliant work spaces. The Hines Rehabilitation R&D center developed a VR application that could be used to test the ADA compliance of architectural designs (Cyberedge, 1993). Such VR applications can help reduce the need for costly building redesign by allowing the designer to implement changes in the early design phase.

Weghorst et al. (1994) demonstrated the utility of a VR based display device as a cueing device for individuals with Parkinson's akinesia. Their system consisted of a head mounted display device which presented cueing stimuli for the user. The presentation of these stimuli was observed to alleviate the shuffle observed in patients suffering from the disease.

Clearly, VR is a technology with tremendous potential in the area of human performance measurement. With the dwindling costs of computing hardware, and progress made in the development of sensor and display technologies, VR can afford a real avenue for training and human performance measurement. VR affords the experimenter a suite of measurement tools and a stable platform for conducting such measurements.

5.2 Therapeutic Robots

In Chapter 38, Reinsmeyer et al. made a case for robots for physical therapy. Within this perspective a reasonable case is made for why such an approach may have a niche, and why understanding the human-technology interface is so important to the process.

6 Assistive Robots for Independent Living

In addition to evaluation and therapy, advanced technologies can also be used to enhance the quality of life, primarily through enhancing a person's capability for independent living and vocational productivity. One class of technology that is of special interest for this book is rehabilitation robotics. Rehabilitation robotics refers to the application of different disciplines that combine to produce an electromechanical aid for a person that has a manipulative disability. A rehab robot is distinct from a prosthesis in that it is not attached to the user, but may reside on a table top, or on the side of a wheelchair, or be in the form of a mobile base. The title is misleading in that most robots are not intended as a manipulative tool (vs rehabilitative therapy). Also, virtually all of the devices that are about to be discussed utilize standard uni-directional control interfaces (i.e., they do no use the concept of EPP).

The field remains relatively small. There are a few groups around the world, comprising mainly of academic based research groups, along with a handful of commercial ventures, that are active in developing the field of rehabilitation robotics. To date, there have only been a handful of robotic devices that have crossed the bridge from a research prototype to commercial product, and even fewer that have demonstrated long term use by a person with a physical disability.

Early attempts at developing rehab robots included the Rancho "Golden" arm (Allen et al. 1970), the Heidelberg arm (Paeslack and Roeslr 1977), the VAPC arm (Mason and Peiser 1979), and the Johns Hopkins arm (Seamone and Schmeisser 1985). Although these devices saw limited use by consumers, they did lay the groundwork for further development in the field.

There are a number of rehab robots currently available or in development. The most prominent device is the MANUS (Verburg et al. 1996), which is a wheelchair-mounted seven-axis (plus gripper) robot. The MANUS, a Dutch project, was designed with the disabled person in mind. It was a unique collaboration between the engineering and rehabilitation worlds, which rendered a well-engineered, quiet, and aesthetic device. The MANUS folds up into an unobtrusive position at the side of the wheelchair and folds out when commanded. Its present inputs include a 16-button keypad, trackball, and joystick, which perform joint or cartesian control. There are currently approximately 50 users of the MANUS, mainly in the Netherlands.

Another project that has had relative success is the Handy I (Topping, 1996). The Handy I uses an inexpensive industrial robot arm (the Cyber 310) to perform programmed tasks. The system is primarily used as a feeding device for children with cerebral palsy. The user uses a chin switch to activate the system and select the food through a scanning selector, which is then automatically brought up to the mouth. The system has enabled a number of children to feed themselves for the first time. Currently, over 40 of these systems have been placed with disabled individuals in the United Kingdom.

The RAID project (Upton 1994) is a collaborative effort on the part of several European concerns. The aim of the project is to develop and demonstrate a prototype workstation for use by the disabled or elderly in a vocational setting. It consists of a six degree of freedom RTX robot placed on a linear track, and a structured workcell. A user may choose his or her preferred input device to control the robot by issuing high-level commands such as "pick up book three." RAID is currently being evaluated in a number of rehabilitation centers in Europe.

In the United States and Canada, there are three commercial projects of note. The first is DeVAR (desktop assistant robot for vocational support in office settings), a Palo Alto VA/Stanford University collaboration that uses a PUMA-260 robot mounted on an overhead rack that performs pre-programmed tasks in a highly structured environment (Van der Loos 1995). The DeVAR system has been evaluated by a number of individuals in various VA centers, and notably by one highly motivated, disabled individual on a two-year trial in his work environment. The project yielded much information on cost/benefit and social issues, however a high price tag has prevented commercial success.

The second project is the Robotic Assistive Appliance (RAA) developed at the Neil Squire Foundation in Vancouver, Canada (Birch et al. 1996). The RAA, which is the result of over ten years of research in rehabilitation robotics, offers a human size manipulator at a workstation with 6 degrees of freedom with either programmed or direct control. The device is currently undergoing testing to assess its advantages over an attendant.

The third commercial prototype is the Helping Hand (Sheredos et al. 1996), which was developed by Kinetic Rehabilitation Instruments (KRI) of Hanover, Massachusetts. It posseses 5 degree of freedom arm, is modular in design, and can be mounted on either side of most powered wheelchairs. The arm comes with its own controller comprising of switches for the joint motors. It does not include a computer which reduces cost and complexity. To date, it has been evaluated in a number of VA centers and has been approved by the FDA. However, it remains to be seen whether the Helping Hand will meet with long term success.

Even though the field of robotics has grown considerably in the last twenty years, from robots operating in the space shuttle to robots used to assist in surgery, the lack of progress of rehabilitation robotics remains enigmatic. The projects alluded to earlier have all had limited success as commercial products. Some of the reasons for this include high cost, lack of understanding of consumer needs and demands, poor interface between a complex electromechanical system and a human with limited capabilities, and social stigma associated with a robot.

These issues and others have been examined at the Rehabilitation Engineering Research Center (RERC) in Rehabilitation Robotics at the Applied Science and Engineering Labs (ASEL) located at the University of Delaware and the duPont Hospital for Children in Wilmington, Delaware. The RERC conducted a number of projects encompassing different aspects of rehabilitation robotics. These are reflected in the themes of the RERC, which include research, application, design, and dissemination. The research area included four projects that emphasized the *user interface* between a robot and human. These are:

• Control of a Powered Upper Extremity Orthosis.
• Proprioceptive User Control of Rehabilitation Robot.
• Multimodal User Direction of Rehabilitation Robot.
• Predictive Robot Planning for Individuals with Severe Disabilities.

The first two (Chen et al. 1994; Rahman et al. 1996) are attempting to make the robot act as an *extension to the person*, whether it be an exoskeleton or a remotely operated robot. It is hoped this intimate link between the person and machine would approach the types of relationships described earlier when we developed the EPP concept. The second two projects (Kazi et al. 1996) are striving to find the right combination of machine and human intelligence that makes the operation of a robot seamless. This includes tools such as predictive planning, object recognition, and voice and laser input.

The application area of the RERC comprise three projects that utilize existing robotics technology in a setting that will have a more short term impact on the lives of people with disabilities (Schuyler et al. 1995; Howell et al. 1996). While the details are not of interest here, of importance is that disabled consumers are involved in the entire design and evaluation process, from initial brainstorming to product development. This is also true for the design area, where the emphasis is placed on developing robotic products that are evolutionary rather than revolutionary. Some researchers in the past have tried to do too much too soon. Here the focus is on simpler, more affordable technology that might not do everything in the manipulative sphere, but concentrate on doing a subset of tasks well. One project (Fee et al. 1995) in the design theme is The Consumer Innovation Laboratory (CIL), a consumer-led, consumer-driven, research effort that is looking at current problems and products specific to people with manipulative disabilities. From the

vast wealth of problems and ideas, consumer researchers explore promising solutions through to a detailed design and prototype stage. These solutions are, in turn, either made available through publication of the designs or passed along to appropriate companies for manufacturing possibilities. A second design-oriented project is the adaptable Integrated Telethesis (Stroud et al. 1996)—a simple, adaptable, wheelchair mounted body-powered robot (the term telethesis is used because of its similarity to a cable operated prosthetic arm). The telethesis has limited degrees of freedom, hence affording the opportunity for direct control of the joints through Bowden cables. The user will have the choice between body-powered, power-assisted, or a hybrid of the two. The adaptability will allow interfacing to a suitable limb in the body depending on the particular disability.

To summarize, for robots to be a viable assistive technology for a person with a disability, two issues must be resolved: (1) development of more "intimate" interfaces (such as use of EPP) that make the robot more user friendly, and more of a extension of the person; and (2) consumers must have more say during the design and evaluation process. Yet there is reason to believe that the future may be bright. It is not difficult to envision a robot with a primary purpose of enhancing independent living, yet also has a virtual interface that facilitates "therapy games," with objective evaluation measures obtained and automatically sent to remote sites.

7   Future Directions

The four perspective chapters that follow have been selected because they provide us with a window toward future directions. Notice the strong focus *on interactive human-technology environments* for all types of consumers, from the person with disability to the clinician making a diagnosis. Also notice the focus on more utilization of objective (sensor-based) measures, on the use of interactive SC techniques to better synthesize data into forms that are more useful to CDMs, and on telerehabilitation.

For science-driven rehabilitation technologies to succeed in the marketplace, the challenges of *product design* (Chapter 41) must become more of a part of the rehabilitation research culture—product design considerations can even influence protocols for basic science experiments.

Finally, we return to the concept of concurrent (just-in-time) research: high returns await rehabilitation research activities that anticipate society-wide technological investment.

Acknowledgments. This work was supported in part by the U.S. Army Medical Research Acquisition Command, contract DAMD17-94-V-4036.

References

Allard, P., Stokes, I.A.F., and Blanchi, J-P. (eds.) (1995). *Three-Dimensional Analysis of Human Movement.* Human Kinetics, Champaign.

Allen, J.R., Karchak, A., and Nickel, V.L. (1970). Orthotic Manipulators. In *Advances in External Control of Human Extremities.* Belgrade.

Ascension Technology. (1995). Flock of birds user's manual.

Barter, J.T. (1957). Estimation of the mass of body segments. *WADC Techn. Report 57-260*, Wright-Pattern A.F.B., Aero. Med. Lab., Ohio.

Barto, A.G., Sutton, R.S., and Watkins, C.J.C.H. (1983). Neuronlike elements that can solve difficult learning control problems. *IEEE Trans. Sys. Man Cybern.*, 13: 835–846.

Bernstein, N.A. (1935). Biodynamics of walking of normal adult man. Moscow, Orignal in Russian, partial translation be B. Bresler, Prosthetics Devices Research Project, University of California, Berkeley, 1947.

Birch et al., G.E. (1996). An assessment methodology and its application to a robotic vocational assistive device. *Tech. Disabil.*, 5:151–166.

Bogner, M.S. (ed.). (1994). Human error in medicine. Lawrence Erlbaum Associates. Hillsdale, New Jersey.

Braune, W. and Fischer, O. (1895). Versuche am unbelasteten und belasteten Menschen. Vol. 21, pp. 151–324.

Bresler, B. and Frankel, J.P. (1950). The forces and moments in the leg during level walking. *Trans. ASME*, 72:27–36.

Bresler, B., Radcliffe, C.W., and Berry, F.R. (1957). Energy and power in the legs of above-knee amputees during normal level walking. *Inst. Eng. Res.*, University of California, Berkeley, Series 11, issue 31, pp. 26.

Budde, J.P. (1990). Independent living centers: a parallel resource. In *Rehabilitation Engineering.* Smith, R.V. and Leslie, J.H. (eds). CRC Press, Boca Raton.

Burdea, G. and Coiffet, P. (1994). *Virtual Reality Technology*, John Wiley & Sons, New York.

Caldwell, D.G. (1993). Natural and artificial muscle elements as robot actuators. *Mechatronics*, 3:269–283.

Card, S., Moran, T., and Newell, A. (1983). The psychology of human-computer interaction. Lawrence Erlbaum Associates, Hillsdale, New Jersey.

Card, S., Moran, T., and Newell, A. (1986). The model human processor. In *Handbook of Perception and Human Performance.* Boff, K., Kaufman, L., and Thomas, J. (eds.), vol 2. John Wiley & Sons, New York.

Carlet, M.G. (1872). Essai experimental sur la locomotion humaine. *Annales des Sciences Naturelles, Section Zoologique*, Series 5, 16:1–93.

Chen, S., Harwin, W., and Rahman, T. (1994). The application of discrete-time adaptive impedance control to rehabilitation robot manipulators. *Proc. of IEEE Int'l Conf on Robotics and Automation*, San Diego, May.

Childress, D.S. (1985). Historical aspects of powered upper-limb prostheses. *Clin. Prosth. Orthotics.*, 9:2–13.

Christensen, J.M., Topmiller, D.A., and Gill, R.T. (1988). Human factors definitions revisited. Human Factors Society Bulletin, 31, pp. 7–8.

Clark, M. and Stark, L. (1975). Time optimal behavior of human saccadic eye movement. *IEEE Trans. Autom. Control* AC-20:345–348.

Cook, A.M. and Hussey, S.M. (1995). Assistive technologies: principles and practice. Mosby, St. Louis.

Cunningham, D.M. (1950). Components of floor reactions during walking. *Prosthetic Devices Research Project, Inst. Eng. Res.*, University of California, Berkeley, Series 11, issue 14.

Cummingham, D.M. and Brown, G.W. (1952). Two devices for measuring the forces acting on the human body during walking. *Proc. Soc. Exp. Stress Analysis*, 9:75–90.

Cyberedge Journal (1993). Product of the year. Sausolito, California, March/April, pp. 3–5.

Davies, A.R., Doyle, M.A., Lansky, D., Rutt, W., Stevic, M.O., and Doyle, J.B. (1994). Outcomes assessment in clinical settings: a consensus statement on principles and best practices in project management. *The Joint Commission on Accreditation of Healthcare Organizations*, pp. 6–16.

Dempster, W.T. (1955). Space requirements of the seated operator. WADC Techn. Report 55-159, Wright-Patterson A.F.B., Ohio.

Doubler, J.A. and Childress, D.S. (1984). An analysis of a prosthesis control system based on the concept of extended physiological proprioception. *J. Rehab. Res. Dev.*, 21:1–18.

Downton, A. (ed). (1991). *Engineering the Human-computer Interface.* McGraw-Hill, London.

Evans, M. (1988). MAGPIE—a lower-limb-operated manipulator. *Engng. Med.*, 17:81.

Fee, J. et al. (1995). The consumer innovation laboratory: an exercise in consumer empowerment. *RESNA Proceedings*, 15: 502–504. Vancouver, June.

Fitts, P.M. (1954). The information capacity of the human motor system in controlling the amplitude of movement. *J. Exp. Psych.*, 47:381–391.

Franier, R., Groleau, N., Hazerlton, L., Colombano, S., Compton, M., Statler, I., Szolovits, P., and Young, L. et al. (1994). PI-in-a-box: a knowledge-based system for space science experimentation. *AI Magazine*, pp. 39–51.

Galvin, J. (1991). The history of rehabilitation engineering. *Assistive Technol.*

Gavrilovic, M.M. and Maric, M.R. (1969). Positional servo-mechanism activation by artificial muscles. *Med. Biol Eng.*, 7:77–82.

General Reality Inc. (1996). Product catalog.

Glanville, A.D. and Kreezer, G. (1937). The characteristics of gait of normal male adults. *J. Exp. Psychol.*, 21:277–301.

Granger, C.V. and Brownscheidle, C.M. (1995). Outcome measurement in medical rehabilitation. *Int. J. Technol. Assess. Health Care*, 11:262–268.

Groleau, N. (1994). ASSET: automation and support system for expert tele-sciences. *Techn. Descrip. Doc., M/S 269–2*, Artif. Intel. Res. Branch, NASA-AMES Research Center, Moffett Field.

Hannaford, B. and Winters, J.M. (1990). Actuator properties and movement control: biological and technological models. In *Multiple Muscle Systems*. Winters, J.M. and Woo, S.L-Y. (eds.), pp. 101–120. Springer-Verlag, New York.

Hannaford, B.,Winters, J.M., Chou, C-P., and Marbot, P-H. (1995). The anthroform biorobotic arm: a system for the study of spinal circuits. *Ann. Biomed. Eng.*, 23:359–374.

Heckathorne, C.W. (1990). Manipulation in unstructured environments: extended physiological proprioception, position control, and arm prostheses. *Proc. Int. Conf. Rehab. Robotics*, pp. 25–40. Wilmington.

Hodges, L.F., Bolter, J., Mynatt, E., Ribarsky, W., and van Teylingen, R. (1993). Virtual environment research at the Georgia Tech GVU Center. *Presence*, 2(3):234–243.

Hogan, N. (1984). Adaptive control of mechanical impedance by coactivation of antagonist muscles. *IEEE Trans. Autom. Control*, AC-29:681–690.

Hogan, N. (1990). Mechanical impedance of single- and multi-articular systems. In *Multiple Muscle Systems*. (Winters, J.M. and Woo, S.L-Y. (eds.), pp. 149–164. Springer-Verlag, New York.

Hogan, N. and Winters, J.M. (1990). Principles underlying movement organization: upper limb. In *Multiple Muscle Systems*. Winters, J.M. and Woo, S.L-Y. (eds.), pp. 182–194. Springer-Verlag, New York.

Howell, R., et al. (1996). Classroom applications of educational robots for inclusive teams of students with and without disabilities. *Technol. Disabil.*, 5:139–150.

Inman, V.T., Ralston, H.J., Saunders, J.B., Feinstein, B., and Wright, E.W. (1952). Relation of human electromyogram to muscular tension. *Electroenceph. Clin. Neurophysiol.*, 4:187–194.

Innman, V. T., Saunders, J.B., and Abbot, L.C. (1944). Observations on the function of the shoulder joint. *J. Bone Joint Surg.*, 26–A:1–30.

Johnston, R.S. (1987). The SIMNET visual system, *Proc. Ninth ITEC Conf.*, pp. 264–273. Washington, DC.

Kazi, Z. et al. (1996). Multimodally controlled intelligent assistive robot. *RESNA Proc.*, vol. 16, pp. 348–350. Salt Lake City, June.

Kijima, R., Shirakawa, K., Hirose, M., and Nihei, K. (1994). Virtual sand box: development of an application of virtual environments in clinical medicine. *Presence*, 3(1):45–59.

Kroemer, K., Kroemer, H., Kroemer-Elbert, K. (1994). *Ergonomics*. Prentice Hall, Englewood Cliffs, New Jersey.

Kruit, J. and Cool, J.C. (1989). Body-powered hand prosthesis with low operating power for children. *J. Med. Eng. & Technol.*, 13:129–133.

LeBlanc, M.A. (1987).

Lamoreux, L.W. (1971). Kinematic measurements in the study of human walking. *Bull. Prosthetic Res.*, BPR 10-15, pp. 3–84.

Lamson, R.J. (1995). Clinical application of virtual therapy to psychiatric disorders. *Virtual Reality and Persons with Disabilities*. San Fransisco.

Levins, A.S., Inman, V.T., and Blosser, J.A. (1948). Transverse rotation of the segments of the lower extremity in locomotion. *J. Bone Joint Surg.*, 30-A:859–872.

Lin, C-T. and Lee, SC.S.G. (1996). Neural fuzzy systems. A neuro-fuzzy synergism to intelligent systems. Prentice Hall, Upper Saddle River, New Jersey.

Liu, A., Tharp, G., French, L., Lai, S., and Stark, L. (1993). Some of what oneneeds to know about using head-mounted displays to improve teleoperator performance. *IEEE Trans Robotics Auto.*, 9:638–648.

Marquardt, E. (1961). Biomechanical control of pneumatic prostheses with special consideration of the sequential control. In *Application of External Power in Prosthetics and Orthotics*. Publ. 874, NAS-RC, pp. 20–31.

Mason, C.P. and E. Peiser. (1979). A seven degree of freedom telemanipulator for tetraplegics. *Conf. Int. sur les Telemanipulators pour Handicapes Physiques*, pp. 309–318.

McDonough, J. (1993). Doorways to the virtual battlefield. *Proc. Virtual Reality, '92*, 104–114.

Michael, J. (1986). Upper limb powered components and control: current concepts. *Clin. Prosth. Orthotics*, 10: 66–77.

Monheit, G. and N. Badler (1991). A kinematic model of the human spine and torso. IEEE Comput. Graphics Appl., Noton, D. and Stark, L. (1971). 11(2):29–38, 1991.

NASA (1993). Virtual Technology for Training. *Technical Report*. Johnson Space Center, Houston, Texas.

Norman, D. (1988). The design of everyday things.

Paeslack, V. and Roesler, H. (1977). Design and control of a manipulator for tetraplegics. *Mech. Machine Theory*, 12:413–423.

Phillips, B. and Chao, (1994).

Pieper, S., Rosen, J., and Zeltzer, D. (1992). Interactive graphics for plastic surgery: a task-level analysis and implementation. In *Proceedings of the 1992 Symposium on Interactive 3D Graphics*, Zeltzer, D., Catmull, E., and Levoy, M. (eds.), pp. 127–134. New York.

Plettenburg, D.H. (1989). Electric versus pneumatic power in hand prostheses for children. *J. Med. Eng. Technol.*, 13:124–128.

Prior, S.D., Warner, P.R., White, A.S., Parsons, J.T., and Gill, R. (1993). Actuators for rehabilitation robots. *Mechatronics*, 3:285–294.

Radcliffe, C.W. (1960). Human engineering: mechanisms for amputees. *Machine Design*, 32:24–28.

Rahman, T. et al. (1996). Task priorities and design of an arm orthosis. *Technol. Disabil.*, 5:197–204.

Ralston, H.J. (1953). Mechanics of voluntary muscle. *Amer. J. Phys. Med.*, 32:166–184.

Reddy, N.P., Sukthankar, S.M., and Gupta, V. (1994). Virtual reality in rehabilitation. *IEEE-EMBS Workshop on Rehabil. Eng.* Baltimore.

Rosen, J.M. (1994). VR and surgery: from simulation to performing complex procedures. *Virtual Reality and Medicine- The Cutting Edge*. New York.

Sanders, M.S. and McCormick, E.J. (1993). *Human factors in Engineering and Design*. McGraw-Hill, New York.

Schulte, R.A. (1961). The characteristics of the McKibben artificial muscle. In *Application of External Power in Prosthetics and Orthotics*, Publ. 874, NAS-RC, pp. 94–115.

Schuyler, J. and Mahoney, R. (1995). Vocational robotics: job identification and analysis. *RESNA Proc.*, 15:542–544. Vancouver.

Seamone, W. and Schmeisser, G. (1985). Early clinical evaluation of a robot arm/worktable system for spinal-cord-injured persons. *J. Rehabil. Res. Dev.*, pp. 38–57.

Sheredos, S.J. et al. (1996). Preliminary evaluation of the helping hand electro-mechanical arm. *Technol. Disabil.*, 5:229–232.

Sheridan, T.B. and Ferrell, W.R. *Man-Machine Systems: Information, Control, and Decision Models of Human Performance*. The MIT Press, Cambridge.

Simpson, D.C. (1974). The choice of control system for the multimovement prosthesis: extended physiological proprioception (e.p.p.). *The Control of Upper Extremity Prostheses and Orthoses*. Herberts, P. et al. (eds.), pp. 146–150. Charles C. Thomas, New York.

Simpson, D.C. and Smith, J.G. (1977). An externally powered controlled complete arm prosthesis. *J. Med. Eng. Tech.*, pp. 275–277.

Stark, L. (1968). *Neurological Control Systems*. Plenum Press, New York.

Stark, L. and Ellis, S.R. (1983). Scanpaths revisited: cognitive models direct active looking. In *Eye Movements: Congnition and Visual Perception*. Fisher, D.F. et al. (eds.). Lawrence Erbaum Assoc.

Stroud, S. et al. (1996). A body powered rehabilitation robot. *RESNA Proc.*, 16:363–365. Salt Lake City.

Suh, C.H. and Radcliffe, C.W. (1967). Synthesis of spherical linkages with use of the displacment matrix. *J. Eng. Ind.*, 89:215–222.

Sukthankar, S.M. (1996). Virtual reality in rehabilitation, *RESNA 1996 Mid-Atlantic Regional Conference*. Philadelphia.

Topping, M. (1996). Handy I, a robotic aid to independence for severely disabled people. *Tech. Disabil.*, 5:233–234.

Upton, C. (1994). The RAID workstation. *Rehab. Robotics Newsletter*, A.I. duPont Institute, 6(1).

Van der Loos, H.F.M. (1995). VA/Stanford rehabilitation robotics research and development program: lessons learned in the application of robotics technology to the field of rehabilitation. *IEEE Trans. Rehab. Eng.*, 3:46–55.

Verburg, G. et al. (1996). Manus: the evolution of an assistive technology. *Technol. Disabil.*, 5:217–228.

Virtual Technologies (1995). CyberGlove User's Manual.

Weghorst, S., Prothero, J., and Furness, T. (1994). Virtual images in the treatment of Parkinson's Disease akinesia. *Medicine meets Virtual Reality II: Visionary Applications for Simulation, Visualization, and Robotics*. pp. 244–246. Aligned Management Associates, San Diego.

Welford, A.T. (1976). Ergonomics: where have we been and where are we going: I. *Ergonomics*, 19(3):275–286.

Werbos, P. (1990). A menu of designs for reinforcement learning over time. In *Neural Networks for Control*. Miller et al., (eds.). The MIT Press.

Wickens, C.D., and Baker, P. (1995). Cognitive issues in virtual reality. In *Virtual Environments and Ad-*

vanced Interface Design. Barfield, W. and Furness, T.A. III (eds.), pp. 514–541. Oxford University Press, New York.

Winters, J.M. (1995a). How detailed should muscle models be to understand multi-joint movement coordination*? Human Mov. Sci.*, 14:401–442.

Winters, J.M. (1995b). An improved muscle-reflex actuator for use in large-scale neuromusculoskeletal models. *Annals Biomed. Eng.*, 23:359–374.

Winters, J.M. (1996). Intelligent synthesis of neuromusculoskeletal signals using fuzzy expert critics. *SPIE Smart Sensing, Processing and Instrumentation: Smart Sensors and Actuators for Neural Prosthesis*, Vol. 2718, pp. 456–468, San Diego.

Winters, J.M. and Sagranichiny, E.S. (1994). Why braided pneumatic actuators in rehabilitation robotics? Principles, properties and suggested applications. *4th Int. Conf. Rehab. Robotics*, pp. 201–208, Wilmington.

Young, L.R. and Stark, L. (1963). Variable feedback experiments testing a sampled data model for eye tracking movements. *IEEE Trans. Human Factors Elec.*, HFE-4:38–51.

Zadah, L.A. (1994). Fuzzy logic, neural networks, and soft computing. *Commun. ACM* 37:77–84.

38

Rehabilitators, Robots, and Guides: New Tools for Neurological Rehabilitation

David J. Reinkensmeyer, Neville Hogan, Hermano I. Krebs, Steven L. Lehman, and Peter S. Lum

1 Introduction

Millions of people in the United States suffer from movement disabilities as the result of neurological injury and disease. Their rehabilitation is labor intensive, often relying on one-on-one, manual interactions with therapists. For many disorders, it is unknown which types of therapeutic manipulations best promote recovery. In addition, patient evaluation is often done subjectively, making it difficult to monitor treatment effects. This chapter demonstrates how appropriately designed machines might be brought to bear on these problems.

The chapter is a cooperative effort from research groups at M.I.T., the Palo Alto Veterans Affairs Medical Center, the Rehabilitation Institute of Chicago, and the University of California, Berkeley, each of which is pursuing new devices for rehabilitation. Each group has focused its initial efforts on stroke, the leading cause of major disability in the United States (American Heart Association, 1993). Other potential target disorders exist, including traumatic brain injury, cerebral palsy, and Parkinson's disease. In this chapter, stroke rehabilitation serves as a specific application illustrating what may be a more general opportunity.

1.1 Overview of Stroke Rehabilitation

More than 500,000 people suffer a stroke each year in the United States; three million stroke victims are alive today (Gresham et al. 1995). Motor deficits persist chronically in approximately one-half of these people (Gresham et al. 1979). Damage to neural areas responsible for controlling

movement and concomitant disuse and persistent abnormal posture of the impaired limb cause a host of centrally and peripherally-based sensory motor impairments (Figure 38.1). Common impairments are decreased passive range of motion, weakness (Gandevia 1993), hyperactive reflexes (Thilmann et al. 1993), and incoordination, manifest in part as an inability to independently coactivate muscles (Dewald et al. 1995). Patients commonly experience some spontaneous recovery, but are also treated with extensive physical and occupational therapy to enhance recovery.

This poststroke therapy is labor intensive, often relying on one-on-one interactions with a trained therapist several hours a day. Therapy extends for at least one or two months poststroke. Typical therapeutic activities include manual manipulations of the patient's limbs, either with the patient passive or attempting movement (Figure 38.1). A therapist may stretch a patient's limb into different positions to help improve passive range of motion, or move it through specific patterns thought to help reduce hyperactive reflexes. Sometimes the therapist will guide or support the limb as a patient attempts to activate desired muscle combinations or to achieve a desired movement or task. Motor activity is also sometimes facilitated by resisting movement, or by applying tactile, thermal, or proprioceptive sensory input to the limb (for a summary of different techniques, see Dewald 1987; Trombly 1995). Patient progress is often evaluated subjectively, with the therapist making hands-on or visual judgments about a patient's movement ability. Sometimes standardized, coarsely discretized scales are used to improve objectivity (Gresham et al. 1995).

516

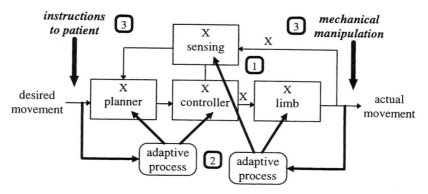

FIGURE 38.1. Systems representation of impaired movement control and manual therapy after stroke. (1) Stroke induces changes (marked by "x's") throughout the human motor control system. In the limb, muscles may shorten, atrophy, and convert to slow fiber types, and connective tissue changes may restrict movement. In supraspinal and spinal controllers, feedforward and feedback processes are typically disrupted. Outflow and in-flow pathways between the controllers and limb may be destroyed. Planning and sensation of movement may also be impaired. (2) Neural and biochemical adaptive processes both contribute to and counteract these changes. (3) Therapists have only indirect access to internal adaptive processes: they can give instructions and feedback to the patient to alter movement intent, and they can mechanically manipulate the limb.

The efficacy of different manual therapy techniques, and of manual therapy in general, has not been established. Also, there is a need for better, more sensitive, more objective methods for measuring movement recovery after stroke. Many fundamental rehabilitation questions remain unanswered (Table 38.1). Pressure for answers is increasing: the health care system is undergoing restructuring for cost effectiveness. Also, stroke frequency doubles with each decade after age 55, and the leading edge of the baby boom will be 55 in a few years (Figure 38.2). This impending wave of new patients is likely to put pressure on a system that does not fully understand how best to treat them.

TABLE 38.1. Some unanswered questions in stroke rehabilitation research.

To what extent does therapy enhance recovery?
What is the role of spontaneous recovery versus therapy?
What is the optimal therapy technique?
What role does manual manipulation play in promoting recovery?
How much therapy should be given for how long?
At what pathophysiological mechanisms should therapy be targeted?
What are the best ways to quantify specific pathophysiological impairments?

1.2   Use of Technology in Stroke Rehabilitation

Currently, technology is rarely used in stroke rehabilitation. Goniometers are sometimes used to measure range of motion of joints, and one degree-of-freedom (DOF) dynamometers to measure strength. There is a lack, however, of clinically viable, instrumented systems for measuring stroke-specific disturbances and determining how they contribute to reduced functional ability. For therapeutic manipulation, passive counterbalancing systems, such as mobile arm supports or overhead slings, are sometimes used to relieve the weight of the arm and thus allow the weakened patient to engage in self-initiated movement. Also, one DOF actuated systems, such as motorized dynamometers, are sometimes used for range of motion exercises or in active-assisted exercises, in which the patient tries to follow the motion of the machine in an attempt to develop volitional control of movement. There is a lack however of systems that replicate the complex spatial patterns of force and movement often used in manual therapy after stroke.

More sophisticated devices that would attach to a patient's limb, measure movement and force generation, and apply therapeutic patterns of forces could make a significant impact on stroke rehabilitation management. Such machines could poten-

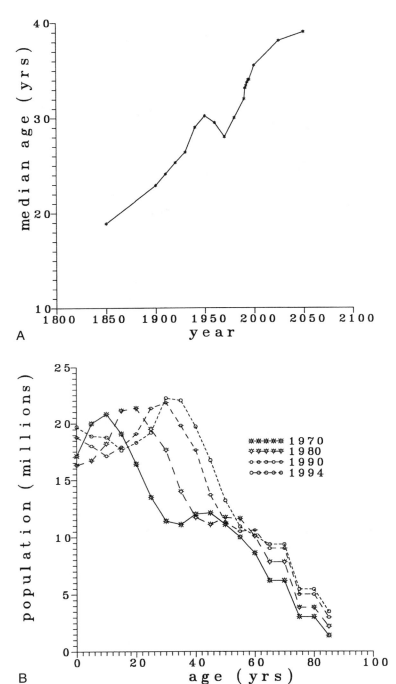

FIGURE 38.2. (A) Median age of the U.S. population. The baby boom caused the only decline in median age in the history of the U.S. census. (B) U.S. population distribution. Baby boomers will soon reach age 55. The frequency of stroke doubles with every decade of age after 55.

tially enhance rehabilitation measurement by quantifying specific pathophysiological mechanisms, spontaneous recovery, functional ability, and therapy dosage more accurately than is now possible. They could also help with therapy itself, replicating key components of current manual therapeutic techniques, or even applying new techniques. They might help answer fundamental scientific questions (Table 38.1), and they might improve cost-efficiency of therapy (Table 38.2).

TABLE 38.2. How rehabilitation machines might improve the cost efficiency of rehabilitation.

Automatic repetitive aspects of therapy.
Facilitate home care (teletherapy: patient at home, therapist at remote site).
Make group sessions more effective.
Help target therapy more accurately.
Help determine appropriate "dosing" and duration for therapy.
Increase therapy "dosage" where appropriate (i.e. more repetitions, more hours per day).

1.3 Overview of New Techniques

Research is just beginning on meeting these goals. This chapter describes initial work on "bimanual rehabilitators," a "reaching guide," "MIME," and "MIT-MANUS," four devices for neurological rehabilitation of the arm (for descriptions of postural, locomotor, and other devices the reader is referred to Flores (1992), Wing et al. (1993), Siddiqi et al. (1994), and Erlandson (1995)). We emphasize that the blueprint for the ideal machine is not obvious from currently available information. Thus, the designs presented in this chapter are reasoned guesses at what might constitute useful machines. They have required distinct design choices regarding manipulation techniques, movement tasks, and actuator, sensor, and software configurations (Table 38.3). One goal is to give concrete examples of ways to approach these choices. The reasoning behind each of the resulting devices, along with illustrative data demonstrating some of each device's capabilities, are presented. The other major goal is to show how these devices are beginning to generate new ideas, information, and potential therapeutic techniques for neurological rehabilitation. The illustrative data in each section shows some of these

TABLE 38.3. Rehabilitation machine design choices.

Tasks for patient to perform.
Manipulations to apply to patient.
Movement parameters to measure.
Linkage geometry and strength.
Number/type/size/location of actuators.
Number/type/size/location of sensors.
Means to physically couple to patient.
Control scheme to implement desired manipulations.
Sensor processing/fusion to derive key parameters.
Feedback to give patient.

results. In addition, the concluding section outlines several basic scientific questions raised by these devices, and possible directions for future technological innovation.

2 Bimanual Rehabilitators

2.1 Design Choices

This section presents devices called "bimanual rehabilitators," which were built to test the applicability of simple machines to stroke rehabilitation. Feasibility experiments with the devices demonstrate how a common therapeutic technique used in stroke rehabilitation can be implemented by the devices, at least for temporarily weakened, unimpaired subjects. This suggests that machine implementation of some therapy techniques might not require complex devices.

The bimanual rehabilitators measure and assist bimanual control—coordination of the two hands to achieve a task. The machines are task specific, and operate under simple feedback control laws. Applied to a hemiplegic stroke patient they would potentially allow the more able hand to train the less able one. Initial tests of the machines (Lum et al. 1993a; Lum et al. 1995) confirmed that that such task-specific machines, operating under simple control laws, could provide adaptive assistance, performing only that part of the task that a weakened person could not perform.

Bimanual control is important for activities of daily living (Washam 1973). Imagine tying shoes, buttoning buttons, or washing dishes without the coordinated use of two hands. Bimanual coordination is so common that people have dedicated motor control systems devoted to two-hand activities: the two hands work together using different control patterns than either one would use separately (Reinkensmeyer et al. 1992); each hand "knows" by feedback mechanisms what the other hand is doing (Lum et al. 1993b); under some circumstances, each hand "knows" by feedforward mechanisms what the other is about to do (Lum et al. 1992). Stroke generally affects one hand more than the other, so that bimanual activities are curtailed, but one hand can still provide motor intention, and perhaps training, for the other. For all these reasons, rehabilitation of bimanual control seems useful and possible.

2.2   Illustrative Data

Two bimanual rehabilitators were built and tested with unimpaired subjects. The Hand-Object-Hand device (Lum et al. 1993a) consists of two handles on a table top, each free to move about an axis coincident with the subject's wrist (Figure 38.3A). Potentiometers attached to the two axes measure position of the hands, and a force transducer mounted in a pencil-like manipulandum measures the grasp force between the outstretched fingers of the subject's hands. A motor under one handle can produce an external torque on one hand. The chief motivation for this device configuration was that it allows performance of bimanual tasks, with the possibility of powered assist for one hand, in a simplified, controlled setting.

In one preliminary experiment, the motor was driven by a simple proportional feedback controller regulating the position of the motor-driven hand. The subject was instructed to squeeze the object between the outstretched fingers. A blood pressure cuff was inflated on the upper arm of the motor-driven hand to occlude blood supply. After a few minutes, ischaemia in this hand and forearm caused a reduction in force and sensation, a crude model of hemiparesis. As the subject's muscles produced less force, the device contributed more by stabilizing the weakened hand, allowing the subject to continue to perform the squeezing task (Figure 38.3B).

In another experiment, subjects were asked to move the object back and forth, maintaining a grasp force adequate to avoid dropping the object. Again, ischaemia was used to create a reduction in force and sensation. Instead of hand position, grasp force was fed back to the motor such that the device assisted the subject in transporting the object back and forth, even though one hand lacked strength and sensation to accomplish the task.

These initial results encouraged us to build a more practical device that could provide adaptive assistance during an activity of daily living. The Bimanual Lifting Rehabilitator (Figure 38.4A) was designed to measure and perturb movements during the lifting of a large object, such as a cafeteria tray or large pot. The device has a handle and force transducer for each hand attached to one rigid link. A second link is connected to this one through a

one DOF bearing, and to a motor. The subject attempts to lift the link by the handles, without tilting it. A potentiometer connected to the bearing measures tilt, and tilt is regulated using a simple control law (Lum et al. 1995). If the object begins to tilt, the motor corrects, assisting the impaired hand.

In an experimental test of the device, it was demonstrated that the device can substitute for

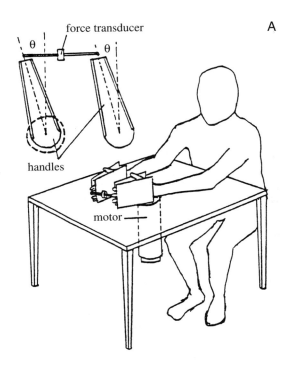

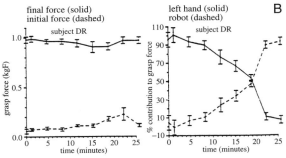

FIGURE 38.3. (A) The Hand-Object-Hand Rehabilitator. (B) Ischemic squeezing by one subject. (Left) Grasp force before (initial) and during (final) squeezing, as ischaemia progresses. (Right) Contribution to final grasp force from subject's hand (solid) and device (dashed). All points are averages over 10 trials. (Adapted from Lum et al. 1993a, © 1993 IEEE.)

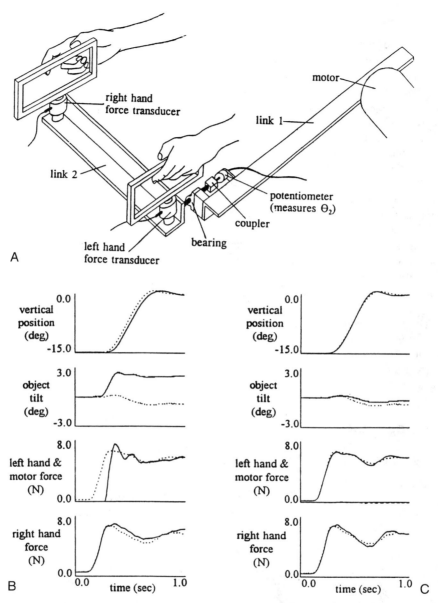

FIGURE 38.4. (A) The Bimanual Lifting Rehabilitator. (B) Lifting and holding performed bimanually (dotted) and with the rehabilitator substituting one hand (solid). Averages of 30 trials. (C) Lifting and holding performed bimanually without assistance (dotted) and bimanually with assistance from the rehabilitator (solid). The third row overlays the normal left hand force (dotted), left hand force while receiving assistance from the rehabilitator (solid), and the motor force (dashed, near zero). Averages of 30 trials. (Adapted from Lum et al. 1995, © 1995 IEEE.)

the left hand of a subject (Figure 38.4B). Using only the right hand, a subject could lift the object with only a small tilt (in an experiment with a cafeteria tray affixed to the device, a large soft drink did not spill). When an unimpaired subject lifted the tray with both hands, the device produced no force. Thus, a simple control law provided appropriate assistance (either full or none) in a bimanual task used in activities of daily living.

2.3 Discussion

These preliminary experiments show how bimanual rehabilitators might replicate adaptive assistance for bimanual therapy. Applied to stroke, the devices could give a measure of the amount of assistance needed in tasks of daily living, providing a means to track improvement and provide motivation to patients. Also, adaptive assistance is a common therapy technique that allows weak and uncoordinated patients to practice functional movement patterns. The simple devices used here have the potential to quantitatively evaluate such assistance as a physical rehabilitation technique. Likewise, bimanual therapy could be rigorously tested using the devices.

The control laws used to implement assistance for movement were simple proportional feedback controllers. In principle, similar assistance could even be provided for some tasks by purely mechanical and purely passive devices. That the bimanual rehabilitators require few actuators would make them more economical. However, the task specificity of the devices is a potential weakness: a therapist might need many such devices if there is little transfer of learning between tasks.

3 Reaching Guide

3.1 Design Choices

The device presented in this section, called a "reaching guide," makes use of a passive constraint to guide one-handed reaching attempts (Figure 38.5). The motivation for this design was that therapists often provide a strategic support or constraint for the arm during patient-initiated movement. The reaching guide is suited to explore the extent to which such guidance is therapeutic. Also, it was hypothesized that the range of motion along the constraint, as well as the extent and direction of guidance forces provided by the constraint during movement, might be used to quantify recovery.

Reaching is fundamental to many activities of daily living and requires sophisticated multimuscle coordination (e.g., see Chapters 25–31). It thus seemed a good target task for a therapeutic device. In addition, because reaching movements typically follow nearly straight-line trajectories, a simple linear bearing can be used to passively steer movement along the desired trajectory. From an engineering viewpoint, passive systems are attractive because they are inherently safe and potentially inexpensive. In the event that actuated assistance is deemed desirable, the guide incorporates a motor for moving the arm along the linear constraint (Figure 38.5B).

For the reaching guide, a linear bearing (LB in Figure 38.5A) is attached to a computer-controlled brake (CB). The bearing can thereby be locked at different elevation angles (θ) in the vertical plane. Movement along the constraint, and the orientation of the constraint, are measured with two optical encoders (OE), one attached to a chain drive. Contact forces and torques against the constraint are measured by a 6-axis force-torque sensor (FT) (For brevity only one force component—the mediolateral

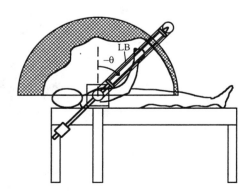
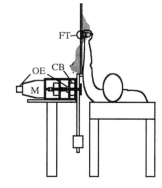

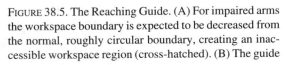

FIGURE 38.5. The Reaching Guide. (A) For impaired arms the workspace boundary is expected to be decreased from the normal, roughly circular boundary, creating an inaccessible workspace region (cross-hatched). (B) The guide can measure contact forces and torques against the constraint during movement. A hypothetical mediolateral contact force pattern against the guide during a reach is shown. See text for description of device components.

contact force shown schematically in Figure 38.5B—is considered here). The device also incorporates a computer-controlled motor (M) and second brake not used in this study but which could be used to give assistance or resistance to reaching along the guide.

3.2   Illustrative Data

As an initial test of the device, we compared guided reaching movements by unimpaired and hemiplegic stroke subjects (see also Reinkensmeyer et al. 1999). We were interested in whether the maximum reach a subject could achieve along the guide at different angles (called the "workspace boundary"), and the contact forces generated against the guide, could be used to quantify movement impairment after stroke.

Six unimpaired subjects and three hemiplegic stroke subjects were tested. The three hemiplegic stroke subjects had sustained hemispheric strokes at least one year prior to testing and were qualitatively classified as mildly, moderately, and severely impaired by an experienced physical therapist. The subjects had no obvious sensory impairment and showed no evidence of neglect of the impaired side.

To measure the workspace boundary, subjects lay supine on an adjustable patient table with the torso along the edge of the bed. The supine position was chosen initially for comfort, but other positions such as sitting and standing are also possible. The height and location of the bed were adjusted to align the approximate center of the head of the humerus, determined by palpation, with the axis of rotation of the guide. The guide was then locked at a series of elevation angles (theta = $[0°, 60°, -10°, 50°, -20°, 40°, -30°, 30°, -40°, 20°, -50°, 10°, -60°, 0°]$; see Figure 38.5). At each angle, the subject started with the elbow flexed against a backstop and reached out as far as possible, 'smoothly and steadily," eight times. Instructions were given not to push to the right or left against the guide, or against the top or bottom of the guide during reaching.

Using this technique, we found that unimpaired subjects reach very consistently to the same maximum reach (avg. SD = 0.5 cm.). Also, the average difference in workspace boundary location for the left and right arms of unimpaired subjects is about 2.0 cm. This number is a sort of minimum "resolution" for workspace measurement with the guide, taking into account the inherent variability between left and right arms, and variability in aligning a subject with the guide.

For the stroke subjects, the repeatability of the maximum reach was approximately the same or decreased only slightly: the average standard deviations of the maximum reach for the hemiplegic arms were 0.5, 2.5, and 1.8 cm., compared with 0.5, 0.7, and 1.1 cm for the corresponding contralateral arms. This suggests the subjects reliably achieved their maximum reach, even after stroke.

Also, for the three stroke subjects, the average differences in workspace boundary location between the hemiplegic and contralateral arms were 5.8, 13.7, and 25.8 cm. (Figure 38.6), exceeding the 2 cm device "resolution" found with unimpaired subjects. For all three subjects, the boundaries for the hemiplegic arms were shrunken particularly in humeral flexion.

Finally, even though all subjects were instructed to minimize contact force against the guide, they generated substantial non-zero medio-lateral contact forces during reaching (Figure 38.7). The pattern of force differed between unimpaired and hemiplegic arms: all unimpaired arms generated a laterally-directed force during reaching (average magnitude 13.5 N). The hemiplegic arms began the movement pushing laterally but changed to pushing medially as they reached outward along the guide (i.e., during elbow extension and shoulder flexion). This spatial pattern of contact force is consistent with the clinically observed abnormal "extension synergy" in which the shoulder is involuntary adducted and internally rotated during efforts at elbow extension (Brunnstrom 1970; Reinkensmeyer et al. 1999).

3.3   Discussion

By accurately measuring range of movement and multiaxial force generation, the reaching guide provides a means to objectively assess functional ability. A key idea in this work is that the forces normal to the desired direction of movement along a constraint contain valuable information about a patient's movement impairment. Measuring such constraint forces allows quantitative assessment of abnormal synergetic control. This is important because currently there are few objective methods available to analyze impaired arm coordination in the clinic.

Future research will evaluate the extent to which workspace and guidance measures correlate with

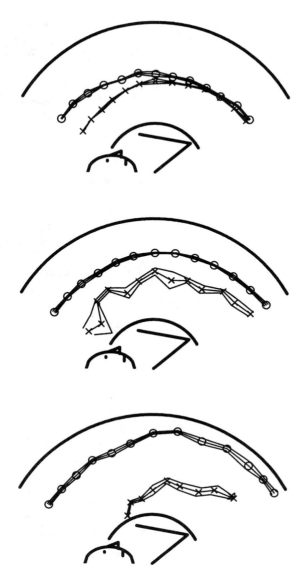

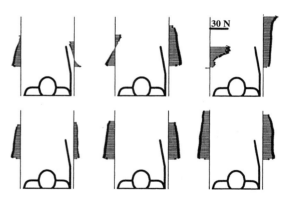

FIGURE 38.7. Mediolateral contact force during reaching. Top row shows spatial averages of mediolateral contact force for three hemiplegic subjects during eight reaches with the guide locked at 0 degree elevation. Bottom row shows contact forces for three unimpaired subjects.

and pattern of constraint forces as a function of workspace position, load, movement velocity, and other key movement variables. Finally, the device provides a simple means to test whether practice of guided reaching across the workspace can induce improved arm function.

4 MIME

4.1 Design Choices

This section discusses a prototype robotic device, called MIME ("Mirror Image Movement Enhancer"), capable of implementing two commonly used therapeutic techniques: passive and active assisted movements. Normally, these techniques are applied by a therapist who moves the paretic limb as the patient either remains passive, or actively attempts to contribute to the movement. MIME can substitute for the human therapist during such exercise tasks.

MIME uses two standard mobile arm supports that limit movement to the horizontal plane, and a 6-DOF robot arm that applies forces and torques to the paretic forearm through the support (Figure 38.8). Use of the arm supports for support against gravity allowed the robot arm to be relatively small, while still allowing performance of coordinated shoulder and elbow tasks. Movements of the arm supports can be controlled with preprogrammed forearm position and orientation trajectories (for

FIGURE 38.6. Workspace boundaries for three chronic hemiplegic stroke subjects. "X's" and "O's" denote the average maximum reach for eight attempts at the given elevation angle for the hemiplegic and contralateral arm, respectively, with the patient supine in the shown position. Bands show plus or minus one standard deviation. Solid arcs are at a radius of 26 and 80 cm.

clinical measures of decreased functional ability, and can be used as more objective, finer-resolution measures of recovery in the clinic. Also, we plan to use the device to better understand impaired coordination after stroke by examining the magnitude

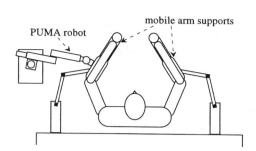

FIGURE 38.8. MIME

which experimental data is presented here), or by a position feedback control system that slaves the trajectories to the movements of the contralateral limb. Optical encoders on the joints of the arm supports measure the position and orientation of the forearms, and a 6-axis force/torque transducer measures the forces/torques applied to the paretic limb.

This approach is different than the other approaches presented in this chapter in that it uses a commercially available industrial robot, and a standard clinical mechanism—the mobile arm support. The design thus builds on established technologies from the robotics and rehabilitation industries, thus benefiting from the many previous design iterations that went into these technologies, for example, in producing a compact yet strong manipulator, or a comfortable yet secure mechanical interface to a patient. Also, use of established technologies allowed construction of a safe system with general movement capability quickly and cost-effectively.

As a first step in evaluating the utility of MIME, we investigated whether improved performance in active-assisted movements (as measured by the interaction forces/torques during robotic assistance) correlates with improved functional recovery of the paretic limb (as measured by an upper extremity Fugl–Meyer examination—a standard clinical exam) in hemiplegic stroke patients (see also Lum et al. 1999). If robot assisted measurements are to yield insight into the mechanisms of lost function, it is important to understand how those measurements correlate with functional ability. Also, it is important to establish that improved performance

ability of a robot exercise task correlates with improved functional recovery, if practicing that exercise task is intended to have functionally significant, therapeutic effects.

4.2 Illustrative Data

Each test session began with an upper extremity Fugl–Meyer examination intended to gauge the subject's arm function with an accepted clinical scale (Fugl-Meyer et al. 1975). Subjects were then strapped into a modified wheelchair, and their forearms were secured in the arm support. Six different point-to-point reaching movements were then tested (Figure 38.9), first with the subject passive and the robot moving, and then with the subject actively moving along with the robot.

The desired movement trajectory for the robot was based on the forearm trajectories of unassisted,

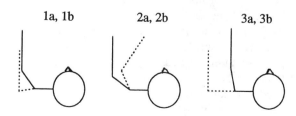

FIGURE 38.9. Top view of the six movement types. Movements 1a, 2a, 3a start from the dashed positions and end at the solid arm positions. Movements 1b, 2b, 3b are in the reverse directions. In all movements the hand moves in a straight line 15 cm.

2.5 second duration, movement trajectories measured from three unimpaired subjects. These trajectories were ensemble averaged across subjects to yield six "normal" trajectories. These trajectories were then programmed into the robot and used as a reference trajectory for a joint-based PID controller supplied with the robot.

For each desired point-to-point movement, the subject was first instructed to remain passive as the robot moved the limb in the programmed trajectory five times. The position and orientation of the forearm, and the interaction forces/torques between the paretic arm and the robot were sampled and stored. Next, the subject was instructed to voluntarily contribute to movement by pushing the robot with approximately one pound of force as it moved along the pre-programmed trajectory. After each trial, the subject was given knowledge of his average force level in the direction of movement and encouraged to increase or decrease his effort accordingly. Ten active-assist trials were collected.

For each desired point-to-point movement, we found the passive force/torque profiles were very consistent. Thus, to estimate the passive force/torque level at each point during the movement, the passive profiles were ensemble averaged. For each active trial, the forces/torques voluntarily generated were then estimated by subtracting the average passive profile from the active profile.

From this estimated voluntary profile we derived several parameters for characterizing performance. The "force magnitude" was defined as the magnitude of the average force vector generated during movement. The "force direction" error of the paretic limb was calculated as the angle between average force vectors generated from the left and right arms. "Work efficiency" was estimated by calculating the work done by the paretic limb during movement, and normalizing it by the "potential work," defined as the work that would have been produced during a trial if at each instant the force magnitude were directed precisely in the desired movement direction, and the torque magnitude were oriented precisely in the desired direction of rotation. Finally, the "normal force per unit work" was defined as the average force magnitude normal to the movement scaled by the positive work done during that movement. The normal torque per unit work was defined similarly.

All these parameters were calculated for each active trial, averaged over the movement duration, averaged across trials, and finally averaged across movement types to reach a scalar value that reflected the subject's average performance during the session. Averaging across movement types was done to compare with the Fugl–Meyer (FM) score, which reflects overall limb ability.

Using these techniques for three unimpaired and seven stroke subjects, we found that the more severely impaired subjects produced higher force magnitudes than less impaired subjects, but these forces were often misdirected relative to the direction of movement, and relative to the force directions generated by their contralateral limb during the same movements (Figure 38.10). There was a negative correlation between the force magnitude error during a trial versus the FM score ($p < 0.02$), with the most impaired subjects (FM scores of 14–17) producing forces of approximately 10 N, and unimpaired arms producing forces of 6 N. The force direction error of the paretic limb also was negatively correlated with the FM score ($p < 0.01$), with the directional errors of the most impaired subjects approximately 70 degrees, and those for the unimpaired arms less than 30 degrees. Finally, the stroke subjects showed a positive correlation between the work efficiency of the paretic limb versus the FM score ($p < 0.02$), and negative correlations between the normal forces and torques per unit work versus the FM score ($p < 0.02$ for forces, $p < 0.01$ for torques).

4.3   Discussion

The experiments with MIME quantified impaired coordination after stroke. During active-assist movements, stroke subjects pushed harder against the robot than unimpaired subjects, but with greater directional errors, creating larger normal forces and torques. They ended up doing increased work. All of these parameters were strongly correlated with the FM score, suggesting that robot-assisted performance has functional relevance.

These results, coupled with the positive results reported from clinical trials of MIT-MANUS presented in the next section, are motivating for a clinical test of the therapeutic efficacy of exercise with MIME. The MIT-MANUS study showed that increased robotic therapy resulted in increased upper limb recovery, an exciting and potentially groundbreaking result. We plan to build upon this result

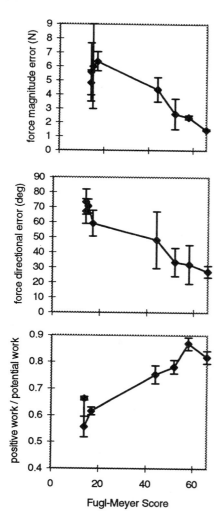

FIGURE 38.10. Force magnitude error, force directional error, and work efficiency vs. Fugl–Meyer score for seven stroke subjects and one composite normal. For the stroke subjects, the mean and standard deviation represents two experimental sessions taken within one week.

by testing if robotic therapy is better than traditional upper limb therapy, and if the active aspect of the therapy is necessary. A test group will exercise each day with our device in addition to their normal therapy, while a control group will receive additional traditional upper limb therapy. A second control group will practice the same movements as the test group in a passive device that constrains the endpoint motion to one direction while providing various levels of loading in that direction, similar to the linear reaching guide presented in the previous

section. This will allow us to determine if an actuated system is absolutely necessary for therapeutic efficacy, or if a passive system would suffice.

5 MIT-MANUS

5.1 Design Choices

As described in the introduction, neurorehabilitation is labor intensive, relying on therapy and evaluation procedures that are typically administered by a single clinician working with a single patient; indeed, this is true of much of the practice of clinical neurology. Labor-intensive procedures suggest that robotics and information technology may be used to improve quality, enhance documentation and reduce cost, and that has been the vision guiding the development of MIT-MANUS. However, at the outset of this research there was no firm evidence that the manipulation of patients' limbs that dominates current neurologic rehabilitation practice (and that might be aided by robotics) has a positive effect on recovery from brain injury. In short, the kind of assistance a robot could provide might not matter.

In a pilot clinical trial, we used MIT-MANUS (Hogan et al. 1995) a robot designed for neurological applications to assess whether manipulation of the impaired limb influences motor recovery in hemiparetic stroke patients (Krebs et al. 1998). Unlike most industrial robots, MIT-MANUS (Figure 38.11.) is configured for safe, stable, and compliant operation in close physical contact with humans. This is achieved using impedance control, a key feature of the robot control system (Hogan 1985). Its computer control system modulates the way the robot reacts to mechanical perturbation from a patient or clinician and ensures a gentle compliant behavior. MIT-MANUS can move, guide, or perturb the movement of a subject's or patient's upper limb and can record motions and mechanical quantities such as the position, velocity, and forces applied. The machine was designed to have a low intrinsic end-point impedance (i.e., be back driveable), with a low and nearly isotropic inertia and friction, and be capable of producing a predetermined range of forces and impedances. We emphasized this design requirement primarily to facilitate control of robot impedance and to ensure that the robot's intrinsic dynamics would be minimally encumbering to the

Portable Robot

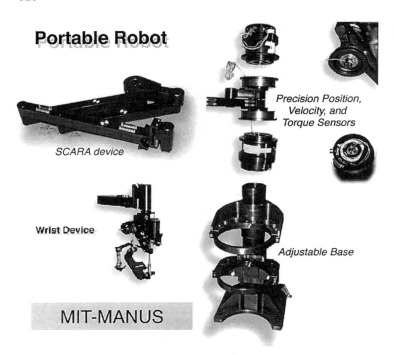

SCARA device

Precision Position, Velocity, and Torque Sensors

Wrist Device

Adjustable Base

MIT-MANUS

FIGURE 38.11. MIT-MANUS. (From Krebs et al. 1998, © 1998 IEEE.)

patient (i.e., we wanted to be sure the robot could "get out of the way" when appropriate).

MIT-MANUS presently has two modules. The 2-DOF module provides two translational degrees of freedom on the horizontal plane for elbow and forearm motion, and it consists of a direct-drive five bar-linkage SCARA (Selective Compliance Assembly Robot Arm) mechanism. The 3-DOF module provides three degrees of freedom for wrist motion (flexion, extension, adduction, abduction, pronation, supination), and it consists of a differential mechanism mounted on a parallelogram linkage driven by three geared actuators. Planar, supported movements are used in activities of daily living and in therapy, with the patient moving the arm across a table top, or with support from the therapist. However, we chose this design primarily to ensure portability, a key requirement of the design to allow us to investigate home-based self-therapy. The present design is portable and meets or exceeds applicable safety standards for operation in a clinical environment.

5.2   Illustrative Data

In a pilot clinical trial with twenty sequential hemiparetic patients, only the 2-DOF module was used.

These twenty sequential patients with a single CT-verified stroke hospitalized on the same acute rehabilitation ward at the Burke Rehabilitation Hospital (White Plains, New York) were enrolled in either a robot-aided therapy group (RT, $N = 10$) or in a standard therapy group plus "sham" robot-aided therapy (ST, $N = 10$). Both groups were comparable in age, physical impairment, and time between onset of the stroke and enrollment in the trial (mean age: RT 58.5, ST 63; mean admission to rehab in weeks since stroke onset: RT 2.8, ST 3.2). All patients were blinded to the treatment group and were assigned to the same blinded clinical team. Both groups received conventional therapy; the RT group received an additional 4 to 5 hours per week of robot-aided therapy typically for 6.5 weeks consisting of peripheral manipulation of the impaired limb correlated with audiovisual stimuli, while the ST group had an hour weekly robot exposure. The sensory-motor training for the RT group consisted of a set of "videogames." Patients were required to move the robot end-effector according to the game's goals, which included drawing circles, stars, squares, diamonds, and navigating through graphical windows. If the patient could not perform the task, the robot assisted and guided the patient's hand.

Menu-driven software allows the clinician to

TABLE 38.4. Change during acute rehabilitation: means, standard deviation, p value. (*) indicates data available for 9 patients. (From Krebs et al. 1998, © 1998 IEEE.)

Group	Experimental (10 patients)	Control (10 patients)	One-T P-value (significance)
Change FIM	25.6 ± 7.2	25.7 ± 12.2	0.51
Change F-M—UE subscore	14.1 ± 9.7	9.9 ± 11.2	0.19
Change MP	3.9 ± 2.9	2.3 ± 2.4	0.10
Change MSS—shoulder/elbow	9.4 ± 5.9 (*)	0.8 ± 3.8	0.00065
Change MSS—wrist	0.9 ± 1.5 (*)	0.5 ± 1.5	0.29
Change MSS—hand	4.6 ± 6.4 (*)	3.5 ± 6.3	0.36

choose different values of impedance (very soft, soft, medium, hard, very hard). However, we opted to use the same soft range (100 N/m, 2 N-sec/m) throughout the trial primarily because of patients' pretrial preferences and to minimize any risk of exacerbating joint or tendon pain. For this trial, the impedance controller was implemented using nonlinear position and velocity feedback structured to produce constant isotropic end-point stiffness and damping. Coupled to our highly back-driveable design, the stability of this controller is extremely robust to the uncertainties because of physical contact (Hogan 1988; Colgate and Hogan 1988).

The training for the ST group was similar to the RT group, except that half of the one hour session consisted of playing the video games with the unimpaired arm and half the session with the impaired arm while the robot passively supported the arm and provided visual feedback. If the patient could not perform the task, he/she used the unimpaired arm to assist and complete the game, or the clinician assisted. The purpose of this "sham" robot therapy was to give the minimum extra "therapy" to the control group yet (1) provide all patients with comparable motivation and attention (2) blind patients and clinicians to which treatment group a patient was in, and (3) familiarize the control group with the machine so that robot-aided evaluation would be feasible and meaningful. All patients were evaluated by the same blinded therapist with standardized clinical assessment procedures and robot-measured assessment of movement kinematics. The standard assessment procedures included measurements of functional status (FIM—Functional Independence Measure) (Dodds et al. 1994), upper extremity motor impairment (F–M—Fugl–Meyer Upper Extremity Subscore, MSS—

Motor Status Score, MP—Motor Power) (Fugl-Meyer et al. 1975; Aisen et al. 1995), as well as pain and shoulder-hand syndrome.

Results of this pilot clinical trial indicated that robot-assisted therapy positively influenced recovery of patients. Mean baseline FIM, F–M, MSS, and MP at admission indicated no statistically significant differences between the control and experimental groups. Nonetheless, at discharge the experimental group outranked the control group in all the clinical assessments of the motor impairment involving shoulder and elbow (limb segments exercised with the robot-aided therapy) (Table 38.4). There was a clear trend in the F–M and MP scores favoring the experimental group ($p < 0.20$ and $p < 0.10$), and a statistically significant improvement in the MSS for shoulder/elbow ($p < 0.05$) (Aisen et al. 1996). At the same time, the measurement of functional status (FIM) indicated no statistically significant difference between groups ($p = 0.98$). We believe that this may be the result of the non-specific nature of the FIM score which broadly assesses overall functional recovery and the fact that both groups had the same standard therapeutic experience aiming at functional recovery. There was no significant difference in frequency between groups for pain in joints and tendons, or shoulder–hand syndrome. Even more encouraging, the MSS results' time history suggests that the control group did not improve after 5 weeks from the stroke onset, while five patients in the experimental group continued to improve up to 8 week's poststroke.

5.3 Discussion

For a number of reasons (most prominently the small sample size) these results should be inter-

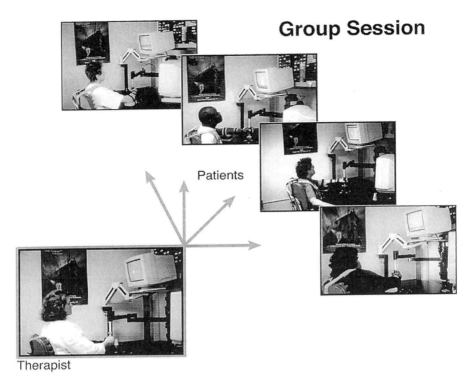

FIGURE 38.12. Group therapy concept. (From Krebs et al. 1998, © 1998 IEEE.)

preted with caution. Nevertheless they are encouraging. To paraphrase one of the Hippocratic principles of medicine: "First, do no harm." At the outset it was unclear whether robot-aided therapy would not impede recovery. At a minimum our results indicate that robot-aided therapy has no adverse effects. On a more positive note, at face value we seem to have answered our first question, "Does robot therapy matter?" in the affirmative. These preliminary results suggest that robotics and information technology may be a promising new tool for neurology and rehabilitation, complementing the clinician's compassion and expertise with repeatable actions, objective quantification and potential cost savings.

A follow-up clinical trial is under way with a larger pool of 60 stroke patients. We are further investigating the potential for cost saving by allowing a clinician to work with more than one patient at a time (Figure 38.12), and by permitting meaningful self-therapy at home by establishing a bilateral teleoperation link between the robot in the home and a robot in the clinic.

6   Conclusion

To summarize, the bimanual rehabilitators showed how simple feedback control laws and machine geometry might replicate seemingly complex therapeutic manipulations. This suggests that simple task-specific devices might suffice for some aspects of therapy. Contact forces between the patient and reaching guide were shown to provide quantitative assessment of impaired coordination. This type of measurement might form the basis for improved assessment and treatment of incoordination. MIME performance measurements were shown to correlate significantly with an accepted clinical measure of function. Machine-based rehabilitation may thus transfer well to more functional contexts. The MIT-MANUS experiments support such a transfer: robot-assisted therapy enhanced recovery according to several clinical measures.

How will these results impact neurological rehabilitation? The questions below indicate possible directions for future research. The questions are divided into three categories: measurement, therapy, and technological innovation. The answers, along with the answers to cost, safety, and ease-of-use questions not discussed here, will play a key role in determining the ultimate pattern of use of this technology.

6.1 Questions Regarding Improved Measurement with Machines

The scientific basis for improved measurement with machines is an important area for future research. Experiments in this chapter suggest that guiding and assisting forces, including those normal to the desired direction of movement, contain useful information. What is the diagnostic and predictive value of these parameters? Other quantitative methods have been developed for measuring weakness (Gandevia 1993), spasticity (Rymer and Katz 1994), and abnormal synergies (Beer et al. 1995; Dewald et al. 1995) in stroke patients. Can these methods be incorporated into machine designs? What is the minimum set of measures needed to adequately characterize impairment and recovery?

A chief motivation for measurement with machines is increased sensitivity in gauging functional recovery. Improved sensitivity could help clinicians to better judge patient progress and the efficacy of different therapy techniques. However, none of the devices in this chapter has yet demonstrated functional measurement at higher resolution than is possible with current clinical scales. There is good reason to question the feasibility of high resolution measurement in neurological rehabilitation: human voluntary movement is inherently variable, impaired movement even more so. To what extent is improved resolution even possible? Can machine-based measurement do better than clinical scales?

6.2 Questions Regarding Therapeutic Manipulation with Machines

Besides machine-based measurement, the scientific bases of therapeutic manipulation are also a key area for future research. Encouragingly, the clinical trial of MIT-MANUS showed that providing additional therapy with a manually assisting robot can improve recovery. This suggests that more therapy can be better, and that machines might help provide this extra therapy. Future research will need to clarify what components of the additional therapy—the peripheral manipulation, the attempts by patients to activate descending pathways, or the enhanced sensory stimulation, to name a few potential candidates—are essential. What are the key components of therapy and how are they best implemented?

It is also important to better understand the nature of transfer from machine therapy to everyday function. In motor learning research with unimpaired subjects transfer of learning between tasks has been found to be small (Schmidt 1988). However, the experiments with MIME and MIT-MANUS suggest that transfer may exist between planar arm movement and overall arm function. What is the nature of this transfer? Could exercise with simple, task-specific devices like the bimanual rehabilitators generate functional transfer? To what extent do performance improvements extend beyond the exercised limb and the specific action used in therapy? What are the mechanisms of transfer and how are they best exploited?

6.3 Questions Regarding Further Technological Innovation

Finally, there is an opportunity for engineering synthesis to expand upon current devices and to create new and better tools for clinical neurology. This raises another set of questions revolving around the possibility of further technological innovation.

Many other device configurations are possible. Each of the devices in this chapter exercises specific degrees of freedom in the arm. Technology applicable to the whole body is technically feasible but would it help? Is there significant benefit to working simultaneously with many degrees of freedom (e.g., the whole arm or even the whole body)? All the machines in this chapter currently implement some sort of guidance or assistance. Other manual therapy manipulation techniques are possible (see Reinkensmeyer et al. 1996, for examples). Can and should these techniques be implemented with mechanical devices? Complex "force fields" have been used to induce motor adaptation in unimpaired subjects (Shadmehr and Mussa-Ivaldi 1994; see also Chapter 24 and Chapter 25). Might such force fields be used for stroke rehabilitation?

Other therapeutic modalities might be combined with manual manipulation. Would tactile stimulation make a difference? What about neuro-muscular stimulation (the focus of Section IX of this book)? Are the emerging "virtual environment" technologies applicable? They may hold the promise of providing tools to address cognitive deficits as well as sensory-motor deficits. Telerobotic technology might enable remote monitoring and intervention (see also Chapter 37). Portable and even wearable sensors, actuators, and information processors (e.g., built into "smart fabrics") are technologically feasible. Can home-based self-therapy be effective?

These are a few of the questions that will help determine the ultimate pattern of use of machine technology in neurological rehabilitation. Answering these questions will ultimately require two types of effort. One is to learn by doing: build the technology and try it out. Much of our work is in this stage now, and more research along these lines is needed. It is also essential to understand the outcomes. Scientific analysis complements design synthesis, and often the two are synergistic. We suggest, and hopefully this chapter has begun to provide support that rehabilitators, robots, and guides will provide a new medium for both rehabilitation innovation and science.

Acknowledgments. David Reinkensmeyer would like to thank Julius P.A. Dewald, PT, Ph.D, and W. Zev Rymer M.D., Ph.D., for their support and assistance, and the NIDRR (Training Grant H133P20016), NIH (NS19331 and National Research Service Award F32HD0806701), and Ralph and Marion C. Falk Medical Research Trust for their support. Neville Hogan and Hermano Krebs would like to thank Mindy L. Aisen, M.D., Bruce T. Volpe, M.D., and Lisa Edelstein, O.T. for their contribution and assistance, and acknowledge the National Science Foundation—grant # 8914032-BCS and the Burke Institute for Medical Research for their support. Peter Lum thanks Charles Burgar, M.D., Machiel Van der Loos, Ph.D., Deborah Kenney, OTR, and the Veteran's Adminstration (grant #B94-846P) for their support.

References

Aisen, M.L., Krebs, H.I., Hogan, N., and Volpe, B.T. (1997). The effect of robot-assisted therapy and reha-

bilitative training on motor recovery following stroke. *Arch. Neurol.*, 54:443–446.

Aisen, M.L., Sevilla, D., Gibson, G., Kutt, H., Blau, A, Edelstein, L., Hatch, J., and Blass, J. (1995). 3,4-Di-aminopyridine as a treatment for amyotrophic lateral sclerosis. *J. Neurol. Sci.*, 129:21–24.

American Heart Association. (1993). *Heart and Stroke Facts and Statistics.*

Brunnstrom, S. (1970). *Movement Therapy in Hemiplegia.* Harper and Row, New York.

Colgate, J.E. and Hogan, N. (1988). Robust Control of Dynamically Interacting Systems. *Int. J. Control*, 48(1):65–88.

Dewald J.P.A. (1987). Sensorimotor neurophysiology and the basis for neurofacilitation therapeutic techniques. In *Stroke Rehabilitation*, Brandstater, M.E. and Basmajian, J.V. (eds.), pp. 109–182. Williams & Wilkins, Baltimore-London-Los Angeles-Sydney.

Dewald, J.P.A., Beer R.F., and Given J.D. (1996). Evidence for task dependent weakness due to abnormal torque synergies in the impaired upper limb of hemiparetic stroke subjects: preliminary results. *Proc. 18th Annual Int. Conf. IEEE Eng. In Medicine and Biology Society*, Oct. 31–Nov. 3, Amsterdam, The Netherlands.

Dewald, J.P.A., Pope, P.S., Given, J.D., Buchanan, T.S., and Rymer, W.Z. (1995). Abnormal muscle coactivation patterns during isometric torque generation at the elbow and shoulder in hemiparetic subjects. *Brain*, 118:495–510.

Dodds, T.A., Martin, D.P., Stolov, W.C., and Deyo, R.A. (1993). A validation of the functional independence measurement and its performance among rehabilitation inpatients. *Arch. Phys. Med. Rehab.* 74:531–536.

Erlandson, R.F. (1995). Applications of robotic/mechatronic systems in special education, rehabilitation therapy, and vocational training: a paradigm shift. *IEEE Trans. Rehab. Eng.*, 3(1):22–34.

Flores, A.M. (1992). Objective measurement of standing balance. *Neurol. Report* 16(1):17–21.

Fugl-Meyer, A.R., Jaasco, L., Leyman, L., Olsson S., and Steglind, S. (1975). The post-stroke hemiplegic patient. *Scand. J. Rehab. Med.*, 7:13–31.

Gandevia, S.C. (1993). Strength changes in hemiparesis: measurements and mechanisms. In *Spasticity: Mechanisms and Management*, Thilmann, A.F., Burke, D.J., and Rymer, W.Z. (eds.), pp. 111–122. Berlin, Springer-Verlag.

Gresham, G.E., Duncan P.W., Stason W.B., et al. (1995). *Post-Stroke Rehabilitation.* Clinical practice guideline, No. 16. Rockville, MD: U.S. Dept. Health and Human Services. Public Health Service, Agency for Health Care Policy and Research. AHCPR Publication No. 95-0662.

Gresham, G.E., Phillips, T.F., Wolf, P.A., McNamara, P.M., Kannel, W.B., and Dawber, T.R. (1979). Epidemiologic profile of long term disability in stroke: the Framingham study. *Arch. Phys. Med. Rehab.*, 60: 487–491.

Hogan, N. (1985). Impedance control: an approach to manipulation: Part I, Part II, Part III. *J. Dynamic Sys., Measurement, and Control—Trans. ASME*; 107:1–24.

Hogan, N. (1988). On the stability of manipulators performing contact tasks. *IEEE J. Robotics Auto.*, 4: 677–686.

Hogan, N., Krebs, H.I., Sharon, A., and Charnnarong, J. (1995). Interactive robotic therapist. U.S. Patent # 5,466,213; Massachusetts Institute of Technology.

Krebs, H.I., Hogan, N., Aisen, M.L., and Volpe, B.T. (1998). Robot-aided neurorehabilitation. *IEEE Trans. Rehab. Eng.*, 6(1):75–87.

Lum, P.S., Burgar, C.G., Kenney, D., and Van der Loos, H.F.M. (1999). Quantification of force abnormalities during passive and active-assisted upper-limb reaching movements in post-stroke hemiparesis. *IEEE Trans. Biomed. Eng.*, 46(6):652–662.

Lum, P.S., Lehman, S.L., and Reinkensmeyer, D.J. (1995). The bimanual lifting rehabilitator: an adaptive machine for therapy of stroke patients. *IEEE Trans. Rehab. Eng.*, 3(2):166–174.

Lum, P.S., Reinkensmeyer, D.J., and Lehman, S.L. (1993a). Robotic assist devices for bimanual physical therapy: preliminary experiments. *IEEE Trans. Rehab. Eng.*, 1:185–191.

Lum, P.S., Reinkensmeyer, D.J., and Lehman, S.L. (1993b). A bimanual reflex during two-hand grasp. *Proceedings of the 15th Annual Int. Conf. of the IEEE Eng. in Medicine and Biology Soc.*, pp. 1163–1164.

Lum, P.S., Reinkensmeyer, D.J., Lehman, S.L., Li, P.Y., and Stark, L.W. (1992). Feedforward stabilization in a bimanual unloading task. *Exp. Brain Res.*, 89:172–180.

Reinkensmeyer, D.J., Dewald J.P.A., and Rymer W.Z. (1996). Robotic devices for physical rehabilitation of stroke patients: fundamental requirements, target therapeutic techniques, and preliminary designs. *Technol. Disabil.*, 5:205–215.

Reinkensmeyer, D.J., Dewald J.P.A., and Rymer W.Z. (1999). Guidance-based quantification of arm impairment following brain injury: a pilot study. *IEEE Trans. Rehab. Eng.*, 7(1):1–11.

Reinkensmeyer, D.J., Lum, P.S., and Lehman, S.L. (1992). Human control of a simple two-hand grasp. *Biol. Cybern.*, 67:553–564.

Rymer, W.Z. and Katz, R.T. (1994). Mechanical quantification of spastic hypertonia. In *Spasticity: Physical Medicine and Rehabilitation State of the Art Reviews*, Katz, R.T. (ed.), 8(3):455–464. Hanley and Belfus, Philadelphia.

Schmidt, R.A. (1988). *Motor Control and Learning.* 2nd ed., Human Kinetics, Champaigne, Illinois.

Shadmehr, R. and Mussa-Ivaldi F.A. (1994). Adaptive representation of dynamics during learning of a motor task. *J. Neurosci.*, 14:3208–3224.

Siddiqi, N.A., Ide, T., Chen, M.Y., and Akamatsu, N. (1994). A computer-aided walking rehabilitation robot. *Am. J. Phys. Med. Rehab.*, 73(3):212–216.

Thilmann, A.F., Burke, D.J., and Rymer W.Z. (ed.). (1993). *Spasticity: Mechanisms and Management.* Springer-Verlag, Berlin.

Trombly, C.A. (ed.). (1995). *Occupational Therapy for Physical Dysfunction.* 4th ed. Williams and Wilkins, Baltimore.

Washam, V. (1973). *The One-Hander's Book: A Basic Guide to Activities of Daily Living.* John Day Co.

Wing, A.M., Allison, S., and Jenner, J.R. (1993). Retaining and retraining balance after stroke. *Baillieres Clin. Neurol.*, 2(1):87–120.

Commentary: Rehabilitators, Robots, and Guides

Dava J. Newman

The devices and data presented in this chapter highlight promising new results and systems for neurological rehabilitation. Most importantly, machine-based techniques were shown to have a significant impact on stroke rehabilitation through different machine designs ranging from simple control and geometry to complex multidegree-of-freedom (DOF) devices. The devices were reported to replicate complex motions, quantify measures of impairment, and even aid recovery according to functional and clinical measures. In the conclusions section the authors pose the question whether their results (and devices) will impact neurological rehabilitation. I agree with them that this is the essential question and very difficult to answer. I will use the three categories (measurement, therapy, and technological innovation) proposed for future research directions as the basis for my commentary.

As suggested, the potential exists for improved measurements using machines and information technology. I would like to suggest that comprehensive experimental studies are necessary to identify the essential parameters that will help quantify the effect that machines have on rehabilitation and then their utility. Is it clear that stroke patients have more debilitation in horizontal than vertical limb motion? All of the machines described in the chapter (except the reaching guide) allowed for only planar horizontal arm motion. Using the same measurements (position, force, and torque) and devices, experiments should be repeated in the vertical plane (and horizontally for the reaching guide device). Only then can the essential judgment of whether specific degrees of freedom rather than the whole limb or whole-body motion is necessary for stroke neurological rehabilitation.

This brings us to the related category of using machines for therapy. Data from the MIME and MIT-MANUS machines suggest that transfer may exist between planar arm movement and overall arm function. However, Schmidt (1988) argues that transfer of learning between tasks is not appreciable. This controversy can be resolved by experiments allowing for limited DOF, in-plane and then multi-DOF, three dimensional arm movements. Statistically significant results between the two paradigms will help determine the necessity of whole limb rehabilitation.

I recommend underwater therapy as a potential for musculoskeletal rehabilitation, and this technique could be coupled with machine rehabilitation. Recovering patients will find relief in the lighter loading and increased mobility of underwater submersion.

Additional technological assessment tools include: noninvasive kinematic analysis (e.g., limb or whole-body motions), piezoelectric force sensors, shape memory alloys to provide resistive or actuation forces, and tactile stimulation (see also Chapter 37). There are many commercially available systems to consider. As alluded to in the chapter, future technological advances might allow us to wear "smart suits", envisioned as lightweight garments with sensors, actuators, and computers woven into the fabric. These technologies could revolutionize rehabilitation methods.

Finally, I believe that the patients' acceptability of the machines as advantageous rather than as intrusive should be given serious consideration.

Reference

Schmidt, R.A. (1988) *Motor Control and Learning*. 2nd ed. Human Kinetics, Champaigne, Illinois.

39
Nonanalytical Control for Assisting Reaching in Humans with Disabilities

Dejan Popovic and Mirjana Popovic

1 Introduction

Restoring reaching and grasping in subjects with spinal cord injury at cervical level (i.e., tetraplegia, is important for improved quality of life) (Triolo et al. 1995). Most subjects who can regain grasping retain elbow flexion to some extent and learn how to compensate for the deficiency of voluntary elbow extension. There is a group of subjects with tetraplegia who retain elbow flexion, and can benefit from functional electrical stimulation (FES) grasping systems (Peckham and Keith 1992), but have paralyzed, yet innervated elbow extensor muscles. These subjects benefit from grasping devices only if, in addition, their reaching movements can be restored. A review of the efforts of three major groups developing assistive systems for upper extremities (Triolo et al. 1995) suggests that there is no practical reaching system available yet for subjects with tetraplegia. This study is a contribution to the development of a new assistive system for restoring reaching. A controller which allows persons to subconsciously and intuitively assist grasping and reaching is essential for the success of such a device (Tomovic et al. 1995).

The traditional approach to the design of controllers starts from the model of the system (e.g., see Chapter 11). Once the parameters of the model and the desired trajectory are known, a controller will generate control signals which, with or without feedback, will drive the system. However, a model of a living system is too complex to be used for real-time control; parameters are typically very difficult to identify; and finally there is an indefinite number of plausible trajectories. In addition,

there are technical and technological difficulties when applying closed-loop control because of inadequate sensors and their positioning, as well as specifics of biological actuators.

Some common principles of motor control organization in organisms (Bernstein 1967) may serve as a useful guideline for the design of control algorithms. The fact that goal-directed movements are reproducible for a given task in different living species implies that optimization of some sort is taking place during learning of such motions (see Chapter 35). This optimization is inherent to the self-organization. In its broad sense, optimization implies the presence of a system goal and at least two feasible options. The role of optimization is to apply the goal function as an ordering criterion of the set of feasible options so that the preferred solutions may be implemented.

When activated to perform a self-paced, nonconstrained functional motion, the neuromuscular system prefers some trajectories over others. The constraint imposed by the task determines the trajectory (e.g., straight-line movements are adequate for fast pointing movements; starting the movement with the proximal joints and adjusting the position of the hand while moving distal segments is effective for precise positioning, etc.).

Skill acquisition processes are optimized in such a way that the open loop control is preferred to the closed-loop control as long as it does not affect the performance of a functional motion. Invariants of functional motions, motor patterns, automatic movements, and reflexes are the means by which the use of the open-loop control mode is extended in the execution of motor acts. Open-loop control sets free the vision system and the conscious level to monitor the environment and prepare the organism for new activities (Tomovic and Bellman 1970). The execution of skilled movements is organized so that the flow of control information from the higher levels is minimized. Lower control levels dispose, therefore, with maximum autonomy while performing a skilled motor act.

The design of a nonanalytical, hierarchical controller for external control of elbow extension in persons with tetraplegia follows some of the principles of biological control (Popovic and Popovic 1994, 1995). The implemented strategy uses the principle of synergy which allows persons with disabilities to plan the movement in terms of the most distal segment, and actually volitionally move the most proximal segment (Bernstein 1967). For reaching, the movement is planned in the external reference frame based only on the visual information of the initial position of the hand and the position of the target (Georgopoulos et al. 1983, 1992; Soechting et al. 1986). The visual information is automatically transferred to the universal, that is, nonuser-specific, constraint between the shoulder and elbow joints (Jeannerod 1988; Soechting and Lacquaniti 1981). Hence, when a person volitionally moves his/her upperarm, the elbow joint angular velocity profile is constrained. An approximation of this constraint is the so-called scaling law (Popovic and Popovic 1994, 1995, 1998). The final step is the usage of the mapping at the lowest, actuator level. The actuator for assisting the elbow extension is Triceps Brachii and its mapping connects the joint angular velocity with the parameters of electrical stimulation. The individual mapping for the Triceps Brachii is expressed as a set of vectors composed of three elements: (1) the elbow joint angular velocity; (2) the increment of the elbow joint angular velocity; and (3) the electric charge delivered to the muscle.

In this algorithm the upper and lower level operate independently, minimizing the need for data transfer. The upper, coordination level operates without feedback (Popovic and Popovic 1994), and the scaling has to be volitionally selected by the user. The lower, actuator level is sensory driven, and the feedback is the elbow joint angular velocity.

Nonanalytical methods are used here for the following tasks: (1) the determination of the scaling coefficients; and (2) the estimation of recruitment characteristics of the elbow extensor muscles. To determine the constraint between joint motions, the system is treated as a black box, and a classification method is applied. The constraint is nonlinear, but the nonlinearity can be approximated with a best linear fit in order to simplify the control. It is noteworthy that the goal is not the copy the biological control, but to find a coercion which can be used to map input and output for the external controller.

The utilization of the hierarchical, nonanalytical controller is based on the assumption that the manipulation can be decomposed into two independent processes (Tomovic et al. 1995): (1) the movement of the hand in a plane due to the shoulder and elbow flexion/extension; and (2) the rotation of that plane around the shoulder joint (shoulder abduction/adduction and humeral rotation). This decomposition is essential because here we only deal with the control of the elbow flexor/extensor muscles, and because potential users have diminished, but still functional, control over their shoulder joints; hence, the abduction/adduction, humeral rotation and flexion/extension of the shoulder joint are expected to be reduced, but still voluntarily controlled.

The trajectory of the hand belongs to three dimensional (3D) space (Figure 39.1), but the elbow

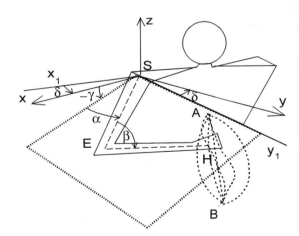

FIGURE 39.1. The mechanical model of the system. (A) and (B) are arbitrarily selected terminal positions of the hand in space. The triangle SEH determines the plane of the elbow (β) and shoulder (α) flexion and extension. γ and δ determine the tilt angle of the arm (SEH) with respect to the horizontal plane, and they are controlled with abduction/adduction and humeral rotation.

flexor and extensor muscles only control the angle $\beta(t)$, that is the movement of the hand in the plane determined with points: S—shoulder, E—elbow, and H—midpoint of the palm (hand). The shoulder flexion angle $\alpha(t)$ is assumed to be voluntarily controlled by the user. The mid point of the palm can follow an infinite number of trajectories if $\alpha(t)$ and $\beta(t)$ are independent variables. A constraint $\beta = f(\alpha)$ imposes the motion at the elbow joint upon the movements at the shoulder joint. We showed that such a constraint between the shoulder and elbow joint angular velocities can be defined with a scaling coefficient C (Popovic and Popovic 1994, 1995). The scaling $C = \dot{\alpha}/\dot{\beta}$ depends only upon the position of points A and B (Popovic 1995).

The scaling implies that the angular velocities are proportional (compare to the joint torque "linear synergy" law described in Chapter 27). The actual synergy $\dot{\beta} = f(\dot{\alpha})$ (Figure 39.2, right panel) obtained from the joint angles (Figure 39.2, left panel) for a movement of the hand in the horizontal plane (Figure 39.2 middle panel) is a nonlinear, propeller like pattern (Popovic 1995). The slope of the long axis of the propeller pattern with respect to the angular velocity at the elbow joint (vertical axis) is equal to the scaling parameter (C). The propeller can be described as a set of two leaf-like shapes. The upper leaf-like shape corresponds to the target

approach phase, while the bottom one relates to the returning movement. Multiple lines shown in Figure 39.2 are for two movements between the same initial and target points.

The structure of the controller (Popovic and Popovic 1994,1995) is depicted in Figure 39.3. The $\dot{\alpha}(t)$, $\dot{\alpha}(t - \Delta t)$ are the angular velocities of the upper arm relative to the body in two consecutive times, and $\dot{\beta}(t)$, $\dot{\beta}(t - \Delta t)$ are the calculated values of the elbow angular velocities obtained after the multiplication of the shoulder angular velocity with C^{-1}. The amplitude of the stimulation pulses (I) and frequency of the stimulation (f) are preselected to allow full range of externally elicited elbow extension. The pulse duration (T) is controlled automatically, and it is responsible for the level of activation of the Triceps Brachii. The incremental pulse duration (ΔT) is the minimal difference between two pulse durations, which generates a noticeable change in velocity, and it is user specific. $\dot{\beta}_m(t)$ is the measured relative angular velocity of the forearm with respect to the upper arm, and $e = \dot{\beta}_m(t) - \dot{\beta}(t)$ is the error used in the sampled feedback control loop.

The paradigm shown in Figure 39.3 decomposes the control to two levels. At a higher level the coefficient C is selected using exclusively the visual information (Popovic and Popovic 1994). This scaling is universal for a given task (Popovic 1995). The

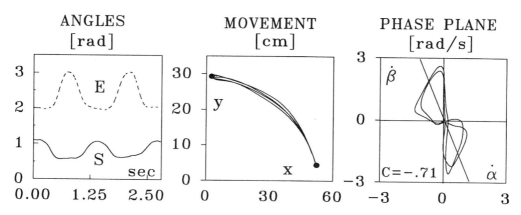

FIGURE 39.2. Original and processed data needed for studying synergies and the determination of the scaling coefficient in able-bodied subjects. The task was to move the hand between two preselected points (middle panel) in both directions. Movements were self-paced and twice repeated. Left panel shows elbow (E-upper trace) and shoulder (S-bottom trace) angular changes during hand movement on the horizontal digitizing board, middle panel. Phase plane presentation is formed of shoulder (x-axis) and elbow (y-axis) angular velocities, right panel. The scaling coefficient for this movement is $C = -0.71$.

COORDINATION LEVEL

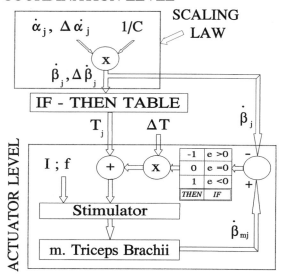

FIGURE 39.3. Rule-based, hierarchical controller for reaching. $\dot{\alpha}_j = \dot{\alpha}(t_j)$ and $\dot{\beta}_j = \dot{\beta}(t_j)$ are the angular velocities at time t_j at the shoulder and elbow joints respectively; $\Delta\dot{\alpha}_j = \dot{\alpha}(t_j) - \dot{\alpha}(t_j - \Delta t)$ and $\Delta\dot{\beta}_j = \dot{\beta}(t_j) - \dot{\beta}(t_j - \Delta t)$ are the increments of the joint angular velocities for the sampling interval; T_j is the pulse duration determined from the look-up table; I and f are the amplitude and frequency of the stimulation pulses; e is the error between the measured $(\dot{\beta}_m(t_j) = \dot{\beta}_{mj})$ and desired $(\dot{\beta}_j)$ elbow angular velocities. The inputs to the controller are: (1) $\dot{\alpha}_j$, $\Delta\dot{\alpha}_j$, voluntary shoulder rotation, (2) C, volitionally selected scaling coefficient based on the visual information, task and zoning method; and (3) ΔT, increment of the pulse duration.

input to the upper control level is the shoulder joint angular velocity. At the lower level, the time-sampled feedback adjusts the stimulation parameters to minimize the error in the velocity space (Popovic and Popovic 1994, 1995). This error-driven control employs a user-dependent rule-based method.

2   Methods

2.1   Determination of Synergies in Reaching

Six able-bodied volunteers participated in the study since: (1) a subject with tetraplegia has reduced reaching abilities and modified strategies of arm

movements; and (2) the biological synergy found for normal reaching movements is to be used as a constraint for control. Subjects were asked to reach and grasp with their dominant arm various objects: glass, snack, fork, bottle, floppy disk, VCR cassette, comb, toothbrush, telephone receiver, books, and papers, following the evaluation methods presented in the literature (Wijman et al. 1990). The reaching movements were in vertical, horizontal and tilted planes, with or without loads, because the controller must adapt for various dynamic and environmental conditions.

The following angles were measured and processed, using the instrumentation and methods described elsewhere (Winter 1990; Popovic and Tepavac 1992): shoulder abduction/adduction, flexion/extension, humeral rotation, and elbow flexion/extension. Each set of the angular velocities (flexion/extension) at the shoulder and elbow joints for a given reaching task (typically 5 movements) was divided into individual reach-return sequences. We disregarded movements which were not typical, and created the phase-space plots from the remaining movements having the shoulder angular velocity at the horizontal axis, and the elbow angular velocity at the vertical axis. The propeller shaped patterns were obtained for all analyzed movements. These propeller patterns were used to determine the best linear fit using the least squares method. The scaling coefficient C, that is the slope of the long axis of a propeller pattern with respect to the elbow angular velocity, was calculated (Popovic and Popovic 1998) for each of the research subjects for many functional movements. The averaged scalings and corresponding standard deviations were determined.

2.2   Determination of Recruitment Parameters

The task was to determine a set of stimulation patterns that, under different dynamic conditions, would generate desired angular velocities at the elbow joint.

Six subjects with spinal cord injuries at C_4 to C_6 levels resulting in paralysis of hand, wrist, elbow extension, and reduced shoulder movements were recruited for experiments. There were no contractures of arm joints, and moderate phasic spasms were treated with the usual doses of antispasticity drugs. All subjects initially responded to the electrical stimulation with elbow extensor contraction, and then received the exercise electrical stimulation of the Triceps Brachii for at least three weeks, with a minimum of five days a week, 45 minutes a day, using an in-house-developed stimulator and surface electrodes.

The dominant arm was instrumented with flexible goniometers at the shoulder and elbow joints. Triceps Brachii was electrically stimulated. Surface electrodes were positioned over the Triceps Brachii in such a way that the cross talk with antagonistic muscles was minimized. The angular velocity was recorded for different movements at both the elbow and shoulder joints. The electrical charge of the compensated, monophasic stimulation was controlled by varying the pulse duration. The frequency was fixed at the level which generated fused contraction of the Triceps Brachii. Frequency was typically at $f \approx 20$ Hz. For each study subject, the current intensity was adjusted at a level for which stimulation started spreading to the Biceps Brachii at a pulse duration of 160 ms while the elbow joint was flexed at 1.5 radian. The current was typically at $I = 60$ mA.

Various movements of the forearm were recorded in several sessions in order to minimize effects of day-to-day variations, muscle fatigue, and other random effects. The pulse duration was increased up to 160 ms, in increments of 10 ms, since the averaged noticeable change of the elbow joint angular velocity ($\Delta \dot{\beta} = 0.05$ rad/s) was found to be $\Delta T = 9.7 \pm 1.4 \ \mu s$. Data was used to produce the angular velocity versus time curves. The curves obtained were then used to create the phase space with the angular velocity and the increment of the angular velocity as the variables. The increment of the angular velocity corresponds to the joint angular acceleration. Both graphs include the pulse duration as a parameter.

Phase plots are displayed on a grid where the distances between the grid lines correspond to the differences of 0.05 rad/s. A method for automatic determination of triplets: (1) angular velocity $\dot{\beta}$; (2) increment of the angular velocity $\Delta \dot{\beta}$ between the previous and its present value, and (3) pulse duration T was developed. A triplet $(\dot{\beta}_j), \Delta(\dot{\beta}_j, T_j)$ denotes the following: *If the angular velocity is* $(\dot{\beta}_j)$, and *if the angular velocity increment is* $\Delta(\dot{\beta}_j)$, *then* the pulse duration T_j is to be applied for stimula-

tion of the Triceps Brachii. These triplets form a look-up table, that is the knowledge base within a rule-based controller depicted in Figure 39.3.

2.3 Implementation of the System

The assistive system comprised of a programmable stimulator with 4×4 keyboard for selection of the scaling coefficient and the task, and two flexible goniometers, was applied in the laboratory conditions to assist reaching of the right arm. Subjects were trained to use the system, after the recruitment parameters had been determined and the look-up table stored in the microcomputer, and Triceps Brachii strengthened through the three weeks of exercise for 45 minutes daily.

Subjects first practiced reaching various objects from the shelf, table and practiced using them (e.g., glass, can, pen, computer disk, VCR tape, etc.). The movements were recorded using the same set of goniometers used for control, and the session was videotaped for later analysis.

The digitizing board (0.3×0.3 m^2) was positioned in front of the subject to assess the precision of reaching. The rectangular reference frame was positioned in such a way that the point $(0,0)$ was in front of the ipsilateral shoulder of the subject, and the vertical axes along the frontal direction. The subject watched his hand moving at the digitizing board, and the movements were shown at the computer display simultaneously.

3 Results

3.1 Determination of Scaling Parameters

Figure 39.4 shows the phase space representation of the synergy patterns for one able-bodied subject when he was asked to reach various objects positioned on the desk and bring the hand to the mouth, simulating drinking. The scaling for the whole workspace varies from -1.65 to 0.54. The negative scaling suggests that the elbow and shoulder joints are both simultaneously flexing or extending, while the positive scaling indicates that when the elbow flexes, the shoulder joint extends, and vice versa.

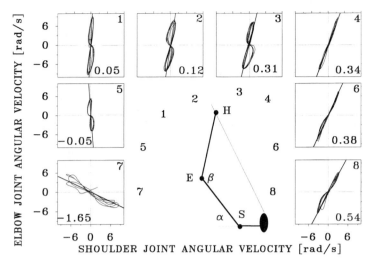

FIGURE 39.4. Phase space diagrams for self-paced movements from/to the mouth in 8 directions recorded in an able-bodied volunteer. The determined scaling coefficients for these movements are between -1.65 and 0.54. The middle sketch shows the target positions.

The averaging results obtained from phase plots $(\dot{\alpha}, \dot{\beta})$ for all six subjects for all recorded movements, and the plan of reducing the number of scalings to simplify the man-FES interface led to the following method: it is possible to divide the workspace into twelve zones, and bring the hand into the vicinity of the target if the scaling law is used as a constraint. The characterization of somewhat overlapping zones is based on two parameters: direction and distance of the target with respect to the body position. The following terms and abbreviations are used to annotate the zones: (1) contralateral (c), ipsilateral (i), ipsilateral-frontal (if) and frontal (f) directions; and (2) distal (d), medial (m) and proximal (p) distances.

After applying the zoning and averaging of the scaling over all six study subjects, the set of 12

scalings to be used within the hierarchical nonanalytical controller for movements to/from the mouth is presented in Table 39.1. Similar tables were determined for other tasks of interest for daily living activities which belong to more transversal and circular movements.

3.2 Determination of the Stimulation Parameters

Figure 39.5 presents the method of determining the required pulse duration using an example. The plots are presented for a series of three pulse durations applied to the m. Triceps Brachii while the hand moved in a tilted without ($\gamma = 1.5$ rad) with no load attached (m = 0). The amplitude and frequency of the stimulation were constant at I = 60 mA, f = 20 Hz. The position of the shaded square (in the middle panel) is determined with two phase coordinates: the angular velocity $\dot{\beta} = 0.7$ rad/s and the increment $\Delta\dot{\beta} = 0.1$ rad/s. This size of the square is 0.05 rad/s $\times$ 0.05 rad/s.

This square is crossed by the recruitment curve obtained at the pulse duration of T = 120 μs (Figure 39.5, left and right panel); hence, the following rule is formed automatically:

IF ($\dot{\beta} = 0.7$ rad/s   AND $\Delta\dot{\beta} = 0.1$ rad/s
                                    THEN $T = 120$ μs   (39.1)

There are squares determined with the angular velocity and its increment that are not crossed with any recruitment curve. In these cases the triplet will

TABLE 39.1. Averaged scalings $C \pm$ the standard deviations determined for the movements from the mouth to various targets. The abbreviations: c—contralateral, f—frontal, if—ipsilateral–frontal, and i—ipsilateral directions; and d—distal, m—medial, and p—proximal distances determine 12 reaching zones.

DIRECTION	DISTANCE		
	d	m	p
c	0.38 ± 0.05	0.35 ± 0.06	0.50 ± 0.09
f	0.30 ± 0.04	0.25 ± 0.05	-0.6 ± 0.12
if	0.13 ± 0.02	0.11 ± 0.02	-1.0 ± 0.14
i	0.05 ± 0.01	-0.05 ± 0.02	-1.5 ± 0.01

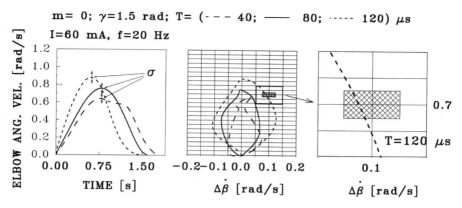

FIGURE 39.5. The elbow joint angular velocity vs. time (left), the elbow joint angular velocity vs. increments of the angular velocity (middle). A magnified detail from the mid-dle panel is shown in the right panel. The presented phase diagram is used for the determination of mapping between the angular velocity and the stimulation parameters.

be formed using the angular velocity $\dot\beta$ and its increment $\Delta\dot\beta$, and the pulse duration T from the nearest phase curve.

Table 39.2 shows a portion of the look-up table which is to be used at the actuator level for control.

3.3 The Reaching Performance with the FES Assistance

Figure 39.6 shows three different reaching tasks (A, B, and C) when a subject with tetraplegia was asked to reach without the assistance. The corresponding elbow and shoulder joint angles are presented at left, while the phase plots created from the joint angular velocities are in the right panels. The best linear fit used for these movements is shown for all three movements. The middle column presents the hand movement with respect to the external reference frame. The upper panels (A) show the movement from the frontal–distal point to the frontal–ipsilateral position, which is a circular motion. The middle panels (B) are for the motion from the ipsilateral-proximal position to the ipsilateral–distal position, and the bottom panels (C) are for the radial movement from the proximal–frontal position to the ipsilateral-distal position. Note that none of the given tasks was performed correctly, mainly because of the lack of controlled elbow extension.

The same subject, assisted with the FES system described in the Methods section performed the same set of movements depicted in Figure 39.6.

Figure 39.7 shows that the subject greatly improved the performance as measured through two elements: precision and speed. It is noteworthy that he now succeeded to fulfill the given tasks, but that the angular velocities, that is the speed of the movement, was still rather slow. The scaling coefficients presented in the right panels of Figure 39.7 show that the scalings are very much different from the scalings without the system.

TABLE 39.2. Part of the look-up table determining triplets to be used for stimulation of Triceps Brachii. Each of the rows is obtained from the phase diagrams formed of the angular velocity vs. the increment of the angular velocity. The pulse duration is the parameter in these plots (Figure 39.5).

IF		THEN
$\dot\beta$ [rad/s]	$\Delta\dot\beta$ [rad/s]	T [μsec]
0.4	−0.20	100
0.4	−0.15	80
0.4	−0.10	60
0.4	−0.05	20
0.4	0.00	0
0.4	0.05	30
0.4	0.10	80
0.4	0.15	120
0.4	0.20	150
0.1	−0.05	80
0.2	−0.05	60
0.3	−0.05	40
0.5	−0.05	40
0.6	−0.05	60
0.7	−0.05	80

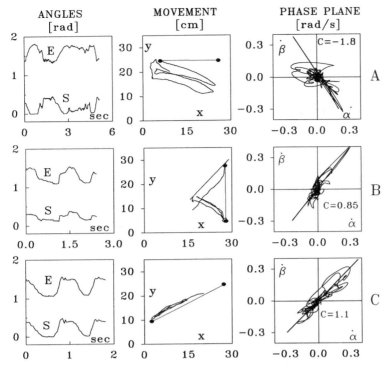

FIGURE 39.6. Joint angles (left), hand position (middle), and phase plots (right) for arm movements of a person with tetraplegia without any assistive system. The tasks were to slide the hand between the marked points (middle panels) on the digitizing board (cases A, B, and C). E and S are notations for the elbow and shoulder joint respectively. Details are described in Figure 39.2. Note that the subject was not able to perform the tasks.

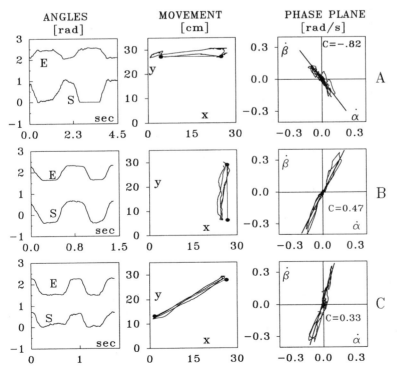

FIGURE 39.7. The same series of recordings presented in Figure 39.6, but with the difference that a FES assistive system was applied. The system is controlled with the non-analytical, hierarchical controller.

4 Discussion

Goal-directed, functional movements belong to the category of dynamical processes, with an infinite number of continuously variable states. The control in this class of tasks can be solved analytically only under definite sets of restrictive conditions. The analytical approach relies on the assumption that the optimal trajectory can be singled out from the infinity of options due to its particular features. The main difficulty arises from the fact that it is almost impossible to single out these particular features.

The requirement for so-called real-time control is essential. Biological plants evolve in time at a rate that is inherent to their nature. Processes in living systems are changing their states according to natural laws. The controller must match its decision speed to the dynamics of controlled objects. The term "real-time control" refers to the matching of decision dynamics and system dynamics. In biological systems, the real-time response to environmental changes can be a matter of survival. As the number of controlled variables increases, the processing and decision time of the controller grows in a nonlinear way. This sets a practical limit to the size of controllable plants, no matter how fast the controller reacts. Such a situation is typical of biological systems where a large inflow of sensory information has to be transmitted and processed by the multilevel nervous structure.

Nature has found nonetheless an elegant way to bypass the so-called curse of dimensionality. The answer to the size challenge is *multilevel control* and the principle of *hierarchy*. Applying such mechanisms, the limits to the growth of control tasks have been increased by many orders of magnitude with respect to the one level approach. The term "large system" refers in the control theory to those plants which generate so much state related information that the controller cannot process it in a straightforward manner.

The properties of *multilevel systems* can be interpreted in terms of general control concepts previously introduced. Multilevel control implies that: (1) the plant must be modeled at each control level in a more and more abstract way (i.e., by a reduced number of synthetic features); (2) each level has its own optimization criterion which encompasses the lower level criteria; (3) criterion functions of sub-systems may be conflicting, in which case the optimal solution is nonexistent; and (4) hierarchy implies that the higher level controller may override the lower level controllers and exercise direct control over the subsystems.

Control theory of large systems must deal with problems not encountered in the design of conventional control structures. How to derive synthetic information relevant for the higher level controllers, or how to synthesize the global optimization criterion out of the local ones, are but a few of the new problems that arise in the control of large systems. Instead of the variational calculus, value judgments are instrumental in the decision processes pertaining to large systems. Heuristics, creativity, domain-oriented experience, are the main approaches used by man in running multilevel organizations. Analytical tools and computer methods appear now not as a means of automatic control, but are subserving the decision maker.

From the control engineering point of view, human extremities appear as the plant involved in the execution of functional motions. Their modeling is thus the prerequisite for the synthesis of control when the dynamics of functional motions is in question. However, human extremities are unlike any other plant encountered in control engineering, especially in terms of joints, actuators and sensors. This fact must be kept in mind when applying the general equations of mechanics to model the dynamics of functional motions. A simple extension of the analytical tools used for the modeling of mechanical plants to the modeling of biomechanical systems may easily produce results sharply at variance with reality.

A basic problem in motor control of human extremities is in *planning* of motions to solve a previously specified task, and then, *controlling* the extremity as it implements the commands. In terms of mechanics, extremities are linked structures with muscles that can set the joint angle to any value in its range. Muscles can also set the desired angular velocity, acceleration, and joint stiffness by their co-contraction and reciprocal inhibition. The motions of extremities are called *trajectories* and consist of a sequence of positions, velocities and accelerations of any part of the system (e.g., hand, fingertip, etc.). It is anticipated that by using the laws of mechanics one can determine the necessary muscle forces to follow the given trajectory.

Biomechanical models of extremities use the equations for the systems of *rigid bodies*, that is, sets of material points with the distances between points being fixed. However, the body segment consisting of a bone, muscles, tendons, and ligaments can hardly be represented as a single rigid body. Inertial properties change in two ways: the center of mass shifts, and the inertia tensor varies with the position of the body part. Joints are modeled as *kinematic pairs*. The theory of mechanisms defines a connection of neighboring segments as a kinematic pair having up to five degrees of freedom (DOF) (see also Chapter 32). A typical human joint is a rotational kinematic pair having up to three DOF forming either a pin (one DOF) or a ball joint. Through evolution, nature evolved many different bone segment interfaces. There are joints that cannot be simplified to a pure pin or ball joint. Some joints are actually multiple joints (e.g., the ankle joint has one joint for internal–external rotation and inversion-eversion, and the second one for movements in sagittal plane (flexion–extension); likewise the elbow is formed as the contact of three bones).

Redundancy is an important feature in biological systems. For example, during rapid movements, some joints will be used, while during some slow postural changes an entirely different strategy will be used. Inverse kinematics involves also a mathematical problem due to the singularity of the model. Inverse kinematics relates to geometry, velocity, and acceleration.

Dynamic analysis needs good *muscle models*. Muscle is not a linear actuator, but exhibits very complex behavior. Many parallel and serial elements contribute to force generation of muscle, and they all depend on the length of the muscle, the velocity of shortening, the firing rate, recruitment and type of the given muscle (e.g., Zahalak 1992; see also Chapters 1, 2, 8, 7, 10, 33).

Dynamic analysis using principles of mechanics is applicable only for open-chain structures, such as an arm when reaching, but it is not applicable for closed-chain structures, such as the arm when contacting an object. Every closed-chain structure is dynamically undeterminable, and the only possible way to determine forces and torques is to use the theory of elasticity. Approximate solution of the dynamics of a closed-chain mechanism assumes how forces and torques are distributed within the structure.

Activation of preserved limb muscles by means of functional electrical stimulation requires a model which will take into account the characteristics of a multijoint, multiactuator system. In this case, the problems arise from the fact that there are many muscles that will not be externally controlled, but will affect the movements (stretch reflexes, cutaneous reflexes, etc.). The activated muscle model itself also has its limitations, and some important elements like muscle fatigue, partial denervation are often omitted completely. A description of agonist/antagonist muscle activity has to be limited to a much simpler system, typically a muscle equivalent determined as a sum of all muscles generating power around a single joint. This simplification is particularly difficult due to existence of biarticular muscles, i.e. muscles acting over two neighboring joints, leading to the model of the muscle activity directly in the form of joint torques (Winters 1990).

It is necessary for the synthesis of control to prescribe the *preferred trajectory* of the limb. Prescription of a trajectory includes that such a trajectory exists, that motion can be fully cyclic or task dependent and that the variation from trial to trial can be neglected.

For reaching and grasping the prescription of the trajectory is complicated, and is highly dependent on the task itself (e.g., see Chapters 23, 25–31). Many motor control studies up to now suggest various strategies, but all of them are applicable only to simple situations, where skills do not play any role (e.g., Soechting and Lacquaniti 1981; Hollerbach and Flash 1982; Moraso 1983; Flash and Hogan 1985; Hogan 1985; Zajac and Gordon 1989).

So far we have emphasized the limitations of analytical control when extended to modeling of biological phenomena. This attitude should not be interpreted, however, as denying the value to analytical methods for studying functional motions. The reason for insisting on inherent constraints of analytical models is to keep in mind that such approaches must be used with due caution.

Considerations of the complexity of models, and multiparameter optimization of the system, in addition to many neurophysiological findings on the role of reflexes and sensory driven control lead to the development of a nonparametric approach, where the model comprises the process of movement, not the plant itself. When speaking of nonanalytical model (Popovic et al. 1991) one regards the process as a black box with multiple inputs (e.g., sensors) and multiple outputs (e.g., joint states, muscle activations, etc.). Specifically, in the case study presented, the model for reaching includes a synergy being the property of the plant as the result of learning and habituation. The model reduces the dynamics and kinematics to a coded description in sensor/actuator space. The nonanalytical model refers to mapping of input–output characteristics for different functional, goal-directed movements. Some computing techniques, such as pattern mapping, are very suitable for designing of a controller. This requires to represent input and output and capture their correlation (Popovic and Popovic 1998). Research efforts to extend the knowledge representation to the domain of *reflex-* and *skill-based* control have been successful, and methods to capture this kind of expertise are now available.

Reflex control can be implemented by using the formalism of *production rules*. A production rule is a situation-action couple, meaning that whenever a certain situation is encountered, given as the left side of the rule, the action on the right side of the rule is performed. There is no a priori constraint on the form of the situation or of the action.

Principles of *rule-based programming* are: (1) each rule is an independent item of knowledge; (2) there is no mechanism anywhere else except in the rule itself that creates conditions that could prevent it from being applied; (3) rules are not ordered in any way and in principle any one can be activated at any moment; and (4) a rule system is able to look first at the established facts and to proceed forward (forward chaining) or to start from the aims and proceed backwards, i.e. from the action part of the rule (backward chaining). (Compare to the Soft Computing approach developed in Chapter 40 and the neurooptimization approach offered in Chapter 35.)

The very same nature of production rules can be found in biology. For many years segmental reflexes were studied in immobile animal preparations or in humans in static postures. This gave the impression that spinally mediated reflexes were invariant. Forssberg et al. (1975) showed that during locomotion in the chronic spinal cat, responses elicited by mechanical stimuli impeding the forward swing of the leg actually reversed when the stimuli were applied during the stance phase, and this finding was

further elaborated by others (e.g., Drew and Rossignol 1985; Dietz et al. 1989). The stumble responses are mainly cutaneously mediated, but the principle of phase-dependent reflex reversal has since been extended to other sensory modalities, including muscle proprioception (Pearson et al. 1992; see also Chapter 17). Phase-dependent reflex reversal may be viewed as a special case of *production rules control*, the rules being of the following type (Prochazka 1993): *IF* swing phase *AND* skin stimulus *THEN* lift and place leg; *IF* stance phase *AND* skin stimulus *THEN* extend leg and prolong stance. Similar rules were identified independently in cat, cockroach, lobster and stick insect (Prochazka 1993).

The rules discussed so far relate largely to sensory input. Centrally generated programs must also be taken into account. Swing phases are triggered even though the last rule given above is not satisfied. This may be explained in terms of the default operation of a central oscillator which cycles autonomously in the absence of sensory input. When the foot contacts the ground, sensory input is restored, and the oscillator is dominated by it, according to the last rule above. Other pattern generators under sensory control include those for respiration, mastication (Lund et al. 1984) and pawshaking (Koshland and Smith 1989). Postural responses to translation of ground support may also be viewed as default programs (MacPherson 1988). IF-THEN rules simply provide a decryption of the associations of inputs and outputs in movements which have multiple phases and are subject to postural constraints.

In the current application, the mappings were used at all three levels of control. At the highest level the visual information is mapped to the scaling between the shoulder-joint angular velocity and the elbow-joint angular velocity. This mapping results in an infinite number of scaling coefficients that are universal (i.e., independent from the subject). This research simplifies the task by taking into account that the shoulder position can be varied volitionally; thus, the fine positioning can be done independently from the shoulder and elbow rotations. The actual mapping, as shown in Figure 39.4, is nonlinear, but in the approximation it was limited to the best linear fit. The reduced mapping model allows the reduction of the number of scaling coefficients to the number which can be learned fast, and for which a simple, intuitive interface can be designed. As shown in Table 39.1, the workspace of interest for a person with tetraplegia is divided in such a way that 12 somewhat overlapping zones with 12 corresponding scalings allow the user of the assistive system to perform daily-living activities. The other mapping defines the stimulation parameters that are control inputs to the elbow-joint angular velocity. This mapping, acting at the actuator level, is individual, and the differences can be found from day to day, or even during one application in the same person with tetraplegia. This variation of the elbow angular velocity to the same stimulation pattern requires some feedback. The averaged mapping serves only as the approximation, and provides a so-called preferred trajectory, but the sampled feedback provides tracking of that trajectory in the velocity space. The mappings at the coordination level are the transferred biological synergies: the adjustments of the elbow-joint angular velocities to match the scaled shoulder-joint angular velocities.

The realization of the knowledge base of both mappings was done by using human expertise and numerous experiments in able-bodied persons and persons with tetraplegia. The implementation of the assistive system is interactive. The usage of the assistive system changes the motor performance for at least these four reasons: (1) new motor schemes are being developed; (2) muscles are strengthened; (3) range of movements is increased; and (4) level of tonic spasticity is changed. These changes demand the modification of the mappings at the lower level, while the scalings remain invariant during the usage of the system.

5   Future Directions

As previously stated, the most important element in the actual organization of a skill-based expert system relates to the design of rules. For many activities there is no knowledge developed in the form suitable for transfer to a machine. Most of the recent skill-based expert systems use hand-crafted rules (i.e., the expertise developed through biomechanical and neurophysiological studies is formalized). Basic problems relate to the fact that a complete understanding of optimization methods is lacking (see Chapter 35, for review), so the designer selects his preferences. The second issue when designing rules is that in many cases there

are no valid methods to assess and compare the result with other methods.

Two methods can be used to help resolve these difficulties. It is feasible to use the artificial neural network or fuzzy logic network approach and resolve the problem of mapping without actually understanding what connectivisms (Arbib 1986) are used for mapping. The real answer to the problem of generating production rules is the use of inductive learning techniques (Jonic et al. 1999; Popovic 1993; Nikolic and Popovic 1994).

Knowledge representation by artificial neural networks (ANNs) is based on completely different principles than the ones built into the operation of production rule systems. In ANNs the connection patterns and the weighing factors are the substrate for knowledge representation. The hardware background for knowledge representation by connectionism is easy to grasp, although the mathematical aspects of the operation of ANNs are quite sophisticated.

There is no general neural network that is able to store arbitrary domains of knowledge. Once the architecture of the dedicated network and the node functions have been fixed, the only adjustable variables for knowledge storage are the weighting factors between the nodes. One way to store the knowledge into the neural network for control of movement has been demonstrated (Jonic et al. 1999; Jovovic et al. 1999; Popovic 1993; Lan et al. 1994; Nikolic and Popovic 1994; Abbas and Chizeck 1995; Kostov et al. 1995; see Chapter 35). There is no guarantee that an iterative minimization procedure will converge. Such a convergence procedure makes no distinction between the local minima and the global minimum.

A neurophysiologist will easily seize the difference between a network of neurons and an ANN. The substrate for neural networks is electrical waveforms, as opposed to electrochemical phenomena in biology. In sharp contrast to the remarkable plasticity of the nervous system, the computer network is a rigid structure. The nodes of the neural network perform a fixed input-output mapping with the weighting factors being the only modifiable elements. This can be compared with the complexity of the electrical and chemical processes taking place in neurons and synapses during every activation event. It was already emphasized that systems with non-repeatable states can be modeled only with great precautions by systems having erasable memory. In spite of these qualitative differences between the biochemical and computer signal processing, an important lesson for the understanding of the organization of motor control can be derived from the study of neural networks. The connectionism and modifications of the interaction strength between network elements are instrumental in knowledge representation.

Comprehensive synthesis methods, as developed in control theory, are not available for ANN since the connectionism and performance requirements have not been related in a general way (Hebb 1949; Fukushima 1980; Hopfield 1982; Kohonen 1982; Rumelhart and McClelland 1986). Heuristic approaches are basically involved in the design of ANN. For instance, one cannot know in advance how many times the weighting factors must be readjusted for a given application so that the input not used for training will be properly classified. In a nonexpert view of this problem, the connectionists' approach is believed to be capable of solving problems where we do not know the model of the system. Although the computer may provide consistent answers, the level of generalization may be very low, because the connectionism does not follow the process, finding a different set of connections satisfying the single training input-output set.

Acknowledgments. The work on this project was partly funded by the Miami Project to Cure Paralysis, University of Miami School of Medicine, Miami, Florida, U.S. and the Ministry for Science and Technology of Serbia, Belgrade, Yugoslavia.

References

Abbas, J. and Chizeck, H.J. (1995). Neural network controller of functional neuromuscular stimulation systems: computer simulation study, *IEEE Trans. Biomed. Eng.*, BME-42:1117–1127.

Arbib, M.A. (1986). *Brain, Machines and Mathematics*. Springer-Verlag, New York.

Bernstein, N.A. (1967). *The Coordination and Regulation of Movements*. Pergamon Press, Oxford.

Dietz, V., Horstmann, G.A., Trippel, M., and Gollhofer, A. (1989). Human postural reflexes and gravity—an under water simulation, *Neurosci. Lett.*, 106:350–355.

Drew, T. and Rossignol, S. (1985). Forelimb responses to cutaneous nerve stimulation during locomotion in intact cats. *Brain Res.*, 329:323–328.

Flash, T. and Hogan, N. (1985). The coordination of the arm movement: an experimentally confirmed mathematical model. *J. Neurosci.*, 5:1688–1703.

Forssberg, H., Grillner, S., and Rossignol, S. (1975). Phase dependent reflex reversal during walking in chronic spinal cats. *Brain Res.*, 85:103–107.

Fukushima, N. (1980). Neocognitron: a self-organization neural network model for a mechanism of pattern recognition unaffected by shift in position. *Biol. Cybern.*, 36:193–202.

Georgopoulos, A.P., Ashe, J., Smyrnis, N., and Taira, M. (1992). The motor cortex and the coding of force. *Science*, 256:1692–1695.

Georgopoulos, A.P., Caminiti, R., Kalaska, J., and Massey, J.T. (1983). Spatial coding of movement: a hypothesis concerning the coding of movement direction by motor cortical populations. *Exp. Brain Res. Suppl.*, 7:327–336.

Hebb, D.O. (1949). *The Organization of Behavior*. John Wiley & Sons, New York.

Hecht-Nielsen, R. (1991). *Neurocomputing*. Addison-Wesley Publishing Co.

Hogan, N. (1985). The mechanics of multijoint posture and movement control. *Biol. Cybern.*, 52:325–332.

Hollerbach, J.M. and Flash, T. (1982). Dynamic interactions between limb segments during planar arm movement. *Biol. Cybern.*, 44:67–77.

Hopfield, J.J. (1982). Neural networks and physical systems with emergent collective computational properties. *Proc. Natl. Acad. Sci. U.S.A.*, 79:2554–2558.

Jeannerod, M. (1988). *The Neural and Behavioral Organization of Goal-directed Movements*. Oxford Physiology Series, No. 15, Clarendon Press, Oxford.

Jonic, S., Jankovic, T., Gajic, V., and Popovic, D. (1999). Three machines learning techniques for automatic determination of rules to control locomotion. *IEEE Trans. Biomed. Eng.*, BME-46:300–310.

Jovovic, M., Jonic, S., and Popovic, D. (1999). Automatic synthesis of synergies for control of reaching—hierarchical clustering. *Med Eng. Phys.*, 21(5):329–341.

Kohonen, T. (1982). Lecture notes in biomathematics. In *Competition and Cooperation in Neural Nets*. Mari, S. and Lara, M.A. (eds.), vol. 45. Springer-Verlag, New York.

Koshland, G.F. and Smith, J.L. (1989). Paw-shake responses with joint immobilization: EMG changes with atypical feedback. *Exp. Brain Res.*, 77:361–373.

Kostov, A., Andrews, B., Popovic, D.B., Stein, R.B., and Armstrong, W.W. (1995). Machine learning in control of functional electrical stimulation (FES) for locomotion. *IEEE Trans. Biomed. Eng.*, BME-42:542–551.

Lan, N., Feng, H., and Crago, P.E. (1994). Neural network generation of muscle stimulation patterns for control of arm movements. *IEEE Trans. Rehabil. Eng.*, TRE-2:213–223.

Lund, J.P., Sasamoto, K., Murakami, T., and Olsson, K.A. (1984). Analysis of rhythmical jaw movements produced by electrical stimulation of motor-sensory cortex of rabbits. *J. Neurophysiol.*, 52:1014–1029.

MacPherson, J.M. (1988). Strategies that simplify the control of quadrupedal stance. II. Electromyographic activity. *J. Neurophysiol.*, 60:218–231.

Morasso, P. (1983). Three dimensional arm trajectories. *Biol. Cybern.*, 48:187–194.

Nikoli, Z. and Popovic, D. (1994). Automatic rule determination for finite state model of locomotion. *Proceedings IEEE Annual Conference EMBS*, vol. 4:1234–1235. Baltimore, Massachusetts.

Pearson, K.G., Ramirez, J.M., and Jiang, W. (1992). Entrainment of the locomotor rhythm by group Ib afferents from ankle extensor muscles in spinal cats. *Exp. Brain Res.*, 90:557–566.

Peckham, P.H. and Keith, M.W. (1992). Motor Prosthesis for restoration of upper extremity function, In *Neural Prostheses: Replacing Motor Function after Disease or Disability*, Stein, R., Peckham, P., and Popovic, D. (eds.), pp. 162–190. Oxford University Press, New York.

Popovic, M. (1995). *A New Approach for Control of Reaching in Quadriplegic Subjects*, Ph.D. thesis, University of Belgrade.

Popovic, D. (1993). Finite state model of locomotion for functional electrical stimulation systems. *Progr. Brain Res.*, 97:397–407.

Popovic, M. and Popovic, D. (1994). A new approach to reaching control for tetraplegics. *J. Electromyog. Kinesiol.*, 4:242–253.

Popovic, D. and Popovic, M. (1995). Control of reaching for tetraplegic subjects. *Proc. II Intern. Symp. on FES*, pp. 166–173, Sendai, Japan.

Popovic, M. and Popovic, D. (1995). Enhanced reaching in tetraplegics by means of FES. *Proc. of the V Vienna Conf. on FES*. pp. 347–350, August, Vienna.

Popovic, D. and Popovic, M. (1998). Tuning of a nonanalytical hierarchical control system for reaching with FES. *IEEE Trans. Biomed. Eng.*, BME-45:203–212.

Popovic, M. and Tepavac, D. (1992). Portable recording unit for gait assessment. *Proceedings 12 IEEE Annual Conference on EMBS*, pp. 1646–1647, Paris, France.

Popovic, D., Tomovic, R., Tepavac, D., and Schwirtlich, L. (1991). Control aspects an active A/K prosthesis. *Intern. J. Man-Machine Studies*, 35:751–767, Academic Press, London.

Prochazka, A. (1993). Comparison of natural and artificial control of movement. *IEEE Trans. Rehabil. Eng.*, TRE-2:7–22.

Rumelhart, D. and McClelland, J.L. (1986). *Parallel Distributed Processing: Exploration in the Microstruc-* *ture of Cognition.* The MIT Press/Bradford Books, Mass.

Soechting, J.F. and Lacquaniti, F. (1981). Invariant characterics of pointing movement in man. *J. Neurosci.*, 1:710–720.

Soechting, J.F., Lacquaniti, F., and Terzuolo, C.A. (1986). Coordination of arm movements in three-dimensional space. Sensorimotor mapping during drawing movement. *Neuroscience*, 17:295–311.

Tomovic, R. and Bellman, R. (1970). Systems approach to muscle control. *Mat. Biosci.*, 8:265–277.

Tomovic, R., Popovic, D., and Stein, R.B. (1995). *Nonanalytical Methods for Control of Movements.* World Sci. Publ., Singapore.

Triolo, R., Nathan, R., Handa, Y., Keith, M., Betz, R.R., Carroll, S., and Kantor, C. (1996). Challenges to clinical deployment of upper limb neuroprostheses. *J. Rehab. Res. Dev.*, 33:111–122.

Wijman, A.C., Stroh, K.C., van Doren, C.L., Thrope, G.B., Peckham, P.H., and Keith, M.W. (1990). Functional evaluation of quadriplegic patients using a hand neuroprosthesis. *Arch. Phys. Med. Rehabil.*, 71:1053–1057.

Winter, D. (1990). *Biomechanics and Motor Control of Human Movements.* John Wiley & Sons, New York.

Winters, J.M. (1990). Hill-based muscle models: a systems engineering perspective. In *Multiple Muscle Systems—Biomechanics and Movement Organization*. Winters J.M. and Woo, S.L.Y. (eds.), pp. 69–73. Springer-Verlag, New York.

Zahalak, G.I. (1992). An overview of muscle modeling. In *Neural Prostheses: Replacing Motor Function After Disease or Disability.* Stein, R.B., Peckham, P.H., and Popovic, D.B. (eds.), pp. 17–57. Oxford University Press, New York.

Zajac, F.E. and Gordon, M.E. (1989). Determining muscle's force and action in multi-articular movement. In *Exercise Sport Sci Rev.* Pandoff, K. (ed.), 17:187–230. Williams and Wilkins, Baltimore.

Commentary:
A Case for Soft Neurofuzzy Controller Interfaces for Humans with Disabilities

Jack M. Winters

The authors' provide an intriguing example of how a rule-based nonanalytic control method can be used to successfully provide what appears to be a robust controller. This approach differs considerably from the traditional control systems approach in which an adaptive controller is based on a linearization around a model operating range (e.g., see Chapter 11). This distinction is important: fundamentally, the controller is based on a set of heuristic, *expert-supplied rules* that are set up as conditional if–then statements; it is an expert system, with its roots in conventional artificial intelligence (AI). The experts in this case are scientists, with the embedded knowledge based on scientific observation (e.g., an observed arm synergy). In their case, a conventional tool—the phase plane—is used to help visualize and refine the production rules, and more explicitly to partition the input space into a set of regions.

A concern is that such rules are inherently crisp—they either fire or they do not (see Table 39.2). This leads to inherent discontinuities at (hard) boundaries (Jang et al. 1997). Optimization search algorithms—including ANNs—tend to have trouble when the search space is inherently discontinuous, as is the case with crisp (hard) rule production algorithms. This seems unfortunate—not only do we lose ties to optimization tools, but rarely are human judgments crisp to begin with. From this perspective fuzzy set approaches are intriguing: rules can partially "fire" (soft boundaries), and systematic approaches are available for synthesis of rules. (Of course, without some care the "curse of dimensionality" many creep up more quickly, at least from a computational perspective.)

Another concern involves extension to more complex systems, where visualization techniques such as the phase-plane plots may not be available to help yield expert insight (e.g., what if the multidimensional shoulder joint also had to be controlled). Additionally, in general many rules can (and for robust performance probably should) be "firing" simultaneously—how does one then synthesize such information? Expert systems such as NASA's CLIPS employ two ad-hoc extensions to help the inference engine handle this problem: rules are given a salience (priority) and a level of certainty. In dealing with this, the authors' suggest a role for hierarchy: higher level controllers may override lower level controllers and exercise direct control. Such gating of some type has often been proposed by neuroscientists studying brain processes (e.g., Chapter 23 and Chapter 34; see also Bullock and

Grossberg 1988). Yet then one may lose the benefit of "the sum being greater than the parts" via the inference process. Interestingly, the authors' note that "value judgments are instrumental in the decision processes pertaining to large systems."

Classic concerns such as these have provided the motivation for fuzzy inference systems development, and what is called Soft Computing (SC) in the next chapter (Chapter 40). Fuzziness differs from probability (which is concerned with the likelihood of nondeterministic, stochastic events) in that in deals with *deterministic plausibility* (i.e., the vagueness (ambiguity, subjectivity) associated with a *concept*). Fuzzy rules have inherently soft boundaries, which has many advantages (e.g., better ties to human inference; gradient calculation; more robust synthesis algorithms).

Thus, this commentator, while agreeing in principal with the "let's wait-and-see and not promise too much" attitude of Tomovic toward new SC approaches in his commentary following Chapter 40, has to respectfully disagree. Why? Because fuzzy (and neurofuzzy) systems have three complementary aspects going for them: (1) a scientifically based, mathematically rigorous *theoretical* framework that is continuously expanding (e.g., crisp rules are a subset of fuzzy rules; probability theory is a subset of fuzzy-based possibility theory; linear control systems are a special case of fuzzy control when certain constraints are placed on the form of membership functions; radial basis functions are a special subset of a certain fuzzy set formulation); (2) a proven *commercial* track record (literally billions of dollars of commercial goods—especially out of Japan—with embedded fuzzy controllers); and (3) the remarkable growth of interest in *integrated neural–fuzzy systems* which take advantage of the strengths of each approach (e.g., fuzzy inference systems with automatic tuning abilities), complemented by an ever-expanding set of engineering textbooks and journals devoted to this topic (suggested, Jang et al. 1997; Lin and Lee, 1996). Even more importantly, SC approaches provide a strong nonanalytic foundation for subjective interpretation of performance. For example, for the reaching with FES assistance tasks of subsection 3.3, the measures that we are asked to consider in Figure 39.7—precision and speed—are somewhat imprecise, and were interpreted by the authors' with the type of fuzzy qualifiers that are part of natural language ("rather slow scalings are very much different").

Nonetheless, it is admitted that creating effective SC systems that embed top-down human reasoning can prove challenging (e.g., rarely do experts fully agree, as seen throughout this book), and involves not only formulating the rules but also determining an appropriate modular neurostructure and choosing how much bottom-up adaptive tuning is allowable (see also Chapter 35). There is also always the issue of "curse of dimensionality"—but here is where scientific insights into synergies (e.g., scaling laws, invariants—see also Chapter 27, Chapter 19, Chapter 22, Chapter 31) can prove useful. Notice that through SC, it is an least possible for top-down hand-crafted expert rules and bottom-up neural learning to coexist. This is an important point. Also, modular, hierarchical structures evolve naturally when fuzzy rules are embedded in ANNs. In essence, through SC, the AI and connectionist approaches can finally converge. Yes, as noted by Tomovic in his commentary on Chapter 40, it is true that to some extent these types of approaches have been around for a while, and SC is in part just repackaging. But now there exists a theoretical foundation, and tremendous academic and commercial momentum outside of neuromotor science to which we can attach.

In their discussion, the authors' propose multilevel hierarchical control, and provide four implications. I agree with three of these, but with one—"criterion functions of subsystems may be conflicting, in which case the optimal solutions is nonexistant"—I would disagree. In real situations conflicting subcriteria happen all the time (e.g., speed versus accuracy or performance versus effort). Indeed, we have found that the most biologically plausible optimization solutions (i.e., neurocontroller signals similar to EMGs) evolve when we use both task-based and effort-based subcriteria that are competing (Seif-Naraghi and Winters 1990; see also Chapter 35). Additionally, multiobjective optimization, aiming for "satisficing" near-optimum solutions, is a rich field within engineering design that ironically includes both rule production algorithms and fuzzy sets among its arsenal of tools (e.g., Sakawa 1993).

With regard to designing assistive systems for humans with disabilities, the nonanalytic foundation based in rules production (and Soft Computing) is very appealing because it is inherently centered on human reasoning (by researchers, clinicians, disabled users) rather than on conventional engineering-centered controllers. This is especially important for technologies in rehabilitation, where for a typical application it is critically important that the controller be robust and dependable for a broad range of applications, with some level of imprecision quite tolerable; certainly this chapter provides an excellent application of this approach.

References

Bullock, D. and Grossberg, S. (1988). Neural dyanmics of planned arm movements: emergent invariants and speed-accuracy properties during trajectory formation. *Psychol. Rev.*, 95:49–90.

Jang, J.-S.R., Sun, C., and Mizutani, E. (1997). *Neuro-Fuzzy and Soft Computing. A Computational Approach to Learning and Machine Intelligence.* Prentice Hall, Upper Saddle River, New Jersey.

Lin, C.-T. and Lee, C.S.G. (1996). *Neural Fuzzy Systems. A Neuro-Fuzzy Synergism to Intelligent Systems.* Prentice Hall, Upper Saddle River, New Jersey.

Sakawa, M. (1993). *Fuzzy Sets and Interactive Multiobjective Optimization.* Plenum Press, New York.

Seif-Naraghi, A.H. and Winters, J.M. (1990). Optimal Strategies for Scaling Goal-Directed Arm Movements. In *Multiple Muscle Systems* (Winters, J.M. and Woo, S.L-Y. (eds.), Chapter 19, pp. 312–334. Springer-Verlag, New York.

40
Soft Computing Techniques for Evaluation and Control of Human Performance

Ron Jacobs and Carole A. Tucker

1 Introduction

The mystical beauty of our nervous system's ability to explore and learn new motor behavior is nicely demonstrated by how newborns and toddlers develop new motor skills. Through extensive periods of "oops" and "wow" learning, the nervous system has learned to control the sensitivities of how different muscles work together in affecting movement at various joints. This learning process starts before birth and continues throughout life. Unfortunately, impairment of the neuromuscular system, whether from injury or disease, may result in a disability of motor performance. Therefore, studying neuromuscular control is eminently important, not only from a scientific point of view to gain better insight into the mysteries of how the nervous system learns and controls movements, but also in studying restoration of movement after injury or disease.

Current applications of knowledge gained from the study of neuromuscular control include:

- *Artificial stimulation for standing and walking control in paraplegics.* By means of functional electrical stimulation (FES), paralyzed muscles can be used functionally (see also Chapters 39–41). To accomplish this, we must know which control strategies should be used for functional standing and walking.
- *Development of adaptive intelligent limb prosthesis.* The design of an ideal prosthetic limb to optimize functional standing and walking must be based on mechanical models as well as models of neuromuscular function (see also Chapter 38).
- *Intervention in rehabilitation, for instance in stroke*

patients. Insight into neural control strategies of standing and walking can be applied to effectively improve evaluation and treatment exercises performed by physicians and physical therapists (see also Chapter 37).

Human movement can be quantified in terms of kinetic, kinematic or electromyographic measures. However, it is still not precisely clear which strategies are utilized by our nervous system for movement control (Chapter 22). Traditional engineering control theory was developed primarily for inanimate electromechanical systems. Application of these traditional techniques, combined with traditional, serial, bivalent-logic based, "hard" computation has been unable to adequately identify and quantify the non-linear behavior and dynamics of the human neuro-musculoskeletal system. Traditional "hard" computation does not resemble human nervous system behavior, especially its temporal adaptation and learning of new motor skills. In addition, human decision making and control of movement does not generally rely on precise inputs and outputs. So the use of "hard" computation to model human neuromuscular performance constrains the model or control system to unhumanlike precise, bivalent inputs and outputs. Thus not surprisingly, even for standing and walking, motor tasks which have been studied extensively for decades with these traditional hard computing techniques, no unified description of control parameters or mechanisms exists yet. What is needed is a technique that is able to:

- Explicitly identify and describe the non-linear behavior of human neuromuscular skeletal dynamics in a meaningful manner.

- Model and simulate the adaptive and learning behavior of a control strategy that resembles human
nervous system behavior.

This chapter proposes the use of soft computing
techniques as a new and unique approach for evaluation and control of human performance. These
techniques are relatively new in comparison to
more traditional approaches, and we suggest that
they offer several advantages in application to human neuromotor control and evaluation of human
motor performance.

2   What Is Soft Computing?

The term applied to traditional mathematical techniques in analysis and modeling is hard computing
(Zadeh 1994). In hard computing, the prime desider-
ata are precision, certainty, and rigor. In contrast,
soft computing does not demand such a high level
of precision and certainty. The point of departure in
soft computing is the thesis that precision and certainty carry a cost and that computation, reasoning,
and decision making should exploit—wherever
possible—the tolerance for imprecision and uncertainty (Zadeh 1994). Soft computing uses the human
system as a role model and at the same time aims at
formalizing the neural processes humans employ so
effectively in the performance of daily tasks.

Soft computing is synonymous with the term
neurocomputation, and these terms are often interchangeable. Neurocomputation is defined as the
technological discipline concerned with parallel,
distributed, adaptive information processing systems
that autonomously develop operational capabilities
in adaptive response to an information environment
(Hecht-Nielsen 1991). For the sake of clarity, the
term soft computing will be used throughout this
chapter.

The principle constituents of soft computing are
fuzzy logic, artificial neural networks, and probabilistic reasoning (e.g., genetic algorithms) (see e.g.,
Zadeh 1994) (Figure 40.1). In general, fuzzy logic
is concerned with imprecision, neural networks with
learning, and probabilistic reasoning with uncertainty. There are substantial areas of overlap, and
these soft computing techniques are complementary
rather than competitive (Zadeh 1994), with each
providing a different set of advantages/disadvan-

Principle Constituents of Soft Computing

Probabilistic reasoning
- *uncertainty* -

Neural networks   Fuzzy logic
- *learning* -   - *imprecision* -

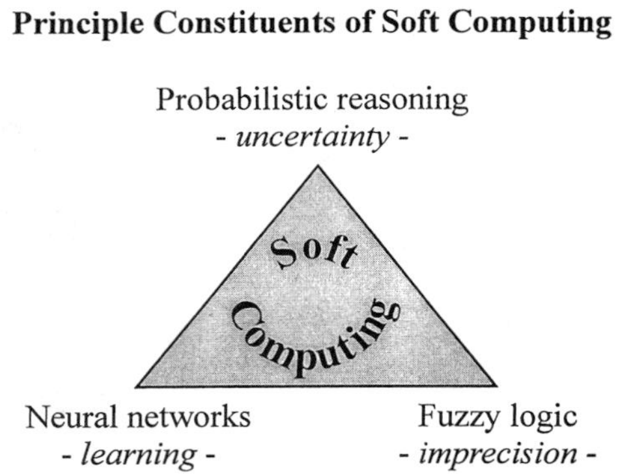

FIGURE 40.1. The constituents of soft computing techniques are relatively new in comparison to the traditional
mathematical hard computing techniques. Soft computing uses the human system as a role model and aims at
formalizing the neural processes human employ so effectively in the performance of daily tasks. Utilizing the
advantages of these soft computing techniques, it is
hoped to adequately identify and quantify the nonlinear
behavior of neuromuscular dynamics.

tages in their performance. For this reason, it is frequently advantageous to employ combinations of
the three techniques rather than use one exclusively.
For instance, a combination of neural networks and
fuzzy logic, called neurofuzzy, has proven its power
in many different complex and highly nonlinear control problems via mainly industrial applications.
Neuro-fuzzy techniques provide explicit and meaningful identification and quantification of control
variables, robustness, and adaptive behavior in systems such as automatic subway control (Sendai,
Japan), car transmission control (e.g., Nissan), dock-
ing control of space shuttle in orbit (Nasa, Berenji
by personal communication), photo camera operation (e.g., Canon, Minolta), and so on (see e.g.,
Kosko 1993; Meier et al. 1994; von Altrock 1995).
These approaches are similar in that fuzzy rule based
systems are combined with neural networks for
adaptations and/or learning of the rules.

Until recently, few reports have been published
utilizing soft computing applied to the field of biomechanics and motor control. Neural networks
have been used for prosthesis control (Kelly et al.
1990; Graupe 1995), for modeling EMG and joint
dynamics during gait (Sepulveda et al. 1993), for
classification of vertical ground reaction forces
(Holzreiter and Kohle 1993; Gioftsos and Grieve

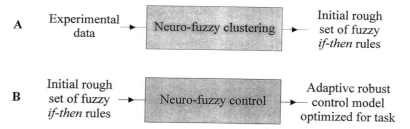

FIGURE 40.2. The combined advantages of hybrid fuzzy logic and neural network systems can be developed into a neurofuzzy technique to study neuromuscular control. Two different phases can be distinguished: (A) Neuro-fuzzy clustering of large, nonlinear experimental data sets that produces the relationship between input and output variables in terms of fuzzy sets and rules; (B) Neurofuzzy control that develops into an adaptive, robust control model by tuning and updating the fuzzy sets and rules.

1995), and for modeling control and learning of voluntary multijoint trajectories (Kawato et al. 1987, 1990; Kawato 1990; Wada and Kawato 1993). Fuzzy logic applied in biomechanics has recently been reported in a few abstracts for clustering of gait analysis data (O'Malley et al. 1995), modeling upper body balance and control (Kubica et al. 1995), and modeling human stance control (Jacobs et al. 1996).

3   New Approach to Study Neuromuscular Control

Using the advantages of these relatively new soft computing techniques, we propose that applications of these techniques for evaluation, quantification, and modeling of human performance will provide additional insights to those gained from more traditional hard computing techniques. Fuzzy logic is a perfect tool to model nonlinear behavior in terms of natural language, which provides explicit interpretation. Neural networks can be used for pattern recognition in large nonlinear data sets and for modeling of the neuromotor control system. The combined advantages of hybrid fuzzy logic and neural network systems can be developed into a neuro-fuzzy technique that is able to:

- *Distill natural groupings of data from a large nonlinear data matrix producing concise representation of the system's control behavior in terms of natural language (i.e., neurofuzzy clustering).* The idea is to lump together data points that populate the multidimensional and nonlinear space of neuromuscular control into a specific number of different control variables. The outcome of the neurofuzzy clustering is a set of fuzzy rules which are constructed in terms of if-then statements resembling a control strategy (e.g., Bezdek 1993; Figure 40.2A).

- *Model the control, learning and adaptive behavior of human neuromuscular skeletal dynamics in a manner that resembles nervous system function during movement (i.e., neurofuzzy control).* The idea is to develop a control model that consists of (1) a fuzzy controller, which is based on the set of fuzzy rules obtained from clustering; and (2) a neural network which acts as a "teacher" to the control model (e.g., Berenji and Khedkar 1992; Jang and Sun 1995) (Figure 40.2B).

The techniques of neurofuzzy clustering and neuro-fuzzy control can then be used in developing combined neural network and fuzzy logic models for quantification and modeling of complex human movements.

4   Principle Constituents of New Approach

What exactly are neural networks and fuzzy logic, and why are these two often integrated into a hybrid neurofuzzy approach? We will briefly discuss the principle constituents that play a role in the neuro-fuzzy approach, and the advantages and disadvantages of these approaches over traditional hard computation approaches.

4.1   Artificial Neural Networks

Artificial neural networks (ANNs) involve large, parallel and adaptive interconnected networks of relatively simple, often nonlinear processing elements (Figure 40.3). Applications of neural networks are often characterized by a high-dimensional feature space, complex interactions between variables, and a solution set that may have a single or multiple solutions (Schalkoff 1992). Neural networks do not require specification of a formal algorithm for classification as in conventional programming. Rather, they require specification of the arrangement of the processing elements and their interconnections. A neural network approach requires no constraints or predetermined rules. The network is adaptive and settles into its own classification scheme based solely on repeated exposures to a representative set of data during the training process. In addition, a neural network approach is capable of handling extensive inputs with ease because of the parallel nature of its computation. This computational style is more similar to our nervous system processing and may therefore be more representative of the human nervous system. In addition, like the human nervous system, neural networks can generalize and often provide correct classification even when faced with an unknown input.

The characteristics of individual units, network's structure or topology, and the training strategies

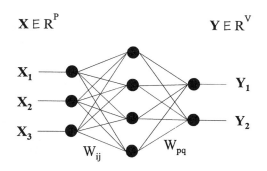

FIGURE 40.3. An example of an artificial neural network. The network is based on collections of "neurons" connected in multiple layers. The neural network learns to map an input X to an output Y by changing the weights W at each neuron. In a neurofuzzy system, the fuzzy sets and rules can be written as a multilayered neural network. Each layer then represents the different phases that underlie a fuzzy system.

need to be defined by the user. The individual processing elements or neurons in an artificial neural network are similar to biological neurons in that they receive inputs, summate or process the inputs for a single net effect, and provide an output based on their activation function. The activation function of a neuron essentially maps the neuron's inputs into an output. These activation functions have a typically linear or sigmoidal relation. The number of inputs and the activation functions, as well as the total number of neurons in a network, are specified by the designer. Individual neurons are often arranged in groups or layers, and the number of neurons and layers affect the classification ability of the network, the training time, and the computational complexity. In biological nervous systems, the arrangement and connections between neurons play a larger role in determining function than the behavior of an isolated neuron. The connectivity between neurons is an important design consideration in artificial neural networks since the connectivity determines much of the performance characteristics of the ANN. Interconnections and the directionality of the connection can be specified between any set of neurons, within or between layers.

In conjunction with the architectural design of the neural network, the training approach used to "teach" the ANN needs to be defined. The values of the interconnections, or weights, are modified from their initial values during the network's training. Training allows the network to correctly classify new or similar relationships between inputs and outputs, and the ANNs knowledge is thus stored in the values of the network's interconnections. Training can be accomplished using supervised, unsupervised, or reinforcement learning (e.g., Barto 1991; Werbos 1991; Berenji and Khedkar 1992; Jang 1992; Kosko 1992). The type of training used in the development of a neural network has a significant impact on the performance characteristics of the neural network. Whatever training type is used, specific learning rules are used to adjust the value of the weights, and the choice learning rule is dependent on the application, training type, desired outcome, and types of inputs and outputs (Zurada 1992).

In supervised learning, knowledge of output classification for each input is used to adjust the interconnection values. Supervised training is analogous to human learning when a "teacher" provides the correct answer as specific, immediate feedback to improve a student's performance. In unsupervised learning, no teacher is available that could provide knowledge of the output classification for each input. When provided with a set of patterns, the network will cluster the patterns based on similarities/differences between the patterns according to a prespecified learning rule. Unsupervised learning can provide a more realistic clustering of patterns than one defined or expected by the user. Self-organizing networks, such as clustering networks and Kohonen feature maps, are examples of unsupervised learning. Reinforcement learning provides the network with either an immediate or delayed evaluative measure of its performance which is maximized through trial and error (Barto 1994). Reinforcement learning requires less extensive sets of data than supervised learning since specific feedback is not provided, and often requires less training time than unsupervised networks since some measure of performance is provided. Reinforcement learning is consistent with biological learning principles and models (Barto 1994), and has recently been used in optimal control and real time dynamic programming applications (Barto et al. 1995).

In comparison to more traditional techniques, these neurocomputational approaches demonstrate the ability to adjust to changes or new conditions, tolerance to imprecise or noisy data, optimal pattern classification, and improved computational efficiency due to the parallel nature of computation (Bezdek and Pal 1992). Neural network software is readily available and can be run on microcomputers making the use of neurocomputational approaches in rehabilitation applications possible in most research or clinical settings.

Neural networks have multiple computational advantages over traditional computational approaches, yet, the neural networks algorithm is not transparent and the neural network acts in essence as a "black box." Although it is possible to gain insight into the ANNs algorithm by inspection of the interconnections' values, this task is near impossible for even moderately large networks where hundreds of interconnections may exist. Another potential disadvantage of ANNs is their bivalent nature being based on traditional logic theory and probability models. The integration of fuzzy logic with neural networks helps to overcome both of these disadvantages.

4.2 Fuzzy Logic

Fuzzy logic is based on reasoning with words and natural language instead of numbers and equations. A fuzzy set is an extension of an crisp set (Zadeh 1965). Fuzzy sets allow partial membership, in contrast to crisp sets which only allow full membership or no membership at all (Figure 40.4). The transition from "belonging to a set" to "not-belonging to a set" is gradual and is characterized by a membership function that gives fuzzy sets flexibility in modeling commonly used linguistic expressions, such as the "muscle force is weak" or "walking speed is comfortable." As Zadeh (1965) pointed out, such imprecisely defined sets play an important role in human thinking, particularly in the domains of pattern recognition, communication of information and abstraction. It is important to note that the fuzziness does not come from the randomness of the constituent members of the sets, but

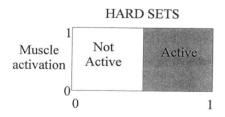

Two-valued membership: 0 OR 1
Muscle is active OR not-active

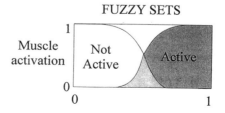

Multi-valued membership: 0 AND 1
Muscle can be active AND not-active

FIGURE 40.4. When does a muscle change from being active to not active? In case of crisp or hard sets only two answers are possible. The muscle is active *or* not active.

In case of fuzzy sets, multiple answers are possible at the same time. The muscle can be active *and* not active (i.e., partially active).

General Idea   Specific Example

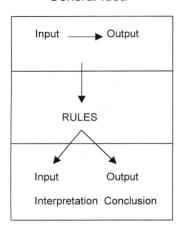 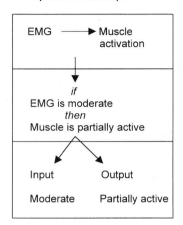

FIGURE 40.5. The idea behind a fuzzy system is to interpret an input, and based on a set of if–then rules, conclude an output.

from the uncertain and imprecise nature of abstract thoughts and concepts. A fuzzy if-then rule assumes the form: *if* x is A *then* y is B, where A and B are linguistic variables defined by fuzzy sets on universes of discourse X and Y, respectively (Figure 40.5). Often "x is A" is called antecedent or interpretation while "y is B" is called the consequence or conclusion. Examples of fuzzy if–then rules are widespread in our daily linguistic expressions. For instance, walking studies often require that subjects walk at a slow pace, comfortable or fast pace. Here a fuzzy rule can be: *If* walking pace is fast, *then* step frequency is high. Using a series of fuzzy if–then rules a control model of a system can be employed. For instance, the fuzzy rules could employ the relationship between input variables (e.g., control variables) and output variables (e.g., muscle torque or stimulation) in a meaningful and explicit manner. An important aspect of using fuzzy logic in our applications is that all rules act in parallel. By acting in parallel, more than one rule can be active with any task demand. Therefore, more than one requirement can be met by various degrees of rule activation.

4.3 Hybrid Neurofuzzy System

Neural networks and fuzzy logic are both soft computing techniques, but differ in their performance characteristics and how knowledge is represented. The algorithm used in a neural network is not transparent, making this process a "black box." The ben-

efits of using fuzzy logic versus neural networks is that fuzzy logic is based on natural language and words and can utilize the expertise of individuals familiar with the data to provide for a more transparent algorithm. In addition to the algorithm being known, the frequency that each rule is used, an indication of the potential importance of the rule, can be quantified (Kosko 1992). However, the definition of multiple fuzzy rules and the computational load required to reason through the bank of rules are two disadvantages to a pure fuzzy-logic approach.

The integration of neural networks and fuzzy logic can be such that the neural network is used within a fuzzy logic system for computational tasks. In this combined neurofuzzy technique, the advantages of a neural network can be used to reduce the computational complexity of a fuzzy logic system. It is also possible that fuzzy logic is incorporated within a neural network for purposes of adaptation and/or learning. In this combined neurofuzzy technique, the advantages of a neural network can be used to train, update or tune a fuzzy rule based systems.

The question arises how data clusters become fuzzy if–then rules. This can be accomplished by approximating any data cluster of numbers and equations into a system of natural language and words (Figure 40.6) (Kosko 1992). In other words, a fuzzy system is a nonlinear mapping between input and output space. A number of fuzzy if–then rules, describing the data, creates a nonlinear mapping that exploits the imprecision common in data. A disadvantage, however, is that the number of

crisp/hard function

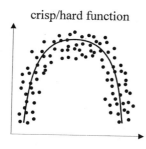

fuzzy graph

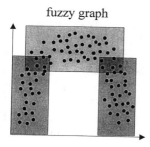

f = small x neg + medium x pos + large x neg

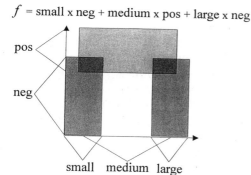

FIGURE 40.6. Experimental data can be approximated into a system of natural language and words. A number of fuzzy sets and if–then rules, describing the data, creates a nonlinear mapping that exploits the imprecision common in data. A fuzzy graph is a functional dependence, $f: A \rightarrow B$, where A and B are linguistic variables. Here the fuzzy graph is equivalent to three fuzzy if–then rules.

rules could grow exponentially as the number of system variables increase.

The concept of linguistic variables serves as a departure point whose use results in data compression (Kosko 1992). The function $f: A \rightarrow B$, with A and B as linguistic variables on universes of discourse X and Y, serves to provide an approximate representation of f in the form of a fuzzy graph (Zadeh 1996):

$$f = \text{small} \times \text{negative} + \text{medium} \\ \times \text{positive} + \text{large} \times \text{negative}$$

Fuzzy if–then rules are employed by combinations of linguistic variables A (small, medium and large) and B (positive and negative). For instance, *if* A is small and B is negative *then* activate knee extensors.

A fuzzy if–then rule is completely defined by the shape of the antecedent variable in input space and the shape of the consequent variable in the output space. A fuzzy rule can be written as a multilayered neural network (Berenji and Khedkar 1992; Bezdek 1993; Jang and Sun 1995). Each layer represents the different processes that underlie a fuzzy system. Training such a neural network dynamically adjusts the fuzzy system parameters so that (1) in case of neurofuzzy clustering a rough and initial set of fuzzy rules will be identified from experimental data describing relations between input space and output space, and (2) in case of neuro-fuzzy control these initial set of rules will be optimized for the specific task requirements in a model of neuromuscular skeletal dynamics of (e.g., standing and walking). In theory, three different types of training can be employed in applying neurofuzzy clustering and neuro-fuzzy control as mentioned previously (i.e., supervised, unsupervised, and reinforcement learning).

Artificial neural networks (ANNs) using supervised learning tune the rules of a fuzzy system as if they were synapses. The user provides the first set of rules. The ANN refines the rules by running through hundreds of thousands of inputs, slightly varying the fuzzy sets each time to see how well the system performs. The ANN tends to keep the changes that improve performance and to ignore the others. The quality of the ultimate rules depends on the quality of the data and therefore on the skills of the expert who generates the data. ANNs using reinforcement training evaluate the state of the system with either an immediate or delayed evaluative measure of its performance which is maximized through trial and error (Barto 1994; see also Chapter 35). Reinforcement learning requires less extensive sets of data than supervised learning because specific feedback is not provided, and often requires less training time than unsupervised networks because some measure of performance is provided.

In summary, neurofuzzy clustering is used to describe the nonlinear behavior in explicit and meaningful manner, with the computational advantages of ANN and fuzzy logic approaches complimenting each other. The result of the initial clustering will be a rough set of fuzzy rules (Figure 40.2A). This initial set of fuzzy rules can then be optimized using a neurofuzzy approach for control where the performance of the fuzzy system will be tuned depending upon the various task requirements (Figure 40.2B). The final developed neuromuscular control strategy can be used as a model for various applications in the field of rehabilitation.

5   Soft Computing Approaches Applied to Rehabilitation: Two Examples

This section presents two specific examples of the potential usefulness of soft computing techniques applied to rehabilitation-oriented areas.

5.1   Synergy Analysis of EMG Data Using ANNs

Electromyography (EMG) is often used in research and clinical settings to study movements and to gain a better understanding of the effect of pathology on the underlying neuromusculoskeletal system. The human body's musculoskeletal and neural control systems have coevolved and evaluation of single muscle, as if it is acting individually and not in concert with each other, may be inadequate to provide useful information about the neuromotor control of the movement. Synergy analysis of EMG data allows for simultaneous analysis of data from multiple muscles and provides more insight into the global differences in muscle activation patterns. Furthermore, synergy analysis provides a better means to describe the overall coordination of muscle activity than analysis from data of a single muscle. Disadvantages of synergy analysis include the additional computation load required to define and analyze each additional muscle channel, as well as the computational complexity required to model the interchannel relationship between multiple muscles.

The ability to objectively identify patterns for synergy analysis of electromyographic data is important for both clinical and research applications (see also Chapter 37). Clinical treatments have become more focused on improving the functional abilities of an individual often through the use of functional activities. The ability to quantify improvement in functional movement for validation of treatment effects is difficult at best, and currently is often described in qualitative terms. Functional movement patterns often require the activation of multiple muscle groups, and the synergy of muscle activation patterns, rather than the activation of a single muscle, often determines the functional outcome. The ability to objectively quantify synergistic muscle activity during gait would be useful both in gait evaluation, and as a feedback tool in the actual process of gait re-education. Previous reports of EMG synergy analysis using traditional hard computing techniques have been only partially successful in providing a quantitative, objective measure of pattern classification of EMG data for gait analysis (Zhang and Shiavi 1991; Chen et al. 1992).

In the following example, the use of ANNs serve two functions. In one case, the ANN is developed for pattern association, to associate an input with the corresponding output (Tucker et al. 1994). In the second case, the ANN is developed as a self-organizing network for unsupervised clustering of the EMG data (Tucker 1996). Surface EMG-linear envelope (LEEMG) data was obtained from 5 lower extremity muscles from a single subject during three different gait conditions: walking at self-selected and fast speeds, and while simulating a pathology (ankle injury). A 51-point feature vector consisting of the normalized LEEMG amplitudes at 2% increments of the gait cycle was then formed for each muscle. The input pattern to the ANNs were 255 point vectors consisting of the ordered series of the individual muscle vectors. These input patterns were classified as belonging to one of the 3 conditions: self-selected, fast or pathological. A total of 39 LEEMG patterns were available, 13 from each of the 3 conditions. The ANNs were programmed using Matlab (The Mathworks Inc, Natick MA).

A pattern association neural network, a feedforward backpropagation network (FF–BP), and a self-organizing map (SOM) were used for synergy analysis of the EMG patterns. After supervised training on a subset of the available patterns, the FF–BP network was able to correctly classify any muscle input pattern in under 1 second. This classification provided a quantitative measure of the

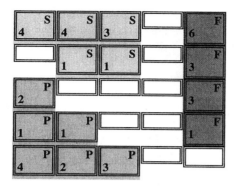

FIGURE 40.7. This Fig graphically represents the 25 neuron self-organizing map (SOM). The physical location of the pattern clusters within the SOM is related to the similarity or differences within the data itself. A total of 39 patterns were clustered from 3 conditions. The letter indicates the condition (S—self-selected, F—fast, P—pathological), while the numbers indicate the number of patterns assigned to that particular neuron. As the figure shows, patterns from each condition cluster close together as expected, with separation noted between the different conditions.

patterns membership in each of the three gait conditions. The SOM network clustered the patterns into a 5 * 5 two-dimensional map consisting of 25 neurons, and as seen in Figure 40.7, with the input patterns from each condition clustered together. As expected, the patterns from the fast condition clustered more closely together because of a greater consistency in patterns obtained during fast gait. The pathological condition was more widely clustered, most likely because of a greater inconsistency of muscle activation when the subject attempted to mimic a pathology. Using the ANN, the self-organization process required less than 5 minutes, after which an unknown pattern could be appropriately clustered in <1 second.

The preceding example demonstrates the feasibility of using ANNs for synergy analysis of EMG data.

5.2 Fuzzy Logic Control Model of Human Stance

Clustering of EMG data into functional muscle groupings is very useful, but contributes only partially to our understanding of human motor control in tasks such as walking and standing. For effective

modeling, we must also determine which control variables and rules are used in motor control and coordination. Insight in these control variables and rules might simplify our understanding of nervous system control and ultimately lead to effective applications. In the process of modeling motor control, we have to determine: (1) how to incorporate the nonprecise experimental data into a control system; (2) how to quantify the relationship between EMG data and functional motor performance.

In the following example, the use of fuzzy logic is demonstrated for the implementation of control variables and rules in a control model of human stance. The design of the control model was stimulated by multiple reports showing several important indications that the nervous system seems to simplify the control of movement in terms of global control variables (see Chapter 22). This is in contrast to the inverse dynamics approach, which involves a computation of the individual joint torques by some complex representation of muscle skeletal dynamics in the nervous.

The proposed model is not based on the control of body's center of mass within the base of support, but rather on the control of *global variables* specific to body orientation and alignment (see also Chapter 22). This is an important difference, since the model can now be applied to changes in environment, for instance stance control in 1G on earth and 0G in space (see also Chapter 20 and Chapter 21). Furthermore, the proposed control model is not based on purely inverted pendulum body mechanics where only motion at one joint is controlled. In the proposed model, the degrees of freedom are controlled by using reciprocal and synergistic muscle actions at multiple joints. The control model is based on control of three sets of different global control variables which act in parallel: (1) *limb length (l)* and its derivative *(l')*, (2) *limb orientation (a)* and its derivative *(a')*, and (3) *trunk attitude (b)* and its derivative *(b')*. In the control strategy, 36 fuzzy *if–then* rules, based on experimental findings, are implemented using Matlab (The Mathworks Inc, Natick MA). Examples of two rules are:

if *l* is **small** and *l'* is **negative** then make **extensor** *muscle action* **large**."

In this example of a fuzzy rule, *l*, *l'* and *muscle action* are fuzzy control variables. *Small*, *negative*, and *large extension* are fuzzy sets.

"if a is **forward** and a' is **zero** then **biarticular posterior** *muscle action.*"

In this example of a fuzzy rule, a, a' and *muscle action* are fuzzy control variables. *Forward, zero,* and *biarticular posterior* are fuzzy sets.

The nature of fuzzy logic allows these two rules to act *in parallel.* In the event that these two rules are activated at the same time, it is interesting to note that coactivation at the knee between vastii and hamstrings and gastrocnemii takes place. Many other combinations of coactivation are possible and will be determined by the instantaneous values of the global control variables in each control mechanism.

The control strategy is implemented using a four-linked segment model consisting of a trunk, thigh, shank and foot. Uni- and biarticular limb muscles and trunk muscles are represented as torque actuators at each individual joint. In the four-linked segment model of a standing human, the three sets of global control variables act in parallel and make corrective and coordinated responses to internal, self-induced perturbations (see Chapter 22, for results). The data shows that the use of fuzzy logic and global control variables successfully enables us to model human standing with sway about a point of equilibrium. Small changes in each of the control variables are similar to those seen during natural sway in human stance. Limb length, limb orientation and trunk attitude controllers are reasonably stable indicating that the selected control variables are appropriate for human stance.

This example demonstrates the feasibility of using fuzzy logic that enables us to model expert knowledge and experimental findings in a meaningful and explicit way. If the goal objectives change, simple changes in the rules will meet the new postural standing requirements. Changes are accomplished by tuning the fuzzy sets and rules in the controller. This tuning utilizes a neural network that can be implemented for optimization of the desired standing task (see also Chapter 19 and Chapter 35).

6   Future Direction

The evaluation and control of human movement performance will become increasingly important in the future. Microelectronics and advanced material technology can provide means for physical substitution of impaired or absent components of human neuromusculoskeletal system. However, until these inanimate components can be provided with a human-like control system, and interfaced with the animate system, their acceptance and widespread use as rehabilitation aids will be slow.

The benefits of using soft computing techniques to elucidate the control parameters by pattern recognition in complex data sets, and computation of control variables has been demonstrated over the past decade in inanimate systems. Application of these techniques to evaluate and control human motor performance should provide similar benefits for humans, reducing disability and improving motor performance. In the analysis and design of human neuromuscular control strategies, our future efforts should include creating soft computing techniques that are able to:

1. Explicitly identify the nonlinear behavior of the neuromuscular system in a meaningful manner,
2. Resemble robust, adaptive and learning behavior of the human neuromuscular system.

It is our belief that the use of the relatively new techniques in Soft Computing (i.e., neural networks, fuzzy logic, and genetic algorithms) offer great potential in the analysis and design of human neuromuscular control. In this chapter, we described an approach that uses these relatively new techniques in soft computing. In this approach, ANNs are a perfect tool for learning patterns in large nonlinear data sets. Fuzzy logic is a perfect tool to model nonlinear behavior in terms of natural language which provides explicit interpretation. The combined advantages of fuzzy logic and neural networks will develop into a neurofuzzy technique. Two phases can be distinguished: (1) neurofuzzy clustering that identifies a rough and initial set of fuzzy rules, and (2) neurofuzzy control that optimizes the performance of this rough and initial set of fuzzy rules that depend upon the task requirements, and leads to robust and adaptive near-optimum behavior.

The techniques of neurofuzzy clustering and neurofuzzy control can be used in developing combined neural networks and fuzzy logic models for control of human movement such as standing and walking. Having these control models, they can be applied to development of applications in FES, prosthetics and orthotics design, and rehabilitation.

References

Altrock, von C. (1995). Fuzzy logic and neuro-fuzzy applications explained. Prentice Hall, Englewoods Cliffs, New Jersey.

Barto, AG (1991). Connectionist learning for control. In *Neural Networks for Control.* Miller, W.T., Sutton, R.S., and Werbos, P.J. (eds.), pp. 5–58. The MIT Press, Cambridge, Massachusetts.

Barto, A.G. (1994). Reinforcement learning control. *Curr. Opin. Neurobiol.*, 4:888–893.

Barto, A.G., Bradtke, S.J., and Singh, S.P. (1995). Learning to act using real-time dynamic programming. *Artif. Intell.*, 72:81–138.

Berenji, H.R. and Khedkar, P. (1992). Learning and tuning fuzzy logic controllers through reinforcements. *IEEE Trans. Neural Networks.*

Bezdek, J.C. (1993). A review of probabilistic, fuzzy, and neural models for pattern recognition. *J. Intell. Fuzzy Sys.*, 1:1–25.

Bezdek, J.C. and Pal, S.K. (1992). Fuzzy models for pattern recognition. IEEE Press, New York, New York.

Chen, J.J.J., Shiavi, R., and Zhang, L.Q. (1992). A quantitative and qualitative description of electromyographic linear envelopes for synergy analysis. *IEEE Trans. Biomed. Eng.*, 39(19):1–18.

Gioftsos, G. and Grieve, D.W. (1995). The use of neural networks to recognize patterns of human movement. *Clin. Biomechan.*, 10(4):179–183.

Graupe, D. (1995). Artificial neural network control of FES in paraplegics for patient responsive ambulation. *IEEE Trans. Biomed. Eng.*, 42(7):699–707.

Hecht-Nielsen, R. (1991). *Neurocomputing.* Addison-Wesley, Reading Massachusetts.

Holzreiter, S.H. and Kohle, M.E. (1993). Assessment of gait patterns using neural networks. *J. Biomech.*, 26(6): 645–651.

Jacobs, R., Koopman, B., Veltink, P., Huijing, P.A., Nene, A., van der Kooij, H., and Grootenboer, H. (1996). A simplified control strategy for postural coordination in human stance. *Engineering Foundation Biomechanics and Neural Control of Movement IX.* Mt. Sterling, Ohio.

Jang, J.S.R. (1992). Self-learning fuzzy controllers based on temporal back propagation. *IEEE Trans. Neural Networks*, 3:714–723.

Jang, J.S.R. and Sun, C.T. (1995). Neuro-fuzzy modeling and control. *Proceedings of the IEEE*, 83:378–406.

Kawato, M. (1990). Computational schemes and neural network models for formation and control of multijoint trajectory. In *Neural Networks for Control*, Miller, T., Sutten, R.S., and Werbos, P.J. (eds.), pp. 197–228. The MIT Press, Cambridge, Massachusetts.

Kawato, M., Furukawa, K., and Suzuki, R. (1987). A hierarchical neural-network model for control and learning of voluntary movement. *Biol Cybern.*, 57:169–185.

Kawato, M., Maeda, Y., Uno, Y., and Suzuki, R. (1990). Trajectory formation of arm movement by cascade neural network model based on minimum torque-change criterion. *Biol. Cybern.*, 62:275–288.

Kelly, N.E., Parker, P.A., and Scott, R.N. (1990). Myoelectric signal analysis using neural networks. *IEEE Trans. Med. Biol.*, 61–64.

Kosko, B. (1992). Neural networks and fuzzy systems. A dynamical systems approach to machnie intelligence. Prentice Hall, Englewoods Cliffs, New Jersey.

Kosko, B. (1993). Fuzzy thinking. Hyperion, New York.

Kubica, E.G., Wang, D., and Winter, D.A. (1995). Modelling balance and posture control mechanisms of the upper body using conventional and fuzzy techniques. Gait and Posture, 3(2):111.

Meier, W., Weber, R., and Zimmerman, H.J. (1994). Fuzzy data analysis—Methods and industrial applications In *Fuzzy Sets and Systems 61.* Negoita, C.V., Zadeh, L.A., and Zimmerman, H.J. (eds.), pp. 19–28. Elsevier, North Holland.

O'Malley, M.J., Abel, M., and Damiano, D. (1995). Fuzzy clustering of temporal-distance and kinematic parameters for cerebral palsy children. Gait and Posture, 3(2):92.

Schalkoff, R. (1992). Pattern recognition: syntactical, structural, and neural approaches. John Wiley & Sons, New York, New York.

Sepulveda, F., Wells, D.M., and Vaughan, C.L. (1993). A neural network representation of electromyography and joint dynamics in human gait. *J. Biomech.*, 26(2): 101–109.

Tucker, C.A. (1996). Artificial neural network approaches to synergy analysis of electromyographic data. In *Engineering Foundation Conference: Biomechanics and Neural Control of Movement IX.* Winters, J. and Crago, P. (eds.), Mt. Sterling, Ohio.

Tucker, C.A., Yack, H.J., and White, S.C. (1994). A neural network approach to synergy analysis of electromyographic data. In *Proceedings of the Tenth Congress of the International Society of Electrophysiology and Kinesiology.* Charleston, South Carolina.

Wada, Y. and Kawato, M. (1993). A neural network model for arm trajectory formation using forward and inverse dynamics models. *Neural Networks*, 6:919–932.

Werbos, P.J. (1991). A menu of designs for reinforcement learning over time. In *Neural Networks for Control.* Miller, W.T., Sutton, R.S., and Werbos, P.J. (eds.), pp. 67–95. MIT Press, Cambridge, MA.

Zadeh, L. (1965). Fuzzy sets. *Infor. Control*, 8:338–353.

Zadeh, L. (1994). Fuzzy logic, neural networks and soft computing. *Comm. Assoc. Computing Machinery*, 3: 77–84.

Zadeh, L. (1996). Fuzzy logic—computing with words. *IEEE Trans. Fuzzy Sys.*, 4:103–111.

Zhang, L.Q. and Shiavi, R. (1991). Clustering analysis and pattern discrimination of EMG linear envelopes. *IEEE-BME*, 38(8):777–784.

Zurada, J.M. (1992). An introduction to artificial neural systems. West Publishing Co., St. Paul, Minnesota.

Commentary:
Soft Computing Techniques for Evaluation and Control of Human Performance

Rajko Tomovic

When a paper advocates the promotion of engineering, mathematical or computer methods, old or new, for the study of the processes in the biological world, it seems to me that the cause in mind will be better served being more of the skeptical rather than optimistic side.

The fact that the engineering control is mediated by electrical waveforms while in biological world this role is assigned to macromolecules should not be overlooked. Neurons are complex biochemical factories, exposed to many chemical and electrical agents, and their relay type responses are the outcome of complex networks connected by the transmitter type nodes (Smith 1989).

It has been also witnessed even in engineering that a settling time is needed to critically evaluate early expectations (dynamic programming, maximum principle, scene analysis, automatic translation, machine learning, etc.). So far, the advances in engineering and computer sciences have proved one thing for sure: that each control method has a limited domain of applications.

With this fact in mind, the statement "we propose the use of soft computing techniques as a new and unique approach for evaluation and control of human performance" seems less convincing. Much more than "two specific examples of the potential usefulness of soft computing techniques" would be needed to support the general claims of the chapter.

By the way, the use of soft computing should not be considered to be a completely new technique. For some time, and for various applications, combined control systems (expert systems, artificial neural networks, fuzzy sets, analytical methods) have been successfully used to solve specific problems. The term human performance covers basically only the control of lower extremities. However, the extension of methods, validated for the control of lower extremities, to the control of upper extremities is by no means straightforward because the spectrum of motor skills of upper extremities is much broader (reaching, tracking, grasping) and combines both dynamic aspects and skill acquisition. It is therefore more appropriate to test the validity of a new approach which pretends to be general for the control of both lower and upper extremities.

Finally, it is questionable if soft computing is the unique approach for control of human performance. It seems that the emphasis in the chapter is on the nonanalytical approaches for motor control versus analytical tools, as in Chapter 39. However, for some time a general nonanalytical approach for the control of functional movements has been developed and successfully applied for the design of advanced assistive systems (Tomovic et al. 1995). This approach relies fully on the biological principles of motor control such as representation of motor patterns, reflexes, and motor skills by expert systems, neural networks, finite automata models, hybrid networks etc. A comparison of relative merits of the soft computing and other nonanalytical approaches for the control performance analysis and advancement of assistive systems would be certainly helpful, both theoretically and for rehabilitation purposes. At least, a reference to other "soft" approaches would help the reader to get a broader insight into the current state of the biologically inspired control methods of human extremities.

References

Smith, C.U.M. (1989). Elements of molecular neurobiology. John Wiley and Sons, New York.

Tomovic, R., Popovic, D., and Stein, R.B. (1995). Nonanalytical methods for motor control. World Scientific, Singapore.

41
From Idea to Product

William K. Durfee

1 Introduction

At some point in their careers, many motor control physiologists have wondered if their research might lead to useful products which could improve the assessment, diagnosis, or rehabilitation of those with motor impairments. Some have gone beyond thought to research, develop, and even commercialize products based on research discoveries. The objective of this perspective chapter is to overview the process of product design and development, to highlight issues specific to rehabilitation products, and to describe some of the pitfalls which face scientists who attempt to commercialize their work.

New products are essential to the survival and growth of any company. With increasing competition, an explosion in new technologies, changing customer needs and shorter product life cycles, it has become more important than ever to design and develop products properly. Developing successful products requires more than just a good idea, more than just good engineering, and more than just good marketing, but rather requires an integrated approach to identify customer needs, create and refine concepts that meet those needs and produce the product at appropriate cost. New product development is still more art than science, but there are structured methodologies that can help the individual decide whether an idea is suitable for turning into a product and which can help an organization through the often confusing and conflicting steps of creating and producing the product. Following new product development best practices is not a guarantee of success, but instead a means for minimizing risks.

Medical products bring special considerations to the process. Products will have multiple customers with competing interests and a complex distribution chain. They are subject to special government regulations and require close scrutiny to minimize liability risk. And most importantly, the success or failure of medical products in recent times is greatly influenced by the rapidly changing health care provider and insurance industries whose policies have a tremendous effect on the way medical products are prescribed, purchased and billed.

2 Why Products Fail

Most product ideas never make it to market and a significant fraction of those that do never make money for the company. This is particularly true for small startup companies relying on a single invention or idea for their product portfolio, and even more true for individual inventors attempting to capitalize on their ideas. The success rate can be improved through informed product management process. A recent survey (PDMA Handbook, Chapter 33) reveals that 75% of new product launches from companies who are members of the Product Design and Management Association (PDMA) were deemed successes by the product manger. Perhaps this is because only one in ten recorded product ideas made it to product launch which meant that poor ideas were stopped relatively early in the development process.

There are many reasons why products fail. First and foremost, products fail because those leading the charge do not understand the market or the needs of the customer. This is particularly prevalent for technology-driven products where the excitement of the idea or novelty of the technology overwhelms all consideration of whether anyone will buy or use or even need it. Picture telephones and motorized orthotics are a good examples of this.

Another common reason for product failure is that the idea is championed and pushed through the development process by a single, typically powerful, person. Although sometimes an aggressive genius personality is required to overcome inertia and capitalize on good ideas (Thomas Edison was good example of this kind of person), usually the result is a poor, or more typically no, process without the usual checks and balances to weed out the weak ideas.

A third reason for failure is exceeding either the technical capability of the organization to refine the product concept, or the manufacturing capability to produce the product in appropriate quantities at reasonable cost, or the sales, distribution and training capability to put the product in the hands of its target market.

3 Generic New Product Development Process

Although each company, small or large follows its own, unique product development process, the essence of a good NPD process can be distilled into a few generic steps. The reader is urged to consult the references for more detailed descriptions of the steps.

3.1 The Ideal Process

The initial stages of the NPD process are often referred to as "the fuzzy front end." Here is where a company (or an individual) determines which products to pursue. In the most forward-looking companies, decisions are made in the context of an overall new products strategy for the company. For example, 3M has a corporate goal that 30% of its revenues must come from products less than five years old, which drives the company to an aggressive front end for product selection. Product areas must be compatible with the organization's core technical and manufacturing competency as well as

satisfying some clear customer need. Sometimes, extensive market research activities take place to help determine whether a product area is worth pursuing. For the biomechanist turned inventor working alone or partnering with a company, this fuzzy front end is the most critical stage in the process because it is here where decisions are made to commit resources to proceed with developing the product or to pursue more lucrative product areas.

After a decision is made to proceed, a product development team is formed with representation from the core areas of marketing, sales, engineering and manufacturing, and depending on the complexity of the project, additional representation from purchasing, quality, finance, and legal. This team is charged with taking the product through to launch acquiring additional resources and people as needed along the way.

The team will first engage in a concept development stage where the target market is defined and customer needs gathered. From there, many ideas are generated and turned into alternative concepts which meet customer needs. A careful, structured selection process should then be used to pick the one or two concepts worthy of proceeding. At this time, a preliminary market assessment is usually conducted to place boundaries on potential market size and profitability of the product for the organization. Truly feasible concepts must have features, benefits, costs, and risks reasonably well defined.

Concepts must then go through a lengthy period of refinement and testing through detail design and concept evaluation. Engineering determines product architecture, defines geometry, selects materials, optimizes cost, prepares designs for manufacture, and produces prototypes in varying levels of functional and aesthetic sophistication. At the same time, marketing is testing the concept through prototype, product use, and market testing using appropriate test sample users from the target market.

Finally, the product moves to the commercialization phase. Plans are put into place for production ramp up, including supplier inventories, sales and advertising and full scale manufacturing systems development. For small-run, one-of-a-kind products, very little planning is required in the commercialization phase. For high-volume products, extensive work is needed during this phase to ensure a smooth launch and subsequent sales.

Following launch, the product must be continually evaluated to determine if sales goals are being met and expected product life cycle being followed. At this time, it is always useful to reflect on the process so that the NPD process can continually improve.

3.2 The Real Process

In reality, very few NPD projects follow the logical, sequential steps outlined above. Time is short, decisions must be made based on incomplete information and development steps inevitably overlap. Many organizations try to survive without any formal plan for product development. Having some kind of structured NPD process is nevertheless important (and required for ISO 9000 accreditation) to provide checkpoints, communicate to others where the project is, help with project management, and allow an organization to continually improve.

The steps in the generic process fit medium-to-large organizations that are fully staffed with representatives from all areas of the company. Many medical product companies, particularly those specializing in biomechanics or assistive technology, are tiny; some have just a handful of employees. For example, *Spinal Designs International* in Minneapolis has only 14 full-time employees and no in-house engineering or manufacturing staff. Their primary product grew from basic research, but they outsourced most of the design to a product development/industrial design company and it is fabricated by a local manufacturing firm. Some researcher/inventors attempt to develop their idea themselves, forming a company with perhaps just one or two full-time employees. It is just as important to follow good product development process for these inventor-owned companies. If one person is in charge, and the product is her "baby," often the search for alternative concepts is overlooked, customer needs not articulated or costs not fully accounted for. Any of these can turn a good idea into an unsuccessful product

Inventors choosing or collaborating with companies to turn their ideas into products should be sensitive to that company's expertise and experience areas, but should also confirm that the company follows good NPD process. If senior managers of the business unit within the company are unable to articulate their NPD process, it might be wise to turn elsewhere with the idea.

4 Customer Needs

Products that do not meet the needs of the target customers fail. Few parts of the NPD process are as important as obtaining reliable, comprehensive customer needs data, and ensuring that the product is focused towards those needs. Although the company developing the product for the inventor should have a logical and comprehensive plan for gathering the "voice of the customer" (see Chapter 11 of the PDMA handbook, or Chapter 3 of Ulrich and Eppinger), the inventor can easily undertake a much smaller scale gathering and assessment of needs data following the suggestions provided here. This will help answer the question of whether the inventor's product has potential for going beyond being a good idea to becoming a viable commercial product. Because accurate assessment of customer needs is such an important part of the product development process, we elaborate on it here.

Customers cannot tell you what your product should be nor how you should develop it. Rather, they are very good at telling you what their problems are, and what the strengths and failings are of existing products and solutions. An administrator in a small dental office will not be able to describe the design for a relational database with direct data output to insurance company forms, links to a patient scheduling program and remote access, but that same person can talk about their paper record-keeping system and how it takes too long to fill out insurance forms, how it is difficult to track when patients should be reminded about appointments, and how he would really like to be able to occasionally do the administrative work from home.

The first step in gathering customer needs data is to define the product and target market. Writing a concise, one paragraph description of what the product is and how it works forces specificity and turns abstract ideas into concrete objectives. Listing the target market forces a context for the product. Is it for physicians? Therapists? Technicians? Patients? Who will buy it and who will use it? The answers may change as the product is refined and a deeper understanding of the market obtained. For example, a rehabilitation product initially thought to have use primarily in the physical therapy clinic, with some variation and refinement, may have more utility (and a bigger market) at home. Once the product is defined, customer needs data gathering can commence. Four methods are common: in-depth interviews, focus groups, on-site observations, and telephone or mail surveys. Of these, interviews and on-site observations are feasible for the inventor without expertise or extensive experience in market research to conduct on his or her own.

Three to ten in-depth interviews should be done with people in the target customer pool. Questions should be written in advance, although the interviewer should be flexible. Start with questions designed to understand what problems the customer has, what the current solutions or standard practices are and the strengths and weaknesses of current products or services. If appropriate, from there move to introducing your product concept and gathering reactions. One of the more difficult aspects of interviewing is keeping your own biases and interpretation out of the questions and your recording of the responses. If possible, include one or two "lead users" among the pool of interviewees. Lead users are customers who anticipate needs before most other users and who often develop unique and innovative solutions to current problems that can be used as a basis for improving your own product concept. Many product development teams have discovered the benefits of integrating lead users into the process to advise in direction and help with decisions.

Observing target customers using existing products or performing current services is invaluable for understanding how a new product might satisfy customer needs. Watching the speed with which a neurologist can assess gross motor skills using simple manual techniques will help one recognize the difficult but very real constraint of designing an automated mechanized product for motor skill evaluation that is fast and simple to use. It is essential that the product inventors and developers be the ones observing, and not relying on second hand information. On-site observations reappear later in the development process when customer tests the initial prototype of a product.

With customer needs raw data in hand, the process of analyzing can begin. In simplest form, this means placing the needs into major and sub-categories and assigning priorities to those needs, and remembering to do this from the customer's rather than the development team's perspective. From there, one can use more formal methods such

as House of Quality (described in the references) to connect customer needs to product attributes.

5 Why Rehabilitation Products Are Different

Rehabilitation is the natural (and often assumed) application area for products arising from research in biomechanics and neural control of movement. Rehabilitation products generally fall into one of three categories: diagnostic and quantitative assessment devices for use in a clinic, therapeutic or function restoration devices for use in the clinic or at home, and assistive technology devices to aid in activities of daily living. A small specialty market also exists for products designed to support rehabilitation or biomechanics research. To develop successful products in any of these categories, one must become an insider and be familiar with current practice, current products and past history. One should also follow trends, read the appropriate professional and trade journals, attend conferences and know standards. Failure to do this will result in solutions to nonproblems or solutions to the wrong problem.

5.1 Decision Makers

The buyer decision chain for rehabilitation products varies depending on which category the product falls into. Diagnostic products such as muscle function testers are usually part of a suite of advanced instruments owned or leased by a clinic. Typically, a committee of clinicians (physicians and physical therapists) will make a decision about which new product, system or service to buy, based on their current practice or on a strategic plan for their clinic. Therapeutic devices such as electrical stimulators for home use are recommended and prescribed by clinicians for individual patients. Here both patient and clinician influence the buy decision. Assistive technology is either prescribed or recommended by the clinician (e.g., an augmentative communication device) or purchased entirely by the end user (e.g., a wheelchair). In some areas, direct marketing of assistive technology to end users is growing and as a consequence, design is much more in tune with customer needs. A good example is the explosion in lightweight wheelchairs (reviewed yearly in *Sports 'N Spokes* magazine) for recreational or sports use.

To a large extent, insurance companies dictate the success or failure of rehabilitation products. If insurers will not pay, physicians will not prescribe and patients will not buy. As the world moves towards a managed care model, cost drives more and more decisions in health care. Before managed care, physicians prescribed treatments and programs based on effectiveness alone and were largely uninfluenced by cost. Modern medical insurance companies not only assess medical effectiveness, but also determine whether the medical outcome leads to a reduction in health care costs. Usually, the quality and utility of a new rehabilitation product concept are appropriate, but a product manufacturer or service provider must also be able to demonstrate cost savings by outcome based performance indices. Vague statements that a stroke rehabilitation tool reduces therapy time are no longer sufficient. A company must back up its claims through documented studies, which show via quantitative assessment, how their product shortens the time for stroke patients to relearn a particular motor skill or how their product reduces the hours of one-on-one interaction with a physical therapist. Convincing documentation must exist showing how the product will save time and money. Companies, such as *Spinal Designs*, sponsor research studies related to their current or future products where the results can be published in peer-reviewed archival journals that have tremendous weight and influence in the medical community.

When developing rehabilitation or biomechanical diagnostic products, it is wise to work directly with the physical or occupational therapists. Because PTs and OTs spend long hours working directly with patients, they sometimes have a better sense of the needs and what products and technologies might work to solve the needs than physicians. Often PTs and OTs are early adopters of new technology, or have used their own creative and inventive skills to solve problems.

To summarize, a successful rehabilitation product must satisfy the needs of the doctors who prescribe, the therapists who instruct or apply, the patients who take or use and the insurance companies who pay. All of these are product customers whose sometimes conflicting needs must be considered (see also Chapter 37).

5.2   Users

Unlike consumer products, each user of a rehabilitation product is different. A single hand grasp orthosis will not be suitable for all potential clients with upper level spinal cord injury. Some may have wrist extension, some wrist flexion, some neither. Some may have sufficient shoulder and elbow control to position their hand at the object to be grasped, some may not. Some may have complete sensory function, others none. Rehabilitation products which interact closely with people generally require customizing to the individual.

Another implication is that products which fill needs for a narrow patient category will not have a big market. Even if there are many patients in a category, often only a fraction of the population could benefit from the proposed technology. For example, a commonly cited statistic justifying continuing development of FES systems for gait restoration is that in the United States there are 200,000 individuals with spinal cord injury with approximately 10,000 new injuries each year. Of those approximately half are paraplegic which leads one to the conclusion that 100,000 individuals could benefit from FES-aided gait technology. In reality, less than 10% of the paraplegic SCI population might be suitable candidates for FES, and of those fewer still might be willing to endure the cost of training, equipment and surgery (for implanted systems) to gain the benefit of limited gait.

5.3   Diagnostic Products

Many biomedical engineers see product opportunities in new technologies for quantitative assessment. Gait analysis systems which capture accurate records of positions and orientations of limb segments during gait are now common in large orthopedic clinics. Isometric muscle strength can be precisely assessed with strain-gage instrumented transducers interfaced to computers for automatic data collection.

At the same time, there have been hundreds of product concepts and prototypes for quantitative assessment of limb and muscle kinematics (e.g., range of motion testers) and dynamics (e.g., single joint power testers) which have failed miserably. The explanation can be found by looking both at the past and the future. Traditionally, orthopedic surgeons, physiatrists and physical therapists are trained to assess limb dynamics manually. From a quick, voluntary push of the patient's forearm against the hand, the clinician can assign a grade of one to five for elbow strength. The test is simple, fast and generally sufficiently informative to enable appropriate treatment decisions. In the future, speed and effectiveness of diagnostic tests are important drivers since patient throughput is essential for a high-efficiency, managed care health system. Contrast this with a technology-based, automated limb tester. The machine takes up space in the clinic (sometimes considerable if whole limbs are being tested), or must be wheeled into the office. The setup time is long, the machine difficult to adjust, and the data, although accurate, is voluminous. It is extremely challenging to develop technology with the speed and sensitivity of the hands of a trained clinician. One area where opportunities may exist, however, are simple to use, but possibly complex diagnostic products that are located remotely, either in satellite clinics or in users homes. Cost savings could be substantial if the product reduces the need for patients located in rural areas to travel to clinics or if it means clinics staffed by nonspecialists can be located remotely.

5.4   Treatment Products

New medical devices for use at home by the patient for long-term or chronic care are being developed at a rapid pace. Along with the requirement that treatment benefits and cost savings must be documented and demonstrated, there are other requirements needed for success.

If the device is complex, then training in its use must occur. The developer must commit significant resources to training clinicians and to seeing that systems are in place for training the user. Good products have failed because the manufacturer did not pay sufficient attention to when and how users were trained with poor training resulting in discouraged (or even angry) patients who reject the device.

Equally important, the company or inventor must decide whether it is selling the product or the whole service. *Spinal Designs* is in the business of low back care. It markets the LTX3000, a product for unweighting the lower spine to improve circulation, reduce pain and aid in healing of acute tissue injuries; but the real value of the company rests not in its product, but rather in the low back programs

it manages. That is, Spinal Designs is a primarily a service company which develops products to support its services.

5.5   FDA

Medical devices are broadly defined as any health care product which does not use chemical action to fulfill its functions. Medical devices include implants, surgical tools, pacemakers, wheelchairs and electrical stimulators. Strict U.S. federal regulations on medical devices were enacted in 1976 as part of the Medical Device Amendments to the federal Food, Drug and Cosmetic Act. These regulations define the requirements for medical device development, testing, approval, and marketing. Because FDA regulations have major influence on the design, safety, cost, and time to market of medical products, every inventor or researcher considering the transfer of their research into a product should be familiar with their basic framework.

Novel devices for biomechanical research designed and constructed in-house and used for research projects are generally covered under an investigational device exemption (IDE) as "nonsignificant risk" (NSR) devices. Here, responsibility for assuring that the device is safe and that good practice is used in test protocols rests with the local institution's human subjects committee of its *Institutional Review Board* (IRB). However, once the device moves beyond the investigation stage, it must go through the FDA 510(k) process, which applies to new products that are substantially equivalent to an existing approved device or through the more rigorous Premarket Approval (PMA) process. Devices are also categorized as Class I, II, or III depending on the level of risk they present to the user or patient, with the complexity and scrutiny of the regulatory process dependent on device class.

The time, cost, and energy to satisfy FDA requirements can be substantial. For PMA devices, the average time from first conception to PMA approval is around ten years and total cost can be in the millions of dollars. For example, the *ParaStep*, a commercial FES-aided gait product marketed by Sigmedics, Inc., is based on technology developed by Daniel Graupe at the University of Illinois at Chicago. The first clinical trials for the *ParaStep* were in 1982, but FDA PMA approval was not received until 1994.

6 Intellectual Property and Technology Transfer

A patent gives its owner an exclusive right to make, use, and sell an invention. Patents provide powerful protection to intellectual property and ensure that innovative companies can profit from their successes. Until recently, academic researchers working in areas that led to technology advances and potential new products did not think much about patents, or actively scorned them as inhibitors to research dissemination and advances. All this changed in the late 1980s when the U.S. Congress passed a law that enabled universities to retain ownership of inventions developed under federally funded research. The result in the United States was the immediate creation or expansion of university technology transfer programs that seek to capitalize on university developments. Because new technology is only useful to a company if it comes with patent protection, most universities have fairly aggressive programs for patenting new technology developed in research laboratories.

Bioengineering researchers working in academia must decide whether to pursue patent protection for potentially useful new technology that they are developing. In general, if the goal of the research is to develop new products or services and it appears that there is real practical value in the work, patent protection should be sought. Usually, the university technology transfer office will assume all costs and administrative responsibilities for the application with little burden on the investigator.

Deciding when to file a patent application can be difficult. If filed too early, the full implications and utility of the technology are not known. If filed too late, others may get there first. One consideration is the rule that U.S. patents must be filed one year after public disclosure of the invention (note that the U.S. Congress is currently considering changes to this policy). For academics, public disclosure includes publication in a journal or presentation at a conference. European patents must be filed before public disclosure. Sometimes, in infringement cases it is important to establish the date of an invention. Disciplined scientists always maintain laboratory notebooks where invention dates can be established, provided the entries are dated and it is clear that entries could not be modified.

Existing patents are useful as a database of ideas for applied research in areas which may lead to products. Much as one might search the scientific literature in *Medline* for prior work in an area, one can also search the patent database for ideas. Although rigorous searches are best left to a patent attorney, initial or more modest searches can easily be accomplished by the research team.

Products can be developed at a university through many channels. Student design teams can tackle practical product design based on research. A local or national company can purchase an exclusive license from the university to develop technology into a product. (Although usually it is up to the researcher to seek out these companies aggressively.) The researcher can start his or her own company to capitalize on research ideas. For small U.S. companies, *the Small Business Innovation Research* (SBIR) *and Small Business Technology Transfer* (STTR) program are effective ways of receiving federal funding for advanced product research and development. By far the most common way for technology transfer of research ideas into the market is through graduate and undergraduate research students who go on to work in industry.

7 Opportunities and Future Directions

There are many opportunities for researchers in biomechanics and rehabilitation to develop new products or product ideas. A particular advantage of products developed through university research is that they are backed by well-documented, peer-reviewed science.

Researchers turning to product development must be aggressive. If you have a good idea, pursue it. Get excited about your work, but be realistic. Follow a good product development process. Finally, remember that not all science and engineering research must have practical outcomes nor must it result in products. The objective of good research is to produce new knowledge that at some point might be useful for new product development; but it is not the obligation of every researcher to make that connection.

References

Rosenau, M., ed. (1996). The PDMA handbook of new product development. John Wiley & Sons, New York.

Ulrich, K. and Eppinger, S. (1995). Product design and development. McGraw-Hill, New York.

Urban, G. and Houser, J. (1993). Design and marketing of new products, 2nd ed. Prentice-Hall, New Jersey.

Commentary:
From Idea to Product

Gerald E. Loeb

There are three aspects of the technology transfer process that bear some additional emphasizing to academicians considering personal involvement (discussed in greater detail in Loeb and Schulman 1992):

- *You will probably get less out it than you expected.* Commercially successful medical devices typically sell for about 10 to 20 times the cost of similar consumer technology, leading one to believe that vast profits will be made and that the inventor deserves a healthy share. Reality is that the costs of highly regulated medical manufacturing processes are many times higher than in consumer technology and the costs of sales, marketing and customer support are staggering (typically 60–75% of selling price). Much of the costs must be committed before the product is successful, so the few successful products must cover the costs of the many failures. For all of these reasons, the inventor can expect only a modest royalty (typically less than 5% of selling price) and little up-front money, except perhaps in the form of consulting and contracting fees to assist in product development.

- *You will probably put more into it than you expected.* Industrial partners typically provide a limited set of highly defined skills that usually do not include knowledge of the specialized science and technology underlying your invention. If you are serious about getting your invention through the technology transfer process, you must convince your partners that you can be trusted to do anything necessary to fix problems rather than fixing blame when problems inevitably arise. This may include unanticipated investments of time and effort at the expense of new research projects and administrative demands in your academic position, at the same time that your academic colleagues are sure that you have sold your scholar's soul for vast personal gain.

- *So why do it?* All the journal articles in the world are not going to make patients healthier. If you really believe that your research is clinically relevant, then there is only one way to prove it. Instead of bemoaning the royalties you are not receiving or the grants you are not getting, add up the money that your industrial partner is investing in your idea. If the idea succeeds, society will be healthier, you will become famous, and your partner will become wealthy. That sounds like a pretty accurate description of the reasons why each of those three parties became involved in research and development in the first place.

Reference

Loeb, G.E. and Schulman, J.H. (1992). The transfer of technology from the laboratory to the real world. In *Neural Prostheses* Stein, R.B., Peckham, P.H., and Popovic, D. (eds.), pp 329–341. Oxford University Press, New York.

Section IX

42
Movement Synthesis and Regulation in Neuroprostheses

Patrick E. Crago, Robert F. Kirsch, and Ronald J. Triolo

1 Introduction

Many movement disorders result from interrupted or disturbed communication from motor centers in the central nervous system to the motorneurons in the spinal cord (i.e., the muscles remain intact, but their activation and control are disturbed). Two prominent examples are spinal cord injury (SCI) and stroke. The ability to activate the muscles electrically via their nerves has opened up the field of motor system neuroprostheses, which has expanded rapidly in the last decade (Peckham and Gray 1996). For a neuroprosthesis to functionally replace part of the damaged nervous system, it must provide appropriate electrical stimuli to the muscles. It must also take on other tasks normally performed by the nervous system to both control and regulate the artificial movements. The control and regulation tasks refer respectively to specification of the temporal patterns of muscle stimulation to produce the desired movements, and the modification of these patterns during use to correct for unplanned changes (disturbances) in the stimulated muscles or in the environment.

Designers of neuroprostheses take on the task of specifying how to perform many of the same functions normally performed by the CNS. The purpose of this chapter is to review the control and regulation of movements in neuroprostheses. These concepts are not unique to engineering; examples can be found throughout physiology (Houk 1988). Many of the techniques have not yet been reduced to clinical practice, but represent the diverse approaches that are likely to be required in future neuroprostheses. The next two sections will review

briefly the input/output properties of stimulated muscle and define some basic control concepts (feedforward, feedback, and adaptive control) that are used in the subsequent sections and in the following Perspective chapters.

2 Electrically Stimulated Muscle As an Actuator

The attraction of neuroprostheses is that they use the body's own muscles to restore movement. The principle difference is that the muscles are controlled artificially (externally). In the absence of any further differences, it would appear that there is a mere substitution of the source of the muscle inputs. This is not exactly true, for there can be significant differences in the properties of electrically stimulated paralyzed muscle. The differences arise from the nonphysiological methods of muscle excitation and from changes in the muscle properties due to the paralysis or denervation. In addition, changes in the controlled system may arise from surgical alterations of anatomy.

2.1 Muscle Excitation and Force Modulation

Muscles are stimulated via electrodes placed either (a) on the skin surface over the motor point, (b) directly within the muscle (intramuscular electrodes), (c) on the surface of the muscle (epimysial electrodes), or (d) on the muscle nerve. In each case, the muscle is activated indirectly via motor axons since the threshold for nerve stimulation is considerably lower than the threshold for activating muscle directly (Crago et al. 1974).

The strength of contraction can by modulated by both recruitment and temporal summation. Individual muscles are typically stimulated via trains of rectangular current pulses delivered through single stimulus channels. Each current pulse activates a number of axons, which in turn activate the muscle fibers innervated by those axons. Thus, varying the amplitude or pulse duration modulates the recruitment level. The frequency of stimulus pulses determines the overlap of successive twitch contractions, thus varying force by temporal summation.

Modulation of recruitment and temporal summation via electrical stimulation does not directly mimic normal control. The recruitment order of axons depends significantly on electrode-nerve geometry and stimulus waveform (Crago et al. 1980). In most cases, the recruitment order should be considered mixed rather than obeying a strict (either increasing or decreasing) axon diameter relationship. Small axons close to the electrode can have the same threshold as larger diameter axons further from the electrode. Temporal summation also differs from the normal case. Individual motor units controlled by one stimulus channel fire synchronously. Thus, a higher average stimulus rate is required to achieve the same force fusion obtained by normal aynchronous firing of individual motor units.

Recruitment modulation by varying stimulus pulse width or amplitude is the primary method used to control muscle force. To reduce fatigue, it is typical to use the minimum fixed stimulus rate that produces a nearly fused contraction. The steady state isometric force as a function of pulse width or amplitude during constant frequency stimulation is used to characterize the recruitment input/output characteristics (a recruitment curve). When the force is normalized to the maximal force produced by that muscle at that length and stimulus frequency, then the recruitment curve can be interpreted as the fraction of the muscle's cross sectional area excited by that stimulus train. Recruitment curves show prominent nonlinearities (threshold, saturation, low and high gain regions), which are similar whether stimulus amplitude or stimulus pulse width is varied (Crago et al. 1980). The exact shape of the recruitment curve is a function of the electrode location

relative to the nerve branches, the size distribution of the axons in the branches, the innervation ratios of the individual axons, and the strengths of the muscle fibers innervated by the axons. Recruitment nonlinearities are observed for all types of electrodes (e.g., skin surface or nerve cuff) since most of the factors are still relevant. The recruitment characteristics can also be time-varying because of fatigue, atrophy or exercise induced strengthening (Durfee and Palmer 1994). Because the shape depends on the relative electrode-nerve location, the recruitment characteristics are different for each electrode activating the same muscle, as well as for different muscles. Since the relative distance between nerves and electrodes can vary with muscle length, the fractional area of the muscle excited by a stimulus can also vary (length dependent recruitment; Crago et al. 1980). Thus, even at a fixed stimulus level, muscle force may not exhibit the expected length-tension properties (Hausdorf and Durfee 1991).

The shape of the recruitment characteristic is important for the specification of functional stimulation patterns because the activation level of the muscle is not simply proportional to the stimulus parameter chosen for modulation. Nearly all neuroprostheses either implicitly or explicitly invert the recruitment nonlinearities, although inversion cannot be perfect as a result of length and time dependent recruitment characteristics. Since most open-loop feedforward systems simply generate a temporally varying (e.g., for locomotion) or parametrically varying (e.g., for hand grasp, Chapter 44) stimulus pattern to produce a movement, the recruitment nonlinearities are accounted for implicitly. However, some systems explicitly compensate for recruitment and other nonlinearities (e.g., the pattern shaper in Chapter 46; Abbas and Chizeck 1995).

Most neuroprostheses require multiple stimulation channels since multiple muscles must be stimulated to produce functional activity. However, each stimulus channel generally does not correspond to a unique muscle, and one stimulus channel may not be able to fully excite a single muscle. The issues of selectivity and completeness of activation are related. As the stimulus current amplitude and pulse width are increased, axons further and further away from the electrode are recruited. If the current density can be concentrated in the region of a whole nerve innervating a single muscle, then it is likely to be able to fully recruit that muscle without recruiting others (Grandjean and Mortimer 1986). Full recruitment of one muscle might not be achieved before "spillover" of activation to other muscles (perhaps even antagonists) takes place if other nerves are nearby, if the current spreads diffusely, or if the innervation of the target muscle is diffuse (see Chapter 44).

The failure to obtain complete recruitment is one factor that limits the force available from the stimulated muscle, and thus limits possible function. The design of electrodes that achieve full recruitment with complete selectivity is an area of active research (Meier et al. 1992; Veraart, Grill, and Mortimer 1993), and it is likely that the future will bring large improvements compared to what can be achieved with present technology. Although it will require great advances in technology, it would even be desirable to subdivide activation of single muscles into smaller units that would allow more natural asynchronous activation.

2.2 Alterations of Muscle Physiological Properties

CNS lesions result in altered muscle properties as a consequence of changes in motor innervation and changes in patterns of use. This generally results in a decrease in the number of muscles available and the forces that the muscles produce, which are further factors limiting the possible functions that can be restored. In stroke, muscles are paralyzed as a result of damage to descending tracts (upper motor neuron lesion). In spinal cord injury, muscles that receive their innervation from motor neurons below the level of the spinal cord damage are also paralyzed by upper motor neuron lesions. Muscles that receive their innervation from cell bodies located in the vicinity of the spinal cord lesion can be paralyzed as a result of direct damage to motor neuron cell bodies in the ventral horn of the spinal cord or to the spinal nerves as they exit the spine (a lower motor neuron lesion). Because most muscles are innervated from two or more spinal cord segments, muscles near the lesion can also have a mixed innervation status: typically a mixture of intact voluntary control and lower motor neuron lesion, or a mixture of both lower and upper motor neuron lesions. The

significance of this is that muscle fibers that lose their spinal innervation eventually atrophy completely, and can not be stimulated to restore function. Therefore, muscles with partial lower motor neuron lesions will produce less force than normal even if they are fully recruited.

The loss of muscles because of denervation can present insurmountable problems to the implementation of neuroprostheses. In paraplegia due to spinal cord injury in the lower thoracic or lumbar regions, denervation can be so extensive that these individuals are eliminated from the pool of candidates. In tetraplegia, individuals classified as having C5 or C6 function have partial denervation of some finger muscles, and extensive denervation of radial wrist muscles. Individuals with C4 function generally have intact hand and wrist muscles, but may have extensive shoulder and elbow muscle denervation. Even prior to the advent of neuroprostheses, surgical procedures were developed to transfer the tendons of muscles with residual voluntary function into the tendons of denervated or paralyzed muscles to restore some function in these individuals (Moberg 1975). For example, transfer of a voluntarily controlled brachioradialis into the denervated extensor carpi radialis brevis can restore wrist extension, and transfer of posterior deltoid into the triceps can restore elbow extension (Lacey et al. 1986). The principles of tendon transfer surgery have been extended to include the transfer of muscles with upper motor neuron lesions into the tendons of denervated muscles so that function can be restored by stimulation. Transfers of both voluntarily controlled and stimulated muscles can be combined to further augment function (Adamczyk et al. 1966). Other surgical reconstruction procedures simplify the implementation of neuroprostheses by imposing kinematic contraints such as tenodesis (Keith et al. 1996).

The loss of activity in muscles with upper motor neuron lesions results in disuse atrophy, with a resulting decrease in strength, increase in contraction speed, and decrease in fatigue resistance (e.g., Lieber 1986). Clinical applications of functional stimulation systems are typically preceded by periods of muscle conditioning (stimulation induced exercise) that can increase strength and endurance (Peckham et al. 1976; Stein et al. 1992). Conditioning is then continued on a regular basis for maintenance.

2.3   Modeling Stimulated Muscles

Muscle (or frequently muscle-joint) models play an important role in neuroprosthesis design/simulation and real-time control/regulation. The two objectives of muscle modeling in the context of neuroprostheses are system identification and parameter estimation (see Section III). System identification is necessary to choose the form of model required to capture important behavior, and parameters must be estimated for specific individuals for implementing clinical systems or for developing simulation models. Because there is no need for the models to represent physical or chemical processes explicitly, most models are black box, and the number of parameters is kept small so that they can be estimated quickly from small amounts of input/output data. The neural input to muscle during stimulation is generated at discrete times (i.e., only at the times of the stimulation pulses). The exact time course of the force during the interstimulus interval is not important if force is adequately fused. For this reason, discrete-time models that give the force only at the stimulus instants are generally used because difference equations are more efficient computationally than continuous time (differential equation) models.

The basic model of stimulated muscle consists of a static nonlinearity representing the recruitment characteristics, in cascade with activation and contraction dynamics (see Figure 42.1). For isometric contractions, models with a static nonlinearity followed by second order linear dynamics, are adequate (Bernotas et al. 1986). Several methods have been developed for estimating the recruitment non-

linearity. The original method of measuring the steady-state force as a function of pulse width or amplitude (Crago et al. 1980) is inefficient with respect to time and stimulation, and requires separate experiments. A ramp response deconvolution method (Durfee and MacLean 1989) requires much less stimulation, but requires separate estimation of the dynamics. Simultaneous estimates of the nonlinearity and the linear isometric dynamics was achieved by a constrained recursive parameter identification method that could be potentially implemented on-line to allow real-time tracking of changes in system properties (Chia et al. 1991). In this case, the nonlinearity was modeled as a polynomial. Nonparametric estimates of the nonlinearity and the dynamics can also be estimated simultaneously from input-output data (Adamczyk and Crago 1996). Mathematical models describing length-dependent recruitment have not yet been developed.

Under nonisometric conditions, where length, velocity, and activation change over wide ranges of values, the dynamics of muscular contraction must be represented by nonlinear models. There is no evidence that different model structures must be used for paralyzed skeletal muscle than for normal muscle, although the models may likely have different parameters than normal. The general structure of the Hill model of muscle has been adopted widely. Hill-based models capture the major nonlinearities in a product relationship: the active force is the product of activation dynamics, force-velocity, and length-tension factors. The product relationship itself is nonlinear, but each factor may also include nonlinearities. Since the formulation of

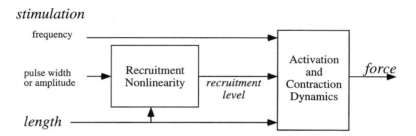

FIGURE 42.1. Model of elecrically stimulated muscle. The neural input to muscle is specified by the stimulus pulse amplitude and pulse width that determine the recruitment level, and stimulus frequency that synchronously acti-

vates all recruited fibers at the same instant. Length affects the contractile dynamics of muscle and can also alter the recruitment nonlinearity.

Hill-type models has been extensively reviewed elsewhere (Zajac 1989; Winters 1990; Chapter 8), we will describe only model variations and parameter estimation methods that were developed specifically for neuroprosthetic applications.

A Hill-based non-isometric model of a muscle-joint system was investigated for the soleus muscle acting at the cat ankle joint by Shue et al. (1995). This model included linear approximations for each of the elements in the multiplicative structure: linear activation dynamics, linear moment-angle, and linear moment-angular velocity factors. The muscle-joint model was chosen to represent the situation that is generally encountered clinically, in which joint angle and moment rather than muscle length and force are all that are accessible to measurement. Parameters were estimated from input–output data collected during random, independent, simultaneous modulation of stimulation and joint angle over a wide range of angles and angular velocities. The ability of this type of model to fit the data from which the parameters were estimated as well as predict the output for other input sequences varied both with the structure of the model and with the range of input conditions. With a total of only four parameters, the model was able to fit the input–output data from which the parameters were derived, as well as predict the output generated by other input sequences, as long as the inputs stayed within a range of angles and velocities. Large increases in angular velocity or changes in mean angle increased the prediction error excessively. Part of the sensitivity to input conditions might be explained by the linear approximations of the moment angle and moment-velocity factors, and by the effects of an angle dependent moment arm on the transformation between muscle dynamic stiffness and joint stiffness. However, the errors with increasing joint velocity could be reduced substantially by including coupling between the activation dynamics and moment-velocity factors that are normally assumed to be independent.

Coupling between activation and velocity was explored further in an isolated cat soleus muscle model rather than a muscle-joint model to eliminate the geometrical effects. Nonlinear force-length and force-velocity models, as well as a nonlinear series elastic element were included in the model (Crago and Shue 1995; Shue and Crago 1998). This expanded model confirmed the benefits of coupling activation dynamics to length and velocity, because the model could predict muscle yielding during stretch, and the length and velocity dependence of the postyielding force trajectory during isovelocity movements. Furthermore, one set of ten free parameters could predict the responses over a large range of lengths, velocities, and activation levels (Crago and Shue 1995; Shue and Crago 1998).

Hill-based models with nonlinear force-length and force-velocity properties have also been tested in isolated stimulated muscles by Durfee (Durfee and Palmer 1994). They did not find it necessary to couple activation to length or velocity to predict the responses to simultaneously modulated length and stimulation input sequences. The seeming disagreement between the findings of Durfee and Palmer and the findings of Shue and Crago can probably be ascribed to differences in stimulation frequency and the different muscles studied. The phenomenon of muscle yielding during stretch is affected greatly by the frequency of stimulation (Joyce, Rack and Westbury 1969), and it is primarily yielding that is accommodated by the coupling in the Shue and Crago model. The Shue and Crago study was carried out at low frequencies of stimulation (10 or 12.5 Hz) in the cat soleus which has uniformly slow muscle fibers. The Durfee and Palmer study employed 40 Hz stimulation in cat medial gastrocnemius and tibialis anterior, which have predominantly fast muscle fibers. Under these conditions, yielding would be expected in the soleus, but not in the other muscles. However, in neuroprostheses, stimulation is generally applied at as low a frequency as possible and if the muscle is fully conditioned, it will likely exhibit the properties of slow fibers. The issue of whether coupling is required in the models of chronically stimulated muscles must be verified experimentally.

Issues regarding modeling of electrically stimulated muscle are not completely resolved. Models of frequency modulation, including the responses to short interval pulse trains, such as doublets, are not available. The work reported in the following perspective (Chapter 43) is a step forward in this regard. Models should also be able to describe time-varying behavior such as potentiation and fatigue in order to be most useful in neuroprostheses. All of the studies described above were carried out under conditions where time variation was minimized to facilitate the experiments. It follows that these models are therefore only suitable for the first few seconds of movement, and that later during movements, the parameters are probably different.

It is unlikely that a single muscle model will be useful in all applications. To fully describe all muscle behavior, a general model would be unnecessarily complex for most applications. What is needed most is a well documented tool box of models that cover many situations. Simulation studies can make use of these models to examine the sensitivity of motor performance to individual parameters and model structure to simplify a model for particular tasks.

3   Feedforward, Feedback, and Adaptive Control

The temporal specification of stimulation patterns (control) for both upper and lower extremity neuroprostheses is difficult because of the complexity of musculoskeletal systems. Specification of the patterns must account for the nonlinear and dynamic relationships between stimulus parameters and muscle output and between muscle output and limb output, as well as the varying load encountered as the subject interacts with the environment. In some systems, only the steady-state (static) input–output properties are considered, but in most systems, both static and dynamic properties are important. For example, in hand grasp neuroprostheses (Keith et al. 1989), grasp control is synthesized by specifying constant relationships (pulse width maps) between a patient-graded command signal and the pulse width applied

to each muscle. Synthesis (reviewed in Chapter 44) is carried out either by following simple qualitative rules or with a quantitative automated process. Grasp synthesis accounts for nonlinear relationships between the stimulus parameters and the grasp output, but does not take into account the dynamic properties of the system being controlled, since the dynamic properties are not very important for grasp-release tasks. Grasp control also does not require a neuroprosthesis generated command signal, since it is under voluntary control by the patient. In contrast, a neuroprosthesis controller for locomotion must specify the basic walking pattern, and the synthesis process must take into account both the nonlinearities of recruitment modulation as well as the nonlinear dynamic properties of the muscles and the limb.

In current neuroprosthetic implementations, locomotion is synthesized by iteratively modifying a basic time varying stimulation pattern to improve the gait of each subject (e.g., Kobetic and Marsolais 1994). Stimulus magnitudes and timing are altered on the basis of walking performance, as assessed visually or by quantitative motion analysis. The iterative modification rules are based on the experience of experts, and compensate for both system nonlinearities and dynamic properties. The dynamic properties that are most important are the inertia of limb segments and the time lag between when a muscle is stimulated and when it actually generates force to accelerate or decelerate the limb. Because of the dynamic system properties, a single pattern is not suitable for all walking speeds.

The control systems described above can be clas-

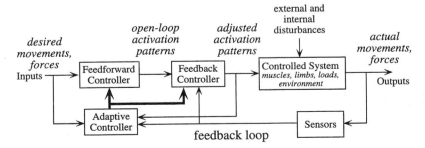

FIGURE 42.2. Generic block diagram illustrating concepts of feedforward (open-loop) control, feedback control and adaptive control. In feedforward control, the desired movements and forces are used to generate the muscle activation patterns that should produce the movement. In feedback control, sensors are used to monitor the actual

movements and forces, and a feedback controller modifies the activation patterns to correct for differences between the desired and actual outputs. In adaptive control, the inputs and outputs are monitored and the feedforward and/or feedback controllers are modified to optimize performance. (From Crago, Lan, Veltink, Abbas and Kantor 1996.)

sified as feedforward control systems (Figure 42.2), since they specify the stimulus parameters (musculoskeletal system inputs) that are expected to be needed to produce the desired movement (system outputs). Feedforward control systems do not make corrections if the actual movement deviates from the desired movement. Deviations are common in neuroprostheses because stimulated muscle properties are time varying (e.g., fatigue) and because the patient operates in a constantly changing environment (e.g., changes in the slope of the walking surface).

Feedback control represents the broad category of control systems that correct for a changing system or environment. In a feedback control system, sensors monitor the output and corrections are made if the output does not behave as desired (Figure 42.2). The corrections are made on the basis of a control law, which is a mathematical prescription for how to change the input to reduce the error between the desired output and the actual output. Much of the work done in automatic control of stimulated muscle in the last 20 years has focused on feedback controllers, with the objectives of assessing how well feedback control can regulate motor activities, and identifying the best control law for the system being controlled and for the type of behavior desired. Feedback control has been successful in regulating hand grasp and standing posture, but it appears that another strategy, adaptive feedforward control, is likely to be required for many dynamic activities such as locomotion.

Adaptation refers to the ability of a control system to change how it responds to inputs or disturbances, based on changes in the properties of the controlled system or the environment. In the case of movement control, the principle of adaptation is to monitor the musculoskeletal system properties by measuring the actual outputs (movements, forces) and the inputs (commands, stimulus parameters) during neuroprosthesis operation. From the inputs and outputs, the feedforward controller and/or the feedback control law are altered to improve performance according to a predetermined optimization criterion. For example in locomotion control, the quadriceps stimulation intensity might be progressively increased in amplitude to compensate for fatigue that would cause the knee to buckle during stance.

There are tradeoffs in the choice of control system for a neuroprosthesis. Even extremely simple feedback control laws can greatly improve performance and can compensate for any source of disturbance. Feedback control also has drawbacks. It requires output sensors and compensation, and is generally slower than feedforward control since an output error must be present to generate a controller response. Thus, feedback control might be best used for slow movements, and for maintaining a steady posture (e.g. hand grasp). On the other hand, feedforward control requires much more detailed information of how the system behaves (i.e. an internal model, see Chapter 24 and Chapter 27) to generate a stimulus pattern that will produce an accurate movement, and may produce poor movements if the system properties change. The most significant advantages of feedforward control are that it can be used for rapid movements (e.g., the swing phase of gait) and may not require sensors. Adaptation requires sensors to monitor output, and can be used to improve the performance of either feedback or feedforward control. The remainder of this chapter will focus on control and regulation methods that have actually been attempted in neuroprosthetic systems.

4 Applications of Control and Regulation in Motor System Neuroprostheses

All of the control and regulation concepts described above have been studied to some extent for their potential applicability to neuroprostheses. The following summary is organized according to the type of control rather than the neuroprosthetic application. Both upper extremity and lower extremity applications are described.

4.1 Feedforward Control

4.1.1 Static Feedforward Control

Limb dynamics have not played a very dominant role in the design of upper extremity neuroprostheses. The most widely deployed neuroprosthesis restores hand grasp in individuals with spinal cord injury. Grasp is synthesized as a series of quasistatic hand postures keyed to different levels of a command signal (Kilgore et al. 1995; Handa et al. 1989;

also see review of grasp synthesis techniques in Chapter 44). For each command level, the stimuli applied to several muscles are adjusted to produce the desired quasistatic posture. The user modulates a continuous command signal by voluntary movements of some part of the body that remains under voluntary control, such as shoulder elevation or wrist extension. As the command signal changes, the stimuli are interpolated between the tuned postures, and the hand moves from one quasistatic posture to the next. The reason that this method has been successful probably lies in three factors: the digits do not have a very high inertia, the command signals are modulated slowly, and errors can be corrected by the person adjusting the command signal.

Another technique of restoring upper limb function by FNS takes advantage of the remaining voluntary function to simplify control. Elbow flexion/extension and forearm rotation have been restored by stimulating the triceps and pronator quadratus respectively (Miller et al. 1989; Lemay et al. 1996; Crago et al. 1996b; Crago et al. 1998). In individuals with C5 and C6 tetraplegia, the elbow extensors and forearm pronators are paralyzed, but their antagonists are not (i.e., the individuals retain elbow flexion and forearm supination). Stimulating the triceps and pronator produces moment biases that can be counteracted by voluntary contractions of their antagonists. Thus, a separate command signal is not needed, and nearly full range of movement can be restored. Stimulation levels are simply chosen to maximize the moment in the given direction, without overpowering the residual voluntary function.

In the case of elbow or forearm rotation control, the muscles can be stimulated at constant levels (Lemay et al. 1996; Crago et al. 1996b), with the potential problem of excessive fatigue. This can be overcome to some extent by using sensors to automatically modulate the stimulation according to the arm posture. Miller et al. (1989) adjusted the stimulation to the triceps according to the computed moment required to overcome gravity. They utilized kinematic sensors on the arm to compute the orientation of the arm in the gravitational field. A somwhat simpler scheme was employed by Crago et al. (1996b) who used an accelerometer on the upper arm to stimulate the triceps whenever the arm was abducted beyond a threshold amount. It is expected that the choice of stimulus level will determine the postural stability of the arm by establishing the joint stiffness, and thus can alter the functional utility. In addition, sensor driven systems are inherently feedback systems, and stability is not guaranteed. Both elbow control schemes would benefit from a dynamic analysis to determine stability requirements.

4.1.2 Dynamic Feedforward Control

The quasistatic techniques typically used to obtain stimulus patterns in upper extremity FNS applications are not appropriate for many lower extremity functions because of the dynamic nature of the motions to be restored. The musculature of the lower extremities must move or stabilize the mass of the entire trunk and upper body in addition to moving the lower extremities themselves, so the mechanical dynamics of the body cannot be ignored. Furthermore, maintaining balance during dynamic tasks such as standing or locomotion is a significantly more complex control problem than is encountered during most upper extremity functions, and the consequences of a control failure (e.g., falling) are quite severe. Thus, muscle stimulation patterns during many tasks cannot be obtained simply by interpolating between a series of static postures as is done successfully in many upper extremity applications. Rather, dynamic stimulation patterns are required to accelerate and decelerate the mass of the body during both ballistic (e.g., sit-to-stand) and cyclical (locomotion) tasks.

Most efforts to restore such dynamic tasks to individuals with paralysis have constructed these dynamic muscle stimulation patterns in a feedforward manner (i.e., the patterns are determined prior to the task and are not modified by performance-related feedback during the task). Initial stimulation levels for the various muscles involved in a task are typically generated on the basis of EMG recordings from the muscles of neurologically intact individuals (Kobetic 1986a,b, 1987; Hoshimiya et al. 1989; Kameyama et al. 1991), the known mechanical requirements of the task (Kagaya et al. 1995), previous experience and intuition (Kobetic and Marsolais 1994), or some combination of all of these. These initial muscle stimulation templates are then modified, again usually based upon previous experience (e.g., Kobetic and Marsolais 1994), to optimize the performance of the task for a particular individual. Typically, initial stimulation levels are chosen to be larger than needed so that the task can continue to be performed for a longer period of time after the onset of fatigue. However, the prefatigue performance of the task is then often exaggerated (e.g., goose-step during locomotion). Thus, the generation of "open-loop" stimulation sequences is often at least partially by trial and error. However, these stimulation sequences implicitly take into account the complex dynamics of body mechanics, the modified properties of paralyzed muscle, and the limited number of stimulated muscles typically available in FNS systems with relatively simple empirical rules.

Open-loop stimulation patterns differ for various tasks in obvious ways. For the sit-to-stand maneuver, maximum stimulation of the knee extensors (quadriceps) is typically used (e.g., Bajd et al. 1982), sometimes in conjunction with hip extensor (e.g., gluteus maximus) stimulation (Kralj et al. 1993), to raise the body into an upright stance. Kagaya et al. (1995) generated knee extensor stimulation amplitudes for the sit-to-stand maneuver based on knee joint moment requirements, while stimulation timing was based on the EMG profiles from able-bodied individuals. Once a standing position is attained, stable balance is typically maintained by high intensity stimulation of the knee extensors to "lock" the knee (Kralj et al. 1983) into full extension, with upper extremity actions (via crutches, a walker, or similar aid) used to hyperextend the hip (to achieve a "C" posture) and to maintain balance. Although a crude approximation of the normal control of standing, this approach has been found to be adequate for standing transfers (Marsolais et al. 1994; Triolo et al. 1996), and the postural stability it provides to individuals with paraplegia has been found to be comparable to that provided by external orthoses (Cybulski and Jaeger 1986), at least for short periods. More extended periods of standing have been achieved by delaying or even preventing the onset of significant fatigue-related force loss in the knee extensors by switching between different stable standing postures which use different combinations of stimulated muscles (Kralj et al. 1986). Fujita et al. (1995) chose muscle stimulation levels for quiet standing which minimized the sum of the gravity moments across the joints of the lower extremity. The stand-to-sit maneuver is typically performed by turning

off all of the stimulation to the lower extremity musculature and using the voluntary upper body musculature to lower the body into a seated position (Kralj 1983; Mulder et al. 1992). Locomotion has been produced by programmed muscle stimulation sequences based on able-bodied EMG patterns that are adjusted in a systematic manner (Kobetic and Marsolais 1994) for each subject to produce reasonably smooth and energy-efficient gait, but final stimulation patterns are usually much different than "normal" EMG. The normal EMG gives a gross starting point, but must be modified because of differences in the number of active muscles. Alternatively, locomotion can be produced using stimulation of fewer individual muscles combined with a stimulated flexion withdrawal reflex (Kralj et al. 1983, 1993a; Kralj and Bajd 1989). Likewise, other functions such as stair climbing (Kobetic et al. 1986b; Chizeck et al. 1988), side-stepping (Kobetic et al. 1986a; Chizeck et al. 1988), and back-stepping (Kobetic et al. 1987) have been restored in individuals with paraplegia using stimulation patterns constructed by experts or from able-bodied EMG patterns.

Dynamic feedforward systems have had much less application in the upper extremity. One of the limits to feedforward control of arm movements is the wide range of possible movements that are required for different tasks (i.e., nearly all movements are novel, rather that repetitive like locomotion). Lan has addressed this issue by developing dynamic model based feedforward systems to generate muscle activation patterns for one and two degree of freedom arm movements that display normal kinematics (Lan, Feng, and Crago 1994; Lan 1996). A model-based approach such as this could also be employed in lower extremity systems to alter stimulation patterns for different speeds of locomotion or for traversing different height steps.

Although open-loop stimulation sequences have been the predominant method for restoring many lower extremity functions via functional neuromuscular stimulation, there are several significant limitations. First, the generation of the stimulation sequences for the various muscles requires highly experienced personnel, especially since each subject is likely to require a unique stimulation template because of differences in muscle properties and upper extremity strength. Second, the developed sequence will be appropriate only if muscle properties remain constant over time. In the short term, fatigue and/or potentiation can change the level of force generated by a predetermined stimulus. In the long term, muscle force capacity can decrease due to disuse or increase due to electrically produced exercise. Third, the energy efficiency of many FNS-restored lower extremity functions is quite low, primarily because of the overstimulation required to help maintain stability. Finally, all of the open-loop approaches rely heavily on the use of the neurologically intact upper extremities (via crutches, walkers, etc.) and/or external orthoses to supply extra power and to provide stability. The need for upper body intervention severely limits the tasks which can be performed by the volitional upper extremities during restored lower extremity functions such as standing.

4.2   Single Variable Continuous Feedback Control

Many of the difficulties with open-loop stimulation sequences described in the previous section can potentially be addressed through feedback control. Conventional closed-loop approaches using continuous feedback have been employed to control the stimulation to paralyzed muscle, primarily during standing in individuals with paraplegia. Chizeck et al. (1985, 1988) used independent PID controllers at each of the ankles and knees of individuals with paraplegia to control sagittal plane stability. This system provided an enhanced disturbance resistance and longer standing durations, but was very sensitive to muscle fatigue. Moynahan and Chizeck (1993) investigated a controller that used open-loop stimulation of the hip and trunk musculature in combination with feedback-controlled stimulation of the quadriceps to maintain stance. The controller modulated both the stimulus period and pulsewidth of the quadriceps in response to changes in knee position, and was tuned to provide a burst-like response to changes in joint position that drove the knee into the stable full extension position. Donner (1986) investigated the use of feedback control of coronal plane hip angle and found that a combination of costimulation and feedback control increased the resistance to imposed disturbances during standing. Abbas and Chizeck (1991) implemented a two-stage controller for coronal plane hip angle which consisted of a

modified PID stage in cascade with a nonlinear single-input, multiple-output costimulation map. This controller was reported to provide enhanced disturbance resistance, indicating that including such control might reduce the reliance on upper body actions to provide stability. Abbas and Chizeck (1987) investigated the control of hip and trunk angles in both the sagittal and coronal planes using skin stretch sensors. This controller also successfully reduced the dependence on upper body actions, and suggested that coordination between the hip and trunk controllers is needed for adequate maintenance of posture in the presence of disturbances.

Continuous feedback control has found far fewer applications for dynamic lower extremity tasks such as walking. The mechanical requirements on the lower extremities change substantially during different phases of gait (see below), so a single fixed controller is unlikely to be appropriate for the whole cycle. Because of delays in force generation due to muscle contraction dynamics and the rapid changes in the gait cycle, it may even be difficult to switch between different control strategies for each phase of gait. In general, control of gait has focused upon predictive mechanisms based on its cyclical nature.

4.3   Multivariable Continuous Feedback Control

Often, something other than control of a single variable such as limb position or contact force is desired. This is particularly important when the limb must interact with objects in the environment, such as during object manipulation. Stiffness regulation, employing a combination of feedback signals from contact force and position sensors, provides a compliant interface with the environment. A stiffness regulator provides position control when the limb is not in contact with an object and force control when it is in contact with a rigid object. This was demonstrated for control of hand grasp (Crago et al. 1986, 1991). Thus, a single control system can effectively switch control modes in response to loads in the environment. The same concept was reported by Willemin and Chizeck (1989) to control the stiffness of the knee joint during the stance phase of FNS-mediated gait where it was implemented with feedback of knee angle and an estimate of knee moment. Such a controller would provide additional knee angle stability which is critical during this phase of gait without requiring the potentially damaging knee hyperextension produced by typical open-loop stimulation patterns. The concept of stiffness regulation has also been extended to the whole limb, where the multi-dimensional relationship between endpoint force and position can be regulated (Lan et al. 1991).

4.4   State-Dependent Control

As mentioned above, many lower extremity tasks can be divided into specific phases or "states," which are mechanically quite distinct. For example, gait is divided into a number of phases (e.g., stance phase and swing phase) that are distinguished by discrete events (e.g., initial contact and toe off) (Chizeck et al., 1988; Popovic 1993; Franken et al. 1994). The nature of the control problem is often quite different during these different phases. For example, during stance phase one of the legs must support the weight of the body while maintaining forward progression, but the goals during swing phase are to prevent the foot from dragging on the ground and to place the foot appropriately to accept weight during its next stance phase. The discrete nature of these phases of a task and the different control needs during each phase has led several groups to view various lower extremity tasks in terms of "finite state" control rather than fixed feedback control of angles or moments. In such systems, event-related (rather than continuous) feedback is used to control the FNS to paralyzed muscle. Most of the control approaches for locomotion rely on its periodic nature and typically target corrective actions on subsequent portions of the gait cycle or the next gait cycle rather than attempting to correct moment-to-moment errors. Most such methods are thus inherently predictive, although more conventional feedback can also be used in parallel. Research has focused both on the detection of the discrete "states" of tasks such as gait and on the development of control methods for each state.

A number of groups have focused on the robust and automatic detection of the different events during the gait cycle, with a goal of reducing or eliminating the cognitive burden on the user to voluntarily initiate the FNS patterns for each individual cycle of gait. Automatic gait event detection algorithms have been based upon signals obtained from body-mounted sensors (Marsolais and Kobetic 1987; Andrews 1989; Ng and Chizeck 1993, 1994) or from sensors mounted on orthoses (Andrews et al. 1989) used in hybrid (i.e., orthoses plus FNS) systems. A number of gait event detection algorithms have been developed both for FNS and hybrid systems. Andrews et al. (1989) and Popovic et al. (1989, 1992) have used "rule-based" methods to control hybrid FNS systems. Ng and Chizeck (1993, 1994) used a fuzzy logic approach to correctly detect the phase of FNS-produced gait of an individual with paraplegia 94% of the time. Kirkwood et al. (1989) introduced a "behavior cloning" method that used sensor signals to automatically predict the learned actions of the user in operating a finger switch that controlled the duration and pattern of stimulation to the knee extensors and to the comon peroneal nerve (which evokes a flexor withdraw). Kostov et al. (1995) subsequently found that similar "machine learning" techniques could use information from foot force and joint angle sensors to accurately predict when FNS users (who had voluntary control over the initiation of stimulation of one or a few channels and presumably had optimized their gait pattern through training) would initiate their stimulation sequences. Finally, the use of "unsupervised" (i.e., exploratory trial and error) machine learning has been proposed (Wang and Andrews 1994; Thrasher et al. 1996) for use in FNS.

The primary use of finite state control for FNS gait has been to simply trigger preprogrammed stimulus sequences following detection of specific events within the gait cycle. Indeed, the first practical clinical systems utilizing FNS used a form of finite state control to prevent foot drop in individuals with hemiplegia. Liberson et al. (1961) described a system that triggered surface electrical stimulation of the peroneal nerve whenever a switch in the sole of the shoe indicated that weight was not being supported by the foot, producing ankle dorsiflexion and thus preventing the foot from dragging on the ground during swing phase (see also Kralj et al. 1993). However, others have developed controllers which adjust stimulation patterns to compensate for detected changes in system properties (e.g., fatigue). Veltink and colleagues (Veltink 1991; Franken et al. 1995) have investigated the control of the swing phase of gait by using a nominal "open-loop" stimulation pattern optimized for the unfatigued muscles, and then adapting the stimulus parameters for a particular cycle based on the performance in the previous cycle. They were able to show that both muscle potentiation and fatigue could be overcome in two individuals with paraplegia during simulated gait in a standing device by a discrete-time PID controller to maintain the hip angle trajectory; the knee flexors and extensors were largely unaffected by fatigue and thus their performance was adequate with open-loop stimulus patterns. Although this approach shows promise for adaptation for slow changes in muscle properties due to potentiation and fatigue, it is not appropriate for more rapid external disturbances (Franken et al. 1994). It is likely that disturbance resistance during FNS gait will have to be provided by a separate mechanism, perhaps via joint stiffness control (Willemin and Chizeck 1989) or by parallel feedback control (Abbas and Chizeck 1995).

Finite state control has also been used for FNS standing up and quiet standing. Mulder et al. (1992b) used an on/off control algorithm to control the stimulation of the quadriceps muscle during experimentally simulated sit-to-stand such that terminal knee angular velocity was reduced (to minimize potential joint damage). Knee angle and angular velocity were monitored during FNS-controlled "standing up," with stimulation switching from maximum to off depending on the relative values of the joint angle and velocity. This relationship effectively produced high initial velocities (by using maximum stimulation levels), but as the knee approached full extension the stimulation was turned off for all but the lowest velocities. Mulder et al. (1990; 1992a) used a finite state controller to control knee joint angle during quiet standing. This controller monitored knee angle but modified the stimulation level to the quadriceps only if it flexed from a locked position more than a certain threshold ($1.8°$). Once this threshold was exceeded, stimulation intensity was switched to its maximum to drive the knee back into full extension (i.e., into a locked position). After this was achieved (detected as an angular velocity of zero), stimulus intensity was exponentially ramped downwards until the knee flexion threshold was again exceeded. This controller was found to significantly to increase standing time relative to open-loop control, presumably because of a reduction of the effects of fatigue.

State dependent control has not found as much application in the upper extremity, probably since the activity is not repetitive, and the user generally has continuous graded control over the stimulation. However, recent successful attempts to record signals from cutaneous afferent nerves in human subjects has made available feedback information that can not presently be obtained from artificial sensors. Whole nerve recordings show bursts of afferent activity when an object slips across the skin. Haugland and Sinkjaer (see Chapter 45) have implemented a system to detect these bursts and automatically increase the stimulation to hand grasp muscles to prevent slip. Slip detection/correction has the potential benefits of a) relieving the neuroprosthesis user from the attention burden of watching their hand continuously (they do not generally have conscious sensation) and b) automatically adjusting the stimulation to a known level above what is required. The latter could help reduce fatigue by minimizing the level of stimulation, and could simplify the adjustment of the stimulation level so that perhaps a Boolean command structure, rather than a continuous command structure could be employed for hand grasp.

5   The Role of Simulations in Neuroprosthesis Development

Model-based simulations have been used by a number of investigators as an alternative to experimental approaches in evaluating the feasibility of controllers for FNS systems. This approach has the advantages of not endangering subjects with untested control systems, while allowing many different types of controllers to be evaluated without requiring long and expensive experimental procedures. Thus, musculoskeletal models have been used in a number of studies to develop and evaluate different FNS controllers for a variety of different tasks.

Jaeger (1986) used an inverted pendulum model of the body in the sagittal plane and a PID controller for the ankle joint to conclude that closed-loop control of standing posture using FNS was feasible if ankle muscle strength was adequate. Khang and Zajac (1989a,b) developed a three joint (ankle, knee, hip) sagittal plane model of the body driven by 13 muscle groups (modeled by Hill-type dynamic muscle models) to design an optimal controller for FNS-restored standing that distributed stimulation to the muscles such that energy consumption was minimized, thus delaying the onset of fatigue. A similar model was used (Kuo and Zajac 1993b) to explore how multijoint mechanical constraints may influence the choice of muscle activation patterns during standing, and to suggest the muscles which should be strengthened to enhance FNS-induced standing (Kuo and Zajac 1993a).

Computer-based models have also been developed to investigate the mechanics and FNS-based control of gait. Yamaguchi and Zajac (1990) developed a 3-dimensional, 8 degree-of-freedom, 14 muscle group forward dynamic model of the body to investigate actuator requirements for restoring FNS-controlled gait. They concluded that FNS-restored gait will probably not require stimulation of all of the muscles of the lower extremities and that full muscle strength is probably needed only in the ankle plantarflexors, and a passive ankle-foot orthosis can be used to augment their strength. Scheiner et al. (1993) developed a 23 degree-of-freedom model of gait which did not include the actions of individual muscles. They subsequently found that stable simulation of the entire gait cycle [rather than the half cycle simulated by Yamaguchi and Zajac (1990)] depended on the inclusion of joint stiffness properties which presumably would arise from the stiffnesses of contracting muscles (Scheiner et al. 1994). Abbas and Chizeck (1995) addressed a different set of issues in FNS-restored gait and other cyclical movements. They used a significantly simpler single degree-of-freedom mechanical model, but used an adaptive neural network-based control system that would be able to adapt to differences in system parameters from subject-to-subject and to adapt to changes in system properties across time (because of fatigue, for example), and to resist external disturbances. As discussed in the perspective by Jayasundera and Abbas (Chapter 46), this model uses a physiologically based central pattern generator to produce nominal stimulation sequences and adaptive "pattern shapers" to modify these patterns relative to the specific properties of the muscles of a given subject. A standard servo feedback controller is used in parallel to the adaptive feedforward controller to improve disturbance resistance

The studies described above have provided valuable insight into the control issues for complex multisegment systems, and indicate that model-based simulations are a potentially powerful tool in the development of control systems for neuroprostheses. However, it should be noted that the models described above were still fairly idealized (most are sagittal plane only, with frictionless rotary joints and both legs assumed identical). Furthermore, system properties (number of available muscles, muscle strength and contraction speed, passive joint properties) have been typically assumed to be similar to those in able-bodied individuals despite clear evidence to the contrary for the targeted paraplegia population. With the exception of the approach of Abbas and Chizeck (1995), none have considered variability in system properties across individual subjects. Finally, none of these models have been validated by experimentation.

6 Future Directions

The fact that current clinical applications of FES for restoring motor functions use only simple feedforward control with a limited number of stimulation channels implies that the control schemes discussed above can be considered as future directions for clinical applications. There are however some other general areas of research that will improve control, and restore new functions to a broader population of the disabled.

- There is a need for better control of more muscles. This will require new and improved methods of exciting individual muscles that can achieve selective, complete, and graded modulation of recruitment and temporal summation. Excitation must be achieved by techniques that are compatible with totally implantable systems with long-term stability.

- Sensory feedback, whether recorded from intact neural sensors (e.g., Chapter 45) or manufactured sensors, will greatly improve the ability to match (adapt) behavior to the constantly changing environment. This is a rich area for future work, as indicated by the above review of current research.

- Future systems will need to incorporate residual voluntary control into the control of stimulated muscles. Complete paralysis of limbs is the exception, rather than the rule, and improved acute care after spinal injury will increase the number as well as the strength of residual voluntarily controlled muscles. In cases where there is sufficient antagonist strength, residual control can simplify the command control process, as in the case of stimulated elbow extension to oppose voluntary elbow flexion [Crago et al. (1998)]. However modulation of the stimulation to account for other voluntary activation patterns is yet to be established.

• Most of the systems described above act at the final motor pathway to the muscles, i.e. excitation takes place at the level of the motor efferents in the periphery. In most motor disorders, significant neural structures are still in place between the site of the lesion and motor efferents. Thus it stands to reason that excitation at higher levels in the neural pathway may make it possible to incorporate residual neural control systems into movement production and regulation. For example stimulation at the level of the spinal cord may produce force fields that can be modulated to coordinate multiple muscles acting across multiple joints (e.g., see Chapters 24 and 25), or may make it possible to produce coordinated grasp with intact slip reflexes or to generate locomotor patterns naturally.

• The population of people who can benefit from FES will be increased significantly when control is adapted to account for, or counteract, abnormal states of residual neural control systems that occur in other motor disorders. Many motor disorders produce undesired contractions as well as paralysis. For example, spasms and spasticity can severely disrupt the motor patterns produced by directly stimulating muscle nerves. Involuntary contractions or synergies of movements are typical in individuals with stroke, complicating the production of even simple movements by FES [e.g., Hines et al. (1995)]. Because the population of individuals with stroke represent a very large pool of individuals who might benefit from FES, these problems represent a significant research challenge with a large potential benefit.

References

Abbas, J.J. and Chizeck, H.J. (1987). Feedback control of the hip and trunk in paraplegic subjects using FNS. *Proceedings 9th IEEE EMBS Soc.*, 1571–1572, Boston.

Abbas, J.J. and Chizeck, H.J. (1991). Feedback control of coronal plane hip angle in paraplegic subjects using functional neuromuscular stimulation. *IEEE Trans. Biomed. Eng.*, 38:687–698.

Abbas, J.J. and Chizeck, H.J. (1995). Neural network control of functional neuromuscular stimulation systems: computer simulation studies. *IEEE Trans. Biomed. Eng.*, 42:1117–1126.

Adamczyk, M.M. and Crago, P.E. (1966). Input-output nonlinearities and time delays increase tracking errors in hand grasp neuroprostheses. *IEEE Trans. Rehab. Eng.*, 4:271–279.

Adamczyk, M.M., Crago, P.E., Gorman, P.H., Keith, M.W., Egelseder, A., Marshall, L., Bryden, A., Kilgore, K.L., and Maher, M. (1966). Stimulated tendon transfers for integrated neuroprosthetic wrist/hand grasp control in tetraplegia. *First Annual Meeting of the International FES Society*. Cleveland, Ohio.

Andrews, B.J., Barnett, R.W., Phillips, G.F., Kirkwood, C.A., Donaldson, N., Rushton, D.N., and Perkins, T.A. (1989). Rule-based control of a hybrid FES orthosis for assisting paraplegic locomotion. *Automedica.*, 11:175–199.

Bajd, T., Kralj, A., and Turk, R. (1982). Standing up of a healthy subject and a paraplegic patient. *J. Biomech.*, 15:1–10.

Bernotas, L.A., Crago, P.E., and Chizeck, H.J. (1986). A discrete-time model of electrically stimulated muscle. *IEEE Trans. Biomed. Eng.*, 33:829–838.

Chia, T.L., Chow, P-C., and Chizeck, H.J. (1991). Recursive identification of constrained systems: an application to electrically stimulated muscle. *IEEE Trans. Biomed. Eng.*, 38:429–442.

Chizeck, H.J. (1992). Adaptive and nonlinear control methods for neural prostheses. In *Neural Prostheses: Replacing Motor Function After Disease or Disability*. Stein, R.B., Peckham, P.H., and Popovic, D.B. (eds.), pp. 298–328. Oxford University Press, New York.

Chizeck, H.J., Kobetic, R., Marsolais, E.B., Abbas, J.J., Donner, I.H., and Simon, E. (1988). Control of functional neuromuscular stimulation systems for standing and locomotion in paraplegics. *Proc IEEE*, 76:1155–1165.

Chizeck, H.J., Lalonde, R., Chang, C.W., Rosenthal, J.A., and Marsolais, E.B. (1985). Performance of a closed-loop controller for electrically-stimulated standing in paralyzed patients. *RESNA 8th Annu. Conf.*, 231–232.

Crago, P.E. and Shue, G-H. (1995). Muscle-tendon model with length-history dependent activation-velocity coupling. *17th Annu. Intl. Conf. the IEEE Eng. in Med. and Biol. Soc.*, Montreal, Canada.

Crago, P.E., Nakai, R.J., and Chizeck, H.J. (1991). Feedback regulation of hand grasp force and position during stimulation of paralyzed muscle. IE*EE Trans. Biomed. Eng.*, 38:17–28.

Crago, P.E., Nakai, R.J., and Chizeck, H.J. (1986). Control of grasp by force and position feedback. *RESNA 9th Annu. Conf.*, Minneapolis.

Crago, P.E., Peckham, P.H., and Thrope, G.B. (1980). Modulation of muscle force by recruitment during intramuscular stimulation. *IEEE Trans. Biomed. Eng.*, 27:679–684.

Crago, P.E., Lan, N., Veltink, P.H., Abbas, J.J., and Kantor, C. (1996a). New feedback control strategies for neuroprosthetic systems. *J. Rehab. Res. Dev.* 33:158–172.

Crago, P.E., Peckham, P.H., Van Der Meulen, J.P., Mortimer, J.T. (1974). The choice of pulse duration for chronic electrical stimulation via surface, nerve, and intramuscular electrodes. *Annals Biomed. Eng.*, 2:252–264.

Crago, P.E., Usey, M., Memberg, W.D., Keith, M.W., Kirsch, R.F., Chapman, G.J., Katorgi, M.A., and Perreault, E.J. (1996b). An elbow extension neuroprosthesis for individuals with tetraplegia. *First Annual Meeting of the Intl. FES Soc.*, Cleveland, Ohio.

Crago, P.E., Memberg, W.D., Usey, M.K., Keith, M.W., Kirsch, R.F., Chapman, G.J., Katorgi, M.A., and Perreault, E.J. (1998). An elbow extension neuroprosthesis for individuals with tetraplegia. *IEEE Trans. Rehab. Eng.*, (in press).

Cybulski, G.R. and Jaeger, R.J. (1986). Standing performance of persons with paraplegia. *Arch. Phys. Med. and Rehab.*, 67:103–108.

Donner, I. (1986). Electrical stabilization of the hip in the lateral plane via a feedback controller. M.S. Thesis, Case Western Reserve University, Cleveland, Ohio.

Durfee, W.K. (1992). Model identification in neural prosthesis systems. In *Neural Prostheses: Replacing Motor Function After Disease or Disability*, Stein, R.B., Peckham, P.H., and Popovic, D.P. (eds.). Oxford University Press, New York.

Durfee, W.K. and MacLean, K.E. (1989). Methods for estimating isometric recruitment curves of electrically stimulated muscle. *IEEE Trans. Biomed. Eng.*, 36:654–667.

Durfee, W.K. and Palmer, K.I. (1994). Estimation of force-activation, force-length, and force-velocity properties in electrically stimulated muscle. *IEEE Trans. Biomed. Eng.*, 41:205–216.

Franken, H.M., Veltink, P.H., and Boom, H.B.K. (1994). Restoration of paraplegic gait by functional electrical stimulation. *IEEE/EMBS Magazine*, 13:564–570.

Franken, H.M., Veltink, P.H., Baardman, G., Redmeijer, R.A., and Boom, H.B.K. (1995). Cycle-to-cycle control of the swing phase of paraplegic gait induced by surface electrical stimulation. *Med. and Biol. Eng. Comput.*, 33:440–451.

Fujita, K., Handa, Y., Hoshimiya, N., and Ichie, M. (1995). Stimulus adjustment protocol for FES-induced standing in paraplegia using percutaneous intramuscular electrodes. *IEEE Trans. Rehab. Eng.*, 3:360–366.

Grandjean, P.A. and Mortimer, J.T. (1986). Recruitment properties of monopolar and bipolar epimysial electrodes. *Annals Biomed. Eng.*, 14:53–66.

Handa, Y., Ohkubo, K., and Hoshimiya, N. (1989). A portable multi-channel FES system for restoration of motor function of the paralyzed extremities. *Automedica*, 11:221–231.

Hausdorff, J.M. and Durfee, W.K. (1991). Open-loop position control of the knee joint using electrical stimulation of the quadriceps and hamstrings. *Med. Biol. Eng. Comput.*, 29:269–280.

Hines, A.E., Crago, P.E. and Billian, C. (1995). FNS for hand opening in spastic hemiplegia. *IEEE Trans. Rehab. Eng.*, 3:193–205.

Hoshimiya, N., Naito, A., Yajima, M., and Handa, Y. (1989). A multichannel FES system for the restoration of motor functions in high spinal cord injury patients: respiration-controlled system for multijoint upper extremity. *IEEE Trans. Biomed. Eng.*, 36:754–760.

Jaeger, R.T. (1986). Design and simulation of closed-loop electrical stimulation orthoses for restoration of quiet standing in paraplegia. *J. Biomech.*, 19:825–835.

Joyce, G.C., Rack, P.M.H., and Westbury, D.R. (1969). The mechanical properties of cat soleus muscle during controlled lengthening and shortening movements. *J. Physiol.*, 204:461–474.

Kagaya, H., Shimada, Y., Ebata, K., Sato, M., Sato, K., Yukawa, T., and Obinata, G. (1995). Restoration and analysis of standing-up in complete paraplegia utilizing functional electrical stimulation. *Arch. Phys. Med. Rehab.*, 76:876–881.

Kameyama, J., Sakuri, M., Handa, Y., Handa, T., Takahashi, H., and Hoshimiya, N. (1991). Control of shoulder movement by FES. *Proceedings of the 13th IEEE EMBS* 13:871–872.

Keith, M.W., Kilgore, K.L., Peckham, P.H., Wuolle, K.S., Creasey, G., and Lemay, M.A. (1966). Tendon transfers and functional electrical stimulation for restoration of hand function in spinal cord injury. *J. Hand Surg.*, 21A:89–99.

Keith, M.W., Peckham, P.H., Thrope, G.B., Stroh, K.C., Smith, B., Buckett, J.R., Kilgore, K.L., and Jatich, J.W. (1989). Implantable functional neuromuscular stimulation in the tetraplegic hand. *J. Hand Surg.*, 14A:524–530.

Khang, G. and Zajac, F.E. (1989a). Paraplegic standing controlled by functional neuromuscular stimulation:

Part I- Computer model and control-system design. *IEEE Trans. Biomed. Eng.*, 36:873–884.

Khang, G. and Zajac, F.E. (1989b). Paraplegic standing controlled by functional neuromuscular stimulation: Part II- Computer simulation studies. *IEEE Trans. Biomed. Eng.*, 36:885–894.

Kilgore, K.L., Peckham, P.H., and Keith, M.W. (1995). An implanted upper extremity neuroprosthesis: a twenty patient follow-up. *J. Spinal Cord Med.*, 18:147.

Kirkwood, C.A., Andrews, B.J., and Mowforth, P. (1989). Automatic detection of gait events: a case study using inductive learning techniques. *J. Biomed. Eng.*, 11:511–516.

Kobetic, R. and Marsolais, E.B. (1994). Synthesis of paraplegic gait with multichannel functional neuromuscular stimulation. *IEEE Trans. Biomed. Eng.*, 2: 66–67.

Kobetic, R., Carroll, S.G., and Marsolais, E.B. (1986). Paraplegic stair climbing assisted by electrical stimulation. *39th ACEMB Conf.*, p. 265.

Kobetic, R., Pereira, J.M., and Marsolais, E.B. (1986). Electromyographic study of the side step for development of electrical stimulation patterns. *RESNA 9th Annu. Conf.*, 372–374.

Kobetic, R., Pereira, J.M., and Marsolais, E.B. (1987). Electromyographic study of the back step for development of electrical stimulation patterns. *RESNA 10th Annu. Conf.*, 630–632.

Kostov, A., Andrews, B.J., Popovic, D.B., Stein, R.B., and Armstrong, W.W. (1995). Machine learning in control of functional electrical stimulation systems for locomotion. *IEEE Trans. Biomed. Eng.*, 42:541–51.

Kotake, T., Nobuyuki, D., Kjiwara, T., Sumi, N., Koyama, Y., and Miura, T. (1993). An analysis of sit-to-stand movements. *Arch. Phys. Med. Rehab.*, 74: 1095–1099.

Kralj, A. and Bajd, T. (1989). *Functional Electrical Stimulation: Standing and Walking After Spinal Cord Injury*. CRC Press, Boca Raton, Florida.

Kralj, A., Acimovic, R., and Stanic, U. (1993). Enhancement of hemiplegic patient rehabilitation by means of functional electrical stimulation. *Prosthet. Ortho. Intl.*, 17:107–14.

Krajl, A., Bajd, T., Turk, R., and Benko, H. (1986). Posture switching for prolonging functional electrical stimulation standing in paraplegic patients. *Paraplegia*, 24:221–230.

Kralj, A., Bajd, T., Turk, R., Krajnik, J., and Benko, H. (1983). Gait restoration in paraplegic patients: a feasibility demonstration using multichannel surface electrode FES. *J. Rehabil. R&D*, 20:3–20.

Kralj, A.R., Bajd, T., Munih, M., and Turk, R. (1993). FES gait restoration and balance control in spinal cord-injured patients. *Prog. Brain Res.*, 97:387–96.

Kuo, A.D. and Zajac, F.E. (1993b). Human standing posture: multi-joint movement strategies based on biomechanical constraints. *Prog. Brain Res.*, 97:349–358.

Kuo, A.D. and Zajac, F.E. (1993a). A biomechanical analysis of muscle strength as a limiting factor in standing posture. *J. Biomech.*, 26:137–150.

Lacey, S.H., Wilber, R.G., Peckham, P.H., and Freehafer, A.A. (1986). The posterior deltoid to triceps transfer: A clinical and biomechanical assessment. *J. Hand Surg.*, 11A:524–527.

Lan, N. (1996). Analysis of an optimal control model of multijoint arm movements. *Biol. Cybern.*, (in press).

Lan, N., Crago, P.E., and Chizeck, H.J. (1991). Control of end-point forces of a multi-joint limb by functional neuromuscular stimulation. *IEEE Trans. Biomed. Eng.*, 38:953–965.

Lan, N., Feng, H-Q., and Crago, P.E. (1994). Neural network generation of muscle stimulation patterns for control of arm movements. *IEEE Trans. Rehab. Eng.*, 2:213–224.

Lemay, M.A., Crago, P.E., and Keith, M.W. (1996). Restoration of pronosupination control by FNS in tetraplegia: experimental and biomechanical evaluation of feasibility. *J. Biomech.*, 29:435–442.

Liberson, W.T., Holmquest, H.J., Scot, D., and Dow, M. (1961). Functional electrotherapy: stimulation of the peroneal nerve synchronized with the swing phase of gait of hemiplegic patients. *Arch. Phys. Med. Rehab.*, 42:101–105.

Lieber, R.L. (1986). Muscle properties following spinal cord injury. *Dev. Med. Child Neurol.*, 28:533–542.

Marsolais, E.B. and Kobetic, R. (1983). Functional walking in paralyzed patients by means of electrical stimulation. *Clin. Ortho. Rel. Res.*, 175:30–36.

Marsolais, E.B. and Kobetic, R. (1987). Functional electrical stimulation for walking in paraplegia. *J. Bone Joint Surg.*, 69A:728–733.

Marsolais, E.B., Scheiner, A., Miller, P.C., Kobetic, R., and Daly, J.J. (1994). Augmentation of transfers for a quadriplegic patient using an implanted FNS system. Case report. *Paraplegia*, 32:573–579.

Meier, J.H., Rutten, W.L., Zoutman, A.E., Boom, H.B., and Bergveld, P. (1992). Simulation of multipolar fiber selective neural stimulation using intrafascicular electrodes. *IEEE Trans. Biomed. Eng.*, 39:122–134.

Miller, L.J., Peckham, P.H., and Keith, M.W. (1989). Elbow extension in the C5 quadriplegic using functional neuromuscular stimulation. *IEEE Trans. Biomed. Eng.*, 37:771–780.

Moberg, E. (1975). Surgical treatment for absent single-hand grip and elbow extension in quadriplegia. Principles and preliminary experience. *J. Bone Joint Surg. [Am]*, 57:196–206.

Moynahan, M. (1995). Postural responses during standing in subjects with spinal-cord injury. *Gait Posture* 3:156–65.

Moynahan, M. and Chizeck, H.J. (1993). Characterization of paraplegic disturbance response during FNS standing. *IEEE Trans. Rehab. Eng.*, 1:43–48.

Mulder, A.J., Boom, H.B., Hermans, H.J., and Zilvold, G. (1990). Artificial-reflex stimulation for FES-induced standing with minimum quadriceps force. *Med. Biol. Eng. Comput.*, 28:483–488.

Mulder, A.J., Veltink, P.H., Boom, H.B.K., and Zilvold, G. (1992a). Low-level finite state control of knee joint in paraplegic standing. *J. Biomed. Eng.*, 14:3–8.

Mulder, A.J., Veltink, P.H., Boom, H.B. (1992b). On/off control in FES-induced standing up: a model study and experiments. *Med. Biol. Eng. Comput.*, 30:205–12.

Ng, S.K. and Chizeck, H.J. (1993). A fuzzy logic gait event detector for FES paraplegic gait. *Proc. 15th IEEE EMBS*.

Ng, S.K. and Chizeck, H.J. (1994). Fuzzy vs. non-fuzzy rule base for gait event detection. *Proc. 16th IEEE EMBS*.

Peckham, P.H. and Gray, D.B. (1996). Functional neuromuscular stimulation. *J. Rehab. Res. Dev.* 33:ix–xi.

Peckham, P.H., Mortimer, J.T., and Marasolais, E.B. (1976). Alteration in the force and fatigability of skeletal muscle in quadriplegic humans following exercise induced by chronic electrical stimulation. *Clin. Orthop.* 114:326–334.

Popovic, D.B. (1992). Functional electrical stimulation for lower extremities. In *Neural Prostheses: Replacing Motor Function After Disease or Disability*. Stein, R.B., Peckham, P.H., and Popovic, D.P. (eds.), pp. 233–251. Oxford University Press, New York.

Popovic, D.B. (1993). Finite state model of locomotion for functional electrical stimulation systems. *Prog. Brain Res.*, 97:397–407.

Popovic, D., Tomovic, R., and Schwirtlich, L. (1989). Hybrid assistive system—Neuroprosthesis for motion. *IEEE Trans. Biomed. Eng.*, 36:729–738.

Rosenthal, J.A. (1984). Control of electrically-stimulated standing for paraplegics. M.S. Thesis, Case Western Reserve University, Cleveland, Ohio.

Scheiner, A., Ferencz, D.C., and Chizeck, H.J. (1993). Simulation of normal and FES induced gait using a 23 degree-of-freedom model. *Proc. 15th IEEE EMBS*.

Scheiner, A., Ferencz, D.C., and Chizeck, H.J. (1994). The effect of joint stiffness on simulation of the complete gait cycle. *Proc. 16th IEEE EMBS*.

Shue, G-H. and Crago, P.E. (1998). Muscle-tendon model with length history dependent activation-velocity coupling. *Annals Biomed. Eng.*, 26:369–380.

Shue, G., Crago, P.E., and Chizeck, H.J. (1995). Muscle-joint models incorporating activation dynam-

ics, torque-angle and torque-velocity properties. *IEEE Trans. Biomed. Eng.*, 42:212–223.

Stanic, U. and Trnkoczy, A. (1974). Closed-loop positioning of hemiparetic patient's joint by means of functional electrical stimulation. *IEEE Trans. Biomed. Eng.*, 21:365–370.

Stein, R.B., Belanger, M., and Wheeler, G. (1993). Electrical systems for improving locomotion after incomplete spinal cord injury: an assessment. *Arch. Phys. Med. Rehab.*, 74:954–959.

Stein, R.B., Gordon, T., Jefferson, J., Sharfenberger, A., Yang, J.F., de Zepetnek, J.T., and Belanger, M. (1992). Optimal stimulation of paralyzed muscle after human spinal cord injury. *J. Appl. Physiol.*, 72:1393–1400.

Thrasher, A., Wang, F., and Andrews, B.J. (1996). Control of neural prostheses III: Self-adaptive neuro-fuzzy control using reinforcement learning. *Proc. RESNA*, 1996:492–494.

Triolo, R.J., Bieri, C., Uhlir, J., Kobetic, R., Polando, G., Scheiner, A., and Marsolais, E.B. (1996). Implanted FNS systems for assisted standing and transfers for individuals with cervical spinal cord injuries: clinical case reports. *Arch. Phys. Med. Rehab.*, 77:1119–1128.

Veltink, P.H. (1991). Control of FES-induced cyclical movements of the lower leg. *Med. Biol. Eng. Comput.*, 29:NS8–12.

Veltink, P.H., Franken, H.M., Van Alste, J.A., and Boom, H.B. (1992). Modelling the optimal control of cyclical leg movements induced by functional electrical stimulation. *Int. J. Art. Organs*, 15:746–755.

Veraart, C., Grill, W.M., and Mortimer, J.T. (1993). Selective control of muscle activation with a multipolar nerve cuff electrode. *IEEE Trans. Biomed. Eng.*, 40:640–653.

Wang, F. and Andrews, B.J. (1994). Adaptive fuzzy logic controller for FES: a computer study. *Proc. Int. Conf. IEEE EMBS*, 16:406–407.

Willemin, D.E. and Chizeck, H.J. (1989). Feedback control of the knee during the stance phase of paraplegic FES gait, using a stiffness regulator. *RESNA 12th Annu. Conf.*, pp. 399–400.

Winters, J.M. (1990). Hill-based muscle models: A systems engineering perspective. In *Multiple Muscle Systems. Biomechanics and Movement Organization*. Winters, J.M. and Woo, S.L-Y. (eds.), pp. 69–93. Springer-Verlag, New York.

Yamaguchi, G.T. and Zajac, F.E. (1990). Restoring unassisted natural gait to paraplegics via functional neuromuscular stimulation: a computer study. *IEEE Trans. Biomed. Eng.*, 37:886–902.

Zajac, F.E. III. (1989). Muscle and tendon: properties, models, scaling and application to biomechanics and motor control. *CRC Crit. Rev. Biomed. Eng.*, 17:359–415.

43
Properties of Artificially Stimulated Muscles: Simulation and Experiments

Robert Riener and Jochen Quintern

1 Introduction

Functional electrical stimulation (FES) has been employed to artificially activate skeletal muscles and so partially restore motor function to patients with upper motor neuron lesions (Kralj and Bajd 1989; Jaeger 1992). However, the control of limb movements induced by FES is complex. The stimulated musculoskeletal system has multiple degrees of freedom with nonlinear, time-varying dynamics. In designing a neural prosthesis, the engineer attempts to develop a mathematical description of the system to predict its mechanical behavior under a variety of specified inputs and external loading. Several investigators have shown that models of the human musculoskeletal system can potentially assist in the design of adequate FES control strategies (e.g., Khang and Zajac 1989; Riener et al. 1996a). The accuracy of the model in predicting the system's behavior is a critical factor in the ability to design effective controllers (Durfee 1992).

The main objective of this chapter is to present a physiologically based model that describes major properties of artificially stimulated muscle. Such features include, for example, the principle of spatial and temporal summation, also known as recruitment and frequency characteristic, respectively (Crago et al. 1980; Mortimer 1981; Carroll et al. 1989). Both properties should be incorporated into the model in order to account for the different effects of frequency and recruitment modulation. Another important property is the doublet phenomenon, also known as nonlinear summation (Cooper and Eccles 1930; Stein and Parmiggiani 1981; Karu et al. 1995). It has been shown that this

effect can be used to obtain effective generation of muscular force during FES (rapid force increase, fatigue reduction). In addition, the model takes into consideration muscle fatigue effects, which are one of the major problems occurring during FES (Mortimer 1981; Garland et al. 1988). A model that accounts for these effects can assist in optimizing open- and closed-loop control strategies, where effective muscle contraction and reduced fatigue is achieved by simultaneously modulating more than one input quantity (e.g., pulse width and pulse frequency). Furthermore, since the model uses available knowledge of the underlying physics, it provides significant insight into muscle processes during FES.

2 Modeling Artificially Stimulated Muscle

2.1 Activation Dynamics

Muscle force is controlled by the number of motor units activated and the amount of force generated in these units. Therefore, we propose to divide the activation dynamics into two submodels (Figure 43.1). A *static recruitment model* determines the motor units recruited (spatial summation). Input to this submodel includes the pulse width $d(t)$ and pulse amplitude pattern $A(t)$, which are generated by the stimulator. A discrete space *dynamic motor unit model* computes the normalized, isometric active state $a(t)$ produced by n motor units. The active state of a single motor unit depends on the input train of nerve potentials and can be adjusted

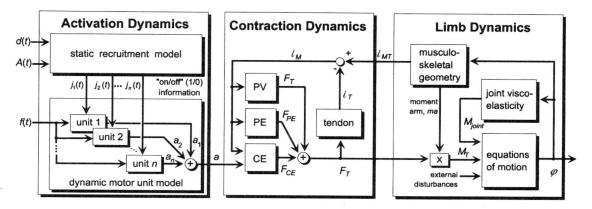

FIGURE 43.1. Model of an artificially stimulated muscle that generates limb motion. The active state of the muscle is calculated from the stimulation signal, which is characterized by pulse width d, pulse amplitude A, and pulse frequency f. The contraction dynamics model con-

siders nonlinear passive viscoelastic and contractile properties of muscle and computes tendon forces. This study modeled the limb dynamics of the hanging shank with a one degree-of-freedom equation of motion.

only by the pulse frequency $f(t)$ (temporal summation). Further input to the dynamic motor unit model is the "on/off-information" $j_i(t)$ of a single unit i determined in the static recruitment model.

We distinguished between slow (corresponding to muscle fiber type I) and fast (corresponding to type II) motor units in both submodels, because many characteristics of artificially stimulated muscle, such as the inverse size-order recruitment (Gorman and Mortimer 1983; Levy et al. 1990) and the shape of the fatigue characteristic (Mortimer 1981; Rabischong and Ohanna 1992), can be attributed to the existence of different motor units with different properties. Muscle fiber types IIa and IIb were ignored for this version of the model. The percentages of fast and slow motor units in the stimulated muscle were estimated on the basis of data from the literature (Johnson et al. 1973).

2.1.1 Static Recruitment Model

One possible way of controlling muscle force is to modulate pulse width or amplitude. Pulse amplitude modulation is, in effect, equivalent to pulse width modulation, since an increase in either recruits more motor units (Crago 1980). For a given pulse width d or pulse amplitude A, the percentage of recruited motor units p_{rec}, also called recruitment level, is computed on the basis of the muscle's recruitment curve. Then, our model assigns a specific selection of recruited motor units to each recruit-

ment level, thus allowing determination of recruited ($j_i = 1$) or nonrecruited ($j_i = 0$) motor units (Riener et al. 1996b).

For constant pulse amplitude the recruitment curve can be approximated by the following equation (Riener et al. 1996a):

$$p_{rec}(d) = c_1\{(d - d_{thr}) \arctan[\kappa_{thr}(d - d_{thr})] - (d - d_{sat})\arctan[\kappa_{sat}(d - d_{sat})]\} + c_2 \quad (43.1)$$

The parameters d_{thr} and d_{sat} denote pulse width values corresponding to threshold and saturation, respectively. The curvatures of the recruitment curve in the area of threshold and saturation can be adjusted by changing κ_{thr} and κ_{sat}, respectively. The constants c_1 and c_2 are global scaling factors.

This submodel can account for the inverse size-order recruitment by distinguishing two different recruitment curves (Riener et al. 1996b): (1) the recruitment curve that represents the recruitment behavior of the small motor units (described by higher threshold and saturation values) and (2) the recruitment curve of the large, fast motor units (lower threshold and saturation values).

2.1.2 Dynamic Motor Unit Model

To describe the force response in a single motor unit as a function of the stimulation frequency f (temporal summation), we developed a model that considers the underlying biophysical processes responsible for the initiation and maintenance of force generation in mammalian muscle. If a motor unit i is recruited ($j_i = 1$), an impulse generator produces a train of action potentials occurring at variable intervals $T = 1/f$. The impulse generator takes into consideration the refractory period as well as the all-or-nothing law: each action potential has the same shape and is approximated by a half-sine wave with a half-period of 1 ms.

The depolarization of the T-tubule membrane in response to a neural action potential has been shown to be well modeled by a second order over-damped system (Hatze 1977). The release of calcium ions from the sarcoplasmic reticulum as a function of the incoming membrane depolarization as well as the reaccumulation process of the calcium ion pump is modeled by another second-order over-damped system (Hatze 1977). In both differential equations different coefficients for fast and slow motor units result in different muscle contraction and relaxation times.

To describe the effect of muscle fatigue and recovery, we introduced a function for muscle fitness, fit, that affects the amount of calcium ions released by the sarcoplasmic reticulum. The muscle fitness value can be computed by the following equation (Riener et al. 1996a):

$$fit(t) = 1/T_{fat}(fit_{min} - fit(t))\overline{\gamma}*(t)$$
$$+ 1/T_{rec}(1 - fit(t))(1 - \overline{\gamma}*(t)) fit(t), \overline{\gamma}*(t) = 0...1$$
$$(43.2)$$

The value $\overline{\gamma}*(t)$ represents the *normalized* concentration of free calcium ions in non-fatigued muscle. T_{fat} and T_{rec} are the time constants for fatigue and recovery, respectively. They differ in fast and slow motor units. The minimum fitness is given by fit_{min}. The fact that the fatigue model reduces the concentration of free calcium ions (it is placed right after the model of the sarcoplasmic reticulum; see Riener 1996b), is justified from a physiological point of view, for it has been demonstrated in the literature that the amplitude of the calcium ion transient decreases as fatigue develops (for review, see Fitts 1994). However, there are some additional effects that lead to fatigue, but are not reflected in the concentration of free calcium ions (e.g., changes in the ATP supply for force generation by the myosin cross-bridges or anaerobic production of lactic acid from glycose and glycogen). These effects were not taken directly into consideration, since the determined fatigue parameters (T_{fat}, T_{rec}, fit_{min}) approx-imate the overall, temporary behavior of muscle fatigue that occurs during fatigue experiments.

To describe the binding process of calcium ions to the troponin-tropomyosin complex in the presence of varying calcium concentration, we applied a static, nonlinear function (Hatze 1978) that uses the concentration of free Ca ions to determine the active state of muscle. Hatze defines the active state a "to be the relative amount of calcium bound to troponin." The active state of the entire muscle is the sum of active states developed in all recruited motor units.

2.2   Contraction and Limb Dynamics

Contraction and limb dynamics are modeled as recently described in Riener et al. (1996a). The stimulated muscle is represented by an active contractile element (CE) in parallel with a nonlinear passive elastic (PE) and a passive viscous (PV) element. The muscle is in series with an elastic tendon. The CE accounts for the force-length and force-velocity property of the muscle, which also depends on the active state of the muscle. The passive effects of the other muscles spanning the same joint as well as connective tissue (in joint, skin, and subcutaneous fat) are summarized by a viscoelastic, joint-angle-dependent moment.

The model used to describe limb dynamics of the accelerated body segment is given by a nonlinear, one degree-of-freedom equation of motion that takes into account gravitational moments, moments of inertia, and the internal moment that acts on the knee joint. This moment is composed of the above-mentioned passive viscoelastic moment and the tendon moment, which is equal to the product of tendon force and joint-angle-dependent moment arm.

3   Methods

3.1   Stimulation Experiments

Stimulation experiments were performed in several patients with complete thoracic spinal cord injury and in healthy subjects. All subjects gave their informed consent to participate in the study. The quadriceps was stimulated with surface electrodes and balanced bipolar stimulation pulses were delivered by a computer-controlled stimulator. Under isometric conditions the moment at the knee joint (knee joint angle 90 degree flexion) was measured with strain-gauge based transducers. Under conditions where the shank swung freely the knee joint angle was measured with an electrogoniometer. In this case only the quadriceps muscle was stimulated and the shank moved against gravity (in neutral position the shank was perpendicular to the supported thigh). Data were digitized and stored in a computer (sampling rate: 200 Hz).

3.2 Model Identification

The model contains a set of subject-specific parameters for each muscle, which have to be adjusted so that the computed muscle force (during isometric conditions) or joint angle (during the condition of a freely swinging shank) agrees with the time course of the measured value. Parameters that do not significantly differ among individuals or are difficult to determine were taken from the literature (e.g., fiber distribution, optimal muscle length, maximum shortening velocity, moment arm).

To show that there exists a set of measurable and model-predictable muscle behaviors that enable the evaluation of subject-specific parameters used in the model, we performed a sensitivity analysis as proposed by Lehman and Stark (1982) (see Appendix). Relative sensitivity coefficients were calculated that reflect the sensitivity of certain model behaviors to changes in specific parameters. If a behavior shows a higher sensitivity to one parameter than to any other, then the measurement of this behavior is a optimal way to evaluate the parameter. The sensitivity analysis provides a better understanding of the model, since it shows how certain model parameters influence model behaviors. Additionally, it is a key to effective design of experiments required to identify and verify the model.

Several isometric and isotonic experiments were derived from this sensitivity analysis (see Appendix) that allow identification of muscle- and subject-specific parameters with trial-and-error adjustments (Riener et al. 1996a; Riener and Quintern 1997). For example, parameters describing the static recruitment model (e.g., threshold and saturation of slow and fast motor units) were adjusted by comparing simulated and measured recruitment characteristics. Fatigue parameters were determined in the model to ensure that the computed muscle force agreed with the muscle force obtained from fatigue/recovery experiments. Model parameters describing impulse generation, T-tubule membrane depolarization, and calcium dynamics were adjusted to match twitch shapes (peak amplitude, contraction and relaxation time of singlets and doublets) as well as measured force outputs at different frequency levels. Peak isometric force was scaled by measuring the knee moment developed during continuous isometric stimulation. Passive viscous properties of muscle could be identified in the model by passive pendulum tests of the shank. In the sensitivity analysis it could be shown that musculotendon elasticity (musculotendon length and stiffness) influence the period of oscillation as well as the neutral position of the shank. Thus, pendulum tests and isometric measurements of passive joint moments at different knee angles served to evaluate tendon slack length and muscle stiffness. Regression equations proposed by Zatsiorsky and Seluyanov (1983) were used to determine anthropometric parameters of the shank and the foot.

Note that changes in a specific model parameter affect all the other simulated force outputs in a more or less distinct way as shown in the sensitivity analysis. Therefore, several iterations of model identifications based on the measured data were necessary to obtain one set of parameters that yielded a satisfactory agreement between measurements and simulations.

4 Muscle Properties: Comparison of Simulation and Measurements

4.1 Spatial Summation

Because motor neurons need a certain threshold value to be excited and because of the spatial separation between fibers and the electrode, the number of active motor units can be graded by the amount of charge injected by the stimulation pulses. The charge of a single stimulus depends on its amplitude (intensity of the stimulus) and its pulse width (duration of the stimulus). Therefore, the number of recruited motor units and, thus, the resulting muscle force can be controlled either by amplitude modulation or by pulse width modulation (Crago 1980). The irregular shape of the recruitment characteristic (muscle force versus pulse width or amplitude) is a result of the nonuniform distribution of nerve fibers within a muscle (Mortimer 1981).

Figure 43.2. shows simulated and measured recruitment characteristics generated by ramping the pulse width up and back down and cross plotting the stimulation input versus the recorded muscle force. The presented simulation correlates well with measured recruitment characteristics of fatigued and nonfatigued muscle. This is not only the result of the application of a nonlinear recruitment characteristic (Eq. 43.1), but also the result of the differentiation of fast and slow motor units with different recruitment and fatigue properties. In most of our experiments

we observed a slight increase of the force at pulse widths around 200 μs and a distinct drop in the slope of the torque at about 350 μs (see large arrows in Figure 43.2). This phenomenon can be explained by the fact that different fiber types have different threshold and saturation values. Our simulations took this effect into account by assuming (according to the inverse size-order recruitment) that the recruitment curve of slow motor units is shifted to higher pulse widths (i.e., $d_{thr,slow} > d_{thr,fast}$ and $d_{sat,slow} > d_{sat,fast}$). Another typical phenomenon observed in simulations and experiments of fatigued muscle is that threshold is increased compared to that of nonfatigued muscle (Figure 43.2C and D). This can be explained by the fact that the activity of fatigue-resistant slow motor

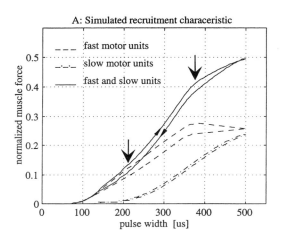

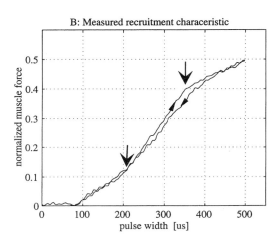

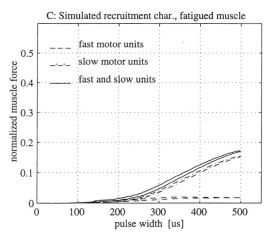

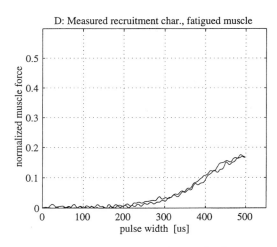

FIGURE 43.2. Isometric recruitment characteristics of nonfatigued (A, B) and fatigued (C, D) quadriceps muscle of a paraplegic patient. Pulse width was linearly increased and decreased (duration: 20 s). Pulse frequency was held constant at 20 Hz. The large arrows mark the often observed change in the slope of the torque.

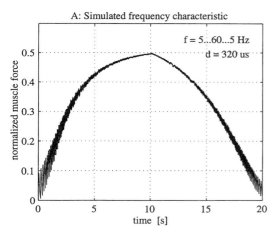
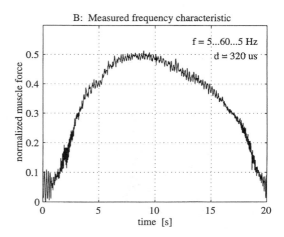

FIGURE 43.3. Frequency characteristic in simulation (A) and experiment (B) of a paralyzed quadriceps muscle. Frequency was linearly increased from 5 to 60 Hz and then again decreased.

units, which have higher threshold values than fast ones (inverse size-order recruitment), dominate in a fatigued muscle with mixed fiber distribution. This is also the reason for the fact that in fatigued muscle the drop in the slope disappears.

4.2   Temporal Summation

A single action potential propagating down an alpha motor neuron produces a single twitch of force in the innervated muscle fibers (motor unit). If two action potentials arrive closely spaced in time, the amount of released calcium ions in the sarcoplasm will increase since not all ions can be sequestered by the sarcoplasmic reticulum during the short interpulse interval. Therefore, the twitch produced by the second pulse will add to the remaining force of the first twitch. With a continuous train of action potentials, muscle force settles at an average force plateau plus some ripple. As the stimulation frequency is increased, the force level increases and becomes smoother, reaching a fused tetanus, i.e. when the individual twitches cannot be distinguished anymore. However, there is a limit of force, since the refractory properties of the nerve and muscle membranes do not allow arbitrarily high frequencies of propagation of action potentials, and after all calcium ions have been released from the sarcoplasmic reticulum, meaning that the muscle has reached its maximal contractile capability.

This property of temporal summation can be used

to control muscle force by frequency modulation alternatively or complementary to recruitment modulation (Mortimer 1981; Carroll et al. 1989; Chizeck et al. 1988). We compared simulated and measured force responses when pulse frequency was linearly increased and decreased. In both simulation and experiment one can see the nonlinear shape of the frequency characteristic, the increased ripple at low frequencies, as well as the saturation of force at high frequencies (Figure 43.3.).

4.3   Doublet Effect

Since the early work of Cooper and Eccles (1930), it has been known that the force produced by a train of several closely spaced stimulation pulses (N-lets) may exceed the force expected from a simple summation of pulse responses. In the last few years several authors have stressed that this behavior of nonlinear summation, also known as doublet effect, can be used to obtain effective generation of muscular force during FES. For example, Karu et al. (1995) proposed to apply N-lets to improve the fatigue-resistance during artificial stimulation. Others describe methods of maximizing the isometric force-time integral per unit pulse for various muscles in order to cause faster force production (Zajac and Young 1980; Karu et al. 1995; Kwende et al. 1995).

The recorded force response to singlets, as well as doublets and triplets with different interpulse intervals shows satisfactory agreement in simulation

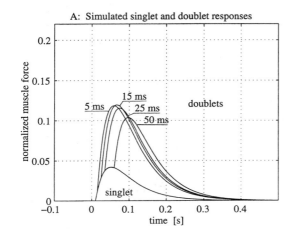

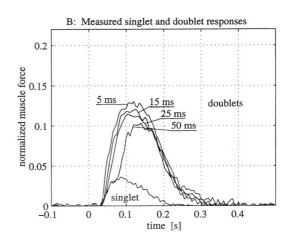

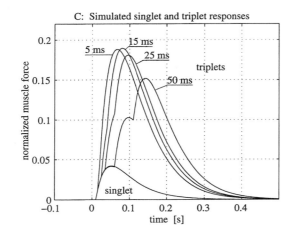

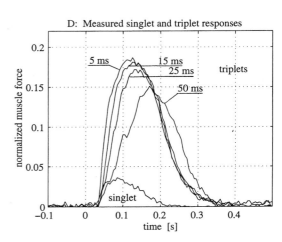

FIGURE 43.4. Simulated and measured twitch responses to doublets (A, B) and triplets (C, D) with different interpulse intervals in comparison to singlets. In the ex- periments, the tibialis anterior muscle of a normal sub- ject was stimulated, which shows the strongest doublet effect compared to the other muscles investigated.

and experimental data (Figure 43.4.). Maximum force time integral per pulse (FTIpP) of the result- ing twitch occurred at higher interpulse intervals in the simulation (9 ms) than in the experiment (5 ms; see also Karu et al. 1995). Good agreement was also obtained during nonisometric condition (Fig- ure 43.5). It was observed in experiments as well as in simulations that the doublet effect can be used to faster accelerate the limb.

4.4 Muscle Fatigue and Recovery

One fundamental problem of FES is that artificially activated muscles fatigue at a faster rate than those

activated by natural physiological processes. The mechanisms responsible for this fast fatigue are nu- merous and include the relative increase of fast- fatiguing muscle fibers that occurs in denervated muscle (Grimby et al. 1976), the inverse size-order recruitment of axons resulting from surface stimu- lation (Gorman and Mortimer 1983; Levy et al. 1990), and the unnaturally high rate of motor unit activation that is required for smooth contractions during synchronous activation of axons (Carroll et al. 1989).

Our model predicts muscle fatigue characteris- tics sufficiently well. Figure 43.6 shows how fast and slow motor units contribute to total muscle

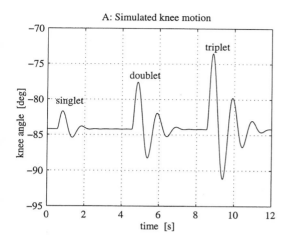

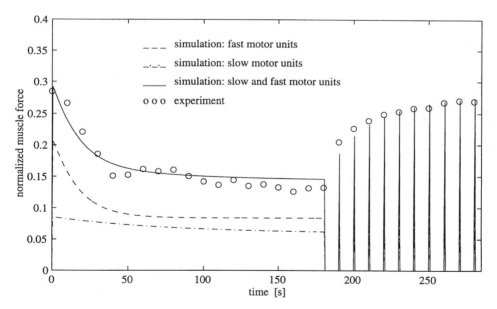

FIGURE 43.5. Simulated (A) and measured (B) knee flex-ion angle of an able-bodied subject under the condition of a freely swinging shank. The quadriceps muscle was stimulated by a singlet, doublet, and triplet pulse. Pulse interval was 5 ms (pulse width: 500μs). A knee angle of 0° means fully extended knee. Graph B is the average of 6 records. Note that doublet and triplet pulses show a significant increase in limb displacement, although in isometric measurements the muscle of this subject showed only a moderate doublet effect (FTIpP of the doublet $\approx$ 1.15).

FIGURE 43.6. Simulated and measured effect of muscle fatigue and recovery on isometric force during stimula-tion of the quadriceps muscle in a normal subject with constant pulse parameters (f = 20 Hz, d = 320 μs). For the first 180 s, the quadriceps was stimulated continu-ously; this leads to a significant decrease of force due to fatigue. The following stimulation pause allowed the muscle to recover. However, to assess the stage of re-covery, test stimuli of 0.5 s duration were given every 10 s (same pulse parameters as during continuous stim-ulation). Note that the normally occurring ripples in force were smoothed in this figure. (Experimental data is shown with permission from Riener et al. 1996a.)

force in order to demonstrate that the shape of the fatigue characteristic can be attributed to the different motor unit types (Mortimer 1981; Rabischong and Ohanna 1992). The model can also predict the frequency dependence of muscle fatigue as observed by others (Garland et al. 1988; Karu et al. 1995): during continuous stimulation, high-frequency stimulation in both simulation and experiment leads to a quicker deterioration of force and to a lower force plateau than low frequency stimulation (see Appendix, Table 43.1, right column; see also Riener et al. 1996b).

5 Discussion

The model presented in this chapter accounts for many relevant properties of artificially stimulated muscle, such as spatial and temporal summation, the doublet phenomenon, and muscle fatigue and recovery. The accuracy of the model in describing these properties is a critical factor in the ability to design effective controllers for neural prostheses. The advantageous muscle properties of temporal and spatial summation, for example, are utilized for the control of muscle force either by recruitment, frequency, or simultaneous recruitment and frequency modulation. Furthermore, it has been shown in this study (Figures 43.4 and 43.5) as well as in other studies (Zajac and Young 1980; Karu et al. 1995; Kwende et al. 1995) that doublets and triplets can considerably enhance force development. For example, single pulses can be added to the stimulation pattern to achieve faster movements (higher accelerations) or to compensate for external (e.g., interaction with environment) or internal disturbances (e.g., spasticity), thus enabling improved control of motion. Another especially significant problem occurring during artificial stimulation is muscle fatigue, which should not be neglected in models used for the design of FES control strategies. Fatigue models in the literature describe how muscle force decreases during continuous stimulation without any change in the stimulation input (e.g., Giat et al. 1993; Rabischong and Ohanna 1992). Our fatigue model involves a more general approach, which has the additional advantage that it can be applied to any input stimulation pattern.

Other authors noted that detailed activation dynamics would not significantly improve the model

prediction results (Franken et al. 1995) and that there would be a large insensitivity of the resulting motion to the perturbations of the neural input (Hatze 1996). However, in our simulation study we paid special attention to the modeling of the activation dynamics and the neural input. As shown in the sensitivity analysis (see Appendix), the resulting muscle force and limb motion strongly depend upon the parameters of the activation dynamics model and the incoming pulse train. It was shown how just one additional stimulus can significantly increase muscle force and resulting joint motion (Figures 43.4 and 43.5). Furthermore, limb motion is affected by muscle fatigue which, in turn, depends on the applied neural input pattern: high frequency modulation shows strong fatigue effects (Garland et al. 1988; Karu et al. 1995; Riener et al. 1996b), whereas irregular stimulation patterns can improve the fatigue behavior (Kwende et al. 1995; Karu et al. 1995).

There are several muscle properties not taken into account in the presented model. Such a property, for example, is the catch-like effect, which was first noted by Burke et al. (1970) as a prolonged increase of cat muscle force, when a subtetanic constant frequency pulse train is preceded by a brief burst (e.g., doublet or triplet) of high-frequency pulses. Some investigators propose to use this property in FES to produce a force output with a more rapid rate of force development than produced by a low-frequency train and a lower rate of fatigue than produced by a high-frequency train (Binder-Macleod and Barker 1991; Karu 1992). However, in a recent study no distinct catch-like effect was observed in paralyzed human muscle (Quintern et al. 1996): the force enhancement when stimulating with a pulse train that starts with a triplet is not significantly longer in duration than the triplet twitch itself. Therefore, this effect was not taken into consideration in this study.

Another effect often observed during electrical stimulation is the potentiation of muscle force either in the sense of post-tetanic-potentiation (increased twitch response after high-frequency stimulation, Brown and Euler 1938), tension creep (slow rise in isometric force production during the initial few seconds of continuous high-frequency stimulation, Morgan 1990; Karu 1992), or staircase rise (after a long period of rest, muscle force steps up until it reaches a plateau, Colomo and Rocchi

1965). Our simulations neglected these phenomena. We believe that this simplification is justified for a tetanized muscle (i.e., muscle has already been "warmed up").

Numerous different models of muscle activation have been developed that describe muscle properties in a realistic way. However, many of these models are developed for natural muscle activation (Hatze 1977). Other models that are used in FES control, have been designed for specific, constrained situations (e.g., constant pulse frequency, no fatigue) or they were identified and verified only in animal experiments (Bernotas et al. 1986; Hannaford 1990; Bobet et al. 1993; Schutte et al. 1993). Unlike black box approaches (Donaldson et al. 1995) our model is more complex. It has the disadvantage of requiring more parameters, which have to be estimated by experiments or taken from the literature. The complexity of the model will increase further when it is extended to a multisegmental model which is needed to describe functional movements induced by stimulating several muscles of the lower extremities simultaneously. However, the greatest advantage of our physiologically based model is the insight it gives into the muscle's biomechanical and physiological connections, thus having the potential of elucidating certain phenomena during FES and studying the effects of different FES control strategies. A current study with a slightly extended version of the presented activation dynamics has shown that significant insight into the complex internal dynamics can be gained by such an approach, thus explaining the underlying reasons for the doublet effect (Dorgan et al. 1997). Simpler models can be easily derived from this comprehensive version, so as to serve for feedforward or adaptive/predictive control techniques.

6   Future Directions

The main objective of this chapter was to present an identified model that has the potential to describe important properties of artificially stimulated muscle. Not shown here are the predictive capabilities of the model. To verify that the model is correct, its output must be compared to experimental output using some input–output sequences which are different from that used to generate and identify the model. If the errors are acceptably small, the results will be of high value in control design applications (Riener and Quintern 1997).

A difficult problem in FES is posed by the influence of increased spinal reflexes because of spasticity. In many paraplegic patients phasic polysynaptic reflexes, also known as spasms, were occasionally observed during the first seconds after onset of the stimulation. These reflexes lead to complex movements with considerable variability between subjects. Because of this variability and because of the occasional occurrence of these reflexes only during the first seconds of stimulation, these phasic polysynaptic reflexes were not considered in this model. Tonic stretch reflexes are more predictable than phasic reflexes. Therefore, in future extensions a tonic stretch reflex model (for review, see Winters 1995) may be added to the presented muscle model.

In the past, numerous different models have been derived to promote an understanding or to predict the behavior of the human neuro-musculo-skeletal system. The models became more complex as it became more important to understand and predict essential properties of the described system. Until now, many models have been identified by a set of generalized parameters taken from different literature sources representing a nominal person. However, in certain future model applications, it will become more and more important to adapt the highly developed models to individuals and individual muscles. Such individually identified models will be applied especially in the fields of rehabilitation and pathophysiology, where they can serve either for the design of individual control strategies (e.g., in neuroprosthetics) or for the estimation and study of internal parameters that cannot be measured non-invasively.

Acknowledgments. The authors thank S. Volz for his support during the measurements. We also thank Prof. G. Schmidt, for the productive discussions on many aspects of this work. This study was supported by the DFG (SFB 462 "Sensomotorik").

References

Audu, M.L. and Davy, D.T. (1985). The influence of muscle model complexity in musculoskeletal motion modeling. *ASME J. Biomech. Eng.*, 107:147–157.

Bernotas, L.A., Crago, P.E., and Chizeck, H.J. (1986). A discrete-time model of electrically stimulated muscle. *IEEE Trans. Biomed. Eng.*, 33:829–838.

Binder-Macleod, S.A. and Barker, C.B. (1991). Use of a catchlike property of human skeletal muscle to reduce fatigue. *Muscle Nerve*, 14:850–857.

Bobet, J., Stein, R.B., and Oguztöreli, M.N. (1993). A linear time-varying model of force generation in skeletal muscle. *IEEE Trans. Biomed. Eng.*, 40:1000–1006.

Brown, G.I. and Euler, U.S. (1938). The after effects of a tetanus on mammalian muscle. *J. Physiol. (London)*, 93:39–60.

Burke, R.E., Rudomin, P., and Zajac, F.E. (1970). Catch property in single mammalian motor units. *Science*, 168:122–124.

Carroll, S.G., Triolo, R., Chizeck, H.J., Kobetic, R., and Marsolais, E.B. (1989). Tetanic responses of electrically stimulated paralyzed muscle at varying interpulse intervals. *IEEE Trans. Biomed. Eng.*, 36:644–653.

Chizeck, H.J., Crago, P.E., and Kofman, L. (1988). Robust closed-loop control of isometric muscle force using pulse width modulation. *IEEE Trans. Biomed. Eng.*, 35:510–517.

Colomo, F. and Rocchi, P. (1965). Staircase effect and post-tetanic potentiation in frog nerve-single muscle fibre preparation. *Arch. Fisiol.*, 64:189–266.

Crago, P.E., Peckham, P.H., and Thrope, G.B. (1980). Modulation of muscle force by recruitment during intramuscular stimulation. *IEEE Trans. Biomed. Eng.*, 27:679–684.

Donaldson, N.N., Gollee, H., Hunt, K.J., Jarvis, J.C., and Kwende, M.K.N. (1995). A radial basis function model of muscle stimulated with irregular inter-pulse intervals. *Med. Eng. Phys.*, 17:431–441.

Dorgan, S.J. and O'Malley, M.J. (1997). A nonlinear mathematical model of skeletal muscle. *IEEE Trans. Rehab. Eng.*, 5:179–194.

Durfee, W.K. (1992). Model identification in neural prosthesis systems. In *Neural Prostheses—Replacing Motor Function after Disease or Disability*. Stein, R., Peckham, H., Popovic, D. (eds.), pp. 58–87. Oxford University Press, New York.

Fitts, R.H. (1994). Cellular mechanisms of muscle fatigue. *Physiol. Rev.*, 74:49–94.

Franken, H.M., Veltink, P.H., Tijsmans, R., Nijmeijer, H., and Boom, H.B.K. (1995). Identification of quadriceps shank dynamics using randomized interpulse interval stimulation. *IEEE Trans. Rehab. Eng.*, 3:182–192.

Garland, S.J., Garner, S.H., and McComas, A.J. (1988). Relationship between numbers and frequencies of stimuli in human muscle fatigue. *J. Appl. Physiol.*, 65:89–93.

Giat, Y., Mizrahi, J., and Levy, M. (1993). A musculo-tendon model of the fatigue profiles of paralyzed quadriceps muscle under FES. *IEEE Trans. Biomed. Eng.*, 40:664–673.

Gorman, P.H. and Mortimer, T. (1983). The effect of stimulus parameters on the recruitment characteristic of direct nerve stimulation. *IEEE Trans. Biomed. Eng.*, 30:407–414.

Grimby, G., Broberg, C., Krotkiewska, M., and Krowieska, M. (1976). Muscle fiber composition in patients with traumatic cord lesions. *Scand. J. Rehab. Eng.*, 8:37–42.

Hannaford, B., (1990). A nonlinear model of the phasic dynamics of muscle activation. *IEEE Trans. Biomed. Eng.*, 37:1067–1075.

Hatze, H. (1977). A myocybernetic control model of skeletal muscle. *Biol. Cybern.*, 25:103–119.

Hatze, H. (1978). A general myocybernetic control model of skeletal muscle. *Biol. Cybern.*, 28:143–157.

Hatze, H. (1996). Sensitivity of human motion to random perturbations of neural control inputs. In *Proceedings of the 9th International Conference on Mechanics in Medicine and Biology*. Ljubljana, July, pp. 211–214.

Jaeger, R.J. (1992). Lower extremity applications of functional neuromuscular stimulation. *Assist. Technol.*, 4:19–30.

Johnson, M.A., Polgar, J., Weightman, D., and Appleton, D. (1973). Data on the distribution of fibre types in thirty-six human muscles: an autopsy study. *J. Neurol. Sci.*, 18:111–129.

Karu, Z.Z. (1992). Optimization of force and fatigue properties of electrically stimulated human skeletal muscle. *Master Thesis*, The MIT Press.

Karu, Z.Z., Durfee, W.K., and Barzilai, A.M. (1995). Reducing muscle fatigue in FES applications by stimulating with N-let pulse trains. *IEEE Trans. Biomed. Eng.*, 42:809–817.

Khang, G. and Zajac, F.E. (1989). Paraplegic standing controlled by functional neuromuscular stimulation: Part I.—computer model and control-system design, Part II—computer simulation studies. *IEEE Trans. Biomed. Eng.*, 36:873–896.

Kralj, A. and Bajd, T. (1989). *Functional Electrical Stimulation: Standing and Walking after Spinal Cord Injury*. CRC Press, Boca Raton, Florida.

Kwende, M.M.N., Jarvis, J.C., and Salmons, S. (1995). The input-output relations of skeletal muscle. *Proc. R. Soc. (London)*, 261:193–201.

Lehman, S.L. and Stark, L.W. (1982). Three algorithms for interpreting models consisting of ordinary differential equations: sensitivity coefficients, sensitivity functions, global optimization. *Math. Biosci.*, 62:107–122.

Levy, M., Mizrahi, J., and Susak, Z. (1990). Recruitment, force and fatigue characteristics of quadriceps muscles of paraplegics isometrically activated by surface functional electrical stimulation. *J. Biomed. Eng.*, 12:150–156.

Morgan, D.L. (1990). New insights into the behavior of muscle during active lengthening. *Biophys. J.*, 57:209–221.

Mortimer, J.T. (1981). Motor prostheses. In *Handbook of Physiology. Nervous System II.* Brooks, V.B. (ed.), vol. 2, pp. 155–187. American Physiological Society, Bethesda, Maryland.

Quintern, J., Volz, S., Riener, R., and Straube, A. (1996). Muscle contraction in normal subjects and patients with upper motor neuron lesions. In *Proceedings of the 9th International Conference on Mechanics in Medicine and Biology.* Ljubljana, July, pp. 175–178.

Rabischong, E. and Ohanna F. (1992). Effects of functional electrical stimulation (FES) on evoked muscular output in paraplegic quadriceps muscle. *Paraplegia*, 30:467–473.

Riener, R. and Quintern, J. (1997). A physiologically based model of muscle activation evaluated by electrical stimulation. *J. Bioelectrochem. Bioenerget.*, 43:257–264.

Riener, R., Quintern, J., and Schmidt, G. (1996a). Biomechanical model of the human knee evaluated by neuromuscular stimulation. *J. Biomech.*, 29:1157–1167.

Riener, R., Quintern, J., Psaier, E., and Schmidt, G. (1996b). Physiologically based multi-input model of muscle activation. In Neuroprosthetics—from basic research to clinical application. Pedotti, A., Ferrarin, F., Quintern, J., and Riener, R. (eds.), pp. 95–114. Springer-Verlag, Berlin.

Stein, R.B. and Parmiggiani, F. (1981). Nonlinear summation of contractions in cat muscles. I and II. *J. Gen. Physiol.*, 78:277–311.

Winters, J.M. (1990). Hill-based muscle models: a systems engineering perspective. In *Multiple Muscle Systems: Biomechanics and Movement Organization.* Winters, J.M. and Woo, S.L.-Y. (eds.), pp. 69–93. Springer-Verlag, New York.

Winters, J.M. (1995). An improved muscle-reflex actuator for use in large-scale neuromusculoskeletal models. *Annals Biomed. Eng.*, 23:359–374.

Zajac, F.E. and Young, J.L. (1980). Properties of stimulus trains producing maximum tension-time area per pulse from single motor units in medial gastrocnemius muscle of the cat. *J. Neurophysiol.*, 43:1206–1220.

Zatsiorsky, V. and Seluyanov, V. (1983). The mass and inertia characteristics of the main segments of the human body. *Biomechanics VIII-B*, University Park Press, 1152–1159.

Commentary: One Muscle Model for All Applications?

Peter H. Veltink

Riener and Quintern indicate that different applications can pose very different demands for muscle models: the loading can be different, the required accuracy, etc. One can question whether it will be possible to apply one general model for all applications. Riener and Quintern show that the sensitivity of different aspects of the model output to the model parameters depends on the loading condition. Such an analysis can indicate which parameters should be identified carefully for each individual, given a certain application, and which parameters can be set to an average value.

The hypothesis that one model can be used for all applications and identified for each individual is very interesting and should be further tested. Especially, the question still exists whether individual identification of a limited set of parameters of this general model indeed yields a suitable model for the individual and for specific applications. Also, it is questionable whether all characteristics relevant to diverse applications can be described by one model. Diverse nonlinear effects have been shown to exist, which may be relevant, but hard to cover by one general model with a limited number of parameters: e.g. the influence of loading, length and activation history on length-force characteristics (Joyce et al. 1996a,b; Roszek et al. 1994; Meijer et al. 1995; Huijng 1996).

References

Huijing, P.A. (1996). Important experimental factors for skeletal muscle modelling: non-linear change of muscle length-force characteristcs as a function of degree of activitation. *Eur. J. Morphol.*, 31:112–120.

Joyce, G.C. and Rack, P.M.H. (1969b). Isotonic lengthening and shortening movements of cat soleus muscle. *J. Physiol.*, 204:475–491.

Joyce, G.C., Rack, P.M.H., and Westbury, D.R. (1969a). The mechanical properties of cat soleus muscle during controlled lengthening and shortening movements. *J. Physiol.*, 204:461–474.

Meijer, K., Grootenboer, H.J., Koopman, H.F.J.M., and Huijing, P.A. (1995). The effect of shortening history on the length-force relationship of the muscle. In *Abstact Book XVth Congress of the International Society of Biomechanics*, Häkkinen, K., Keskinen, K.L., Komi, P.V., and Mero, A. (eds.), pp. 618–619.

Roszek, B., Baan, G.C., and Huijing, P.A. (1994). Decreasing stimulation frequency-dependent length force characteristics of rat muscle. *J. Appl. Physiol.*, 77:2115–2124.

Appendix: Sensitivity Analysis

Relative sensitivity coefficients of the presented model were calculated as proposed by Lehman and Stark (1982). Sensitivity coefficients are ratios of changes in normalized model behavior to changes in normalized parameter values at a certain operating level. For each behavior b_i and each parameter p_j, we defined a sensitivity coefficient

$$S_{ij} = \frac{\Delta b_i / b_{i0}}{\Delta p_j / p_{j0}} \quad (43.3)$$

where p_{j0} and b_{i0} are nominal values (e.g., estimated, initial values before parameter variation) of the parameter and behavior. For example, a sensitivity value of 1.0 means that the value describing the behavior increases by the same percentage as the parameter value was changed. If a behavior is not sensitive to a parameter, then the sensitivity coefficient is 0.

TABLE 43.1. Sensitivity coefficients for the presented musculoskeletal model.

		Activation dynamics										
		t-tubule system			Calcium dynamics				Fatigue		Recruitment	
		c_4	c_5	c_6	c_1	c_2	c_3	ρ	T_{fat}	T_{rec}	d_{thr}^{*}	d_{sat}^{*}
I	Peak twitch force	−.05	−.87	1.11	−.70	−.30	1.11	1.11	0	0	−.09	−.81
S	Contraction time	0	−.20	.20	0	−.40	.20	.20	0	0	0	0
O	Relaxation time	−.27	−.07	.13	.80	−.40	.13	.13	0	0	0	0
M	Doublet FTIpP	−.13	.48	−.28	.32	.01	−.28	−.28	.01	0	0	0
E	Tetanic force	0	−.03	.02	0	.31	.02	.02	0	0	−.09	−.81
T	Tetanus to twitch rat.	.04	1.48	−.70	1.07	.31	−.70	−.70	0	0	0	0
R	Percent ripple	−.04	1.46	−.95	−.69	3.01	−.95	−.95	0	0	0	0
I	Recruit, threshold	0	0	0	0	0	0	0	0	0	1.0	0
C	Recruit, saturation	0	0	0	0	0	0	0	0	0	0	1.0
	Fatigue index	.02	.54	−.45	0	.56	−.45	1.41	1.53	−.91	0	0
	$T_{fat/2}$	.01	.31	−.26	.01	.31	−.26	.93	1.28	−.11	0	0
	$T_{rec/2}$	0	.58	−.55	.10	.55	−.55	−.61	−.03	1.13	0	0
I	$\Delta\varphi_{max}$ (after triplet)	−.19	−.71	.99	−.26	−.61	.99	.99	0	0	−.09	−.85
S	v_{max} (after triplet)	−.17	−.65	.85	−.39	−.45	.85	.85	0	0	−.09	−.83
O	a_{max} (after triplet)	−.12	−.51	.46	−.48	−.15	.46	.46	0	0	−.09	−.81
T	Doublet PApP	−.14	.40	−.29	.36	−.03	−.29	−.29	0	0	0	−.01
O	Neutral position	0	0	0	0	0	0	0	0	0	0	0
N	Oscillation period	0	0	0	0	0	0	0	0	0	0	0
I	Settling time	0	0	0	0	0	0	0	0	0	0	0
C												

Calculated coefficients for 50% parameter variations are given for each behavior (rows) and each parameter (columns). Isometric behaviors are peak twitch force, contraction and relaxation time of a single twitch, normalized force-time-integral per pulse (FTIpP) after doublet stimulation (5 ms pulse interval), tanic force ($f = 100$ Hz), tetanus to twitch ratio, force ripple as percentage of average force ($f = 20$ Hz), threshold and saturation of the recruitment characteristic, fatigue index (final to initial force value), time when force is 50% of initial value due to fatigue ($T_{fat/2}$), and time when force is recovered to 50% of initial value ($T_{rec/2}$). Isotonic behaviors are maximum knee angle, velocity and acceleration, when stimulating with a triplet pulse (5 ms pulse interval), the normalized peak angle per pulse (PApP) after doublet stimulation (5 ms pulse interval), the neutral position of the shank (deviation from vertical position), and the oscillation period and the settling time (5% initial value) in a passive pendulum test. Note that all forces are tendon forces.

Parameters: c_4, c_5, c_6, and c_1, c_2, c_3 are coefficients for the membrane depolarization and sarcoplasmatic reticulum dynamics, respectively (see Hatze, 1977), ρ is a calcium-binding constant (Hatze, 1978), T_{fat} and T_{rec} are fatigue and recovery time constants,

Table 43.1 shows the sensitivity coefficients for the musculoskeletal model presented above. For simplicity, homogeneous fiber distribution was assumed in this analysis. We varied parameters p_j from the nominal value to 1.5 times nominal value and calculated the sensitivity coefficients for many different behaviors that were defined for isometric and isotonic conditions. Winters (1990) describes several additional input–output combinations for skeletal muscle (e.g., isokinetic, external perturbed, "free" against mass) that can be used to define other behaviors.

Note that the calculated coefficients have to be interpreted carefully, because they represent the sensitivity of the behaviors only for the defined parameter variation and the actual operating level. As a result of the nonlinearity of the model, changes in the parameter variations or any other model parameters and input values can yield significantly different sensitivity coefficients.

A glance at a row of a certain behavior reveals how a specific parameter influences this behavior. If, for example, only one coefficient in a row is large, then the model behavior is strongly influenced by the parameter corresponding to that coefficient; we say the "resolution" is high. In the model presented the resolution of most behaviors is low (i.e., most behaviors are influenced by many parameters). Inspection of the table as a whole shows that isometric behaviors are sensitive to pa-

TABLE 43.1. (*Continued*).

| | Contraction dynamics | | | | | | Limb dynamics | | | | | Pulse input | |
| PE, PV, tendon | | | | | CE | | Scale | Joint | Anthropometry | | | | |
k_e^M	k_e^T	l_{opt}^M	l_s^{T**}	k_d^M	v_m^M	a_{fl}	F_s	k_d^J	m	I	l_{COG}	d, A	f
.04	0	2.44	3.14	−.19	.07	.38	1.0	0	0	0	0	.38	—
−.13	−.19	.94	2.23	.50	−.19	0	0	0	0	0	0	0	—
−.05	−.03	.63	.75	.25	.13	−.08	0	0	0	0	0	0	—
.01	.01	−.17	−.25	−.06	.01	−.01	0	0	0	0	0	0	—
−.03	−.06	2.40	3.45	0	0	.25	1.0	0	0	0	0	.38	.37
−.06	−.06	−.03	.26	.22	−.07	−.11	0	0	0	0	0	0	.37
.03	.59	−.63	1.53	−.52	.32	−.06	0	0	0	0	0	0	1.70
0	0	0	0	0	0	0	0	0	0	0	0	—	0
0	0	0	0	0	0	0	0	0	0	0	0	—	0
0	0	0	0	0	0	0	0	0	0	0	0	0	−.54
0	0	0	0	0	0	0	0	0	0	0	0	0	−.31
0	0	0	0	0	0	0	0	0	0	0	0	0	0
−.01	−.02	2.60	3.65	−.12	.09	.27	.86	−.05	−.55	−.54	−.73	.40	—
.04	−.02	2.18	3.08	−.08	.08	.27	.94	−.03	−.57	−.97	−.92	.39	—
.05	.09	1.96	2.57	−.09	.13	.25	1.06	−.01	−.63	1.14	−1.05	.36	—
.01	.02	−.32	−.62	−.02	.03	−.04	.03	0	−.02	−.01	−.02	.01	—
.64	.05	6.04	8.64	0	0	0	.71	0	−.56	0	−.56	—	—
−.07	−.01	.72	.76	−.04	0	0	−.11	0	.03	.06	.45	—	—
.01	−.03	.28	.66	−.50	0	0	−.53	−.31	.87	.10	1.94	—	—

respectively (Eq. 2), d_{thr} and d_{sat} are recruitment parameters (Eq. 1), k_e^M is a factor for muscle elasticity (corresponding to k_1 of Audu and Davy, 1985), k_e^T is a factor for tendon elasticity (corresponding to the constant slope of the tendon characteristic), l_{opt}^M is the optimal muscle length, l_s^T is the tendon slack length, k_d^M is the muscle damping coefficient (Riener et al., 1996a), v_m^M is the maximum muscle contraction velocity, a_{fl} is a shape parameter of the force-length relation (Winters, 1995), F_s is a parameter that scales the dimensionless tendon force to physiologic tendon force, k_d^J is a damping coefficient for joint viscosity (Riener et al., 1996a), and m is the mass, I the moment of inertia, and l_{COG} is the distance to the center of gravity of shank and foot. Large coefficients in each row and each column are underlined.
*Coefficients in these two columns depend significantly on actual d_{thr}, d_{sat}, and d.
**Note that for the calculation of the sensitivity coefficients of l_{opt}^M and l_s^T variations of only 10% were performed (because 50% variations led to unrealistic musculotendon lengths).

rameters describing the activation and contraction dynamics. Muscle fatigue and recovery depend only on parameters of activation dynamics. Passive behaviors such as the neutral position of the shank, oscillation period, and settling time, are influenced by contraction and limb dynamics, but not by activation dynamics. Note that some isotonic behaviors, for example, the peak joint angle, velocity, acceleration, and the peak angle per pulse (PApP) after doublet stimulation significantly depend on many parameters of activation dynamics.

A glance at one parameter and its column reveals which behaviors are most sensitive to the selected parameter. If, for example, a coefficient in the column of a specific parameter is large, then the corresponding model behavior is strongly influenced by this parameter. Thus, the measurement of this behavior is a optimal way to evaluate the parameter.

Some interesting examples for behaviors and their sensitivity to certain model parameters are listed below (see also Table 43.1).

Row-by-Row View

- FTIpP and PApP, two measures for the observable doublet effect during isometric and isotonic conditions, respectively, are most sensitive to the parameters c_5 (T-tubule membrane depolarization) and c_1 (calcium dynamics).
- The threshold and saturation behavior have high resolution, because they are influenced only by d_{thr} and d_{sat}, respectively.

- Fatigue behavior is most sensitive to the fatigue parameters T_{fat} and $T_{rec.}$
- Shank motion ($\Delta\varphi_{max}$, v_{max}, a_{max}) is influenced mainly by musculotendon length. Note that there is also a remarkable sensitivity to certain parameters of activation dynamics.
- The neutral position of the hanging shank depends mostly on musculotendon length (l_{opt}^M, l_s^T).
- Oscillation period during a passive pendulum test (no stimulation) is sensitive to musculotendon length (l_{opt}^M, l_s^T) and center of gravity (l_{COG}), but not significantly to mass.

Column-by-Column View

- Muscle and tendon elasticities (k_e^M, k_e^T) have only a slight influence on most isometric and isotonic behaviors presented. k_e^M influences mainly the neutral position of the shank, and k_e^T the ripple in the isometric force output.
- Changes in musculotendon length (l_{opt}^M and l_s^M) have a significant influence on isometric force production as well as most isotonic behaviors. Thus, identification of these parameters have to be performed carefully to avoid erroneous model predictions.
- Muscle damping has a strong influence on the ripple in the isometric force output, twitch contraction time, and settling time.
- Pulse frequency has the most significant influence on the ripple in the force output and fatigue behavior of muscle.

44
Synthesis of Hand Grasp

Kevin Kilgore

1 What Is Grasp Synthesis?

Functional electrical stimulation (FES) has been used to provide grasp and release for individuals with a cervical level spinal cord injury (Nathan and Ohry 1979; Peckham et al. 1980; Handa et al. 1989; Keith et al. 1989; Nathan 1993; Perkins et al. 1994). A block diagram of the typical neuroprosthesis providing hand grasp is shown in Figure 44.1. Stimulation is applied to the muscles of the forearm and hand via surface, percutaneous, or implanted electrodes. Patterned stimulation is used to provide functional grasp movement (Thrope et al. 1985; Kilgore et al. 1989; Handa et al. 1992). The resultant movement is controlled by remaining voluntary movements from the quadriplegic individual. Typically, this takes the form of a signal proportional to grasp opening and closing (Johnson and Peckham 1990). A controller/stimulator is responsible for converting the proportional command signal into the appropriate stimulus levels for each electrode so that the desired grasp pattern is achieved (Buckett et al. 1988). The relationship between the control signal generated by the patient, and the stimulus levels delivered to each muscle, is referred to as the stimulus map (Kilgore et al. 1989; Kilgore and Peckham 1993a). The process of developing the stimulus map is referred to as grasp synthesis. The stimulus map must be customized for each subject because the output of each electrode varies from subject to subject (Crago et al. 1980; Thrope et al. 1985). Typically the stimulus maps are stored as lookup tables in the controller/stimulator unit.

This chapter reviews the current methods of grasp synthesis and describes ongoing research into

new methods (see also Chapter 30 for basic research studies of grasp and manipulation tasks). The types of grasp patterns provided by the neuroprosthesis are described in Section II. The factors which affect grasp output and the grasp synthesis process are presented in Section III. Section IV reviews the methods of grasp synthesis that are currently in clinical use. These methods rely on qualitative investigator observations and the grasp patterns are modified by trial and error. Section V reviews investigations into more biomechanically based automated grasp synthesis procedures. This method relies on accurate measures of joint moments and angles as inputs to a process performed mathematically by a computer. Finally, the future role of the automated method in clinically deployed neuroprostheses is discussed.

2 Grasp Patterns and Templates

Grasp patterns in FES are defined by the moments and angles of the thumb and fingers as a function of the proportional command signal. The complete set of these functions for a particular grasp pattern is known as the grasp template (Kilgore et al. 1989). Two basic grasp patterns are generally provided for functional activities: lateral pinch and palmar prehension (Peckham et al. 1983). The grasp templates for these two grasps are shown in Figure 44.2. Other grasp patterns have been described for use in FES, including a "pinch grip" between the index and thumb (Nathan 1993) and the "parallel extension grasp" with finger extension and thumb adduction (Handa et al. 1992). These patterns are

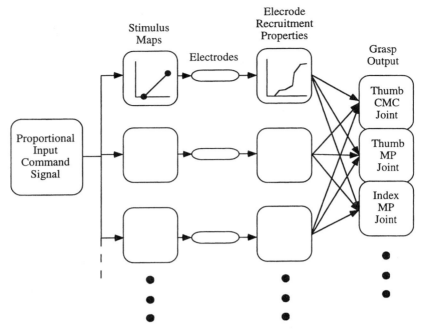

FIGURE 44.1. Block diagram of a typical neuroprosthesis for control of grasp opening and closing. A single proportional signal controls the stimulus level to electrodes implanted in various muscles of the hand and arm. The relationship between the input control signal and the output stimulus level to each electrode is determined by the stimulus map. Typically, there is a single map associated with each electrode. The actual grasp output depends on the recruitment properties of the electrode, as well as the physiology and biomechanics of the joints involved. A single electrode-muscle combination may develop moments about several joints. Current systems incorporate as many as 30 electrodes and have an effect on all 22 degrees-of-freedom of the hand and wrist.

modifications of the palmar grasp template as shown in Figure 44.2. At present, grasp templates are derived empirically from clinical experience. The templates shown in Figure 44.2 are not necessarily the optimum templates for any particular task or for function across several activities, although they have been shown to provide a functional benefit to patients (Wijman et al. 1990; Wuolle et al. 1994). More research is needed to determine what these optimum templates should be for the type of tasks performed by the user population. Potentially, templates could be optimized according to the individual's specific functional goals.

3 Factors Influencing Grasp Synthesis

The goal of grasp synthesis is to develop grasp patterns that match as closely as possible the desired grasp template (Kilgore and Peckham 1993a).

Therefore, the optimum stimulus map is defined as the map that results in a minimization of the error between the grasp template and the stimulated grasp movement and force. Optimization of the stimulus map is difficult due to several factors that influence the electrically elicited grasp output. These factors can be placed into five categories according to the manner in which they affect the grasp output, as shown in Table 44.1.

They are: (1) the controlled variables, (2) the electrode recruitment properties, (3) the static physiological and biomechanical properties of the extremity, (4) the dynamic physiological properties, and (5) the task dependent variables. The controlled variables are the stimulus pulse duration, amplitude, and frequency. These are the only variables under direct control by any neuroprosthesis. The second category consists of the input–output recruitment properties of each electrode: threshold, nonlinear force recruitment, and changes in electrode-nerve geometry with muscle length. These

FIGURE 44.2. Grasp templates for the lateral and palmar grasp patterns. These curves relate the proportional command input to the desired grasp output. The dashed lines show the "parallel extension" grip. The dotted lines show the template for the tip pinch grip, where the long, ring and small fingers remain extended, while the index finger follows the same pattern as the palmar grasp.

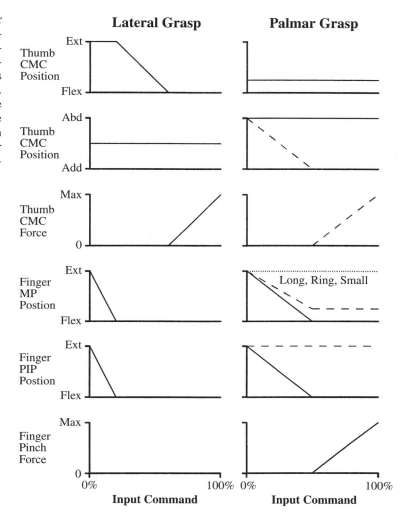

properties are determined by the position of the electrode relative to the muscle innervation, as well as size and type of the nerve and muscle fibers. Although these properties may be affected by exercise, they should be relatively constant over time as long as the electrode remains in the same position within the muscle. A large change in these characteristics probably indicates movement of the electrode. The third category includes the length-tension property of muscle, moment arm changes as a function of joint angle changes, the passive properties of each joint and reflex responses (if present). In addition to these rather static or slowly changing properties, there are also factors such as fatigue and potentiation that result in changes in the muscle properties from moment to moment. The passive properties of the joint can also vary about

a baseline, as exhibited by the warm-up effect (Esteki and Mansour 1996). Finally, there are factors which change whenever the subject attempts to grasp an object, including object weight, size and shape; the orientation of the upper extremity with respect to gravity; and externally applied forces. The goal of stimulus map synthesis is to utilize the controlled variables (category 1) to compensate for the effect of the static properties (categories 2 and 3), and minimize the effect of the dynamic variables, on the grasp output.

Another difficulty in grasp synthesis optimization is to determine what measure(s) should be used to quantify the error between the grasp output and the grasp template. This is difficult because of the complex relationship between the grasp output and function. A measure such as a simple mean squared

TABLE 44.1. Variables that affect the grasp output.

TABLE 44.1. Variables that affect the grasp output.

I. Controlled Variables
  1. Stimulus amplitude
  2. Stimulus pulse duration
  3. Stimulus period
II. Electrode Recruitment Properties
  1. Threshold for force recruitment
  2. Nonlinear force output as a function of stimulus input
  3. Changes in the electrode/nerve geometry with muscle contraction
III. Static Physiological/Biomechanical Properties
  1. Length-tension property of active muscle
  2. Changes in tendon moment arms with joint angle
  3. Passive properties of each joint
  4. Reflex responses
IV. Dynamic Physiological Properties
  1. Fatigue and potentiation
  2. Time varying passive properties
V. Task Dependent Variables
  1. Orientation of upper extremity
  2. Size and shape of object
  3. Externally applied forces

The slope error characterized the modulation of the force and position over the entire command range. It was defined as the magnitude of the difference between the slope of the desired grasp output and the slope of the actual grasp output. The slopes were compared at 5% increments of the command range and the magnitude of each difference was summed over the entire command range. The slope error represents the degree to which force and position can be modulated. A slope error of 100% indicates that neither force nor position can be modulated (that is, the grasp output can achieve only two states): maximum hand opening and maximum pinch force, with no modulation between these states. The slope error is based on the assumption that subjects map relative command input to grasp output rather than mapping absolute command input (i.e., subjects expect that a certain change in shoulder angle results in a consistent

error would not capture the functional aspects of the grasp. For example, if the grasp output achieves only full opening and full closing force, with an infinite gain between those two extremes, the mean squared error would be very high. However, such a grasp pattern is fairly useful for many tasks. In contrast, a grasp output in which the joint movement follows a perfectly linear trajectory, but does not generate any closing force, will have a lower mean squared error but will be useless for any task requiring pinch force.

To attempt to quantify grasp output error, two parameters were developed to allow comparisons between different grasp outputs (Kilgore and Peckham 1993a). These parameters were termed the range error and slope error, as shown in Figure 44.3. The range error characterized the grasp output at the minimum and maximum extremes of the command input. These extremes directly relate to the size and weight of the objects that can be acquired. The range error was defined as the percentage difference between the desired maximum force (and maximum position), and the actual maximum force (and maximum position). For example, if the grasp output achieved 80% of the maximum desired force, then the range error was 20%. The range error was zero if the desired ranges were met or exceeded. There were separate error terms for force and position.

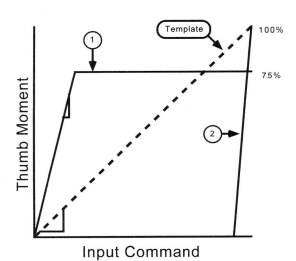

FIGURE 44.3. A method for quantifying the error between grasp output and grasp template. The desired grasp output is the line labeled "template". The line labeled "1" shows a grasp output with a range error of 25%. The slope error is calculated by comparing the slope of the desired grasp output with the slope of the actual grasp output. The line labeled "2" shows a grasp output with a range error of 0% (maximum moment is matched) but with a maximum slope error, i.e. 100% of the moment is attained within one segment (5 % of the command range). Grasp output "2" is defined as having 100% slope error. (From Kilgore KL, Peckham PH, Medical & Biological Engineering & Computing, 31, 607–614, 1993, with permission)

change in grasp output, rather than expecting that a specific shoulder angle results in a specific level of pinch force). This is, at present, an unproven assumption. These simple measures of grasp output error await replacement by more carefully evaluated measures, which take into account the type of tasks performed.

4 Existing Methods for Stimulus Map Synthesis

The grasp synthesis methods currently used in open loop neuroprosthetic systems have been described by researchers in Cleveland, Ohio (Peckham and Mortimer 1977; Peckham et al. 1980; Peckham et al. 1983; Keith et al. 1989; Kilgore et al. 1995) and Sendai, Japan (Handa et al 1989; Handa et al. 1992). Historically, grasp patterns were developed in early neuroprosthetic systems by an experienced investigator through trial and error. As the use of these systems spread to new clinical sites, it became necessary to clearly define the process of grasp synthesis so that it could be performed by less experienced investigators. This process relies on investigator observation of the electrode recruitment properties and the grasp output, and therefore does not require any special instrumentation. Synthesis of the stimulus map generally involves determining minimum and maximum stimulus levels for each electrode and linearly modulating the stimulus between these levels. These methods are reviewed in the paragraphs that follow.

4.1 Cleveland Percutaneous Neuroprosthesis

The neuroprosthesis developed in Cleveland employs a rule-based procedure to synthesize the grasp stimulus map (Kilgore et al. 1989). A "muscle group" template was defined for each grasp pattern that described the necessary activation levels for the following five functional muscle groups: (1) thumb flexors, (2) thumb extensors, (3) thumb abductors, (4) finger flexors, and (5) finger extensors (Table 44.2). The muscle group template for the lateral grasp is shown in Figure 44.2. Each electrode was characterized in an "electrode profile" and placed into one of the five functional muscle groups. The parameters of the profile included two pulse duration values: threshold, and spillover; and a number

of other descriptive ratings of the electrode output (e.g., dependence of the output on the position of other joints). The threshold was defined as the minimum stimulus necessary to elicit a muscle contraction. It was determined by increasing the pulse duration from 0 μS (amplitude constant at 20 mA, stimulus period constant at 80 ms) until a visible movement was observed in the arm or hand as a result of muscle contraction. The muscle recruited at the lowest stimulus level was defined as the primary muscle. Muscles recruited at higher stimulus levels were referred to as secondary muscles. Each primary and secondary muscle was assigned to a functional muscle group as shown in Table 44.2. The spillover pulse duration was defined as the minimum stimulus at which contraction of a muscle not a member of the primary muscle group occurred.

The information provided by the electrode profile served two purposes: (1) to establish the minimum and maximum stimulus levels for each electrode (using the threshold and spillover parameters) and (2) as a way to select the "best" electrodes (using the descriptive ratings). Selecting the "best" electrode was important for a percutaneous system where multiple electrodes were implanted in the same muscle, but only a subset were used in a functional grasp at any point in time.

Once the muscle group, minimum and maximum stimulus levels were determined, the stimulus values were assigned to the appropriate points in the muscle group template, as shown in Figure 44.4.

TABLE 44.2. Functional muscle groups.

| *Thumb Flexors* |
| Adductor pollicis |
| Flexor pollicis longus |
| Flexor pollicis brevis |
| *Thumb Extensors* |
| Extensor pollicis longus |
| Extensor pollicis brevis |
| *Thumb Abductors* |
| Abductor pollicis brevis |
| Abductor pollicis longus |
| Opponens pollicis |
| *Finger Flexors* |
| Flexor digitorum profundus |
| Flexor digitorum superficialis |
| *Finger Extensors* |
| Extensor communis |
| Extensor indicis proprius |
| Extensor digiti minimim |

Lateral Grasp

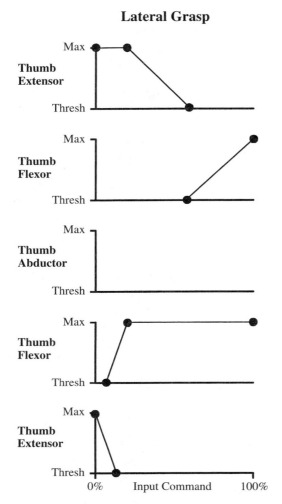

FIGURE 44.4. Muscle group template for the lateral grasp. Each circle represents a breakpoint in the curve. Linear interpolation is used to determine points between each breakpoint. The threshold (Thresh) and maximum (Max) stimulus levels are determined for each electrode in each muscle group. The thumb abductor is not used in the lateral grasp.

Linear interpolation was used between these points. Because the relationship between stimulus level and grasp output is nonlinear, the grasp output will also be nonlinear (Crago et al. 1980; Kilgore et al. 1989). Therefore, the grasp output that resulted from grasp patterns developed in this manner were subject to the variables described earlier. The grasp patterns could be refined by the investigator after qualitative observation of the grasp output. This process was later modified for use with an implanted neuroprosthesis (Smith et al. 1987; Keith et al. 1989; Kilgore and Peckham 1993b; Kilgore et al. 1995; Kilgore et al. 1997).

4.2   Sendai Percutaneous Neuroprosthesis

The Sendai neuroprosthesis is a 30-channel percutaneous system (Handa et al. 1989; Handa et al. 1992). The grasp patterns are developed using a procedure similar to the Cleveland group except that the templates are organized by individual muscles rather than by muscle groups. EMG analysis of normal muscle during grasping movements was used to develop these templates (Handa et al. 1989). The EMG recordings were made through electrodes implanted in each of the muscles of the thumb and fingers of a nonparalyzed subject. The recordings were made as the hand performed different grasp functions. The relative EMG signals were then rectified and filtered and approximated by a trapezoidal envelope. The grasp patterns were then developed by finding the minimum and maximum stimulus levels for each electrode, and assigning those levels in the stimulus map for each muscle. The minimum and maximum stimulus levels could be adjusted for each muscle until the desired movement was achieved. The stimulus amplitude was varied linearly between the minimum and maximum levels. In some cases the subjects were provided with a means for limited adjustment of the maximum stimulus levels to account for day-to-day variability.

5   Automated Grasp Synthesis Procedure

5.1   Significance

The existing methods for grasp synthesis described above do not account for the nonlinear recruitment properties of muscle. The major advantage of these methods is that they can be performed with a minimum of instrumentation; typically all the measurements are made by investigator observation. These methods are probably adequate for simple grasp patterns as evidenced by the success of the current systems (Wuolle et al. 1994; Wijman et al. 1990; Kilgore et al. 1997). However, as grasp patterns become more complex, and more and more electrodes and muscles are included in the grasp patterns, qualitative methods will become extremely difficult to implement on a widespread basis. Therefore, a method of automating and optimizing this process has been pursued (Kilgore and Peckham 1993a). This method relies on instrumentation to make quantitative measures of input/output characteristics and computer optimization to develop the stimulus map parameters. Joint moments and angles are measured externally under static conditions. The procedure for determining the stimulus level necessary to generate the desired grasp output is mathematically defined so that it can be performed by computer. This method removes the need for specially trained individuals and can result in improved grasp patterns.

5.2   Objective and General Description

The goal of the external moment grasp synthesis procedure (GSP) was to minimize the error between the grasp output and the grasp template. The GSP compensates for the influence of the electrode recruitment properties and the static physiological properties on the grasp output, and attempts to minimize the influence of the dynamic physiological properties and the task dependent variables on the grasp output (see Table 44.1). The recruitment properties of each electrode can be measured as a function of the input stimulus level and joint angle. Compensation for these properties can be achieved by determining the stimulus levels applied to each electrode which will yield the desired moments and joint angles at a series of discrete points in the command range. The curves relating the joint moments as a function of joint angle and stimulus level are called

the joint moment recruitment curves (JMRC) (Kilgore and Peckham 1993 a,b). An external moment model (EMM), described below, uses the JMRCs for each electrode, the passive moments (including tendon, ligament, soft tissue and gravity moments), and the grasp template, to synthesize the stimulus map. The grasp output can then be measured and compared with the grasp template. Differences that remain can be further modified if necessary. The entire procedure, shown in Figure 44.5, can be performed automatically by computer because the optimization process is mathematically defined.

5.3   External Moment Model

The EMM describes the interaction between the active moments produced by electrical stimulation, the passive joint moments, and the total joint position and moment output. It is called the external moment model because it incorporates only the externally measurable active and passive moments about the joints of the hand, with no assumptions regarding the forces developed on each tendon and ligament. All measurements are made isometrically. The EMM assumes that the moments produced by stimulation of each electrode/muscle combination can be summed with the passive moments to predict the total moment, and that joint angle can be predicted as the angle at which the sum of all moments acting across a joint are zero.

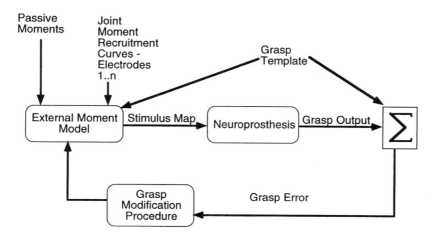

FIGURE 44.5. Block diagram of the automated grasp synthesis procedure (GSP). The inputs to the model are the joint moment recruitment curves (JMRC) of each electrode, the passive moments and the grasp template. The output of the model is the stimulus map for each electrode. The measured grasp output is used as the input to the grasp modification procedure. (From Kilgore KL, Peckham PH, *Medical & Biological Engineering & Computing*, 31, 607–614, 1993, with permission)

The objective of the EMM is to find the combination of stimulus levels to each electrode which minimizes the difference between the desired grasp output and the predicted grasp output (i.e., minimizes the grasp error). This objective can be stated as:

(1) $\Phi = G^T E$
  Φ = Objective function to be minimized
  E = Vector of grasp error terms
  G = Vector of grasp error term gains

The error vector terms (E) are the grasp error values for each moment component of the grasp output, for example, thumb carpometacarpal (CMC) extension/flexion moment, thumb CMC abduction/adduction moment, index finger metacarpalphalangeal (MP) extension/flexion moment, and so on. The gain vector (G) indicates the relative importance of each moment component to the overall grasp output.

The stimulus values which minimize the value of the objective function can be found by searching all combinations of pulse width values for each electrode. The EMM will produce a stimulus map that will compensate for the influence of the electrode recruitment properties as measured in the JMRCs. The influence of the dynamic variables (e.g., fatigue) on the grasp output cannot be eliminated in the open loop system. However, the appropriate choice of grasp template parameters, which include the assumed contact point and degree of cocontraction, can minimize their influence.

5.4   Validation of Basic EMM Principles

The EMM is based on two assumptions regarding the biomechanics of the hand. First, it assumes that the individual static moment vectors resulting from the stimulated muscles, and the passive joint properties, can be mathematically summed to predict the total moment output when these components are combined. Second, it assumes that joint position can be predicted as the joint angle at which the sum of moments about the joint is zero. These assumptions should be true if the system is linear and not time varying, if all the moments and joint angles are accurately measured, and if the dynamic effects are small. Of course, in the practical situation, these conditions are not met. Force vectors resulting from electrical stimulation of two electrodes may not sum if the stimulus fields overlap one an-

other. The system is known to be time varying due to fatigue and potentiation. In addition, it is usually impractical to measure all of the moments and joint angles of the hand (since there are 22 degrees of freedom). However, it is possible to evaluate the sensitivity of the grasp output to these different assumptions. This has been performed on a limited basis for the thumb (Kilgore and Peckham 1993a; Kilgore et al. 1995) and is reviewed here.

To test the first assumption, the moments generated about the carpometacarpal joint of the thumb were measured using a two degree-of-freedom force transducer. Flexion/extension and abduction/adduction moments were measured as a function of stimulation applied to electrodes placed in the muscles of paralyzed subjects. Two muscles were stimulated individually, and then simultane-

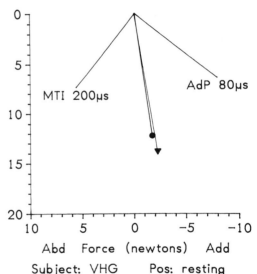

FIGURE 44.6. Example of vector summation involving muscles from the median thenar intrinsic group (MTI—includes abductor pollicis brevis, opponens pollicis and flexor pollicis brevis) and the adductor pollicis (AdP). The triangle represents the resultant vector from the addition of the two individual force vectors. The circle represents the actual force output when the two muscles are stimulated simultaneously. The predicted vector is about 15 % greater but in the same direction as the actual vector. (From Kilgore KL, Peckham PH, Keith MW, Thrope GB, *IEEE Transactions on Biomedical Engineering*, 37, 12–21, 1990, with permission)

ously. The moment vector obtained with simultaneous stimulation was compared to the mathematical sum of the two individual vectors. An example of the experimental results is shown in Figure 44.6. In half of the cases (22 of 44), the predicted vector was within 10% of the actual vector.

Predicting joint position based on the second assumption of the EMM was found to be rather inaccurate for the thumb CMC joint. This is due to small changes in the passive moment over the full range of the joint, and to hysteresis in joint movement. Therefore, it was found to be necessary to measure the joint position, and use the errors between the actual and desired joint position to modify the stimulus levels. It was found that the position errors could be minimized within three iterations.

5.5 Application of the GSP to a Simplified Lateral Grasp

The GSP was successfully used to develop simplified lateral grasp patterns in three C5/C6 quadriplegic subjects. In this grasp pattern, only the CMC joint of the thumb and the MP joint of the index finger were controlled. Only two electrodes, a thumb extensor and thumb flexor, were used in each grasp. In all cases a successful grasp was generated. The range and slope errors were minimized within three iterations. An example of the grasp output obtained by this process is shown in Figure 44.7. The grasp output developed by the GSP was compared with the grasp output developed by the Cleveland rule based method (subsection 4.1) for

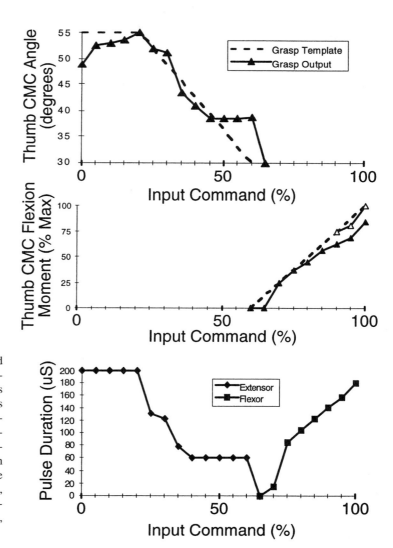

FIGURE 44.7. Grasp output developed using the GSP for a quadriplegic subject with an implanted neuroprosthesis for grasp function. Three iterations were performed on the initial grasp output developed by the GSP. The open triangles show the grasp output after compensation for fatigue. The bottom graph shows the stimulus map used to produce the grasp output. (From Kilgore KL, Peckham PH, *Medical & Biological Engineering & Computing*, 31, 607–614, 1993, with permission)

all three subjects. Both the position and force slope errors were lower for the automated procedure in all cases. Therefore, this method shows promise but has yet to be successfully demonstrated in the general case where there are many electrodes acting on muscles of both the thumb and fingers.

5.6 Practical Implementation of Automated Grasp Synthesis Procedures

Accurate measurement of joint moments and joint angles is important in the automation of the grasp synthesis process. There are no readily available devices for accurate measurement of all joints involved in grasp synthesis. Simple devices such as pinch meters and goniometers do not provide enough information to be usable. Pinch meters typically measure contact force normal to the surface of the pinch meter. This information does not accurately capture the true vector of force, which is especially important for the wrist and thumb. In addition, because force is measured, the moment arm between the force application and the center of joint rotation must be accurately measured. This can be a source of considerable error. Goniometers, although providing an accurate non-invasive measure of joint angle, are slow to use because they typically act across one joint at a time. Some transducers described in the literature for force/moment measurement (Nathan 1979; Kilgore et al. 1990; Kilgore and Peckham 1993a) and joint angle measurement (Thomas and Long 1964; Wise et al. 1990; Kilgore and Peckham 1993a) may provide solutions in the future.

Research is needed to determine the minimum amount of information necessary to perform these automated procedures. The entire process should not take more time than the current methods (typically two to three hours). The potential for an automated process to be cost effective is great if it can reduce the amount of clinician time and training necessary to develop grasp patterns.

6 Summary

Stimulus map synthesis for the development of grasp patterns in upper extremity FES remains a somewhat subjective process, but the rules governing this process have been well defined. This has allowed the transfer of technology to a variety of clinical sites, resulting in the successful implementation of upper extremity systems, as evidenced by the functional improvements reported by these patients. As the grasp patterns become more complex, and as the systems are deployed by less experienced personnel, it will probably be necessary to have more quantitative methods of grasp synthesis if additional function is to be obtained. Although methods exist for automating the process and performing it completely by computer, these methods have not yet been implemented practically because of the need for specialized instrumentation.

7 Future Directions

Future research in the area of grasp synthesis should be focused in two primary directions. First, sensors must be developed that allow clinicians to make measurements of joint moments and angles quickly and easily. There will be a trade-off between the need to obtain a large number of accurate measurements and the need to minimize the time and instrumentation necessary to make these measurements. Initially, measurements might be made using specialized instrumented objects that can be placed in the hand to obtain gross grasp information (Memberg and Crago 1997). It will be important to determine the relationship between the number and accuracy of the measurements to be made and the quality of the resulting grasp pattern. Ultimately, it will be the functional needs for improved grasp performance that will determine the appropriate balance.

The second area of future research is to determine the grasp output parameters for the optimum grasp patterns. These will probably be determined through the analysis of task performance. It will be particularly important to determine the necessary accuracy and resolution of the grasp output. For example, what is the allowable variability in the direction of an applied pinch force vector before an object begins to slip or is twisted out of the grasp? What is the relative importance of the MP, PIP and DIP joint angles of the fingers? The answers to these questions will allow us to predict functional ability and may indicate a need for improved control of the stimulated musculature. The optimum grasp patterns for paralyzed individuals may not be identical to the optimum grasp patterns for the normal population because of the weakness in their proximal musculature.

Finally, grasp patterns will eventually be developed so that they can be changed dynamically (i.e., can be adjusted to conform to an object). This will require grasp synthesis methods that are more directly based on biomechanical information and models. It is likely that, in the future, grasp patterns will be developed through accurate prediction of tendon forces and passive joint properties.

References

Buckett, J.R., Peckham, P.H., Thrope, G.B., Braswell, S.D., and Keith, M.W. (1988). A flexible, portable system for neuromuscular stimulation in the paralyzed upper extremity. *IEEE Trans. Biomed. Eng.*, 35:897–904.

Crago, P.E., Peckham, P.H., and Thrope, G.B. (1980). Modulation of muscle force by recruitment during intramuscular stimulation. *IEEE Trans. Biomed. Eng.*, 27:679–684.

Esteki, A. and Mansour, J.M. (1996). An experimentally based nonlinear viscoelastic model of joint passive moment. *J. Biomech.*, 29:443–450.

Forsythe, G.E., Malcolm, M.A., and Moler, C.B. (1977). *Computer Methods for Mathematical Computations.* Prentice-Hall, Inc., Englewood Cliffs, New Jersey.

Handa, Y., Ohkubo, K., and Hoshimiya, N. (1989). A portable multi-channel FES system for restoration of motor function of the paralyzed extremities. *Automedica*, 11:221–231.

Handa, Y., Handa, T., Ichie, M., Murakami, H., Hoshimiya, N., Ishikawa, S., and Ohkubo, K. (1992). Functional electrical stimulation (FES) systems for restoration of motor function of paralyzed muscles—versatile systems and a portable system. *Frontiers Med. Biol. Eng.*, 4:214–255.

Johnson, M.W. and Peckham, P.H. (1990). Evaluation of shoulder movement as a command control source. *IEEE Trans. Biomed. Eng.*, 37:876–885.

Kailath, T. (1977). *Linear Systems.* Prentice-Hall, Inc., Englewood Cliffs, New Jersey.

Keith, M.W., Peckham, P.H., Thrope, G.B., Buckett, J.R., Stroh, K.C., and Menger, V. (1988). Functional neuromuscular stimulation neuroprostheses for the tetraplegic hand. *Clin. Orthop. Related Res.*, 233:25–33.

Keith, M.W., Peckham, P.H., Thrope, G.B., Stroh, K.C., Smith, B., Buckett, J.R., Kilgore, K.L., and Jatich, J.W. (1989). Implantable functional neuromuscular stimulation in the tetraplegic hand. *J. Hand Surg.*, 14A:524–530.

Kilgore K.L. and Peckham P.H. (1993a). Grasp synthesis for upper extremity FNS—Part I: An automated method for synthesizing the stimulus map. *Med. Biol. Eng. Comput.*, 31:607–614.

Kilgore, K.L., Peckham, P.H., Thrope, G.B., Keith M.W., and Stone K.A. (1989). Synthesis of hand movement using functional neuromuscular stimulation. *IEEE Trans. Biomed. Eng.*, 36:761–770.

Kilgore, K.L., Peckham, P.H., Keith, M.W., and Thrope, G.B. (1990). Electrode characterization for functional application to upper extremity FNS. *IEEE Trans. Biomed. Eng.*, 37:12–21.

Kilgore, K.L., Peckham, P.H., and Keith, M.W. (1995). An implanted upper extremity neuroprosthesis: a twenty patient follow-up. *J. Spinal Cord Med.*, 18:147.

Kilgore, K.L., Peckham P.H., Keith M.W., Thrope G.B., Wuolle K.S., Bryden A.M., and Hart R.L. (1997). An implanted upper extremity neuroprosthesis: a five patient follow-up. *J. Bone Joint Surg.*, 79A:533–541.

Memberg, W.M. and Crago, P.E. (1997). Instrumented objects for quantitative evaluation of hand grasp. *J. Rehab. Res. Dev.*, 34:82–90.

Nathan, R.H. (1979). Functional electrical stimulation of the upper limb: charting the forearm surface. *Med. Biol. Eng. Comput.*, 17:729–736.

Nathan, R.H. (1993). Control strategies in FNS systems for the upper extremities. *Crit. Rev. Biomed. Eng.*, 21:485–568.

Nathan, R.H. and Ohry, A. (1990). Upper limb functions regained in quadriplegia: a hybrid computerized neuromuscular stimulation system. *Arch. Phys. Med. Rehabil.*, 71:415–421.

Peckham, P.H. and Keith, M.W. (1992). Motor prostheses for restoration of upper extremity function. In *Neural Prostheses: Replacing Motor Function After Disease or Disability.* Stein, R.B., Peckham, P.H., and Popovic, D.B. (eds.), Oxford University Press, New York.

Peckham, P.H. and Mortimer, J.T. (1977). Restoration of hand function in the quadriplegic through electrical stimulation. In *Functional Electrical Stimulation: Applications in Neural Prosthesis*, Reswick, J.B. and Hambrecht, F.T. (eds.), pp. 83–95. Marcel Dekker, New York.

Peckham, P.H., Mortimer, J.T., and Marsolais, E.B. (1980). Controlled prehension and release in the C5 quadriplegic elicited by functional electrical stimulation of the paralyzed forearm musculature. *Annals Biomed. Eng.*, 8:369–388.

Peckham, P.H., Thrope, G.B., Buckett, J.R., Freehafer, A.A., and Keith, M.W. (1983). Coordinated two mode grasp in the quadriplegic initiated by functional neuromuscular stimulation. *IFAC Control Aspects of Prosthetics and Orthotics.* Campbell, R.M. (ed.), pp. 29–32. Pergamon Press, Oxford.

Perkins, T.A., Brindley, G.S., Donaldson, N.N., Polkey, C.E., and Rushton, D.N. (1994). Implant provision of key, pinch and power grips in a C6 tetraplegic. *Med. Biol. Eng. Comput.*, 32:367–372.

Smith, B., Buckett, J.R., Peckham, P.H., Keith, M.W., and Roscoe, D.D. (1987). An externally powered, multichannel, implantable stimulator for versatile control of paralyzed muscle. *IEEE Trans. Biomed. Eng.*, 34:499–508.

Thomas, D.H. and Long, C. (1964). An electrogoniometer for the finger: a kinesiologic tracking device. *Am. J. Med. Electronics*, 3:96–100.

Thrope, G.B., Peckham, P.H., and Crago, P.E. (1985). A computer controlled multichannel stimulation system for laboratory use in functional neuromuscular stimulation. *IEEE Trans. Biomed. Eng.*, 32:363–373.

Wijman, C.A., Stroh, K.C., Van Doren, C.L., Thrope, G.B., Peckham, P.H., and Keith, M.W. (1990). Functional evaluation of quadriplegic patients using a hand neuroprosthesis. *Arch. Phys. Med. Rehabil.*, 71:1053–1057.

Wise, S., Gardner, W., Sableman, E., Valainis, E., Wong, Y., Glass, K., Drace, J., and Rosen, J. (1990). Evaluation of a fiber optic glove for semi-automated goniometric measurements. *J. Rehab. Res. Dev.*, 27:411–424.

Wuolle, K.S., Van Doren, C.L., Thrope, G.B., Keith, M.W., and Peckham, P.H. (1994). Development of a quantitative hand grasp and release test for patients with tetraplegia using a hand neuroprosthesis. *J. Hand Surg.*, 19A:209–218.

45
Control with Natural Sensors

Morten Haugland and Thomas Sinkjær

1 Introduction

In Functional Electrical Stimulation (FES), a para-lyzed person's remaining part of the motor system is attempted to be used as a functional benefit to the user. In the case of muscle paralysis from spinal cord injury or trauma to the brain, many of the muscles connected to the central nervous system below the injury level can be made to contract by apply-ing electrical impulses to the nerves innervating those muscles. Such electrical stimuli can mimic the neural impulses which would normally origi-nate from the brain, but which are prevented from reaching the intended muscles by the injury. De-pending on each person's situation, the muscles benefiting from this artificial activation could in-clude those used in picking up objects (Peckham and Keith 1992), in standing and walking (Krajl and Bajd 1989; Marsolais et al. 1991), in control-ling bladder (Brindley 1994) and bowel function and even for breathing (Glenn and Phelps 1985; Talonen et al. 1990).

Despite substantial progress over nearly three decades of development, many challenges remain to provide a more efficient functionality of FES systems; the most important of these is an improved control of the activated muscles (see Chapter 42). This has been attempted by applying a variety of external and implanted sensors to provide feedback on the parameters which one wishes to control (Crago et al. 1986). However, instead of trying to develop artificial sensors, with all the practical problems involved in the application of these, it might be a good idea to try using the body's own sensors, which are already "installed" and opti-mized through millions of years of natural evolu-tion. In able-bodied persons a variety of sensors embedded in the muscles, tendons, joint capsules and skin send proprioceptive information about muscle length, limb loading and joint position to the central nervous system (CNS) via peripheral nerves (Gandevia 1995). If information from these sensors could be extracted, it might be used to pro-vide feedback in clinical FES assistive devices. The natural sensors are an attractive alternative to arti-ficial sensors because they are available and func-tioning in virtually all the individuals who would use a motor prosthesis (Johnson et al. 1995). If nat-ural sensors are used instead of artificial sensors, it would also complement FES in the sense that FES uses the body's own motors (the muscles) to pro-duce movement rather than artificial motors. A sys-tem using electrical stimulation of motor nerves to produce movement and using the natural sensors to correct for disturbances and perhaps to provide cognitive feedback to the user might be able to replicate some of the functions of the spinal cord and its communication with the brain (Figure 45.1).

If a recording method is to be clinically applica-ble, the nerve activity must be nearly stable and in-variant over time and the recording electrodes must be robust. Of the different approaches developed for long-term recording of peripheral nerve activity, only the nerve cuff electrodes (Figure 45.2) have been shown in several animal studies to be able to provide a stable interface between a nerve and an electronic device (Hoffer 1990). This type of elec-trode records the summed activity of all the axons in the nerve. The recorded nerve activity reflects both the recruitment of receptors and changes in the

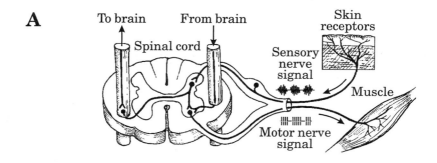

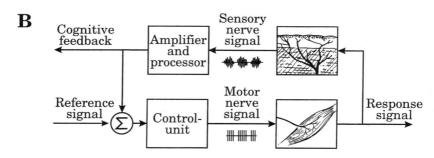

FIGURE 45.1. Schematic illustration showing how information from natural sensors is integrated into the activation of the muscle system in (A) natural automatic spinal cord control and (B) a neural prosthesis using functional electrical stimulation and processed signals recorded from the natural sensors.

firing patterns of active receptors responding to changing stimuli (Stein et al. 1975; Hoffer 1990).

Much is known about the function of the many different receptors both in various animals as well as humans (for a review, see Gandevia 1995). It is not the purpose of this chapter to go into details about these receptors. Instead we will concentrate on the practical and clinical issues regarding the use of signals recorded with a nerve cuff electrode implanted on a peripheral nerve in a human subject as feedback to an already existing clinical FES system. By this we hope to show that even though it is not yet possible to record very detailed information from many single receptors in the way which the central nervous system does, it is still possible and worth while to use the body's own sensors to improve performance of practical FES systems. In the same manner it is still possible and worth while to apply electrical stimulation and produce movement in paralyzed humans although it is not yet possible to stimulate every single motor unit selectively.

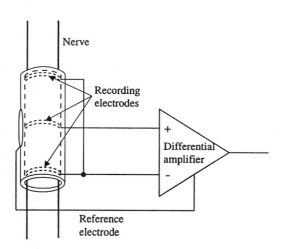

FIGURE 45.2. Principle for recording of neural signals with implanted 3-polar nerve cuff electrode.

2   Methods

The design, fabrication, and surgical installation of nerve cuff recording electrodes were reviewed in detail by Hoffer (1990). The electroneurographic (ENG) activity recorded from a nerve cuff electrode depends on the number of active fibers, is usually dominated by the activity of the largest axons, and is biased in favor of superficial axons. Cuff impedances usually range between 1 and 2 KOhm when a 1 kHz sinusoidal test signal is used. As the signal amplitude is small, an ultralow noise differential preamplifier should be used. Preferably the amplifier noise should be below $4nV/\sqrt{Hz}$ (corresponding to $0.25\mu V_{RMS}$ at a 4KHz bandwidth), which is the theoretical thermal noise of a 1 KOhm resistor at a temperature of 37°C. This is a value that can be difficult to obtain in practical situations meaning that often amplifier noise will be the dominating noise source.

The nerve signal is recorded by shorting the end electrodes in the cuff and amplifying the difference between the center electrode and the two end electrodes as shown in Figure 45.2. The advantage of using a tripolar cuff electrode instead of a monopolar or bipolar electrode is that it gives a reduction of artifacts from the surroundings (Stein et al. 1975).

It has been proposed that the reason for the artifact reduction in tripolar recording is that there can be no potential gradient across the cuff in the longitudinal direction when the end electrodes are shorted (Stein et al. 1975). Later it has been shown that the linearization of external fields within the cuff (because of the insulating wall) has an effect on reducing artifacts (Struijk and Thomsen 1995). If the impedances of the end electrodes are equal and the cuff is symmetrical, the shortened end electrodes give the same potential as the center electrode with respect to the external field. When these potentials are subtracted in a differential amplifier, the artifacts will be reduced. It does, however, require an amplifier with a high common-mode rejection ratio (CMRR) preferably higher than 120dB, which can be difficult to achieve at the frequencies of interest (typically 1–5Khz).

One of the drawbacks of using a cuff electrode is the low signal amplitude (typically below 5 μV when caused by activity from skin receptors in the human applications; Sinkjær et al. 1994; Haugland et al. 1995). The signal amplitude can be increased by increasing the length of the cuff up to an optimal length. Thomsen et al. (2000) have recently shown that the optimal length for a tripolar recording electrode in a typical human application is more than 6 cm. This contradicts earlier studies (e.g., Stein et al. 1975; Marks and Loeb 1976) showing the optimal length to be below 4 cm. In some ap-

plications such as hand grasp with a cuff placed on a nerve in the palm of the hand, it is impossible to implant a 6 cm long cuff.

The nerve cuff and the amplifier configuration reject a large fraction of the external noise. However, when nearby muscles are stimulated, the cuff will still pick up some stimulation artifact. We have tried to show this in Figure 45.3. These data were recorded from a digital nerve in a human subject (will be described later), while the flexor pollicis longus muscle (FPL) was stimulated at a fixed rate of 20 Hz. Stimulation artifacts appear as large spikes saturating in both positive and negative direction. We did not attempt to blank the artifacts by shorting or disconnecting the input to the amplifier as it is often done when recording EMG dur-

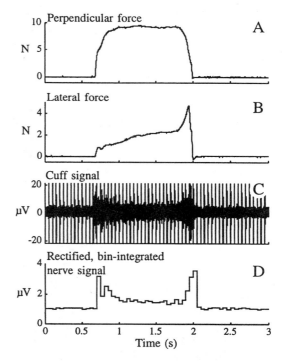

FIGURE 45.3. Signal recorded from cutaneous nerve while stimulating nearby muscles electrically. Force is being applied to the skin within the innervation area of the nerve by a hand-held force probe. (A) Force applied perpendicularly on the skin. (B) Lateral force applied along the skin. In this case the lateral force was increased so that the probe slipped across the skin at the end of the trial. (C) Signal recorded from the nerve. (D) The same signal after rectification and bin-integration of the artifact-free periods. Notice that bin-integration introduces a delay of half a bin-width, on average.

ing stimulation (e.g., Knaflitz and Merletti 1988) because if this is done to an amplifier with this high gain, it can easily cause even larger artifacts because of switching noise and changes in source impedance. Instead, we used an amplifier that did not "hang" when driven to saturation. In this way the artifacts could be removed after amplification by simply ignoring the periods in the signal when the artifacts were present.

If the stimulated muscles are very close to the cuff, evoked EMG responses may also be picked up. These responses can selectively be suppressed by highpass filtering as they contain frequencies mainly below 1 KHz and the nerve signal (depending on the exact dimensions of the cuff and the type of nerve fibers in the nerve) usually contains frequencies above 1 KHz.

The information contained in the nerve signal appears to be stored in the amplitude of the signal rather than in the frequency content. A simple way of measuring amplitude is to rectify and integrate the signal. If this is done in bins containing all the valid data between two stimulation artifacts, a signal, like the lower trace in Figure 45.3, can be obtained. The method of highpass filtration and bin-integration of nerve cuff signals has been described in (Haugland and Hoffer 1994a). More sophisticated signal analysis, such as use of higher order statistics (Upshaw and Sinkjær 1996) has been used and shown to be able to extract more reliable information from the nerve signal.

3   Results

We have used the nerve signal recorded from cutaneous nerves in two different human applications: to replace the external heel-switch of a system for correction of drop foot by peroneal stimulation, and to provide an FES system for restoration of hand grasp with feedback from the fingertip. These two applications will be described in the following after a brief description of cutaneous mechanoreceptors.

3.1   Skin Receptors

The largest fibers in cutaneous nerves are group II fibers (A_β fibers) with conduction velocities between 35 and 75 m/s, and these are generally conducting impulses from the mechanoreceptive units (Burgess and Perl 1973). In human glabrous skin, four types of units have been identified based on their rate of adaptation when the mechanical stimulus is kept constant (SA = slow adaptation, FA = fast adaptation) and area of receptive field (type I = small, II = large) (reviewed by Vallbo and Johansson 1984). SA I units adapt slowly and have small receptive areas. The receptors are Merkel cells. SA II units adapt slowly and have large receptive areas. The receptors are Ruffini endings. FA I units adapt rapidly and have small receptive areas. The receptors are Meissner corpuscles and are comparable to RA (Rapidly Adapting) units found in cats. FA II units adapt rapidly and have large receptive areas. The receptors are Pacinian corpuscles.

The innervation density of cutaneous mechanoreceptors in the finger tips ranges from 10 to 140 units/cm^2 (Johansson and Vallbo 1979) compared to 10 to 20 units/cm^2 in human sural nerve innervating the foot (Buchtal 1982).

3.2   Natural Sensory Information Used in Drop Foot Prosthesis

Electrical stimulation of the peroneal nerve used for correction of gait has proven to be a potentially useful means for enhancing dorsiflexion in the swing phase of walking in lower extremities of individuals with hemiplegia (Strojnik et al. 1987). The stimulation is applied during the swing phase of the affected leg and prevents drop-foot so that the person walks faster and more securely. The stimulator is often located distal to the knee on the lateral part of the tibia. The stimulator can be either external or partly implantable. In both cases the stimulator is triggered by an external heel-switch linked to the stimulator through a wire running from the switch under the heel up to the stimulator.

An application with a large clinical impact would be to replace the external heel switch with a cuff electrode that records the neural activity from the skin of the foot and uses this as a trigger to turn on and off the electrical stimulation, which enhances the dorsiflexion in the swing phase of walking (Sinkjær et al. 1992).

The rationale for implanting a cuff electrode on a cutaneous nerve innervating the foot is to remove the external heel switch used in existing systems for foot drop correction and thereby making it possible to use such systems without footwear and preparing it to be a totally implantable system.

Two individuals with spastic hemiplegia and foot drop were each instrumented with a nerve cuff electrode on a cutaneous nerve innervating the foot of the affected leg. In one person a tripolar cuff electrode was implanted on the sural nerve (Figure 45.4) which innervates the lateral side of the foot (Haugland and Sinkjær 1995). The other person had a cuff electrode implanted on the calcaneal nerve (Figure 45.4) which innervates the heel area of the foot sole (Upshaw and Sinkjær 1996). By percutaneous lead wires, the nerve cuff electrode was connected in a tripolar configuration to an external amplifier providing an overall gain of approximately 110,000. Initially the electronics to correct the drop foot was based on an analogue device (Haugland and Sinkjær 1995), but in later trials the output of the amplifier has been connected to a portable battery operated Digital Signal Processing (DSP) system (Upshaw and Sinkjær 1996). The output from the DSP system was fed to a microprocessor controlled stimulator that activated the ankle dorsiflexor muscles through bipolar surface stimulation of the peroneal nerve (Figure 45.4).

Without stimulation the results showed that the human cutaneous nerves responded in a similar way to pushes and slips applied on the skin as earlier demonstrated in cats (Hoffer and Sinkjær 1986; Haugland et al. 1994). These recordings are consistent with the recordings from single fibers and the distribution of receptors in the skin of the human foot (Haugland and Sinkjær 1995). From modeling studies (Haugland et al. 1994) and more controlled experiments (Haugland and Sinkjær 1995), the contributions of FAI receptors (Meissner corpuscles) seem to dominate the recorded nerve signals. During walking, the nerve signal modulated strongly and gave a clearly detectable response at foot contact and a silent period when the foot was in the air through the swing phase of the walking cycle (Figure 45.5).

After processing the signal as described above, the bin integrated signal was passed into a peak detector, which indicated heel strike. A timer was used to estimate heel-off. An analysis of over 1100 steps (25 minutes of walking) showed that 85% ($\pm6\%$ SD) of heel contacts could be detected us-

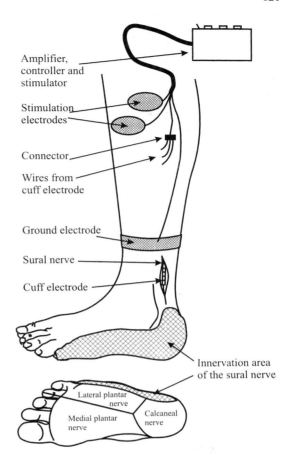

FIGURE 45.4. Natural sensory information used as replacement of heel-switch in peroneal nerve stimulation for compensation of drop foot. Hatched areas show the innervation areas of the sural nerve and the innervation area of the three branches (lateral plantar nerve, medial plantar nerve and calcaneal nerve) of the tibial nerve. Modified from Haugland and Sinkjær (1995).

ing the afferent nerve signal information alone. One of the most significant problems in correlating nerve signal activity to the actual heel contact during walking is noise originating from nearby muscle activity (Haugland and Sinkjær 1995). Fortunately, the erroneous detections can be reduced further by applying more advanced signal processing methods and by making some assumptions about the gait cycle (Kostov et al. 1996; Upshaw and Sinkjær 1996).

One obvious effect of the stimulation was that the steps became more secure as the foot was lifted further off the floor. Without stimulation the heel

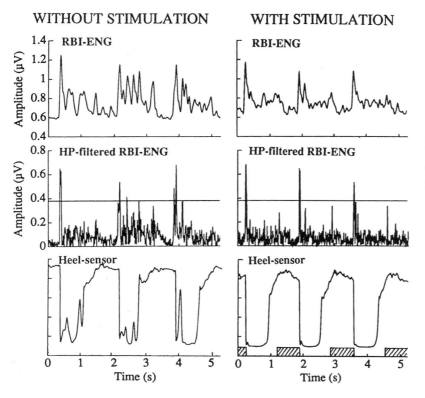

FIGURE 45.5. Sural nerve signal compared to heel switch during walking for a hemiplegic volunteer subject. Left traces show walking without stimulation of the peroneal nerve. Right traces show walking when the nerve signal is used for turning on and off the stimulation of the peroneal nerve. Heel sensor signal was low during heel contact. The stimulation was carried out during the periods shown by the hatched bars.

did not always contact the ground firmly. This is clearly reflected as several distinct peaks in the nerve signal and the external heel sensor during the stance phase (Figure 45.5, left). With stimulation the stance phase was more stable and well defined (Figure 45.5, right).

3.3 Natural Sensory Information Used in Hand Prosthesis

A very interesting application of signals from cutaneous receptors is the hand neuroprosthesis. In able-bodied individuals, the skin receptors play an important role in the control mechanisms when, for example, lifting an object in a precision grip. The fingers of the human hand are subserved by an estimated 17,000 touch sensing receptors within the skin. Such tactile sensors are required to signal the amount of grasp effort needed to secure an object with sufficient force to prevent slippage, but with an economy of effort to avoid undue muscle fatigue (Johannson and Westling 1990). If the initial muscle activity in a hand grasping an object only leaves a minute safety margin against slips, small fric-

tional slips might elicit brief burst responses in FAI, FAII, and SAI units. These afferent volleys trigger an upgrading of the grip force about 70 ms after the onset of the slip (Johansson and Westling 1988). Figure 45.6 shows a schematic illustration of a simple lifting task, including the forces and receptor activity related to the task. Able-bodied subjects are able to control the grip force when holding a given object, independent of the weight and surface texture of the object. This is possible because the mechanoreceptors shown in Figure 45.6 give information about small slips and skin deformation.

We have implemented an algorithm that makes an FES system able to mimic this function based on the compound information from the mechanoreceptors (lower trace in Figure 45.6) as recorded by a nerve cuff electrode. The algorithm was initially developed in an animal preparation (Haugland and Hoffer 1994b) and later we have implemented it in two spinal cord injured subjects.

Results from a 25-year-old tetraplegic male with a complete C5 spinal cord injury (two years postinjury) are presented here. The person had no vol-

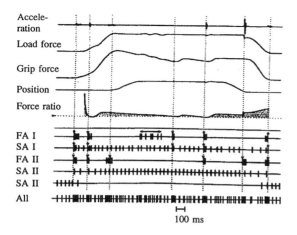

Accele-ration
Load force
Grip force
Position
Force ratio

FA I
SA I
FA II
SA II
SA II
All

⊢⊣
100 ms

FIGURE 45.6. Schematic illustration of the various tactile afferent responses during lifts and some of their motor effects. The data show an able-bodied person lifting an object from a table, holding it in the air for a moment and then replacing it. (Adapted from Johansson 1991). The lower trace shows the added activity from all four (five) types of receptors as they would be recorded with a nerve cuff.

untary elbow extension, no wrist function and no finger function. He used a splint for keeping the wrist stiff. He had partial sensation in the thumb, but no sensation in 2nd to 5th finger.

During general anesthesia, the individual was implanted with a tripolar nerve cuff electrode on the palmar interdigital nerve to the radial side of the index finger branching off the median nerve.

The cuff was 2 cm long and had an inner diameter of 2.6 mm. Eight intramuscular wire electrodes were placed in the following muscles: Extensor Pollicis Longus (EPL), Flexor Pollicis Brevis (FPB), Adductor Pollicis (AdP) and Flexor Digitorum Superficialis (FDS) (See Figure 45.7).

A functional handgrasp was produced by a template mapping a linear command signal to stimulation activity of each of the involved muscles so that 0% corresponded to an open grasp (extended thumb and inactive finger flexors) and 100% corresponded to closed grasp (flexed/adducted thumb and flexed fingers) similar to the method developed by Peckham and coworkers (1992), (see, e.g. Kilgore et al. 1989 and Chapter 44).

The stimulator was controlled by a computer, which also sampled the nerve signals and performed the signal analysis. When stimulation was turned on, the object could be held in a lateral grasp (key grip) by the stimulated thumb. If the object slipped, either because of decreasing muscle force or increasing load, the computer detected this from the processed nerve signal and increased the stimulation intensity with a fixed amount so that the slip was arrested and the object again held in a firm grip. Two examples of this are shown in Figure 45.8.

When comparing the performance of the system to able-bodied subjects doing the same task, it was shown that the FES-generated force could automatically adjust to levels close to able-bodied subjects' levels. When extra weight was added, the slipped

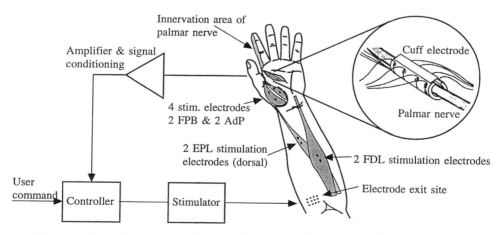

FIGURE 45.7. Schematic illustration of system for restoration of lateral key grip in spinal cord injured volunteer subject, including natural sensory feedback from cutaneous digital nerve.

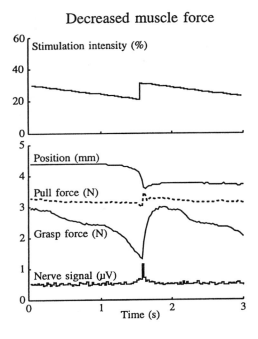

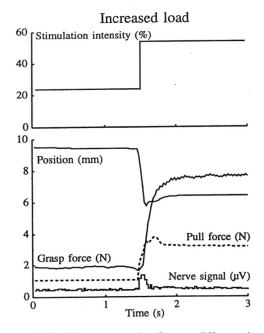

FIGURE 45.8. Slip compensation for two different situations. In (A) the slip was caused by the stimulation intensity being slowly decreased, causing the grasp force to become too low to hold the object. This could either simulate fatigue or an intended action to probe the necessary level of stimulation intensity. In (B) the slip was caused by a sudden increase in external load.

distance was also comparable to the performance of able-bodied subjects (Lickel et al. 1996).

In a preliminary experiment we asked our subject to grasp and lift an object with the slip-compensation algorithm running. An example is shown in Figure 45.9. The algorithm proved to be very helpful in the aquisition of a grasp. If the grasp force produced by the initial stimulation intensity was not sufficient to hold the object, the slips caused by the attempt to lift it made the stimulation intensity increase, until the object was held in a secure grasp. The algorithm then continuously decreased the stimulation intensity to probe which intensity was necessary to hold the object. This caused a couple of slips to occur during the hold phase. The function of the system during both the aquisition phase and the hold phase could be compared to the schematic representation of a lifting task in able-bodied subjects (Figure 45.6). During the replacement of the object on the table, the algorithm increased the stimulation intensity when neural activity was caused by the object touching the table. This was of course not an intended reaction, and should be avoided by turning off the slip compensation when the users wish to replace/release the object.

4   Conclusion

We believe that the two human studies presented here show that it is indeed possible to make functional use of natural skin sensors in FES systems with the techniques available today. An important area not being dealt with in this chapter is the multitude of different electrode-designs which have been used and are being developed for recording (and stimulation) of peripheral nerves. Future research will show whether cuffs and other types of electrodes can be used to reliably extract signals from the large number of other receptors in the body to improve and expand on the use of natural sensors in clinical FES systems.

5   Future Directions

This chapter has focused mainly on the use of information from cutaneous mechanoreceptors, as they are presently the type of receptors being used

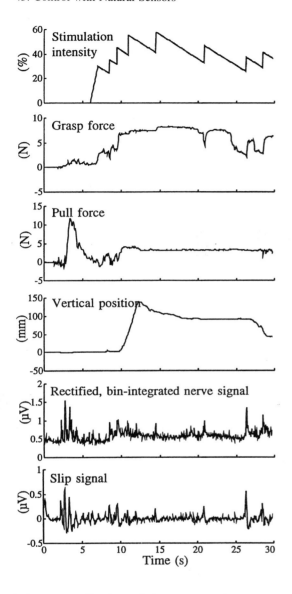

FIGURE 45.9. An individual with C5 tetraplegia grasping and lifting an instrumented object by means of FES with natural sensory feedback. During the first six seconds, the object is entered into the grasp causing some erroneous force readings and some nerve activity. Then the stimulation was started by increasing linearly to 30%, and the person attempted to lift the object. In this situation the grasp force was not sufficient, and the object stayed on the table causing a slip to occur between the fingers and the object. This made the computer increase the stimulation intensity to give a better grip. This happened again after approximately 10 seconds, and by then the force was high enough and the object was lifted off the table. The stimulation intensity was continuously decreased (top trace) causing new slips later on in the trial (bottom trace). Again the detected slips automatically increased the stimulation intensity. After 28 seconds the object was replaced at the table (at a different place causing the position signal to be different at the end compared to the start).

fibers (only the muscle spindles) with conduction velocities of 35 to 75 m/s. This means that it is technically possible to record the activity with a cuff electrode in man.

Basic studies in muscle receptor properties have demonstrated that muscle spindle receptors in passive muscles can signal joint angle (Gandevia 1995). If similar information can be obtained from cuff electrodes placed around muscle nerves, muscle afferent feedback signals can perhaps be used as feedback signals in clinical FES control (Hoffer et al. 1996) of joint position during, for example, standing. Yoshida and Horch (1996) demonstrated closed loop FES control of paralyzed muscles by intrafascicular recordings of muscle spindle signals in an acute animal model.

In a rabbit model of human peripheral nerves Mosallaie et al. (1996) use cuff electrodes to record neural activity simultaneously from tibial nerve and peroneal nerve during passive trapezoidal motion of the ankle joint. The descending branches of tibial and peroneal nerves innervating the foot and the ankle are transected eliminating proprioceptive and cutaneous inputs from ankle joint and foot. In this way, the model simulates some of the muscle nerve branches which are accessible for cuff electrode implants in the human lower extremity. In agreement with muscle spindle firing properties they observe pronounced hysteresis in recorded activity from both

in human applications. Other receptors in the body are also found that might be of interest with respect to clinical FES systems (Figure 45.10).

5.1 Muscle

The proprioceptors signal the body's own movement. Any receptor that can signal position or movement about joints qualifies as a proprioceptor. The two major proprioceptors are the muscle spindles and the tendon organs. They are innervated by the group I nerve fibers with conduction velocities of 80 to 120 m/s and by the group II nerve

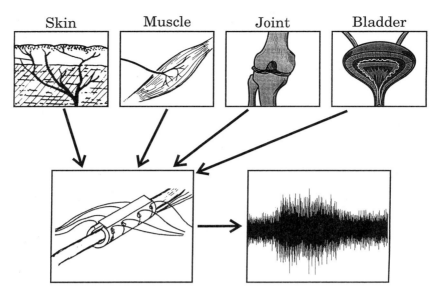

FIGURE 45.10. Illustration of the natural sensors that might be interesting to investigate for control of neural prostheses.

the tibial and the peroneal nerves (where there is little or no activity during the relaxation phase). In an FES application, the hysteresis effects can be overcome by using activity only during muscle stretching, and then during shortening by using activity from an antagonist muscle (e.g., by using activity from the tibial nerve during plantar flexion and activity from the peroneal nerve during extension). However, because afferent fibers in a muscle nerve also contain fibers from the Golgi tendon organs, it may be difficult to determine the origin of nerve activity when the muscle is changing length and force at the same time. It still needs to be investigated to what extent this will be a problem or whether it will actually be possible to use this information for force estimation, given it is known that the muscle is shortening (which could perhaps be determined from the antagonist spindle activity).

5.2   Joint

The joint receptors are mechanoreceptors located in the intraarticular structures (joint ligaments, discs, and menisci). The receptors are slowly adapting Golgi and Ruffini endings, rapidly adapting Pacinian and paciniform endings with myelinated axons and a large number of free nerve endings. Usually, their afferents run in special articular nerves, but only for short distances before they join nerves from, e.g. muscles and skin. The distribution of the receptors seems to reflect the location of stresses (Gandevia 1995).

The joint receptors are activated by the movement of the joint. When the joint moves, some of the receptors are compressed while others are stretched. This results in changes in the discharge rates of the concerned receptors. The Golgi organs respond to compressive forces normal to the capsular surface while the Ruffini afferents respond to planar forces (Proske et al. 1988).

It seems generally accepted that joint receptors are capable to act as limit detectors and to detect movement at the limits of the joint range (Gandevia 1995).

Hines et al. (1995, 1996) studied the feasibility of recording from articular afferents in the human knee joint, which may be able to provide a feedback signal in a lower extremity FES system. A surgical approach was developed to access articular branches of the tibial nerve in human cadavers. Since the main articular branch of the tibial nerve contains a branch projecting distal to the knee joint capsule, the best location for a neural recording interface (e.g., a nerve cuff electrode) appears to be the well-defined ramifications of the articular branch penetrating the joint capsule. The most appropriate branch for a neural recording interface contains approximately 650 myelinated fibers. Fiber diameter distributions computed for nerves from eight cadavers displayed a peak at 3 or 4 μm, and in one specimen an additional peak at 9 μm was found. As the fibers are predominantly group III (diameter $= 1-6$ μm) with fewer group II fibers (diameter $= 6-12$ μm) and virtually no group I fibers (diameter >12 μm), the ability of a nerve cuff electrode to record from these fibers is unknown. Another neural recording interface, such as a microelectrode array (e.g., Kovacs et al. 1994) or intrafascicular electrodes (Yoshida and Horch 1996), might be more suitable for this application.

5.3 Bladder

A sacral root stimulator as developed by Brindley (1994) has proven to be a useful device to control the emptying of the bladder. This and other bladder emptying devices do, however, not tell the person when the bladder is full. The individual often makes his or her decision based on other peoples' visits to the rest room or simply by emptying the bladder at regular intervals. A full bladder increases the pressure on the kidneys, which can be a life-threatening situation in a chronic state because the kidneys may not continue to function properly. Ideally a bladder evacuation prosthesis would derive bladder pressure signals from the natural bladder wall receptors (Crago et al. 1986). Häbler et al. (1993) demonstrate that a good relation exists between neural activity from individual small myelinated group III nerve fibers innervating the bladder wall and the pressure in the bladder. If one can estimate the pressure level in the bladder from the processed nerve signal, it could be used to reduce the risk of impairments in the bladder and kidneys.

To investigate if cuff electrode recordings can be used to detect neural activity from myelinated nerve fibers innervating the bladder, we instrumented anesthetized female pigs (approximately 40 kg) with 22-mm long cuff electrodes on the left sacral roots S2, S3 and S4 (inner diameter 2.2, 1.8, and 1.3 mm respectively) where bladder afferents from the pelvic nerve terminate in the sacral cord.

All three roots showed clear modulation when the skin close to the anus of the pig was touched, but recordings from S2 and S4 showed practically no modulation with bladder pressure. S3 did, however, increase its activity when the bladder was filled.

The change in amplitude of the nerve signal during bladder filling was very small, which was to be expected as the bladder afferent fibers are very small. However, because the filling of the bladder takes place over a long period, suitable filtering and averaging might be able to reduce noise and discriminate (e.g., breathing and movement artifacts). To avoid pickup from nerve fibers not related to the bladder, the pelvic nerve in the human might be superior to the sacral root S3 recordings if surgically accessible.

If the neural signal is to be used as a "warning" signal of when the bladder gets full, one can imagine that when the neural signal has exceeded a threshold corresponding to a critical bladder pressure (approximately 40 cm H_2O in humans) for several minutes, the individual is reminded to empty his/her bladder by a warning signal.

Acknowledgments. The Danish National Research Foundation, The Danish Research Councils, The Danish Foundation for Physical Disability and The Obel Family Foundation are kindly acknowledged for financial support.

References

Brindley, G.S. (1994). The first 500 patients with sacral anterior root stimulator implants: general description. *Paraplegia*, 32:795–805.

Buchtal, F. (1982). Human nerve potentials evoked by tactile stimuli. I. Maximum conduction and properties of compound potentials. *Acta Physiol. Scand Suppl.*, 502:5–18.

Burgess, P.R. and Perl, E.R. (1973). Cutaneous mechanoreceptors and nociceptors. In *Handbook of Sensory Physiology, vol II: Somatosensory System.* Iggo, A. (ed.), pp. 28–78. Springer-Verlag, New York.

Crago, P., Chizeck, H.J., Neuman, M.R., and Hambrecht, F.T. (1986). Sensors for use with functional neuromuscular stimulation, *IEEE Trans. Biomed. Eng.*, 33:256–268.

Gandevia, S.C. (1995). Kinesthesia: roles for afferent signals and motor commands. In *Handbook, Integra-*

tion of Motor, Circulatory, Respiratory and Metabolic Control During Exercise. American Physiological Society.

Glenn, W.W.L. and Phelps, M.L. (1985). Diaphragm pacing by electrical stimulation of the phrenic nerve. *Neurosurgery,* 17:974–984.

Häbler, H.-J., Jänig, W., and Koltzenburg, M. (1993). Myelinated primary afferents of the sacral spinal cord responding to slow filling and distension of the cat urinary bladder. *J. Physiol.,* 463:449–460.

Haugland, M. and Hoffer, J.A. (1994a). Artifact-free sensory nerve signals obtained from cuff electrodes during functional electrical stimulation of nearby muscles. *IEEE Trans. Rehabil. Eng.,* 2:37–39.

Haugland, M. and Hoffer, J.A. (1994b). Slip information provided by nerve cuff signals: application in closed-loop control of functional electrical stimulation. *IEEE Trans. Rehabil. Eng.,* 2:29–36.

Haugland, M. and Sinkjær T. (1995). Cutaneous whole nerve recordings used for correction of footdrop in hemiplegic man. *IEEE Trans. Rehabil. Eng.,* 3:307–317.

Haugland, M., Hoffer, J., and Sinkjær, T. (1994). Skin contact force information in sensory nerve signals recorded by implanted cuff electrodes. *IEEE Trans. Rehabil. Eng.,* 2:18–28.

Haugland, M., Lickel, A., Riso, R., Adamczyk, M.M., Keith, M., Jensen, I.L., Haase, J., and Sinkjær, T. (1995). Restoration of lateral hand grasp using natural sensors, *Proc. of the 5th Vienna Int. Workshop on FES,* Vienna, pp. 339–342.

Hines, A., Birn, H., Stubbe Teglbjærg, P., and Sinkjær, T. (1995). Evaluation of human knee joint articular nerves for nerve cuff recordings. *5th Vienna International Workshop on Functional Electrostimulation,* Vienna, Austria, pp. 247–250.

Hines, A., Birn, H., Teglbjærg, P.S., and Sinkjær, T. (1996). Fibre type composition of articular branches of the tibial nerve at the knee joint in man. *Anatomical Record,* 246:573–578.

Hoffer, J.A. (1990). Techniques to record spinal cord, peripheral nerve and muscle activity in freely moving animals. In *Neurophysiological Techniques: Applications to Neural Systems.* Neuromethods 15, Boulton, A.A., Baker, G.B., and Vanderwolf, C.H. (eds.), pp. 65–145. Humana Press, Clifton, New Jersey.

Hoffer, J.A. and Sinkjær, T. (1986). A natural "force sensor" suitable for closed-loop control of functional neuromuscular stimulation. In *Proceedings of the 2nd Vienna International Workshop on Functional Electrostimulation,* pp. 47–50.

Hoffer, J.A., Stein, R.B., Haugland, M.K., Sinkjær, T., Durfee, W.K., Schwartz, A.B., Loeb, G.E., and Kantor, C. (1996). Neural signals for command control and feedback in functional neuromuscular stimulation. *J. Rehabil.,* 33:145–157.

Johansson, R.S. (1991). How is grasping modified by somatosensory input? In *Motor Control: Concepts and Issues.* Humprey, D.R. and Freud, H.J. (eds.), pp. 331–355.

Johansson, R.S. and Vallbo, Å.B., (1979). Tactile sensibility in the human hand: relative and absolute densities of four types of mechanoreceptive units in glabrous skin. *J. Physiol.,* 286:283–300.

Johansson, R. and Westling, G. (1988). Programmed and triggered actions to rapid load changes during precision grip. *Exp. Brain Res.,* 71:72–86.

Johansson, R. and Westling, G. (1990). Tactile afferent signals in the control of precision grip. In *Attention and Performance.* Jannerod, M. (ed.), XIII:677–713. Erlbaum, Hilldale, New Jersey.

Johnson, K.O., Popovic, D., Riso, R.R., Kors, M., Van Doren, C., and Kantor, C. (1995). Perspectives on the role of afferent signals in control of motor neuroprostheses. *Med. Eng. Phys.,* 17:481–496.

Kilgore, K., Peckham, P.H., Thrope, G.B., Keith, M.W., and Gallaher-Stone, K.A. (1989). Synthesis of hand grasp using functional neuromuscular stimulation. *IEEE Trans. Biomed. Eng.,* 36:761–770.

Knaflitz, M. and Merletti, R. (1988). Suppression of stimulation artifacts from myoelectric-evoked potential recordings. *IEEE Trans. Biomed. Eng.,* 35:758–763.

Kostov, A., Sinkjær, T., and Upshaw, B. (1996). Gait event discrimination using ALNs for control of FES in foot-drop problem. *18th Annual International Conference IEEE Engineering in Medicine and Biology Society,* Amsterdam, The Netherlands, (only available on CD-ROM).

Kovacs, G.T.A., Storment, C.W., Halks-Miller, M., Belczynski Jr., C.R., Santina, C.C.D., Lewis, E.R., and Maluf, N.I. (1994). Silicon-substrate microelectrode arrays for parallel recording of neural activity in peripheral and cranial nerves. *IEEE Trans. Biomed. Eng.,* 41:567–577.

Krajl, A. and Bajd, T. (1989). *Functional Electrical Stimulation, Standing and Walking after Spinal Cord Injury.* CRC Press, Boca Raton, Florida.

Lickel, A., Haugland, M., and Sinkjær, T. (1996). Comparison of catch responses between a tetraplegic patient using an FES system and healthy control subjects. *18th Annual International Conference IEEE Engineering in Medicine and Biology Society.* Amsterdam, The Netherlands, (only available on CD-ROM).

Marks, W.B. and Loeb, G.E. (1976). Action currents, internodal potentials and extracellular records of myelinated mammalian nerve fibres derived from node potentials. *Biophys. J.,* 16:655–668.

Marsolais, E.B., Kobetic, R., Chizeck, H.J., and Jacobs,

J.L. (1991). Orthoses and electrical stimulation for walking in complete paraplegia. *J. Neurol. Rehabil.*, 5:13–22.

Mosallie, K., Riso, R.R., and Sinkjær, T. (1996). Muscle afferent activity recorded during passive extension-flexion of rabbit's foot. *18th Annual International Conference IEEE Engineering in Medicine and Biology Society*, Amsterdam, The Netherlands, (only available on CD-ROM).

Peckham, P.H. and Keith, M.W. (1992). Motor prostheses for restoration of upper extremity function. In *Neural Prostheses Replacing Motor Function After Disease or Disability*. Stein, R.B., Peckham, P.H., and Popovic, D.B. (eds.), pp. 162–190. Oxford University Press, New York.

Proske, U., Schaible, H.-G., and Schmidt, R.F. (1988). Joint receptors and kinaesthesia. *Exp. Brain Res.*, 72:219–224.

Sinkjær, T., Haugland, M., and Haase, J. (1992). The use of natural sensory nerve signals as an advanced heel-switch in drop-foot patients. In *Proc. 4th Vienna International Workshop on Functional Electrostimulation*, pp. 134–137.

Sinkjær, T., Haugland, M., and Haase, J. (1994). Natural neural sensing and artificial muscle control in man. *Exp. Brain Res.*, 98:542–545.

Stein, R.B., Charles, D., Davis, L., Jhamandas, J., Mannard, A., and Nichols, T.R. (1975). Principles underlying new methods for chronic neural recording. *Canad. J. Neurol. Sci.*, 2:235–244.

Strojnik, P., Acimovic, R., Vavken, E., Simic, V., and Stanic, U. (1987). Treatment of drop-foot using an implantable peroneal underknee stimulator. *Scand. J. Rehab. Med.*, 19:37–43.

Struijk, J.J. and Thomsen, M. (1995). Tripolar nerve cuff recording: stimulus artifact, EMG, and the recorded nerve signal. *17th Annual International Conference IEEE Engineering in Medicine and Biology Society*. Montreal, Quebec, Canada, (only available on CD-ROM).

Talonen, P.P., Baer, G.A., Häkkinen, V., and Ojala, J.K. (1990). Neurophysiological and technical considerations for the design of an implantable phrenic nerve stimulator. *Med. Biol. Eng. Comput.*, 28:31–37.

Thomsen, M., Struij, J.J., and Sinkjær, T. (2000). Nerve cuff recording with a combined monopolar and bipolar electrode. *IEEE, Trans. Rehab. Eng.*, (submitted).

Upshaw, B. and Sinkjær, T. (1998). Digital signal processing algorithms for the detection of afferent activity recorded from cuff electrodes. *IEEE Trans. Rehab. Eng.*, 6:172–181.

Vallbo, Å.B. and Johansson, R.S. (1984). Properties of cutaneous mechanoreceptors in the human hand related to touch sensation. *Human Neurobiol.*, 3:3–14.

Yoshida, K. and Horch, K. (1996). Closed-loop control of ankle position using muscle afferent feedback with functional neuromuscular stimulation. *IEEE Trans. Biomed. Eng.*, 43:167–176.

Commentary: Control with Natural Sensors?

Dejan Popovic

Technology to restore movement by FES has advanced substantially. Implantable electrodes are available for safe and selective percutaneous or implantable stimulation. However, adequate control remains a major problem for multiple reasons, including (1) muscles can not be controlled in the same way that the central nervous system controls them; (2) central nervous system injury results in modified reflexes (e.g., spasticity), so inappropriate contractions may be produced; and (3) the changed patterns of activity are responsible for modifying the contractile properties of different muscle groups. Overcoming these problems requires a hierarchical controller which combines feedforward with closed-loop control with reliable, reproducible sensory feedback (see Chapter 42).

Both feedforward and closed-loop control techniques require information on limb position, slip, force, foot or finger contact, and changes in load. Manufactured transducers provide such signals, but they are plagued with reliability, positioning, and mounting problems. Thus, they have not yet proven suitable for practical assistive systems. Natural sensors (i.e., the mechanoreceptors found in the skin, joint capsules, and muscles), which normally carry afferent signals to the central nervous system, are an attractive alternative to artificial ones because they are available and functional in virtually all patients who would use a motor prosthesis as it is nicely demonstrated by Haugland and Sinkjær. In order to use the information provided by natural sensors, one must develop techniques that will (1) permit long term recording of peripheral nerve activity; (2) allow the separation of wanted from unwanted information; and (3) enable the determination of a relationship between neural activity and the physical variables required for control of the motor prosthesis.

This commentary aims to highlight some difficulties in the application of the peripheral electrodes with regard to the three requirements listed above.

1 Technology for Long-Term Recordings

This chapter describes in detail the efficacy of cuff electrodes. There have been minor improvements in the design of the cuff over the last 20 years, and the new electrode patented recently by Hoffer and colleagues seems to resolve some of the problems. This electrode is closed around the nerve minimizing EMG pickup; it is made of thin silastic being flexible and compliant; and the closure method is greatly improved. However, there are still numerous problems associated with the wiring. The wires have to travel a long way along the muscles, over the joints and then exit through the skin. There is no amplifier integrated into the electrode.

An additional problem is the invasive procedure necessary to implant the electrode. The success rate in putting the electrode into position has to be very high, possibly 100%. The preparation of the nerve is complicated, and needs individual approaches because of the variability from human to human in the size, position of blood vessels, and so on. The longest experience with cuff electrodes by my best knowledge at this point is about four years (unpublished data). Once the electrode is in place, the recordings are relatively stable, and they do not deteriorate with time. Considerable connective tissue is built up around the electrode, and its removal is very difficult. A small number of electrodes failed, and in some experiments the nerve was damaged and recordings were not obtained. The experience with percutaneous wires is not very encouraging and this is one of the main reasons for the development of fully implantable systems. If this system is to become practical and useful better electrodes will have to be developed.

2 Separation of Wanted from Unwanted Information

This is one of the most complicated questions because the nerve consists of many afferent and effer-

ent fibers, and it may transmit various signals that are of no relevance for the control. It is relatively easy to instrument mostly afferent nerves such as the sural nerve, tibial nerve, or digital nerves in the hand. However, most of the nerves that might be of interest contain both efferent and afferent fibers and their contributions must be discriminated. The complexity arises also because in a real application many muscles will be stimulated at various rates asynchronously, introducing artifacts.

Our results from semichronic recordings in human hands were encouraging and effective, but pointed to the problem of distinguishing wanted from unwanted information (Popovic and Raspopovic 1992). Able-bodied subjects were instrumented with cuff electrodes (four digital nerves) and asked to perform different tasks. The neural activity was largest when force changed (increasing force, sliding, decreasing force). Activity decayed exponentially when a constant force was exerted. Relative motion of fingers and rubbing of adjacent fingers could not be distinguished from contact with the object or slipping, because a number of receptors lateral and medial on index, middle and ring fingers innervate the digital palmar nerves.

3 Determination of a Relationship Between Neural Activity and the Physical Variables

In a series of studies of the precision (pinch) grip in able-bodied individuals, investigators showed that the simple act of lifting an object from a supporting surface, holding it in the air, and then replacing it on the support involves a series of stereotyped stages. Specifically, following the application of the fingers to the object, the grip force builds linearly to a level that is preprogrammed according to the expected weight of the object. In order to complete the lifting activity as expediently as possible, the grasper applies the grip force at a faster rate when lifting slippery objects. In addition, the grasper applies and increases the lifting force in a parallel manner with the grip force rather than delaying its onset until the full grip force level is reached. This parallel buildup of the grip and load forces is carefully coordinated so that the grip level is always sufficiently greater then the lifting force to prevent the object from slipping. The ratios of the grip and lifting forces during the object "pick up" and "hold" phases are matched to the friction conditions present between the fingers and the object. If slip does occur, the natural grasp control system provides for an automatic increase in this ratio until the slip is arrested (Johnson et al. 1995).

The natural sensor based in cuff electrodes and processing techniques described by Haugland and Sinkjær at this point are not capable of discriminating different functions. The sensor at this time is used basically as a switch. A possible solution to this problem is to use a biological (i.e., a connectionist approach). Sensory channels have been generally thought of as providing a signal, that is a measure of some relevant variable, transformed through the sensory transfer function. The term relevant variable refers to the general hypothesis that the brain has to quantify and to code some "relevant variables" in order to program and control movements or to build an invariant representation of the external world. The fact that individual sensors provide reliable information only on a limited and sometimes variable working range, in both spatial and temporal domains, and the fact that the degradation of the measure by noise can depend on the context and the existence of multiple interpretations suggest that sensory signals cannot be processed as a set of measures but a set of constraints upon the internal estimates. This is to propose that sensory signals are not used to directly estimate the relevant variables, but to estimate the mismatch between internal estimates and measures. This biological organization can be replicated by the usage of multiple recording sites, and correlating those to motor functions (Popovic and Raspopovic 1992, Kostov et al. 1993, Popovic et al. 1993).

Finally, it is important to mention that there is an extensive effort to develop better electrodes and allow recordings not only from peripheral nerves, but also from the brain and integrate these signal into the control.

References

Johnson, K.O., Popovic, D., Riso, R.R., Koris, M., Van Doren, C., and Kantor, C. (1995). Perspectives on the role of afferent signals in control of motor neuroprostheses. *Med. Eng. Phys.*, 17:481–496.

Kostov, A., Popovic, D., Stein, R.B., and Armstrong, W.W. (1993). Learning of EMG patterns by adaptive logic networks. *IEEE Annu. Conf. on EMBS*, pp. 1135–1136. San Diego, California.

Popovic, D. and Raspopovic, V. (1992). Afferent signals in palmar digital nerves. In *Proc. IV Vienna Intern. Workshop on FES*, September, 22–26, Vienna, Austria, pp 105–108.

Popovic, D., Stein, R.B., Jovanovic, K., Dai, R., Kostov, A., and Armstrong, W. (1993). Sensory nerve recording for closed-loop control to restore motor functions. *IEEE Trans. Biomed. Eng.*, 40:1024–1031.

46
Control of Rhythmic Movements Using FNS

Srinath P. Jayasundera and James J. Abbas

1   Abstract

It is widely accepted that neural systems for controlling cyclic movements such as walking, running, chewing, and breathing utilize pattern generating neural circuitry. A network of coupled pattern generator units may be used to control such movements in multisegmented skeletal systems. In the work presented here, we use a network of coupled pattern generators as one component in a control system for generating movements using electrical stimulation. The control system is intended for eventual use in Functional Neuromuscular Stimulation (FNS) systems to restore locomotor function to individuals with neurological disorders. Results of computer simulation studies are presented to demonstrate that: (1) the pattern generator component can generate oscillatory patterns over a wide range of model parameter values; (2) the coupling amongst individual pattern generator components can be automatically adjusted to generate specified intersegmental phase lags; and (3) the control system can automatically adjust stimulation parameters to generate a specified movement of a multisegmented skeletal system.

2   Introduction

FNS is a technique used to stimulate muscles to restore lost functionality to individuals with central nervous system disorders, such as spinal cord injury, stroke, or head injury. FNS has been successfully used to restore stance and locomotion (Marsolais and Kobetic, 1987; Kralj and Bajd, 1989), hand function (Peckham, 1987; Kilgore et al. 1989), and bladder control (Janknegt et al. 1992). Success in the laboratory, however, has proven to be easier to achieve than success in a clinical environment. In particular, there are many formidable issues to be addressed in the development of a practical lower extremity FNS system to restore locomotor function (Marsolais and Kobetic 1987; Chizeck et al. 1988; Jaeger 1992; Popovic 1992). This work is directed at the development of the control system component of the lower extremity FNS system.

2.1   Control of Lower Extremity FNS Systems

Several control strategies have been used in the past to control rhythmic movements, such as gait, using FNS (see Figure 46.1 and Chapter 42). These control strategies can be divided into two broad categories: feedforward and feedback. The distinction between the two is somewhat important in dealing with different control schemes, but it should be noted that most applications utilize a combination of feedforward and feedback strategies. The simplest and most widely used control strategy is a feedforward controller that uses a predetermined stimulation pattern. Here, a stimulation pattern for one cycle of movement is stored in a look-up table and the same stimulation pattern is provided for each cycle of the movement. Typically, the trigger to initiate each step is provided by the user via a finger switch (Kralj and Bajd 1989; Graupe and Kohn 1994; Kobetic and Marsolais 1994). Examples of more advanced approaches to open-loop control (Crago et al. 1996) include the use of in-

FIGURE 46.1. Classification of FNS control strategies for loco-motion.

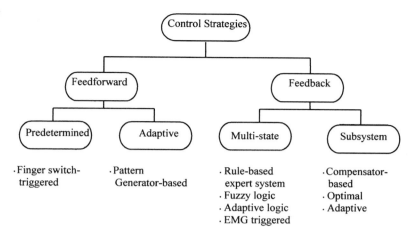

verse dynamic models (Hausdorff and Durfee 1991), dynamic programming techniques on computer simulated models (Yamaguchi and Zajac 1990), and the use of adaptive "cycle-to-cycle" control of swing phase in experimental studies (Franken et al. 1995).

Much of the FNS control system research has been directed at the design and development of feedback controllers, which utilize measurements of system outputs in order to determine the stimulation to be sent to the muscles. Chizeck et al. (1988) described a hierarchical structure that included an upper level, or *multistate*, controller and a lower-level, or *subsystem*, controller. The multistate controller uses sensor measurements (and possibly other signals or timers) to track movements of the system through a finite number of states (e.g. single support, double support, etc.). The various multistate strategies that have been investigated include rule-based expert system controllers (Andrews et al. 1989; Chizeck et al. 1988; Kobetic and Marsolais 1985), fuzzy logic controllers (Skelly et al. 1994), adaptive logic controllers (Kostov et al. 1995), and EMG-triggered multistate control (Graupe and Kordylewski 1995). The subsystem controller, which may be one of several active in a given state, uses sensor measurements (and possibly other signals or timers) to determine the stimulation for a subset of the muscles to be stimulated. For example, a joint tracking subsystem controller would compare measurements of joint angle to a reference trajectory in order to calculate an error signal that would be used to determine muscle stimulation values according to a control law. In general,

the approaches used for lower extremity applications are similar to, and/or have been borrowed from, the approaches used in upper-extremity FNS applications. The various subsystem feedback controllers that have been investigated or proposed for use in FNS systems for locomotion include compensator-based controllers (for reviews, see, Chizeck et al. 1988; Jaeger 1992; Franken et al. 1994), optimal controllers (Veltink et al. 1992), and adaptive controllers (Hatwell et al. 1991; Sepulveda and Cliquet 1995).

To determine an appropriate control strategy, several factors must be considered. Among the several practical issues encountered in FNS control of gait, we have focused on addressing two: intersubject variability and muscle fatigue. Intersubject variability arises from the fact that there is high variability in muscle responses as well as variability in body segment parameters such as length and mass. Muscle fatigue is an especially important issue due to limited muscle use in neurologically impaired individuals and due to the fact that electrical stimulation recruits the more fatiguable fibers first (i.e., "reverse recruitment order") (Mortimer 1981). Hence, each person using an FNS walking system requires an *individualized stimulation pattern*, and the stimulation pattern *requires altering over time*. To address the problems of individualizing and adapting the stimulation pattern, we have developed an adaptive feedforward control system that is based upon the concept of a central pattern generator (see below). The overall control system structure provides the capability of utilizing feedback for triggering state transitions and/or for subsystem control, but the feedback component is not discussed here.

2.2 Neural Pattern Generators: Physiological Evidence and Models

A central pattern generator (CPG) is a collection of neurons with intrinsic properties and/or interconnectivity that can give rise to rhythmic patterns in the absence of any extrinsic phasic inputs. Existence of such neuronal networks as an underlying control system has been conclusively demonstrated in lower vertebrates (e.g. cats, lamprey) (Grillner and Dubuc 1988; Pearson et al. 1992; Sigvardt 1993; Cohen 1994). In humans, although conclusive evidence of CPG circuitry is not available, CPG-based models have been hypothesized (Grillner 1981; Grillner and Dubuc 1988).

Two basic strategies have been used to model central pattern generator circuitry. One strategy, the network oscillator, relies on the interconnectivity between neurons to elicit oscillations (Cohen et al. 1992; Sigvardt and Williams 1992; Buchanan 1992; Jung et al. 1996). The other, the pacemaker oscillator, relies on the intrinsic membrane properties of individual oscillators. The pacemaker category includes models based upon membrane biophysical properties (Brodin et al. 1991; Ekeberg et al. 1991) and models that use equations which do not explicitly model neural membrane properties (Skinner et al. 1993; Ermentrout and Kopell 1994).

2.3 Pattern Generator-Based Control

Models of pattern generator circuitry have primarily been used to investigate the viability of various motor control theories. Such studies have included investigations of the role of intracellular mechanisms in invertebrate neural circuits (Marder et al. 1993) to the role of interoscillator coupling in generating coordinated multisegmental movements (Williams 1992; Ermentrout and Kopell 1994). Other studies have incorporated models of peripheral dynamics (i.e., the dynamics of the muscular and/or skeletal system) to investigate the role of neuro-mechanical coupling (Ekeberg et al. 1991; Beer et al. 1993; Taga 1995; Cruse et al. 1995). The issues around central pattern generation and the interaction with peripheral dynamics are discussed in detail in Section IV.

In our work, we have used models of pattern generator circuitry to design systems for controlling cyclic movements using FNS. Our approach has been to utilize a combination of neural models at the block diagram level, neural models at the cellular level, artificial neural network algorithms, and engineering control systems techniques. Figure 46.2 shows the two-stage structure of the PG/PS control system. The pattern generator (PG) is responsible for generating the basic oscillatory rhythm, while the patter shaper (PS) adaptively filters the signal from the PG before it is sent to the muscles. This two-stage structure is loosely based on a hypothesized organizational structure of the spinal cord, with one set of neurons (the pattern generator) generating the basic rhythm and a different, although possibly overlapping, set (the spinal segmental circuitry) modifying the signals before they reach the muscles. The design of the PG component is based on biophysical models of neural pattern generator circuitry, the PS component uses an adaptive artificial neural network learning algorithm, and traditional engineering feedback controllers can be used to supplement the action of the PG/PS control system.

The PG/PS control system was first evaluated in a set of simulation studies where the outputs of the controller were used as inputs to a pair of agonist/antagonist muscles that acted on a one-segment planar system. Results of those studies indicated that the control system could automatically customize the stimulation parameters to generate a specified movement in a given individual (Abbas and Chizeck 1995) and that the pattern generator component was capable of generating reflex-type

PG/PS Control System

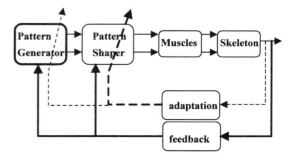

FIGURE 46.2. Block diagram of the control system.

behavior that would be desirable in an FNS system (Abbas 1996). In a series of experiments using spinal cord injured human subjects, the control system was used to generate cyclic isometric contractions. Results of these studies indicated that the control system could rapidly adapt to customize the stimulation parameters for a particular individual and then adapt the stimulation parameters in order to account for muscle fatigue (Abbas and Triolo 1993).

2.4   Objectives

In the work presented here, we sought to enhance the PG/PS control system for the purpose of controlling multisegmented musculoskeletal systems in a coordinated manner. Our first objective was to further characterize the dynamics of the PG circuit model. Here, we sought to investigate the oscillatory behavior of the PG circuitry over a wide range of parameter values. Our second objective was to develop a mechanism for coordinating the activity of a set of pattern generator modules. Towards this objective, we incorporated an adaptive algorithm to automatically adjust specific synaptic weights in order to learn a specified set of intersegmental phase lags. The third objective was to investigate the possibility of using the on-line PS learning algorithm to automatically adjust stimulation parameters for controlling multisegmental movements. Here, we used a straightforward extension of our previous PS learning algorithm.

3   Methods

The properties and intersegmental coupling of the PG component were investigated in a series of computer simulation studies using only the PG circuit model. The multi-segment movement controller was evaluated in computer simulation studies that used the PG circuit model, the PS component, and a mechanical model that included three pairs of muscles acting on a three segment planar skeleton. The simulation studies and analysis were performed on a Unix-based workstation using in-house developed software, Xppaut [a public-domain package for analysis of differential equations (Ermentrout 1994)], and Matlab™.

3.1   The Pattern Generator

Three models of CPG circuits were used to generate activity patterns required for cyclic movements. In all the models, rhythmic neural activity is due to the interaction of depolarizing NMDA channel currents, repolarizing calcium-dependent potassium channel currents and synaptic connections between the neurons. In the first model, the oscillator consisted of a pair of mutually inhibitory pacemaker neurons. The second model consisted of two pairs of mutually inhibitory neurons with excitatory connections between neurons of opposite pairs. The third model, which was based upon a model of the neural circuitry controlling locomotion in the lamprey, consisted of a chain of three unit pattern generators; each unit pattern generator consisted of a six-neuron network (two groups of three neurons in mutual inhibition). The set of differential equations used in all models to describe the dynamics of an individual PG neuron (Abbas 1996) has been adapted from models of neural control of locomotion in the lamprey (Brodin et al. 1991; Ekeberg et al. 1991).

3.2   The Musculoskeletal System Model

The model of the musculature (Abbas and Chizeck 1995) included nonlinear recruitment properties, second order linear dynamics with an input time delay, length-tension properties and velocity-tension properties. The skeletal system model describes the dynamics of three rigid segments in a swinging pendulum configuration with linear and nonlinear stiffness and damping at each joint.

4   Results

This section reports results from three different studies. The first study, demonstrates the use of nonlinear dynamical systems techniques to characterize the behavior the PG network in different regions of parameter space. The next section addresses the problem of intersegmental coordination by learning an appropriate set of phase lags between different PG components. The final section presents results from studies where the PG is used in conjunction with the PS to control movements of a musculoskeletal system model.

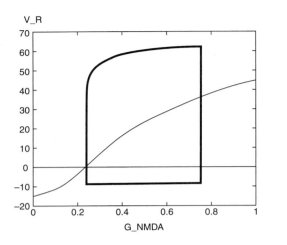
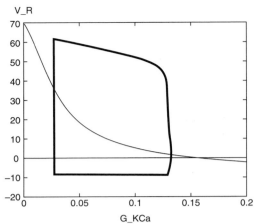

FIGURE 46.3. One parameter bifurcation diagram for a two neuron network. Control parameters were g_{NMDA} (left) and g_{KCa} (right).

4.1 Dynamics of the Pattern Generator

The PG is responsible for generating a pattern of oscillatory signals to drive a given movement. In the PG model used for this system, each neuron in the circuit includes many parameters (Brodin et al. 1991; Abbas 1996). Changes in any model parameter may result in profound changes in system output. We have investigated the sensitivity to such changes in model parameters using nonlinear dynamical systems analysis techniques. An example of one type analysis used is shown in Figure 46.3. Here we have qualitatively shown the effects of varying a single parameter at a time (NMDA conductance, g_{NMDA}, and calcium dependent potassium conductance, g_{KCa}) on the outputs of a two-neuron pattern generator system. In this plot, regions of the parameter space where oscillations were not observed are indicated by a single thin line (g_{NMDA} <0.22; g_{NMDA} >0.72 and g_{Kca} <0.03 and g_{KCa} >0.14). These regions have a stable fixed point—a voltage to which the membrane voltage would settle. Regions where oscillations were observed are indicated by a set of three lines (0.22 < g_{NMDA} <0.72 and 0.03 < g_{KCa} <0.14); the top and bottom line indicate the maximum and minimum membrane voltages, respectively, and the middle line indicates the unstable fixed point about which the system oscillates. (Note that, in theory, if the membrane were to be placed on the unstable fixed point, it would stay there until perturbed). The point at which the system undergoes such a qualitative change in behavior is call a *bifurcation point*. For example, the transition from a stable fixed point to a limit cycle oscillation at g_{NMDA} = 0.22 is a *Hopf bifurcation*.

The results shown in Figure 46.3 indicate that system exhibits stable oscillations over a wide range of parameters and that the oscillations can be terminated by adjusting a single control parameter. Selecting a parameter value that is in the center of the oscillatory region may be a suitable choice to ensure stable oscillations; selecting a value near the border of the oscillatory region may provide a means to rapidly and conveniently terminate an oscillation. It should be noted, however, each diagram exhibits the effects of varying only one parameter of several in the model. There may be strong interdependencies among the various model parameters that must also be considered. In ongoing work, we are studying the effects of changes in parameters such as synaptic conductances and the tonic drive current.

4.2 Learning Phase Lags

The second issue that we will address is how to generate oscillatory patterns with specified phase delays between individual component oscillators. In this work, we seek to develop a technique to automatically adjust model parameters to result in patterns of neural activity with an appropriate set of phase relationships. This feature will play a role

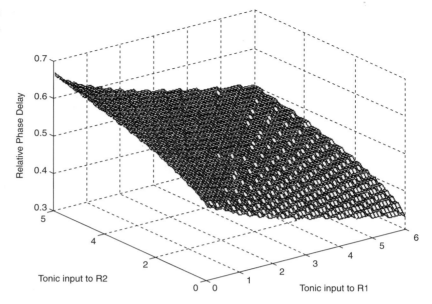

FIGURE 46.4. Phase delays as a function of tonic input for a two-neuron network.

in an adaptive control system that learns to generate coordinated movements of multi-segment systems. We have focused this work on investigating the influence of two model parameters (the tonic drive to individual oscillator components and the weights on the synaptic connections between oscillator components) on the resulting phase relationships. Figure 46.4 shows results from a study where the tonic drive to each neuron of a two-neuron oscillator (i.e., two neurons in mutual inhibition) was varied. These results demonstrate that the phase delay between the two neurons (plotted on the vertical axis) can be manipulated over a wide range. Figure 46.5 shows the results from a study where the synaptic strengths among oscillator components of a four-neuron network were automatically adjusted using a neural network-type learning algorithm to achieve a desired phase lag. The figure on the left shows a phase portrait (output of one oscillator vs. output of another) before learning occurs, where the network is symmetric and the phase lag equals 0.5 (expressed as a fraction of the cycle period). Note that the slope of the system trajectory is near −1, indicating that the output variables

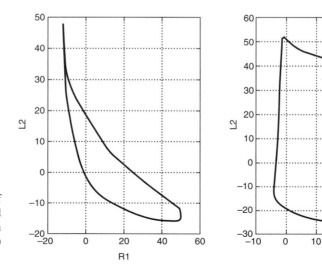

FIGURE 46.5. Phase portrait of membrane voltages of R1 and L2 neurons of a four-neuron network: before learning (left) and after learning (right).

are changing in opposite directions (one is going up while the other is coming down). The figure on the right shows the phase portrait where the oscillation is asymmetric and it has learned to oscillate with the new desired phase lag (0.7). Note that the system follows a trajectory along near-vertical and near-horizontal lines, indicating that only one of the two output variables is changing at a given time. These results demonstrate that phase lags can be manipulated by changing system parameters in a pre-specified manner or through an on-line adaptation algorithm.

4.3 Controlling Movements of Multisegment Systems

The third issue to be addressed is that of adaptively filtering the signals from the PG in order to fine tune the stimulation parameters for a given individual. Our approach here assumes that the basic features of the stimulation pattern will be the same for all individuals, but the details will be different. Therefore, the same PG circuit can be used for everyone, but the PS would adapt to fine-tune the pattern for a specific individual. Figure 46.6 shows the results of a simulation run in which the set of oscillatory signals from the PG was adaptively filtered by the PS to generate a specified movement in a planar three-segment skeletal model. Here, the PG consisted of three unit PGs (Grillner et al. 1987; Abbas 1996), one for each joint to be controlled. Each unit PG consists of a six-neuron oscillator with two output signals. A desired joint angle trajectory (dashed line) was specified and the tracking error signal was used in an on-line learning algorithm (Abbas and Chizeck 1995) to adjust the parameters of the PS. Results (e.g., Figure 46.6) have indicated that the on-line learning is capable of adapting PS filtering properties to fine tune the stimulation for a particular individual. One particularly important feature of this learning paradigm is that it is not necessary to explicitly account for the intersegmental coupling, for example, updates to stimulation patterns for muscles at the knee can be based on measurements of knee angle tracking error without regard for the segmental accelerations caused by the activation of muscles that cross the hip. It should be noted that the data shown in Figure 46.6 were generated by a stimulation pattern that was adapted over 200 cycles from an initial PS

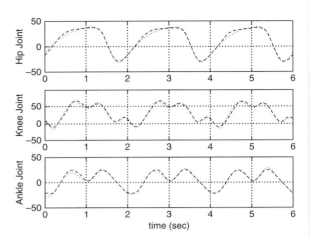

FIGURE 46.6. Time traces of joint trajectories after 200 s of learning: desired trajectory (dashed line), actual trajectory (dotted line).

parameter set equal to 0. In actual implementation, the adaptation time could be significantly improved by pretuning the PS parameters on a system model, although a very accurate model would not be required.

5 Discussion and Future Directions

To control cyclic movements using FNS, we have used an adaptive feedforward control system. We believe that a feedforward strategy is necessary because a purely reactive feedback controller would perform poorly for locomotion control due to the slow response time of electrically stimulated muscle and due to the unavoidable noise in sensor measurements. Although a feedback controller may play an important role as a component in the overall system, our approach seeks to utilize feedback only as a supplement to the feedforward control system to enhance disturbance rejection.

The adaptive nature of the control system provides a very important feature that can greatly enhance the practicality of the control system by simplifying the process of "fitting" the neural prosthesis to the user. We seek to develop a control system that can be initially pre-tuned off-line using musculoskeletal models and then automatically fine tuned in a short time during operation on a specific individual. One way to utilize this fine tuning would be in an intermittent mode where a user comes to the clinic for fine tuning of the stimulation pattern that would then be utilized on a regular basis in open-loop mode. This would avoid the need to wear sensors all of the time; they would only be required during fine tuning sessions. A second way to utilize the fine tuning would be to continually adapt the control system parameters during operation. This approach offers the advantage of providing a means to adapt stimulation parameters to account for muscle fatigue, but unfortunately, it would require that sensors be worn continuously.

Our implementation of the adaptive feedforward control system utilizes models of neural pattern generator circuitry (PG) in cascade with an adaptive filter (PS). We believe that the PG component will play an important role in generating coordinated movements as well as in generating reflexes that are phase dependent (Abbas and Chizeck 1994; Abbas 1996). The PS provides the adaptive capability that may be a critical component to a practical FNS control system.

In future work, we will seek to further characterize the oscillatory properties of the PG circuit and to develop additional methods for automatically adjusting system parameters. The paradigm used in the phase lag learning studies uses a localized learning algorithm. That is, a few network parameters were adjusted based upon a specific characteristic of the output pattern (a synaptic weight was adjusted based upon a measurement of interoscillator phase lag). We believe that this approach to learning may provide a rapid and efficient means of making adjustments to complex neural models such as the one used in the PG. We will investigate the use of such localized learning algorithms for pattern generator based control systems.

Two general issues are faced in the development of this (and many other) FNS control systems: (1) how will this approach scale when used on more complex musculoskeletal system models, and (2) how will this approach fare when stepping out of the world of computer simulation studies to experimental trials (and hopefully on to clinical use). We plan to enhance the design of the control system through an iterative process that includes simulation studies and experiments in a wide variety of situations.

Acknowledgments. This work was supported by the National Science Foundation (NSF-BCS-9216697).

References

Abbas, J.J. (1996). Using neural models in the design of a movement control system. In *Computing Neuroscience.* Bower, J. (ed.), pp. 305–310. Academic Press, New York.

Abbas, J.J. and Chizeck, H.J. (1994). Phase-dependent reflexes in a neural network control system. *Proc. Southern Biomed. Conf.*, pp 494–497.

Abbas, J.J. and Chizeck, H.J. (1995). Neural network control of functional neuromuscular stimulation systems: computer simulation studies. *IEEE Trans. Biomed Eng.*, 42(11):1117–1127.

Abbas, J.J. and Triolo R.J. (1993). Experimental evaluation of an adaptive feedforward controller for use in functional neuromuscular stimulation systems. *Proc. IEEE/EMBS Conf.*, 15:1326–1327.

Andrews, B.J., Barnett, R.W., Phillips, G.F., Kirkwood, C.A., Donaldson, N., Rushton, D.N., and Perkins, T.A. (1989). Rule-based control of a hybrid FES orthosis for assisting paraplegic locomotion. *Automedica*, 11: 175–199.

Bajd, T., Andrews, B.J., Kralj, A., and Katakis, J. (1985). Restoration of walking in patients with incomplete spinal cord injuries by use of surface electrical stimulation—preliminary results. *Prosthetics Orthotics Intl.*, 9:109–111.

Beer, R.D., Ritzmann, R.E., and McKenna, T. (1993). Biological neural networks in invertebrate neuroethology and robotics. Academic Press, Boston.

Brodin, L., Traven, H.G.C, Lansner, A., Wallen, P., Ekeberg, O. and Grillner, A. (1991). Computer simulations of N-Methyl-D-Aspartate receptor-induced membrane properties in a neuron model *J. Neurophysiol.*, 66(2):473–483.

Buchanan, J.T. (1992). Neural network simulations of coupled locomotor oscillators in the lamprey spinal cord. *Biol. Cybern.*, 66:367–374.

Chizeck, H.J., Kobetic, R., Marsolais, E.B., Abbas, J.J., Donner, I.H., and Simon, E. (1988). Control of Functional Neuromuscular Stimulation Systems for Standing and Locomotion in Paraplegics. *Proc. IEEE*, 76(9):1155–1165.

Cohen, A., Ermentrout, G., Kiemel, T., Kopell, N., Sigvardt, K., and Williams, T. (1992). Modeling of intersegmental coordination in the lamprey central pattern generator for locomotion. *TINS*, 15(11):434–438.

Cohen, A.H. (1994). Evolution of the vertebrates central pattern generator for locomotion. In *Neural Control of Rhythmic Movements in Vertebrates.* Cohen, A.H.,

Rossignol, S., and Grillner, S. (eds.), pp. 129–166. John Wiley & Sons, New York.

Crago, P.E., Lan, N., Veltink, P.H., Abbas, J.J., and Kantor, C. (1996). New Control strategies for neuroprosthetic Systems. *J. Rehab. Res. Dev.*, 33(2):158–172.

Cruse, H., Bartling, C., Cymbalyuk, G., Dean, J., and Dreifert, M. (1995). A modular artificial neural net for controlling a six-legged walking system. *Biol. Cybern.*, 72:421–430.

Ekeberg, O., Wallen, P., Lansner, A., Traven, H., Brodin, L., and Grillner, S. (1991). A computer based model for realistic simulations of neural networks; I. The single neuron and synaptic interaction. *Biol. Cybern.*, 65:81–90.

Ermentrout, B. (1994). XPPAUT1.2-the differential equation tool. (http://mrb.niddk..nih.gov/xpp/).

Ermentrout, B., and Kopell, N. (1994). Learning of phase lags in coupled neural oscillators. *SIAM J. Appl. Math.*, 54(2):478–507.

Franken, H.M., Veltink, P.H., and Boom, H.B.K. (1994). Restoration of paraplegic gait by functional electrical stimulation. *IEEE/EMBS Magazine.*

Franken, H.M., Veltink, P.H., Baardman, G., Redmeijer, R.A., and Boom, H.B.K. (1995). Cycle-to-cycle control of swing phase of paraplegic gait induced by surface electrical stimulation. *Med. Biol. Eng. Comput.*, 33:440–451.

Graupe, D. and Kohn, K.H. (1994). *Functional Electrical Stimulation for Ambulation by Paraplegics.* Krieger Publishing Co., Melbourne, Florida.

Graupe, D. and Kordylewski, H. (1995). Artificial neural network control of FES in paraplegics for patient responsive ambulation. *Trans. Biomed. Eng.*, 42(7): 699–707.

Grillner, S. (1981). Control of locomotion in bipeds, tetrapods and fish. In *Handbook of Physiology, Sect. 1: The Nervous System II, Motor Control.* Brooks, V.B. (ed.), pp. 1179–1236. American Physiological Society, Waverly Press, Bethesda, Maryland.

Grillner, S. and Dubuc, R. (1988). Control of locomotion in vertebrates: spinal and supraspinal mechanisms. *Adv. Neurol.*, 47:425–453.

Grillner, S., Wallen, P., Dale, N., Brodin, L., Buchanan, J., and Hill, R. (1987). Transmitters, membrane properties and network circuitry in the control of locomotion in lamprey *TINS*, 10:34–41.

Hatwell, M.S., Oderkerk, B.J., Sacher, C.A., and Inbar, G.F. (1991). The development of a model reference adaptive controller to control the knee joint of paraplegics. *IEEE Trans. Auto. Cont.*, 36:683–691.

Hausdorff, J.M. and Durfee, W.K. (1991). Open-loop position control of the knee joint using electrical stimulation of the quadriceps and hamstrings. *Med. Biol. Eng. Comput.*, 29:269–280.

Jaeger R. (1992). Lower extremity applications of functional neuromuscular stimulation. *Assist. Technol.*, 4:19–30.

Janknegt, R.A., Baeten, C.G.M.I., Weil, E.H., and Spaans, F. (1992). Electrically stimulated gracilis sphincter for treatment of bladder sphincter incontinence. *Lancet*, 340:1129–1130.

Jung, R., Kiemel, T., and Cohen, A.H. (1996). Dynamical behavior of a neural network model of locomotor control in the lamprey. *J. Neurophysiol.*, 75(3):1074–1086.

Kilgore, K.L., Peckham, P.H., Thrope, G.B., and Keith, M.W. (1989). Synthesis of hand movement using functional neuromuscular stimulation. *IEEE Trans. Biomed. Eng.*, 36(7):761–770.

Kobetic, R. and Marsolais, E.B. (1985). Automated electrically induced paraplegic gait. *Proc. 38th ACEMB*, 293.

Kobetic, R. and Marsolais, E.B. (1994). Synthesis of paraplegic gait with multichannel functional neuromuscular stimulation. *IEEE Trans. Rehab. Eng.*, 2(2):66–79.

Kostov, A., Andrews, B.J., Popovic, D.B., Stein, R.B., and Armstrong, W.W. (1995). Machine learning in control of functional neuromuscular stimulation systems for locomotion. *IEEE Trans. BME*, 42(6):541–551.

Kralj, A. and Bajd, T. (1989). *Functional Electrical Stimulation: Standing and Walking After Spinal Cord Injury.* CRC Press, Boca Raton, Florida.

Marder, E., Abbott, L.F., Buchholtz, F., and Epstein, I.R. (1993). Physiological insights from cellular and network models of the stomatogastric nervous system of lobsters and crabs. *American Zoologist*, 33(1):29–39.

Marsolais, E.B. and Kobetic, R. (1987). Functional electrical stimulation for walking in paraplegia. *J. Bone Joint Surg.*, 69-A(5):728–733.

Mortimer, J.T. (1981). Motor prostheses. In *Handbook of Physiology, Section I: The Nervous System*. Brooks, V.B. (ed.), Vol. 2, pp. 155–187. American Physiological Society, Bethesda, Maryland.

Pearson, K., Ramirez, J., and Jiang, W. (1992). Entrainment of the locomotor rhythm by group Ib afferents from ankle extensor muscles in spinal cats. *Exp. Brain Res.*, 90:557–566.

Peckham, P.H. (1987). Functional electrical stimulation: current status and future prospects of applications to the neuromuscular system in spinal cord injury. *Paraplegia*, 25:279–288.

Popovic, D.B. (1992). Functional electrical stimulation for lower extremities. In *Neural Prostheses: Replacing Motor Function After Disease or Disability*. Stein, R.B., Peckham, P.H., and Popovic, D.P. (eds.), pp. 233–251. Oxford University Press, New York.

Sepulveda, F., and Cliquet, Jr., A. (1995). An Artificial neural system for closed loop control of locomotion produced via neuromuscular electrical stimulation. *Artif. Organs*, 19(3):231–237.

Sigvardt, K.A. (1993). Intersegmental coordination in the lamprey central pattern generator for locomotion. *Neurosciences*, 5:3–15.

Sigvardt, K.A. and Williams, T.L. (1992). Models of central pattern generators as oscillators: the lamprey locomotion CPG. *Semi. Neurosci.*, 4:37–46.

Skelly, M.M., Chizeck, H.J., and Ferencz, D.C. (1994). Pattern switching and a fuzzy rule base for control of paraplegic walking, *Proc. Neural Prosthesis: Motor Systems IV*, Mt. Sterling, Ohio, p. 34.

Skinner, S.K., Turrigiano, G.G., and Marder, E. (1993). Frequency and burst duration in oscillating neurons and two-cell networks. *Biol. Cybern.*, 69:375–383.

Taga, G. (1995). A model of the neuro-musculo-skeletal system for human locomotion: I. Emergence of basic gait. *Biol. Cybern.*, 73:97–111.

Veltink, P.H., Franken, H.M., Van Alste, J. A., and Boom, H.B.K. (1992). Modeling the optimal control of cyclic leg movements induced by functional electrical stimulation. *Int. J. Artif. Organs*, 15(12):746–755.

Williams T. (1992). Phase coupling in simulated chains of coupled oscillators representing the lamprey spinal cord. *Neural Comput.*, 4:546–558.

Yamaguchi, G.T. and Zajac, F.E. (1990). Restoring unassisted natural gait to paraplegics via functional neuromuscular stimulation: a computer simulation study. *IEEE Trans. BME*, 37(9):886–902.

Commentary

Peter H. Veltink

Jayasundera and Abbas describe a very interesting approach to the artificial control of cyclical human movements assisted by FES. They propose a control strategy based on the concept of a central pattern generator, which has been demonstrated to exist in lower vertebrates (e.g., see Chapter 12 and Chapter 15).

The application of such a physiological control concept in artificial control systems is in itself a challenging and interesting problem. The authors demonstrate the feasibility in a model study and in isometric experiments. Beyond this, the question arises whether such approaches can yield a superior performance in comparison to more classical approaches to adaptive feedforward control of cyclical movements, as e.g. investigated by our group (e.g., Franken et al. 1995). Jayasundera and Abbas do not answer this question.

Superior performance with the central pattern generator approach may be possible, since the strategies described are highly non-linear, which potentially results in a superior adaptation and learning response. Also, the automatic tuning of the system to the characteristics of the individual user is an important characteristic, although it remains to be seen how much tuning and setting of parameters is still required. I would like to encourage the investigation of the benefits of the central pattern generator strategy in comparison to alternative strategies, in order to demonstrate its strength and importance.

Reference

Franken, H.M., Veltink, P.H., Baardman, G., Redmeijer, R.A., and Boom, H.B.K. (1995). Cycle-to-cycle control of swing phase of paraplegic gait included by surface electrical stimulation, *Med. Biol. Eng. Comput.*, 33:440–451.

Section X

Appendix 1
Morphological Data for the Development of Musculoskeletal Models: An Update*

Frans C.T. van der Helm and Gary T. Yamaguchi

This appendix describes morphological studies. Several methodologies and type of data are compared and combined. The actual morphological data can be found through a website:

http://www.wbmt.tudelft.nl/mms/morph_data/

1 Introduction

Probably the most used chapter of *Multiple Muscle Systems*, edited by Winters and Woo (1990), was the appendix on morphological data sets written by Yamaguchi, Sawa, Moran, Fessler, and Winters (1990). They compiled a huge database of morphological data from 24 original sources. These types of data bridge the gap between the development of a model structure and the "real" world (i.e., comparison with experimental results and applications of all sorts). This chapter serves the same purpose and intends to encourage those involved in computer simulations to use real data and to compare predictions with experimental results.

The form of the data presented by Yamaguchi et al. (1990) is static. Once they are derived, there will be hardly any updates on them. Instead of repeating this data, it has been decided to put all morphological data on a website. For data published since 1990, and data not covered by Yamaguchi et al. (1990), there will be links to websites with the original data, organized by the authors of the study.

Major advantages of this procedure is that the original authors remain responsible for their data and can be contacted by E-mail in case of questions. Data are only available for noncommercial use. It seems fair to pay the authors a reasonable fee if their data are used to make money!

Data are categorized according to their anatomical location:

1. Lower extremity musculature.
2. Trunk musculature.
3. Upper extremity musculature.
4. Hand musculature.
5. head and neck musculature.

For a complete musculoskeletal model, the following data should all be available (see also Yamaguchi et al. 1990):

- Mass and rotational inertia of the segments, center of gravity.
- Geometrical data of joints, ligaments, and muscles.
- Muscle parameters.

Unfortunately, few studies provide all necessary data for a specific region. Most models are build by "assembling" data from various sources. This will be the task and responsibility of the reader.

2 Parameters

In general, all parameters are presented in the local coordinate systems of the original studies. Local coordinate systems are usually defined with respect to bony landmarks (preferred!), but sometimes using such vague notions as the longitudinal

---

*For an earlier version of this appendix, see Yamaguchi et al. (1990), pp. 717–773. Used with permission.

axis or perpendicular to bony surfaces. To combine data, the same bony landmarks must have been used, and the relation between the local coordinate systems must be known.

In the parameter list, inertial data are only incorporated if they are accompanied by other muscle data, because the purpose is to build *musculoskeletal* systems in the first place. Many more inertial data can be found in original studies as Dempster (1955), with further data processing by Chandler et al. (1975), Hinrichs (1985), and Yeadon and Morlock (1989). In these studies, typical regression equations are used to estimate inertial properties from anthropometric data.

Muscle data can be split into two types: geometrical data (attachment sites, moment arms) and "physiological" data (related to magnitude of force exertion and/or muscle dynamics). These types of data can easily be combined.

3   Geometrical Data

A full set of *geometrical* data consists of:

1. *Bony landmarks.* Too often forgotten, bony landmarks are usually needed firstly to determine the local coordinate systems of the bone, and secondly to measure motions. If two studies use the same set of bony landmarks, at least the local coordinate systems can be rotated into each other, and all data can be compared. If some bony landmarks are different (and missing), it *is very difficult* to compare the results of two studies. In addition, if one wants to combine two data sets of separate joints from different studies (e.g., combine shoulder data and elbow data), it is only possible if the two data sets share enough common information, such as the same bony landmarks for the humerus.
2. *Joint rotation center-rotation axis.* Obviously this is the first step into modeling, because the joints are almost always represented as idealized geometrical figures, such as spherical joints or hinge joints. The orientation of the articular surfaces might be important if the stability of the joint is of concern (e.g., large shear forces in the should could dislocate the joint).
3. *Ligament attachments and/or moment arm.* Ligaments are usually represented by a straight line. Extracapsular ligaments can easily be recorded.

The definition of intracapsular ligaments is sometimes difficult because they are in continuation with the capsule. The mechanical properties of the ligaments (stress/strain data, or collagen content and cross-sectional area) are important.

4. *Muscle attachments and/or moment arms.* Most attention is focused on the location of the muscles. In the modeling approach, muscles are typically represented by a single or multiple lines of action, despite having a certain thickness, a sometimes large attachment site, fiber pennation, and parts that may contract independently. Thus, many of the muscle properties might already be lost in the first stage of the modeling process. The following items should be considered:

- *Architecture*: The architecture of a muscle deals with the fiber direction between origin and insertion. If the origin and/or insertions attach over a certain region, the muscle may not be adequately described by just one muscle line of action, but more are needed. Most studies a priori divide the muscle in a number of parts, each described by a muscle line of action. Van der Helm and Veenbaas (1991) presented a methodology for assessing the appropriate number of muscle lines of action a posteriori, by analyzing the resulting mechanical effect of the muscle.

- *Excursion* of the muscle line of action. For many muscles a straight line between origin and insertion is not the correct representation of the muscle line of action. Basically, there are two solutions: (1) use geometric representations of bony contours to wrap the muscle around (e.g., Högfors et al. 1987; Van der Helm et al. 1992), or (2) use prespecified via-points for the muscle line of action (Delp et al. 1990). In the latter approach, the effective moment arm will be between the last via-points attached to the proximal bone and the first via-point attached to the distal bone.

- *Moment arms.* Obviously, if the origin and insertion are given, the moment arm of the muscle can be calculated as a function of joint angle. Some studies however record directly the muscle moment arm, without explicitly measuring the attachments, either by using CT scans or MRI images to determine the centroid line (Jensen and Davy 1975) or by measuring muscle length versus joint angle (e.g., Spoor et al. 1990). The derivative of this function will be the muscle moment arm. A musculoskeletal model can be developed with just the muscle moment arms, neglecting the origin-insertion attachments and thereby the force direction. However, it will be impossible to reconstruct the joint reaction forces and bending moments of hinge axes.

4 Physiological Data

The describing parameters for the *dynamic behavior* of muscles depend on the type of muscle model used. Most muscle models are Hill-type models, with a mathematical description of the force-length and force-velocity relation. Some authors employ rather complex muscle models (e.g., Hatze 1981, 1997); others attempt to simplify models as much as possible without loosing important features (e.g., Winters and Stark 1985; Otten 1988; Zajac 1989). The more complex a muscle model, the more difficult it will be to obtain a sufficient set of parameters. Huxley-type cross-bridge models (Huxley 1957; Zahalak 1981) are not yet used for large-scale musculoskeletal models. These models require some detailed parameters for cross-bridge attachment and detachment rates, which can only be approximated through a fit of the force-velocity relation. Other parameters like force-length relation, cross-sectional area, tendon properties need to be obtained similarly to Hill-type models.

Winters and Stark (1988) used estimates of seven basic parameters (several interdependent), that is, muscle length, tendon length, fiber length, pennation angle, cross-sectional area, fiber type, and muscle mass, plus estimated muscle moment arms to arrive at muscle moments. Relations between muscle parameters and basic parameters are weakly based on muscle experiments reported in literature. Zajac (1989) mentioned five basic parameters needed for his muscle model, that is, tendon slack length, peak isometric muscle force (~cross-sectional area), optimal muscle fiber length, optimal muscle fiber pennation angle and maximum shortening velocity. Delp et al. (1990) used the geometry data of Brand et al. (1981) and heuristically adapted the basic parameters for Zajac's model to fit a majority of experimental data for the human leg. In Yamaguchi et al. (1990) the following parameters have been mentioned:

- Muscle length.
- Fiber length.
- Volume.
- Muscle cross-sectional area (volume/fiber length).
- Pennation angle.
- Fiber type.
- Tendon length.
- Tendon cross-sectional area.

It is clear that muscle length and fiber length depend on the joint angle. Fiber length itself is a not well-defined quantity. Optimal fiber length is needed to derive a force-length relationship and can be estimated by meticulous measurements of sarcomere length, using for example laser diffraction. Klein Breteler et al. (1997) are the only ones to measure optimal fiber length for all muscles of one joint (i.e., the shoulder joint).

Muscle volume is most often (and most accurately) obtained by measuring muscle mass, and dividing by the muscle density (between 1.02 and 1.05 kg/liter). Subsequently, muscle cross-sectional area can be obtained by dividing volume by optimal fiber length. Though as mentioned before, data on optimal fiber length are lacking, and optimal fiber length is replaced by *measured* fiber length, so the approximation of muscle cross-sectional area is inaccurate.

Pennation angle is the angle between the *muscle* force acting along the muscle line of action and the muscle fiber force acting along the fiber direction. Maximal muscle force is calculated from the muscle cross-sectional area, which is taken perpendicular to the muscle *fibers*. The maximal muscle force should be scaled by the cosine of the pennation angle. However, pennation angles below 10° can be neglected (cosine is 0.985 or higher). A quick scan through the tables of Yamaguchi et al. (1990) shows that most muscles have pennation angles below 10°.

5 Principal Sources

Sources *[1]* to *[24]* are reprinted from Yamaguchi et al. 1990, while sources *[25]* to *[31]* are new.

1. *Alexander and Vernon (1975)*

Muscles were dissected from the right leg of an embalmed male cadaver (166 cm, 64 kg, 48 year old) who "seemed superficially similar to the bodies of reasonably healthy middle-aged men." Muscles were weighed, then strategically cut for pennation angle and fiber length estimation. Pennation angles were measured in two places for ext. digitorum longus and peroneus longus; both are given.

2. Brand, Crowninshield, Wittstock, Pederson, Clark, and van Krieken (1982)

Three fresh cadavers (163 cm female, 172 cm male, 183 cm male) were used in this study. Muscles were dissected from their origins and insertions so that radiographically visible markers (single or multiple nails) could be attached in their place. Where muscle origin or insertion sites occurred over broad areas, the centroids of the attachment areas (estimated) were used to specify origin/insertion "points." Some muscles (e.g. gluteus maximus) which exert forces upon exceptionally large regions of bone were considered to be composed of two or three components; thus, such muscles can be seen to have multiple numbered listings in the table. For muscles that do not act in a straight line, "effective" origin and/or insertion points were marked by a single point located where the estimated centroid of the cross section of the muscle or tendon crossed the joint and had the most realistic effect on moment arm predictions.

X-ray imaging was used to determine the Cartesian coordinates of the embedded metal markers. Locations of these points are specified with respect to tibial, femoral, and pelvic reference frames defined as follows: "The pelvic reference frame has its origin at the center of the acetabulum with axis directions defined as:

$$Y^P = P_3^P - P_2^P$$
$$X^P = Y^P \times (P_1^P - P_2^P) \qquad \text{(A1.1)}$$
$$Z^P = X^P \times Y^P$$

Where (P_1^P) is the right anterior superior iliac spine, (P_2^P) is the midpoint between the pubic tubercles, and (P_3^P) is the midpoint between the anterior superior iliac spines. The femoral reference frame has its origin at the midpoint between the medial and lateral epicondyle with axes directions defined as:

$$Y^F = P_3^F - P_2^F$$
$$X^F = Y^F \times (P_1^F - P_2^F) \qquad \text{(A1.2)}$$
$$Z^F = X^F \times Y^F$$

where (P_1^F) is the lateral epicondyle, (P_2^F) is the midpoint of the medial and lateral epicondyle, and

(P_3^F) is the femoral head center. The tibial reference frame has its origin at the midpoint between the medial and lateral malleolus with directions defined as:

$$Y^T = P_3^T - P_2^T$$
$$X^T = Y^T \times (P_1^T - P_2^T) \qquad \text{(A1.3)}$$
$$Z^T = X^T \times Y^T$$

where (P_1^T) is the lateral malleolus, (P_2^T) is the midpoint between the medial and lateral malleolus, and (P_3^T) is the tibial tuberosity."

3. Friedrich and Brand (1990)

The legs of two embalmed cadavers were dissected: (S1) 37 year old, 183 cm, 91 kg male, and (S2) a 63 year old, 168cm, 59 kg female. Musculotendon (total) lengths and average pennation angles were measured with the bodies placed in a supine position ("with hips close to full extension and neutral abduction-adduction and rotation, knees extended, and feet in a neutral position neither dorsi- nor plantar-flexed"), prior to dissection. Excised muscle lengths were used instead of origin to insertion distances for muscles that wrapped around joints and hence followed curving, instead of straight, pathways.

Muscle volumes were obtained after excision via water displacement. Average fiber lengths were measured directly from single fibers, which were prepared from macerated fiber bundles obtained by sampling different parts of each muscle. Physiological cross-sectional areas were calculated by dividing the muscle volume by average muscle fiber length. Separate measurements of tendon were not included in these studies. The reader is cautioned that muscle volumes, and hence PCSAs, appear to shrink following death. Therefore PCSAs as measured in cadavers may be smaller than those in living subjects (Brand 1990; Edgerton 1990).

4. Pierrynowski and Morrison (1985)

Muscle properties were obtained via a combination of dry-bone measurements (from a dried, disarticulated, male Caucasian skeleton), scaling, and literature surveys. A measurement device capable of 1 mm accuracy in 3D was built to collect data on the 38 muscles deemed to be important. Some muscles were partitioned into "two or more distinct structures due to functional considerations," so that data on 47 "muscles" were provided. Data includes useful measures such as tendon and fiber lengths as a percentage of total length, a shape factor E, the "maximum anatomical cross-sectional area divided by its mean anatomical cross-sectional area, tendon cross-sectional area, pennation, and muscle mass. To obtain muscle volumes, Eycleshymer and Shoemaker's cross-section anatomy text (1970) was used to define muscle cross-sections, from which areas could be measured via plaimetry, and volumes estimated by summing the products of area and slice thickness. Area representations of three fiver types (percentages of SO, FO, FG) are also provided. However, many of these percentile values were estimated, for example, the subdivision of the fast twitch fibers into the oxidative and glycolytic group.

Dry-bone measurements of muscle-tendon origin and insertion locations, as well as up to four intermediate points describing a nonlinear pathway are reported to have been measured. Unfortunately, this data and the measured values of musculotendon lengths, were not available from this publication.

5. *Dostal and Andrews (1981)*

Musculotendon origins and insertions were determined for the muscles of the pelvis. Points representing the centers of muscle attachment were marked on the bony pelvis and right femur of an adult, male, dry-bone specimen using anatomical texts as a guide (Schaeffer 1953; Gray 1959; Grant and Basmajian 1965; Hollinshead 1967; Gardner et al. 1969). Each long, two-joint hip muscle that crosses the knee was given a fictitious distal attachment point on the femur. Likewise, two muscles that wrapped around the underlying bony pelvis prior to crossing the hip were also assigned fictitious pelvic attachments. These origins were taken as the last point of muscle contact with the bony pelvis.

Coordinates of bony landmark and muscle attachment points were measured in a convenient right-handed orthogonal laboratory reference system which was subsequently transformed to embedded pelvic and femoral coordinate systems. The origins of each of these coordinate systems are located at the center of the acetabulum/femoral head. The axes of each system were defined such that they would be coincident when the pelvis and femur were in the position referred to as the zero joint

configuration, a position similar to the anatomical position. In such a position, the plane containing the two anterior superior iliac spines and the more anterior of the two pubic tubercles would be parallel to the frontal (y,z) plane. The straight line connecting the two anterior superior iliac spines was defined to be parallel to the z-axis. Hence with the positive x-axis directed anteriorly, and with the positive z-axis directed laterally for this right hip specimen, the positive y-axis would therefore be defined superiorly from the origin.

6. *White, Yack, and Winter (1989)*

Musculotendon origins and insertions measured from several dry-bone specimens (six disarticulated pelvises, nine femurs and combined tibia-fibula, one reconstructed skeletal foot, and one dissected cadaver foot) are provided in scaled form. All coordinates were scaled homogeneously and mapped to a subject of 66.5 kg mass and 1.77 m height. Measurements were made using graphical techniques (graph paper and drawing tools) to a precision of 2 mm. Four coordinate systems were used to measure bony landmarks and the origins and insertions of 40 muscle-tendons. The pelvis coordinate system used the right anterior superior iliac spine as the origin with left anterior superior iliac spine and right pubic tubercle as markers for defining axis orientation (x-axis points anteriorly, y-axis points superiorly, and the z-axis points laterally). The origin of the femoral coordinate system was located at the greater trochanter and orientation was defined using the medial lateral epicondyle. The three axes of the femoral system are all parallel to their respective axes in the pelvic system. The tibial coordinate system used the tibial tuberosity as its origin with the same orientation as the other two systems. The origin of the foot segment is located at the heel with the same orientation as the other three systems.

7. *Spoor, Van Leeuwen, De Windt, and Huson (1989)*

All anatomical measures were taken from one embalmed male specimen (a 48 cm baby) who had died during partus. Values of pennation angle, muscle length from origin to insertion, and physiological cross sectional area were determined for both the fetal and neonatal posture. The fetal posture, designated "fet" in the table, is described as hyper flexion of hip and knee, lateral rotation and very little abduction of the femur, the foot touches the trunk. The neonatal posture, designated "neo" in the table, is described as abduction, lateral rotation, and slight flexion of the hip.

8. Seireg and Arvikar (1989)

Musculotendon parameter and coordinate data are based on (1) direct measurements made on a skeleton of average size, (2) several approximately same-size (171 cm, 68 kg) cadavers (Seireg, 1990) and (3) scaled (2D) diagrams based on Braus (1954). Data sets from this reference are used in all five appendixes. The coordinate systems used to define the lower extremity muscles are defined as follows: pelvis, femur, tibia, calcaneus, retinaculum, talus, clavicle, cuboid, cuneiforms one through three, metatarsals one through five, and finally the toes. The coordinate systems used to define the trunk muscles are thorax, pelvis, femur, and lumbar vertebrae (L1 through L5). The upper extremity musculature utilized five coordinate systems as follows: ulna, radius, humerus, scapula, and hand. The hand musculature used twenty different coordinate systems: four for each finger, three for the thumb and one for the ulna/radius. They are ulna/radius, five metacarpals (MC1 through MC5), five proximal phalanxes (PP1 through PP5), four middle phalanxes (MP2 through MP5), and five distal phalanxes (DP1 through DP5). Finally, the head musculature utilized ten reference frames. They include the first seven cervical vertebrae (C1 through C7), the skull, the thorax, and the clavicle.

In general, coordinate axes are located at the joint centers and aligned along bony axes. For vertebrae, reference frames are located at the centers of the inferior intervertebral disks, oriented such that Z points superiorly, and X points anteriorly. For the upper-extremity limbs, the reference frames appear to be located at the centers of the proximal joints, and oriented similarly to the vertebral axes in the anatomical position. Lower-extremity reference frames are located a distal joint centers. Finger-segment reference frames are located at the centers of the proximal joints, and oriented such that X points distally along the bone axes, and Y points laterally. The skull reference is located at the "level of the atlanto occipital joint, midway between the two occipital condyles." The clavicular frame is located at its junction with the sternum; orientations of the skull and clavicular frames are similar to that for the vertebral frames. The reader is referred to the original source for further definition.

9. Wickiewicz, Roy, Powell, and Edgerton (1983)

Architectural features of the major knee flexors/extensors and ankle plantarflexors/dorsi-flexors were measured in three human cadavers (as denoted by S1, S2, and S3 in Table A.1). The hemipelvectomy sections were fixed in formalin with the hip and knee in maximum extension and the ankle in maximum dorsiflexion. Muscle lengths were measured as "the distance between the most proximal and the most distal" muscle fibers during dissection. Muscles were cleaned of fat and excessive connective tissue before weighing and preparation for architectural determinations (see Sacks and Roy, 1982). PCSA was computed as

$$PSCA = \frac{m}{\rho l}cos(\alpha) \qquad \text{(A.4)}$$

where m was the wet weight of the fixed muscles and ρ was assumed to be 1.056 (g/cm^3).

Their findings are characterized by a "marked uniformity of fiber length throughout a given muscle," similarity of fiber lengths within muscles of a synergistic group, "generally consistent" pennation angles throughout a given muscle, and "remarkably similar" ratios of muscle length to fiber length.

10a. Delp, Loan, Hoy, Zajac, Topp, and Rosen (1990); [10b] Delp (1990)

These data sets are modifications of data presented originally by Brand et al. (1982, 1986) and Friederich and Brand (1990). Musculotendon pathways have been defined relative to 3D bone surface models. To acquire the bone surface data, bone surfaces were marked with a mesh of polygons, and digitized using a Polhemus electromagnetic digitizer to determine coordinates of the vertices. Bone surface coordinates [those measured by Delp et al. (1990) were used to define the pelvis and femur; other skeletal segments were derived by Stredney (1989)] were displayed on a computer graphics workstation (Silicon Graphics, IRIS 2400T) as Gouraud-shaded surfaces. Based on the anatomical landmarks of the bone surface models, the paths (i.e., the lines of action) of 43 musculotendon actuators were defined. Each musculotendon path is represented as a series of line segments. Origins and insertions alone, in some cases, were sufficient for describing the muscle path (e.g., soleus). In other cases, where muscle wraps over bone or is constrained by retinacula, intermediate "via-points" were introduced to represent the muscle path more accurately. The number of via-points was dependent on the body position. For example, the quadriceps tendon contacts the distal femur when the knee is flexed beyond about 80 degrees (Yamaguchi and Zajac, 1989), but does not when the knee is extended. Thus, additional "wrapping points," were introduced for large knee flexion angles so that the quadriceps tendon would wrap over, rather than pass through, the bone.

Seven different coordinate systems were used to define the lower extremity muscles in this paper. In the anatomical position, each of these reference frames are oriented such that the x-axes point anteriorly, the y-axes superiorly, and the z-axes point laterally.

11.   White (1989)

The average pennation angle was obtained from data from several sources. Fiber type compositions are provided as being of either type I (slow twitch, or SO), or type II (fast twitch, or FG), and were assigned using data from the literature. Fiber resting lengths were calculated from average numbers of sarcomeres per fiber (Wickiewicz et al. 1983), and an average sarcomere rest length of 2.6 μm. See also White et al. (1989).

12.   Hoy, Zajac, and Gordon (1989)

Grouped musculotendon actuators were constructed from the data sets of Brand et al. (1982) and Friedrich and Brand (1990), according to sagittal-plane function. The coordinate systems used were consistent with those utilized by Brand et al. (1982), except that the longitudinal coordinate axis of the shank reference frame (the "Y^T" axis) was rotated 5 degrees about a medial to lateral axis and displaced 1.4 cm distal to the tibial frame of Brand et al. (1982). By doing so, the local coordinate axes were more nearly placed at the rotational axes of the ankle joint, and the y-axes of the shank frame pointed toward the origin of the femoral coordinate frame instead of the tibial tuberosity. Curved musculotendon pathways were approximated by one to three straight-line segments defined in Cartesian space. Tendon slack lengths were determined by comparing the joint angles at which (1) passive mo-ments initiate and (2) active moments peak to in vivo measurements. Origins and insertions were defined relative to the pelvic, femoral, tibial and calcaneal reference frames.

13.   Freivaldes (1985)

Musculotendon origin/insertion coordinates were measured from scaled photographs obtained from anatomical texts (among others, McMinn and Hutchings 1977), and are given for the right side of the body. The reference frames relating to 10 of the 15 body segments are referred to by: pelvis, center torso, upper torso, neck, head, femur, tibia, foot, upper arm, and lower arm. Origins of the segmental reference frames are defined relative to the center of mass of the segment identified, however, orientation was not specified. Joints are defined as: head junction, neck junction, waist, L5/S1, hips, knees, ankles, shoulders, and elbows. Muscle cross-sectional areas were determined by an independent investigator (C.S. Davis, not referenced) and are taken primarily from the Schumacher and Wolff (1966) reference.

14.   Reid and Costigan (1985)

Cross-sectional areas and average volumes for the rectus abdominis and erector spinae muscles were calculated based on computer tomography scans of 28 living subjects (16 normal males and 12 normal females). Transverse (coronal) scans were taken at 0.5 to 2.0 cm intervals between the xiphoid process and the symphysis pubis. The subjects ranged from 17 to 75 years old (mean 53.5 years), 147 to 185.4 cm in height (mean 166.1 cm) and 47 to 100.9 kg (mean 69.25). Absolute muscle moment arms about the spinal column were also calculated, though this data is not presented in the tables.

15.   Johnson, Polgar, Weightman, and Appleton (1973)

The method of Jennekens et al. (1971) was used to measure the distribution of Type I and II muscle fibers in samples of 36 human muscles obtained from 6 normal, male adults at autopsy. The subjects ranged in age from 17 and 30 years (mean age 21.8 years), in height from 183 to 198 cm (mean 186.3), and in weight from 61 to 98 kg (mean 78.5). Muscle specimens were removed, placed in solution, and frozen within 24 hours after death in preparation for later measurement.

16. Lehmkuhl and Smith (1983)

This source contained data on muscle cross-sectional area as adapted from Fick (1911); however, methods by which values were obtained were not stated in the current text. The cross-sectional areas of the upper extremity muscles were given for several positions of the arm (e.g., supination, midposition, pronation). The values of the arm in "midposition" were used in the table.

17. Wood, Meek, and Jacobsen (1989a,b)

The torso of an embalmed, male cadaver (from a person having good muscle definition and estimated to be 180 cm tall and 91 kg total weight) was used to define muscle parameters. The eviscerated torso was filled with expanding polyurethane foam and secured to a frame. Distributed sets of points defining musculotendinous attachment areas were collected using an electromechanical digitizer (accuracy estimated at better than 0.5 mm) and averaged to yield the single-point origin and insertion data. Computer surface topographies (bicubic patches) of the muscles were also defined with the aid of the digitizer. Muscle lengths, surface areas, volumes, and cross-sectional areas were defined from these computerized representations. Reference frames ulna are shown graphically and appear to be defined with respect to the eleven revolute axes of the arm linkage. These reference frames thus appear to be located along the elbow flexion/extension axis, at the humeral head (the intersection of the humeral flexion/extension, axial rotation, and ab/adduction axes), and at the approximate kinematic centers of the thorax/clavicle and clavicle/scapula joints. The reader is referred to Table 3 and Section VI.A. of the original reference for further definition.

18. Hogfors, Sigholm, and Herberts (1987)

Three dissected cadavers (range of subject ages: 55–71 yrs) and one human skeleton were used to determine the insertions of 21 muscles spanning the shoulder. Measurements were made using a ruler; accuracy was estimated to be within ** 4 mm in worst-case situations. Coordinates of the corresponding muscle insertions were defined using the following coordinate systems: thorax, scapula, clavicle, and humerus. Coordinates of the origin points, however, were not included.

The sternum system has its origin in the center of the sternoclavicular joint. It is such that the x-axis goes through the middle of both articular surfaces (i.e., it is normal to the sagittal plane and directed away from it). The entire system is orthonormal and the x-y plane contains the midpoint of the first thoracic vertebra. The y-axis and z-axis are directed forwards and upwards, respectively. The clavicular system has the same origin as the sternum system. Its x-axis goes through the center of the acromial articular surface. The y-axis is orthogonal to the x-axis and parallel to the upper planar surface on the lateral end of the clavicula. The y-axis is directed forwards and the z-axis upwards. The scapula system has its origin in the point common to the clavicula and the scapula. The x-axis of the system is directed through the inferior angle. The x-y plane contains the superior angle in its first quadrant. The last coordinate system is that fixed in the humerus. The gliding surface of the humeral head has an essentially spherical shape. The center of this sphere is used as the origin of the humerus system. The x-axis is directed along the humerus through the end of the ridge between the coronoid fossa and the radial fossa. The y-axis lies in the plane of the x-axis and the angular mobility direction of the ulna, with the ulna having a positive y-coordinate.

19. Daru (1989)

Muscle parameters (lengths, fiber compositions, tendon lengths, pennation angles, and PCSAs) of 38 muscles in the human head and neck are presented. Muscle lengths were measured from an anthropomorphic model representing a "small man." Origin/insertion points of the anthropomorphic model were based on data presented in Warfel (1973). Masses for the majority of the muscles were estimated from consideration of cross-sectional slices from Koritke and Sick (1983); other muscle masses were estimated based on their sizes relative to larger ones. Pennation angles were conservatively estimated at 10 degrees for lack of better data. Thickness estimates for PCSAs were estimated based primarily on measurements made from Toldt and Hochstatter (1957). Percentages of slow muscle fibers were estimated based primarily on measurements made from Toldt and Hochstatter (1957). Percentages of slow muscle fibers were estimated (and scaled toward 50 percent) based on data giving the ranges of slow fiber distributions in cat muscle (Richmond, 1988). Tendon lengths were estimated from inspection of anatomy sources.

20.   Brand, Beach, and Thompson (1981)

Here, the ratios of muscle mass and PCSA were investigated, based on the observation that while muscle strength varies enormously with time, and from subject to subject, the ratios of muscle strengths within the same limb vary much less.

To develop a list of muscle qualities independent of time or the overall strength of an individual, the fleshy part of each of the muscles below the elbow (including portions of brachioradialis and extensor carpi radialis longus that extend above the elbow, but not including the anconeus) in 15 cadavers was weighed. Muscle "mass fraction" was defined as the percentage of individual muscle mass to total muscle mass. These enable comparisons to all arms as long as the total muscle weight is known or can be estimated. Mass fractions in atrophied and non-atrophied arms were found to be comparable.

Mean fiber lengths were measured in five of the cadavers. These specimens were of normal build and were placed with the hands at a position of "physiologic rest" (wrist in 10 degrees of extension, finger joints flexed at about 45 degrees, thumb opposite the middle segment of the middle finger). Finally, the mass of each muscle was converted to volume and divided by the mean fiber length of the muscle to get the physiological cross-sectional area (PCSA) of the muscle. The PCSAs of the hand and forearm were then totaled. The percentage of each muscle in comparison to the total PCSA was termed "tension fraction." The average total PCSA of the muscles in each arm was 141 cm^2.

It must be clearly understood that these values for tension fraction are only proportional and involve no numerical statement of force, tension or PCSA. However, their figures do suggest that a muscle with a tension fraction of three is probably capable of 50 percent more tension than one with a value of two.

21.   Amis, Dowson, and Wright (1979)

Four limbs of stout muscular build were removed from embalmed cadavers, complete with shoulder girdle and pectoral muscles. Alexander's formula (1975) was used to compute each muscle's PCSA. To apply the formula, they measured the following upper-limb parameters: muscle mass, fiber length, and layer thickness of muscle pennation. For bipennate muscle fibers, each pennation angle was computed independently, as was each PCSA, then the PCSAs were added to obtain the total PCSA for the bipennate muscle. Muscle density was assumed to be 1050 kg/m^3.

22.   Kleweno (1987)

Parameters for muscles crossing the elbow joint were estimated based on the data of An et al. (1981), scaled for an "average sized male." Additionally, origin-insertion values were checked and in a few cases modified slightly based on attachment of strings to an "average sized male" anthropomorphic replica with plastic bones (bones obtained from Carolina Biological Supply Co.). Not presented here are computer model results which estimate muscle moment arms and muscle length changes as a function of shoulder, elbow, and wrist angles and the different assumptions utilized to curve muscle paths.

23.   An, Hui, Morrey, Linscheid, and Chao (1981)

The volumes, fiber lengths and physiological cross-sectional areas of muscles spanning the elbow joint were obtained from four fresh, unembalmed cadaver specimens "comparable in size." The specimens were dissected with the elbows in resting positions (70 degrees flexion) following the technique developed by Brand (Brand et al. 1981). Volumes of all the muscles were measured by water displacement. The physiological cross-sectional area was calculated by dividing the muscle volume by its fiber length.

24.   An, Chao, Cooney, and Linscheid (1979)

Ten normal hand specimens were examined to develop a normative model of the hand. Small segments of each tendon and muscle near each of three finger joints (metacarpal-phalangeal (MP), proximal interphalangeal (PIP), and distal interphalangeal (DIP) joint in the finger; CMC, MCP, and IP joints in the thumb) were exposed via surgical incision. Surgical wire markers were inserted into the centers of each tendon and muscle at sites "immediately proximal and distal to the joint." Intermediate points between the tendon origins/insertions were defined "with reference to the pulley systems, retinacular ligaments and transverse bands which constrain the tendons and muscles at the joint" and "minimize bowstringing with flexion and extension," so as to best represent the direction of tendon force. Measurements were taken from anterior-posterior and lateral X-ray films.

Coordinates for each finger are normalized by

the length of the middle phalanx. The length of the proximal phalanx normalized data for the thumb. The normative data presented was determined by averaging the "force" and "moment potentials" of each muscle-tendon as determined by the X-ray measures (see original source for definitions). Data is presented with respect to six Cartesian coordinate systems for each finger, numbered from distal (1) to proximal (6). Primary coordinate systems (numbers 2, 4, and 6) are "located at the approximate center of rotation of the phalangeal and metacarpal heads," and the secondary systems (numbers 1, 3, and 5) are located at "the centers of the concave articular surfaces." Therefore, reference frames are located immediately distal and proximal to each joint (i.e., odd-numbered frames are located immediately distal to each joint, while even-numbered frames are located immediately proximal). X-axes are directed proximally along the phalangeal or metacarpal shafts (from the origin of one frame to the origin of the next). Y-axes are projected dorsally from each segment, and the z-axes are projected radially for the right hand and ulnarly for the left hand.

25. Veeger, Van der Helm, Van der Woude, Pronk, and Rozendal (1991); Van der Helm, Veeger, Pronk, Van der Woude, and Rozendal (1992); Van der Helm and Veenbaas (1991)

In these three papers, a morphological study is described in which the right and left shoulders of seven cadavers have been measured. The shoulder mechanism is defined as a configuration of four bones (thorax, clavicle, scapula, humerus) and the joints in between. These papers provide a fairly complete data set. Veeger et al. (1991) provide anthropometric measurements, from which the inertial properties of the upper and lower arm and the hand are derived. In addition, the mass and cross-sectional areas of all muscles of the shoulder are provided, in combination with a literature review. In Van der Helm et al. (1992), a methodological framework is worked out for measuring geometrical data. Joint articular surfaces, ligament and muscle attachments and bony contours have been recorded using a four-bar linkage, the so-called palpator (accuracy 0.93 mm standard deviation per coordinate). The data for each morphological structure is fitted by geometrical shapes, that is, (curved) lines, planes, spheres, cylinders, ellipsoids. Pa-

rameters for the fits are provided for one cadaver. Mean residual error is reasonably low. Van der Helm and Veenbaas (1991) show how many muscle lines of action are needed for an appropriate model of the muscle geometry, taking into account the shape of the attachment site (curved line, plane) and architecture (i.e., direction of muscle fibers between origin and insertion). A final set of data is given which has been used for a musculoskeletal model of the shoulder (Van der Helm, 1994a,b). No physiological data except for muscle cross-sectional area and muscle mass are provided. All data are presented in the same global coordinate system with the Incisura Jugularis (suprasternal notch) as origin, and x-, y- and z-axis along the medial–lateral, caudal–cranial, and ventral–dorsal axis, respectively. Data on bony landmarks are lacking in the papers, but are provided on the website.

26. Klein Breteler, Spoor, and Van der Helm (1997)

In a conference proceeding, the authors mention a morphological study of the shoulder joint of one cadaver. It is a sort of follow-up on the previous study of Veeger et al. (1991) and Van der Helm et al. (1992). This study is unique since it provides data on *optimal muscle length* for all shoulder muscles, in addition to geometrical and inertial data. Geometrical data consist of joint rotation centers (from articular surfaces and from a "pivot" point approximation), ligament and muscle attachments, bony contours, and bony landmarks. Muscles have been divided a priori into muscle bundles, of which the optical centroid of the origin and insertion site have been measured using a four-bar linkage, the so-called palpator (accuracy 0.93 mm standard deviation per coordinate). In accordance with Van der Helm and Veenbaas (1991), a sufficient number of separate muscle bundles is measured to account for the mechanical effect of the muscle. Focus of the study was on deriving *physiological* data: Muscle cross-sectional area (by direct measurement and by dividing volume by optimal fiber length), muscle optimal fiber length and tendon length. For each muscle bundle the sarcomere length is measured using laser diffraction. From the recorded fiber length, sarcomere length and optimal sarcomere length (= 2.7 μm), the optimal fiber length could be calculated. Simulations are provided to show the excursion of muscles through their force-length curves. Data are not provided in the proceedings, but can be found on the website.

27.   Johnson, Spalding, Nowitzke, and Bogduk (1996)

Modeling of the shoulder mechanism is dependent upon reliable data on the morphometry and points of attachment of the relevant muscles. In this paper, the authors present coordinate data that have been derived from the radiography of dissected cadavers. The measurement method, previously described by the authors (Johnson et al. 1992) relies upon the use of uncalibrated biplanar radiography in which coordinate data is derived by digitization of pairs of radiograph images. Validation of the data was achieved by ensuring good agreement of the measurements of the "common coordinate" measured on each of the radiographs. It was found that for all muscles except those of the rotator cuff, the variations between cadavers exceed the differences in measurements of the common coordinate. Fascicles were defined according to their bony attachments and physiological cross section areas were calculated from measurements of length (using a ruler) and volume (using a measuring cylinder). The measurements were taken from three adult embalmed cadavers (6 shoulders).

Data were collected for the following muscles: Deltoid, the muscles of the rotator cuff, teres major, trapezius, levator scapulae, rhomboid (major and minor), serratus anterior, pectoralis minor. All coordinates were calculated in the anatomical frames originally defined by Hogfors et al. (1987). Physiological cross sectional areas were normalised according to the PCSA of deltoid. The measurements of PCSA were found to agree well with those of Veeger et al. (1991). Similarly, the majority of descriptions of morphometry concur with accepted texts, the authors have ascribed rather different properties to the fascicles of trapezius, pointing out that they achieve the production of lifting forces at the shoulder by applying predominantly medial forces to the mechanism created by the clavicle and scapula.

28.   Murray, Delp, and Buchanan (1995)

This paper describes a morphological study of the elbow of two specimen. Muscle moment arms for flexion/extension and pro/supination were determined by measuring the muscle length dis-

placement during varying angles. Five muscles were included in the study: Biceps long head, brachialis, pronator teres, brachioradialis and triceps medial head. The muscles were partly replaced by wires. Results are presented as moment arm vs. angle curves. These curves were also approximated by a model developed using polyhedra bones to obtain the muscle attachments. The model approximation to the data set was fair. A thorough literature review on moment arm data for the elbow is provided in the paper. There are no data of this study available on the website.

29.   Stokes and Gardner-Morse (1995)

The database now consists of the locations and orientations of 12 thoracic and 5 lumbar vertebrae relative to the sacrum, the shape and positions of the 24 ribs, and the origins, insertions and physiological cross sectional area (PCSA) of 180 muscles which have attachments to the lumbar spine and/or the pelvis.

The bony anatomy (lumbar and thoracic spine and rib cage) was obtained from stereoradiographs of 4 healthy adults and digitized to give 3D coordinates defining the local axis system of each of the 17 vertebrae and the midlines of each of the 24 ribs.

Dorsal muscles and the psoas were obtained from various publications of Bogduk, (e.g. Bogduk & Twomey 1987) These authors express the muscle anatomy descriptively in terms of locations of attachments to the vertebrae and quantitatively by the physiological cross-sectional area. The attachments were referred to the coordinate system defined by the above bony anatomy. The same geometric information was inferred for the quadratus lumborum from serial CT sectional data published by Han et al.

Geometry of the abdominal muscles was obtained by averaging the digitized outlines of serial MRI scans of 15 healthy males. The resulting volumetric reconstructions for the right and left sides were expressed as an equivalent line of action for the rectus abdominus, and 6 equivalent lines for each of the internal and external oblique muscles. These lines of action were based on the predominant fiber directions visible in anatomical specimens. Equal area was assigned to each line. The division of muscle between the internal and external obliques was based on observations of the cross-sections of the National Library of Medicine's "Visible Human," and by using the weights of abdominal muscles obtained from embalmed anatomical specimens.

An earlier version of the database is listed in Stokes and Gardner-Morse (1995). The improved database (described above) should be completed shortly.

30. Kepple, Sommer, Siegel, and Stanhope (1997)

Three-dimensional musculoskeletal databases of the lower extremities were developed for use in human musculoskeletal models. Three-dimensional muscle origin and insertion locations as well as selected landmarks were digitized from 52 dried skeletons. Specimens were selected for equal representation within four gender/race categories with stature spanning the 5th and 95th percentile for each gender. Data were collected on seven anatomical segments per skeleton (pelvis, two femurs, two tibia-fibula, and two feet). Distributed muscle origin and insertions were approximated by one or more centroid locations. When a muscle line-of-action would be expected to wrap over bony structure, connector points were selected as effective intermediate origins and insertions. A total of 43 lower extremity muscles and 60 muscle units per side were included in the database.

Affine scaling was used to accumulate normative models. Data from left and right segments were pooled. Multivariate statistical methods were then used to compare and statistically group the normative models.

With over 1200 muscle origin and insertion locations digitized these new databases greatly enhance the ability to provide accurate estimates of muscle kinematics from surface landmarks.

31. Veeger, Yu, and An (1997); Veeger, Yu, An, and Rozendal (1997)

Veeger et al. (1997) recorded morphological data of the elbow and forearm joint on five fresh cadaver specimens (four right, one left). They used a Polhemus electromagnetic tracking device to record the data. The flexion/extension and prosupination rotation axes were calculated as the instantaneous helical axes from motion data (Veeger, 1997). All origin and insertion attachment sites of the muscles, and underlying bony contours, were recorded with respect to bony landmarks of the upper arm and forearm. Bony contours were represented by cylinders. Moment arms could be derived for all combinations

of flexion/extension and pro/supination axes. The moment arm derived from muscle length-joint angle relations was compared with the moment arm calculated from the attachment sites, connected by a straight line or curved around bony contours. They concluded that there were no significant differences. The elbow data of Veeger et al. (1997) have been connected with the shoulder data obtained by Veeger et al. (1991) and Van der Helm et al. (1992). Unpublished data showed a good resemblance with the moment arms of Murray et al. (1995), and with other moment arms reviewed by these authors.

References

Alexander, R.McN., and Vernon, A. (1975). The dimensions of knee and ankle muscles and the forces they exert. *J. Human Movement Studies* 1:115–123.

Amis, A.A., Dowson, D., and Wright, V. (1979). Muscle strengths and musculo-skeletal geometry of the upper limb. *Engineering in Medicine* 8:41–48.

An, K.N., Chao, E.Y., Cooney, W.P. III, and Linsheid, R.L. (1979). Normative model of human hand for biomechanical analysis. *J. Biomech.* 12:775–788.

An, K.N., Hui, F.C., Morrey, B.F., Linsscheid, R.L. and Chao, E.Y. (1981). Muscles across the elbow joint: a biomechanical analysis. *J. Biomech.* 14:659–669.

Bogduk, N. and Twomey, L.T. (1987). *Clinical Anatomy of the Lumbar Spine.* New York, Chuchill Livingstone.

Brand, P.W., Beach, R.B., and Thompson, D.E. (1981). Relative tension and potential excursion of muscles in the forearm and hand. *J. Hand Surgery* 6:209–219.

Brand, R.A., Crowninshield, R.D., Wittstock, C.E., Pedersen, D.R., Clark, C.R., and Van Krieken, F.M. (1982). A model of lower extremity muscular anatomy. *J. Biomech. Engrg.* 104:304–310.

Brand, R.A., Pedersen, D.R., and Friederich, J.A. (1986). The sensitivity of muscle force predictions to changes in physiologic cross-sectional area. *J. Biomech.* 19: 589–596.

Braus (1954). *Anatomic Der Menschen.* Springer-Verlag, Berlin.

Chandler, R.F., Clauser, C.E., McConville, J.T., Reynolds, H.M., and Young, J.W. (1975). Investigation of inertial properties of the human body. DOT HS-801 430, Wright Patterson Air Force Base, Ohio

Delp, S.L. (1990). A computer-graphics system to analyze and design musculoskeletal reconstructions of the lower limb. Ph.D. Thesis, Dept. of Mechanical Engineering, Stanford University.

Delp, S.L., Loan, J.P., Hoy, M.G., Zajac, F.E., Topp, E.L., and Rosen, J.M. (1990). An interactive graphics-based model of the lower extremity to study orthopaedic surgical procedures. *IEEE Trans. Biomed. Eng.* (in press).

Daru, K.R. (1989). Computer simulation and static analysis of the human head, neck, and upper torso. M.S. Thesis, Dept. of Chemical, Bio, and Materials Engineering, Arizona State University, Tempe, AZ.

Dempster, W.T. (1955). Space requirements of the seated operator. WADC-TR-55-159. Wright Patterson Air Force Base, Ohio.

Dostal, W.F. and Andrews J.G. (1981). A three dimensional biomechanical model of hip musculature. *J. Biomech.* 14:803–812.

Eychleshymer, A.C. and Schoemaker, D.M. (1911). *A Cross-Section Anatomy.* Appleton-Century-Crofts, New York.

Fick, R. (1911). Anatomie und Mechanik der Gelenke: Teil III, Spexielle Gelenk und Muskel Mechanik. Fischer, Jena.

Freivalds, A. (1985). Incorporation of active elements into the articulated total body model. Paper #AAMRL-TR-85-061, Armstrong Aerospace Medical Research Laboratory, Wright-Patterson Air Force Base, OH.

Friederich, J.A. and Brand, R.A. (1990). Muscle fiber architecture in the human lower limb. Technical Note, *J. Biomech.* 23:91–95.

Gardner et al. (1969). *Anatomy*, W.B. Saunders, Philadelphia.

Grant, J.C. and Basmajian, J.V. (1965). *Grants Method of Anatomy*, Williams and Wilkins, Baltimore.

Gray, H. (1959). *Anatomy of the Human Body* (Edited by Goss, C.), Lea and Febinger, Philadelphia.

Högfors, C., Sigholm, G., and Herberts, P. (1987). Biomechanical model of the human shoulder—I. Elements. *J. Biomech.* 20:157–166.

Hollinshead, W.H. (1967). *Textbook of Anatomy*, Harper and Row, New York.

Hoy, M.G., Zajac, F.E., and Gordon, M.R. (1989). A musculoskeletal model of the human lower extremity: The effect of muscle, tendon, and moment arm on the knee and ankle. *J. Biomech.* 22:157–169.

Jennekens et al. (1971). Data on the distribution of fiber types in five human limb muscles. *J. Neurol. Sci.* 14:245–257 (as quoted in Johnson et al. (1973).

Johnson, G.R., Buxton, T., House, D., and Bogduk, N. (1992). The use of uncalibrated biplanar radiography for the measurement of skeletal coordinates around the shoulder girdle. *J. Biomed. Eng.*, 14:490–494.

Johnson, G.R., Spalding, D., Nowitzke, A., and Bogduk, N. (1996). Modelling the muscles of the scapula: morphometric and coordinate data and functional implications. *J. Biomech.*, 29:1039–1051

Johnson, M.A., Polgar, J., Weightman, D., and Appleton, D. (1973). Data on the distribution of fiber types in thirty-six human muscles. *J. Neurol. Sci.* 18:111–129.

Hatze, H. (1981). Myocybernetic control models of skeletal muscles. University of South Africa.

Hatze, H. (1997). Progression of musculoskeletal models toward large cybernetic myoskeletal models. In *Biomech. Neural Control Movement.* Winters, J.M. and Crago, P.E. (eds.), Springer Verlag, New York

Hinrichs, R.N. (1985). Regression equations to predict segmental moments of inertia from antrhopometric measurements: an extension of the data of Chandler et al. (1975). *J. Biomech.*, 18:621–624.

Huxley, A.F. (1957). Muscle structure and theories of contraction. *Prog. Biophys. Biophys. Chem.* 7:257–318.

Jensen, R.H., and Davy, D.T. (1975). An investigation of muscle lines of actions about the hip: a centroid approach vs. the straight line approach. *J. Biomech.*, 8:103–110.

Kepple, T.M., Sommer, H.J., Siegel, K.L., and Stanhope, S.J. (1997). *J. Biomech.*, (provisionally accepted).

Klein Breteler, M.D., Spoor, C.W., and Van der Helm, F.C.T. (1997). Measuring muscle and joint geometry parameters for a shoulder model. *Proc. First Conference of the International Shoulder Group.* Veeger, H.E.J., Van der Helm, F.C.T., and Rozing, P.M. (eds). Delft University of Technology, The Netherlands.

Kleweno, D.G. (1987). Physiological and theoretical analysis of isometric strength curves of the upper limb. M.S. Thesis, Dept. of Chemical, Bio & Materials Engineering, Arizona State University.

Koritké, J.G., and Sick, H. (1983). *Atlas of Sectional Human Anatomy, Frontal, Sagittal, and Horizontal Planes, Vol. 1, Head, Neck Thorax.* Urban & Scharzenburg, Baltimore.

Lehmkuhl, L.D. and Smith, L.K. (1983). *Brunnstrom's Clinical Kinesiology.* F.A. Davis Co., Philadelphia, pp. 391–401.

McMinn, R.M.H., and Hutchings, R.T. (1977). *Color Atlas of Human Anatomy.* Yearbook Medical Publishers, Chicago.

Murray, W.M., Delp, S.L., and Buchanan, T.S. (1995). Variation of muscle moment arms with elbow and forearm position. *J. Biomech.*, 28(5):513–525.

Pierrynowski, M.R. and Morrison, J.B. (1985). A physiological model for the evaluation of muscular forces in human locomotion: Theoretical aspects. *Math. Biosci.* 75:69–101.

Otten, E. (1988). Concepts and models of functional architecture in skeletal muscle. *Exerc. Sports Sci. Rev.*, 89–137.

Reid, J.G. and Costigan, P.A. (1985). Geometry of adult rectus abdominis and erector-spinae muscles. *J. Orthop. and Sports Phys. Ther.* 5:278–280.

Richmond, F.J. (1988). The motor system: joints and muscles of the neck. In *Control of Head Movements*, Chapter 1, pp. 1–22.

Sacks, R.D., and Roy, R.R. (1982). Architecture of the hind limb muscles of cats: Functional significance. *J. Morphol.* 173:185.

Schaeffer, J. (1953). *Morris' Human Anatomy*. Blakiston, New York.

Schumacher, G.H. and Wolff, E. (1966). Trockengewicht and Physiologischer Querschnitt der Menschlichen Skelettmuskulatur, II, Physiologische Querschnitte, *Anat. Anz.* 119:259–269.

Seireg, A. (1990). University of Wisconsin, Madison—personal communication.

Seireg, A. and Arvikar, R.J. (1989). *Biomechanical Analysis of the Musculoskeletal Structure for Medicine and Sports*. Hemisphere Publishing Corporation, New York.

Spoor, C.W., Van Leeuwen, J.L., De Windt, F.H.J., Huson, A. (1989). A model study of muscle forces and joint-force direction in normal and dysplastic neonatal hips. *J. Biomech.*, 8:873–884.

Spoor, C.W., Van Leeuwen, J.L., Meskers, C.G.M., Titulaer, A.F., and Huson, A. (1990). Estimation of instantaneous moment arms of lower-leg muscles. *J. Biomech.*, 23:1247–1259.

Stokes, I., Gardner-Morse, M. (1995). Lumbar spine maximum efforts and muscle recruitment patterns predicted by a model with multijoint muscles and flexible joints. *J. Biomech.*, 28(2):173–186.

Stredney, D. (1989). Ohio State University—personal communication.

Toldt, C., and Hochstetter, F. (1957). *Anatomischer Atlas für Studierende und Ärtze*, Vol. 1. Heinrich Hayek, ed., 23rd edit., Urban & Scharzenburg, Wien.

Van der Helm, F.C.T. (1994a). A finite element musculoskeletal model of the shoulder mechanism. *J. Biomech.*, 27(5):551–569.

Van der Helm, F.C.T. (1994b). Analysis of the kinematic and dynamic behavior of the shoulder mechanism. *J. Biomech.*, 27(5):527–550.

Van der Helm, F.C.T. and Veenbaas, R. (1991). Modelling the mechanical effect of muscles with large attachment sites: application to the shoulder mechanism. *J. Biomech.*, 24(12):1151–1163.

Van der Helm, F.C.T., Pronk, G.M., Veeger, H.E.J., Van der Woude, L.H.V., and Rozendal, R.H. (1992). Geometry parameters for musculoskeletal modelling of the shoulder mechanism. *J. Biomech.*, 25:129–144.

Veeger, H.E.J., Yu, B., An, K.N., and Rozendal, R.H. (1997). Parameters for modeling the upper extremity. *J. Biomech.*, 30:647–652.

Veeger, H.E.J., Yu, B., and An, K.N. (1997). Orientation of axes in the elbow and forearm for biomechanical modelling. *Proc. First Conference of the International Shoulder Group*. Veeger, H.E.J., Van der Helm, F.C.T., and Rozing, P.M. (eds.). Delft University of Technology, The Netherlands.

Veeger, H.E.J., Van der Helm, F.C.T., Van der Woude, L.H.V., Pronk, G.M., and Rozendal, R.H. (1991). Inertia and muscle contraction parameters for musculoskeletal modelling of the shoulder mechanism. *J. Biomech.*, 24(7):615–629.

Warfel, J.H. (1973). *The Head, Neck, and Trunk Muscles and Motor Points*. 4th ed. Lea & Febiger, Philadelphia.

White, S.C. (1989). A deterministic model using EMG and muscle kinematics to predict individual muscle forces during normal human gait. Ph.D. Thesis, Dept. of Kinesiology, University of Waterloo.

White, S.C., Yack, H.J., and Winter, D.A. (1989). A three-dimensional musculoskeletal model for gait analysis. Anatomical variability estimates. *J. Biomech.* 22:885–893.

Wickiewicz, T.L., Roy, R.R., Powell, P.L., and Edgerton, V.R. (1983). Muscle architecture of the human lower limb. *Clin. Orthop. Rel. Res.* 179:275–283.

Winters, J.M. and Stark, L. (1985). Analysis of fundamental human movement patterns through the use of in-depth antagonistic muscle models. *IEEE Trans. Biomed. Eng.*, 32(10):826–839.

Winters, J.M. and Stark, L. (1988). Estimated mechanical properties of synergistic muscles involved in movements of a variety of human joints. *J. Biomech.*, 21:1027–1042.

Winters, J.M. and Woo, SL-Y. (1990). Multiple muscle systems, biomechanics and movement organization. Springer Verlag, New York.

Wood, J.E., Meek, S.G., and Jacobsen, S.C. (1989a). Quantitation of human shoulder anatomy for prosthetic arm control—I. Surface modeling. *J. Biomech.* 22:273–292.

Wood, J.E., Meek, S.G., and Jacobsen, S.C. (1989b). Quantitation of human shoulder anatomy for prosthetic arm control—II. Anatomy matrices. *J. Biomech.* 22:309–325.

Yamaguchi, G.T., Zajac, F.E. (1989). A planar model of the knee joint to characterize the knee extensor mechanism. *J. Biomech.* 22:1–10.

Yamaguchi, G.T., Sawa, A.G.U., Moran, D.W., Fessler, M.J., and Winters, J.M. (1990). A survey of human-musculotendon actuator paramaters. In *Multiple Muscle Systems* Winters, J.M. and Woo, S.L-Y. (eds.), Springer Verlag, New York.

Yeadon, M.R. and Morlock, M. (1989). The appropriate use of regression equations for the estimation of segmental inertia parameters. *J. Biomech.*, 22:683–689.

Zahalak, G.I. (1981). A distribution-moment approximation for kinetic theories of muscular contraction. *Math. Biosc.*, 55:89–114.

Zajac, F.E. (1989). Muscle and tendon: properties, models, scaling and application to biomechanical and motor control. *CRC Crit. Rev. Bioimed. Engn.*, 17:359–411.

Appendix 2
Move3d Software

Tom M. Kepple and Steven J. Stanhope

Move3d software was designed for use in the Rehabilitation Medicine Department at the National Institutes of Health. The software uses data obtained from a 3D motion measurement system to provide quantitative analysis of human musculoskeletal motion including three-dimensional graphical displays. The software is designed for clinical flexibility and can be used to study a variety of movement disorders. The software, which runs on VMS, Silicon Graphics Irix, and PC DOS operating systems, has been become one of the most widely used clinical motion analysis software tools.

The Move3d model can include up to 15 anatomical segments that are customized to the subject based on geometric primitives and anthropometric measurements. The user can elect to build a default model that includes the whole body or elect to build custom models that may provide a more detailed analysis of specific anatomical regions. The position and orientation of the anatomical segments are measured using sets of noncolinear tracking targets (minimum of three, maximum of eight). Move3d operates on three input files: a motion file, a subject calibration file, and a model file. The motion file contains data collected during the movement including the coordinates of the tracking targets and the force platform vectors. The subject calibration file, which contains data collected on stationary segments, relates the tracking targets to the segmental anatomy. This file contains coordinate data of targets placed on the anatomical landmarks as well as the coordinate data of the tracking targets. The model file contains the subject specific model to be used by Move3d.

The kinematic parameters that can be evaluated include joint angular displacements, velocities and accelerations as well as segmental translations, velocities, and accelerations. Kinetic parameters that can be evaluated include net joint reaction forces and moments, joint powers, segmental powers and segmental energies. Move3d also incorporates a musculoskeletal model to allow muscle lengths and velocities to be estimated. The Move3d musculoskeletal model is based on a anatomical database generated from 52 dry skeletal specimens (Kepple et al. 1995). The musculoskeletal model can be scaled to each subject based on the location of key surface landmarks.

Move3d has been instrumental in the improvement of motion analysis algorithms. Move3d algorithms led to an improved method for calculating net joint power (Buczek et al. 1994) and to an improved agreement between the foot segmental power and rate of energy change calculations (Lohmann Siegel et al. 1995). Move3d was also used to assess the previously developed musculoskeletal geometric models (Kepple et al. 1994) and to aid in the development of a new lower-extremity musculoskeletal database (Kepple et al. 1995).

Move3d contains a graphical display component that can be used to animate the measured movement as either geometric primitives, stick figures, or 3D skeletons. In addition Move3d can generate files that can be directly entered into the ADAMS dynamic analysis software (Mechanical Dynamics Inc., Ann Arbor, MI) allowing the ADAMS user to produce computer simulations based on the measured movements.

All major components of the Move3d software have been written and maintained at the National Institutes of Health's biomechanics laboratory. This has enabled the software to be made available at no cost to most research and clinical facilities worldwide.

References

Buczek, F., Kepple, T., Lohmann Siegel, K., and Stanhope, S. (1994). Translational and rotational joint power terms in a six degree-of-freedom model of the normal ankle complex. *J. Biomech.*, 27:1447–1457.

Kepple, T., Arnold, A., Stanhope, S., and Lohmann Siegel, K. (1994). Assessment of a method to estimate muscle attachments from surface landmarks: a 3d computer graphics approach. *J. Biomech.*, 27:365–371.

Kepple, T., Sommer H.J., Lohmann Siegel, K., and Stanhope, S. (1995). Three dimensional normative databases of muscle origins and insertions for the lower extremities. *Proc. Am. Soc. Biomech.*, 37–38.

Lohmann Siegel, K., Kepple, T., and Caldwell, G. (1995). Improved agreement of foot segmental power and rate of energy change during gait: inclusion of distal power terms and the use of three dimensional models. *J. Biomech.*, 29:823–827.

Appendix 3
Simulation of an Antagonistic Muscle Model in Matlab

Bart L. Kaptein, Guido G. Brouwn, and Frans C.T. van der Helm

1 Goal of the Muscle Model

An antagonistic muscle model has been built using Matlab. This model describes the behaviour of a single joint controlled by an antagonistic muscle pair. The goal of this model is to introduce a basic understanding of muscle dynamics, in combination with segmental inertia. Because the model has a complete graphical user interface, it can be used especially for educational purposes. Input to the model is a user-defined neural input signal to the agonist and the antagonist muscle. This signal can be generated using the mouse. The graphical output of the model shows a number of responses (e.g., muscle force and joint angle) of the model to the input signals. The user can also determine an external moment that acts on the joint.

The software for the antagonistic muscle model can be found at the following website: http://www.wbmt.tudelft.nl/mms/antagonistic-muscle-model/.

2 Description of the Muscle Model

A macroscopic Hill-type muscle model is implemented (Winters and Stark 1985). This model computes muscle force (F_m) from neural input (u) and muscle length (l_m). The model structure consists of an excitation/activation part and a contractile part. The model describes the following muscle properties:

- The excitation dynamics describe the time-constant that determines the conversion from the supra-spinal command signal u to muscle excitation signal e, resembling the alpha-motoneuron signal.
- The activation dynamics describe the electro-chemical time-constant that determines the conversion from the electrical excitation signal e to the muscle active state a, resembling intramuscular calcium concentration.
- The contractile element CE of muscle fibers, described by the nonlinear force-length and force-velocity relations. The CE is activated by a, which causes active muscle contraction:
- Concentric contraction (positive work done by muscle): $a > 0$ and $l_{ce} < 0$.
- Eccentric contraction (negative work done by muscle): $a > 0$ and $l_{ce} > 0$.
- The elastic SE (lumped crossbridges and tendon) translates CE-contraction into active muscle force ($l_{se} > 0$). The SE is described by a nonlinear force-length relation, whereby the force increases exponentially with SE-stretch. Maximal muscle force is reached at 5% SE-stretch.
- The elastic PE (passive tissue within and surrounding muscle) is described by a nonlinear force-length relation, whereby the force increases exponentially with PE-stretch. Maximal muscle force is reached at 40% PE-stretch. The PE also works as a smooth mechanical stop for extreme muscle extensions and adds as passive force to total muscle force.

In the model muscle force yields $F_m = F_{se} + F_{pe}$ and muscle length yields $l_m = l_{ce} + l_{se}$. The following model parameters can be altered by the user: joint inertia, passive damping of the joint, maximal isometric force of the muscles, and maximal shortening velocity of the muscles. Additionally the user can specify duration of the simulation and goal angle of the joint. The following model signals are shown on the screen: joint angle, muscle force, muscle length, series elastic force, series elastic length, neural inputs of agonist and antagonist, external moment. Experienced users may want to alter other muscle parameters and run the simulation program for their own purposes. This is possible as the Matlab software is completely accessible.

3   Application of the Muscle Model

The antagonistic muscle model makes it possible to simulate different kinds of physical joints: shoulder, elbow, wrist, neck, eye, leg, and so on. Shoulder muscles control a much larger inertia than eye muscles. Eye muscles behave more viscous than shoulder muscles. These properties have a significant effect on optimal muscle control signals, for instance to make a time-optimal movement.

For shoulder muscles a tri-phasic neural input pattern is required, whereas for a time-optimal eye movement a biphasic pattern suffices. Besides joint movement, also joint stabilization under external disturbances can be simulated. Co-contraction can decrease the effect of external moment disturbances, because joint visco-elasticity increases with co-activation of antagonistic muscles.

Reference

Winters, J.M. and Stark, L. (1985). Analysis of Fundamental Movement Patterns Through the Use of In-Depth Antagonistic Muscle Models. *IEEE Trans. Biomed. Eng.*, BME-32:826–839.

Appendix 4
SPACAR: A Finite-Element Software Package for Musculoskeletal Modeling

Frans C.T. van der Helm

An important tool for biomechanical research is computer simulation. In inverse dynamic simulations, the recorded motions are input to the model, and muscle forces and/or muscle activations are calculated by evaluation of the motion equations, which are in an algebraic format since the position, velocities and accelerations are given. In forward dynamic simulations, typically the neural input to the muscles is input to the model, and the resulting motions are calculated by integrating the (differential) motion equations. To solve the motion equations, universal computer packages such as SPACAR have been developed.

SPACAR was developed at the Engineering Mechanics Group at Delft University of Technology (Van der Werff 1977; Jonker 1988) and still exists as an experimental program. It is based on the finite element approach, in which *each* morphological structure is represented by *one* appropriate element, of which the mechanical behavior is known. By connecting the elements through shared endpoint nodes, the mechanical behavior of the whole mechanism can be derived. A complex mechanism can be built with reasonably simple means. The system is very flexible for adding or removing elements, and changing coordinates. For application in biomechanics, some special purpose elements have been developed (Van der Helm 1994a). Bones are represented as rigid bodies (with local coordinate systems), with attachment sites fixed in the bodies. Joints are represented by three perpendicular HINGE elements, muscles and ligaments by sliding TRUSS elements that are capable of lengthening and shortening. If the muscle is wrapped around a bony contour, a CURVED-TRUSS ele-

ment is used. Each time step the shortest path around a bony contour (sphere, cylinder, ellipsoid) is calculated. For application in the shoulder region, the sliding of the scapula over the thorax is represented by specially developed SURFACE elements, defining the compression forces between the medial border of the scapula and the thorax. For each element the dynamical behavior can be described with user-supplied subroutines.

For all elements, the mathematical relation between the *deformation vector E* and the (position and orientation) *coordinates X* is calculated. New elements can be developed by specifying a certain relation between the deformation of the element and its coordinates. Using the principle of virtual work (relating stresses associated with the deformations to the forces associated with the coordinates), and d'Alembert principle for including inertial forces, the motion equations are derived for the whole mechanism. The methodology is quite equivalent with the derivation of Kane's equations.

Being a more experimental package consisting of a number of *Fortran* subroutines, the performance is not very fast. Major advantages are the unified framework in which the elements are treated, allowing for user-specified elements, the full access to the subroutines and all variables, and the flexibility in building mechanisms. For example, adding an elbow model structure to the existing shoulder model took only a few hours.

Recorded motions are supplied as the time-course of the generalized variables. The post-processing consists of Matlab files with all variables, and Matlab subroutines for selecting the simulation results by pull-down menus. Input-files for a 3D musculoskeletal visualization package SIMM (Musculographics, Inc.) are also generated, so that the motion can be viewed from any angle and magnification (see Figure A4.1).

Applications have been built for some small-scale 1-DOF and 2-DOF models, but also for large-scale models of the shoulder and elbow. Van der Helm (1994b) used a shoulder model with 95 muscle lines of action for the inverse static analysis of abduction and forward flexion. Happee and Van der Helm (1996) used the model for inverse dynamic analysis of goal-directed arm movements, including the application of an inverse muscle model to account for the muscle dynamics. The flexible structure of SPACAR made it very suitable for quasi-static stability analysis (Van der Helm and Winters 1994). Rozendaal (1997) developed a (quasistatic) linearized state-space representation, and included proprioceptive feedback (length, velocity, force) with time-delays (modeled by second-order Padé filters). Forward dynamic optimization simulations have been tested, but abandoned because of the large amount of CPU time expected for optimization.

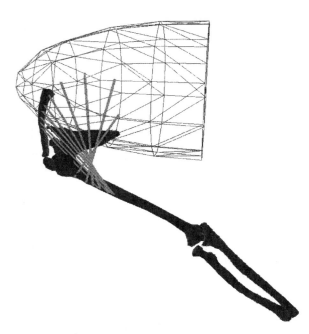

FIGURE A4.1. Visualization of the muscle lines of action of the shoulder model. (Visualization by SIMM, Musculographics, Inc.)

Ergonomic applications were in the analysis of manual wheelchair propulsion (Van der Helm and Veeger 1996) and an unpublished study of bricklayers. Clinical applications were in the sensitivity analyses of a shoulder arthrodesis (Van der Helm and Pronk 1994) and a glenohumeral endoprosthesis (Leest et al. 1996). Current research topics involve the forward dynamic simulation of the musculoskeletal model including feedback loops, and connecting the model to motion recordings in a clinical setting, to be used by physicians.

References

Happee, R. and Van der Helm, F.C.T. (1995). The control of shoulder muscles during goal directed movements, an inverse dynamic analysis. *J. Biomech.*, 28: 1179–1191.

Jonker, B. (1978). A finite element dynamic analysis of flexible spatial mechanisms and manipulators. Doctoral thesis, Delft University of Technology, The Netherlands.

Leest, O de., Rozing, P.M., Rozendaal, L.A., and Van der Helm, F.C.T. (1996). The influence of glenohumeral prosthesis geometry and placement on shoulder muscle forces. *Clin. Orthopaed. Related Res.*, 330:222–233.

Rozendaal, L.A. (1997). Stability of the shoulder: intrinsic muscle properties and reflexive control. Doctoral thesis, Delft Univerrsity of Technology, The Netherlands.

Van der Helm, F.C.T. (1994a). A finite element musculoskeletal model of the shoulder mechanism. *J. Biomech.*, 27(5):551–569.

Van der Helm, F.C.T. (1994b). Analysis of the kinematic and dynamic behavior of the shoulder mechanism. *J. Biomech.*, 27(5):527–550.

Van der Helm, F.C.T. and Pronk, G.M. (1994). The loading of shoulder girdle muscles in consequence of a glenohumeral arthrodesis. *Clin. Biomech.*, 9:139–148.

Van der Helm, F.C.T. and Winters, J.M. (1994). Optimizing postural stability in large-scale upper limb system: kinematic and actuator redundancy? *Abstr. II World Congr. on Biomechanics*. Blankevoort, L. and Kooloos, J.G.M. (eds.). St. World Biomechanics (publisher), II-113.

Van der Helm, F.C.T. and Veeger, H.E.J. (1996). Quasistatic analysis of the shoulder forces in wheelchair propulsion. *J. Biomech.*, 29:39–52.

Van der Werff, K. (1977). Kinematic and dynamic analysis of mechanisms: a finite element approach. Doctoral thesis, Delft University of Technology, The Netherlands.

Appendix 5
DataMonster

E. Otten

1 Introduction

DataMonster, so named in analogy with Cookie Monster for its ability to eat massive amounts of data, is a software package running on an Apple Macintosh computer for signal analysis and modeling for research in motor control. It was grown in the laboratory of Medical Physiology in Groningen. The group is devoted to the study of motor control, both in humans and animals. DataMonster was developed with eight purposes in mind:

1. To perform a large set of common mathematical operations and signal processing (also found in programs like LabVIEW from National Instruments).
2. To perform some very specific signal processing linked to the study of motor control (like signal recognition and spike detection).
3. To be able to solve numerically large sets of differential equations.
4. To be able to handle large sets of datafiles with a similar structure in the same way with a minimum of mouse clicks and keyboard strokes.
5. To have a large variety of graph types (included shaded 3D surface graphs, etc.) with up to 25 panels on a page.
6. To be able to produce animations of graphs or a large collection of different graphs in one Quicktime movie file.
7. To do some modeling and simulation of motor control problems with an icon lay out interface.
8. To be able to quickly compare measurements and simulation results.

Data Monster was developed within Macintosh Programmers Workshop in Pascal and runs on a Power Macintosh.

2 Overview

DataMonster works with a number of calculation passes that transform the data to required results. Apart from the possibility to use formulas in which whole columns of data are used and constants defined in each pass, column operators can be called inside formulas, such as:

- A set of frequency filters (low pass, high pass, band reject, band pass).
- An unclip operator that reconstructs peaks that have been cut off by an amplifier.
- Interpolator that tries to recover a very short piece of signal that was missing (like in 3D movement registration when a marker is covered for a short while).
- Click removal from a signal.
- Signal recognition based on a signal template (useful in recognizing motor units from a needle recording).
- Forward and inverse fast fourier transforms and frequency spectrum.
- Convolution and deconvolution of signals.
- Cross correlation of signals and calculation of similarity of signals.
- Removal of cross talk from signals.
- Calculation of angles from 3D or 2D data.

- Many geometrical operators, such as projection of points on lines,
- 3D rotation of points.
- Automatic stick diagram generation of sets of 3D datapoints.
- Polynomial fitting of data.
- Random number generators.
- Set of signal generators (including signals with variable amplitude and frequency.
- Piecewise linear line segment generator.
- Four-dimensional function generator (producing a series of files with rows and columns in which the function value is stored).
- Solving a set of up to 256 coupled differential equations.
- Solving a set of up to 256 linear equations.
- Integration and differentiation.
- Sorting of columns with association of other columns.
- A large set of operators that distribute, move, collect or assemble data.
- Calculation of statistical distributions.

There is also a window in which simple models of control can be programmed using icons, connected by lines (very much like the programming technique in LabVIEW). The models that can be used are:

- A solid body.
- A linear spring.
- A skeletal muscle.
- A tendon.
- A muscle spindle.
- A golgi tendon organ.
- A neuron.

With these elements, simple neural networks can be programmed and control loops from spindles and Golgis back to muscles. Only movements along a single line of simulated solid bodies can be calculated at present.

3 Graphics

The graphics output (on screen, printer, or PICT file) possibilities are:

- Point graphs.
- Line graphs.
- Mixed point and line graphs.
- Bar graphs.
- Area graph.
- Contour plots.
- Scaled charts (every signal optimally scaled).
- Stick diagrams (with animations).
- A field of line graphs with colour coding.
- Intensity graphs (value coded in colour or gray tone).
- 3D surface graph (both from function value only and real 3D surfaces).

The view is controlled by a mouse, with choice of shading, position of lamp, and so on.

4 Absolute Limits

DataMonster can handle files with up to 2,000,000 rows of data and 256 columns when given enough memory (4 Gb). At 8 Mb of memory, DataMonster handles 3000 rows and 64 columns. The number of columns can be reduced when more rows are needed (their product is the limit given).

Formulas can have 8 levels of brackets and in each level 8 blocks of brackets. In each block 35 constants can be used and 30 operators.

Most operators (like a low pass filter) can be used within a formula, two levels deep and 10 operators in each formula.

5 World Wide Web

More information, images and facts on how to obtain DataMonster can be found on internet page:

coo.med.rug.nl/fmw/c3111.htm

Index